Lecture Notes on Data Engineering and Communications Technologies

Volume 95

The aim of the book series is to present cutting edge engineering approaches to data technologies and communications. It will publish latest advances on the engineering task of building and deploying distributed, scalable and reliable data infrastructures and communication systems.

The series will have a prominent applied focus on data technologies and communications with aim to promote the bridging from fundamental research on data science and networking to data engineering and communications that lead to industry products, business knowledge and standardisation.

Indexed by SCOPUS, INSPEC, EI Compendex.

All books published in the series are submitted for consideration in Web of Science.

More information about this series at https://link.springer.com/bookseries/15362

Mohammad Shamsul Arefin · M. Shamim Kaiser ·
Anirban Bandyopadhyay ·
Md. Atiqur Rahman Ahad · Kanad Ray
Editors

Proceedings of the International Conference on Big Data, IoT, and Machine Learning

BIM 2021

Springer

Editors
Mohammad Shamsul Arefin
Chittagong University of Engineering and Technology (CUET)
Chittagong, Bangladesh

M. Shamim Kaiser
Jahangirnagar University
Dhaka, Bangladesh

Anirban Bandyopadhyay
National Institute for Materials Science
Tsukuba, Japan

Md. Atiqur Rahman Ahad
University of Dhaka
Dhaka, Bangladesh

Kanad Ray
Amity University
Jaipur, India

ISSN 2367-4512 ISSN 2367-4520 (electronic)
Lecture Notes on Data Engineering and Communications Technologies
ISBN 978-981-16-6635-3 ISBN 978-981-16-6636-0 (eBook)
https://doi.org/10.1007/978-981-16-6636-0

This Springer imprint is published by the registered company Springer Nature Singapore Pte Ltd.
The registered company address is: 152 Beach Road, #21-01/04 Gateway East, Singapore 189721, Singapore

Organization

General Chairs

V. R. Singh, National Physical Laboratory, New Delhi, India
Yasuhiko Morimoto, Hiroshima University, Japan
Nazmul Siddique, University of Ulster, UK

General Co-chairs

M. Ibrahim Khan, CUET, Bangladesh
K. Hasan Talukder, KU, Bangladesh
Rashed Mustafa, University of Chittagong, Bangladesh

International Advisors

Muhammad H. Rashid, University of West Florida, USA
Manzur Murshed, Federation University Australia
M. Tariqul Islam, Universiti Kebangsaan Malaysia
Tarek M. Sobh, Lawrence Technological University, USA
Fakhrul Alam, Massey University, New Zealand
Issam W. Damaj, Beirut Arab University, Lebanon
Saidur Rahman, Sunway University, Malaysia
Santanu Behera, NIT Rourkela, India
M. Julius Hossain, Research Scientist, EMBL, Heidelberg, Germany
Stefano Vassanelli, University of Padova, Italy
Amir Hussain, Edinburgh Napier University, UK
Ning Zhong, Maebashi Institute of Technology, Japan

Mufti Mahmud, Nottingham Trent University, UK
Yang Yang, Maebashi Institute of Technology, Japan
Wolfgang Maas, Technische Universität Graz, Austria
S. M. Riazul Islam, Sejong University, South Korea

National Advisors

M. Kaykobad, Brac University, Bangladesh
B. Chandra Ghosh, RUET, Bangladesh
S. Prasad Majumder, BUET, Bangladesh
M. Rafiqul Alam, CUET, Bangladesh
R. Alam Seikh, RUET, Bangladesh
M. H. Azad Khan, EWU, Bangladesh
H. M. Hasan Babu, DU, Bangladesh
M. Shorif Uddin, JU, Bangladesh
A. S. M. Latiful Hoque, BUET, Bangladesh
M. Saidur Rahman, BUET, Bangladesh
M. Mahfuzul Islam, BUET, Bangladesh
M. Rafiqul Islam, KU, Bangladesh
M. Shahadat Hossain, CU, Bangladesh
M. Mahbubur Rahman, MIST, Bangladesh
Nasim Akhtar, DUET, Bangladesh
M. Abdur Razzaque, DU, Bangladesh
Kaushik Deb, CUET, Bangladesh
M. Quamruzzaman, CUET, Bangladesh
M. Sohel Rahman, BUET, Bangladesh

Organizing Chair

M. Shamsul Arefin, CUET, Bangladesh

Organizing Co-chairs

M. R. Tanvir Hossain, CUET, Bangladesh
M. Mokammel Haque, CUET, Bangladesh
M. Obaidur Rahman, DUET, Bangladesh

Organizing Secretary

Abdur Rouf, DUET, Bangladesh

Organizing Committee Members

K. M. Rokibul Alam, KUET, Bangladesh
M. Obaidur Rahman, CUET, Bangladesh
M. Kamal Hossain, CUET, Bangladesh
M. Hoq Chowdhury, CUET, Bangladesh
G. M. Atiqur Rahaman, KU, Bangladesh
M. A. Rahman Khan, NSTU, Bangladesh
M. M. Kamal Bhuiya, CUET, Bangladesh
S. Md. Galib, JSTU, Bangladesh
Juel Sikder, RMSTU, Bangladesh
S. M Taohidul Islam, PSTU, Bangladesh
Rezaul Karim, CU, Bangladesh
Md. Ahsan Habib, MVSTU, Bangladesh
Sajeeb Saha, JnU, Bangladesh
Tahmina Khanam, CUET, Bangladesh
Lamia Alam, CUET, Bangladesh
Shayla Sharmin, CUET, Bangladesh
A, Chandra Roy, CUET, Bangladesh
M. Sabir Hossain, CUET, Bangladesh
M. S. Alam Forhad, CUET, Bangladesh
S. Chandra Tista, CUET, Bangladesh
M. Mynul Hasan, CUET, Bangladesh
Annesha Das, CUET, Bangladesh
Ashim Dey, CUET, Bangladesh
Omar Sharif, CUET, Bangladesh
Billal Hossain, CUET, Bangladesh
M. A. Islam Rizvi, CUET, Bangladesh
Sabiha Anan, CUET, Bangladesh
M. Rashadur Rahman, CUET, Bangladesh
M. Fazlur Rahman, CUET, Bangladesh
M. Rashedul Islam, CUET, Bangladesh
M. Kamrul Hossain, CUET, Bangladesh
M. K. I. Zinnah Apu, CUET, Bangladesh
M. Akber Hossain, CUET, Bangladesh
M. Moyeen Uddin, CUET, Bangladesh
K. Hassan Shakib, CUET, Bangladesh
M. Shafiqul Islam, CUET, Bangladesh

M. M. M. Hussain, UCTC, Bangladesh
Nishu Chowdhury, UCTC, Bangladesh

TPC Chairs

K. M. Azharul Hasan, KUET, Bangladesh
M. Al Mamun, RUET, Bangladesh
A. S. M. Kayes, La Trobe University, Australia

TPC Co-chairs

M. Zahidul Islam, Islamic University, Bangladesh
A. Wasif Reza, EWU, Bangladesh
M. A. Akber Dewan, Athabasca University, Canada

Track Chairs

M. Shamim Kaiser, JU, Bangladesh
P. K. Dhar, CUET, Bangladesh
M. S. Chowdhury, University of Chittagong, Bangladesh
M. Chowdhury, CUET, Bangladesh
A. Kumar Bairagi, KU, Bangladesh
W. Rahman Miah, DUET, Bangladesh
Kamruddin Nur, AIUB, Bangladesh
Firoz Mridha, BUBT, Bangladesh
M. J. Alam Patwary, Shenzhen University, China
Syed Galib, JUST, Bangladesh
G. M. Atiqur Rahaman, KU, Bangladesh
S. Islam Khan, IIUC, Bangladesh
A. K. M. Masum, IIUC, Bangladesh

TPC Members

John H. L. Hansen, The University of Texas at Dallas, USA
C. K. Roy, University of Saskatchewan, Canada
Paul Watters, La Trobe University, Australia
Zubair Fadlullah, Lakehead University, Canada

M. M. Hassan, King Saud University, KSA
Alex Ng, La Trobe University, Melbourne, Australia
Indika Kumara, Jheronimus Academy of Data Science, Netherlands
Tarique Anwar, Macquarie University, Australia
Omaru Maruatona, Aiculus Pty Ltd, Melbourne, Australia
A. R. Bin Shahid, Concord University, USA
Shahriar Badsha, University of Nevada, Reno, USA
Kanad Ray, Amity University, Jaipur, India
Abdur Rouf, DUET, Bangladesh
M. Kamal Hossain, CUET, Bangladesh
Mohammad Hammoudeh, Manchester Metropolitan University, UK
Tauhidul Alam, LSU Shreveport University, USA
M. J. Alam Patwary, Shenzhen University, China
P. K. Dhar, CUET, Bangladesh
Nursadul Mamun, The University of Texas at Dallas, USA
M. Ahmed, Edith Cowan University, Perth, Australia
M. Azad Hossain, CUET, Bangladesh
Adnan Anwar, Deakin University, Australia
A. Bandyopadhyay, NIMS, Japan
M. Moshiul Hoque, CUET, Bangladesh
A. K. M. Masum, IIUC, Bangladesh
M. Hanif Seddiqui, CU, Bangladesh
Asaduzzaman, CUET, Bangladesh
M. Osiur Rahman, CU, Bangladesh
M. S. Chowdhury, CU, Bangladesh
Rashed Mustafa, CU, Bangladesh
Nur Mohammad, CUET, Bangladesh
J. Chakrabartty, CUET, Bangladesh
G. M. Sadiqul Islam, CUET, Bangladesh
N. Karim Chowdhury, CU, Bangladesh
R. Uddin Faruqui, CU, Bangladesh
Ahmed Imteaj, Florida International University, USA
S. M. Rafizul Haque, Canadian Food Inspection Agency, Canada
M. Fazlul Kader, CU, Bangladesh
S. Chandra Banik, CUET, Bangladesh
S. M. Galib, JSTU, Bangladesh
M. H. Chowdhury, City University of Hong Kong, Hong Kong
S. Majumder, Texas A&M University, USA
Kafi Rahman, Truman State University, USA
M. Mizanur Rahman, CUET, Bangladesh
M. Sanaul Rabbi, CUET, Bangladesh
Monjurul Islam, Canberra Institute of Technology, Australia
Rahma Mukta, University of New South Wales, Australia
A. S. Muhammad Sayem, CUET, Bangladesh
M. A. Rahman Ahad, Osaka University, Japan

M. Golam Hafez, CUET, Bangladesh
M. Kamruddin Nur, AIUB, Bangladesh
M. Nurul Huda, UIU, Bangladesh
Aseef Iqbal, CIU, Bangladesh
E. Hoque Prince, York University, Canada
M. Saiful Islam, Griffith University, Australia
M. Firoz Mridha, BUBT, Bangladesh
S. Kibria, SUST, Bangladesh
A. Ahmad, SUST, Bangladesh
S. I. Khan, IIUC, Bangladesh
M. Manirul Islam, AIUB, Bangladesh
A. Das Antar, University of Michigan, USA
I. H. Sarkar, CUET, Bangladesh
M. H. Chowdhury, CUET, Bangladesh
Nilanjan Dey, JIS University, Kolkata, India
A. A. Mamun, JU, Bangladesh
A. K. M. Mahbubur Rahman, IUB, Bangladesh
A. B. M. Aowlad Hossain, KUET, Bangladesh
M. Saiful Islam, CUET, Bangladesh
A. S. M. Sanwar Hosen, JNU, South Korea
Antesar Shabut, Leeds Trinity University, UK
Antony Lam, Mercari Inc., Japan
Banani Roy, University of Saskatchewan, Canada
Atik Mahabub, Concordia University, Canada
Aye Su Phy, Computer University Kalay, Myanmar
Belayat Hossain, Loughborough University, UK
Manohar Das, Oakland University, USA
M. Hoque-Tania, Oxford University, UK
Md. Abu Yousuf, JU, Bangladesh
M. B. Alam Miah, UPM, Malaysia
M. I. Miah, CUET, Bangladesh
M. Sanaul Haque, University of Oulu, Finland
M. Abu Layek, Jagannath University, Bangladesh
M. Ahsan Habib, MBSTU, Bangladesh
M. Golam Rashed, Rajshahi University, Bangladesh
M. Habibur Rahman, MBSTU, Bangladesh
Farah Deeba, DUET, Bangladesh
G. D. Bashar, Boise State University, USA
H. Liu, Wayne State University, USA
Hishato Fukuda, Saitama University, Japan
Imtiaz Mahmud, Kyungpook National University, Korea
Khoo Bee Ee, Universiti Sains Malaysia, Malaysia
Linta Islam, Jagannath University, Bangladesh
Lu Cao, Saitama University, Japan
M. Rashidul Hasan, Rajshahi University, Bangladesh

Ryote Suzuki, Saitama University, Japan
Saiful Azad, Universiti Malaysia Pahang, Malaysia
Sajjad Waheed, MBSTU, Bangladesh
Surapong Uttama, Mae Fah Luang University, Thailand
Tabin Hassan, AIUB, Bangladesh
Tomonori Hashiyama, The University of Electro-Communications, Japan
Wladyslaw Homenda, Warsaw University of Technology, Poland
Tushar Kanti Shaha, JKKNIU, Bangladesh

Preface

We are in the era of the Fourth Industrial Revolution that is a new chapter in human development enabled by extraordinary technological advances and making a fundamental change in the way we live, work, and relate to one another. It is an opportunity to help everyone, including leaders, policy-makers, and people from all income groups and nations, to harness converging technologies in order to create an inclusive, human-centered future. Big data, IoT, and machine learning are three important components of 4.0 Industrial Revolution. To do this, we must prepare our graduates and researchers to conduct their research using Industry 4.0-related technologies such as big data, machine learning, Internet of things, robotics, augmented reality, virtual reality, 3D printing, and so on. As part of our efforts to achieve sustainable development, we must develop and put into effect policies that are focused on the components of 4.0 Industrial Revolution. Considering this fact, we organized the International Conference on Big Data, Internet of Things (IoT) and Machine Learning (BIM 2021) on September 23–25, 2021. Initially, we planned to organize BIM 2021 at Cox's Bazar, Bangladesh. However, due to the COVID-19 pandemic situation, BIM 2021 took place in full virtual mode. Although we had to arrange BIM 2021 virtually, the research community reacted amazingly well at this challenging time. The support partners of BIM 2021 were CUET Intelligent Computing Lab, IEEE Computer Society Bangladesh Chapter, and the Center for Natural Science and Engineering Research (CNSER).

There were three main tracks at BIM 2021. These are data science and big data, Internet of things, and machine learning. There were a total of 263 submissions from fourteen different countries at BIM 2021. The submitted papers underwent a double-blind review process soliciting expert opinion from at least three experts: at least two independent reviewers and the respective track chair. After the rigorous review reports from the reviewers and the track chairs, the technical committee has selected 59 high-quality papers for presentation in the conference and possible inclusion in Lecture Notes on Data Engineering and Communications Technologies. We hope that the papers published in this volume will help researchers, professionals, and students to enrich their knowledge to continue their research with cutting-edge technologies.

We are thankful to authors who have made a significant contribution to the conference and have developed relevant research and literature in data science, IoT, and machine learning. We would like to express our gratitude to the members of international and national advisory committees, general chairs, general co-chairs, organizing committee members, and technical committee members for their unconditional support to make BIM 2021 a grand success. We are highly grateful to the faculty members of the Department of Computer Science and Engineering, Chittagong University of Engineering and Technology, for their wholehearted support for BIM 2021. We are grateful to Mr. Aninda Bose, Mr. Nareshkumar Mani, and other members of Springer Nature for their continuous support in coordinating this volume publication. Last but not least, we thank all of our volunteers for their tremendous support during this challenging time to make BIM 2021 a successful one.

Chittagong, Bangladesh — Mohammad Shamsul Arefin
Dhaka, Bangladesh — M. Shamim Kaiser
Tsukuba, Japan — Anirban Bandyopadhyay
Dhaka, Bangladesh — Md. Atiqur Rahman Ahad
Jaipur, India — Kanad Ray
July 2021

Contents

Artificial Intelligence for Imaging Applications

Data Science and Big Data

Internet of Things for Smart Applications

Internet of Things for Wellbeing

About the Editors

Professor Dr. Mohammad Shamsul Arefin is affiliated with the Department of Computer Science and Engineering (CSE), Chittagong University of Engineering and Technology, Bangladesh. Earlier he was the Head of the Department. Prof. Arefin received his Doctor of Engineering Degree in Information Engineering from Hiroshima University, Japan with support of the scholarship of MEXT, Japan. As a part of his doctoral research, Dr. Arefin was with IBM Yamato Software Laboratory, Japan. His research includes privacy preserving data publishing and mining, distributed and cloud computing, big data management, multilingual data management, semantic web, object oriented system development and IT for agriculture and environment. Dr. Arefin has more than 110 referred publications in international journals, book series and conference proceedings. He is a senior member of IEEE, Member of ACM, Fellow of IEB and BCS. Dr. Arefin is the Organizing Chair of BIM 2021; TPC Chair, ECCE 2017; Organizing Co-Chair, ECCE 2019; and Organizing Chair, BDML 2020. Dr. Arefin visited Japan, Indonesia, Malaysia, Bhutan, Singapore, South Korea, Egypt, India, Saudi Arabia and China for different professional and social activities.

Dr. M. Shamim Kaiser is currently working as Professor at the Institute of Information Technology of Jahangirnagar University, Savar, Dhaka-1342, Bangladesh. He received his Bachelor's and Master's degrees in Applied Physics Electronics and Communication Engineering from the University of Dhaka, Bangladesh, in 2002 and 2004, respectively, and the Ph.D. degree in Telecommunication Engineering from the Asian Institute of Technology, Thailand, in 2010. His current research interests include data analytics, machine learning, wireless network and signal processing, cognitive radio network, big data and cyber security, renewable energy. He has authored more than 100 papers in different peer-reviewed journals and conferences. He is Associate Editor of the *IEEE Access Journal*, Guest Editor of *Brain Informatics Journal* and *Cognitive Computation Journal*. He is Life Member of Bangladesh Electronic Society; Bangladesh Physical Society. He is also a senior member of IEEE, USA, and IEICE, Japan, and active volunteer of the IEEE Bangladesh Section. He is the founding Chapter Chair of the IEEE Bangladesh Section Computer Society

Chapter. He organized various international conferences such as ICEEICT 2015–2018, IEEE HTC 2017, IEEE ICREST 2018 and BI2020.

Anirban Bandyopadhyay is Senior Scientist in the National Institute for Materials Science (NIMS), Tsukuba, Japan. He received Ph.D. from Indian Association for the Cultivation of Science (IACS), Kolkata, 2005, December, on Supramolecular Electronics. From 2005 to 2007, he was ICYS Research Fellow NIMS, Japan, and, 2007, is now a permanent Scientist in NIMS, Japan. He has ten patents on building artificial organic brain, big data, molecular bot, cancer and Alzheimer drug, fourth circuit element, etc. From 2013 to 2014, he was a visiting scientist in MIT, USA, on biorhythms. He worked in World Technology Network, as WTN Fellow, (2009 continued); he received Hitachi Science and Technology Award 2010, Inamori Foundation Award 2011–2012, Kurata Foundation Award, Inamori Foundation Fellow (2011), Sewa Society International SSS Fellow (2012), Japan; SSI Gold Medal (2017).

Md. Atiqur Rahman Ahad, SMIEEE, SMOSA; Professor, University of Dhaka (DU); Specially Appointed Associate Professor, Osaka University. He studied at the University of Dhaka, University of New South Wales, and Kyushu Institute of Technology. His authored/edited 10 books in Springer, e.g., *IoT-sensor based Activity Recognition*; *Motion History Images for Action Recognition and Understanding*; *Computer Vision and Action Recognition*. He published 180+ journal/conference papers, chapters, 120+ keynote/invited talks, 35+ Awards/Recognitions. He is an Editorial Board Member of Scientific Reports, Nature; Associate Editor of *Frontiers in Computer Science*; Editor of *International Journal of Affective Engineering*; Editor-in-Chief: *IJCVSP*; Guest-Editor: *PRL*, Elsevier; *JMUI*, Springer; *JHE*, Hindawi; *IJICIC*; Member: ACM, IAPR.

Kanad Ray (Senior Member, IEEE) received the M.Sc. degree in physics from Calcutta University and the Ph.D. degree in physics from Jadavpur University, West Bengal, India. He has been Professor of Physics and Electronics and Communication and is presently working as Head of the Department of Physics, Amity School of Applied Sciences, Amity University Rajasthan (AUR), Jaipur, India. His current research areas of interest include cognition, communication, electromagnetic field theory, antenna and wave propagation, microwave, computational biology and applied physics. He has been serving as Editor for various Springer book series. He was Associate Editor of the *Journal of Integrative Neuroscience* (The Netherlands: IOS Press). He has been visiting Professor to UTM and UTeM, Malaysia, and visiting Scientist to NIMS, Japan. He has established MOU with UTeM Malaysia, NIMS Japan and University of Montreal, Canada. He has visited several countries such as Netherlands, Turkey, China, Czechoslovakia, Russia, Portugal, Finland, Belgium, South Africa, Japan, Singapore, Thailand and Malaysia for various academic missions. He has organized various conferences such as SoCPROS, SoCTA, ICOEVCI and TCCE as General Chair and Steering Committee Member.

Machine Learning for Disease Detection

Performance Analysis of Classifier for Chronic Kidney Disease Prediction Using SVM, DNN and KNN

Md. Omaer Faruq Goni, Abdul Matin, Tonmoy Hasan, and Md. Rafidul Islam Sarker

Abstract Disruption of the regular operation of the kidney is named chronic kidney disease (CKD). CKD is widespread, and the death rate due to this increases rapidly. To reduce the amount of death, early detection of CKD is necessary. This paper aims to help medical practitioners to diagnose CKD patients by applying machine learning (ML) techniques. We have applied several ML algorithms to the chronic kidney disease dataset which is archived at the machine learning repository of the University of California Irvine (UCI). The classification approaches have been analyzed in this study including deep neural network(DNN), support vector machine (SVM) and K-nearest Neighbor (KNN). To fulfill this study, the missing values have been imputed with different techniques according to the characteristics of the features and relations among them. Hyperparameters of each algorithm have been tuned through experiments. The proposed approach has been evaluated with the best-tuned parameter. The assessment has done based on different performance metrics such as train–test sensitivity, accuracy, f-measure, specificity and Matthews correlation coefficient (MCC). The empirical result shows that SVM and KNN have enhanced accuracy, and DNN shows the most optimistic result with 100% accuracy compared to the existing.

Keywords Chronic kidney disease · Support vector machine · Deep neural network · K-nearest neighbors

1 Introduction

It is called chronic kidney disease when it fails to perform its regular operation. It is also called kidney failure. It refines our blood by removing wastes and spare fluid as urine. It has some very critical effects such as damage to the nerve and immune system that adversely reduce living standards and the chance of living. The number

Md. O. F. Goni (✉) · A. Matin · T. Hasan · Md. R. I. Sarker
Department of Electrical and Computer Engineering, Rajshahi University of Engineering and Technology (RUET), Rajshahi, Bangladesh

M. S. Arefin et al. (eds.), *Proceedings of the International Conference on Big Data, IoT, and Machine Learning*, Lecture Notes on Data Engineering and Communications Technologies 95, https://doi.org/10.1007/978-981-16-6636-0_1

of CKD patients increases due to diabetes, high blood pressure and unsound habit [1]. Gradual failure of the kidney leads to death. The number of CKD patients increases globally [2]. Hence, an early screening system is required.

In the recent world, there is a large number of available data that requires proper usage, otherwise it will be useless. Data mining is the technique to make the appropriate use of large data. Using statistical methods and machine learning (ML) algorithms, it discovers exciting patterns and interactions in data which helps in making useful predictions [3]. In this paper, we use different ML methods that include SVM, KNN and DNN for early diagnosis of CKD in CKD dataset recorded at the machine learning repository of the UCI.

In real-world data, there is a probability of missing and erroneous values that can fail the actual purpose. Adequate data cleaning and missing value imputation techniques are required. It has a crucial impact on model accuracy. CKD dataset is imputed using different techniques. Again, the best feature subset is selected to make the model more efficient. Performance of all the models is assessed with some performance metrics including train–test precision, accuracy, recall, specificity, sensitivity, f-measure and MCC.

This study is structured as follows. In Sect. 2, existing methods on CKD patients classification are discussed. Section 3 represents the detailed explanation of CKD dataset as well as the proposed methods. Section 4 illustrates the result and discussion of this study. Section 5 implies the conclusion.

2 Literature Review

Charleonnan et al. [4] have performed a comparative study of SVM, KNN, decision tree (DT) and logistic regression (LR) classifiers for predicting CKD. SVM has achieved the highest accuracy of 98.3%.

Salekin and Stankovic [5] have analyzed SVM, KNN and ANN with and without imputing the missing values for classification. Feature selection techniques have also been employed. Among them, KNN has achieved the highest prediction accuracy of 0.993 in terms of $F1$-measure with missing values imputation.

Polat et al. [6] have analyzed the effect of feature selection on the performance of a ML classifier for CKD patients classification. To conduct the analysis, SVM has been examined with different feature subset searching methods and evaluators. Among all the combinations, SVM obtained a great accuracy of 98.5% where the filter method and the best first search have been considered as feature subset evaluator and the feature selector, respectively.

Sara and Kalaiselvi [7] have introduced a hybrid feature selection (FS) technique HWFFS, which combines the warper and the filter method of FS. With and without HWFFS, a performance comparison of NB, SVM and ANN has been conducted. SVM-HWFFS has achieved the highest accuracy of 90%.

Almansour et al. [8] have performed a comparative study of not only the most prominent but also strong classifiers, ANN and SVM in the diagnosis of CKD patients. All the missing values have been filled with the mean value of associated features. ANN has achieved the highest accuracy of 99.75% while SVM has also attained a great accuracy of 97.75%.

3 Materials and Methods

This study has been conducted into four stages, data imputation, feature subset selection, training & testing of ML model and assessment using performance metrics on CKD dataset recorded at the machine learning repository of the UCI. All the stages are discussed in brief in the next sections. Figure 1 represents the steps of this study.

3.1 Dataset Description

The chronic kidney disease dataset used in this research is openly accessible from the machine learning repository of UCI. It has 400 instances, 24 features, 11 numerical and 13 nominal features. The nominal features are albumin, anemia, appetite,

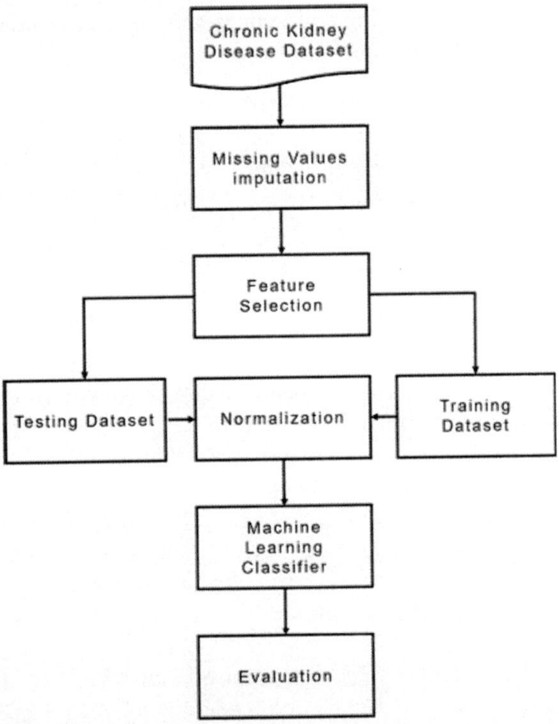

Fig. 1 Proposed methodology

bacteria, coronary artery disease, diabetes mellitus, hypertension, pus cell, pus cell clumps, pedal edema, red blood cells, specific gravity and sugar. The numerical features are age, blood glucoses, blood pressure, blood urea, hemoglobin, packed cell volume, potassium, red blood cell count, serum creatinine, sodium and white blood cell count. These 400 samples contain 250 (62.5%) CKD patients and 150 (37.5%) non-CKD patients. The dataset has 1012 (10.1%) missing values.

3.2 Dataset Cleaning

It is very important to prepare data before implementing the data for a classification model. CKD dataset has only 160 instances that do not have any missing attributes. That means more than 50% of instances have missing values. There are three things to do with missing values.

(a) By using a strong and rapid nonlinear classifier that can manage missing values as well as noisy data together. But data cleaning should be such that it can be applied to any model.
(b) Eliminate the missing values, but there is a problem of losing a large amount of data that may contain important data patterns.
(c) Imputation of missing value . But due to poor imputation, a classifier may be biased to the imputed value. If missing values are imputed properly, it can overcome all the problems.

Missing values can be imputed in different ways.

Imputation using mean value
All the missing values can be imputed with the mean value of the corresponding feature. In this experiment, age, Bp, Sg, Bgr, Bu, Sc, sod and pot are imputed using this method.

Imputation using maximum occurrence
This method can be used for missing nominal data types. In this study, this technique is used for all the missing data of the nominal type using the most frequently occurred value of the corresponding feature.

Imputation using regression model
To apply this technique, it is required to study the interaction between features and find relation among the features if there is any. Then fit a regression model using the related features and predict the missing values. In Fig. 2, it shows the interaction of bu, hemo and pcv with rc. We can fit a regression model to impute the missing values.

In this study, at first, some of the missing values of rc are imputed with the predicted value of the linear regression model fitted by hemo and rc. Then the second part of

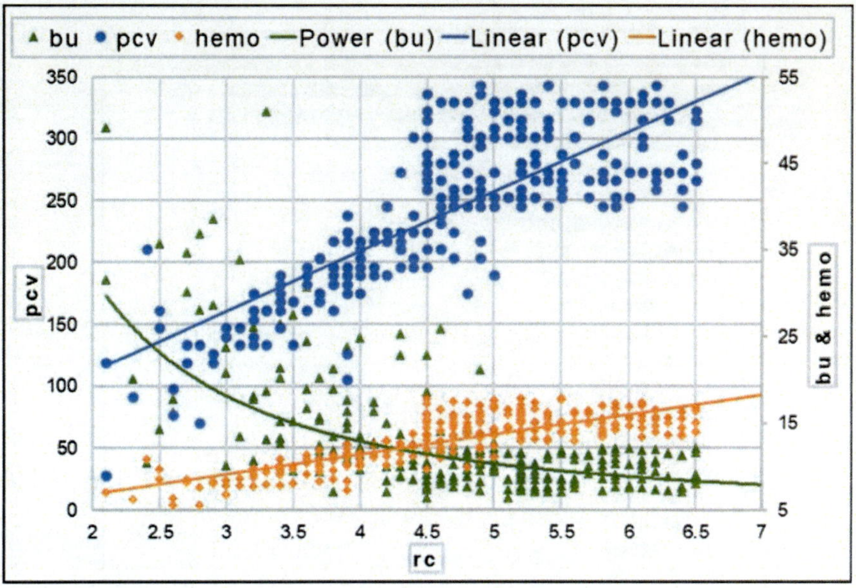

Fig. 2 Interaction of bu, hemo and pcv with rc

the missing values is imputed in the same way, and the model is fitted by rc and pcv. The rest of the missing values of rc are fitted with the mean of rc. And finally, missing values of pcv, hemo and bu are imputed using a regression model fitted using rc and corresponding features.

3.3 Feature Selection

All the dataset features do not have enough impact on decision making. It is better to remove less important and irrelevant features, and this technique is called feature selection [9]. The efficiency of a classifier along with its effectiveness is greatly impacted by feature selection. It reduces the time consumption of a model and makes it faster and efficient [10].

In this study, to select the best features, different feature subsets are created according to the feature importance and tested with the models. The best feature subset consists of 11 features. Feature importance can be calculated in many ways. In this paper, a random forest model is used to do this job, and the best feature subset is selected using the feature importance shown in Fig. 3.

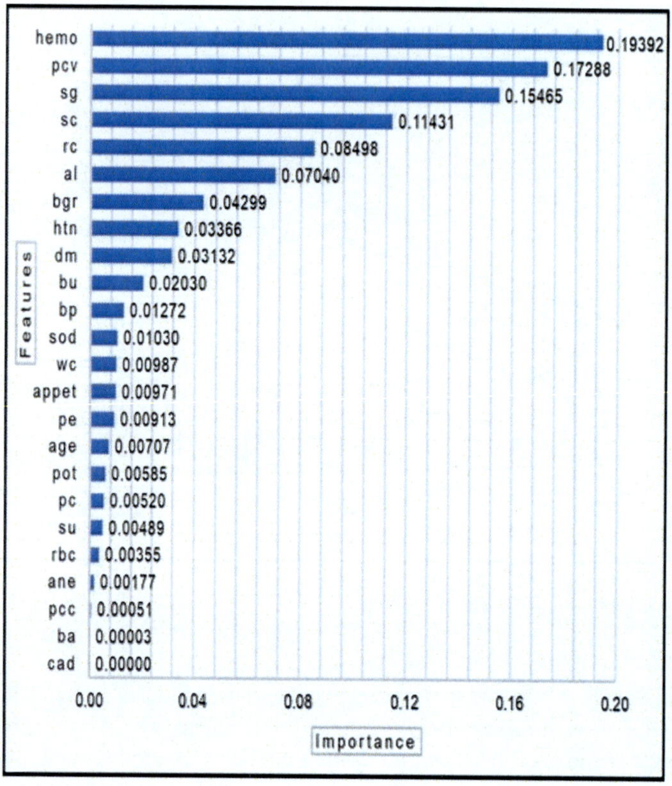

Fig. 3 Feature importance

3.4 *Normalization*

As part of data preparation for machine learning, technique of normalization is used sometimes to adjust the numeric column values to a standard scale in the dataset, without altering the differences in the value ranges [11]. Each dataset does not need normalization for machine learning. It is applied only when dataset features have different ranges. There are several approaches to normalization that include z-score, logistic, min–max, tah, lognormal, etc. To conduct this study, min–max normalization has been used. Mathematically, it can be expressed as follows:

$$\text{reshaped: } Z = \frac{Z - \min(Z)}{\max(Z) - \min(Z)} \tag{1}$$

where Z is the column required to normalize.

3.5 Performance Metrics

Mathematical expression of each performance metrics is given below.

$$\text{Accuracy} = \frac{(\text{TP} + \text{TN})}{(\text{TP} + \text{TN} + \text{FP} + \text{FN})} * 100\% \tag{2}$$

$$\text{Sensivity} = \frac{\text{TP}}{(\text{TP} + \text{FN})} * 100\% \tag{3}$$

$$\text{Specificity} = \frac{\text{TN}}{(\text{FP} + \text{TN})} * 100\% \tag{4}$$

$$\text{Presission} = \frac{\text{TP}}{(\text{TP} + \text{FP})} \tag{5}$$

$$\text{Recall} = \frac{\text{TP}}{(\text{TP} + \text{FN})} \tag{6}$$

$$F - \text{measure} = \frac{2 * (\text{Presission} * \text{Recall})}{(\text{Presission} + \text{Recall})} \tag{7}$$

$$MCC = \frac{(\text{TP} * \text{TN}) - (\text{FP} * \text{FN})}{\sqrt{(\text{TP} + \text{FP})(\text{TP} + \text{FN})(\text{TN} + \text{FP})(\text{TN} + \text{FN})}} \tag{8}$$

where, true positive (TP) = counts CKD as CKD, true negative (TN) = counts not CKD as not CKD, false negative (FN) = counts CKD as not CKD, false positive (FP) = counts not CKD as CKD.

3.6 Classification Models

Three classification models (e.g., DNN, SVM and KNN) have been applied in our approach.

Deep Neural Network (DNN) A deep neural network is a feed-forward neural network with multiple layers [12]. Data travels only one direction, input layer to the output layer and between them there is hidden layer containing the different or same number of neurons [11]. All the neurons of one layer are linked to the neurons of the forward layer. DNN does not contain any loops or cycles. Learning the weights of each neuron is done using the backpropagation method [13]. Each neuron has an activation function.

Support Vector Machine (SVM) For both classification and regression, the support vector machine is a strong supervised machine learning algorithm [14]. It is widely used due to its simplicity and flexibility. Using hyperplanes, the dataset is divided into classes. It can be applied for both cases: linearly separable data and linearly non-separable data. The major objective of SVM is to locate a maximum marginal hyperplane (MMH) by using a different kernel function.

K-Nearest Neighbors (KNN) K-nearest neighbor (KNN) is a supervised machine learning algorithm. It can be used for both the classification problem and the regression problem. It is called the lazy Learning algorithm because instead of learning a discerning function, it memorizes the training data [15]. It is also a nonparametric method because the number of parameters increases with the size of the training dataset [16]. Here, only the unknown parameter is k, which is a small integer number.

4 Result and Discussion

The classification task is performed using three ML models including KNN, SVM and DNN with data cleaning and selecting the best feature subset. After observing the type, characteristics and relation between each other, the missing values are filled in three different ways including mean value, maximum occurred value and using a regression model. Best feature subset is selected using the feature importance shown in Fig. 3 which consists of the first 11 features. The dataset has been split into a ratio of 70:30 for training and testing purposes.

For better outcomes, hyperparameter tuning has been performed for the DNN model. The best parameters obtained in the DNN model consist of 11 input nodes, while the number of hidden layer is 5 having each 50 nodes. ReLu has been used for each node as activation. HeNormal has been used as initializer for the weight initialization of hidden layers. Adam has been used as model optimizer and MSE as loss function.

KNN and SVM, both of them, have achieved improved accuracy of 100% train and 99.17% test accuracy. They show a specificity of 97.92% and sensitivity of 100%. DNN has achieved the highest accuracy of 100%. Figure 4 shows the training accuracy, test accuracy, specificity and sensitivity of KNN, SVM and DNN. Figure 5 shows the recall, precision, f-measure and mcc score of them.

In Table 1, a comparative investigation of proposed approaches with the existing methods is represented. The DNN model outperforms the existing methods. The empirical result and statistical analysis indicate the effect of adequate data cleaning tasks.

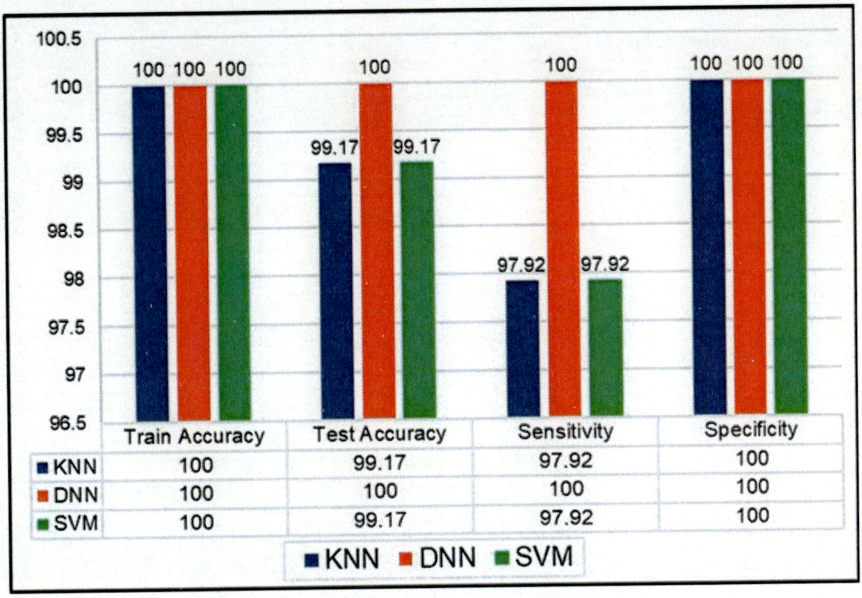

Fig. 4 Graphical comparison of classifiers in terms of train accuracy, test accuracy, sensitivity and specificity of KNN, DNN and SVM

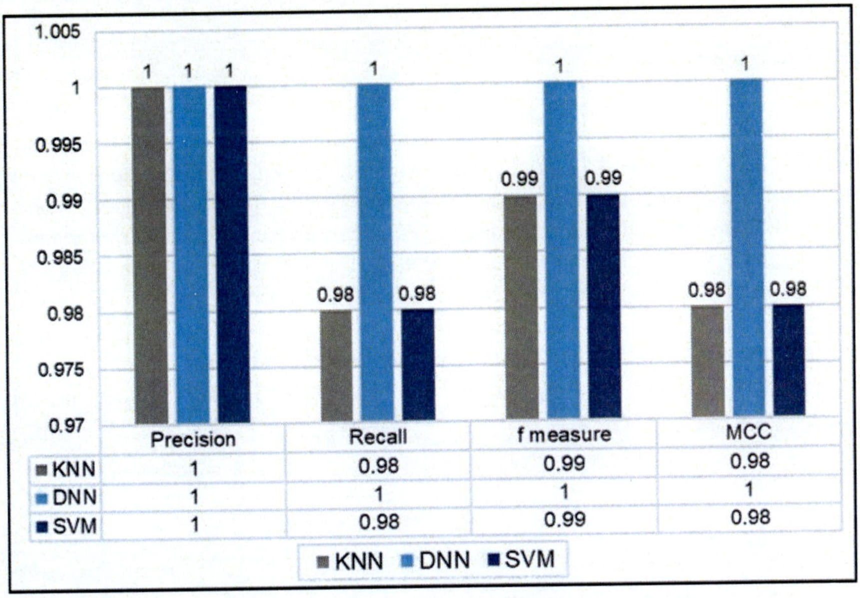

Fig. 5 Graphical comparison of classifiers in terms of precision, recall, F-measure and MCC of KNN, DNN and SVM

Table 1 Comparison of proposed approaches with existing methods

Author	Model	Accuracy (%)
Charleonnan et al. [4]	SVM	98.30
	LR	96.55
	DT	94.80
	KNN	98.10
Sara & Kalaiselvi [7]	NB	66.67
	SVM	73.33
	ANN	70
	NB-HWFFS	76.67
	ANN-HWFFS	78.79
	SVM-HWFFS	90
Almansour et al. [8]	ANN	99.75
	SVM	97.75
Polat et al. [6]	SVM	98.5
Proposed approach	**DNN**	**100**
	SVM	99.17
	KNN	99.17

As the proposed model achieved the highest accuracy in this CKD dataset, the result has been marked as bold in the table

5 Conclusion

An automatic system that serves early detection with precision is crucial in medical diagnosis and very helpful for medical personnel. It can reduce the mortality rate of CKD patients. Again, the effectiveness of such a system depends on proper preprocessing (like missing value imputation, outlier detection, etc.). This paper represents a robust and effective deep neural network with decent imputation of missing values. The best feature subset is selected based on their importance to make the model more efficient. SVM and KNN provide improved accuracy. DNN has achieved 100% accuracy and also outperformed the existing methods. It can be used to classify CKD patients more accurately. In future, a GUI can be designed to use this model in real life.

References

1. Chen Z, Zhang X, Zhang Z (2016) Clinical risk assessment of patients with chronic kidney disease by using clinical data and multivariate models. Int Urol Nephrol 48(12):2069–2075
2. Zhang L, Wang F, Wang L, Wang W, Liu B, Liu J, Chen M, He Q, Liao Y, Yu X et al (2012) Prevalence of chronic kidney disease in china: a cross-sectional survey. Lancet 379(9818):815–822

3. Sarker MRI, Matin A (2021) A hybrid collaborative recommendation system based on matrix factorization and deep neural network. In: 2021 international conference on information and communication technology for sustainable development (ICICT4SD), pp 371–374. https://doi.org/10.1109/ICICT4SD50815.2021.9397027
4. Charleonnan A, Fufaung T, Niyomwong T, Chokchueypattanakit W, Suwannawach S, Ninchawee N (2016) Predictive analytics for chronic kidney disease using machine learning techniques. In: 2016 management and innovation technology international conference (MITicon), IEEE, pp MIT–80
5. Salekin A, Stankovic J (2016) Detection of chronic kidney disease and selecting important predictive attributes. In: 2016 IEEE international conference on healthcare informatics (ICHI). IEEE, pp 262–270
6. Polat H, Mehr HD, Cetin A (2017) Diagnosis of chronic kidney disease based on support vector machine by feature selection methods. J Med Syst 41(4):55
7. Sara SBV, Kalaiselvi K (2018) Ensemble swarm behavior based feature selection and support vector machine classifier for chronic kidney disease prediction. Int J Eng Technol 7(2.31):190–195
8. Almansour NA, Syed HF, Khayat NR, Altheeb RK, Juri RE, Alhiyafi J, Alrashed S, Olatunji SO (2019) Neural network and support vector machine for the prediction of chronic kidney disease: a comparative study. Comput Biol Med 109:101–111
9. Shaikh R (2018) Feature selection techniques in machine learning with python. https://towardsdatascience.com/feature-selection-techniques-in-machine-learning-with-python-f24e7da3f36e. Accessed 30 July 2020
10. Blum AL, Langley P (1997) Selection of relevant features and examples in machine learning. Artif Intell 97(1–2):245–271
11. Goni MOF, Matin A, Hasan T, Siddique MAI, Jyoti O, Hasnain FMS (2020a) Graduate admission chance prediction using deep neural network. In: 2020 IEEE international women in engineering (WIE) conference on electrical and computer engineering (WIECON-ECE). IEEE, pp 259–262
12. Goni MOF, Hasnain FMS, Siddique MAI, Jyoti O, Rahaman MH (2020b) Breast cancer detection using deep neural network. In: 2020 23rd international conference on computer and information technology (ICCIT). IEEE, pp 1–5
13. Nahiduzzaman M, Nayeem MJ, Ahmed MT, Zaman MSU (2019) Prediction of heart disease using multi-layer perceptron neural network and support vector machine. In: 2019 4th international conference on electrical information and communication technology (EICT). IEEE, pp 1–6
14. Hasan T, Matin A, Kamruzzaman M, Islam S, Goni MOF (2020) A comparative analysis of feature extraction methods for human opinion grouping using several machine learning techniques. In: 2020 IEEE international women in engineering (wie) conference on electrical and computer engineering (WIECON-ECE). IEEE, pp 272–275
15. Guo G, Wang H, Bell D, Bi Y, Greer K (2003) KNN model-based approach in classification. In: OTM confederated international conferences on the move to meaningful internet systems. Springer, pp 986–996
16. Zhang H, Berg AC, Maire M, Malik J (2006) SVM-KNN: Discriminative nearest neighbor classification for visual category recognition. In: 2006 IEEE computer society conference on computer vision and pattern recognition (CVPR'06). IEEE, vol 2, pp 2126–2136

Comparative Analysis of Machine Learning Techniques in Classification Cervical Cancer Using Isolation Forest with ADASYN

Fariha Iffath, Sabrina Jahan Maisha, and Maliha Rashida

Abstract Cervical cancer is a form of cancer that forms in the cervix area. Majority of this form of cancer are related to human papillomavirus infection. Cervical cancer is linked with a number of risk factors. It is important to consider the importance of cervical cancer test factors when categorizing patients based on the findings. In this paper, we have performed a comprehensive analysis on cervical cancer prediction using various machine learning algorithms. We have applied seven different machine learning algorithms which are artificial neural networks (ANN), logistic regression (LR), decision tree (DT), random forest (RF), gradient boosting classifier (GBC), K-nearest neighbor (KNN), and AdaBoost (ADA) for predicting cervical cancer. Adaptive synthetic sampling method was used in this paper for filling out the missing data. For reducing the dimension of the dataset, linear discriminant analysis was used. In addition, we have used isolation forest outliers detection method for detecting outliers. From the comprehensive analysis, we have concluded that attains the maximum accuracy of 95%. We have performed our experiments on publically available cervical cancer dataset. This dataset is available in UCI machine learning repository. Overall, decision tree algorithm with outlier detection approach has performed comparatively well with comprehensive accuracy for predicting women exhibiting clinical symptoms of cervical cancer.

Keywords Cervical cancer · Machine learning · Outlier detection · Isolation forest

1 Introduction

The most common form of cancers that occur in the reproductive tract of a woman are gynecological cancers. Cervical cancer is one of the most critical gynecological

F. Iffath (✉) · S. J. Maisha · M. Rashida
Department of Computer Science and Engineering, Chittagong University of Engineering and Technology, Chittagong 4349, Bangladesh
e-mail: fariha@bgctub.ac.bd

S. J. Maisha
e-mail: sabrina@bgctub.ac.bd

M. S. Arefin et al. (eds.), *Proceedings of the International Conference on Big Data, IoT, and Machine Learning*, Lecture Notes on Data Engineering and Communications Technologies 95, https://doi.org/10.1007/978-981-16-6636-0_2

cancers among them. Cervical cancer develops in the cervix cell—the lower portion of the uterus that connects to the vagina. Various forms of sexually transmitted virus, the human papillomavirus (HPV), are a crucial factor for causing cervical cancer [7]. The mortality risk among women for cervical cancer is high due to a lack of awareness for early detection of cervical cancer [17]. According to a study, cervical cancer is the cause of 7.5% of female cancer deaths worldwide [14]. Early diagnosis of this life-threatening cancer can potentially reduce the death rate. Machine learning algorithms, for instance, applied to medical science data [16], are growing significantly due to their high efficiency in early predicting outcomes and making real-time life-saving decisions. Machine learning model can be used to predict the outcome of the biopsy by indicating the likelihood of the presence/absence of cervical cancer in patients. However, it is very difficult to predict the findings of the biopsy because there are a number of missing values in the dataset due to privacy concerns. In addition, the high dimension in the attributes and the presence of outliers in the dataset result in poor accuracy in the prediction result.

An outlier is an sample that lies in an abnormal distance from other samples provided in a dataset [1]. One probable reason of outlier is due to the experimental error, data collection error, etc. Outliers present in the dataset influence the mean and median values of the dataset [2]. It can also induce over-fitting of classification models. Removing outliers can also increase the accuracy of prediction result [5].

In this paper, we applied seven machine learning algorithms, namely random forest (RF), logistic regression (LR), artificial neural networks (ANN), gradient boosting classifier (GBC), decision tree (DT), K-nearest neighbor (KNN), and AdaBoost (ADA) in predicting cervical cancer. At first, we performed necessary data preprocessing steps including label encoding, missing value imputation, data balancing using ADASYN [10], dimensionality reduction using linear discriminant analysis (LDA) [20], outlier detection using isolation forest [8]. Then, we applied ML models after cleaning data and determined most efficient model for predicting cervical cancer. The contribution of our work can be summarized as follows:

- We have shown a higher prediction accuracy of cervical cancer after missing value imputation.
- We have implemented linear discriminant analysis (LDA) for dimensionality reduction and Adaptive Synthetic sampling approach (ADASYN) to properly balance the dataset and got better outcome.
- Using isolation forest algorithm, we have effectively identified outliers in cervical cancer datasets and removed them from the dataset.
- We have performed a variety of cancer dataset studies to evaluate the efficacy of our outlier detection method.
- We have developed an effective classifier model for cervical cancer prediction using different ML techniques and evaluated the accuracy of the models with different evaluation approach to obtain higher performance efficacy.

The rest of the paper is organized as follows: Sect. 2 reviews related works of cervical cancer prediction. We present our proposed method of outlier detection in cervical cancer dataset and deployment of ML models in Sect. 3. Our experimental result is discussed in Sect. 4. The last Sect. 5 concludes the paper and outlines the future work.

2 Related Work

Various researches have been accomplished so far on outlier detection in dataset as well as cervical cancer prediction. Parikh and Menon [15] applied three distinct ML techniques, K-nearest neighbor (KNN), decision tree classifier (DT), and random forest (RF) to the cervical cancer dataset. They aimed at the variable number of features for each algorithm. Hence, observed that KNN outperforms the other two algorithms and, therefore, has the best AUC and F1-values of 88% and 94%, respectively. In this study, however, separate numbers of training and test data samples from cervical cancer datasets were used for each algorithm.

Abdullah et al. [4] developed a predictive model using two ML algorithms, random forest (RF), and support vector machine (SVM). The dataset used gene expression profiling, with 58 samples and 714 features. Their predictive model for cervical cancer diagnosis is 94.21% accuracy in case of a random forest algorithm. RF algorithm is, therefore, performed better than SVM in cervical cancer prediction.

Wu and Zhou [21] proposed two extended SVM methods, support vector recursive feature elimination (SVM-RFE), and support vector machine principle component analysis (SVM-PCA) to analyze malignant cancer samples. They examined cervical cancer dataset that included 32 risk factors and 4 target variables. The four targets were analyzed and labeled using the three SVM-based approaches, respectively. Hence, SVM-PCA was shown to have the highest accuracy in cancer prediction.

Sobar et al. [18] analyzed a dataset that was obtained by designing the questionnaire. Eight features were included, and nine questions were included on each feature. The questionnaire was approached to 72 respondents. Later, the data obtained from the respondent were analyzed using two ML methods, naive Bayes (NB), and logistic regression (LR), to estimate the probability of cervical cancer dependent on behavior and its determinant with accuracy of 91.67% and 87.5%, respectively. Therefore, they found naive Bayes (NB) had a higher rate of accuracy over logistic regression (LR) in prediction of cervical cancer risk factor.

Abdoh et al. [3] have used synthetic minority oversampling technique (SMOTE) in addition to random forest (RF) algorithm with two feature reduction methods, (i) recursive feature elimination (RFE) and (ii) principal component analysis (PCA), to predict risk factors for cervical cancer. They used a sample composed of four target variables and 32 risk factors. The four target variables used in their work were cytology, Schiller, Hinselmann, and biopsy. After comparing the results, they fond that the combination of a random forest classification technique with SMOTE increases the efficiency of the classification. They also found that RFE and PCA can be used for accurate prediction for diagnosis of patients with cervical cancer. But, SMOTE-RF gives better accuracy than RFE and PCA in cervical cancer predictions.

Muhammad et al. [11] developed a model of cervical cancer prediction for early cervical cancer prediction. Two outlier detection methods used such as density-based spatial clustering of noise applications (DBSCAN) and isolation forest (iForest). As oversampling method, they used synthetic minority oversampling technique (SMOTE) and SMOTE with Tomek link (SMOTETomek). At the end, random forest (RF) classifiers were deployed for prediction. They also contrasted their proposed model with other ML classifier techniques and found that RF had the maximum accuracy.

A great deal of research has been conducted on outlier identification in datasets. For instance, Mascaro et al. [12] presented an outlier detection method, using dynamic and static Bayesian network models to augment the efficiency of the detection. Sun et al. [19] identified numerous outlier detection methods that could be differentiated into univariate versus multivariate procedures along with parametric versus non-parametric procedures. Aggarwal and Yu [6] formulated a process capable of identifying low-dimensional representations that are locally sparse. As a consequence, their approach is ideal for outlier identification in high-dimensional data. Mohamed and Kavitha [13] implemented a model to identify the data of the sensor node in different categories, such as a local or cluster outlier or a network outlier, using a conventional support vector machine (SVM). Hautamaki et al. [9] proposed an indegree number (ODIN) algorithm based on K-nearest neighbor for outlier detection. A certain improvements are also made to the current KNN distance-based system.

Cervical cancer dataset is imbalanced with a substantial amount of missing values. However, recent findings have demonstrated that the cervical cancer dataset considered in different studies has eliminated instances of missing values and provided less priority in evaluating important attributes. Eventually, it is difficult to process with missing values in the dataset, to evaluate the exact features and to obtain better accuracy of estimation. Therefore, our proposed work is designed to face all these challenges.

3 Methodology

In this section, overall methodology of our proposed model will be described. Initially, we preprocessed the raw dataset for normalization of the features. Secondly, we impute missing values in features using decision tree (DT) classifier. Next, we apply an oversampling method to reduce data imbalancing. After that, we perform linear discriminant analysis (LDA) for dimentionality reduction. Following that, we use an approach for outlier detection and elimination using isolation forest tree algorithm. Hence, we evalute the efficacy of our proposed model in predicting cervical cancer. The step by step approach of our model is presented in Fig. 1.

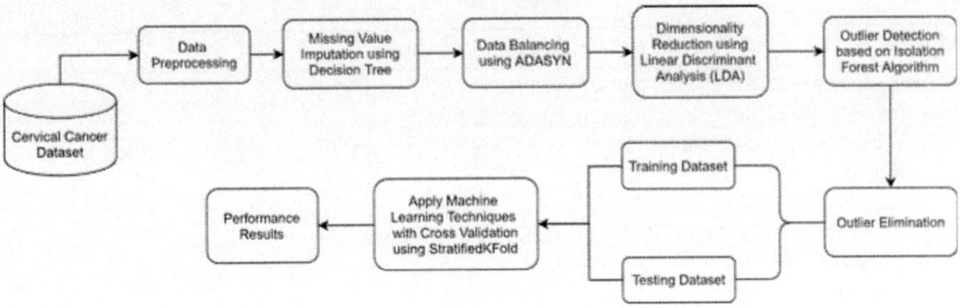

Fig. 1 Proposed methodology for cervical cancer prediction

3.1 Dataset and Data Preprocessing

In our work, we used a dataset collected from the "Hospital Universitario de Caracas" in Caracas, Venezuela. The dataset consists of 858 patients' demographic statistics, activities, and historical health records. This dataset includes 35 features as well as a target feature. The target feature is cervical biopsy prediction, which is typically performed due to detection of an abnormality during cytology (detect abnormal cells in the cervix). As a result, the biopsy prediction acts as an indicator/diagnosis of cervical cancer.

In the data preprocessing approach, initially, it is observed that the features "STDs: time after first diagnosis" and "STDs: time after last diagnosis" have approximately 85% null values. As a result, we remove those columns. Besides this, the features "smokes" and "first sexual intercourse" have a few missing value; therefore, we eliminate those missing records from the two columns. Finally, we split the features into numerical and categorical sections.

3.2 Missing Value Imputation

Medical sector being the most sensitive arena playing significant role in human lives needs proper handling of data values. Moreover in real world, missing of real data values is a common well as challenging factor to deal with. Traditional statistical methods such as mean, median, or mode technique implementation in case of data imputation method have not shown satisfactory results. So in the process of filling up the fields of missing data in our proposed system, machine learning models have been used. Algorithm 1 shows the procedure for missing value imputation.

Algorithm 1 Imputation of data values in columns

procedure MISSING VALUE IMPUTATION
    Identify the columns with missing values as Y
    Create independent column list X excluding the columns that requires imputation Y
    Repeat
    for $i \leftarrow 0$ to X **do**
        if row == null **then** Fill up row using $Median$ or $Mode$
        end if
    end for
Set TestData := Missing values of Y and TrainData := Filled values of Y
Build a ML model:
    $training \rightarrow$ filled values of Y & $prediction \rightarrow$ missing values of Y
Implement ML model according to column data type:
    $DecisionTreeRegressor \rightarrow (type)$Numerical Columns
    $DecisionTreeClassifier \rightarrow (type)$Categorical Columns
end procedure

3.3 *Oversampling Technique for Imbalanced Data*

Dataset sometimes includes patient data that impact the output to a greater extent not being an important factor to consider resolving that issue. Furthermore, often in real world, there can be certain instances having unequal distribution of number of data per class; i.e, values of some classes can be fewer in comparison to other existing classes with more amounts of data. This phenomenon in dataset is stated as "imbalanced." To fix such kind of disparity, oversampling measures must be undertaken. Usually, the actual data values of classes having lesser data are duplicated till being equivalent to the higher number of data value classes. Though duplication of actual sample values may bring out models with higher accuracy but it can question out reliability of such a system whether it is able to predict complex and different types of sample cases. In this purpose, an advanced sampling approach named as "ADASYN" can outperform to show better results solving oversampling issue.

ADASYN is an Adaptive Synthetic algorithm that generates complex, rare, and different artificial data rather not representing replica of the actual dataset in minority classes. The core strategy of this method is to determine the weighted distribution based on the level of difficulty to train each existing instances per minority classes. Using ADASYN, our proposed system can achieve more efficiency over data learning resolving two important factors (a) scaling down of the impact of biasness introduced due to imbalance and (b) alteration of the classification decision boundary by means of adaptive learning of rare cases. According to the data distribution of the samples, this approach processes by simultaneously and automatically fine-tuning the weights and adaptive learning mechanism.

3.4 Linear Discriminant Analysis (LDA)

Linear discriminant analysis is a technique that is used for dimensionality reduction of supervised multi-class classification problems. The prime concern is to reduce the "curse of dimensionality" by transforming the higher space features into a lower dimensional space assuring good separability among the labeled classes. LDA basically focuses on two points: (a) to ascertain the parameters that associates a certain class more precisely and (b) to build up a good model for separating the groups. The features that distinguish this kind of modeling technique is its ability to perform data categorization, computation of direction through the axes, and to outline class separability in unambiguous and explicit measure to solve more accurately multi-classification cases.
Through the execution of LDA for the datasets of cervical cancer, feature redundancy and dependency can be eliminated to a greater extent. Moreover, it analyzes the labeled classes to a deeper form in order to extract features to define the output classes more definitely.

3.5 Isolation Forest-Based Outlier Detection

Isolation forest [19] is an outlier detection technique which aims on segregating outliers. Usually, outlier detection algorithms concentrate on creating a profile of "normal" instances, after which outliers are reported in the dataset as those that do not comply to the normal profile. Whereas, isolation forest directly isolates outlier points in the dataset. Isolation forest is based on two properties of outlier data points in a sample. They are made up of fewer instances and have feature values that vary greatly from those of "normal" instances. They constitute the minority.

The outlier score of an instance is calculated by observing that the arrangement of "iTrees," which is similar to binary search trees. The estimated average $h(x)$ for the termination of the outlier node is the computed using the following equation $c = Z\ m = x\ n = y\ x = m$

$$Z(x) = \begin{cases} 2H(x-1) - \frac{2(x-1)}{x} & \text{for } x > 2 \\ 1 & \text{for } x = 2 \\ 0 & \text{otherwise} \end{cases} \quad (1)$$

In the above equation, n refers to the size of testing data, m refers to the size of sample set, and H denotes the harmonic number. This can be estimated by

$$H(i) = \ln(i) + 0.5772156649. \quad (2)$$

Here, the value of $c(x)$ means the average $h(m)$ for a given x. Later, it is used to normalize $h(m)$ and obtain an estimation of the outlier score for a given instance x:

$$s(m, x) = 2^{\frac{-E(h(m))}{c(x)}} \tag{3}$$

where $E(h(m))$ denotes the average value of $h(m)$ from a set of "iTrees". Finally, if the value of s instance m is assigned to outlier is near to one, else if the value of s is smaller than 0.5, then m is considered to be in normal instance. After identifying outliers, isolation forest is used to remove all the outliers to evaluate the system's improved performance in terms of classification accuracy. We employed seven traditional machine learning classification methods, which are gradient boosting classifier (GBC), artificial neural network (ANN), K-nearest neighbor (KNN), random forest (RF), logistic regression (LR), decision tree (DT), and AdaBoost classifier (ADC) to predict cervical cancer.

4 Experimental Evaluation

In this section, we will describe the experimental analysis of this work. We have used performance metrics precision, recall, accuracy, and F1-score to test our model.

4.1 Result Analysis on the Original Dataset

In the first experiment, we have tested all the mentioned method on the original dataset. From Table 1, we can see that ANN gives the highest accuracy, precision, and F1-score than other used algorithms, and the percentages are 90%, 93%, and 95%, respectively, whereas LR gives the highest recall rate of 98%. LR algorithm gives better accuracy after ANN, and the percentage is 72%. GBC gives the accuracy of 71%, in case of KNN which is 70%. Apart from this, ADA and RF give accuracy rate of 68% and 62%, respectively, and hence, DT gives the lowest accuracy rate (58%) on original dataset. The precision rate is observed between 70 and 75% in case of remaining six algorithms. Hence, ANN gives the better result than other algorithms.

4.2 Result Analysis on the Resampled Dataset

In the methodology section, we described that we resampled the dataset and imputed missing values using ADASYN algorithm. From Table 1, we can see that all the mentioned algorithms give accuracy, precision, recall, and F1-score above 90%. Among the mentioned algorithms, GBC and ADA achieve the highest accuracy rate which is 95%, and in case of LR, RF, and DT, the accuracy rate is 94%, and finally, KNN has the lowest accuracy of 93%. LR, GBC, and DT obtain the best precision

Table 1 Result analysis of different machine learning algorithms

Algorithm name	Type of dataset	Accuracy	Precision	Recall	F1-score
ANN	Original dataset	0.90	0.93	0.97	0.95
	After resampling	0.93	0.97	0.96	0.96
	After outlier detection	0.94	0.98	0.96	0.97
LR	Original dataset	0.72ty	0.72	0.98	0.83
	After resampling	0.94	0.99	0.95	0.97
	After outlier detection	0.95	0.97	0.98	0.97
RF	Original dataset	0.62	0.7	0.82	0.76
	After resampling	0.94	0.98	0.96	0.97
	After outlier detection	0.95	0.99	0.96	0.97
GBC	Original dataset	0.71	0.72	0.98	0.83
	After resampling	0.95	0.99	0.98	0.97
	After outlier detection	0.94	0.99	0.95	0.97
KNN	Original dataset	0.7	0.74	0.92	0.82
	After resampling	0.93	0.95	0.96	0.97
	After outlier detection	0.94	0.94	0.97	0.97
DT	Original dataset	0.58	0.72	0.67	0.69
	After resampling	0.94	0.99	0.95	0.97
	After outlier detection	0.96	0.99	1.00	0.98
ADA	Original dataset	0.68	0.73	0.88	0.8
	After resampling	0.95	0.98	0.97	0.97
	After outlier detection	0.94	0.97	0.97	0.97

rate of 99%. RF and ADA attain 98% precision, and in case of ANN and KNN, the rate is 97% and 95%, respectively. The highest recall rate is achieved by GBC, and the percentage is 98%. After GBC, ADA obtain the better recall rate of 97%. The recall rate in case of ANN, RF, KNN is similar, and its percentage is 96%. Hence, LR and DT give the lowest recall score of 95%. Finally, it can be observed that all the algorithms give similar F1-score of 97% except ANN. In case of ANN, the rate is 96%. Hence, we can say that after resampling, GBC performs better than remaining six algorithms.

4.3 Result Analysis on the Dataset After Removing Outliers

Using isolation forest algorithm, we have removed all the outliers to improve the classification accuracy of the machine learning models. We can observe that the classification accuracy of cervical cancer detection gained substantial improvement after removing outliers except for ADA and GBC. The proposed method shows improved performance in terms of accuracy, precision, recall, and F1-score values. Table 1 shows that DT classifier attains the highest accuracy, precision, recall, and F1-score, and the percentages are 96%, 99%, 100%, 98%, respectively. The accuracy rate of ANN, GBC, KNN, and ADA is 94%. LR and RF give accuracy of 95%. After analyzing the precision score, it can be observed GBC, RF also give the highest precision rate of 99% as the DT classifier. From the remaining classifiers, ANN gives better precision rate of 98%. LR and ADA obtain similar precision rate of 97%. Lastly, KNN gives the lowest precision score of 94%. From the recall score, we can observe that LR classifier achieves the highest recall rate of 98% after DT. After LR, ADA and KNN obtain the highest percentage of recall, and it is 97%. Apart from these, ANN and RF give recall rate of 96%, and GBC obtains the recall rate of 95% which is the lowest rate till now. Lastly, apart from DT, all the classifiers give F1-score of 97%.

Hence, we can come to a conclusion that all the algorithms are giving the highest accuracy, precision, recall, and F1-score after detection of outlier except GBC and ADA, and the overall performance of DT is better than the remaining classifiers (Fig. 2).

5 Conclusion

In this paper, we present an approach to predict biopsy result of cervical cancer. In this approach, at first, we imputed decision tree, a machine learning algorithm, for missing value imputation. Then, we have done data balancing using an oversampling method called Adaptive Synthetic algorithm for data balancing, and following that, we have reduced the dimension of data using linear discriminant analysis method to eliminate data redundancy and dependency. After that, we have used isolation

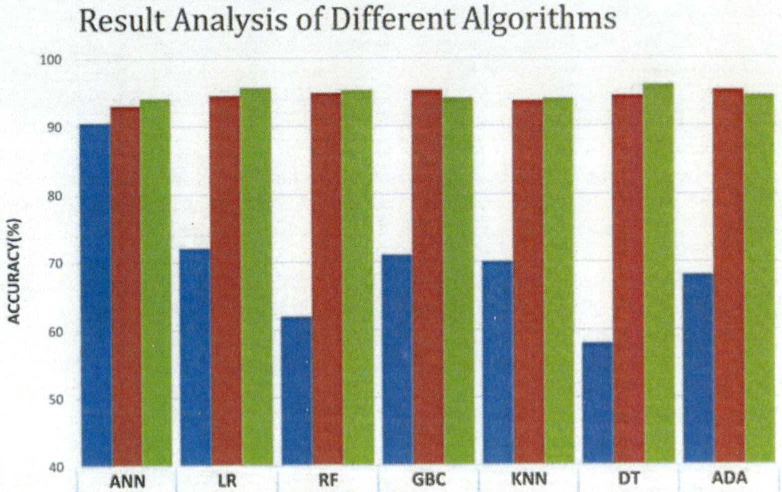

	ANN	LR	RF	GBC	KNN	DT	ADA
Original	90.48	72	62	71	70	58	68
After Sampling	93	94.44	94.84	95.23	93.65	94.44	95.23
After Outlier	94	95.63	95.23	94.05	94.04	96.03	94.44

Fig. 2 Performance analysis of various machine learning algorithm in original dataset, resampled dataset, and outlier detected dataset

forest algorithm for the detection of outliers in the features and removed outliers for the better performance of our method. We have used seven machine learning classifiers for the prediction of cervical cancer and have found that the performance of decision tree classifier is superior while considering the overall evaluation metrics. Our model would be a cost-effective solution for low- and middle-income people. Apart from that, this model will help medical experts predict cancer more accurately than traditional approaches. Furthermore, this model may help in the faster diagnosis of cancer in its early stages. We will try to use more social, cultural, and eating-habit-related features in the future. For performance enhancement, our future work will include utilizing CT-scan images of cervix and extracting the features using deep learning architectures.

References

1. In: Intelligent computing & optimization, conference proceedings ICO 2018. Springer, Cham. ISBN 978-3-030-00978-6
2. In: Intelligent computing and optimization, proceedings of the 2nd international conference on intelligent computing and optimization 2019 (ICO 2019). Springer International Publishing. ISBN 978-3-030-33585-4
3. Abdoh, S.F., Rizka, M.A., Maghraby, F.: Cervical cancer diagnosis using random forest classifier with smote and feature reduction techniques. IEEE Access 6, 59475–59485 (2018)

4. Abdullah A, Sabri N, Khairunizam W, Ibrahim Z, Mohamad Razlan Z, Abu Bakar S (2019) Development of predictive models for cervical cancer based on gene expression profiling data. IOP Conf Ser Mater Sci Eng 557:012003. https://doi.org/10.1088/1757-899X/557/1/012003
5. Acuna E, Rodriguez C (2004) On detection of outliers and their effect in supervised classification
6. Aggarwal CC, Yu PS (2001) Outlier detection for high dimensional data. In: Proceedings of the 2001 ACM SIGMOD international conference on management of data, pp 37–46
7. Asadi F, Salehnasab C, Ajori L (2020) Supervised algorithms of machine learning for the prediction of cervical cancer. J Biomed Eng 10. https://doi.org/10.31661/jbpe.v0i0.1912-1027
8. Abu Bakar Z, Mohemad R, Ahmad A, Mat Deris M (2006) A comparative study for outlier detection techniques in data mining, pp 1–6. https://doi.org/10.1109/ICCIS.2006.252287
9. Hautamaki V, Karkkainen I, Franti P (2004) Outlier detection using k-nearest neighbour graph. In: Proceedings of the 17th international conference on pattern recognition, 2004. ICPR 2004, vol 3. IEEE, pp 430–433
10. He H, Bai Y, Garcia E, Li S (2008) ADASYN: adaptive synthetic sampling approach for imbalanced learning, pp 1322–1328 . https://doi.org/10.1109/IJCNN.2008.4633969
11. Ijaz M, Attique M, Son Y (2020) Data-driven cervical cancer prediction model with outlier detection and over-sampling methods. Sensors (Basel, Switzerland) 20
12. Mascaro S, Nicholso AE, Korb KB (2014) Anomaly detection in vessel tracks using bayesian networks. International Journal of Approximate Reasoning 55(1):84–98
13. Mohamed MS, Kavitha T (2011) Outlier detection using support vector machine in wireless sensor network real time data. Int J Soft Comput Eng 1(2)
14. Nithya B, Ilango V (2019) Evaluation of machine learning based optimized feature selection approaches and classification methods for cervical cancer prediction. SN Applied Sciences 1:1–16
15. Parikh D, Menon V (2019) Machine learning applied to cervical cancer data. Int J Math Sci Comput 5:53–64. https://doi.org/10.5815/ijmsc.2019.01.05
16. Purba JH, Ratodi M, Mulyana M, Wahyoedi S, Andriana R, Shankar D, Nguyen P (2019) Prediction model in medical science and health care. Int J Eng Adv Technol 8:815–818. https://doi.org/10.35940/ijeat.F1158.0986S319
17. Purnami SW, Khasanah PM, Sumartini SH, Chosuvivatwong V, Sriplung H (2016) Cervical cancer survival prediction using hybrid of smote, cart and smooth support vector machine. AIP Conf Proc 1723(1):030017. https://doi.org/10.1063/1.4945075. https://aip.scitation.org/doi/abs/10.1063/1.4945075
18. Sobar, Machmud R, Wijaya A (2016) Behavior determinant based cervical cancer early detection with machine learning algorithm. Adv Sci Lett 22:3120–3123. https://doi.org/10.1166/asl.2016.7980
19. Sun L, Versteeg S, Boztas S, Rao A (2016) Detecting anomalous user behavior using an extended isolation forest algorithm: an enterprise case study. arXiv preprint arXiv:1609.06676
20. Shereena VB, David J (2015) Comparative study of dimensionality reduction techniques using PCA and LDA for content based image retrieval. Comput Sci Inf Technol 41–55. https://doi.org/10.5121/csit.2015.50905
21. Wu, W., Zhou, H.: Data-driven diagnosis of cervical cancer with support vector machine-based approaches. IEEE Access 5, 25189–25195 (2017)

Computer-Aided Cataract Detection Using Random Forest Classifier

Tasmina Tasin and Mohammad Ashfak Habib

Abstract Cataract is one of the most common causes of vision impairment, particularly in aged people. Early diagnosis and treatment of cataracts will prevent vision impairment from progressing to blindness. Medical facilities in remote areas are limited and ophthalmologists use a slit lamp to diagnose cataracts which is costly. As a result, a simple and effective auxiliary diagnostic method is suggested here. As our main focus is to detect cataracts from iris images, the iris area has been extracted using a contour detection process from the binary mask image. Two types of texture feature Gray-Level Co-occurrence Matrix (GLCM) and Histogram texture features are extracted from the images. Random Forest classifier is used for the automatic detection of cataracts. By achieving an overall accuracy of 97.92%, the proposed method classifies cataracts effectively.

Keywords Cataract detection · Image processing · Texture features · Machine learning · Random forest classifier

1 Introduction

One of the most essential sensory organs in the human body is the eye. Eye disease is a serious health problem that affects people all over the world. One of the most common eye diseases is cataracts. Since the dawn of civilization, cataract has been recognized as the most common cause of blindness. About 16–20 million people suffering from blinding cataracts. The clouding of the eye lens is known as cataract. Proteins in our lens tend to break down as we get older, and the lens becomes cloudy. Since cataracts normally grow slowly and do not obstruct vision at first, people may not even know they have one. About 90% of people have a cataract by the age of 65, and half of those aged 75–85 have lost any vision due to a cataract. The leading causes of blindness are

T. Tasin (✉) · M. A. Habib
Department of Computer Science and Engineering, Chittagong University of Engineering and Technology, Chattogram 4349, Bangladesh
e-mail: ashfak@cuet.ac.bd

M. S. Arefin et al. (eds.), *Proceedings of the International Conference on Big Data, IoT, and Machine Learning*, Lecture Notes on Data Engineering and Communications Technologies 95, https://doi.org/10.1007/978-981-16-6636-0_3

refractive error (123 million), cataract (65 million), glaucoma, corneal opacities, and diabetic retinopathy. According to the 2001 World Health Report, there are 20 million people worldwide who are bilaterally blind due to age-related cataracts. By the year 2021, the figure would have risen to 40 million. Cataract has a significant economic and public health impact, especially in developing countries. According to a World Health Organization (WHO) study on global blindness, developing countries account for roughly 90% of global cataracts, with more than 82% of all blindness occurring in people aged 50 and up. Seva and the International Agency for the Prevention of Blindness (IAPB) released a situational study in June 2018 to provide a snapshot of the eye care background in Bangladesh's Cox's Bazar district in Rohingya refugee camps and eye care suggests that they have a high rate of cataracts and need eye glasses. In refugee camps, 30% of those seeking eye care had cataracts. In 2013, Bangladesh had 1193 cataract surgeries per million, compared to 6353 in the United States.

There are several reasons of cataract formation. Most of the cataracts are age-related. Some inherited genetic disorders, other eye problems, previous eye surgery or medical conditions such as diabetes, long-term use of steroid medications, can cause cataracts to develop. To detect cataracts, ophthalmologists use different tools such as a slit lamp camera or an ophthalmoscope. However, there are some drawbacks to use these machines, such as the need for specialized training. The contribution of image processing techniques and machine learning approaches is more prevalent in the modern medical domain. Machine learning methods can automatically learn critical features and incorporate feature learning into the model-building process, resulting in a more accurate model. We have proposed an effective method for cataract detection using image processing and machine learning techniques which can be used as a backend model of an android application of cataract detection.

The rest of the paper is organized as follows: Sect. 2 presents the related works and the contributions of this study. Section 3 describes the methodology. Experimental outcomes are discussed in Sect. 4. Finally, Sect. 5 concludes the paper and discusses some considerations for future work.

2 Related Works

Previously, several works have been done on cataract detection. All of the studies applied different feature extraction and classification methods. In Zhang et al. [1] proposed a solution that aims to develop a system by using the Deep Convolutional Neural Network (DCNN) to detect and grade cataracts automatically. It also displays some of the feature maps at the pool 5 layer along with their high-order empirical semantic sense. The effect of the G-filter on removing unequal illumination of fundus images and the classification accuracy of DCNN were verified separately in this paper. The accuracy they got is 86.69%. Kaur et al. [2] developed an android application system for detecting cataracts. They used a microscopic lens in the mobile camera for capturing retinal images. For the rooted method implementation, they used a modified Neural Network. Network is a Java application for neural training that runs on a personal computer. In [3], Dong et al. provided a solution for the classification of cataract fundus image using deep learning. For feature extraction, a CNN model was trained. Five convolution layers were used. Following the feature extraction step, SVM and Softmax were used to classify the features extracted from the CNN model. They stated that because of the image's high quality, training the current sample took a long time, and if the number of images grows rapidly, they will face significant challenges. Harini et al. [4] developed a system of automatic cataract classification using the SVM classifier, and the fundus image was graded as non-cataract or cataract. The RBF network was used to grade the cataract image as mild or extreme. The dataset they used was not so large(60 images). In Fuadah et al. [5] used K-Nearest Neighbor (k-NN) as a classification model, which was implemented on an android smartphone. Their findings show that dissimilarity, contrast, uniformity are the best texture features to combine. The system's highest level of accuracy is 97.5% using a dataset of 160 images. Pathak et al. [6] presented a texture-based algorithm for detecting cataracts in adult human subjects from digital eye images. Experiments were performed on true color images obtained from a compact digital camera. Ik et al. [7] aimed to develop an alternate cataract screening solution with a flash enabled smartphone. In this method, a red reflex flash method was used. Eye image of the patient was captured in a dark room. Before capturing images, eyes were dilated using a drop and all of these steps are done by an ophthalmologist. They used a smartphone that must have greater light intensity ensuring no light sources are present when an image is captured. In Li et al. [8] established a new system for automatic grading of nuclear cataracts (AGNC) by merging clinical and image analytic knowledge. The grades of nuclear cataracts were predicted using support vector machine (SVM) regression and features taken from the lens anatomy.

In Zhang et al. [9] proposed a solution for cataract classification that used an approach dependent on residual focus. The B-Scan Eye ultrasound image dataset was used. An object detection network, three pretrained classification networks: DenseNet-161, ResNet-152, ResNet-101, and a model ensemble module made up the proposed model. Nayak et al. [10] developed a solution that aims to grade normal, cataract, and post-cataract images of the eye. Edge Pixel Count (EPC) and Object Perimeter of the optical eye images were extracted. For automatic classification, the same features were used in an automatic classifier such as SVM. They used a small dataset (174 images) and trained their model using NMR images. In Jagadale et al. [11] presented a system that detects cataracts at an earlier stage by combining slit lamp images from an ophthalmologist at an eye hospital with computer-aided image processing. The use of the Hough circle detection transform for lens detection and SVM for categorization improves overall accuracy. Miguel et al. [12] proposed a system that used ultrasound images to diagnose the nuclear cataract early, classify its intensity, and extract features from the collected signals, which were then used to train and test various classifiers to reliably distinguish normal and cataractous lenses. Jindal et al. [13] used image processing methods which were used on eye images to assess the presence and intensity of cataracts. On a dataset of images of eyes with differing degrees of cataract, two separate image processing algorithms

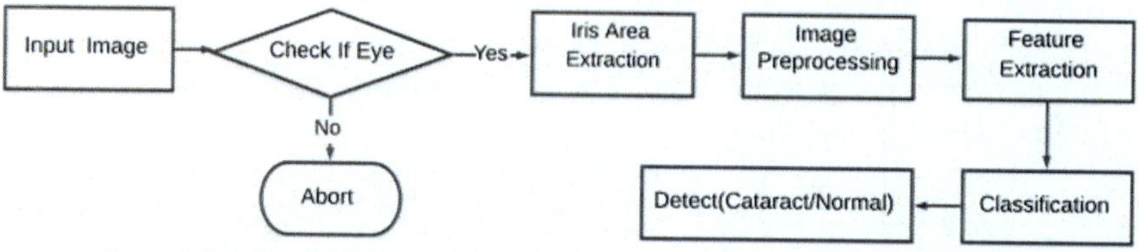

Fig. 1 Block diagram of the proposed system

were applied. In Hu et al. [14] proposed an algorithm whose major objective was to automatically classify the cataract severity based on the photometric appearance of the crystalline lens using smartphone-based slit lamp pictures from individuals with various cataract severity. The study provided a framework for automated nuclear cataract severity grading which is inspired by grafting, combines deep learning and classical feature extraction approaches, and achieved 93.48% accuracy. Hossain et al. [15] proposed an automatic cataract detection method based on Deep Convolution Neural Network that can distinguish between cataract and non-cataract images in the retinal fundus. For cataract detection, a trained classifier model based on Res-Net50 was utilized. They used 4000 retinal fundus images to train the cataract detection system and tested it on 1418 images obtaining a 95.77% accuracy.

Most of the discussed works are based on retinal-based fundus images. A fundus camera, also known as a retinal camera, is a specialized low-power microscope with an attached camera for photographing the eye's internal surface. In the medical field, cataract detection methods rely on either a fundus camera or a Digital Single-Lens Reflex (DSLR) camera, both of which are highly expensive.

In this paper, we have developed a computer-aided cataract detection technique based on iris images. We have used 3004 iris images which are captured by a normal camera. Several images preprocessing and machine learning techniques have been applied to the system, and we have obtained a satisfactory result.

3 Methodology

A framework for automatic cataract detection is shown in Fig. 1. Initially, data of eye images for both cataracts and normal eyes are collected. Then, the images are processed for formatting and removing the anomalies. Next, data is augmented for increasing the quantity of the dataset. After that, several features are extracted from the augmented data. Lastly, the Random Forest classifier is used for classifying the cataract or normal eye.

3.1 Collection of Images

We have collected images from google and Kaggle [16]. More than 300 images of cataracts and normal eyes were collected. The images were checked one by one manually, and 19 images from a total of 332 images were removed, because of blurriness and poor quality. We have collected 150 cataract images (positive samples) and 150 normal eye images (negative samples). Then, we augmented the data using three data augmentation techniques-random rotation, horizontal flip, change brightness and made a whole of 3004 images (cataract and normal).

3.2 Eye Detection

We have detected eye using Haar cascade Classifier. OpenCV provides a pretrained Haar Cascade model to detect eye from an image. The required XML file is loaded using the cv::CascadeClassifier::load method after a cv::CascadeClassifier has been developed. The detection process is then completed, yielding boundary rectangles for the detected eyes. Now, if the system detects eye from the image, we will proceed for the further steps to detect cataract. Otherwise, the procedure will be stopped.

3.3 Iris Area Extraction and Image Preprocessing

Since the images were unconstrained, some data preprocessing and data cleaning steps were applied in the first phase to prepare the input images and make them available for model input. To preprocess our images, we have followed several steps. The steps are demonstrated in Fig. 2. At first, the true color RGB images are converted to the grayscale image for reducing complexity. Then, the region of interest is extracted in two steps. In the first step, an optimal threshold value is selected, then a binary masked image is found from that. After that, morphological closing has been applied. To extract areas inside or near the iris, contours are used. In the second step, with the coordinates of the contour, the region of interest from the original image frame is extracted. All the images are resized into 256×256 so that every image has the same number of pixels. It means the images have 256 pixels in horizontal and 256 pixels in vertical dimensions. In order to denoise the images, the Gaussian filter has been applied. A 5×5 kernel is used for Gaussian blurring. After that, histogram equalization is done for enhancing the contrast of each of the images. All of this steps are shown in Fig. 3.

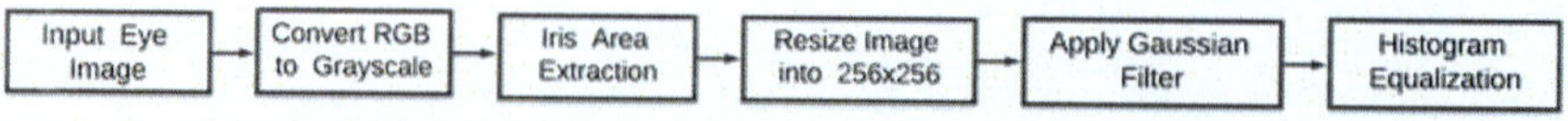

Fig. 2 Image preprocessing steps

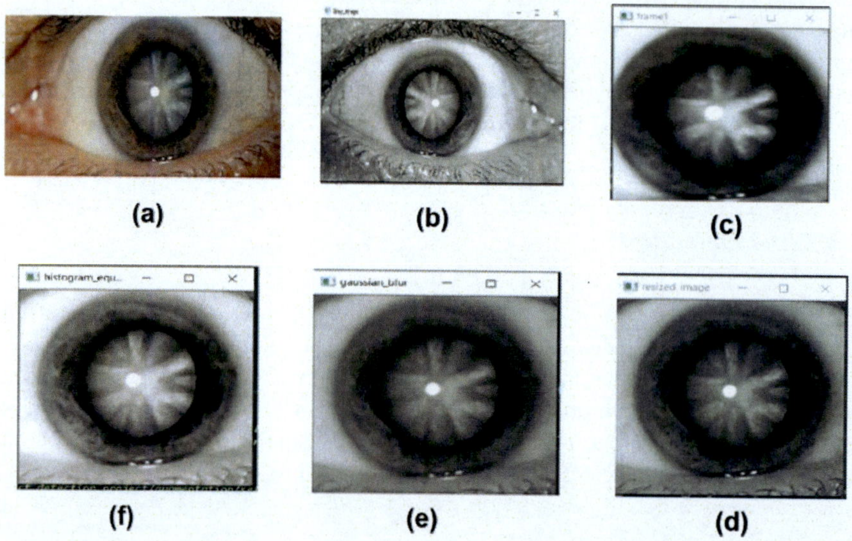

Fig. 3 Image preprocessing steps applied on a cataract eye image. **a** True color image; **b** grayscale image; **c** extracted iris area; **d** resize image (256 × 256); **e** Gaussian blur; **f** histogram equalization

3.4 Feature Extraction

Feature extraction is a dimensionality reduction method that reduces a large collection of raw data into smaller groups for processing. Two types of texture features have been used in our method. The first-order statistical texture method and the second-order statistical texture method are the two methods of statistical texture analysis. The first-order statistical texture method produces a function dependent on the histogram image's characteristics. In certain cases, the first-order statistical method cannot distinguish between images. So, we have applied both to extract features from our dataset. We have used two histogram features-mean intensity and standard deviation. Since the whitish color of cataract eyes comes from the lens, so cataract eyes have higher intensities than normal eyes. A low standard deviation value means that pixel values appear to be very similar to the average value and vice-versa. Figure 4 shows that cataract eyes have higher intensity values than the normal ones. The co-occurrence matrix is a second-order histogram that looks at the gray-level distribution of pixels in pairs. We have extracted five GLCM features named Contrast, Dissimilarity, Homogeneity, Correlation and Entropy.

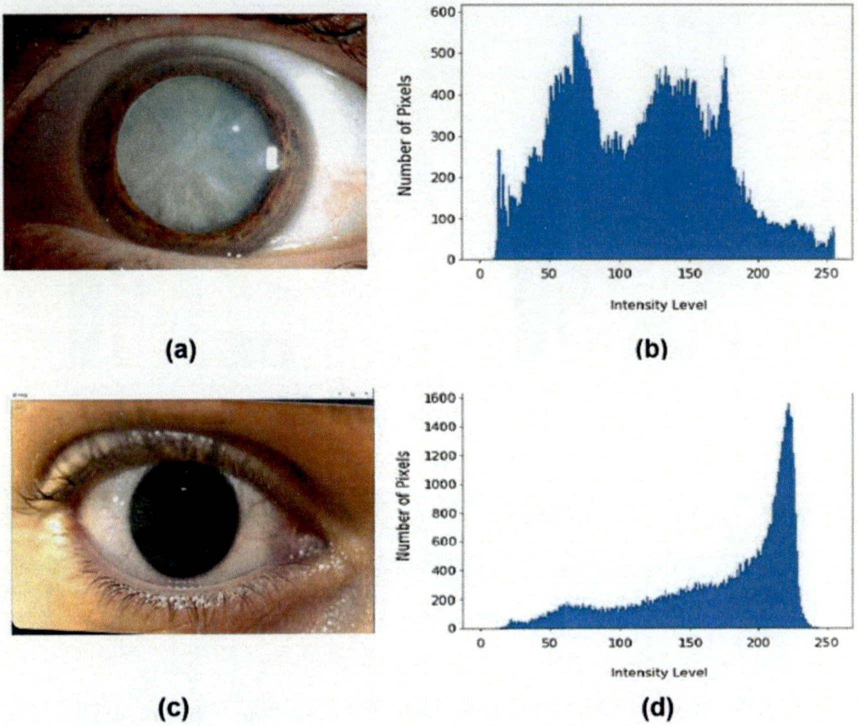

Fig. 4 Histogram analysis of both cataract and normal eye. **a** Cataract eye; **b** histogram of cataract eye; **c** normal eye; **d** histogram of normal eye

3.5 Classification by Random Forest Classifier

Random Forest classifier is a supervised learning algorithm. Random Forest classifier generates decision trees from randomly chosen data samples, obtain predictions from each tree, and vote on the best solution. Space is divided into classes depending on the training data classification; in this case, there are two classes: normal and cataract. The testing images are classified as Cataract or Normal based on their resemblance to two groups. When comparing the accuracy of the Random Forest classifier to that of the other classifiers, it has been found that the Random Forest classifier gives higher accuracy than the other classifiers.

4 Experimental Outcomes

The system is implemented on a computer having operating system windows 10 with a 2.50 Core i3-4100 processor and 4GB RAM. Python 3.6.7 (version) was used to develop the system. The dataset consists of 1502 cataract images and 1502 normal

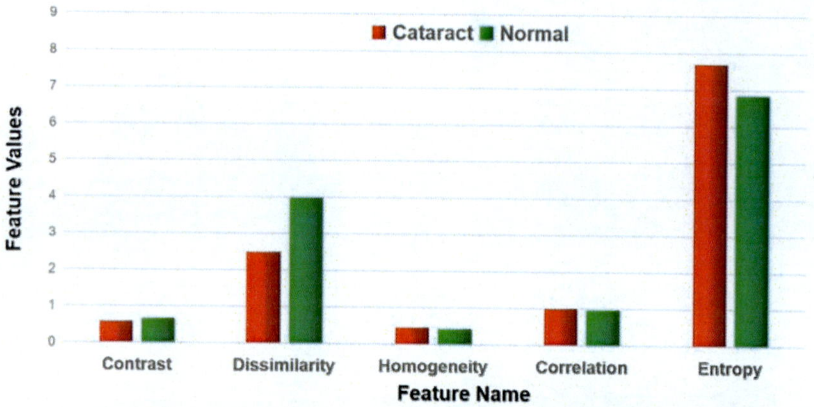

	Contrast	Dissimilarity	Homogeneity	Correlation	Entropy
Cataract	0.5835	2.5732	0.4619	0.9987	7.7213
Normal	0.6623	3.9815	0.4331	0.9638	6.8540

Fig. 5 GLCM texture features of both cataract and normal eye

images. 3004 image's feature values are saved in the Random Forest model file. The whole dataset is split into two parts, where 75% of the dataset is used for training the model and the remaining 25% is used for testing.

4.1 Analysis of Extracted Features

The related features of each cataract and normal images are extracted. Then, the extracted features are combined to create a matrix, which is stored as a matrix format. This matrix file is the classifier's input, the relevant features of the test image are also extracted and combined to create a matrix format, which is then used to test the image. Normal and cataract images are used to train the random forest classifiers. The classifiers classify the input images into normal or cataract based on the extracted features.

Figure 5 shows the distinct values of GLCM feature value for cataract and normal eye images. It is observed that cataract and normal class have distinct values for dissimilarity' (2.57 and 3.98) and entropy' (7.72 and 6.85) features, respectively. Whereas the other three features have almost the same values for both categories. The mean intensity value for a normal eye fluctuates between 40.83 and 108.33 and the standard deviation is found to be less than 70.65 and more than 21.32. Whereas the mean intensity value for a cataract eye is greater than 124.62 and the standard deviation is above 70. Figure 6 shows the distinct values of histogram features for cataract and normal eye images.

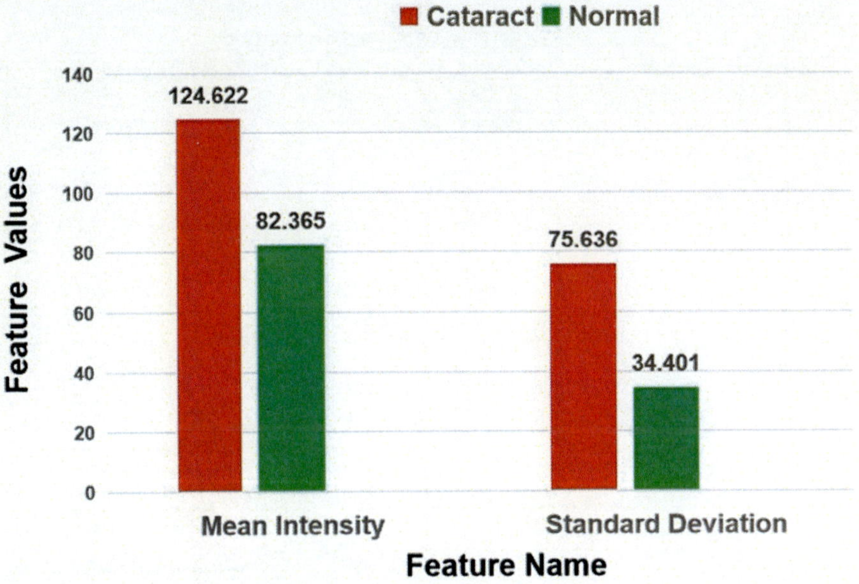

Fig. 6 Histogram texture features of both cataract and normal eye

Table 1 Confusion matrix of random forest model

		Target	
		Cataract	Normal
Testing result	Cataract	374	19
	Normal	7	350

4.2 Performance Analysis

True Positive, True Negative, False Positive, and False Negative are the four parameters that determine the accuracy of any machine learning algorithm. Confusion Matrix displays (Table 1) all four parameters, and other parameters such as recall, precision, F-Score, and Receiving Operating Characteristic (ROC) evaluated based on the confusion matrix which is shown in Fig. 7. Table 1 illustrated that our model classified 724 samples correctly among 750 samples. The number of the true positive and true negative are 374 and 350, respectively.

ROC Curve: We have applied k-fold stratified cross-validation for determining the actual accuracy. Data were folded five times. Then, the mean of this five-fold accuracy is counted. The ROC curve of this framework is represented in Fig. 7. From the curve, we can visualize that our true positive rate is very high and accuracy is around 97%.

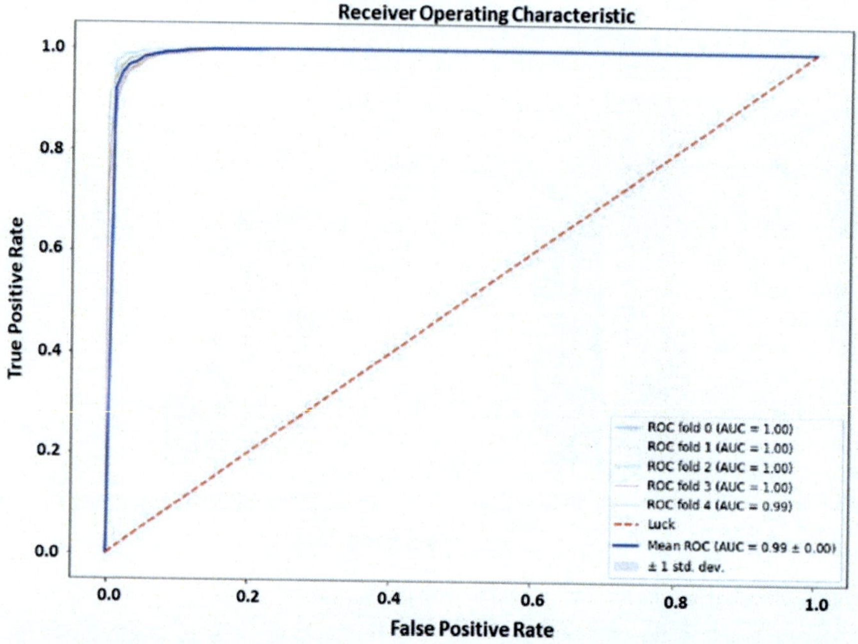

Fig. 7 ROC curve of the proposed model

Table 2 Comparison with existing method

Method	Precision (%)	Recall (%)	$F1$ score (%)	Accuracy (%)
GLCM + KNN [5]	90.46	91.76	90.07	90.23
Statistical parameter + SVM [11]	88.29	87.06	87.42	88.83
GLCM + histogram + random forest (proposed)	97.16	98.72	97.93	97.92

4.3 Comparison with Recent Works

The analysis of the result revealed that our model is an effective one to classify cataracts from iris images. Thus, we compared the performance of our model with some existing techniques to assess its effectiveness. We implemented previous methods on our dataset and recorded the accuracy. Table 2 shows a summary of the comparison.

The result shows that our model outperforms the previous techniques by achieving the highest accuracy (97.92%).

5 Conclusion and Future Works

Cataract is one of the most common diseases associated with aging, and as a result, many people suffer from it. The described system is based on a combination of machine learning and image processing algorithms. This technique can be used to diagnose cataracts in a user-friendly manner. The proposed methodology is compared to SVM and KNN, with the results indicating that the Random Forest-based methodology outperforms both. Due to the Covid-19 pandemic situation, we couldn't collect real-time images of cataract patients so, in the future, we will try to collect real-time images, and for further research, we will implement the model on a smartphone to develop an android application to make it portable and simple to use.

References

1. Zhang L, Li J, Zhang I, Han H, Liu B, Yang J, Wang Q (2017) Automatic cataract detection and grading using deep convolutional neural network. In: 2017 IEEE 14th international conference on networking, sensing and control (ICNSC), pp 60–65 (2017). https://doi.org/10.1109/ICNSC.2017.8000068
2. Kaur M, Kaur J, Kaur R (2015) Low cost cataract detection system using smart phone. In: 2015 international conference on green computing and internet of things (ICGCIoT), pp 1607–1609 (2015). DOIurl10.1109/ICGCIoT.2015.7380724
3. Dong Y, Zhang Q, Qiao Z, Yang J (2017) Classification of cataract fundus image based on deep learning. In: 2017 IEEE international conference on imaging systems and techniques (IST), pp 1–5 (2017). https://doi.org/10.1109/IST.2017.8261463
4. Harini V, Bhanumathi V (2016) Automatic cataract classification system. In: 2016 international conference on communication and signal processing (ICCSP), pp 0815–0819 (2016). https://doi.org/10.1109/ICCSP.2016.7754258
5. Fuadah YN, Setiawan AW, Mengko TLR, Budiman: mobile cataract detection using optimal combination of statistical texture analysis. In: 2015 4th international conference on instrumentation, communications, information technology, and biomedical engineering (ICICI-BME), pp. 232–236 (2015). https://doi.org/10.1109/ICICI-BME.2015.7401368
6. Pathak S, Gupta S, Kumar B (2016) A novel cataract detection algorithm using clinical data mapping. In: 2016 IEEE region 10 humanitarian technology conference (R10-HTC), pp 1–5 (2016). https://doi.org/10.1109/R10-HTC.2016.7906816
7. Ik ZQ, Lau SL, Chan JB (2015) Mobile cataract screening app using a smartphone. In: 2015 IEEE conference on e-learning, e-management and e-services (IC3e), pp. 110–115 (2015). https://doi.org/10.1109/IC3e.2015.7403496
8. Li H, Lim JH, Liu J, Mitchell P, Tan AG, Wang JJ, Wong TY (2010) A computer-aided diagnosis system of nuclear cataract. IEEE Trans Biomed Eng 57(7):1690–1698. https://doi.org/10.1109/TBME.2010.2041454
9. Zhang X, Lv J, Zheng H, Sang Y (2020) Attention-based multi-model ensemble for automatic cataract detection in b-scan eye ultrasound images. In: 2020 international joint conference on neural networks (IJCNN), pp 1–10 (2020). https://doi.org/10.1109/IJCNN48605.2020.9207696
10. Nayak J (2013) Automated classification of normal, cataract and post cataract optical eye images using SVM classifier. Lecture Notes Eng Comput Sci 1:542–545

11. Jagadale AB, Sonavane S, Jadav D (2019) Computer aided system for early detection of nuclear cataract using circle Hough transform. In: 2019 3rd international conference on trends in electronics and informatics (ICOEI), pp 1009–1012 (2019). https://doi.org/10.1109/ICOEI.2019.8862595
12. Caixinha M, Amaro J, Santos M, Perdigo F, Gomes M, Santos J (2016) In-vivo automatic nuclear cataract detection and classification in an animal model by ultrasounds. IEEE Trans Biomed Eng 63(11):2326–2335. https://doi.org/10.1109/TBME.2016.2527787
13. Jindal I, Gupta P, Goyal A (2019) Cataract detection using digital image processing. In: 2019 Global Conference for Advancement in Technology (GCAT) (2019). https://doi.org/10.1109/gcat47503.2019.8978316, https://www.jncet.org/Manuscripts/Volume-7/Issue-10/Vol-7-issue-10-M-07.pdf
14. Hu S, Wang X, Wu H, Luan X, Qi P, Lin Y, He X, He W (2020) Unified diagnosis framework for automated nuclear cataract grading based on smartphone slit-lamp images. IEEE Access 8:174169–174178. https://doi.org/10.1109/ACCESS.2020.3025346
15. Hossain MR, Afroze S, Siddique N, Hoque MM (2020) Automatic detection of eye cataract using deep convolution neural networks (dcnns). In: 2020 IEEE region 10 symposium (TENSYMP), pp 1333–1338 (2020). https://doi.org/10.1109/TENSYMP50017.2020.9231045
16. Mohammed A (2020) Cataract image (Jul 2020). https://www.kaggle.com/alexandramohammed/cataract-image

COV-Doctor: A Machine Learning Based Scheme for Early Identification of COVID-19 in Patients

Ferdib-Al-Islam and Mounita Ghosh

Abstract The COVID-19 pandemic brought about by the SARS-CoV-2 keeps on representing a critical danger to worldwide wellbeing. The most approved indicative test for Coronavirus, utilizing reverse transcriptase-polymerase chain response (RT-PCR) kit has deficiency sometimes in low-income countries. This adds to expanded disease rates and defers basic preventive measures. Successful screening empowers fast and effective analysis of Coronavirus and can relieve the burden on medical care services. Machine learning (ML) models are being used to anticipate the presence of COVID-19 in patients to support clinical staff worldwide, particularly with regards to restricted medical services assets. In this research, machine learning models have been developed to identify COVID-19 in the early stage of sickness using the information of symptoms and exterior activities of patients. Among the four machine learning classifiers, the Decision Tree and Extreme Gradient Boosting (XGBoost) performed equally better with 98% of accuracy, precision, and recall. The feature importance scores have been calculated also to understand the feature's impact on the development of the machine learning model. The proposed work has outperformed the existing works with better execution.

Keywords COVID-19 · Decision tree · XGBoost · Correlation · Feature importance score

1 Introduction

The spread of COVID-19, a respiratory illness triggered by the SARS-CoV-2, is a severe and burning universal concern. The World Health Organization (WHO) has stated a pandemic caused by the COVID-19 virus. According to WHO figures as of

Ferdib-Al-Islam (✉)
Northern University of Business and Technology Khulna, Khulna, Bangladesh

M. Ghosh
Khulna University of Engineering & Technology, Khulna, Bangladesh

M. S. Arefin et al. (eds.), *Proceedings of the International Conference on Big Data, IoT, and Machine Learning*, Lecture Notes on Data Engineering and Communications Technologies 95, https://doi.org/10.1007/978-981-16-6636-0_4

14 May 2021, it had caused over fifteen million cases and 3,499,979 deaths worldwide [1]. Initially, coronavirus infections were thought to cause harmless respiratory human symptoms that were not lethal. However, the beta variant of coronavirus was later linked to the emergence of serious and fatal respiratory disorders such as the "Severe Acute Respiratory Syndrome (SARS)" and the "Middle East Respiratory Syndrome (MERS)". SARS and MERS were responsible for about 9.7% and 35.8% of all deaths correspondingly. Machine learning has been adapted to classify the most relevant and meaningful clinical signs that will forecast real COVID-19 positive events [2].

The most frequent COVID-19 symptoms, such as fever and cough, are close to those of a variety of other communicable illnesses making rapid identification difficult for health practitioners. According to reports, results from the RT-PCR, which is at present the most accurate indicative test, frequently take more than a week to become usable [3]. On the other hand, the latest growth in the use of modern quick diagnostic tests, which are susceptible to certain accuracy problems, may raise the threat of inadequate health resource allocation [3]. However, the process of RT-PCR tests is difficult, and it normally takes 5–6 hours or lengthier to obtain the data. Furthermore, due to little virus counts in initial COVID-19 sufferers, RT-PCR studies yielded false-negative findings in a variety of cases. It has significantly hampered global pandemic prevention and management. As a result, it is critically important to develop a fast indicative model to monitor high-threat patients for COVID-19 contagion [4]. This pandemic threatens to strain medical services around the world in a variety of ways, comprising sharp spikes in demand for clinic beds and acute scarcities of medicinal supplies, as many healthcare staff has been contaminated. As a result, the ability to make immediate treatment decisions and use healthcare services effectively is critical [5]. For these difficulties, COVID-19 improved cases are frequently restricted to data accumulated at the country level, are obtained only from verified cases, and can vary depending on the concept of "recovery" or form of authorization [6].

Models focused on symptoms such as deficiency of smell and taste have been suggested as useful instruments for predicting COVID-19 detection as well as early markers of the efficacy of containment strategies in new outbreaks. Researchers have also been interested in new methods focused on ML algorithms, which have recently been utilized to study and forecast olfactory dysfunction [7]. As opposed to traditional statistical approaches, ML methods rely on algorithms that can be extended to a population-based systems approach, modeling dynamic relationships and correlations between multiple variables [7].

In this research, machine learning techniques (Logistic Regression, Random Forest, Decision Tree, and XGBoost) have been utilized in the COVID-19 symptoms dataset to predict the patient is suffering from COVID-19 or not in the early stage and to show the most relevant symptoms (features) of patients by calculating feature importance scores that impact the creation of the machine learning model.

The organization of this paper is as follows—Section 2 expresses the recent works on COVID-19 detection, Sect. 3 explains the implementation of the work in detail,

Sect. 4 represents the results obtained from this study and finally, Sect. 5 presents the conclusion of the paper.

2 Literature Review

In this section, the recent researches on COVID-19 prediction utilizing machine learning, and other related techniques have been explained.

Reddy et al. [2] developed a model that used supervised machine learning algorithms to recognize specific features to predict COVID-19 accurately. These features included gender, breathing difficulty, observation of fever, as well as clinical details like the cough as well as lung infection and congestion. Some machine learning techniques had been implemented and the precision was computed. For both age groups, the highest precision was greater than (50%) of actual patients. The information was gathered from COVID-19 supportive patients, an anonymous study, and a social survey conducted at research centers. Following that, they used different approaches such as data pre-processing, model validation, and statistical analysis. For a greater understanding, the probability and precision of a patient were shown using different techniques of Machine learning.

Batista et al. [3] used machine learning to forecast the likelihood of a positive COVID-19 diagnosis by using the data from crisis attention tests. The data was obtained from 235 mature patients of which 102 (43%) obtained a positive result of COVID-19 via RT-PCR tests. On an arbitrary sample of 70% of the data, 5 ML techniques were applied. The support vector machines got the best predictive results (AUC: 85%; Sensitivity: 68%; Specificity: 85%; Brier Score: 16%). The quantity of lymphocytes, leukocytes, and eosinophils were the three most significant variables for the algorithm's predictive efficiency.

Sun et al. [4] developed a system for predicting early COVID-19. That study aimed to derive threat elements from medical data of primary COVID-19 diseased patients using 4 conventional machine learning methods for quick COVID-19 identification. The findings showed that the LR model has a higher accuracy rate of 95%, a region under the AUC of 97%, and an increased sensitivity rate of 82%, making it ideal for early COVID-19 infection screening. Zoabi et al. [5] developed a machine learning method that was applied on 51,831 tested individuals and tested on data from the following week. Using just eight features, their model accurately predicted COVID-19 test results: Gender, age over 60, as well as a history of concussions were all factors to consider. Mackey et al. [6] adapted a study in which a total of 4,492,954 tweets containing words similar to COVID-19 indications were received. A total of 3465 (1%) tweets contained user-produced discussions about interactions that users shared with potential COVID-19 symptoms and other illness involvements after using BTM to classify related subject clusters and eliminating redundant tweets. These tweets were divided into 5 groups: first and second-hand accounts of indications, indication identification simultaneous with a lack of testing, discussion of healing, clarification of a negative COVID-19 analysis after testing, and users remembering indications and wondering if they had previously been affected with COVID-19.

Wu et al. [8] developed a study where 11 principal blood indices were found out using a random forest method to create the ultimate associate perception method from 49 clinically obtainable blood test data obtained using commercial blood test equipment. The approach demonstrated a robust performance in reliably identifying COVID-19 from several suspicious patients with identical CT information or signs, with the precision of 97.95% and 96.97% for the cross-validation and test set correspondingly. The tool also performed admirably on an exterior validation range that was entirely free of the modeling procedure, achieving sensitivity, precision, and overall accuracy of 95.12%, 96.97%, and 95.95% correspondingly. Furthermore, 24 samples from COVID-19 contaminated patients from other countries were used to perform an in-depth clinical examination with a precision of 91.67%.

Yan et al. [9] built an analytical model based on the XGBoost technique which was used to test 29 patients. Their model identified three main clinical characteristics namely lactic dehydrogenase, and lactic de, lymphocytes, and high-sensitivity C-reactive protein (hs-CRP), lactic dehydrogenase, lactic dehydrogenase, and lactic dehydrogenase from more than 300 features. The prognostic prediction model based on three indices was developed and it was able to calculate the death chance and provide a medical pathway for distinguishing critical cases from serious cases and certain instances.

3 Methodology

The methodology of the proposed work has been separated into the following steps:

- Data Collection and Preprocessing
- Exploratory Data Analysis
- ML Algorithms for Classification

The architecture of the proposed work has been demonstrated in Fig. 1.

3.1 Data Collection and Preprocessing

The dataset that has been utilized in this research is available on Kaggle [10]. The dataset contains the symptoms of COVID-19 which was seen in patients provided by the "World Health Organization (WHO)" [11]. This dataset contains 5434 instances of 20 features and a target variable. The target variable is "COVID-19". The data distribution of each attribute has been listed in Table 1.

In machine learning, most of the algorithms require numerical values as input because computing machines cannot deal with categorical variables. So, it is needed to convert the categorical features into numerical format before feeding them into the

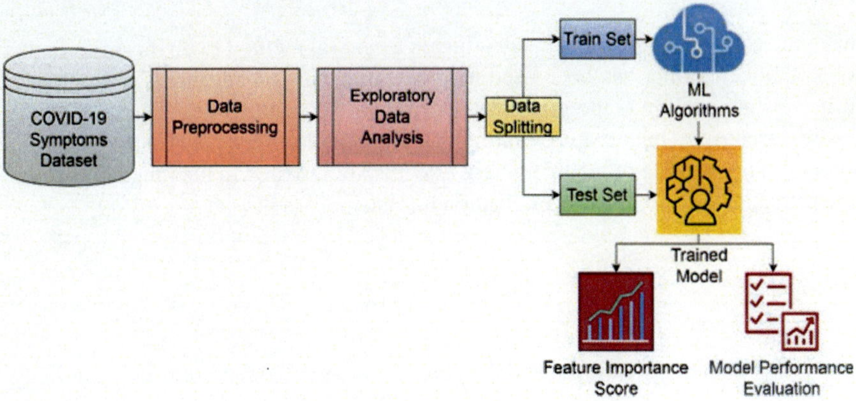

Fig. 1 Proposed system architecture

Table 1 Dataset details

Attribute name	Data distribution
"Breathing problem"	Yes (67%), No (33%)
"Fever"	Yes (79%), No (21%)
"Dry cough"	Yes (79%), No (12%)
"Sore throat"	Yes (73%), No (27%)
"Running nose"	Yes (54%), No (46%)
"Asthma"	Yes (46%), No (54%)
"Chronic lung disease"	Yes (47%), No (53%)
"Headache"	Yes (50%), No (50%)
"Heart disease"	Yes (46%), No (54%)
"Diabetes"	Yes (48%), No (52%)
"Hyper tension"	Yes (49%), No (51%)
"Fatigue"	Yes (52%), No (48%)
"Gastrointestinal"	Yes (47%), No (53%)
"Abroad travel"	Yes (45%), No (55%)
"Contact with COVID patient"	Yes (50%), No (50%)
"Attended large gathering"	Yes (46%), No (54%)
"Visited public exposed places"	Yes (52%), No (48%)
"Family working in public exposed places"	Yes (42%), No (58%)
"Wearing masks"	Yes (0%), No (100%)
"Sanitization from market"	Yes (0%), No (100%)
"COVID-19"	Yes (81%), No (19%)

machine learning models. Label encoding is a general method to do this task. In this work, label encoding has been used to convert the categorical inputs into numeric form. Feature scaling is the principle to bring all the features to the same scale. In this research, min–max feature scaling or normalization for every feature has been applied. It is a scaling procedure where esteems are rescaled in the range of 0 and 1. The principle for applying normalization has been given in (1):

$$A\prime = \frac{A - A_{\min}}{A_{\max} - A_{\min}} \quad (1)$$

where $A_{\max}$ and $A_{\min}$ are the top and the bottom values of the feature correspondingly.

3.2 Exploratory Data Analysis

Exploratory Data Analysis (EDA) is a method of analyzing or interpreting the data and exploring insights or fundamental attributes of the data. In this step, at first, the correlation among the variables has been computed. Correlation has control over feature significance. As two features/variables are related, a variation in one will create variation in another. So there is no reason to keep each of them. In Fig. 2, the correlation of the variables has been illustrated. It can be seen that no input variables have strong correlations among them, so, no feature is eliminated from the input feature list. "Wearing Masks" and "Sanitization from Market" contain only one type of category ("No"), so, these variables do not correlate with others. However, symptoms like "Breathing Problem", "Fever", "Dry Cough", "Sore throat" and external activities like "Abroad travel", "Contact with COVID Patient" and "Attended Large Gathering" are the main cause to get affected with COVID-19.

3.3 ML Algorithms for Classification

Before applying ML algorithms for classification, the dataset has been split using the "percentage split" concept. 70% of data has been used in the training set for constructing the model and the leftover 30% of the data has been used in the test set for testing. There were 3803 instances in the training set and 1631 instances in the test set. "Grid Search CV" algorithm has been utilized to obtain the finest parameters of the classifiers [12].

Logistic Regression Logistic regression is one of the most basic and generally utilized machine learning algorithms [13]. Logistic regression isn't a regression method yet a probabilistic classification algorithm. The motivation in Logistic regression is to solve the problem as a summed up linear regression algorithm as in (2):

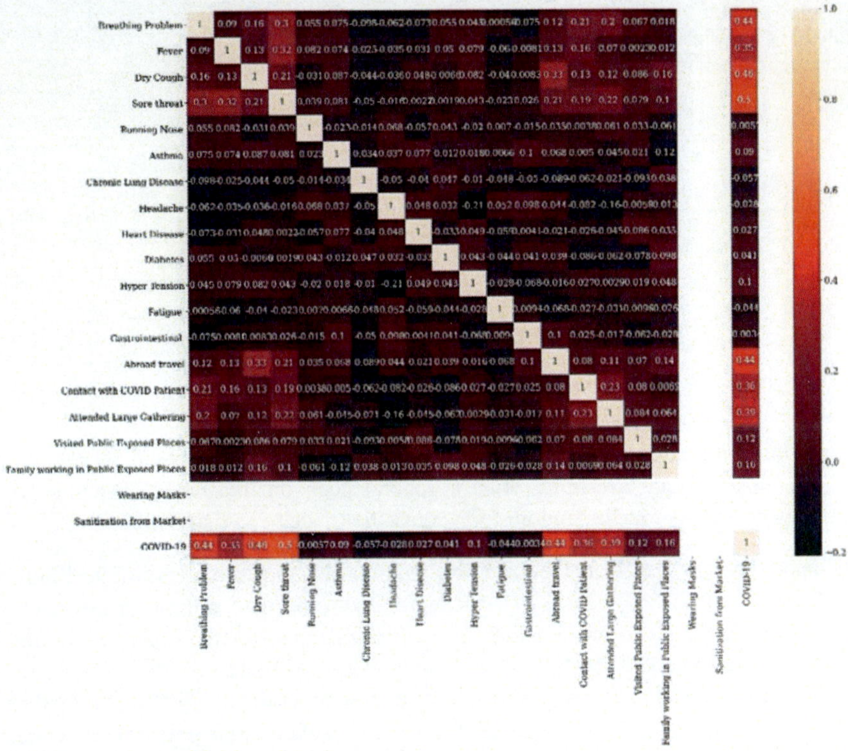

Fig. 2 Correlation among variables

$$\hat{y} = \beta_0 + \beta_1 x_1 + \ldots + \beta_n x_n \quad (2)$$

where $\hat{y}$ = predicted esteem, x = independent features, and the β = learning coefficients. The selected parameters' values for the logistic regression classifier have been listed in Table 2.

Random Forest It is a supervised learning technique. The "forest" it accumulates, is a collection of decision trees, normally trained with the "bagging" principle [13]. The general motivation of the "bagging" principle is that an arrangement of learning models builds the general consequence. Accordingly, in a random forest, just an irregular subset of the features is selected to part a node. Even it can be made trees

Table 2 Chosen parameter values for logistic regression classifier

Parameter name	Chosen value
"C"	0.01
"penalty"	"l2"
"solver"	"lbfgs"

Table 3 Chosen parameter values for random forest classifier

Parameter name	Chosen value
"n_estimators"	100
"random_state"	5
"max_depth"	30

Table 4 Chosen parameter values for decision tree classifier

Parameter name	Chosen value
"criterion"	"gini"
"max_depth"	200
"random_state"	5

more uninformed by also developing thresholds for each feature instead of searching for the most perfect thresholds (like a general decision tree does). The selected parameters' values for the random forest classifier have been listed in Table 3.

Decision Tree A Decision Tree is a non-parametric supervised learning technique applied for classification tasks [14]. It learns from the data sample to estimate a sine arc with a collection of if-then-else decision rules. As the depth of the tree increases, the rules and the fitter of the model becomes more complex. A decision tree is a hierarchical tree configuration where an inner node demonstrates a feature, the branch demonstrates a system, and every leaf node demonstrates the outcome. The selected parameters' values for the decision tree classifier have been listed in Table 4.

XGBoost Among the gradient boosting (ensemble) methods in tree-based machine learning algorithms, Extreme Gradient Boosting (XGBoost) is one of the mainstream algorithms that possesses improved and quick execution [15]. In the collection of ensemble learning methods, XGBoost represents the boosting method set. A set of classifiers which are the combination of several models that are used for delivering superior classification performance is the concept of ensemble learning. XGBoost algorithm is the advancement of gradient boosting methods that have regularization factors. The selected parameters' values for the XGBoost classifier have been listed in Table 5.

4 Result and Discussion

It was mentioned earlier that, Logistic Regression, Random Forest, Decision Tree, and XGBoost classifier has been utilized to predict COVID-19 in patients. The implemented system performance was measured using different performance metrics—accuracy, precision, and recall using the principles presented in (3), (4), and (5) respectively.

Table 5 Chosen parameter values for XGBoost classifier

Parameter name	Chosen value
"colsample_bytree"	0.3
"learning_rate"	0.1
"max_depth"	50
"alpha"	10
"n_estimators"	1000
"objective"	"binary: logistic"
"booster"	"gbtree"

$$\text{Accuracy} = \frac{\text{TP} + \text{TN}}{\text{TP} + \text{FP} + \text{FN} + \text{TN}} \quad (3)$$

$$\text{Precision} = \frac{\text{TP}}{\text{TP} + \text{FP}} \quad (4)$$

$$\text{Recall} = \frac{\text{TP}}{\text{TP} + \text{FN}} \quad (5)$$

The detailed classification report of each classification model has been represented in Table 6. Among the 4 machine learning models, the decision tree and the XGBoost model performed equally better than the other models with 98% of accuracy, precision, and recall.

The feature importance scores of both the DT and XGBoost model have been represented in Figs. 3 and 4 respectively. "Sore throat", "Breathing Problem", and "Abroad travel" is the top-3 significant feature from DT model, where "Hyper Tension". "Asthma", and "Heart Disease" is the top-3 significant feature from XGBoost model. A contrast between the proposed work and other existing works has been represented in Table 7.

Table 6 Classification report of machine learning models

Model	Class	Accuracy (%)	Precision (%)	Recall (%)
Logistic regression	Yes	97	98	98
	No		92	92
Random forest	Yes	88	87	100
	No		100	41
Decision tree	Yes	98	100	97
	No		91	99
XGBoost	Yes	98	99	98
	No		92	98

Fig. 3 Feature importance scores from DT model

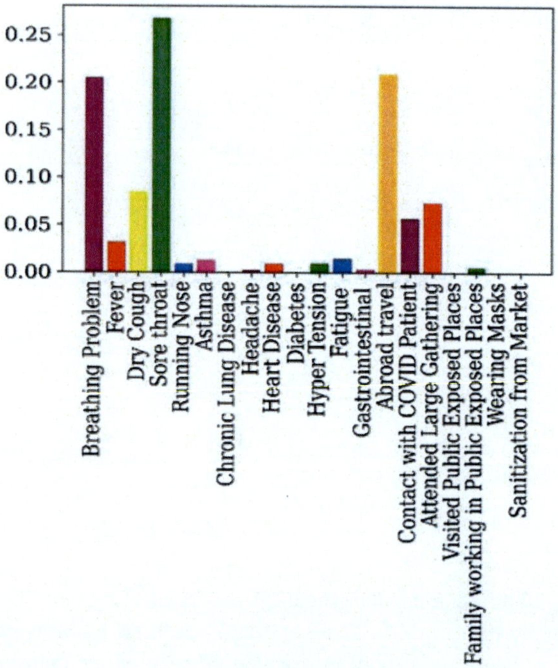

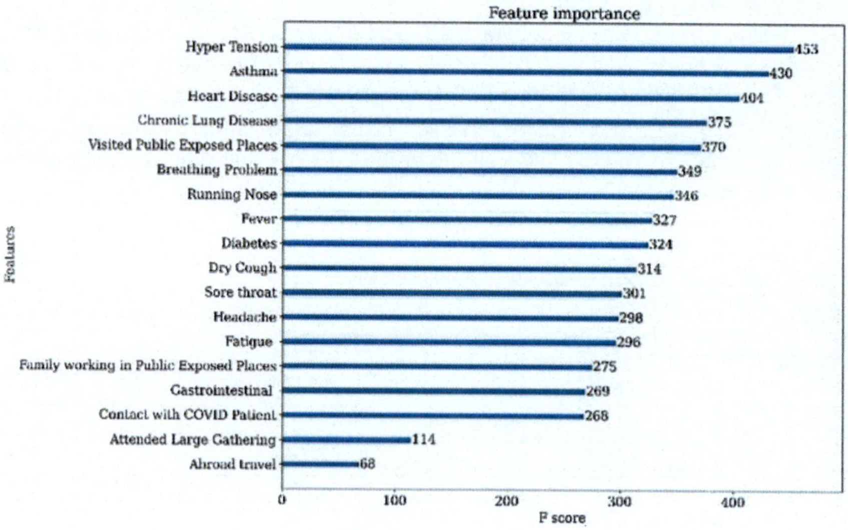

Fig. 4 Feature importance scores from the XGBoost model

Table 7 Proposed work comparison with other works

Author	Model used	Accuracy (%)	Precision/PPV (%)	Recall/Sensitivity (%)	Feature importance score calculation
Reddy et al. [2]	Logistic regression	50	N/A	N/A	No
Batista et al. [3]	Support vector machine	N/A	77.8	68	No
Sun et al. [4]	Logistic regression + Feature selection	91	N/A	87	No
Zoabi et al. [5]	Gradient boosting	N/A	N/A	87.3	No
Proposed work	Decision tree	98	98	98	Yes
	XGBoost	98	98	98	

It has been shown that the proposed work obtained better performance than the existing works in different performance metrics along with the feature importance score, which was not computed in the existing works.

5 Conclusion

COVID-19 is a contagious disease and so it has been turned into a pandemic rapidly. Early detection of COVID-19 in patients can prevent further damages in the body, help to reduce mortality, and also allow to spread of infection in other humans. In this study, COVID-19 symptoms and people's exterior activities have been considered. Only symptoms can't be used for detecting COVID-19 in asymptotic patients. Machine learning models have been built to anticipate the status of COVID-19 in the patient. Among the models used in this research, the Decision Tree and XGBoost model outperformed other models and also existing works with superior performances in different metrics as—accuracy (98%), precision (98%), and recall (98%). This research can help clinicians to identify COVID-19 rapidly and also when the test kits are limited.

References

1. COVID live update: 162,539,539 cases and 3,499,979 deaths from the Coronavirus—Worldometer. https://www.worldometers.info/coronavirus/

2. Reddy K et al (2021) A machine learning approach to analyse the symptoms of COVID-19 for the initial diagnosis of a patient. Int J Sci Res Comput Sci Eng Inf Technol 34–40
3. Moraes Batista A et al (2020) COVID-19 diagnosis prediction in emergency care patients: a machine learning approach
4. Sun N et al (2020) A prediction model based on machine learning for diagnosing the early COVID-19 patients
5. Zoabi Y et al (2021) Machine learning-based prediction of COVID-19 diagnosis based on symptoms. NPJ Digital Med 4(1)
6. Mackey T et al (2020) Machine learning to detect self-reporting of symptoms, testing access, and recovery associated with COVID-19 on twitter: retrospective big data infoveillance study. JMIR Public Health Surveill 6(2):e19509
7. Callejon-Leblic M et al (2021) Loss of smell and taste can accurately predict COVID-19 infection: a machine-learning approach. J Clin Med 10(4):570
8. Wu J et al (2020) Rapid and accurate identification of COVID-19 infection through machine learning based on clinical available blood test results
9. Yan L et al (2020) A machine learning-based model for survival prediction in patients with severe COVID-19 infection
10. Symptoms and COVID presence. https://www.kaggle.com/hemanthhari/symptoms-and-covid-presence
11. Coronavirus disease 2019 (COVID-19)—Symptoms. https://www.cdc.gov/coronavirus/2019-ncov/symptoms-testing/symptoms.html
12. Paper D (2020) Hands-on scikit-learn for machine learning applications. Apress, Berkeley, CA
13. Couronné R et al (2018) Random forest versus logistic regression: a large-scale benchmark experiment. BMC Bioinformatics 19:1
14. Charbuty B, Abdulazeez A (2021) Classification based on decision tree algorithm for machine learning. J Appl Sci Technol Trends 2(01):20–28
15. Feng Y et al (2020) An XGBoost-based casualty prediction method for terrorist attacks. Complex Intell Syst 6(3):721–740

Ovarian Cancer Prediction from Ovarian Cysts Based on TVUS Using Machine Learning Algorithms

Laboni Akter and Nasrin Akhter

Abstract Ovarian Cancer (OC) is type of female reproductive malignancy which can be found among young girls and mostly the women in their fertile or reproductive. There are few number of cysts are dangerous and may it cause cancer. So, it is very important to predict and it can be from different types of screening are used for this detection using Transvaginal Ultrasonography (TVUS) screening. In this research, we employed an actual datasets called PLCO with TVUS screening and three machine learning (ML) techniques, respectively Random Forest KNN, and XGBoost within three target variables. We obtained a best performance from this algorithms as far as accuracy, recall, f1 score and precision with the approximations of 99.50%, 99.50%, 99.49% and 99.50% individually. The AUC score of 99.87%, 98.97% and 99.88% are observed in these Random Forest, KNN and XGB algorithms. This approach helps assist physicians and suspects in identifying ovarian risks early on, reducing ovarian malignancy-related complications and deaths.

Keywords Ovarian cancer · Transvaginal ultrasonography (TVUS) · KNN imputer · Smote · Machine learning · Feature correlation

1 Introduction

Ovarian cancer (OC) has been the world largest seventh leading causes of death and disability in women [1]. In the year 2018, 295,414 women were diagnosed with OC, and 184,799 women died as a result of the condition [2]. Ovarian cancer is a disease of the ovaries, the woman reproductive structures that generate eggs and generate estrogen and progesterone. Ovarian cancer treatment is developing, and the best results are usually shown whenever the cancer is detected early. Epithelial ovarian cancers (EOC) or OC are the most typical types of ovarian cancer [3]. Ovarian cysts are fluid-filled sacs or pockets that form within or on the ovarian membrane. Despite the fact that post-menopausal females have a higher chance of malignancy to

---

L. Akter (✉) · N. Akhter
Department of Biomedical Engineering, Khulna University of Engineering and Technology, Khulna, Bangladesh

M. S. Arefin et al. (eds.), *Proceedings of the International Conference on Big Data, IoT, and Machine Learning*, Lecture Notes on Data Engineering and Communications Technologies 95, https://doi.org/10.1007/978-981-16-6636-0_5

premenopausal women, the most of ovarian cysts in postmenopausal women with no unequivocal malignancy markers, such as solid regions, papillary features, or thick uneven members to develop, are benign [4]. The growing utilization ultrasonography in ovarian screenings and new progress in image analysis have boosted the detection of ovarian cysts in untreated post—menopausal females.

Analyzing a vast number of data acquired via several actual patients' databases yields a wealth of data for providing high-quality healthcare at lower prices. The key to lowering the fatality rate from ovarian cancer is early identification. The physicians will use an efficient and trustworthy screening technique to make an initial prognosis. Conventional diagnostic approaches for ovarian cancer include serum cancer antigen 125 (CA-125) screening and transvaginal ultrasonography (TVUS). The significance of categorizing ovarian cancer patients between low— and high categories has prompted numerous healthcare and bioinformatics research groups to investigate the use of machine learning (ML) technologies. As a result, these methods have been used to model the progression and therapy of malignant diseases. Furthermore, the capability of ML algorithms to find essential variables in complicated dataset demonstrates their value. In cancer research, a range of these strategies, such as KNN, Random Forest, and XGBoost, have been widely used to construct prediction algorithms resulting in efficient and precise making decisions.

In this study, the main approach is that many researchers have done a lot of work on ovarian cancer but no author has done the work of predicting ovarian cancer from cysts using ML. So it can be said that this work is the principal to predict ovarian cancer from ovarian cysts. From an unprocessed set of data, this paper applies machine learning algorithms to create an understanding method for early ovarian cancer diagnosis.

2 Related Work

Yasodha and Ananthanarayanan [5] analyzed big datasets to develop an experience and understanding method of OC for earlier diagnosis. They had used three algorithms for this work that was Multiclass SVM, ANN, and Naïve Bayes. To properly categorize data either normal or abnormal, PGSO is utilized to improve the rough set feature minimization. The performance for SVM, ANN Naïve Bayes of accuracy were 98%, 95%, 93%, specificity 96.7%, 92.9%, 91.4%, sensitivity 99%, 97%, 96% respectively. Guan et al. [6] approached to predict OC from metabolomics liquid chromatography spectrometry data SVM was used. The SVM algorithm being tested on LC/TOF MS metabolomics data, with the goal of identifying pairings of putative metabolic diagnostic biomarkers. With 90% accuracy, 37 OC patients. Alqudah [7] classified a evaluation of ML and feature selection systems for OC utilizing serum proteome profiling and wavelet features. There are 207 no cancers and 262 ovarian cancers in the given dataset. With 44 features, they employed ANN, SVM, KNN, and ELM ML methods. The accuracy 99%, 99.45% sensitivity 93.21% precision which combining PCA with SVM. Lu et al. [8] performed to predict ovarian cancer using three machine learning algorithms which are DT, LR, and ROMA. They used 235 patients' data where 89 BOT and 146 OC for DT model and 114 patients where 89 BOT and 25 OC for ROMA and LR model with 49 variables. The training data gained the accuracies 79.6%, 87.2%, 84.7% for DT, LR, ROMA algorithms and test data gained 92.1%, 95.6%, 97.4% DT, LR, ROMA algorithms respectively. For training data highest for sensitivity DT was 82.2% and specificity 100% for LR whereas for test data the sensitivity 100% for ROMA and specificity 97.8% for LR. Wang et al. [9] accomplished the HE4 seems significant to identifying OC, particularly in the post—menopausal community, according to a meta-analysis consisting on 32 reports that looked at the prognostic significance of HE4, CA125, and ROMA. ROMA and CA125 remain better tests for detecting OC in post—menopausal women. Zhang et al. [10] developed the dual markers that indicated the quantity of epidemiological data in the onset also progress of OC, therefore a linear multi-marker system incorporating CA125, HE4, estradiol, and progesterone was developed. While associated to CA125 or HE4, their multi-marker approach was much better at distinguishing BPM from EOC patients. Chen et al. [11] designed a model of earlier demise in individuals with left-testis malignant tumor with accuracy of 76.1 percent AUC 0.621, sensibility 0.130, positive 0.659 and F1 score 0.216. In this work clinical variables were obtained since a unit of 273 ovarian cancer patients with phase I and II and a ML algorithm for L2 Regression was developed because of number of patients with mortality forecast issue under 20 months, the twenty-fifth percentile of total survival.

3 Methodology

The methodology part has been divided into several sections.

- Data Collection
- Data Preprocessing
- Imbalanced Data Handling by SMOTE Analysis
- Dataset Splitting
- Feature Scaling
- Machine Learning Algorithm for Classification\
- Model Result and
- Performance Evaluation Methods

The work's procedure flow-diagram is shown in Fig. 1.

3.1 Data Collection

The data collected in this research work from PLCO dataset of National Cancer Institute (NCI), United States [12]. This dataset has a number of

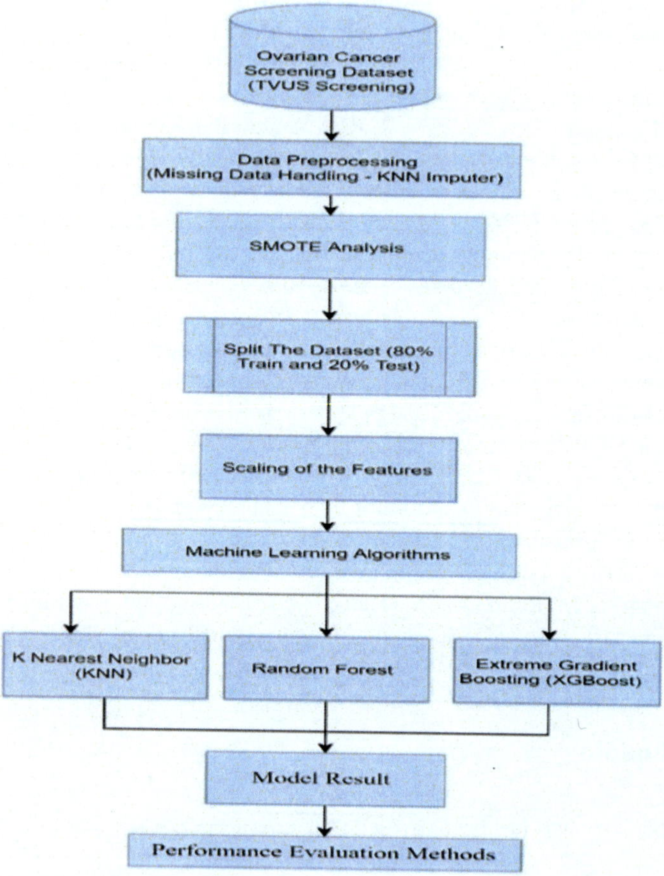

Fig. 1 Flowchart of methodology

attributes that aren't required for this purpose. As a result, have selected eighteen separate features from the dataset, as well as a parameter with a target. In this work, taken a number of approaches which are numcystl, numcystr, ovary_diaml, ovary_diamr, ovary_voll, ovary_volr, ovcyst_diaml, ovcyst_diamr, ovcyst_morphl, ovcyst_morphr, ovcyst_outlinel, ovcyst_outliner, ovcyst_solidl, ovcyst_solidr, ovcyst_suml, ovcyst_sumr, ovcyst_voll, ovcyst_volr, "ovar_result" as the class.

3.2 *Data Preprocessing*

Missing Data Handling by KNN Imputer

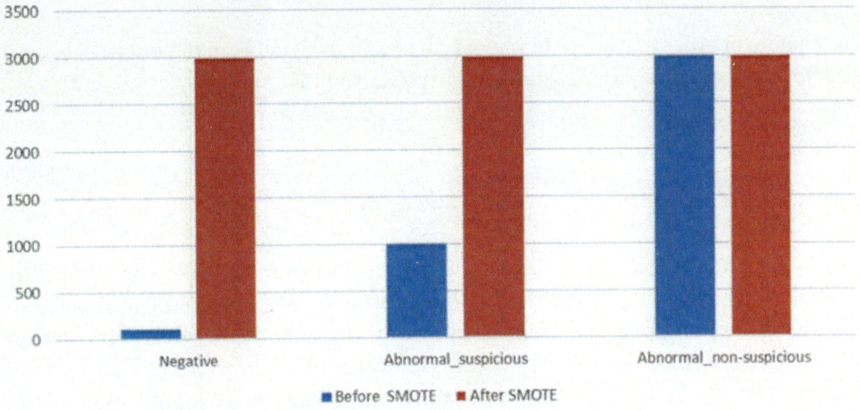

Fig. 2 Dataset distributions by class before and after SMOTE

In this dataset there have lots of missing value. First of all, we have selected a feature that has the lowest missing data value. However, a small amount of missing data remains. So, then we used KNN Imputer. Several techniques are available to counter missing values in a dataset. In this research work, the KNN imputer has been applied. In scikit-learn, KNN Imputer is a popular approach for imputing missing data. It is largely accepted as a viable alternative to typical impute methods. Inhere $k = 5$ number of neighbors to eliminate missing data contained in the dataset [13].

3.3 Imbalanced Data Handling by SMOTE Analysis

We found data imbalances for several categories when doing the categorization function. Due to the fact that this was an actual medical datasets, unbalanced classifications were unavoidable. When faced with unbalanced datasets, traditional machine learning system analysis methods fail to accurately characterize proposed system. SMOTE (synthetic minority oversampling method) is one of the most often used oversampling approaches for dealing with the imbalanced class problem [14]. Figure 2 shows the class wise imbalanced data before the SMOTE and the data set has balanced after SMOTE.

3.4 Dataset Splitting

The dataset was split into binary parts: a training set and a test set. The training dataset contained 80% of the whole data, while the test set contained 20% of the total data. The ML establishes a relationship with the independent and dependent parameters in order to foresee or choose an alternative, and then the test data is used to determine whether the ML approach is effective [15].

3.5 Feature Scaling

Feature scaling is a technique for bringing most of the features to the same scale. We used min–max feature scaling (normalization) for all of the features in this study. It's a rescaling approach in which estimations are shifted and resized till they're someplace between 0 and 1. Therefore, it's known as normalization. This is a method for normalizing the level of data's self-sufficient features. It is usually done as part of the data pre-treatment procedure [16]. The Standard Scalar Python Library was applied in this research.

3.6 Implemented Machine Learning Algorithms

KNN (K Nearest Neighbor)
The KNN classifier is a common basic ML method which is being utilized to categorize images. It is dependent on the feature vector separation, and we have labeled data to categorize and recover the image's exact class. The Euclidean distance was employed as the similarity function. Model performance is linked to determining the optimal number of neighbors. KNN had a neighbor number of 9 in this study [17].

Random Forest
Because of classification, Random Forest (RF) is a supervised ML technique, is used. We concluded that a forest is made up of trees, and that the additional trees there are, the more powerful the forest. Similarly, the RF algorithm proposes DT on data tests and then receives the urge to every one of these before selecting the optimum configuration using voting form projection procedures. It is a way of dressing that reduces over-fitting by averaging the results [14].

XGBoost (Extreme Gradient Boosting)
XGBoost is a DT based ensemble ML approach that uses gradient boosting. In unorganized dataset forecasting, ANN outperform all established techniques or systems. The XGBoost algorithm was used to do classifying in the dataset. We split the data using percent split methodology, with 80% of the data in the training data and 20% of the data in the test data, and have used classification algorithms. On classification problems, XGBoost works effectively on small data sets. Boosting is a grouping strategy in which newer versions are introduced to resolve current models' combination [16].

3.7 *Performance Evaluation Methods*

The ML outcomes depending on confusion matrix findings were analyzed to determine the efficiency of the deployed ML methods. Accuracy, recall, $f1$ score and precision are the performance metrics. Four arithmetical keys, like false positive (FP), false negative (FN), true positive (TP), and true negative (TN) were generated to compute the accuracy, precision, recall, and $f1$ score of this methodology. The accuracy, recall, precision, and $f1$ score were calculated as follows:

$$\text{Accuracy} = (\text{TP} + \text{TN})/(\text{TP} + \text{TN} + \text{FP} + \text{FN}) \tag{1}$$

$$\text{Recall} = \text{TP}/(\text{TP} + \text{FN}) \tag{2}$$

$$\text{Precision} = \text{TP}/(\text{TP} + \text{FP}) \tag{3}$$

$$F1 \text{ score} = 2 \times (\text{Precision} \times \text{Recall})/(\text{Precision} + \text{Recall}) \tag{4}$$

4 Result and Discussion

Through this portion, we analyze and evaluate the outcomes obtained by the ML methods. Figure 3 shows the feature correlation matrix. This matrix calculates the correlation across two features in order to reveal their link. The correlation value runs ranging from -1 to $+1$, with ±1 denoting negative/positive correlation is a optimal and 0 denoting no connection at all. The diagonal components of this symmetrical matrix are all $+1$. We clearly see a strong positive correlation between the numcystr and a strong negative connection between the ovary_voll in the matrix. That means that the ovarian cancer class is heavily impacted by the characteristics.

Figure 4 shows the results of KNN, Random Forest, XGBoost algorithms for predict the ovarian cancer with three classes "Negative", "Abnormal, suspicious", "Abnormal, non-suspicious". The accuracy of the KNN, Random Forest, and XGBoost were 93.82%, 99%, 99.50% respectively. The precision was gained of KNN, Random Forest, and XGBoost were 93.83%, 98.99%, 99.49% respectively. The recall values which was achieved of KNN, Forest, and XGBoost 93.75%, 99.01%, 99.50% respectively. The F1 Score were obtained of KNN, Random Forest, and XGBoost were 93.73%, 98.99%, 99.50%.

Table 1 shows the comparison between proposed work and the previous study of some papers to predict the ovarian cancer. From this table we can see that some of the author's was not calculated the precision, recall and $f1$ score and the number of features was also a lot where in this work the number of features is less and the accuracy is high than others.

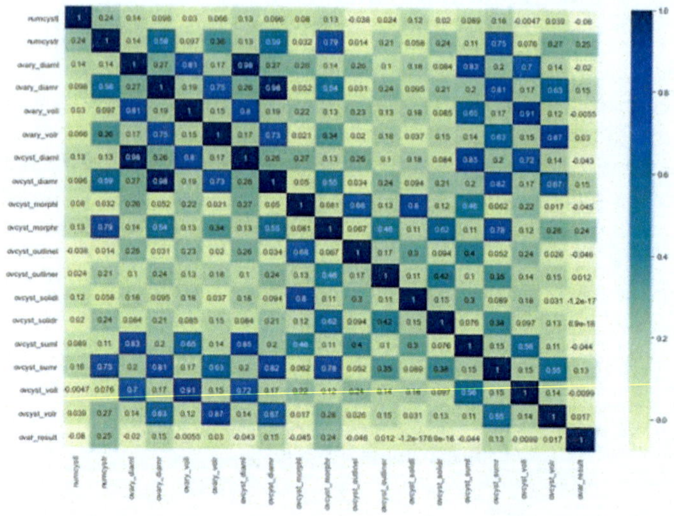

Fig. 3 Feature correlation matrix

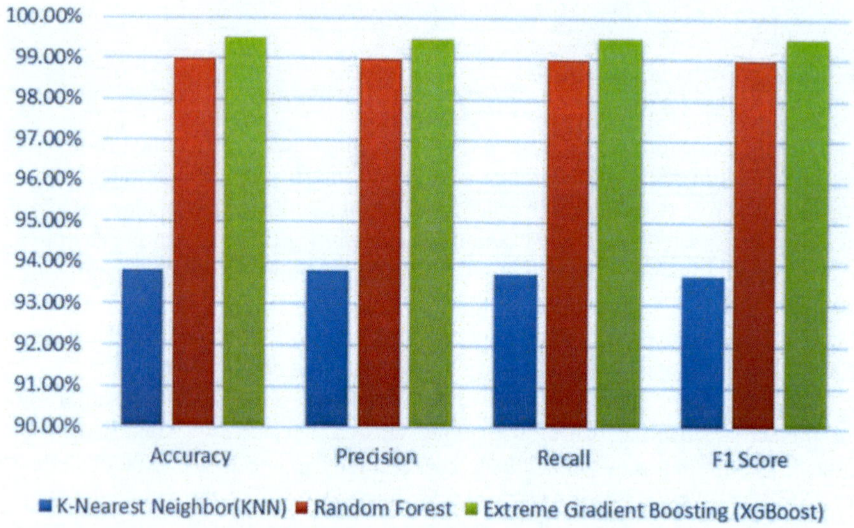

Fig. 4 Accuracy, precision, recall and *f*1 score of KNN, random forest and XGBoost algorithms

We utilize the Area Under the Curve (AUC) and the Receiver Operating Characteristics (ROC) curve to graphically examine the outcomes of the three classes categorization. AUC is a measurement of distinction over classifications obtained by a given classifier while ROC is a probability curve. In this situation, the algorithms performed well in terms of categorization accuracy. The ROC and AUC for the KNN, Random Forest, and XGBoost method in categorizing the three classes are shown in

Table 1 Comparison table with existing work

References	Number of features	Algorithms	Accuracy (%)	Precision (%)	Recall (%)	*F*1-Score
[6]	N/A	Multiclass SVM ANN NB	98	96.7	99	N/A
[7]	N/A	SVM	90	N/A	N/A	N/A
[8]	44	ANN SVM KNN	99	93.21	99.45	N/A
[9]	49	ROMA DT RF	84.7 (Train) 97.4 (Test)	100	82.2	N/A
Proposed result	18	KNN RF XGBoost	99.50	99.49	99.50	99.50%

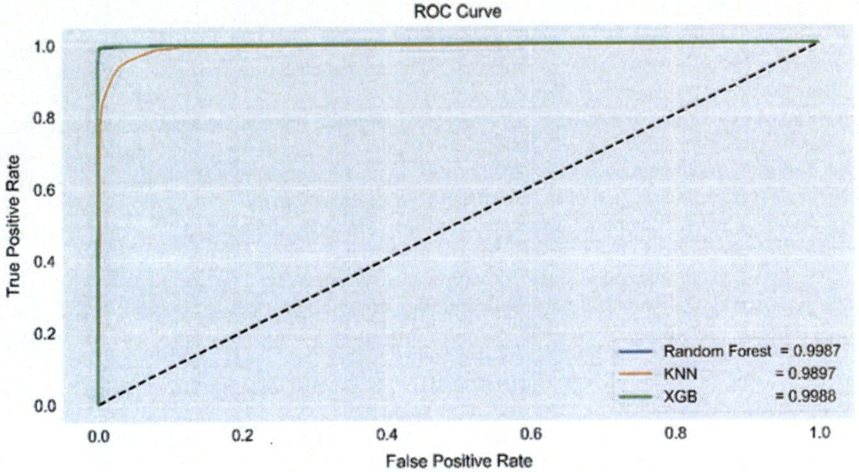

Fig. 5 The ROC curve of random forest, KNN, XGBoost algorithms

Fig. 5. In the three classes' organizations, for Random Forest, KNN, XGBoost the AUC score was determined to be 99.87%, 98.97%, 99.88% correspondingly.

5 Conclusion

In this study, we dispensation ML techniques a forecast of ovarian cancer from ovarian cysts of TVUS screening. We obtained high accuracy 99.50%, *f*1 score

99.50%, recall 99.50%, precision 99.49% from this KNN, RF, XGBoost algorithms respectively with "Negative", "Abnormal, suspicious", "Abnormal, non-suspicious" classes. In this work, the missing is handle by KNN Imputer and the imbalanced data is handle by SMOTE. The use of as an approach will result in a much more precise assessment of system forecasting accuracy. Further study will combine the suggested approach with additional techniques like as ultrasonic imaging recognition and merge all utilizing machine learning and deep learning approaches to improve selection process effectiveness. The implementation of this work for early detection of ovarian cancer may beneficial of this deadly disease.

References

1. Shabir S, Gill P (2020) Global scenario on ovarian cancer—its dynamics, relative survival, treatment, and epidemiology. Adesh Univ J Med Sci Res 2:17–25
2. Bray F, Ferlay J, Soerjomataram I et al (2018) Global cancer statistics 2018: GLOBOCAN estimates of incidence and mortality worldwide for 36 cancers in 185 countries. CA Cancer J Clin 68(6):394–424
3. Ovarian cancer: early signs, detection, and treatment, Healthline 2021. [Online]. Available: https://www.healthline.com/health/cancer/ovarian-cancer-early-signs. Accessed: 22-May-2021
4. Guraslan H, Dogan K (2016) Management of unilocular or multilocular cysts more than 5 centimeters in postmenopausal women. Eur J Obstet Gynecol Reprod Biol 203:40–43
5. Yasodha P, Ananthanarayanan N (2015) Analysing big data to build knowledge based system for early detection of ovarian cancer. Indian J Sci Technol 8(14)
6. Guan W, Zhou M, Hampton C, Benigno B, Walker L, Gray A, McDonald J, Fernández F (2009) Ovarian cancer detection from metabolomic liquid chromatography/mass spectrometry data by support vector machines. BMC Bioinformatics 10(1):259
7. Alqudah AM (2019) Ovarian cancer classification using serum proteomic profiling and wavelet features a comparison of machine learning and features selection algorithms. J Clin Eng 44(4):165–173
8. Lu M, Fan Z, Xu B, Chen L, Zheng X, Li J, Znati T, Mi Q, Jiang J (2020) Using machine learning to predict ovarian cancer. Int J Med Inf 141:104195
9. Wang J et al. (2014) Diagnostic accuracy of serum HE4, CA125 and ROMA in patients with ovarian cancer: a meta-analysis. Tumor Biol 35(6):6127–6138
10. Zhang P et al. (2016) Development of a multi-marker model combining HE4, CA125, progesterone, and estradiol for distinguishing benign from malignant pelvic masses in postmenopausal women. Tumor Biol 37(2):2183–2191
11. Chen R, Rosado AM, Zhang J (2020) Machine learning for ovarian cancer: lasso regression-based predictive model of early mortality in patients with stage I and stage II ovarian cancer. medRxiv
12. Ovarian—Datasets—PLCO—The cancer data access system. In: Cdas.cancer.gov. https://cdas.cancer.gov/datasets/plco/23/. Accessed 15 May 2021
13. Kaushik (2021) (Scikit-learn), KNNImputer | Way to impute missing values. Analytics Vidhya. Available: https://www.analyticsvidhya.com/blog/2020/07/knnimputer-a-robust-way-to-impute-missing-values-using-scikit-learn/. Accessed: 15 May 2021
14. Akter L, Akhter N (2020) Detection of ovarian malignancy from combination of CA125 in blood and TVUS using machine learning. Advances in intelligent systems and computing, pp 279–289

15. Raihan MMS, Shams AB, Preo RB (2020) Multi-class electrogastrogram (EGG) signal classification using machine learning algorithms. In: 2020 23rd International Conference on Computer and Information Technology (ICCIT), pp 1–6, https://doi.org/10.1109/ICCIT51783.2020.9392695
16. Akter L, Ferdib-Al-Islam (2021) Dementia identification for diagnosing Alzheimer's disease using XGBoost algorithm. In: 2021 international conference on information and communication technology for sustainable development (ICICT4SD), pp 205–209. https://doi.org/10.1109/ICICT4SD50815.2021.9396777
17. Ferdib-Al-Islam, Akter L (2020) Early identification of Parkinson's disease from hand-drawn images using histogram of oriented gradients and machine learning techniques. In: 2020 emerging technology in computing, communication and electronics (ETCCE), pp 1–6. https://doi.org/10.1109/ETCCE51779.2020.9350870

A Comprehensive Analysis of Most Relevant Features Causes Heart Disease Using Machine Learning Algorithms

Faria Rahman and Md. Ashiq Mahmood

Abstract Coronary artery disease is a very known word to us because day by day the number of affected people is increasing dramatically, where equally occurs almost both men and women. Globally this disease is the number one causes of death. Low-income and middle-income country's people suffer most. By analyzing biological data we can extract the possible reason for it so in our proposed model, we feed the data and try to mark the most relevant possible outcome of early heart disease. In this paper, we have used two methods for focusing on features: Pearson's Correlation Heatmap and Chi Squared Test, three algorithms of classification: SVM (Support Vector Machine), Decision Tree and K-Nearest Neighbor, three boosting techniques: Ada boost, Gradient boost, XG Boost and two ensemble techniques: Stacking and Voting. 10-fold cross validation has been used by us. By using the top rank features which are coming out through Pearson's Correlation Heatmap run over Stacking ensemble technique where our proposed model has given 97.00% accuracy on it.

Keywords Heart disease · SVM · Decision tree · K-nearest neighbor · Stacking · Voting

1 Introduction

CVD which means cardiovascular disease is an important phrase that encloses a number of fields which affect the heart. Irregular heartbeats, disease of the heart muscle, blood vessel issues, chest pain and so on are related to this [1]. Heart disease is such a leading public health concern because the whole body suffers when the heart traces some problems. Cardiovascular disorders, such as heart attacks and strokes, affect 17 million people worldwide each year [2]. Heart disease caused by rheumatic and different cardiovascular diseases are examples of CVDs and other conditions where heart attacks and strokes cause four out of every five CVD deaths, with one-third of these deaths occurring before the age of 70 [3]. It's more important to find

F. Rahman (✉) · Md. Ashiq Mahmood
Institute of Information and Communication Technology (IICT), Khulna University of Engineering and Technology, Khulna, Bangladesh

M. S. Arefin et al. (eds.), *Proceedings of the International Conference on Big Data, IoT, and Machine Learning*, Lecture Notes on Data Engineering and Communications Technologies 95, https://doi.org/10.1007/978-981-16-6636-0_6

out what's causing the issues, as well as the risk factors and conditions that come with heart disease.

Heart fails to do normal functional work because of irregular responses from its several organs. But by maintaining a healthy lifestyle and diet many problems can be preventable. Many symptoms vary from different heart states, but these are very known alarming signs include shortness of breath, fatigue, light-headedness, chest pain or pressure and numb or cold extremities etc. To maintain our required essential nutrition and oxygen in body parts heart plays an important role on it [4]. The proper working of the heart ensure the life of any organism because the other organs are fully depend on its proper functionality [5]. The health system can be modernized with the new technology, improving the overall population's life expectancy. Leading diseases such as Heart Disease and Cancer are major causes of death worldwide, and every year, the risk of death from coronary heart disease rises at a shocking rate [6].

From the survey of WHO (2016), cardiovascular disease causes 31% of all deaths worldwide including heart attacks and stroke which accounts 85% of all deaths [7]. Disease diagnosis is the process in the medical field which can be taken as a way of decision making, where unknown and new cases are analyzed from the medical databases [8]. To make the analyzing process easy, more accurate and faster we used some techniques of machine learning in our proposed model. Techniques of Machine learning have the ability to learn the input from the system and generate cases based on the previous output and try to improve the final result with the help of previous experiences [9]. In our experiment, we first identified the features that are most closely related and generated a good impact on Cleveland heart disease dataset by performing two feature selection methods, then for each of the feature selection methods KNN, SVM, Decision Tree, Stacking and Voting has been used. The better efficiency comes out with nine extracted features from Pearson's Correlation Heatmap applying Stacking.

2 Related Work

In the previous paper [10] for successful calculation and to raise and improve the performance, LMT techniques, Hoeffding Tree, Random Forest, SVM and Gaussian NB were used. Each algorithm was put to the test on the dataset, and the outcomes were measured in terms of accuracy. They identified that Random forest suits better and the initial and final accuracy was 88.52% and 95.08% respectively. Prakash et al. [11] they used Information gain to discover the best features which plays important role to predict the output from it they extracted age, blood sugar (fasting) > 120, chest pain type and sex mark as top features. Then applied Random Forest, XG-Boost, Support Vector Machine, Logistic Regression and Naive Bayes. Where, SVM (79%) and LR (79%) both has given better accuracy. Patra and Khuntia [12] has proposed a model where they applied Information gain concept for best attribute selection then used J48, KNN, RBF, Naive Bayes with the support of weka tool and KNN, SVM, Decision Tree through the support of python tool. Where, decision tree classification technique holds good accuracy 93.4%. Naseer et al. [13] in this paper they has described a Mamdani Fuzzy Inference based expert system for diagnosing heart illness that effectively has recognized heart disease. The current study has taken six favorable factors in an exploratory pattern for the goal of fuzzy logic technical advancement in the analysis of heart disease and overall performance of the proposed DHD-MFI expert system is 94.00%. Kasbe and Pippal [14] this article has employed a fuzzy expert system for heart disease diagnosis. The Cleveland heart disease dataset has run in this study where 13 input and 1 output parameters has used in the proposed fuzzy expert system. As a development tool, MATLAB is utilized. The proposed system has given 93.33% accuracy.

Chauhan et al. [15] in this paper they used data mining methods to predict heart disease and had a 60% success rate. Jesmin Nahar et al. [16] here they used UCI Cleveland dataset and apply three algorithms for generating rules Ttertius, Predictive Apriori and Apriori. Anooj [17] a weighted fuzzy rule-based CDSS for risk estimation of heart disease patients was introduced in this paper.

2.1 *Justification*

In the previous work [10], the authors applied LMT techniques, Hoeffding Tree, Random Forest, SVM and Gaussian NB over Cleveland dataset that contain 303 cases. The results pointed out that Random forest works better and the initial and final accuracy was 88.52% and 95.08% respectively. In our experiment, we used two methods for marking features to identify the most related features that have a strong effect on the Cleveland heart disease dataset, and then we used KNN, SVM, Decision Tree, Stacking and Voting on each of the feature selection methods. Stacking with nine extracted features from the Pearson's Correlation Heatmap generated better results, with an accuracy of 97.00%.

3 Proposed Work

The method of extracting useful information from raw data is known as data mining. The all process is done by following some steps for analyzing properly the desired output. The steps are like data collection, data preprocessing, handling the data and applying machine learning algorithms etc.

In our proposed work, at first we applied data preprocessing, due to this we worked on missing values, handle the noisy data because the existence of these values make the process complex for analyzing a dataset. In the Cleveland dataset some values are absent, so many ways to handle absent values but here we just replace it with zero. Then we use Pearson's Correlation Heatmap and Chi-squared test two effective feature selection methods. Those methods at first make a ranking by calculating independent variables with respect to target value and Pearson's Correlation Heatmap rated attributes by calculating one another relations. The dataset is analyzed using tenfold cross validation through K-Nearest Neighbors, SVM, Decision Tree classification algorithms. Beside this, Stacking and Voting ensemble methods similarly give an outcome to predict the diseases. Two rank based collection of features used after preprocessing where the rank value indicates how tightly each variable is attached in the dataset. These feature selection algorithms are capable of dealing with large amounts of information. Classification is described as the procedure of identifying each target class while also preparing the dataset to return a desired state of boundary.

Here we used K-Nearest Neighbor, Decision Tree and SVM and applied ensemble techniques Stacking and Voting also, it built up the thought that not to depend only on a single decision, take many and come to a decision on the basis of overall outputs. It gathers all outputs from poor learners and decides on the final one based on those. A different model can be used for each base learner. We run XG Boost models, Gradient Boosting, Ada Boost and among them select the best performance by Voting and KNN, SVM, Decision Tree are run and among them Stacking allows us to choose the best results. Recall, ROC accuracy, F-measure and Precision were also measured to see how well the predictive model performed (Fig. 1).

3.1 Dataset

We use Cleveland datasets for early heart disease prediction in our model. This is the original dataset of heart disease presented by UCI Machine Learning Repository. There are 303 samples and 14 attributes in total, as well as a target attribute. This Cleveland datasets contain 76 attributes but among them a subset of 14 attributes used, based on the reference of all published experiments. Patients without heart disease have a target attribute of 0 and patients with heart disease have a target attribute of 1. In Cleveland dataset 164 samples belong to class 0 and 139 belong to class 1.This data helps to predict heart disease early. The features that exist inside the dataset are shown in Table 1.

3.2 Experimental Design

To find out the most related features, we picked features using the Pearson's Correlation Heatmap and Chi-squired test methods. This improves the accuracy of our model for predicting diseases. We have collected six features by using Chi-squired test; we get nine features from another method. Then, ruled KNN, SVM, Decision Tree and ensemble techniques (Voting and Stacking).

Beside this, for raising the accuracy of the model ensemble techniques are applied. When we are performing any of the machine learning techniques to predict the target variable, bias and variance are the key difference in predicted and actual values. To minimize these factors here we try to use ensemble methods. By measuring multiple

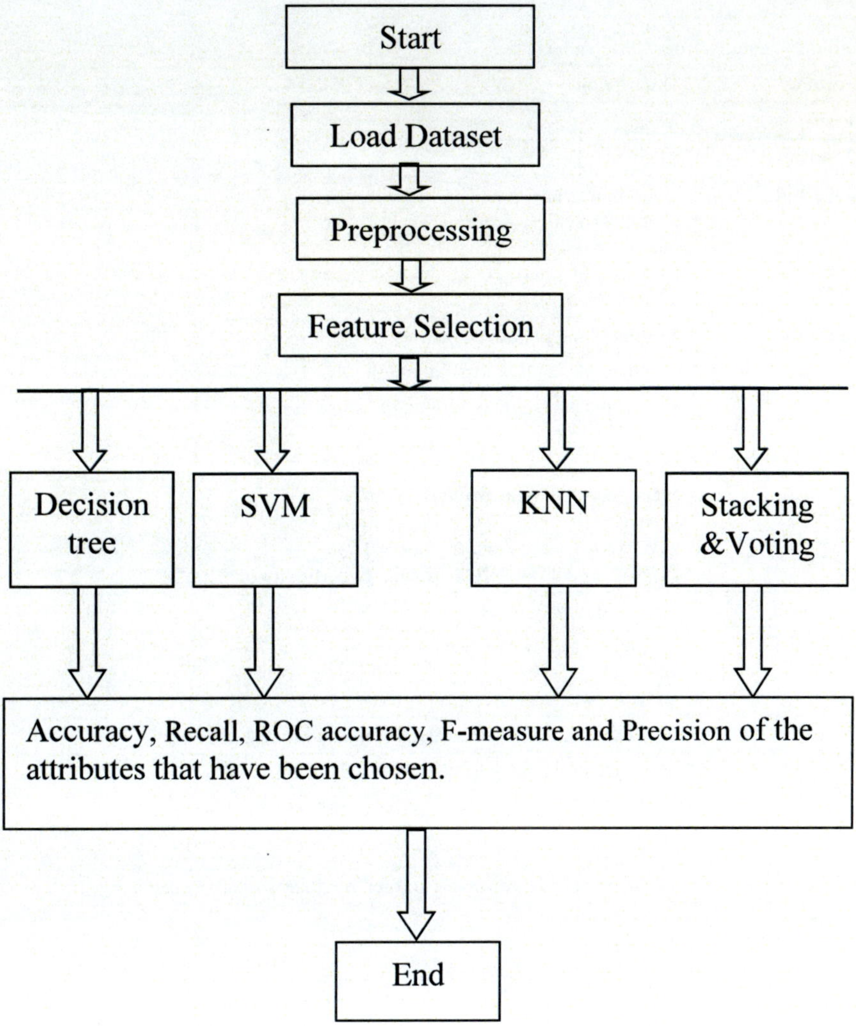

Fig. 1 Flowchart of proposed model

models, ensemble learning aims to improve machine learning performance. Table 2 shown all experimental schemas,

4 Outcome

For the implementation purpose in this work, 'python 3' programming language and scikit library has been used. We applied the criterion value as 'gini' in the decision

Table 1 Features from Cleveland datasets

Features	Description
1. age	Years
2. sex:	Male = 1; Female = 0)
3. cp	Chest pain type angina (typical) = 1 angina (atypical) = 2 pain (non-anginal) = 3 asymptomatic = 4
4. trestbps	Resting blood pressure
5. chol	Serum cholestoral in mg/dlin mg/dl
6. fbs	Blood sugar (fasting) > 120 mg/dl true = 1; false = 0
7. restecg	Electrocardiographic results (Resting) normal = 0 having ST-T wave Abnormality = 1 Possible or definite left ventricular hypertrophy by Estes' criteria = 2
8. thalach	Achieved maximum heart rate
9. exang	Angina caused by exercise yes = 1 no = 0
10. oldpeak	Exercise-induced ST depression compared to rest
11. slope	The slope of the ST portion of the peak exercise up sloping = 1 flat = 2 down sloping = 3
12. ca	The number of large vessels colored by flourosopy (between 0 and 3)
13. thal	Normal = 3; fixed defect = 6; reversable defect = 7
14. num	Without heart disease = 0 with heart disease = 1

tree algorithm. In our proposed model at first we pre-processed our data by removing noisy values then applied two methods Chi Squared Test and Pearson's Correlation Heatmap for deciding features. After that we applied three classification algorithms. Then we used stacking ensemble technique where we run KNN, SVM and Decision Tree algorithms and stacking ensemble technique calculated the most relevant output from those algorithms. We also applied three boosting techniques and used voting ensemble methods then generated the possible output. When we loaded the Cleveland datasets in our model after pre-processing we applied two feature selection methods Chi-squired test and Pearson's Correlation Heatmap. Chi squired test extracted six features from the dataset based on the top rank. The extracted attributes are shown in Table 3.

Table 2 Experimental design

Schema No.	Feature selection	Schema initial	No. of features
1	Chi squires test	DT-6	6
2	Chi squires test	SVM-6	6
3	Chi squires test	KNN-6	6
4	Chi squires test	Stacking-6	6
5	Chi squires test	Voting-6	6
6	Correlation heatmap	DT-9	9
7	Correlation heatmap	SVM-9	9
8	Correlation heatmap	KNN-9	9
9	Correlation heatmap	Stacking-9	9
10	Correlation heatmap	Voting-9	9

Table 3 Selected features from Cleveland dataset

Chi squired test	Correlation heatmap
1. thalach	1. thalach
2. ca	2. ca
3. oldpeak	3. oldpeak
4. thal	4. thal
5. exang	5. exang
6. chol	6. cp
	7. slope
	8. sex
	9. age

After the chi squared test method we run extracted features over three classifiers, stacking ensemble method, three boosting and voting ensemble method. Where the accuracy of K-Nearest Neighbor, SVM and Decision Tree 73.00%, 93.33% and 77.00% respectively. Accuracy from the boosting techniques Ada boost given 87.10%, Gradient boost 87.10% and XG boost 90.30%.The Ensemble techniques Stacking and Voting has given accuracy 93.00% and 90.00%. Using this feature selection methods SVM generated best output and the accuracy 93.33%, precision 100%, recall 92.00%, f1_value 92.00% and roc_auc 95.00% (Fig. 2).

From Pearson's Correlation Heatmap we selected nine features from the dataset based on the top rank. Next Pearson's Correlation Heatmap method, in the same way we run selected features over three classifiers, stacking ensemble method, three boosting and voting ensemble method. Where the accuracy of K-Nearest Neighbor,

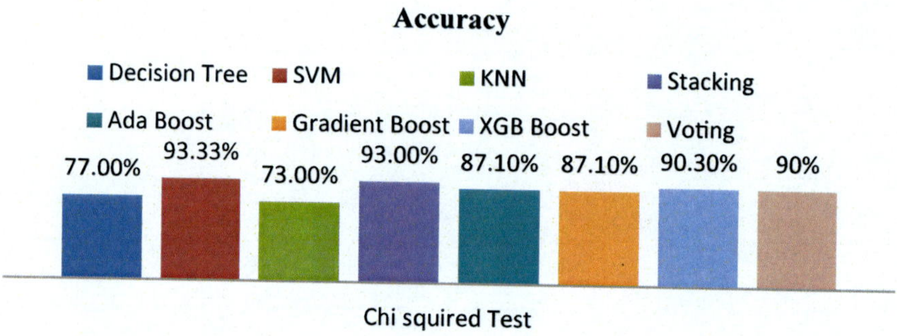

Fig. 2 Accuracy for the Chi squired test feature selection

SVM and Decision Tree 77.00%, 96.66% and 90.32% respectively. Accuracy from the boosting techniques Ada boost given 86.90%, Gradient boost 86.90% and XG boost 88.50%. The Ensemble technique Stacking and Voting has given accuracy 97.00% and 88.50%. Using this feature selection methods Stacking generated best output and the accuracy 97.00%, precision 100%, recall 100%, f1_value 96.00% and roc_auc 96.00% (Fig. 3).

From above all results we can define easily that using Pearson's Correlation Heatmap with Stacking ensemble technique has given better result where the accuracy 97.00% (Figs. 4 and 5).

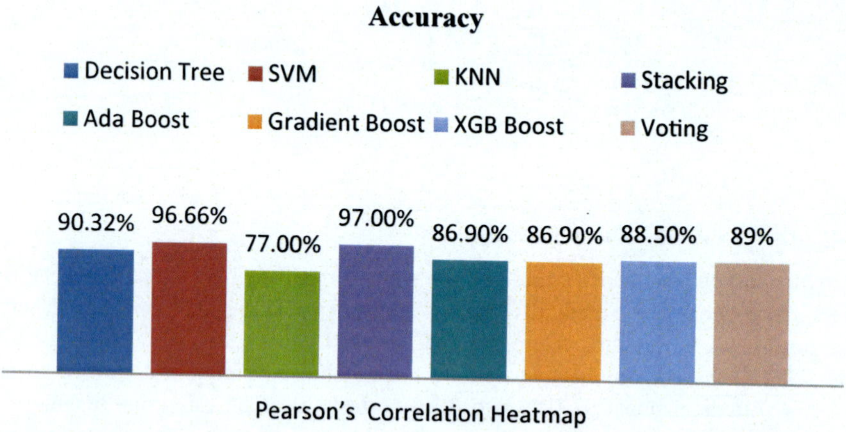

Fig. 3 Accuracy for the pearson's correlation heatmap feature selection

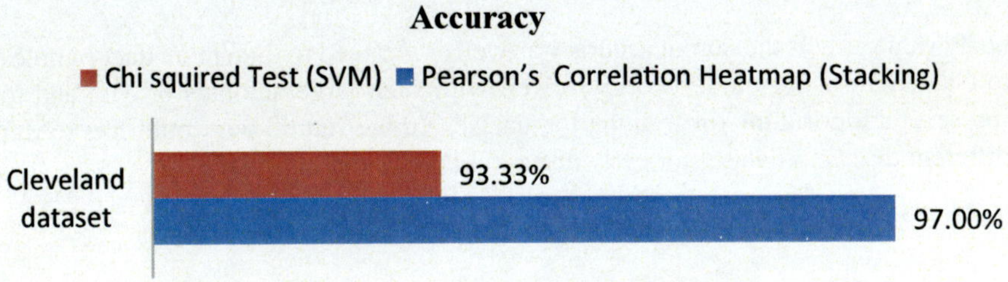

Fig. 4 Best accuracy from pearson's correlation heatmap and chi squired test feature selection

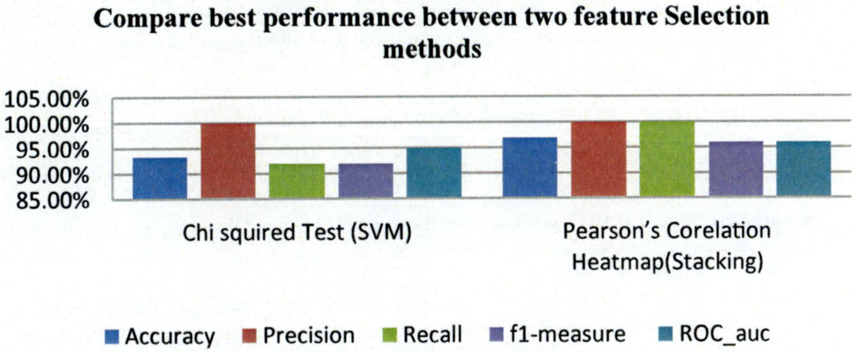

Fig. 5 Compare best performance between pearson's correlation heatmap and chi squired test

5 Conclusions

The model we proposed here has four phases: first, we gathered relevant data, and then we used two methods for selecting features to identify the far more common characteristics that were traced with a decent sign over the output. Utilizing 10-fold cross validation, the third stage consists of K-Nearest Neighbors (KNN) , SVM (SVM) and Decision Tree, after applying those algorithms, we have further used Stacking and Voting ensemble techniques for better results. Using chi squired test the accuracy of K-Nearest Neighbor, SVM and Decision Tree is 73.00%, 93.33% and 77.00% respectively with ensemble techniques Stacking and Voting has given accuracy 93.00% and 90.00%. On the other side using Pearson's Correlation Heatmap the accuracy of K-Nearest Neighbor, SVM and Decision Tree is 77.00%, 96.66% and 90.32% respectively, with ensemble techniques Stacking and Voting has given accuracy 97.00% and 88.50%. Our model performed better when we have used Pearson's Correlation Heatmap with Stacking, giving us 97.00% accuracy. Here we have used machine learning algorithms because it has the capability to make possible decisions based on the input data. Some limitations exist in every model. Our model has also some limitations that it takes more time to generate outcomes due to large amount of data where processing the raw data is also an important part because noisy value affects the output accuracy as well as we need to train more data samples to classify how our model works so well with such large amounts of data and to run several algorithms for training the dataset. In the future, we would like to run different disease's dataset for early disease prediction in our model.

References

1. Normawati D, Winarti S (2018) Feature selection with combination classifier use rules-based data mining for diagnosis of coronary heart disease. In: 2018 12th international conference on telecommunication systems, services, and applications (TSSA). https://doi.org/10.1109/tssa.2018.8708849
2. Krishnaiah V, Srinivas M, Narsimha G, Chandra N (2014) Diagnosis of heart disease patients using fuzzy classification technique. Int Conf Comput Commun Technol. https://doi.org/10.1109/iccct2.2014.7066746
3. Cardiovascular diseases (2021) Retrieved 21 April 2021. From https://www.who.int/health-topics/cardiovascular-diseases/#tab=tab_1
4. Junaid M, Kumar R (2020) Data science and its application in heart disease prediction. In: 2020 international conference on intelligent engineering and management (ICIEM). https://doi.org/10.1109/iciem48762.2020.9160056
5. Sonawane J, Patil D (2014) Prediction of heart disease using multilayer perceptron neural network. In: International conference on information communication and embedded systems (ICICES2014). https://doi.org/10.1109/icices.2014.7033860
6. Pawlovsky A (2018) An ensemble based on distances for a KNN method for heart disease diagnosis. 2018 international conference on electronics, information, and communication (ICEIC). https://doi.org/10.23919/elinfocom.2018.8330570
7. Cardiovascular diseases (2021) Retrieved 24 April 2021. From https://www.who.int/health-topics/cardiovascular-diseases/#tab=tab_3
8. Nassif A, Mahdi O, Nasir Q, Talib M, Azzeh M (2018) Machine learning classifications of coronary artery disease. In: 2018 international joint symposium on artificial intelligence and natural language processing (Isai-NLP). https://doi.org/10.1109/isai-nlp.2018.8692942
9. Abu Yazid M, Haikal Satria M, Talib S, Azman N (2018) Artificial neural network parameter tuning framework for heart disease classification. In: 2018 5th international conference on electrical engineering, computer science and informatics (EECSI). https://doi.org/10.1109/eecsi.2018.8752821
10. Motarwar P, Duraphe A, Suganya G, Premalatha M (2020) Cognitive approach for heart disease prediction using machine learning. In: 2020 international conference on emerging trends in information technology and engineering (IC-ETITE). https://doi.org/10.1109/ic-etite47903.2020.242
11. Prakash C, Madhu Bala M, Rudra A (2020) Data science framework—heart disease predictions, variant models and visualizations. In: 2020 international conference on computer science, engineering and applications (ICCSEA). https://doi.org/10.1109/iccsea49143.2020.9132920
12. Patra R, Khuntia B (2019) Predictive analysis of rapid spread of heart disease with data mining. In: 2019 IEEE international conference on electrical, computer and communication technologies (ICECCT). https://doi.org/10.1109/icecct.2019.8869194
13. Naseer I, Khan BS, Saqib S, Tahir SN, Tariq S, Akhter M S (2020) Diagnosis heart disease using Mamdani fuzzy inference expert system. EAI Endorsed Trans Scalable Inf Syst 7(26). https://eudl.eu/doi/10.4108/eai.15-1-2020.162736
14. Kasbe T, Pippal R (2017) Design of heart disease diagnosis system using fuzzy logic. In: 2017 international conference on energy, communication, data analytics and soft computing (ICECDS). https://doi.org/10.1109/icecds.2017.8390044

15. Chauhan A, Jain A, Sharma P, Deep V (2018) Heart disease prediction using evolutionary rule learning. In: 2018 4th international conference on computational intelligence and communication technology (CICT). https://doi.org/10.1109/ciact.2018.8480271
16. Nahar J, Imam T, Tickle KS, Chen YPP (2013) Association rule mining to detect factors which contribute to heart disease in males and females. Expert Syst Appl 40(4):1086–1093. https://dl.acm.org/doi/abs/10.1016/j.eswa.2012.08.028
17. Anooj P (2011) Clinical decision support system: risk level prediction of heart disease using weighted fuzzy rules and decision tree rules. Open Comput Sci 1(4). https://doi.org/10.2478/s13537-011-0032-y

Early Stage Detection of Heart Failure Using Machine Learning Techniques

Zulfikar Alom, Mohammad Abdul Azim, Zeyar Aung, Matloob Khushi, Josip Car, and Mohammad Ali Moni

Abstract With a devastating health impact, heart attack prediction is an essential aspect of human health due to well understood early heart attack symptoms. The recent advancement of Artificial Intelligence (AI) and Machine learning (ML) provides a significant part in illness detection as well as prediction upon many phenomena. This makes AI and ML great techniques to predict heart attack prediction. This research chose the well-known Logistic Regression (LR), Naive Bayes (NB), Random Forest (RF), Decision Tree (DT), Support Vector Machine (SVM), and k-Nearest Neighbor (k-NN) algorithms to predict heart attacks. A comparative study of the algorithmic performances is performed to identify the best algorithm that could be useful in the clinical decisions system.

Z. Alom · M. A. Azim
Department of Computer Science, Asian University for Women (AUW), Chattagram, Bangladesh
e-mail: zulfikar.alom@auw.edu.bd

M. A. Azim
e-mail: azim@ieee.org

Z. Aung
Department of Electrical Engineering and Computer Science, Khalifa University, Abu Dhabi, United Arab Emirates
e-mail: zeyar.aung@ku.ac.ae

M. Khushi
School of Computer Science, The University of Sydney, Sydney, Australia
e-mail: matloob.khushi@sydney.edu.au

School of EAST, University of Suffolk, Ipswich, UK

J. Car
Center of Population Health Sciences, Lee Kong Chian School of Medicine, Nanyang Technological University, Singapore, Singapore

Global eHealth Unit, Department of Primary Care and Public Health, School of Public Health,Imperial College London, London, UK

M. A. Moni (✉)
School of Health and Rehabilitation Sciences, Faculty of Health and Behavioural Sciences, The University of Queensland, St Lucia, QLD 4072, Australia
e-mail: m.moni@uq.edu.au

M. S. Arefin et al. (eds.), *Proceedings of the International Conference on Big Data, IoT, and Machine Learning*, Lecture Notes on Data Engineering and Communications Technologies 95, https://doi.org/10.1007/978-981-16-6636-0_7

Keywords Artificial intelligence · Machine learning · Heart disease · Biomedical · Healthcare

1 Introduction

With 17.9 million deaths due to cardiovascular disease globally, it is take out the proper take out priority to pay medical attention to patients with severity and potential subjectivity. A heart attack or Myocardial infarction (MI) is the utmost medical emergency where blood towards the heart is instantaneously occluded [4, 9]. This kind of unavailability from the blood in the center may grimly impair one's heart muscle and be frequently life-threatening.

The three types of heart episodes are ST-elevation myocardial infarction (STEMI), non-ST-elevation myocardial infarction (NSTEMI), as well as coronary spasm [8]. However, these leading causes of death, i.e., heart attack [16] early symptoms, are well understood. However, this knowledge is little utilized to save enough souls and minimize damages by detecting and preempting, and caring for the patients.

Artificial Intelligence (AI) is one of the advancements of technology that has emerged as a feasible tool in medical treatment in recent years. Most medical professionals and experts still depend on the primordial way of medical treatment. A part of the medical professionals seems that irremediable diseases might be cured flawlessly by using AI techniques.

Utilizing the AI diagnosing and finding the severity of the risk of the heart disease of a person help both an individual and the hospital decision support system. Consequently, this research focuses on finding the potential treat based on some small set of essential features and employing machine learning techniques.

The assembly of this paper is structured as follows. In the following Sect. 2, a literature assessment on the topic of concern has been discoursed. Section 3 represents the material and methods of the study used to develop the proposed model. Section 4 illustrates the experimental results, and finally, Sect. 5 concludes the paper.

2 Related Work

A number of literature has proposed the detection and prediction system for heart disease and heart attack probabilities [1, 24, 25]. Reference [18] studied machine understanding classification techniques in line with the Naive Bayes as well as Support Vector Machines in which the system showed accuracy as well as predicts attributes, for example, age, intercourse, blood stress, and blood sugar levels and the likelihood of a diabetic patient obtaining a heart illness.

A heart disease warehouse for heart attack prediction is proposed where the information is pre-processed, and data mining methods are in place [19]. Here, the K-means algorithm is used to cluster the relevant data related to a heart attack. Next, the MAFIA algorithm [7] is employed to find frequent patterns mined from the clustered data. Reference [22] proposed utilizing the Tanagra tool [10] to classify data while evaluating the system by N-fold cross ($N = 10$) validation. The NB, k-NN, decision list [23] methods are used as a classifier in this study.

Reference [17] applied the adaptive neuro-fuzzy inference system (ANFIS) [12], where heartbeat, exercise, bloodstream pressure, grow older, cholesterol, upper body pain kind, blood sugars, and sex would be the input towards the fuzzy techniques. The output provides four different outputs (small, low, substantial, and extremely high) of the heart attack probability. A random forest-based ML algorithm is used to classify the heart disease in reference [5]. Reference [5] proposed a mobile device-based heart disease detection system using the camera, mobile stethoscope. A fuzzy-based data mining is adopted as a decision system.

Reference [6] proposed a decision system for the predicted heart attack risk by utilizing the Bagging method, an ensemble machine learning classifier. Reference [15] proposes an IoT-based heart rate monitoring system and intelligent blood pressure system to accommodate heart attack detection. Reference [21] proposed a primary cardiac disease risk prediction method utilizing Classification Tree, NB, RF, and SVM algorithms. A majority voting ensemble method for heart disease prediction technique is presented in [3]

Reference [13] utilized RF, DT, and Hybrid models (Hybrid of RF and DT) on Cleveland heart failure dataset to predict heart illness. Reference [26] provided a comparative performance analysis of DT, NB, SVM, k-NN, LR, and RF. Reference [8] developed a wearable sensor subsystem and a smart heart assault detection as well as a warning subsystem. The sensor subsystem information the heart's electric activity utilizing an electrocardiogram (ECG) find, subsequently employing a portable decision-making subsystem, center attack signs and symptoms are discovered.

3 Proposed Methodology

In this section, we present the framework of the heart disease detection method the central architecture of the proposed system, as shown in Fig. 1. The proposed system consists of several steps, including data acquisition, pre-processing, feature/attribute selection, classifications, and performance evaluation, briefly described below.

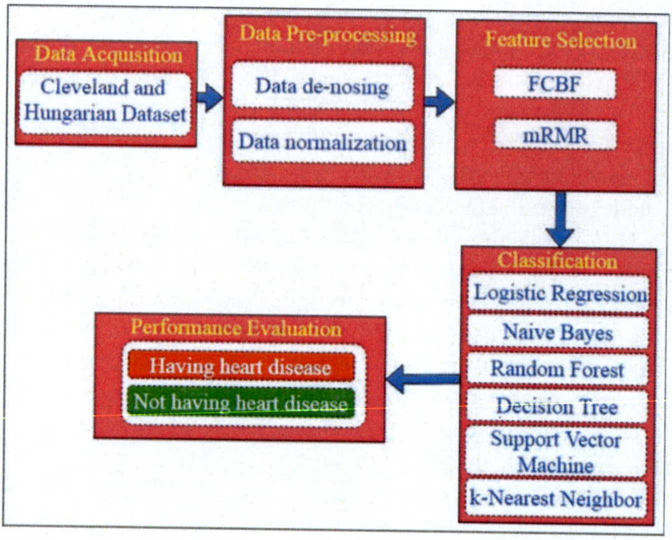

Fig. 1 The proposed architecture of heart disease detection model

3.1 Data Acquisition

We collected two datasets, specifically, the Cleveland cardiovascular disease dataset (Dataset I) and Hungarian cardiovascular disease dataset (Dataset II), available online in the University of California Irvine (UCI)[1] and Kaggle[2] repository. The datasets have two classes: having heart disease or not having heart disease (1 = yes, 0 = no). The Dataset I consists of 303 patients information, with pre-classified 165 patients with heart disease and 138 patients who do not have heart disease. Likewise, Dataset II consists of 1025 instances in which 525 instances belong to the heart disease class. In contrast, the rest of the 500 instances belong to not having a heart disease class. It is noteworthy to mention that both datasets have similar attributes. The properties of the datasets' attributes are shown in Table 1.

3.2 Data Pre-processing

Data pre-processing is a process associated with transforming uncooked data into meaningful patterns. In this step, we exploit the data de-noising and normalization techniques. Denoising is the task of removing noise (i.e., missing values, outliers) from the data. However, machine learning techniques provide the best overall per-

[1] https://archive.ics.uci.edu/ml/index.php.

[2] https://www.kaggle.com/ronitf/heart-disease-uci.

Table 1 Characteristics of the dataset

Feature	Attribute	Feature name	Type	Description	Values
f_1	*age*	Age	Numeric	Patient age	26–88
f_2	*sex*	Sex	Nominal	Male	1
				Female	0
f_3	*cp*	Chest muscles pain variety	Nominal	Atypical angina	0
				Normal angina	1
				Asymptotic	2
				Non-anginal ache	3
f_4	*trestbps*	Relaxing blood force	Numeric	mmHg	94–200
f_5	*chol*	Serum cholesterol	Numeric	in mg/dl	120–564
f_6	*fbs*	Fasting blood glucose levels	Nominal	True	1
				False	0
f_7	*restecg*	Resting electrocardiographic benefits	Nominal	Normal state	0
				Abnormality in $ST-T$	1
				LV hypertrophy	2
f_8	*thalach*	Maximum pulse rate achieved	Numeric	heart per minute (bpm)	71–202
f_9	*exang*	Exercise induced angina	Nominal	Yes	1
				No	0
f_{10}	*oldpeak*	ST depressive disorder induced by simply exercise in accordance with rest	Numeric	–	0–6.2
f_{11}	*slope*	The slope in the peak exercising ST message	Nominal	Up sloping	0
				Flat/no slope	1
				Down slope	2
f_{12}	*ca*	Number of major vessels colored by flourosopy	Nominal	–	0
					1
					2
					3
f_{13}	*thal*	Thalium stress result	Nominal	Normal	0
				Fixed defect	1
				Reversible defect	2

formance once the input information have 0 mean and 1 variance. Therefore, we normalize[3] the dataset to make it far better for the classification.

3.3 Feature Selection

The feature selection technique selects the most beneficial features among each of the features within a dataset. It is noteworthy to mention that due to irrelevant features in the dataset, the classification performance degrades. Therefore, we used the attribute selection technique to enhance classification accuracy. In this step, Fast Correlation-Based Filter (FCBF) [28] and Minimal Redundancy Maximal Relevance (mRMR) [20] methods are used for the selection of optimal features to send the classifications step.

3.4 Classifications

In this subsection, we present the machine learning techniques used to detect heart disease. In the experiment, we used six classification models, namely, Logistic Regression (LR), Naive Bayes (NB), Random Forest (RF), Decision Tree (DT), Support Vector Machine (SVM), and k-Nearest Neighbor (k-NN), briefly described in what follows [14].

Logistic Regression. LR is a statistical technique, that discovers a formula that forecasts an outcome for any binary variable (i.e., Y) in one or several input parameters (i.e., X). Throughout logistic regression, linear regression productivity is passed over the activation function called your Softmax function. This function enables to calculation of the possibilities of the events. However, the output on this function is actually in the stove [0,1], and the sum of the productivity values is adequate to 1. More technically, LR classifies insight to type 1 if the output on this function can be closed to 1 along with classifies to class 2 if the output can be closed to 0.

Naive Bayesian. NB classifier is dependant on the likelihood theory (i.e., Bayes theorem) [14]. This design is popular because this gives a great performance as well as requires much less computational period for instruction the design. It calculates some probabilities through counting the actual frequency as well as combinations associated with values inside a dataset. The probability of the feature within the dataset comes by determining the rate of recurrence of function value inside a class of the training information set. Usually, the instruction dataset is a subset of the

[3] To accomplish this, first, we estimate the necessarily mean value along with standard deviation of each one feature. Up coming, we take away the necessarily mean value via each attribute. Finally, we divide the significance of every single feature by simply its normal deviation. Mathematically: $x' = \frac{x-\overline{x}}{\sigma}$, where x will be the original attribute vector, $\overline{x}$ will be the mean value of these feature vector, along with σ can be its normal deviation.

dataset accustomed to train the actual classifier model by utilizing known ideals for forecasting unknown ideals.

More technically, the Naive Bayesian classifier can be described as follows [11]: $P(A|B) = \frac{P(B|A)*P(A)}{P(B)}$, where, $P(A)$ as well as $P(B)$ would be the probability associated with A and B, respectively. They are called the last probability, and their own values could be computed in the training information. $P(B|A)$ is known as the conditional likelihood, which indicates the likelihood of B because A occurs. $P(A|B)$ indicates the likelihood of A because B occurs. It is known as the posterior likelihood.

However, the NB classifier algorithm works as follows, where each patient's information is modeled as a vector D with its feature values:

1. Let TD function as the training dataset along with class labeling. Each person's record is actually represented through an n-dimensional vector, $D = (d_1, d_2, \ldots\ldots, d_n)$.
2. Consider that there are m classes (in our case, $m = 2$) $c_1, c_2, c_3, \ldots., c_m$. Let U be an unlabeled patient's record which we want to classify, the classifier will forecast that U belongs to the class with the highest posterior likelihood. More particularly, the NB classifier assigns patient U to the class *having the disease* if and only if $P(\text{disease}|U) > P(\text{not-disease}|U)$.

Random Forest. RF is a supervised algorithm used in machine learning to do regression and classification tasks. This algorithm has a large number of small decision trees, each one of them called an estimator. Each estimator generates its predictions, and then the random forest generates an accurate prediction from the combination of them. However, compared to other algorithms, a random forest algorithm has the following advantages: It avoids the overfitting[4] problem; it might be used intended for both classification and regression chores, and it might be used intended for large datasets.

The RF algorithm includes two stages. The first example may be to generate a random forest, and the second reason is to come up with a prediction through the random forest classifier. However, the randomly selected RF classifier algorithm works as follows.

1. In the first stage, we start selecting the data points randomly from the total data points. Then, we use the best split point method to calculate the node and split it into smaller nodes. We repeat those steps until we get the total number of trees we want.
2. In the second stage, we assign a number for each decision tree and then find the predictions for each decision tree. After that, we calculate the votes for each prediction to find the winner category.

[4] Overfitting is a modeling error within the machine understanding methods, it occurs whenever a classifier fits working out (training) data as well tightly and does not generalize well to test data.

Decision Tree. DT is a compelling and popular machine understanding classifier due to the simplicity, also it gives great results by utilizing less storage. The steps performed by the DT classifier are as follows:

1. Set the most effective attribute (according to information acquired and entropy) with the dataset because of the root with the tree.
2. Divide working out (training) dataset into subsets, where every subset consists of data using the same value to have an attribute.
3. Repeat step 1 and step 2 on every single subset until eventually finding leaf nodes as well as terminal nodes to all the branches in the tree.
4. Repeat step 1 and step 2 for every subset till setting the actual class content label (i.e., leaf node) in most the branches from the decision sapling.

Support Vector Machine. SVM is the plane-based category algorithm which constructs the discrete hyperplane, which maximizes the actual margin between two classes. The primary objective associated with SVM would be to reveal the very best hyperplane within training data between two classes. A perfect SVM creates a hyperplane that completely sets apart the vectors into two non-overlapping courses. However, perfect separation might not be possible, so in this instance, SVM discovers the hyperplane that maximizes the margin as well as minimizes the mis-classifications.

***k*-Nearest Neighbors**. k-NN algorithm is a supervised algorithm that is widely used in pattern recognition and statistical estimation. This algorithm was proposed by Thomas Cover and is still being used to solve both classification and regression problems. In k-NN, an entity is classified by a plurality vote of its neighbors. This particular voting assigns the actual entity the class that's most typical among its k closest. Note which k this is a small integer. If $k = 1$, then the object is merely assigned toward the class of this single closest neighbor. However, the steps performed by the k-NN classifier are as follows.

1. Determine the parameter K, i.e., the quantity of nearest others who live nearby.
2. Calculate the Euclidean, Manhattan, or Hamming distance involving the query illustration and every one of the training trials.
3. Sort the length to look for the nearest neighbors concerning the K_{th} minimal distance.
4. Collect the closest neighbor class. The simple most of the group of nearest neighbors may be the defined category.

3.5 *Performance Evaluation*

To be able to evaluate the actual performance in our approach, we used four metrics as accuracy, precision, recall, and f1-score, briefly described in Sect. 4.1. However, according to the experimental results (see Sect. 4), we can see that SVM and RF gives the highest accuracy over Dataset I and Dataset II, respectively.

4 Experimental Results

We present the experimental evaluation benefits we accomplished to show the potency of the recommended approach. Over the experiments in two real-world coronary disease datasets, we tried to discover answers to the following pair of questions:

- how effective could the proposed strategy be for discovering heart failure?
- which machine learning algorithm is more effective to detect heart failure?

To be able to answer these types of questions, first of all, we expose the overall performance metrics. After that, we existing the fresh setup for that experiments. Finally, we living the overall performance evaluation in addition to analysis as well as discussion from the results.

4.1 Performance Metrics

In order to evaluate the performance of our approach, we consider the evaluation matrix illustrated in Table 2. It shows four variables: True Positives (TP), False Positives (FP), True Negatives (TN), and False Negatives (FN). Here, TP means the patient has the disease, and the models have correctly classified it as a disease. FP represents the patient does not have the disease, but the models wrongly misclassified as a disease. TN refers to the patient who does not have the disease, and the models correctly classified as Not-disease. FN expresses that the patient has the disease, but the models wrongly misclassified it as Not-disease.

According to the evaluation matrix, we used four standard metrics, namely accuracy, precision, recall, and f1-score, briefly described in what follows [2].

- **Accuracy (A)** is the ratio of the correctly classified observations over the total number of observations in the dataset and is expressed by $A = \frac{(\text{TP+TN})}{(TP+FN+FP+TN)}$.
- **Precision (P)** refers to the ratio of correctly predicted observations to the total number of predicted observations and is expressed by $P = \frac{\text{TP}}{(\text{TP+FP})}$.
- **Recall (R)** defines the ratio of correctly predicted observations to the total observations in actual class and is expressed by $R = \frac{\text{TP}}{(\text{TP+FN})}$.
- **F1-score (F1)** is measured to be a weighted average on the precision in addition to recall, defined as $F1 = \frac{2P*R}{(P+R)}$.

Table 2 Evaluation matrix

		Predicted	
		Disease	Not-disease
True	Disease	TP	FN
	Not-disease	FP	TN

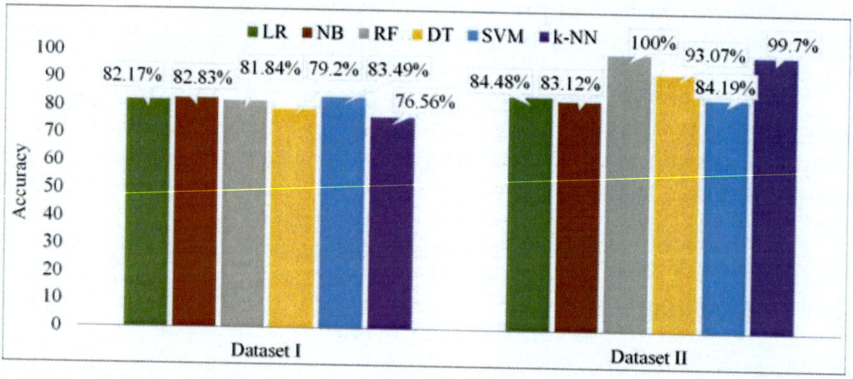

Fig. 2 Accuracy of the LR, NB, RF, DT, SVM and k-NN algorithms for Dataset I and Dataset II

4.2 Experimental Setup

On the experiments, we put into use Google Colab,[5] that is a product right from Google Explore. It will allow for anybody for you to and conduct arbitrary python code on the browser. A great deal more technically, Colab is mostly a hosted Jupyter portable service that requires no setup to try while featuring free permission to access computing strategies, including GPUs.

4.3 Evaluation

Here, we evaluate the performance of the different ML algorithms on Dataset I and Dataset II, respectively.

Over Dataset I, we can see that the accuracy of the SVM model is greater than the others, as shown in Fig. 2. It reaches the highest accuracy of 83.49% for the SVM classifier. Similarly, in terms of $f1$-score SVM gives better results (i.e., 85.9%), as shown in Fig. 5. Likewise, the precision and recall value of the SVM model is 80.4% and 92.1%, respectively, which are also more significant than the other machine learning approaches, as shown in Figs. 3 and 4.

[5] https://colab.research.google.com/.

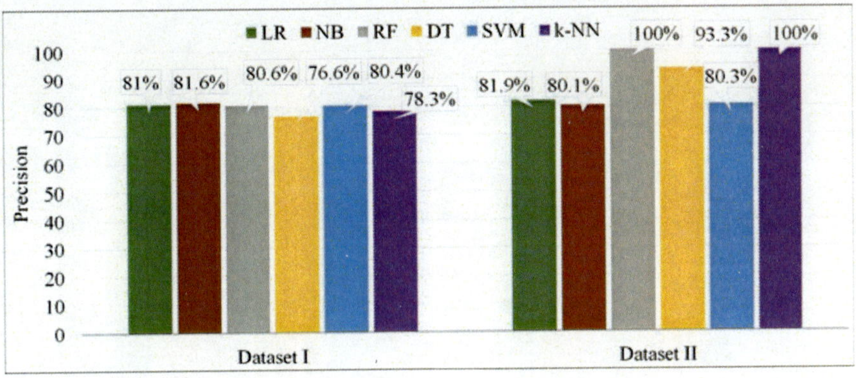

Fig. 3 Precision values of the LR, NB, RF, DT, SVM and k-NN algorithms for Dataset I and Dataset II

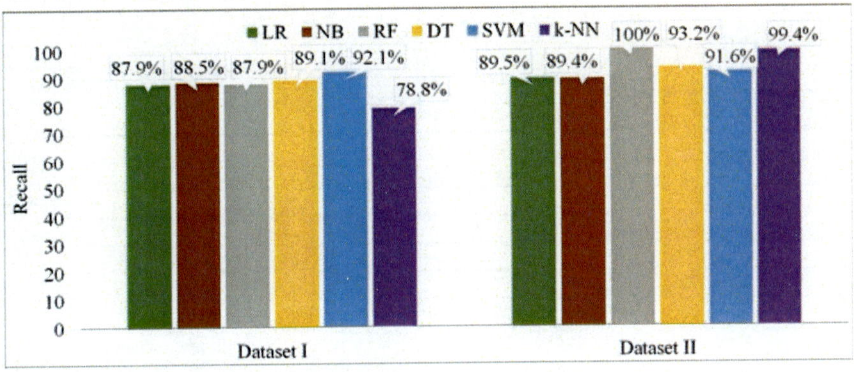

Fig. 4 Recall values of the LR, NB, RF, DT, SVM and k-NN algorithms for Dataset I and Dataset II

Over Dataset II, from Fig. 5, it can be seen that the $f1$-score of RF is 100%, which is greater than the other ML approaches. Similarly, from Figs. 2, 3 and 4, we see that RF gives accuracy, precision, and recall values are the same that is 100%.

Feature ranking: To be able to verify the significance of the actual features, everyone used the actual feature ranking method. To get this done, we used the info gain function selection method that's available on Weka [27]. Weka facilitates feature choice via info gain while using $Info\_Gain\_Attribute\_Eval$ feature evaluator. It calculates the entropy (i.e., info gain) from each attribute. This value range from 0 to 1. The benefits that contribute additional information will enjoy a higher knowledge gain value and that can be chosen, whereas individuals that do not likely add a whole lot of information are going to have a smaller score and that can be cleaned up and removed. Information gain is calculated the following: $IG(M, P_i) = H(M) -$

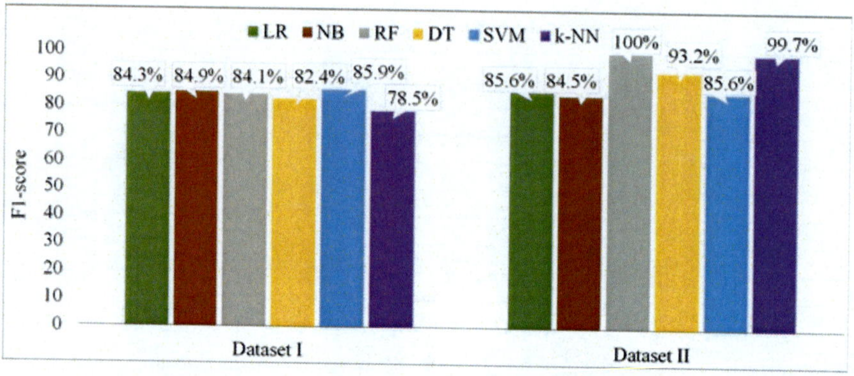

Fig. 5 $F1$-score of the LR, NB, RF, DT, SVM and k-NN algorithms for Dataset I and Dataset II

Table 3 Top 10 features

Rank	Information gain
1	Thalium stress result (*thal*)
2	Chest muscles pain variety (*cp*)
3	Number of major vessels colored by flourosopy (*ca*)
4	ST depressive disorder induced by simply exercise in accordance with rest (*oldpeak*)
5	Maximum pulse rate achieved (*thalach*)
6	Exercise induced angina (*exang*)
7	The slope in the peak exercising ST message (*slope*)
8	Age (*age*)
9	Serum cholesterol (*chol*)
10	Sex (*sex*)

$H(M|P_i)$, where M certainly is the output quality, P_i and additionally H certainly is the entropy.

Table 3 listed the best 10 valuable attributes within all features from Dataset I and Dataset II.

However, according to the experimental results (as shown in Table 4), it can be observed that the SVM algorithm gives the highest accuracy (about 83.49%) on Dataset I, and the RF algorithm provides the highest accuracy (about 100%) on Dataset II, to predict whether the patient has heart disease or not.

Table 4 Performance comparison of different learning algorithms

		LR (%)	NB (%)	RF (%)	DT (%)	SVM (%)	k-NN (%)
Dataset I	Accuracy	82.17	82.83	81.84	79.2	83.49	76.56
	Precision	81	81.6	80.6	76.6	80.4	78.3
	Recall	87.9	88.5	87.9	89.1	92.1	78.8
	$F1$-score	84.3	84.9	84.1	82.4	85.9	78.5
Dataset II	Accuracy	84.48	83.12	100	93.07	84.19	99.7
	Precision	81.9	80.1	100	93.3	80.3	100
	Recall	89.5	89.4	100	93.2	91.6	99.4
	$F1$-score	85.6	84.5	100	93.2	85.6	99.7

5 Conclusion

Heart disease is one of the deadliest and fatal chronic diseases on the rise, even though its early symptoms are well understood. The diagnosis of the disease is the key to reduce the damage considerably in the early stages. Subsequently, we have developed an intelligent predictive system based on contemporary machine learning algorithms to diagnose heart disease. The contributing steps of this research consist of data acquisition, data pre-processing, feature selection, classifications, and performance evaluation. The feature selection algorithms were FCBF, and mRMR. The LR, NB, RF, DT, SVM and k-NN algorithms were the models of concern of this research.

In future, we intend to improve the performance of the predictive system for the diagnosis of heart disease. We believe if we can apply a machine learning-based system similar to our proposed framework as a part of a decision support system in health care and clinic it will help to detect heart failure at the early stage.

References

1. Alexander CA, Wang L (2017) Big data analytics in heart attack prediction. J Nurs Care 6(393):2167–1168
2. Alom Z, Carminati B, Ferrari E (2018) Detecting spam accounts on twitter. In: 2018 IEEE/ACM international conference on advances in social networks analysis and mining (ASONAM). IEEE, pp 1191–1198
3. Atallah R, Al-Mousa A (2019) Heart disease detection using machine learning majority voting ensemble method. In: 2019 2nd international conference on new trends in computing sciences (ICTCS). IEEE, pp 1–6
4. Bhopal RS (2019) Epidemic of cardiovascular disease and diabetes: explaining the phenomenon in south Asians worldwide
5. Buettner R, Schunter M (2019) Efficient machine learning based detection of heart disease. In: 2019 IEEE international conference on e-health networking, application and services (HealthCom). IEEE, pp 1–6
6. Bulut F (2016) Heart attack risk detection using bagging classifier. In: 2016 24th signal processing and communication application conference (SIU). IEEE, pp 2013–2016

7. Burdick D, Calimlim M, Flannick J, Gehrke J, Yiu T (2005) Mafia: a maximal frequent itemset algorithm. IEEE Trans Knowl Data Eng 17(11):1490–1504
8. Chowdhury ME, Alzoubi K, Khandakar A, Khallifa R, Abouhasera R, Koubaa S, Ahmed R, Hasan A (2019) Wearable real-time heart attack detection and warning system to reduce road accidents. Sensors 19(12):2780
9. Gibson AL, Wagner D, Heyward V (2018) Advanced fitness assessment and exercise prescription, 8E. Human Kinetics
10. Hasim N, Haris NA (2015) A study of open-source data mining tools for forecasting. In: Proceedings of the 9th international conference on ubiquitous information management and communication, pp 1–4
11. Jadhav SD, Channe H (2016) Comparative study of KNN, Naive Bayes and decision tree classification techniques. Int J Sci Res 5(1)
12. Jang JS (1993) Anfis: adaptive-network-based fuzzy inference system. IEEE Trans Syst Man Cybern 23(3):665–685
13. Kavitha M, Gnaneswar G, Dinesh R, Sai YR, Suraj RS (2021) Heart disease prediction using hybrid machine learning model. In: 2021 6th international conference on inventive computation technologies (ICICT). IEEE, pp 1329–1333
14. Kotsiantis SB (2007) Supervised machine learning: a review of classification techniques
15. Manisha M, Neeraja K, Sindhura V, Ramaya P (2016) IoT on heart attack detection and heart rate monitoring. Int J Innov Eng Technol (IJIET)
16. Murray CJ, Lopez AD (1997) Global mortality, disability, and the contribution of risk factors: global burden of disease study. Lancet 349(9063):1436–1442
17. Opeyemi O, Justice EO (2012) Development of neuro-fuzzy system for early prediction of heart attack. Inf Technol Comput Sci 9(9):22–28
18. Parthiban G, Srivatsa S (2012) Applying machine learning methods in diagnosing heart disease for diabetic patients. Int J Appl Inf Syst (IJAIS) 3(7):25–30
19. Patil SB, Kumaraswamy Y (2009) Extraction of significant patterns from heart disease warehouses for heart attack prediction. IJCSNS 9(2):228–235
20. Ponsa D, López A (2007) Feature selection based on a new formulation of the minimal-redundancy-maximal-relevance criterion. In: Iberian conference on pattern recognition and image analysis. Springer, pp 47–54
21. Puyalnithi T, Viswanatham VM (2016) Preliminary cardiac disease risk prediction based on medical and behavioural data set using supervised machine learning techniques. Indian J Sci Technol 9:31
22. Rajkumar A, Reena GS (2010) Diagnosis of heart disease using datamining algorithm. Global J Comput Sci Technol 10(10):38–43
23. Rivest RL (1987) Learning decision lists. Mach Learn 2(3):229–246
24. Roopa C, Harish B (2017) A survey on various machine learning approaches for ECG analysis. Int J Comput Appl 163(9):25–33
25. Sharma H, Rizvi M (2017) Prediction of heart disease using machine learning algorithms: a survey. Int J Recent Innov Trends Comput Commun 5(8):99–104
26. Sujatha P, Mahalakshmi K (2020) Performance evaluation of supervised machine learning algorithms in prediction of heart disease. In: 2020 IEEE international conference for innovation in technology (INOCON). IEEE, pp 1–7
27. Witten IH, Frank E, Hall MA, Pal CJ (2016) Data Mining: Practical machine learning tools and techniques. Morgan Kaufmann
28. Yu L, Liu H (2003) Feature selection for high-dimensional data: a fast correlation-based filter solution. In: Proceedings of the 20th international conference on machine learning (ICML-03), pp 856–863

Artificial Intelligence for Imaging Applications

Automatic License Plate Recognition System for Bangladeshi Vehicles Using Deep Neural Network

Syed Nahin Hossain, Md. Zahim Hassan, and Md. Masum Al Masba

Abstract The goal of Automatic License Plate Recognition (ALPR) is localizing the license plate of a vehicle from an image and extracting text from it to recognize and track the vehicle. Each year, the amount of vehicles in Bangladesh is increasing at a significant rate. With the increasing number of vehicles, the intelligent transport system (ITS) has become essential. The automatic license plate recognition system (ALPRS) is a key part of ITS. The ALPRS can also help monitor traffic, surveillance of certain areas, crime investigations, etc. This paper has proposed an optimal end-to-end approach for the ALPR system for Bangladeshi vehicles by experimenting with the various deep neural network (DNN) models. These models have been trained and evaluated on our rich datasets of Bangladeshi vehicles and license plates. We have also introduced an algorithm that eliminates the need for the typical segmentation phase and generates properly formatted output efficiently. The final proposed system offers 99.37% accuracy in license plate localization and 96.31% accuracy in text recognition from the license plate (LP)s.

Keywords ALPR · YOLO · SSD · DNN · Bangla license plate recognition

1 Introduction

ALPRS has been widely used by many smart cities worldwide, and there is an amazing opportunity for developing an end-to-end optimal system for this sector. ALPR is one of the major key concerns for building an ITS. Bangladesh is a fast-growing country. Its economy is increasing rapidly. As a result, the number of vehicles is also increasing in the big cities both in the private and trading sectors. As a result, traffic is also increasing on the roads and becoming difficult to manage. Moreover, increasing the number of vehicles also increases the tendency of criminal activities related to vehicles such as drunk and drive, hit and run, kidnapping, car stealing, driving in the wrong lane as manually keeping track of these vehicles is very difficult.

S. N. Hossain (✉) · Md. Z. Hassan · Md. M. A. Masba
Khulna University of Engineering & Technology, Khulna, Bangladesh

M. S. Arefin et al. (eds.), *Proceedings of the International Conference on Big Data, IoT, and Machine Learning*, Lecture Notes on Data Engineering and Communications Technologies 95, https://doi.org/10.1007/978-981-16-6636-0_8

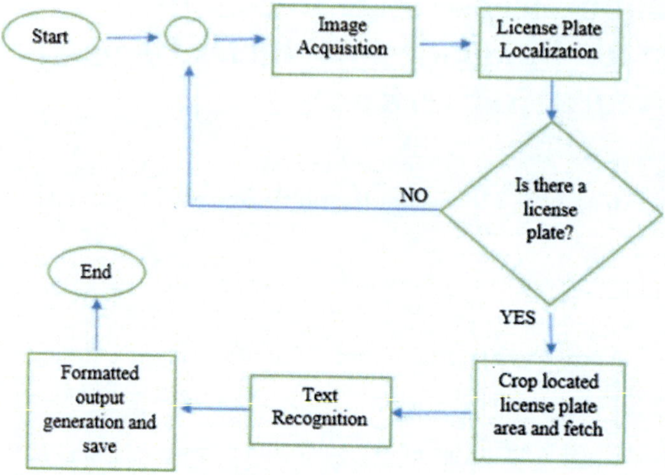

Fig. 1 Workflow diagram of our proposed system

ALPR can reduce these problems and be helpful in toll collections, border securities, investigation of crimes, etc. ALPR can be also helpful in an automatic parking space management system, in surveillance of any place or region.

An ALPRS has three main challenges as a whole. These are properly detecting a vehicle's license plate, segmenting the texts in the license plate, and recognizing the texts or characters or digits. Apart from these three challenges, another important challenge is to generate properly formatted output which is ready to use. This paper focuses on various ways of overcoming these challenges with the help of deep neural networks and presents the most feasible solution for a complete ALPR system. Experimenting with different DNN models, we have come up with the most efficient and robust solution in every stage. Figure 1. demonstrates our full workflow. We have merged the segmentation phase with the recognition phase to make this process easy. By doing such, we can save a lot of time and computational effort. Finally, we have presented our custom algorithm, which is computationally efficient and generates a properly formatted output for our system.

Contributions of this article are as follows:

1. We have shown the performance comparison of various DNN models on the Bangladeshi license plates.
2. We have introduced rich datasets of images of Bangladeshi license plates from major cities of Bangladesh.
3. We have presented an optimal end-to-end solution for the ALPR system.
4. We have preprocessed the raw images in different DNN model-specific formats.
5. We have created a dataset containing almost 2800 images for localization and around 4000 images for text recognition stages.
6. Our proposed model offers one of the most promising results with 27 classes.
7. We have created a custom algorithm to emphasize properly formatted output generation.

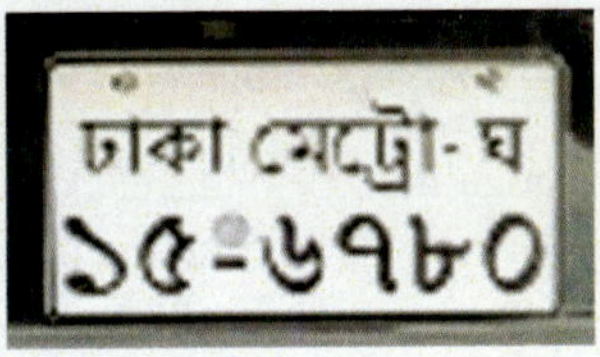

Fig. 2 BRTA standard license plate format

The rest of the paper is organized as follows, Sect. 2 is about introduction to the Bangladeshi standard license plate, Sect. 3 discusses the literature review, discussion about dataset is in Sect. 4, Our proposed methodology is discussed in Sect. 5, and Sect. 6 is about results and discussion. Finally, Sect. 7 has the conclusion followed by references.

2 Overview of Bangladeshi Standard License Plate

Bangladesh Road Transport Authority (BRTA) is a regulatory body to control, manage, and ensure discipline as well as to maintain safety in the road transport sector of Bangladesh. In 2012, as a part of digitalization, BRTA introduced a new vehicle license plate system called retro-reflective license plate mostly known as vehicle digital license plate. Since then, it has become mandatory for vehicles to attach this license plate to the rear side. The new digital license plate is of mainly two categories: private vehicle's license plate and trading vehicle's license plate. The color combinations are a white background with black text for private vehicles and green background with black text for trading vehicles.

The license plate has two separate rows of texts, characters, and digits as shown in Fig. 2. From the first row, the first word indicates the district name where the vehicle was registered. The second word is optional, if it is under the metropolitan area, then it is used to indicate the area. The only character in the first row separated by a hyphen indicates the vehicle category.

Coming to the second row, the first two digits in this row is the class registration number of the vehicle, and the next four digits again separated by a hyphen as a whole represent the serial number of the vehicle. It is mandatory to use Bangla language on the license plate.

3 Literature Review

The ALPR system has been a key sector of research for many years. Researchers worldwide have tried to develop this system in many ways. Quadri et al. [1] used a smearing algorithm for extracting plate region, then used row and column segmentation for OCR to recognize text from it. Shidore et al. [2] used the Sobel filter, morphological operations, connected component analysis along with vertical projection analysis. They also used the support vector machine(SVM) for character recognition. Lekhana et al. [3] have shown an approach using spectral analysis along with connected component analysis and SVM to recognize characters in the license plate. Astari et al. [4] have achieved significant accuracy according to their paper, where they proposed a system based on color features and a hybrid classifier that comprises a decision tree and an SVM. Wang et al. [5] have used image processing techniques for localization and segmentation parts and a CNN model for character recognition. Jain et al. [6] used image processing techniques with Sobel edge detection, then OCR to recognize the license plate. Lin et al. [7] used the YOLOv2 model for vehicle and the license plate localization, classic image processing operations for segmentation, and a custom LPR-CNN model for character recognition.

Kumari et al. [8] proposed to use image prepossessing operations on the image, then contour tracing and edge detection for LP localization. Then, they used neural network models for character segmentation and recognition. Ahmed et al. [9] and Choudhary et al. [10] mainly focused on the recognition part. In [9], they have used horizontal, vertical projection and gray-level occurrence to extract edible text. On the other hand, [10] have used the CNN-LSTM model combined for character segmentation and recognition. They claimed a 99.64% success rate of their approach. Venkateswari et al. [11] concentrated on LP localization. To do so, they have used horizontal and vertical highest histogram value for extracting the region of interest (ROI). In [12], Surekha et al. claim to get 97% accuracy. They have done several image preprocessing operations, then showed a comparison between morphological processing and edge processing for LP area extraction. They extracted the characters with connected component analysis and recognized them using a supervised learning model.

Most of these proposed systems are not properly applicable to Bangladeshi vehicle license plates. Because, most of them are specific to a region, language, and type of license plate. Some previous work has been done for Bangladeshi vehicle license plates also. Nooruddin et al. [13] proposed the use of color features with MinPool and MaxPool features to detect license plates. Amin et al. [14] proposed a system combination of edge detection, binary thresholding, and Hough transformation for plate localization, and OCR for Bangla language to recognize text. Their accuracy is not noteworthy as well as that is not a generalized process. Baten et al. [15], in their paper, proposed a method that uses a special feature of the Bangla language called "matra" and connected component analysis for detection and segmentation of text, then they used template matching for the recognition phase. However, they did not reveal much about their dataset and accuracy. Abedin et al. [16] proposed using contour properties for both license plate detection and character segmentation. They proposed to use a CNN model for the character recognition part. They claim to get an accuracy of the total procedure 92% within 0.11s. However, their dataset mostly includes private vehicles, they did not consider all the categories of vehicles, and they did not focus on night conditions. Rahman et al. [17] only focused on the recognition task, so they had to manually cut the license plate then the characters from it, then they used the characters in a CNN model to recognize it. They used a dataset containing 1750 images. They had to do huge work to get the dataset.

Abdullah et al. [18] used YOLOv3 and then ResNet-20 in their paper. Their dataset contains 1500 images and 6400 character images for the localization model and recognition model, respectively. They claimed to get 92.7% accuracy. But, they have not extracted all texts from the plate as it is only for Dhaka Metropolitan Area. So, it fails to generalize for other cities. Dhar et al. [19] proposed a shape validation technique to detect license plates, then tilt correction and connected component analysis to segment different texts, characters, and digits. They used an AdaBoost classifier with two main features which are histogram of gradient(HOG) and local binary pattern(LBP). They introduced a dataset of 2800 images of only 14 different classes for the recognition task. They achieved 97.2% accuracy. Sarif et al. [20] proposed a system that uses YOLOv3 for the plate localization and a custom segmentation algorithm to segment the texts, characters, digits from the plate which they later fed into a CNN model to recognize them. They gained 97.5% recognition accuracy. However, they only tested on 16 different classes which are not sufficient for the real scenarios of Bangladeshi vehicle license plates. Moreover, they tested mostly on private vehicles of Dhaka city only which makes their claim more vulnerable in the case of the trading vehicle license plates. Saif et al. [21] proposed to use the YOLOv3 model in number plate localization and the recognition stage. They used a small dataset containing only 1050 images of private vehicles. They claim to get 99.5% accuracy. Their claim completely fails in the case of trading vehicle license plates which is absent in their dataset. Moreover, they measured accuracy in the binary fashion of the license plate as a whole.

In [22], Azam et al. mainly focused on removing noise from images for detecting LP regions. They gained 94% detection accuracy. They used a method with frequency domain mask to remove rain stroke, contrast enhancement method, Radon transform for tilt correction, and image entropy-based method to filter LP regions. Hossain et al. [23] proposed a system depending on various image processing operations. They proposed to use the Sobel edge operator, dilation, erosion, boundary features, and horizontal and vertical projection to extract LP regions. Then, dividing the extracted LP region into two halves, they used boundary features to segment the character and template matching to recognize them. However, their system fails in case of ambiguous character recognition and more than 10°C tilted image. They claimed 90% accuracy. Chowdhury et al. [24] extracted the LP region based on color information, segmented that in two halves using centroid information, and extracted characters using bounding box parameters. Then, they used SVM to recognize characters. They claimed 99.3% recognition accuracy. But they only used private vehicle images, their system struggles when LP is not in focus and image is not ideal and tested only on 14 classes. In [25], after preprocessing, horizontal and vertical projection with geometric properties was used by Islam et al. to extract LP regions. Then, character localization is done with connected component analysis and bounding box technology. SVM is used to recognize characters using the features extracted with HOG. They got very good recognition accuracy, but they ignored non-ideal conditions. Their system fails when the image resolution is not high and struggles to detect trading vehicle's LP. Ahsan et al. [26] proposed a system that uses template matching to localize LP region, spatial super resolution technique to enhance the image, bounding box method for segmenting characters, and AlexNet to recognize them. They attained 98.2% accuracy which seems to be currently one of the highest. However, they did not reveal much about the number of classes that AlexNet was trained on. Also, the template matching technique often finds it hard to detect a target when that is tilted in an image.

4 Dataset

One of the main contributions of this paper is the rich datasets for both localization and recognition of the Bangladeshi license plate. Our first dataset contains almost 2800 images for localization shown in Fig. 3. The second dataset contains around 4000 license plate images cropped from our first dataset shown in Fig. 4 which are the most so far in this sector. We have split our datasets into 70:15:15 and 85:10:5 for training, validation, and testing purpose in license plate localization and text recognition stage, respectively. Our datasets contain images from four different cities of Bangladesh: Dhaka, Khulna, Chattogram, Jashore including 12 different vehicle categories license plates from both private and trading vehicles. Our datasets are diverse enough and cover almost every possible condition, angle, and environment. To create our datasets more diverse, we have gathered images from different sources. From Nooruddin et al. [13], we are given their dataset of only trading vehicles. Most of the private vehicle images are used from this paper Rahman et al. [27], and the rest of them are collected by us.

Fig. 3 Images in plate localization dataset

Fig. 4 Images in text recognition dataset

5 Our Proposed Methodology

Our full system is divided into three major parts. First, license plate localization from an image. Next, recognizing text, characters, digits from the license plate, and finally, presenting a formatted output.

5.1 License Plate Localization

Many have proposed to use color features [4, 13], Sobel edge detection [2, 6, 14], image processing techniques [5], deep learning models [7, 18, 20, 21]. Generally, image processing operations are time-consuming and computationally expensive. Apart from that, image processing technique does not generalize for every situation. So, we prefer DNN models over them. However, some researchers also used DNN models previously those are not state-of-the-art models anymore. Here, we have implemented some new pre-trained DNN models and tried to compare their performances in Sect. 6.

YOLOv4 is one of the latest versions of you only look once (YOLO), a real-time object recognition system that can recognize multiple objects in a single frame. YOLO is a one-stage object detector. YOLOv4 has surpassed its ancestors in terms of accuracy and speed [28]. Single shot detector (SSD) [29] is designed for real-time object detection, and it uses VGG-16 [30] to extract feature maps and small convolutional filters to detect objects. It achieves accuracy close to that of faster R-CNN [31] in terms of lower resolution images.

We have implemented SSD MobileNetV2 FPNLite 320×320, SSD MobileNetV1 FPN 640×640, and SSD ResNet-50V1 FPN 640×640. EfficientDet [32] family is the successors of the EfficientNets. EfficientDet detectors are single shot detectors much like SSD. Their backbone networks are ImageNets pre-trained EfficientNets. They use BiFPNs to create bidirectional feature fusion to detect objects. We have implemented EfficientDet-D0 512×512, EfficientDet-D1 640×640, EfficientDet-D2 768×768 in this stage.

5.2 Text Recognition

Recognition of texts in the license plate region plays a crucial role in building an ALPR system. Most of the previous researchers devoted their effort to segmenting different texts in the license plate area in different ways and then recognizing those segmented parts separately. However, this is a somewhat lengthy process, and we have proposed an idea to merge this two parts: segmentation and recognition into one.

From the previous stage, after localization, we get the desired license plate region cropped from our model. This cropped image is then used in this stage to recognize texts on the license plate. We have considered each separated text as an object and then tried to detect them in the license plate. By doing so, we have eliminated the need for any other segmentation algorithm. In this stage, we have implemented various models, and their performances are discussed in Sect. 6. Implemented models are YOLOv4, SSD MobileNetV2 FPNLite 320 × 320, SSD MobileNetV1 FPN 640 × 640, SSD ResNet-50V1 FPN 640 × 640, EfficientDet-D0 512 × 512, EfficientDet-D1 640 × 640, EfficientDet-D2 768 × 768.

5.3 Formatted Output Generation

In this part, we have proposed a unique solution that plays a key role in eliminating the need for any segmentation part and yet generates the standard BRTA license plate number form.

From the text recognition stage, the output is provided for each image in three different lists. One list contains the coordinates of the individual characters and words in the image, the second list is of confidence score for each of the coordinates, and the third list is the class ID of each of the predicted characters and words. Using these lists, we have developed an algorithm that generates output from each image in BRTA standard format and stores them in a CSV file.

Algorithm 1: Formatted Output Generation

```
Result: Set of characters from the license plate as a string
Push 0-9 numbers as a string in a list NBR;
Push class names in a list CN where indexes are according to class ID;
Applying non-max suppression on the predictions and get the indexes into a
list IDX;
while i from 0 to length(box_coordinates) do
    if i in IDX then
        Get (x,y,w,h) from box_coordinates list;
        Push (x,y,w,h,i) into OBJ list;
    else
        Continue;
    end
end
Sort OBJ;
while obj in OBJ do
    Index_no = obj [4];
    if CN[Index_no] in NBR then
        Append CN[Index_no] in PN string;
    else
        Append CN[Index_no] in PP string;
    end
end
Store PN and PP into a CSV file
```

6 Results and Discussion

We have used Google Colaboratory which provides a single 12GB Nvidia Tesla K18 GPU RAM, for training and evaluating our models. However, we have labeled the images of our dataset for different models in a local machine. Figure 5a shows the output from the license plate localization stage. Figure 5b shows the output from the text recognition stage, and finally, Fig. 5c shows the formatted output. For the evaluation of the models, we have considered two main metrics which are mean average precision(mAP) for 0.5IoU and recall.

Table 1 shows the result from different models that we have trained and evaluated for the license plate localization. From the table, we can see that in the license plate localization stage, both YOLOv4 and EfficientDet-D1 640 × 640 have performed very well. But as the recall of YOLOv4 is much higher than that of EfficientDet-D1 640 × 640, YOLOv4 is surely a better choice here. Similarly from Table 2, we can conclude that YOLOv4 has performed significantly better than other models in terms of both metrics.

From the above discussion, it is clear that using the YOLOv4 model in both the license plate localization and text recognition stage can be combined to provide the best solution to the automatic license plate recognition for Bangladeshi vehicles. Moreover, we have compared our system with previously existing systems, and it has proven to be one of the best solutions for this type of system keeping in mind that no existing system has been evaluated on 27 different classes before.

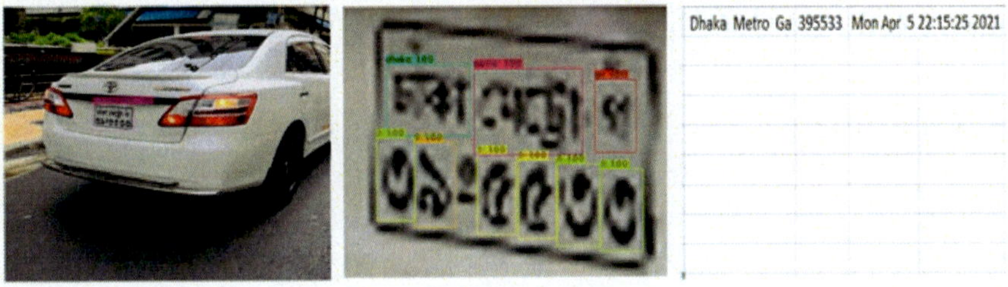

(a) (b) (c)

Fig. 5 **a** Output from license plate localization stage, **b** Output from text recognition stage, and **c** Formatted output in a CSV file

Table 1 Performances of different models on localizing license plate from images

Metric	Model						
	Efficient Det-D0	Efficient Det-D1	Efficient Det-D2	SSD MobileNet 320 Lite	SSD MobileNet 640	SSD ResNet-50 640	YOLOv4
mAP	0.9089	0.9184	0.8940	0.9254	0.9448	0.9359	0.9631
Recall	0.6869	0.7334	0.7127	0.7407	0.7477	0.7437	0.9800

Table 2 Performances of different models on text recognition from images

Metric	Model						
	Efficient Det-D0	Efficient Det-D1	Efficient Det-D2	SSD MobileNet 320 Lite	SSD MobileNet 640	SSD ResNet-50 640	YOLOv4
mAP	0.9089	0.9184	0.8940	0.9254	0.9448	0.9359	0.9631
Recall	0.6869	0.7334	0.7127	0.7407	0.7477	0.7437	0.9800

Table 3 Accuracy comparison with other approaches

Method	Accuracy (%)
[23]	90
[17]	92.7
[18]	97.2
[19]	97.5
[26]	98.2
[24]	99.3
Our proposed method	96.31

From Table 3 and considering our diverse datasets, we can confidently conclude that our proposed method can perform significantly better than most other method previously proposed in a real-life scenario.

7 Conclusion

This paper is mainly focused on finding the best solution for an automatic license plate recognition system for Bangladeshi vehicles. To do so, we have used seven different methods in two different stages and successfully come up with the optimal and efficient solution to this problem. Using the YOLOv4 model in both the license plate localization and text recognition stage offers the most promising result so far. Besides, we have successfully generated the BRTA standard format from the result and are able to store it in a CSV file. Moreover, this paper also suggests a computationally efficient way to eliminate the segmentation part using our custom algorithm along with the YOLOv4 model. Apart from that, here, we have also introduced two rich and diverse datasets. These datasets contain images from four major cities of Bangladesh and the 12 common vehicle category license plates in Bangladesh. With the YOLOv4 model, this system has managed to gain 99.37 and 96.31% license plate localization and text recognition accuracy, respectively. However, in the license plate localization stage, the YOLOv4 model is mostly trained on daylight condition images. So, this system may not perform as expected in the night condition until more of those conditions images are included in the dataset for training. The whole system can be made more generalized by adding more images to both datasets. Moreover, running the system on a GPU will provide the best frame per second (FPS) in case of any video footage. There is a scope of research for the future researchers to prevent this system from storing similar license plate numbers from consecutive frames of video footage. Considering the whole methodology, performance, and least limitations of this system, we believe that we have found the state-of-the-art approach, and it shows satisfying results for an end-to-end Bangladeshi license plate recognition system.

References

1. Qadri MT, Asif M (2009) Automatic number plate recognition system for vehicle identification using optical character recognition. In: 2009 international conference on education technology and computer. IEEE, pp 335–338
2. Shidore M, Narote S (2011) Number plate recognition for Indian vehicles. IJCSNS Int J Comput Sci Netw Secur 11(2):143–146
3. Lekhana G, Srikantaswamy R (2012) Real time license plate recognition system. Int J Adv Technol Eng Res 2(4):5–9
4. Ashtari AH, Nordin MJ, Fathy M (2014) An Iranian license plate recognition system based on color features. IEEE Trans Intell Transp Syst 15(4):1690–1705
5. Wang CM, Liu JH (2015) License plate recognition system. In: 2015 12th international conference on fuzzy systems and knowledge discovery (FSKD). IEEE, pp 1708–1710
6. Jain K, Choudhury T, Kashyap N (2017) Smart vehicle identification system using OCR. In: 2017 3rd international conference on computational intelligence and communication technology (CICT). IEEE, pp 1–6
7. Lin CH, Lin YS, Liu WC (2018) An efficient license plate recognition system using convolution neural networks. In: 2018 IEEE international conference on applied system invention (ICASI). IEEE, pp 224–227
8. Kumari S, Gupta L, Gupta P (2017) Automatic license plate recognition using OpenCV and neural network. Int J Comput Sci Trends Technol (IJCST) 5(3):114–118
9. Ahmed AK, Taha MQ, Mustafa AS (2018) On-road automobile license plate recognition using co-occurrence matrix. J Adv Res Dyn Control Syst 10(7)
10. Choudhary N, Tech Scholar M, Jain K (2019) License plate recognition using combination of CNN-LSTM. Int J Analyt Exp Modal Anal
11. Venkateswari P, Steffy EJ, Muthukumaran DN (2018) License plate cognizance by ocular character perception. Int Res J Eng Technol 5(2):536–542
12. Surekha P, Gurudath P, Prithvi R, Ananth V (2018) Automatic license plate recognition using image processing and neural network. ICTACT J Image Video Process 8(4)
13. Nooruddin S, Sharna FA, Ahsan SMM (2020) A Bangladeshi license plate detection system based on extracted color features. In: 2020 23rd international conference on computer and information technology (ICCIT), pp 1–6
14. Amin MR, Mohammad N, Bikas MAN (2014) An automatic number plate recognition of Bangladeshi vehicles. Int J Comput Appl 93(15)
15. Baten RA, Omair Z, Sikder U (2014) Bangla license plate reader for metropolitan cities of Bangladesh using template matching. In: 8th international conference on electrical and computer engineering. IEEE, pp 776–779
16. Abedin MZ, Nath AC, Dhar P, Deb K, Hossain MS (2017) License plate recognition system based on contour properties and deep learning model. In: 2017 ieee region 10 humanitarian technology conference (R10-HTC). IEEE, pp 590–593

17. Rahman MS, Mostakim M, Nasrin MS, Alom MZ (2019) Bangla license plate recognition using convolutional neural networks (CNN). In: 2019 22nd international conference on computer and information technology (ICCIT). IEEE, pp 1–6
18. Abdullah S, Hasan MM, Islam SMS (2018) Yolo-based three-stage network for Bangla license plate recognition in Dhaka metropolitan city. In: 2018 international conference on bangla speech and language processing (ICBSLP). IEEE, pp 1–6
19. Dhar P, Abedin MZ, Karim R, Hossain MS et al (2019) Bangladeshi license plate recognition using adaboost classifier. In: 2019 joint 8th international conference on informatics, electronics and vision (ICIEV) and 2019 3rd international conference on imaging, vision and pattern recognition (icIVPR). IEEE, pp 342–347
20. Sarif MM, Pias TS, Helaly T, Tutul MSR, Rahman MN (2020) Deep learning-based Bangladeshi license plate recognition system. In: 2020 4th international symposium on multidisciplinary studies and innovative technologies (ISMSIT). IEEE, pp 1–6
21. Saif N, Ahmmed N, Pasha S, Shahrin MSK, Hasan MM, Islam S, Jameel ASMM (2019) Automatic license plate recognition system for Bangla license plates using convolutional neural network. In: TENCON 2019-2019 IEEE region 10 conference (TENCON). IEEE, pp 925–930
22. Azam S, Islam MM (2016) Automatic license plate detection in hazardous condition. J Vis Commun Image Represent 36:172–186
23. Hossain MJ, Uzzaman MH, Saif AS (2018) Bangla digital number plate recognition using template matching for higher accuracy and less time complexity. Int J Comput Appl 975:8887
24. Chowdhury MBU, Dhar P, Guha S (2020) Detection and recognition of Bangladeshi license plate. Int J 9(3)
25. Islam R, Islam MR, Talukder KH (2020) An efficient method for extraction and recognition of Bangla characters from vehicle license plates. Multimedia Tools Appl 79(27):20107–20132
26. Ahsan M, Based M, Haider J et al (2021) Intelligent system for vehicles number plate detection and recognition using convolutional neural networks. Technologies 9(1):9
27. Rahman M. A study on Bangladeshi car name plate detection and recognition. In: Thesis/project No. CSER-19-24, KUET
28. Bochkovskiy A, Wang CY, Liao HYM (2020) Yolov4: Optimal speed and accuracy of object detection. arXiv:2004.10934
29. Liu W, Anguelov D, Erhan D, Szegedy C, Reed S, Fu CY, Berg AC (2016) SSD: Single shot multibox detector. In: European conference on computer vision. Springer, pp 21–37
30. Qassim H, Verma A, Feinzimer D (2018) Compressed residual-vgg16 CNN model for big data places image recognition. In: 2018 IEEE 8th annual computing and communication workshop and conference (CCWC). IEEE, pp 169–175
31. Ren S, He K, Girshick R, Sun J (2015) Faster r-CNN: Towards real-time object detection with region proposal networks. arXiv:1506.01497
32. Tan M, Pang R, Le QV (2020) Efficientdet: Scalable and efficient object detection. In: Proceedings of the IEEE/CVF conference on computer vision and pattern recognition, pp 10781–10790

Vulnerability Analysis and Robust Training with Additive Noise for FGSM Attack on Transfer Learning-Based Brain Tumor Detection from MRI

Debashis Gupta and Biprodip Pal

Abstract Deep learning-based high-precision computerized brain tumor diagnosis helps to obtain significant clinical features for proper treatment. Research also revealed that medical deep learning systems are easily compromised by several small imperceptible perturbation strategies and resultant adversarial attacks. Medical deep learning systems for brain MRI-based tumor classification has been unexplored for susceptibility to adversarial attack except some abstract description of the vulnerability. In this research, the vulnerability of a highly accurate pretrained deep learning model has been studied in presence of adversarial samples. The potential risk associated with this model has been illustrated in terms of performance drop for misclassification, correct classification, and visual perceptibility. It is found that a very small perturbation variation of 0.0001–0.0007 can cause the performance to drop from 97 to 82%. Finally, a Gaussian additive noise-based robustness improvement strategy has been presented to overcome the drop of correct classification probability criteria. The results has been validated with publicly available dataset. These findings can be useful to raise safety concerns and design more robust medical deep learning systems.

Keywords Adversarial attack · FGSM · Brain tumor · Deep learning · Robustness

1 Introduction

Brain tumors classified as either metastatic or primary consist of abnormal growth of tissue as a result of uncontrolled multiplication of brain cells. Gliomas are kind of brain tumors originating from glial cells. Chemotherapy, surgery, and radiotherapy are the techniques used, often in combination, to treat low-grade gliomas like astrocytomas or oligodendrogliomas as well as most aggressive glioblastoma multiforme (GBM) [16]. Before any therapy-based treatment, the segmentation of the tumor is very crucial to protect healthy tissues while damaging and destroying tumor cells

D. Gupta · B. Pal (✉)
Rajshahi University of Engineering & Technology, Rajshahi, Bangladesh

M. S. Arefin et al. (eds.), *Proceedings of the International Conference on Big Data, IoT, and Machine Learning*, Lecture Notes on Data Engineering and Communications Technologies 95, https://doi.org/10.1007/978-981-16-6636-0_9

during the therapy. The main goal of computerized brain tumor diagnosis is to obtain important clinical information regarding the tumor presence, position, and type accurately. To inspect the infected brain sections, magnetic resonance imaging (MRI) is widely used. MRI records various sections of the brain and later reconstructs these as a three-dimensional image (3D) or a two-dimensional image (2D). It produces high-quality segmentation images with distinct views of brain structures like axial, coronal, and sagittal.

Deep learning methods are especially used to assess the brain MRI to identify a class of cognitive disorders. Deep learning (DL) models, particularly, pretrained models or a customized convolutional neural network (CNN) is found to be very efficient to segment and classify the abnormality in the growth of human tissue in the brain. These researches range from developing CNN with custom simple architecture, cascading multipath architectures to deeper CNN architectures with more nonlinearities and have less filter weights [1, 4, 11]. Apart from customized models, pretrained models are also widely used and achieved remarkable accuracies [10] . Multi-level features extraction and concatenation from two pretrained models Inception v3 and DensNet-201 that came out with reliable performance. Among VGG-16, Inception v3, and ResNet-50 models transfer learning from ResNet-50 obtained the highest accuracy of 95% [14]. These works illustrate that pretrained models are used frequently, and the results are promising.

In parallel to the progress of medical DL, adversarial examples have uncovered vulnerabilities in many state-of-the-art DL systems [3]. Adversarial examples typically tend to attempt to reduce the prediction confidence of the target machine learning model and change the output of classification of some sample to any different class from the original class. The need for automated diagnosis can make the diagnosis process vulnerable to adversarial attack. Ma and others analyzed the susceptibility of AI-based medical image processing approaches to adversarial attack [9]. Li et al. have illustrated the susceptibility of brain MRI to adversarial attack for age prediction task [8]. Wang and others demonstrated [17] practicality of particular type of attack called backdoor attacks in transfer learning models for brain MRI classification. Cheng et al. showed vulnerability of CNN-based semantic segmentation for brain tumor from MRI [2]. Hence, brain MRI images are found to be vulnerable to different types of attacks. Most of the successful brain tumor detection deep learning-based methods consist of pretrained weights and architecture description that are publicly available because of research transparency and reusability. Hence, attacks like FGSM attack can be easily crafted for these models. Jai et al. have shown brain MRI classification approaches are vulnerable to FGSM attack but lack quantification of attack effects [6]. Although significant research on developing defensive models has been considered like network distillation, adversarial training, or input reconstruction, most heuristic methods are often vulnerable to adaptive attacks [18]. Moreover, most methods fail to obtain non-trivial robustness for large data and models. However, additive noise for training can provide certified bounds to adversarial robustness [7] which has not been analyzed previously for FGSM attack on MRI-based brain tumor detection technique.

Fig. 1 Brain MRI dataset

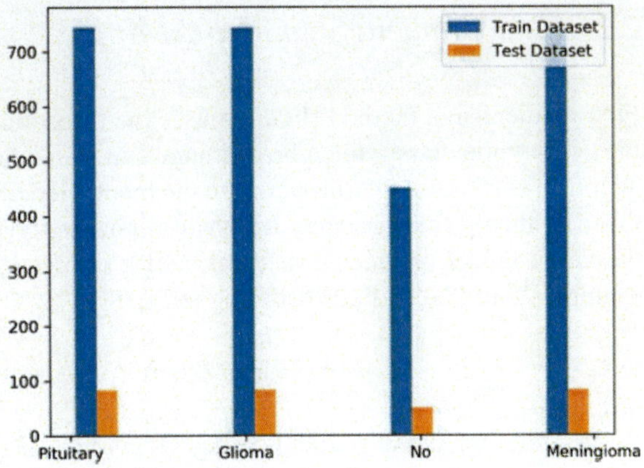

In this paper, the vulnerability to FGSM attack of a pretrained ResNet-50 DL model that obtained the best performance among some frequently used DL models for MRI-based brain tumor classification has been explored [14]. FGSM attack has been crafted followed by quantification of corresponding misclassification and intuitive analysis of visual perceptibility as well as correct classification probability drop for the mentioned model for brain tumor detection. Finally, an additive Gaussian noise-based robustness improvement has been empirically studied. The findings have been validated using standard dataset available publicly as described in the dataset description.

2 Materials and Methods

2.1 Dataset Description

The dataset has been taken from the Kaggle competition [13]. It contains 2975 images with four different labels. This dataset has been split into 90% for training and 10% for the testing. There are a total of 743, 740, 450, 744 MRI samples in training dataset and 83, 82, 50, and 83 MRI samples in the testing dataset for "glioma," "meningioma," "no tumor," and "pituitary" types, respectively. Figure 1 shows the distribution. In both scenarios, i.e., without the additive noise training and with the additive noise training, the training and testing datasets were kept alike along with different image augmentation techniques like rotation_range, fill_mode, shear_range, zoom_range, width_shift_range, height_shift_range, brightness_range, horizontal_flip, and vertical_flip.

2.2 First Gradient Sign Method Attack

First gradient sign method (FGSM) uses the direction of the calculated gradient of the CNN while classifying a brain image and creates an adversarial brain image by adding crafted noise (perturbation) to the input. Goodfellow et al. [3] first came with this new theory for producing adversarial images. For an original image x, label y, the target model parameter vector θ, and loss function $J(\theta, x, y)$, the adversarial example X' can be crafted according to Eq. (1).

$$x' = x + \epsilon \text{sign}(\nabla_x J(\theta, x, y)) \tag{1}$$

First, the sign of the gradients using a loss function for the input image calculated by the targeted model is taken. Here, sign $\nabla_x J(\theta, x, y)$ represents the sign of the calculated gradients for the corresponding inputs by the model. In this case, the sign can be both positive or negative relying on the applied loss function. Moreover, the positive sign indicates any enlargement in pixel intensity increases the loss, i.e., the error that the targeted model makes. On the contrary, the negative sign denotes a reduction in pixel intensity that raises the overall loss calculated by the model. FGSM attempts to increase the loss in a systematic manner and fool the diagnostic model or any inspector. This causes the failure of the model which deals with a relationship between an input pixel intensity and the class score for correct classification. A small crafted noise called perturbation represented by ϵ is multiplied with the signed value calculated from the gradient vector which is shown by $\epsilon * \nabla_x J(\theta, x, y)$. Finally, the obtained result is added to the original image turning it into an adversarial image which forces the model to misclassify the original label.

$$x' = x + \eta \tag{2}$$

Here, η represents $\epsilon * \text{sign}(\nabla_x J(\theta, x, y))$ and x' indicates the adversarial image which is imperceptible to the human visual system for smaller perturbation [18].

2.3 Transfer Learning with Additive Noise

Residual neural network ResNet-50 is a variant of residual network containing a stack of 50 convolutional layers. The feature extractor layers are kept frozen having the same weight during ImageNet classification. For brain MRI classification, the classifier part of ResNet-50 consists of a two-dimensional global average pooling layer followed by a dropout of 50%. In addition, the last fully connected layer is connected to four output units to classify brain MRI images into one of the four classes with a softmax activation function. Categorical cross entropy $-\sum_{c=1}^{M} y_{o,c} \log(p_{o,c})$ is used as a loss function where $M = 4$ for brain image classification to four different classes, y is a binary indicator if class label c is the correct classification for observa-

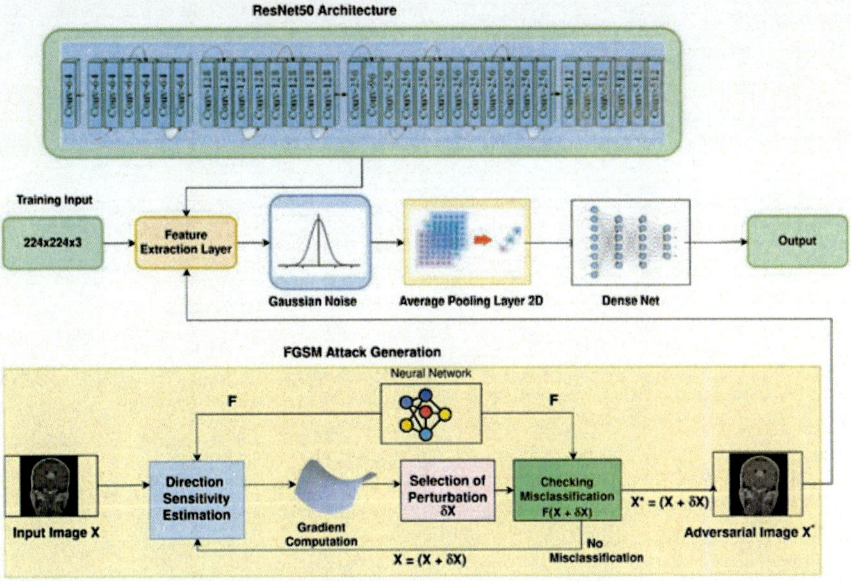

Fig. 2 Proposed architecture with additive noise training

tion o and p is the predicted probability observation. Adam is used as an optimizer to minimize the loss function [5]. The structure of the targeted model does not deal with external noise; on the other hand, additive noise provides theoretical bound to robustness. Therefore, an extra hidden layer of Gaussian noise has been added into the architecture. Gaussian noise $N(0, \sigma^2)$ is added to each pixel of features extracted from x and applies the classifier f on it. The output $c_i = f(x + N(0, \sigma^2 I))$ is further used for fine-tuning the classifier weights (Fig. 2).

3 Experimental Study

The TensorFlow deep learning library and Python programming language were used to implement the program of DL models and FGSM attacks. The experiment was held from two different viewpoints. First, it had been analyzed the performance of ResNet-50 for brain tumor classification from MRI images by an in-depth analysis of the drop of performance of this model in the presence of the FGSM attack. Second, the model was trained with a Gaussian noise for different standard deviation, and performance improvement was quantified with extensive experiment.

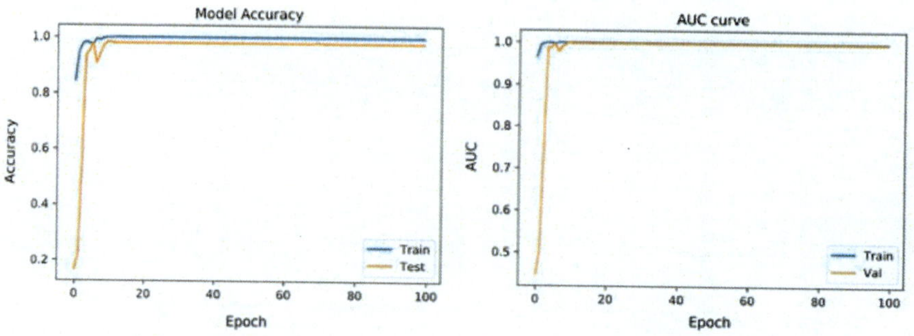

(a) Accuracy Curve (b) AUC Curve

Fig. 3 Accuracy and AUC curve for ResNet-50 model for brain tumor classification

Table 1 Confusion matrix of ResNet-50 model for brain tumor classification

	Glioma	Meningioma	No tumor	Pituitary
Glioma	83	0	0	0
Meningioma	0	78	0	0
No tumor	0	3	50	1
Pituitary	0	1	0	82

3.1 *Baseline Classification Performance of ResNet-50*

To start with, the performance of the ResNet-50 model for proper diagnosis in an attack-free environment was analyzed. The accuracy and AUC curve in Fig. 3 and confusion matrix in Table 1 show the outcome of brain tumor classification using the ResNet-50 pretrained model without the presence of any adversarial image. In the presence of sufficient data, the network training performance saturates quickly, and test outcomes reflect the reliability of ResNet-50 for MRI-based brain tumor classification. The test performance is close to 100% for this multiclass classification scenario.

3.2 *Perturbation Effect on Visual Perceptibility of Adversarial Images*

To visualize and discuss the potential risk and performance drop and effect on visual perceptibility, analysis of misclassification performance and correct classification performance drop was illustrated through intuitive analysis.

For this experiment, testing images with smaller (0.0001) to higher (0.01) perturbation (ϵ) were generated, and corresponding performance was enlisted. Figure 4 clearly describes that subtle perturbation of 0.0005 is sufficient to generate adver-

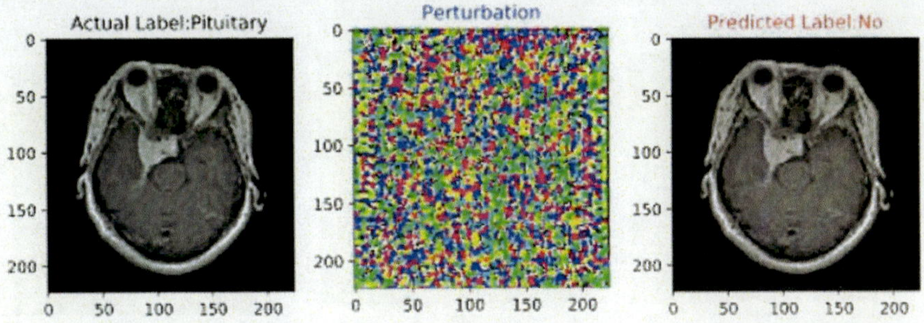

(a) Original Label
Prediction: **Pituitary**

(b) Perturbation
Epsilon (ϵ)=0.0005

(c) Adversarial Image
Prediction: **No Tumor**

Fig. 4 Misclassification of adversarial images by ResNet-50 with visually imperceptible perturbation

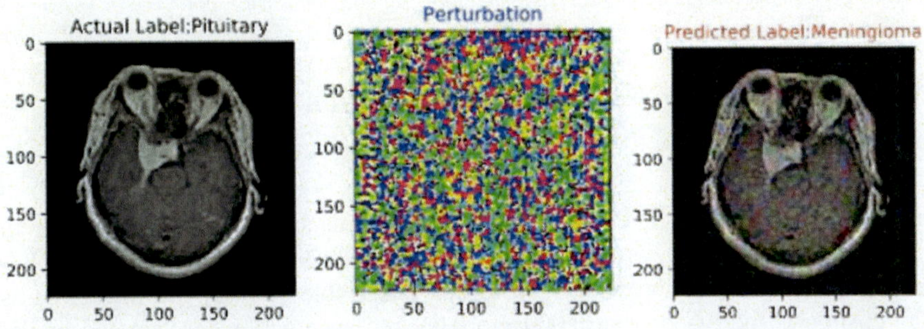

(a) Original Label
Prediction: **Pituitary**

(b) Perturbation
Epsilon (ϵ)=0.01

(c) Adversarial Image
Prediction: **Meningioma**

Fig. 5 Misclassification of adversarial images by ResNet-50 with visually perceptible perturbation

sarial images that can cause misclassification, and corresponding adversarial images can easily fool the model without being recognized by the human eye. But, despite successful misclassification, Fig. 5 illustrates that adversarial image can easily be detected by the human eye as a tempted image with increased ϵ.

Since ResNet-50 is a complex architecture well-trained with ImageNet feature extraction weights, it can still classify some of the tumor adversarial images correctly. However, the confidence or in other words the probability of correct prediction decreases over a small variation of perturbation on successful classification. Figure 6 depicts that for a meningioma image, although the ResNet-50 correctly predicts the type, for a perturbation of 0.0003, the probability of this image to belong to Meningioma class drops from 0.99 to 0.86.

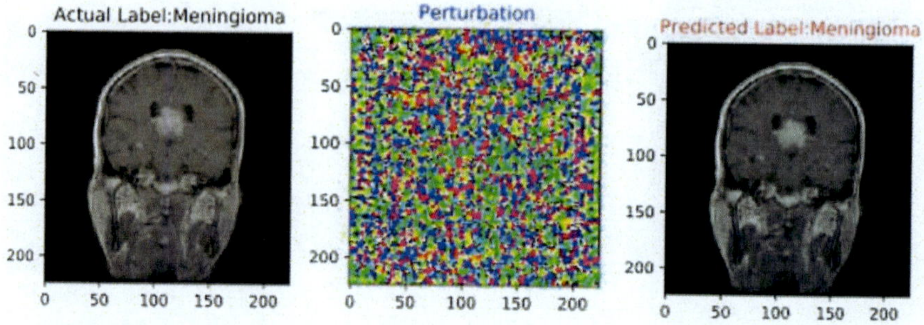

(a) **Status**: Without Attack
Orginal Image
Prediction:Meningioma
Prediction Probability: 95.89%

(b) Perturbation
Epsilon (ϵ)=0.0003

(c) **Status** : Under FGSM Attack
Adversarial Image
Prediction : Meningioma
Prediction Probability: 86.95%

Fig. 6 Performance degradation of ResNet-50 model for correct classification

3.3 *Perturbation Effect on Misclassification and Correct Classification*

This study attempts to clearly quantify the rate of misclassification and the correct classification probability drop for FGSM attack on ResNet-50 mode for brain MRI-based tumor classification with rigorous analysis. Table 2 clearly illustrates that as the epsilon is increased from 0.0001 to 0.01, the accuracy of the targeted model drastically falls from 97 to 29% and the number of misclassification increases dramatically. In case of correctly predicted images, the probability of the correct class also drops. Table 2 shows that average original probability for correct prediction (for correctly predicted original images) also drops while classifying corresponding adversarial images correctly. Often, a significant drop of 0.99 to 0.91 is seen for perturbation like 0.003. Catastrophic drop of such diagnostic performance shows the vulnerabilities of well-trained pretrained models on large dataset.

3.4 *Effect of Additive Noise on Correct Classification Confidence Improvement*

The robustness can be improved for performance drop of misclassification as well as correct classification. Experiment has been carried out by training the model with additive Gaussian noise for standard deviation of noise 0.2, 0.4, 0.6, and 0.8. Figure 7 shows that the performance of ResNet50 remains stable for noisy training while classifying attack free images.After training the model with additive noise, the result shows much improvements in presence of adversarial images as shown in Fig. 8. With 0.8 standard deviation of noise, the target model accuracy was reduced from 95 to 37% for perturbation ranged from 0.0001 to 0.01. But for the other standard

Table 2 Performance drop of targeted model for perturbation variation

Epsilon	Correct predict	Wrong predict	Accuracy	Average original probability	Average adversarial probability	Probability drop
0.0001	289	9	0.96	0.99	0.99	0.3
0.0003	274	24	0.91	0.99	0.98	0.8
0.0005	264	34	0.88	0.99	0.96	0.11
0.0007	246	52	0.82	0.99	0.95	0.17
0.0009	234	64	0.78	0.99	0.95	0.21
0.001	232	66	0.77	0.99	0.94	0.22
0.003	159	139	0.53	0.99	0.91	0.46
0.005	116	182	0.38	0.99	0.91	0.61
0.007	97	201	0.32	0.99	0.95	0.63
0.009	89	209	0.29	0.99	0.96	0.70
0.01	89	209	0.29	0.99	0.95	0.70

Table 3 Confusion matrix for brain tumor classification with additive noise

	Glioma	Meningioma	No tumor	Pituitary
Glioma	83	0	0	0
Meningioma	0	78	0	0
No tumor	0	3	50	1
Pituitary	0	1	0	82

noise deviation, the targeted model performance falls to almost 29%. Based on the selected = 0.8, the performance of ResNet was analyzed in details.

Based on the misclassification accuracy, using additive noise for training, the standard deviation of 0.8 has been chosen. Table 3 shows the confusion matrices of the targeted model under the training with 0.8 additive noise. The performance is almost similar to classification performance of original attack-free images. Table 4 shows the comparative analysis of proposed approach with some traditional ML approach and deep learning approach on brain tumor detection.

While the correct classification probability drops significantly with increase in perturbation, proposed training approach reduces the probability drop of ResNet-50 to certain extent for brain tumor detection.

It has been shown how the probability changes on average for correctly classified adversarial images in terms of proposed training approach and traditional transfer learning approach in Fig. 9. In spite of some sharp drop for lower perturbation, the performance of the proposed approach outperformed significantly as perturbation increased in the range of 0.002–0.008. For larger μ, the adversarial images can easily be detected in the human eye. Therefore, improvement of performance for lower μ is significant. Since noisy training helps the network to understand noise effects,

Table 4 Comparison with existing works

Authors	Applied architecture	Model accuracy (%)
Afshar et al. [1]	One convolutional layer with 64 feature maps	86.56
Noreen et al. [10]	8-combined_densenet_block	99.51
Rathi et al. [12]	SVM	97.82
Seetha et al. [15]	CNN	97.50
	Proposed architecture (ResNet-50 with additive noise)	99.45

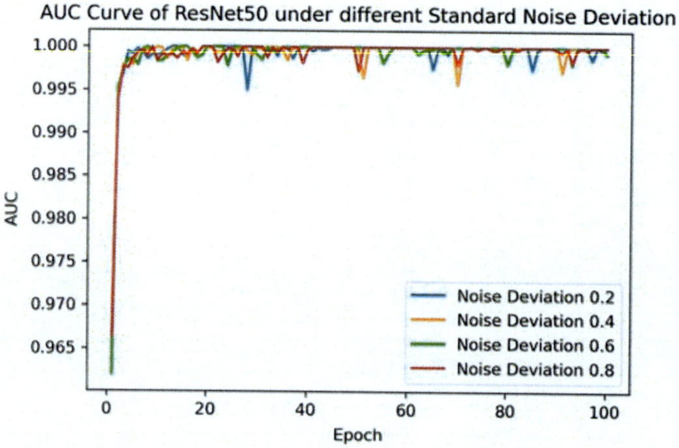

Fig. 7 AUC curve of ResNet-50 under different standard noise deviation

adversarial images with very subtle noise have not been detected by ResNet-50. Also, the model can successfully classify some images for very low perturbation.

As perturbation is basically a crafted noise, it can be clearly seen that addition of the Gaussian noise layer as the hidden layer of the targeted model can nicely deal with external noise added to the images using perturbation and, hence, is more robust to this type of attack.

4 Conclusion

Deep transfer learning-based brain tumor classification from MRI image obtained promising outcomes. Some research on adversarial attacks has explored the vulnerability of similar DL models on several tasks including medical imaging tasks. This research clearly illustrates the performance drops in case of the best perform-

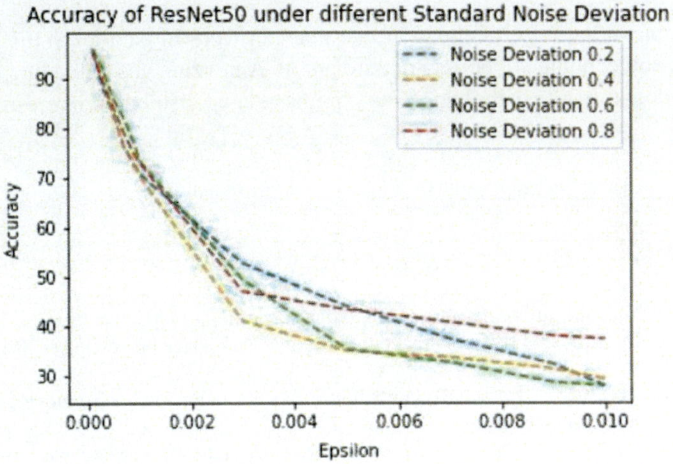

Fig. 8 Performance of ResNet50 model for dierent standard deviation of gaussian noise

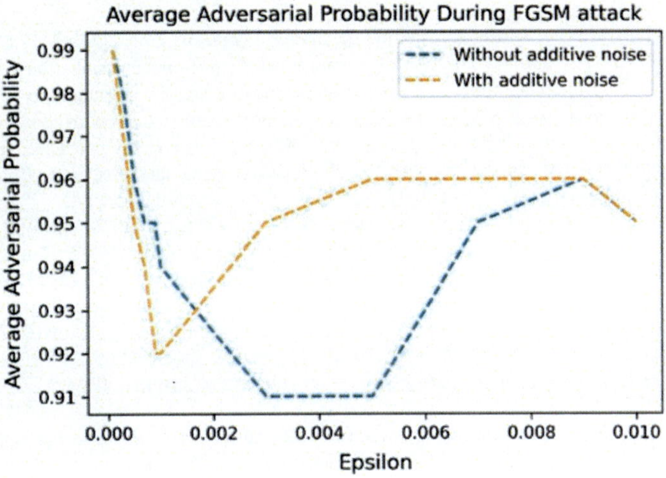

Fig. 9 Correct classification probability score of ResNet-50 model for noisy and noiseless trainings

ing pretrained model ResNet-50 for very subtle perturbation in adversarial images. The research also illustrates that additive noise can improve the correct classification confidence of such weakly performing models for lower perturbations. This research provides an intuitive analysis of vulnerability to attacks like FGSM attack for a common and reliable brain tumor classification DL model. This proposed scheme can be used in different use cases, on availability of necessary data and performing fine-tuning the model accordingly. Regardless of the performance improvement, there are scopes to enhance performance for lower perturbation like 0.001 and higher than 0.008 as shown in figure. Future works can include designing robust training strategies to improve the misclassification rate that can work in parallel to the proposed improvement technique of correct classification. Analyzing and extending this work for other domains like pathological or CT images can also be an interesting future work.

References

1. Afshar P, Mohammadi A, Plataniotis KN (2018) Brain tumor type classification via capsule networks. In: 2018 25th IEEE international conference on image processing (ICIP). IEEE, pp 3129–3133
2. Cheng G, Ji H (2020) Adversarial perturbation on MRI modalities in brain tumor segmentation. IEEE Access 8:206009–206015
3. Goodfellow IJ, Shlens J, Szegedy C (2014) Explaining and harnessing adversarial examples. arXiv:1412.6572
4. Havaei M, Davy A, Warde-Farley D, Biard A, Courville A, Bengio Y, Pal C, Jodoin P, Larochelle H (2016) Brain tumor segmentation with deep neural networks. Cornell University Library. arXiv:1505.03540 (2016)
5. Kingma DP, Ba J (2014) Adam: a method for stochastic optimization. arXiv:1412.6980 (2014)
6. Kotia J, Kotwal A, Bharti R (2019) Risk susceptibility of brain tumor classification to adversarial attacks. In: International conference on man–machine interactions. Springer, pp 181–187
7. Li B, Chen C, Wang W, Carin L (2019) Certified adversarial robustness with additive noise. In: Wallach H, Larochelle H, Beygelzimer A, d'Alché-Buc F, Fox E, Garnett R (eds) Advances in neural information processing systems, vol 32. Curran Associates, Inc. https://proceedings.neurips.cc/paper/2019/file/335cd1b90bfa4ee70b39d08a4ae0cf2d-Paper.pdf
8. Li Y, Zhang H, Bermudez C, Chen Y, Landman BA, Vorobeychik Y (2020) Anatomical context protects deep learning from adversarial perturbations in medical imaging. Neurocomputing 379:370–378. https://www.sciencedirect.com/science/article/pii/S0925231219315279
9. Ma X, Niu Y, Gu L, Wang Y, Zhao Y, Bailey J, Lu F (2021) Understanding adversarial attacks on deep learning based medical image analysis systems. Pattern Recogn 110:107332
10. Noreen N, Palaniappan S, Qayyum A, Ahmad I, Imran M, Shoaib M (2020) A deep learning model based on concatenation approach for the diagnosis of brain tumor. IEEE Access 8:55135–55144
11. Pereira S, Pinto A, Alves V, Silva CA (2016) Brain tumor segmentation using convolutional neural networks in MRI images. IEEE Trans Med Imaging 35(5):1240–1251
12. Rathi VP, Palani S (2012) Brain tumor MRI image classification with feature selection and extraction using linear discriminant analysis. arXiv:1208.2128
13. Sartaj N, Chakrabarty SK (2020) Brain tumor classification (MRI). https://www.kaggle.com/sartajbhuvaji/brain-tumor-classification-mri/activity
14. Saxena P, Maheshwari A, Maheshwari S (2019) Predictive modeling of brain tumor: a deep learning approach. In: Innovations in computational intelligence and computer vision. Springer, pp 275–285
15. Seetha J, Raja SS (2018) Brain tumor classification using convolutional neural networks. Biomed Pharmacol J 11(3):1457
16. Stupp R, Roila F (2009) Malignant glioma: ESMO clinical recommendations for diagnosis, treatment and follow-up. Ann Oncol 20(Suppl. 4):126–128
17. Wang S, Nepal S, Rudolph C, Grobler M, Chen S, Chen T (2020) Backdoor attacks against transfer learning with pre-trained deep learning models. IEEE Trans Services Comput
18. Yuan X, He P, Zhu Q, Li X (2019) Adversarial examples: attacks and defenses for deep learning. IEEE Trans Neural Netw Learn Syst 30(9):2805–2824

Performance Evaluation of Convolution Neural Network Based Object Detection Model for Bangladeshi Traffic Vehicle Detection

S. M. Sadakatul Bari, Rafiul Islam, and Syeda Radiatum Mardia

Abstract Vehicle detection has numerous applications in modern day like smart toll plaza, parking, traffic management, etc. Many Convolution Neural Network (CNN) based object detection models, designed to train specific object detection and test them in real world. However, the challenge is preparing these models for vehicle detection according specific custom datasets and train them with minimal devices such as low GPU, memory, and test those models with embedded devices. Both accuracy and speed are matter for real-time vehicle detection and counting. In general, real-time object detection model based on CNN are 2 types—One stage method (YOLOv3, SSD) and two stage method (Faster-RCNN resnet50, resnet101). Both of the methods have complex network architecture which makes the real-time detection slow. But two stage method is slower than one stage method and in particular YOLO series is faster as it requires less GPU than other models. This paper has shown the performance evaluation of YOLOv3, YOLOv3-tiny, and YOLOv4-tiny in terms of precision, recall, F1-Score, mAP (Mean Average Precision), IoU (Intersection over Union), and Average FPS (Frame per second) for moving traffic vehicle detection and count them using the dataset named 'Dhaka-AI (Dhaka traffic detection challenge dataset). Experiments show that YOLOv4-tiny performs better compared to other two models in terms of recall, F1-Score, AVG_FPS and mAP.

Keywords Deep learning · Real-time object detection · Vehicle detection · Convolutional neural network (CNN) · YOLOv3 · YOLOv3-tiny · YOLOv4-tiny

1 Introduction

Because of the swift improvement of the economy and the gradual progress of the quality of life of people, almost everyone wants to lead a better life. For this reason, people develop cities, industrial zones, and business areas at different places. So,

S. M. Sadakatul Bari (✉) · R. Islam · S. R. Mardia
Department of CSE, Bangladesh Army University of Science and Technology, Saidpur, Bangladesh
e-mail: sadakatul@baust.edu.bd

M. S. Arefin et al. (eds.), *Proceedings of the International Conference on Big Data, IoT, and Machine Learning*, Lecture Notes on Data Engineering and Communications Technologies 95, https://doi.org/10.1007/978-981-16-6636-0_10

the only way to merge these cities, industrial zones, and businesses together are through the transportation system. As the importance of the transportation system is increasing, the number of vehicles on the road is also increasing. More and more of the population own private cars, for industrial and business purposes. Due to transportation needs between cities people have adopted the heavy vehicle. This has been conducted to the quick outgrowth of city traffic vehicles and highway traffic vehicles, which is greatly challenging for traffic management. Additionally, in urban areas, parking area is limited so nowadays private parking areas for business purposes have been provided. However, it is difficult to record the number of vehicles that are entering and which position is free. Consequently, quick detection, identification, and counting of vehicles have turned into a crucial assignment in city traffic management, highway traffic management, parking area surveillance and restricted area surveillance and an experiment focus in the sector of Artificial Intelligence. However, there are still a number of challenges, such as a lack of availability of proper datasets, lack of images, lack of resources, not enough GPU, etc. Researchers have developed models to address these issues. However, they are difficult to train and do not solve these problems. Currently, the 4th industrial revolution is ongoing and there have been great improvements in tyers, engines, the aerodynamic body shape, increasing HP of the engine, etc. For this reason, the speed of vehicles is increasing day by day. If our detection speed is low then AI has lost their ability. Theoretical research was started on this technology in the 1970s, and it has been used for commercial purposes since 1990s. Before deep learning was utilized, some mathematical models were applied. Basically, the conventional object detection method is splitting within three stages, namely (1) select candidate areas, (2) extract the characteristics of these candidate regions, (3) the trained classifier is used for classification. This method has faced a number of issues [1]. To solve this convolutional neural network (CNN) was introduced. Due to the remarkable performance of deep learning, deep neural network has replaced the conventional feature extraction method. This network can independently acquire necessary features by a huge volume of data training [2]. There are two types of CNN models, the (1) one stage method (YOLO [3]) and the (2) two stage method (Fast RCNN [4] and Faster RCNN [5]). The two stage method has provided a very high level of accuracy but has a very low speed. On the other hand, the one stage method has high speed but low accuracy. Regarding speed, the two stage method does not fulfil the requirement of state of art real-time object detection. Even though the one stage is fast compared to the two stage method. It still does not fulfil the requirements of real-time object detection. To address this problem, light weight models have been introduced. These light weight models have been invented from their core model, such as YOLOv3-tiny comes from YOLOv3 and YOLOv4-tiny comes from YOLOv4. These two light weight models perform very well in their category, are quite popular, easy to use require less GPU, provide good accuracy and remarkable speed.

This paper has conducted a cumulative comparison between YOLOv3, YOLOv3-tiny, and YOLOv4-tiny with a custom dataset. After that, they have been tested using real life examples with embedded devices such as the smartphone and the result and performance has been counted and briefly analyzed.

2 Related Work

In this section, we have discussed different types of convolutional neural networks (CNN) and compared them regarding their complex network size, working process, accuracy, and speed. We have also discussed why we select YOLO series and light weight model.

Typically, object detection model in deep learning is divided into two categories, one stage, and two stage. The one stage method directly extracts features from the input image and sends it to the next convolutional layer such as the YOLO series. YOLO means you only look once and was first proposed by Redmon et al. [3], It is the first regression-based method. After that came YOLOv2, in which the fully connected layers were removed and introduced pooling layers [6] and worked with a new basic network named DarkNet-19. It has more layers than V1 and improved accuracy. The last approach from Redmon was YOLOv3, which works with darknet 53 [7]. YOLOv3 was the advanced method compared to the previous versions and it introduced fully connected convolutional layers. It has increased accuracy but decreases speed and it works with a total of 106 layers. YOLOv3-tiny [8] has come from YOLOv3. In YOLOv3-tiny, the fully connected layers are removed and the pooling layer is used. Also, there is a decreased the number of layers which made it a lot faster. After two years YOLOv4 model was introduced by Alexey et al. [9]. It has given the highest accuracy among all the YOLO models. However, based on the YOLOv4, a lightweight method was released and it was YOLOv4-tiny [10]. In place of CSPDarknet53 backbone network which is using in YOLOv4, it uses the CSPDarknet53-tiny backbone network and it also uses two scale predictions. YOLOv4-tiny is also very fast. On the other hand, the two stage method first selects the region, finds out the probability of object-ness, pulls the ROI (region of interest) with the first feature map, and then passes it to the convolutional layer such as Fast RCNN [4] and Faster RCNN [5]. These two methods have given the highest accuracy compared to all other models but, because of the complex network, have very low speed and also required very powerful GPU, which is not possible in embedded devices. Embedded devices (such as smartphone camera and video surveillance camera) have limited computing power and limited memory. It is very difficult to maintain constant performance. So, for better performance, a simple model which provides moderate accuracy and high speed is needed because in the present system speed does matter.

Thus, in this research YOLOv3 and the light weight models, YOLOv3-tiny, and YOLOv4-tiny were selected, because researchers have detected moving vehicles and at the same time counted and categorized them. Also, this is why a very high FPS (frame per second) rate was required. The YOLO series is faster than the two stages method but still does not fulfil the speed requirements. Among the YOLO core models, YOLOv3 has given quite good accuracy and speed compared to others. Also, light weight models were selected because of their remarkable speed and good accuracy. YOLOv4 was not considered because it has a large network. YOLOv5 was also not considered because it is still in the developing process.

Some related works done by different researchers are given below:

- Vision-based vehicle detection and counting system using deep learning in highway scenes [11]. The researchers used the vehicle dataset with a total of 57,290 annotated instances in 11,129 images. As for the CNN model they used YOLOv3 and improved it.
- Real-time object detection method for embedded devices [12]. YOLOv4-tiny was used for the embedded system.
- An Improved Vehicle Detection Algorithm based on YOLOv3 [1]. The YOLOv3 model was improved and achieved 6% higher accuracy.

3 Materials and Methods

3.1 Network Architecture

YOLOv3: YOLOv3 has a very big network, it contains 106 layers and all these layers are fully connected. It has more layers than previous versions and this has improved the accuracy. YOLOv3 has used a regression classifier instead of softmax. YOLOv3 predicts bounding boxes at three different scales. The three scale predictions are 52×52, 26×26, and 13×13. As a feature extraction backbone, YOLOv3 has used Darknet-53. It has 53 convolutional layers and a total of 106 fully connected convolutional layers. Because of its large amounts of layers, it is slow (Fig. 1).

YOLOv3-tiny: YOLOv3-tiny comes from YOLOv3. It is a light weight model and researchers proposed it to reduce the network layers to increase speed. Due to the decrease in the layers, it lost accuracy. In place of ResBlock structure in the Darknet53 network, YOLOv3-tiny conducts seven convolution layers and six max-pooling layers. It has conducted two scale prediction (26×26 and 13×13) in place of three scale,, if the input image size is 416×416 and (40×40 and 20×20) for 640×640 (Fig. 2).

YOLOv4-tiny: YOLOv4-tiny is a very light weight object detection model. It is based on the regression problem. It contains only 37 layers and the detection layers are the 30th and 37th layers. It uses the six max-pooling layers instead of the convolution connected layer. CSPDarknet53-tiny is the backbone network of YOLOv4 tiny. CSPDarknet53-tiny helps to extract features. YOLOv4 uses three scale predictions but instead of three scales, it conducts two scale predictions (26×26 and 13×13). For this reason, it has decreased accuracy but increased speed. For reducing detection time, it has needed feature extraction. For this, it has used FPN (feature pyramid networks). That's why it is so fast. The CSPDarknet53-tiny network has used the CSPBlock module in place of the ResBlock module in its residual network. The CSPBlock module can enhance the learning ability of the convolution network compared with the ResBlock module.

Fig. 1 Darknet-53 [7]

	Type	Filters	Size	Output
	Convolutional	32	3 × 3	256 × 256
	Convolutional	64	3 × 3 / 2	128 × 128
1x	Convolutional	32	1 × 1	
	Convolutional	64	3 × 3	
	Residual			128 × 128
	Convolutional	128	3 × 3 / 2	64 × 64
2x	Convolutional	64	1 × 1	
	Convolutional	128	3 × 3	
	Residual			64 × 64
	Convolutional	256	3 × 3 / 2	32 × 32
8x	Convolutional	128	1 × 1	
	Convolutional	256	3 × 3	
	Residual			32 × 32
	Convolutional	512	3 × 3 / 2	16 × 16
8x	Convolutional	256	1 × 1	
	Convolutional	512	3 × 3	
	Residual			16 × 16
	Convolutional	1024	3 × 3 / 2	8 × 8
4x	Convolutional	512	1 × 1	
	Convolutional	1024	3 × 3	
	Residual			8 × 8
	Avgpool		Global	
	Connected		1000	
	Softmax			

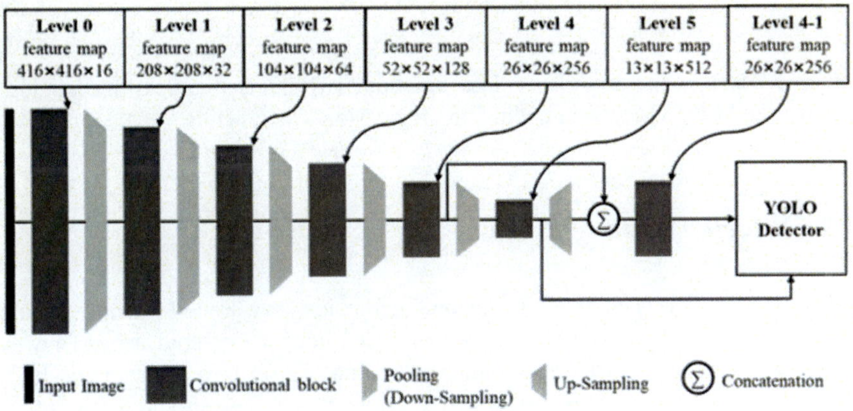

Fig. 2 Network architecture of YOLOv3-tiny [13]

3.2 Prediction Process

The prediction process is same for all of these three models. At first, the input image is divided into the $S * S$ grid. After that, the depth of the feature map $B * (5 + C)$ is calculated, where B represents the number of bounding boxes, which is predicted by each grid and C is the total class number. 5 represents each bounding box has 5 predicted values: x, y, w, h, and confidence score. The (x, y) coordinates represent the center of the box relative to the bounds of the grid cell. The width and height are predicted relative to the whole image. If an object center point falls in a grid, the model will predict it. In this prediction process, the confidence threshold is used to check redundancy and decrease it. The confidence score function is shown below:

$$\text{Confidence score} = P * \text{IOU} \tag{1}$$

where P is a function of Object and IOU is the intersection over the union. IOU is calculated as the percentage of predicted bounding box overlapping on actual bounding box.

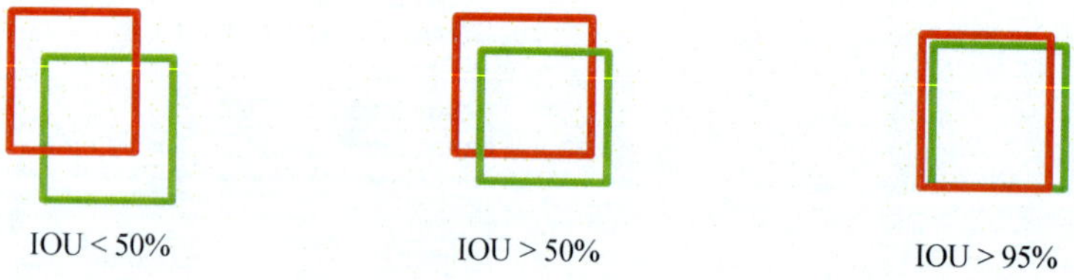

YOLOv4 tiny uses three loss functions.

$$\text{LOSS} = \text{LOSS1} + \text{LOSS2} + \text{LOSS3} \tag{2}$$

where LOSS1 is the confidence loss function, LOSS2 is the classification loss function and LOSS3 is the bounding box regression loss function.

3.3 *Vehicle Counting Process*

When an object has been detected, a bounding box is drawn around this object. This bounding box is not permanent, it has been drawn frame by frame and also has moves when the object moves frame by frame. So, at first, we selected a specific position on the viewing window. Suppose the window height is 500 pixels and 500 – 200 = 300 pixels is selected. When a vehicle has been detected, the bounding box moves down. When the bounding box center point has come to this pixel point it checks a condition and counts it (Fig. 3).

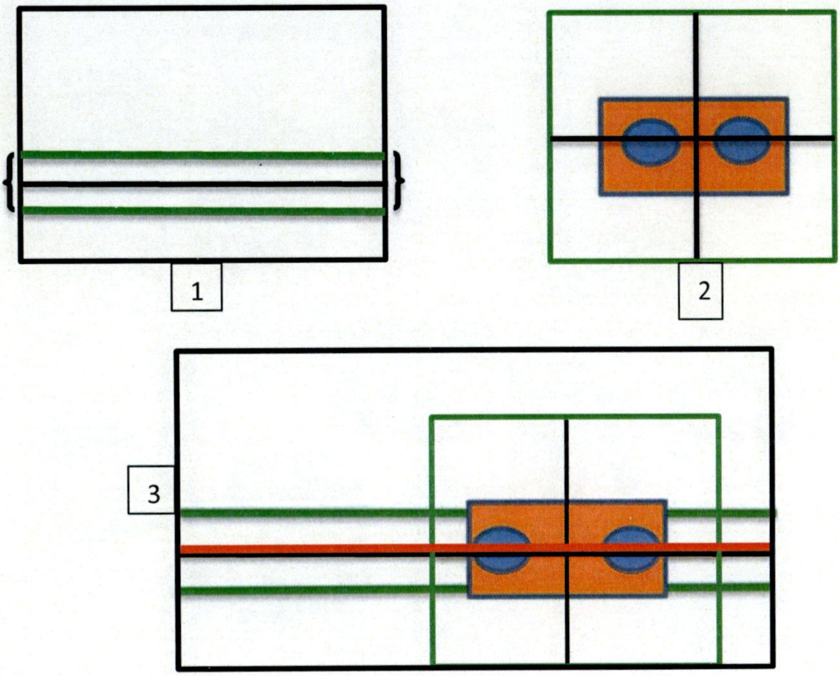

Fig. 3 Counting process

4 Experimental Results and Analysis

4.1 Dataset

The dataset that we have collected is from Kaggle. Its name is 'Dhaka-AI (Dhaka traffic detection challenge dataset) [14], which contained 3006 images for training with annotations and 500 for test without annotations. We did not modify this dataset so therefore we did not label the test images. Instead of using test images, we used train images for both train and test at a ratio of 70:30. This dataset contains total 21 classes and all the vehicles found in Bangladesh. This dataset, which was its first version, was used in the "Dhaka-AI" competition. We have removed 117 of the 3006 images because those ones had some issues. We used the remaining 2889 images instead. 70% of them, which means 1929, were used for training, and 30% of them, which means 960 of them were used for testing (Fig. 4).

Experimental setup: For the training and evaluation process we have used Google Co-laboratory and the local machine (laptop) (Table 1).

Fig. 4 Dataset

Table 1 Experimental setup (hardware and Software)

Name	Configuration
Local machine CPU	Intel core i5
Co-lab GPU	NVIDIA Tesla T4
Local machine GPU	NVIDIA GEFORCE 940 MX
GPU memory Google Co-lab/local machine	16 GB/2 GB
RAM Google Co-lab/local machine	13 GB/16 GB
GPU software	CUDA 10.1, CUDNN 7.6.5
Language/environment software	PYTHON/Anaconda navigator
Software library	Tensor flow, OpenCV
Object detection API	Darknet API

4.2 Experimental Results

We have tested the models with motion videos as well as embedded devices such as smartphone cameras (Fig. 5).

In this evaluation process, the types of measurements that we have used are precision, recall, mAP@0.50, mAP@0.75 (mean average precision), average precision of each class, F1-score, and FPS (frames per second. In order to visualize the results, we used "PASCAL VOC detection metrics", which had been used in the PASCAL VOC competition. It looks the same as a normal confusion matrix but they count the mAP instead of the accuracy since it is better to count the precision and recall rather than accuracy when FN and FP are high.

- **Precision**: Also called Positive Predictive Value. It is the ratio of correct positive predictions to the total of predicted positives. It is a measure of exactness.

$$\text{Precision} = \text{TP}/(\text{TP} + \text{FP}) * 100\% \tag{3}$$

- **Recall**: Also called Sensitivity, Probability of Detection or; True Positive Rate. It is the ratio of correct positive predictions to the total of positives examples.

$$\text{Recall} = \text{TP}/(\text{TP} + \text{FN}) * 100\% \tag{4}$$

- **F1-score**: It conveys the balance between precision and recall.

$$\text{F1-score} = 2 * ((\text{precision} * \text{recall})/(\text{precision} + \text{recall})) \tag{5}$$

- **mAP (mean average precision)**: The mean average precision is the average value of each category, which is used to evaluate the accuracy of prediction mAP is calculated as follows:

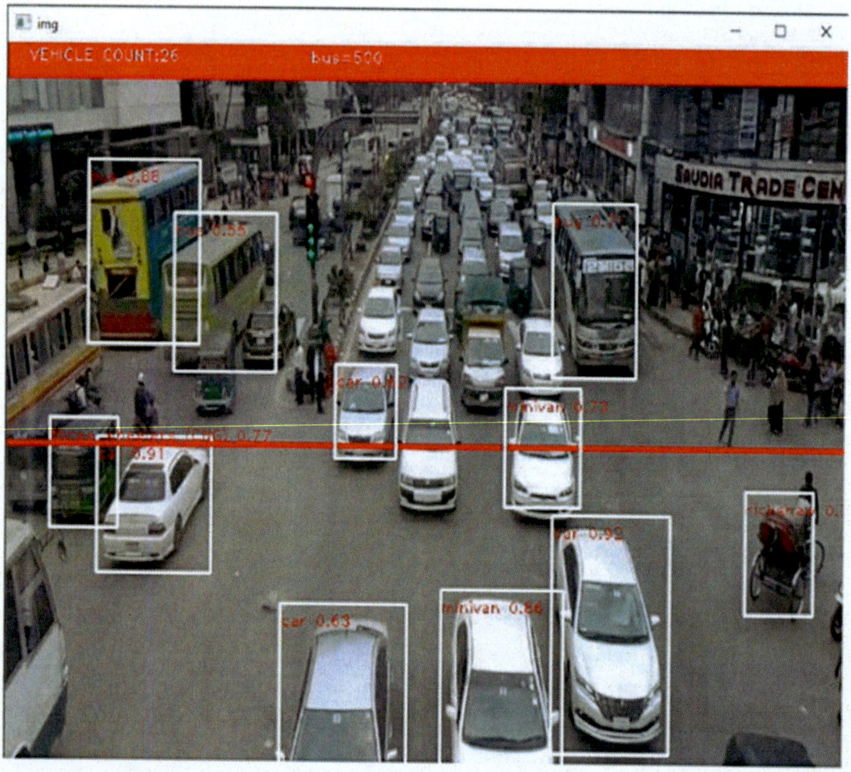

Fig. 5 Test results

$$\text{mAP }(\%) = \left(\sum \text{AP}\right)/n * 100\% \tag{6}$$

- **IoU**: IoU is the overlap rate of candidate boxes and ground truth, which are both generated during the target detection process, It is calculated like this:

$$\text{IoU} = (\text{Area of Overlap})/(\text{Area of Union}) \tag{7}$$

Accuracy mAP@0.5 = 18.50% in the console—this indicator is better than Loss, as train occurs while mAP increases [10]. For training, we have set a fixed size for input, which is at 640 * 640. And the total features are 409,600. We ran 42,000 steps for each model, 2000 steps for every class (which is the minimum number of steps, and if you want to run training with bigger steps, you can do it with higher GPU), the size of each batch was 2 learning rate was 0.001 for every model.

In Fig. 6, we can see that YOLOv3 has the highest precision, which is 76%, while YOLOv3-tiny and YOLOv4-tiny have both achieved 71%. However, when it comes to Recall as shown in Fig. 7, YOLOv4-tiny is the highest, which is 44%, while YOLOv3 achieved 41% and YOLOv3-tiny achieved 37%.

Fig. 6 Precision

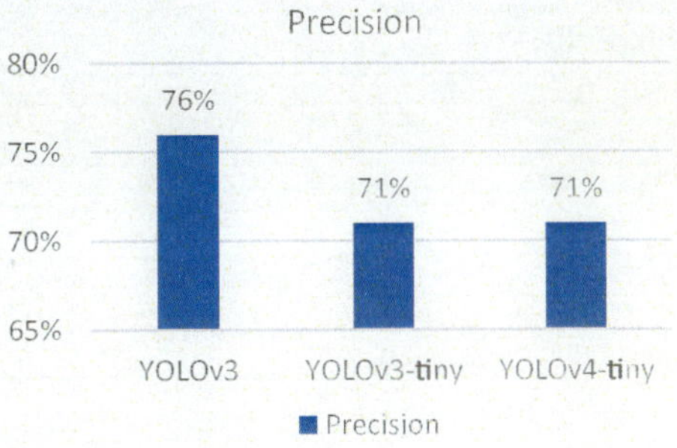

Fig. 7 Recall

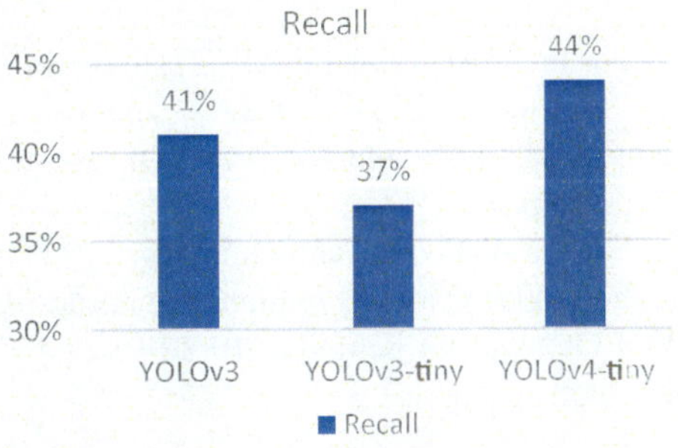

In Fig. 8, YOLOv3 has the highest IoU, which is 60%, while YOLOv3-tiny and YOLOv4-tiny were 53% and 59% respectively. In the matter of the F1-score shown in Fig. 9, those three have achieved 53%, 48%, and 54% respectively. YOLOv4-tiny has the highest F1-score and is almost equal to the IoU of YOLOv3.

mAP@0.50: It means mAP with an IoU of 50%. It is a standard measurement in the field of object detection.

mAP@0.75: It means mAP with an IoU of 75%. It is another standard measurement in the field of object detection.

Fig. 8 IoU

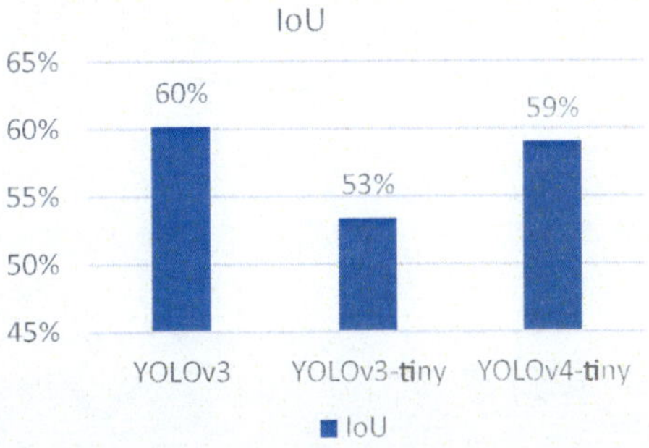

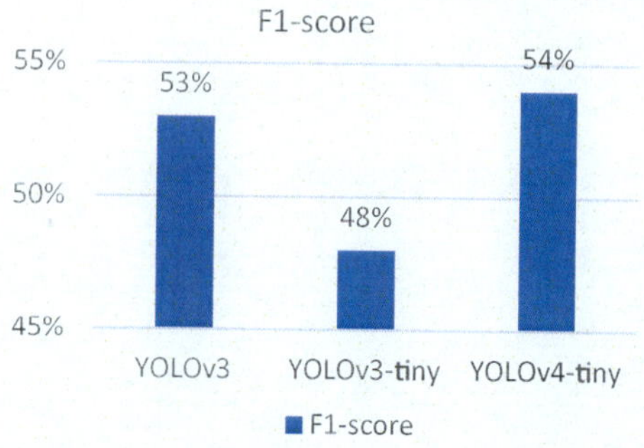

Fig. 9 F1-score

For the mAP.50 and mAP.75, shown in Figs. 10 and 11, YOLOv4-tiny achieved 30.25% and 17.09%, YOLOv3 achieved 26.56% and 14.53% and YOLOv3-tiny achieved 27.56% and 10.30%. YOLOv4-tiny achieved the highest in both of these measurements.

For the AVG_FPS shown in Fig. 12, YOLOv3-tiny was 111.9 FPS, which is the highest. YOLOv4-tiny, which was not too far behind, achieved 102 FPS. However, YOLOv3 was far behind at 19.2 FPS and it was not able to detect vehicles moving at a high speed.

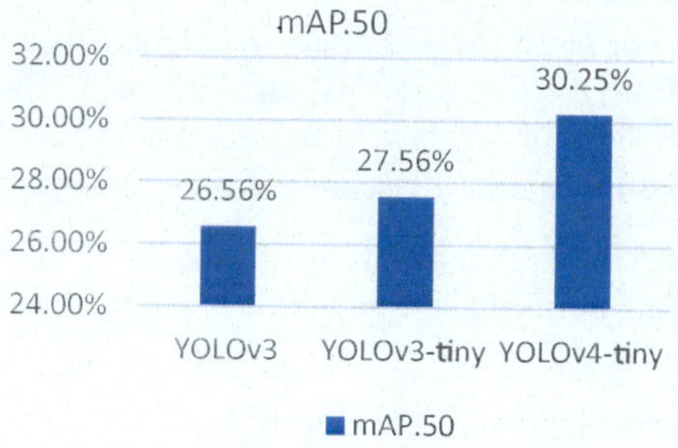

Fig. 10 mAP.50

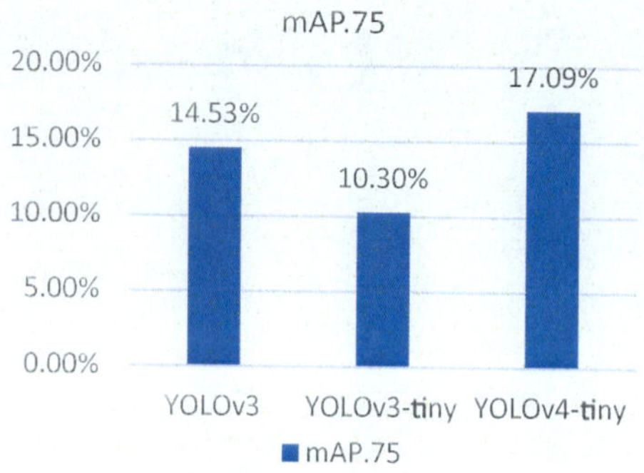

Fig. 11 mAP.75

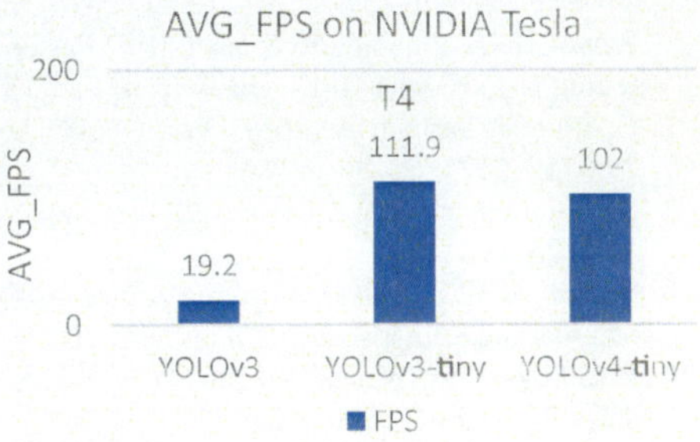

Fig. 12 AVG_FPS

From the experimental results depicted above, it is found that YOLOv4-tiny performed better compared to YOLOv3 and YOLOv3-tiny in Recall, F1-Score, mAP, and AVG_FPS. Therefore, YOLOv4-tiny could be considered the most suitable for object detection in real time with an embedded system.

5 Conclusion

The main motive of this research is to find real-time object detection models with speed and accuracy that can be used for Bangladeshi traffic vehicle detection and counting. We have trained three object detection models namely YOLOv3, YOLOv3-tiny, and YOLOv4-tiny to achieve these goals. On average, YOLOv4-tiny performed better than both YOLOv3-tiny and YOLOv3. It has very good speed, which is the key factor for embedded systems when detecting and counting moving vehicles. It also has good accuracy when compared to the other two.

Acknowledgements We would like to acknowledge Department of Computer Science and Engineering, Bangladesh Army University of Science and Technology to support this works.

References

1. Sun X, Huang Q, Li Y, Huang Y (2019) An improved vehicle detection algorithm based on YOLOV3. In: IEEE international conference on parallel & distributed processing with applications, big data & cloud computing, sustainable computing & communications, social computing & networking. IEEE, Xiamen, China, pp 1445–1450
2. Zhou Y, Liu L, Shao L (2018) Fast automatic vehicle annotation for urban traffic surveillance. IEEE Trans Intell Transp Syst 19(6):1973–1984
3. Redmon J, Divvala S, Girshick R, Farhadi A (2016) You only look once: unified, real-time object detection. In Proceedings of the IEEE conference on computer vision and pattern recognition. IEEE, Las Vegas, NV, USA, pp 779–788
4. Girshick R (2015) Fast R-CNN. In: Proceedings of the IEEE international conference on computer vision. IEEE, Santiago, Chile, pp 127–135

5. Ren S, He K, Girshick R, Sun J (2017) Faster R-CNN: towards realtime object detection with region proposal networks. IEEE Trans Pattern Anal Mach Intell 39:1137–1149
6. Redmon J, Farhadi A (2017) YOLO9000: better, faster, stronger. IEEE Trans Pattern Anal 29:6517–6525
7. Redmon J, Farhadi A (2018) YOLOv3: an incremental improvement. IEEE Trans Pattern Anal 15:1125–1131
8. Redmon J (2020) Darkent: open source neural networks in C, 2013–2016. https://pjreddie.com/darknet/. Last accessed 02 Nov 2020
9. Bochkovskiy A, Wang CY, Liao HYM (2020) YOLOv4: optimal speed and accuracy of object detection. arXiv 2020, arXiv: 2004.10934
10. Alexey B (2020) Darknet: open source neural networks in python (2020). https://github.com/AlexeyAB/darknet. Last accessed 02 Nov 2020
11. Song H, Liang H, Li H (2019) Vision-based vehicle detection and counting system using deep learning in highway scenes. Eur Transp Res Rev 11:51
12. Mao QC, Sun HM, Liu YB (2019) Mini-YOLOv3: real-time object detector for embedded applications. IEEE Access 7:133529–133538
13. Han B-G, Lee J-G, Lim K-T, Choi D-H (2020) Design of a scalable and fast yolo for edge-computing devices. Sensors 20(23):6779
14. Shihavuddin ASM, Rifat M, Rashid A (2020) DhakaAI. Harvard Datavers, V1. https://doi.org/10.7910/DVN/porexf

Hyperspectral Image Classification Using Factor Analysis and Convolutional Neural Networks

A. F. M. Minhazur Rahman and Boshir Ahmed

Abstract Hyperspectral image sensors can provide valuable data for land covers, oceans, and the earth atmosphere at various spatial and spectral scales. Rich spectral and spatial information of a location makes hyperspectral image (HSI) an excellent way to work with materials, identify them, or define their properties. However, computer-automated analysis and classification of hyperspectral image is a challenging problem. Most of the spectral information in hyperspectral image is correlated, containing redundant information. High number of bands in input image contributes to the curse of dimensionality problem that reduces classifier performance. In many applications, the amount of labelled hyperspectral data that can be acquired is minimal. The complexities associated with HSI motivate us to propose a method named FA-CNN. We have used factor analysis (FA) dimension reduction technique to remove band correlation while maintaining useful spectral information in a lower number of bands. Then, we have applied convolutional neural network (CNN) for combining spectral and spatial features of HSI. Finally, multilayer perceptron classifier is used for classifying each of the input pixels in HSI. Our proposed method achieved 99.59% overall accuracy and 99.75% average accuracy on Indian Pines dataset; 99.95% overall accuracy and 99.90% average accuracy on Pavia University dataset while requiring a lower number of trainable parameters and training data compared to other methods.

Keywords Hyperspectral image classification · Dimension reduction · Factor analysis · Spectral and spatial feature extraction · Convolutional neural networks

1 Introduction

Hyperspectral remote sensing aims to collect spectral and spatial data from objects on the earth's surface based on their reflectance property acquired by airborne or spaceborne sensors. It has finer wavelength resolution, contiguous wavelength bands, and

A. F. M. Minhazur Rahman (✉) · B. Ahmed
Department of Computer Science and Engineering, Rajshahi University of Engineering & Technology, Rajshahi, Bangladesh

M. S. Arefin et al. (eds.), *Proceedings of the International Conference on Big Data, IoT, and Machine Learning*, Lecture Notes on Data Engineering and Communications Technologies 95, https://doi.org/10.1007/978-981-16-6636-0_11

a higher range than traditional image acquisition techniques. It makes precise analysis of soils, waterbody, and materials possible [1]. Because of the large amount of detailed information available, hyperspectral image classification of ground objects has become a common research topic among researchers. However, high dimensionality [2] of hyperspectral images combined with limited training samples leads to Hughes phenomenon [3]. Hughes phenomenon has the potential to significantly reduce classifier performance. To overcome the problem of high dimensionality, various dimension reduction techniques like PCA [4], LDA, etc. are commonly used. Spectral features extracted from these dimension reduction techniques are insufficient for providing satisfactory classification performance using SVM [5], Multilayer Perceptron (MLP) etc.

Deep learning-based frameworks have become a popular method for hyperspectral image classification task. Support vector machine, random forest, decision tree, and other conventional machine learning approaches require laborious feature engineering. In comparison, deep learning-based methods automatically extract features that are effective in minimizing task errors [6]. On the other hand, combining spatial context with spectral features has shown great promise [7]. Rather than treating hyperspectral image cubes as a simple pixel array, designing a classifier that integrates spectral and spatial features is an efficient way to enhance classification efficiency. Spatial features provide additional information correlated with the shape and structure of a material [8]. The rise of deep learning techniques and the effectiveness of spectral–spatial classification framework have led to broader adoption of convolutional neural networks (CNN) [9–12]. Major drawbacks of CNN include increased training time, large training data, and high computational power requirements compared to traditional machine learning techniques. However, hyperspectral images typically do not have a large amount available training samples [10]. This constraint forces the development of a robust CNN framework that can capture the underlying discriminant features of the HSI using a minimal amount of training samples.

One of the recent CNN-based method—Makantasis et al. [12] used Randomized-PCA to reduce dimension and represent spectral information in a compressed form while keeping spatial information intact. This dimension reduced image was fed into convolutional neural network (CNN) using 2D convolution, which conducts high-level feature construction and a multilayer perceptron, which classifies each pixel of the input image. This method performed extremely well using 30 PCA bands. Haque et al. [11] used a similar method to [12]. However, they considered multiple spatial contexts like 3×3, 5×5 and 7×7 at the same time to extract different level of spatial features and passed it to CNN using 2D convolution, 2D max-pooling, and MLP classifier. This method performed well but increased the number of trainable parameters compared to other methods. Roy et al. [13] proposed a joint 2D and 3D-CNN-based method called HybridSN. This method utilizes 3D-CNN for spectral–spectral feature learning from dimensional reduced spectral bands. 2D-CNN follows the 3D-CNN framework. 2D-CNN learns deep spatial features, which further ensures proper utilization of available spatial information. Usage of both 2D and 3D-CNN and a MLP classifier resulted in high accuracy in HSI classification task. One downside of this method is the added complexity of joint 3D and 2D convolution and high number of trainable parameters.

In this research work, we have proposed a method called FA-CNN, which uses factor analysis dimension reduction technique to overcome the Hughes phenomenon by finding the original image bands' underlying factors and representing spectral information of the original image using those factors. After that, we have used a convolutional neural network to combine spectral and spatial information of the image in a single step and a multilayer perceptron classifier to classify each pixel in the input hyperspectral image. Our proposed method overcomes some of the drawbacks in the hyperspectral image classification mentioned above.

2 Methodology

Our research methodology is comprised of three major steps. In the first step, the input hyperspectral image cube is projected into a new subspace using factor analysis dimension reduction technique. After that, multiple overlapping three-dimensional patches are created from the dimensionally reduced image for providing spatial–spectral information to the CNN framework. Finally, fully connected layers in the CNN are used to classify each input pixel fed into our proposed framework. The following subsections elaborate on these main components of our research methodology.

2.1 Dataset Description

Indian Pines (IP) hyperspectral dataset (Fig. 1) was captured using AVIRIS sensor in Northwest Indiana, USA. There are 200 spectral bands in this dataset. The spatial dimension of the hyperspectral image cube is 145×145. After removing background samples, 10249 samples from 16 classes are left for experimentation.

Pavia University (PU) hyperspectral dataset (Fig. 2) was captured using ROSIS sensor in Northern Italy. This dataset contains 103 spectral bands. The spatial dimension of the hyperspectral image cube is 610×340. After removing background samples, 42776 samples from 9 classes are left for experimentation.

2.2 Dimension Reduction Using Factor Analysis

Factor analysis is an unsupervised feature extraction method that finds the underlying factors that explain the original spectral information of hyperspectral data in fewer dimensions while keeping the spatial information intact. This dimension reduction

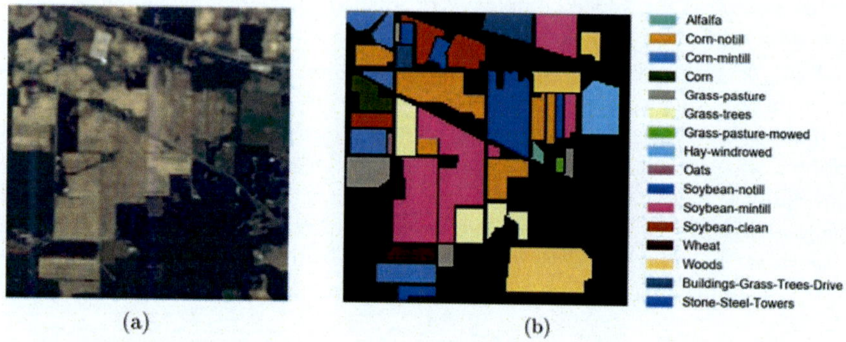

Fig. 1 Indian Pines dataset. **a** False colour composite image; **b** ground truth

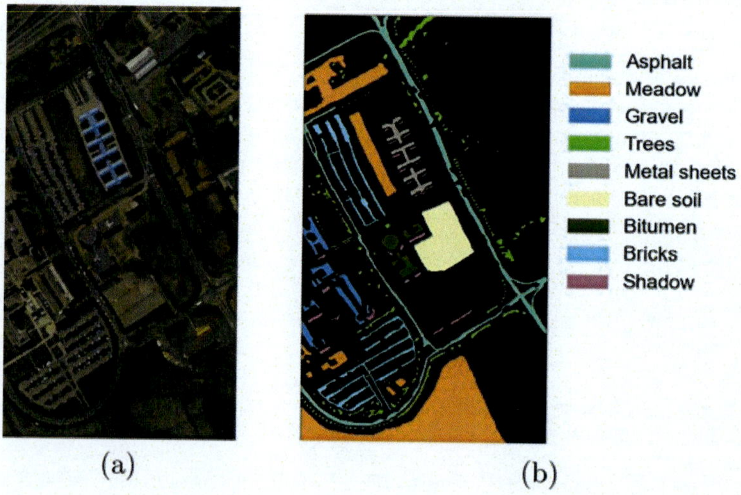

Fig. 2 Pavia University dataset. **a** False colour composite image; **b** ground truth

process alleviates the curse of dimensionality problem, reduces training parameters, and compresses the original data for memory efficiency.

Factor analysis assumes that all p observed variables X_1, X_2, ..., X_p in a dataset is linearly dependent on m unobservable, common factors $F_1, F_2, \ldots, F_m$. Observed variables also have unique variances $\epsilon_1, \epsilon_2, \ldots, \epsilon_p$ [14]. For a single observation,

$$X_i - \mu_i = l_{i1}F_1 + l_{i2}F_2 + \cdots + l_{im}F_m + \epsilon_i \tag{1}$$

l_{ij} is loading or weight of the ith variable on the jth factor. μ_i is the mean of X_i. For n observation, the matrix form of factor model,

$$X - \mu = LF + E \tag{2}$$

A maximum likelihood estimation technique will find factors that increase the likelihood of generating the covariances matrix of the original data. Assumption is made that the data are independently sampled from a multivariate normal distribution with mean vector μ, and covariance matrix of the form $LL^T + \psi$, where ψ is the covariance matrix of E [14]. The log likelihood function to find MLE estimators μ, L, ψ is,

$$\mathcal{L}(\mu, L, \psi) = -\frac{np}{2}log2\pi - \frac{n}{2}log|LL^T + \psi| - \frac{1}{2}(X_i - \mu)^T(LL^T + \psi)(X_i - \mu) \tag{3}$$

According to Bartlett method [15], we can obtain the factor score matrix F,

$$F = (L^T\psi^{-1}L)^{-1}L^T\psi^{-1} \tag{4}$$

Finally, projection of X into new subspace is obtained by,

$$Y = F^T(X - \mu) \tag{5}$$

By following the above method, a hyperspectral image cube X with $W \times H \times \lambda$ dimension can be reduced to $W \times H \times B$ where $B <= \lambda$.

2.3 Neighbourhood Patch Creation

The input of the hyperspectral image classification pipeline is normally a single image cube. This differs from conventional classification problems, where many images are used to train the model.

To create multiple image cubes for CNN model, HSI cube Y of dimension $W \times H \times B$ needs to be split into many neighbourhood patches(also called spatial context) of dimension $S \times S \times B$. This can be achieved by considering a neighbourhood window of $S \times S$ size for each pixel in the input HSI (Fig. 3). This procedure creates $(W - S + 1) \times (H - S + 1)$ number of 3D-patches from Y [13]. We added padding of size $\frac{S-1}{2}$ to the original image before creating the patches to ensure that the number of samples remains the same after creating the 3D-patches.

2.4 CNN Architecture

We will use CNN for constructing deep high-level spectral–spatial features. Spectral and spatial feature extraction is performed by using a 2D convolution layer followed by one 2D max-pooling layer and finally another 2D convolution layer. Since input

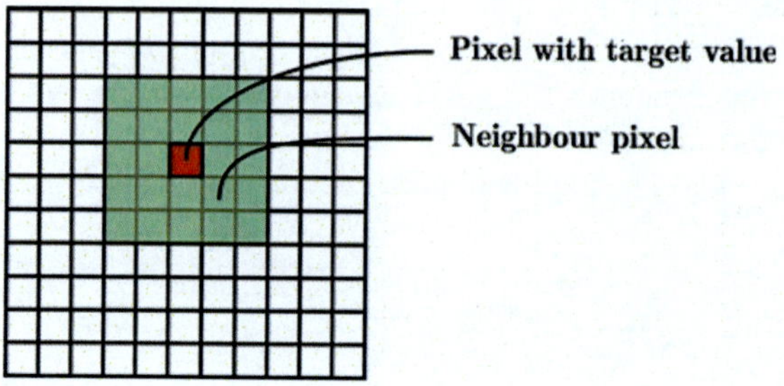

Fig. 3 Creating neighbourhood patch from input image (2D view)

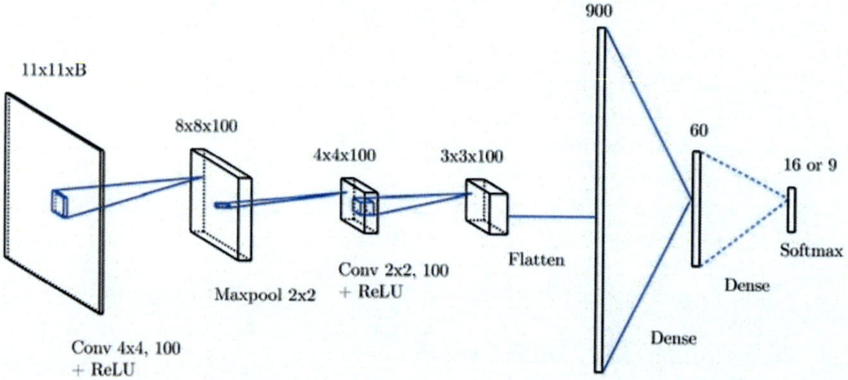

Fig. 4 Proposed CNN architecture

image cubes have extremely low spatial dimension, filter size is restricted to 4×4 (100 filters) and 2×2 (100 filters). The activation function of choice for all convolution layers is ReLU. ReLU has proven to converge faster in HSI classification [10]. Max-pooling layer accounts for spatial invariance. Performing convolution and pooling operation generates invariant, nonlinear, and discriminant features that will facilitate effective classification [9].

Multilayer perceptron with one hidden layer and softmax output layer is used for classification of each pixel in the input image cube. The output layer has 16 units (Indian Pines) or 9 units (Pavia University), each representing the likelihood of an input image cube belonging to a certain class. The choice of loss function for the network is categorical cross-entropy (Fig. 4).

The proposed CNN architecture is shallow because using deep network with small training data available could lead to overfitting problem [10].

3 Experimental Analysis

3.1 Experimental Setup

We have trained our deep learning model using Keras. Keras is a framework for creating deep neural networks that provides an abstraction on top of TensorFlow library.

We used Google Colab Jupyter Notebook powered by Intel(R) Xeon(R) CPU@2.20 GHz, 12 GB RAM, NVIDIA Tesla T4 GPU running on Linux-based operating system for training and testing our architecture.

Indian Pines and the Pavia University datasets are divided into training and test sets. Only 40% of the initial dataset is used as the training sample, while 60% is used as the test set. For hyperparameter tuning, 60% of the training set is used as the validation set. Training on such a small number of samples is critical in developing reliable classification method because hyperspectral images seldom have a large number of training data available for the classification task [10, 16].

The model was trained for 130 epochs using mini batch gradient descent. Selected batch size was 16 and the learning rate was empirically determined to be 0.001.

3.2 Result and Performance Analysis

We considered different numbers of spatial context or neighbourhood window sizes $S = \{7, 9, 11, 13\}$ and extracted bands $B = \{1, 3, 5, 7, 9, 11\}$ for our experiment to find the optimal configuration that would give maximum overall accuracy and average accuracy.

Figure 5 shows that increasing the number of extracted band and spatial context positively affects accuracy. However, increasing the spatial context increases trainable parameters and consequently increases training time. Also, a higher value of S can lead to overlapping problem [17]. So we settled on $S = 11$. A lower number of extracted band does not contain enough discriminating information to classify HSI correctly. Increasing the number of extracted bands up to a certain extent adds useful spectral information for the classifier to make correct decisions without introducing the dimensionality problem. Based on the above results, the optimal value is $S = 11$ and $B = 11$. So we have used, $11 \times 11 \times 11$ neighbourhood patches, constructed from original hyperspectral image cube, as input to our CNN framework for best performance.

Tables 1 and 2 demonstrate that FA-CNN performs well on overall accuracy, average accuracy, per-class accuracy, Cohen's Kappa(κ), and $F1$ score metrics on both Indian Pines and Pavia University dataset. The classifier only makes 17 misclassification out of 4100 test samples on Indian Pines dataset and 9 misclassification out of 17111 samples on Pavia University dataset.

Table 1 Performance of FA-CNN using $S = 11$ and $B = 11$ on Indian Pines

#	Class	Accuracy (%)	Training samples	Testing samples
1	Alfalfa	100	16	19
2	Corn-notill	99.47	514	571
3	Corn-mintill	99.40	299	332
4	Corn	100	85	95
5	Grass-pasture	98.96	174	193
6	Grass-trees	100	263	292
7	Grass-pasture-mowed	100	10	11
8	Hay-windrowed	100	172	191
9	Oats	100	7	8
10	Soybean-notill	98.97	350	389
11	Soybean-mintill	99.59	884	982
12	Soybean-clean	100	213	237
13	Wheat	100	74	82
14	Woods	99.60	455	506
15	Buildings-Grass-Trees-Drives	100	139	155
16	Stone-Steel-Towers	100	34	37
Average accuracy		99.75		
Overall accuracy		99.59		
κ score		99.53		
$F1$ score		99.66		

Table 2 Performance of FA-CNN using $S = 11$ and $B = 11$ on Pavia University

#	Class	Accuracy (%)	Training samples	Testing samples
1	Asphalt	100	2387	2652
2	Meadows	100	6173	7460
3	Gravel	99.40	756	839
4	Trees	100	1103	1226
5	Metal sheets	100	484	538
6	Bare Soil	100	1810	2012
7	Bitumen	100	479	532
8	Bricks	99.73	1326	1473
9	Shadows	100	341	379
Average accuracy		99.90		
Overall accuracy		99.95		
κ score		99.93		
$F1$ score		99.91		

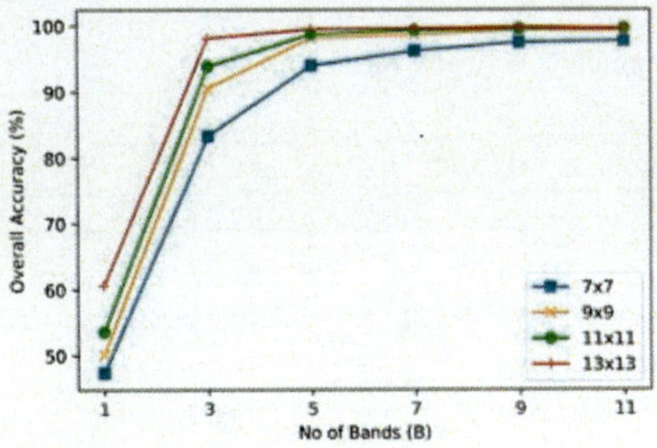

(a) Indian Pines overall accuracy.

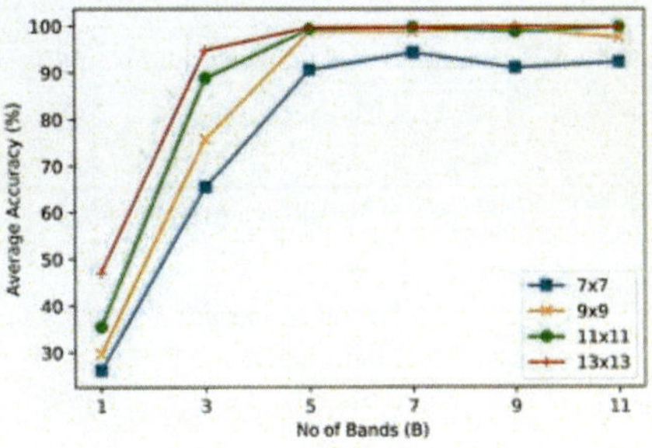

(b) Indian Pines average accuracy.

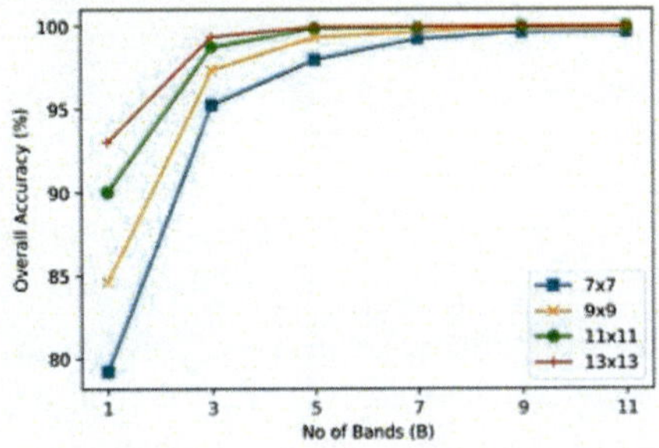

(c) Pavia University overall accuracy.

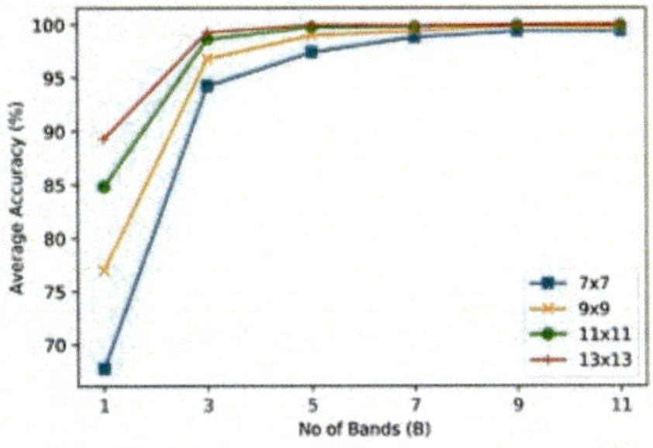

(d) Pavia University average accuracy.

Fig. 5 Accuracy results for different values of extracted bands (B) and spatial context (S)

3.3 Comparison with Other Methods

We compared our proposed FA-CNN with two recent spectral–spatial classification methods—HybridSN [13] and PCA-MS-CNN [11]. For Indian Pines dataset, 30 PCA bands accounting for 99.24% variance of the original dataset were extracted, as suggested by those papers. For Pavia University dataset, 15 PCA bands accounting for 99.90% variance of the original dataset were extracted for both methods. Spatial context for HybridSN and PCA-MS-CNN is, respectively, $S = 25$ and $S = \{3, 5, 7\}$, as recommended by those papers.

We have implemented all of these methods on our machine and used the same training split of 36% from the original dataset to train the models and the same test split for evaluating performance.

From Table 3, we can see that our method FA-CNN compares favourably to HybridSN and PCA-MS-CNN on accuracy metrics while requiring a vastly lower number of trainable parameters. A lower number of trainable parameters indicates that our network can be trained faster compared other two approaches.

Table 3 Comparison of accuracy and trainable parameters between various methods

Dataset	Measurements	Methods		
		Our method	HybridSN	PCA-MS-CNN
IP	Overall accuracy	**99.59**	98.37	97.61
	Average accuracy	**99.75**	96.15	97.79
	Total trainable parameters	**112,836**	5,122,176	10,035,746
PU	Overall accuracy	**99.95**	99.78	99.68
	Average accuracy	**99.90**	99.41	99.40
	Total trainable parameters	**112,409**	4,844,793	5,680,283

Bold indicates the best result among all the compared methods for a specific dataset and measurement

4 Conclusion

This research work proposed a hyperspectral image classification method called FA-CNN. FA-CNN extracted spectral features from the hyperspectral data using an unsupervised dimension reduction technique called factor analysis. Factor analysis extracted the underlying factors that explain the original spectral information in fewer dimensions. Combining effective spectral feature extraction with spatial information using CNN, FA-CNN achieved high accuracy using a relatively small training set and lower number of trainable parameters. Our method compared favourably to several state-of-the-art methods. In future, we want to apply this method to other hyperspectral datasets to leverage the power of hyperspectral imaging in solving various problems.

References

1. Bioucas-Dias JM, Plaza A, Camps-Valls G, Scheunders P, Nasrabadi N, Chanussot J (2013) Hyperspectral remote sensing data analysis and future challenges. IEEE Geosci Remote Sens Mag 1(2):6–36
2. Donoho DL et al (2000) High-dimensional data analysis: the curses and blessings of dimensionality. AMS Math Challenges Lecture 1(2000):32
3. Hughes GF (1968) On the mean accuracy of statistical pattern recognizers. IEEE Trans Inf Theory IT 14(1):55-63
4. Rodarmel C, Shan J (2002) Principal component analysis for hyperspectral image classification. Surveying Land Inf Sci 62(2):115–122
5. Melgani F, Bruzzone L (2004) Classification of hyperspectral remote sensing images with support vector machines. IEEE Trans Geosci Remote Sens 42(8):1778–1790
6. Audebert N, Le Saux B, Lefèvre S (2019) Deep learning for classification of hyperspectral data: a comparative review. IEEE Geosci Remote Sens Mag 7(2):159–173
7. Li S, Song W, Fang L, Chen Y, Ghamisi P, Benediktsson JA (2019) Deep learning for hyperspectral image classification: an overview. IEEE Trans Geosci Remote Sens 57(9):6690–6709
8. Ahmad M (2020) A fast 3D CNN for hyperspectral image classification. arXiv:2004.14152

9. Chen Y, Jiang H, Li C, Jia X, Ghamisi P (2016) Deep feature extraction and classification of hyperspectral images based on convolutional neural networks. IEEE Trans Geosci Remote Sens 54(10):6232–6251
10. Gao Q, Lim S, Jia X (2018) Hyperspectral image classification using convolutional neural networks and multiple feature learning. Remote Sens 10(2):299
11. Haque M, Zaman S (2019) Spectral-spatial feature extraction using PCA and multi-scale deep convolutional neural network for hyperspectral image classification, pp 1–6
12. Makantasis K, Karantzalos K, Doulamis A, Doulamis N (2015) Deep supervised learning for hyperspectral data classification through convolutional neural networks. In: 2015 IEEE International geoscience and remote sensing symposium (IGARSS). IEEE, pp 4959–4962
13. Roy SK, Krishna G, Dubey SR, Chaudhuri BB (2019) HybridSN: exploring 3-D-2-D CNN feature hierarchy for hyperspectral image classification. IEEE Geosci Remote Sens Lett 17(2):277–281
14. Kassim S, Hasan H, Mohd Ismon A, Muhammad Asri F (2013) Parameter estimation in factor analysis: maximum likelihood versus principal component. AIP Conf Proc 1522(1):1293–1299
15. Bartlett MS (1937) The statistical conception of mental factors. British J Psychol 28(1):97
16. Huang L, Chen Y (2020) Dual-path siamese CNN for hyperspectral image classification with limited training samples. IEEE Geosci Remote Sens Lett
17. Lange J, Cavallaro G, Götz M, Erlingsson E, Riedel M (2018) The influence of sampling methods on pixel-wise hyperspectral image classification with 3d convolutional neural networks. In: IGARSS 2018-2018 IEEE international geoscience and remote sensing symposium. IEEE, pp 2087–2090

A Convolutional Neural Network Model for Screening COVID-19 Patients Based on CT Scan Images

Md. Fazle Rabbi, S. M. Mahedy Hasan, Arifa Islam Champa, Md. Rifat Hossain, and Md. Asif Zaman

Abstract The novel coronavirus (COVID-19) spread all over the world within a few months and turned into a pandemic. Early diagnosis is the only way to combat this pandemic by isolating the affected cases from healthy ones for refraining it from further spreading. At present, RT-PCR is extensively used for screening coronavirus cases, however, WHO stated that it suffers from low sensitivity and low specificity in the early-stage patients. Recent studies advise that the CT scan image embraces key features for detecting this disease. The application of deep learning approaches combined with CT images could be useful for the precise diagnosis of positive coronavirus patients. In this research, we have employed the Convolutional Neural Network (CNN) architecture of deep learning on publicly accessible CT images dataset to build a prediction model for classifying positive COVID-19 from other pulmonary diseases and healthy patients. Moreover, this prediction model has also been utilized for identifying COVID-19 cases from other pulmonary diseases, which is a binary classification. In ternary classification, our proposed CNN model has obtained an accuracy of 98.79%, a precision of 94.98%, a sensitivity of 98.85%. In contrast, for binary classification, it has acquired an accuracy of 98.79%, a precision of 94.98%, a sensitivity of 98.85%. Therefore, this proposed model can be a faster and alternative tool or even a supportive tool along with RT-PCR in rural areas of many countries where there is a scarcity of testing kits and expert physicians.

Keywords Coronavirus (COVID-19) · Deep learning · Convolutional neural network · CT scan images

Md. Fazle Rabbi (✉) · A. I. Champa · Md. Rifat Hossain
Department of Computer Science and Engineering, Bangladesh Army International University of Science & Technology, Cumilla, Bangladesh
e-mail: fazle@baiust.edu.bd

S. M. Mahedy Hasan · Md. Asif Zaman
Department of Computer Science and Engineering, Rajshahi University of Engineering & Technology, Rajshahi, Bangladesh

M. S. Arefin et al. (eds.), *Proceedings of the International Conference on Big Data, IoT, and Machine Learning*, Lecture Notes on Data Engineering and Communications Technologies 95, https://doi.org/10.1007/978-981-16-6636-0_12

1 Introduction

Coronavirus disease 2019 or in short COVID-19 is an exceedingly infectious disease that was originated from Wuhan, Hubei Province, China, in December 2019. The disease spread all over China from Wuhan within 30 days [1]. Eventually, more than 127 million people have been affected and causing more than 2.7 million deaths over 215 countries, by May 19, 2020 [2]. The diagnosis of COVID-19 is of immense importance because that's the only way to identify infected persons and quarantine them so that they cannot spread the disease. Also, early diagnosis is a must to treat the patients because it can be fatal as 2% of affected people died.

At present, the reverse transcription-polymerase chain reaction (RT-PCR) technique is amply applied for the diagnosis of COVID-19 disease. Nonetheless, the main drawback of RT-PCR is that it suffers from low sensitivity, low specificity, and reproducibility [3]. RT-PCR test is expensive, time-consuming, and requires skilled medical expertise [4, 5]. Moreover, due to the shortage of RT-PCR test kits, doctors, or medical practitioners urge to utilize radiological images to diagnose COVID-19 patients [6, 7]. Computed Tomography (CT) scan image contains fruitful information that can be helpful to diagnose COVID-19 patients [8, 9]. Chest CT images are widely used to detect COVID-19 cases because of the shortage of huge number of testing kits in different countries such as Turkey. For early diagnosis of COVID-19 patients, CT images with human observations make the diagnosis indistinguishable from other types of viral and bacterial pneumonia. Here, Artificial Intelligence (AI) with deep learning techniques plays a vital role to distinguish COVID-19 patients of different kinds of diseases. In recent years, deep learning technologies are widely used to diagnose various medical disorders such as skin cancer detection, breast cancer detection, lung cancer detection, heart disease prediction, brain tumor classification, pneumonia detection from chest CT/X-ray images, and outperformed other conventional methods [10]. Deep learning smoothly extracts features and constructs a robust model that can be used for effective classification. Numerous researches have been done using AI with deep learning techniques for ascertaining COVID-19 patients using chest CT or X-ray images [11, 12].

In this research, a deep learning-based CNN model is proposed for automatic detection of COVID-19 cases. This proposed model is applied to a comparatively large dataset consists of 3791 images. The experimental findings imply that our suggested model could be utilized for the confirmation of COVID patients and, therefore, help people around the globe to fight against this pandemic.

The remaining segments of this article are arranged in a systematic way as follows: Sect. 2 includes related studies conducted in recent times. Section 3 elaborates on the dataset and methods adopted for this study. Section 4 analyzes the findings of the suggested CNN model. Section 5 concludes the paper.

2 Literature Review

From the beginning of the outbreak, various researchers have been focusing on the construction of a model for automatic detection of COVID-19 cases. Li et al. [13] developed a deep learning model named COVNET for detecting COVID-19. To build the model, they have utilized 4356 CT images of the chest to differentiate between COVID-19 and pneumonia. To evaluate the model performance, they considered sensitivity, specificity, and Area Under Curve (AUC) and obtained 87%, 92%, and 0.95%, respectively. Bai et al. [14] designed an Efficient Net B4 architecture based on deep learning for identifying COVID-19 patients using chest CT images. They collected CT images of 1121 patients, where 521 patients were COVID-19 positive, and 665 non-COVID pneumonia patients to build their proposed model. For performance evaluation, they used different statistical assessment techniques, i.e., accuracy, sensitivity, specificity, and Receiver Operating Characteristics (ROC AUC), and acquired performances of 87%, 89%, 86%, and 0.9% respectively. Ardakani et al. [15] compared ten different CNN architectures to diagnose COVID-19 cases fast. CT images of 108 COVID-19 positive patients and 86 viral pneumonia patients were examined. Among those architectures, ResNet-101 achieved the highest accuracy of 99.51%. Kang et al. [16] proposed a multi-view representation learning framework to distinguish COVID-19 patients from pneumonia patients. They implied 2522 chest CT images for training and testing their suggested framework. Their proposed technique obtained an accuracy of 95.5%, 96.6% sensitivity, and 93.2% specificity. Xu et al. [17] used ResNet-18 for extracting features and differentiate between COVID-19, pneumonia, and normal cases. Their proposed model was built based on 618 CT samples and obtained an accuracy of 86.7%. Butt et al. [18] employed Resnet-18 and Resnet-23 based deep CNN architecture to build deep learning frameworks to differentiate coronavirus from non-coronavirus cases. A set of 618 thoracic CT samples were utilized for training and testing. Their proposed architecture attained 98.2% sensitivity, 92.2% specificity, and 0.996 AUC. Tang et al. [19] combined quantitative feature analysis Random Forest (RF) classifier for screening coronavirus patients. In this research, they collected CT images from 176 patients and extracted 63 quantitative features to build the RF model. To evaluate the model performance, they examined different statistical measures and obtained an accuracy of 87.5%, a true positive rate of 93.3%, a true negative rate of 74.5%, and AUC of 0.91. Shi et al. [20] proposed a machine learning-based framework for screening coronavirus patients using location-specific features. A group of CT images from 2685 patients was considered to identify COVID-19 from pneumonia. Their experimental analysis showed that the proposed RF model achieved 87.9% accuracy, 90.7% sensitivity, 83.3% specificity. Alom et al. [21] used both CT scans and X-ray images to create an Inception Residual Recurrent Convolutional Neural Network (IRRCNN) model to detect COVID-19. They exerted 420 CT images where 247 for normal cases and 178 for COVID-19 cases. The overall accuracy of their IRRCNN model was 98.78%. Özkaya et al. [22] proposed a transfer learning-based CNN model where ResNet-50, Google Net, and VGG-16 architectures were utilized for extracting deep features.

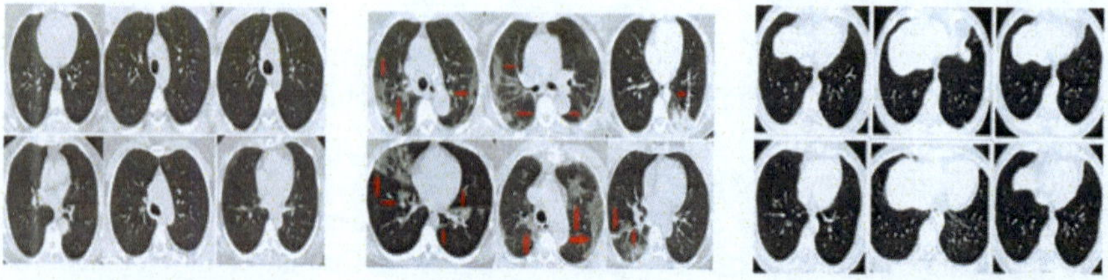

(a) Healthy cases (b) COVID-19 cases (c) Pulmonary cases

Fig. 1 Samples of CT scan images of the dataset

Those pre-trained CNN architectures were then trained to use transfer learning on two subsets of images. They acquired an overall classification accuracy of 95.60% in subset-1, and 98.27% in subset-2. Jin et al. [23] suggested an AI system based on ResNet-152 for screening COVID-19 positive and negative cases. In this study, they have considered CT images of 496 COVID-19 positive patients and 260 COVID-19 negative patients. The obtained accuracy, AUC, sensitivity, and specificity are 94.98%, 97.91%, 94.06%, and 95.47% respectively. Ying et al. [24] developed a deep learning-based model named DeepPnumonia to differentiate between COVID-19 from bacterial pneumonia. They collected CT images of 88 COVID-19 positive patients and 101 bacterial pneumonia patients. They obtained 95% of AUC and 96% sensitivity.

3 Materials and Methods

3.1 Dataset Description

In this study, publicly available CT scan images are collected from Kaggle and China National Center for Bio information for the detection of COVID-19 cases [25, 26]. This dataset comprises 3791 chest CT scan images that were accumulated from a public hospital of Sao Paolo, Brazil, and (CC-CCII) the China Consortium of Chest CT Image Investigation. Among those images, 1252 chest CT scan images are of positive COVID-19 cases, 1230 chest CT scan images are of other pulmonic diseases, and the rest of the 1309 chest CT scan images are healthy patients. Some samples of CT images from this dataset are displayed in Fig. 1.

3.2 Proposed CNN Architecture

In Fig. 2, the structure of the proposed CNN model is delineated.

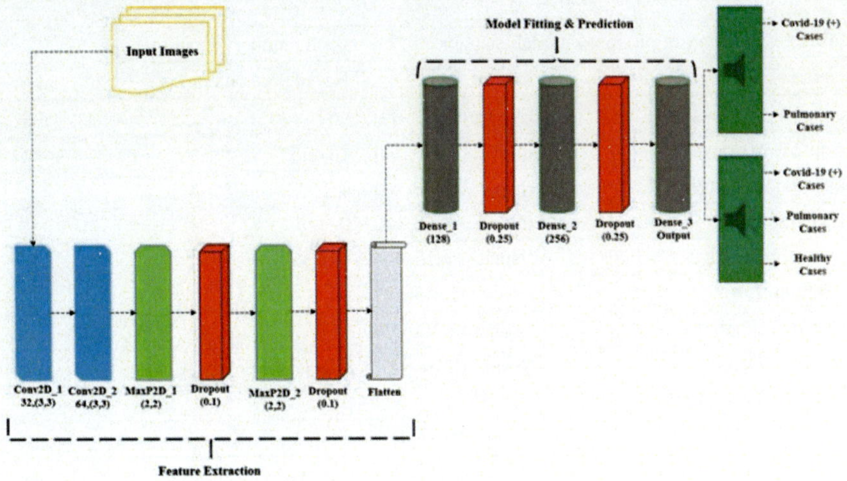

Fig. 2 The structure of proposed CNN model

The experimental dataset has images of different shapes. Therefore, those images are reshaped into a uniform shape of 64 × 64 pixels. To extract features from the input images, two convolution layers are utilized in our proposed CNN model. In the first convolution layer, there are 32 filters of 3 × 3 shapes applied. To increase non-linearity, ReLU is widely used as an activation function. Moreover, to the first convolution layer, a second convolution layer is applied to the convolved features with 64 filters of 3 × 3 shapes and the ReLU activation function. Subsequently, 2 × 2 size max pooling is utilized to minimize the size of the feature map. To confront the overfitting problem, a dropout layer is utilized. In the Keras module, the dropout rate ranges from 0 to 1, where it refers to the fraction of the number of neurons to drop. After performing max pooling, 10% of the input neurons are dropped. On top of the dropout layer, the second max-pooling layer is added with 2 × 2 size to shorten the activation map's size. Furthermore, a second dropout layer is applied with a rate of 0.1.

After the completion of convolution and max-pooling steps, pooled features are transformed into 1-Dimensional array by flattening, which is further fed into the ANN. Two hidden or dense layers have also been used. The number of hidden units is 128 in the first dense layer, and ReLU is employed as an activation function. Then, 25% of the neurons are excluded from the dropout layer. The second dense layer of 256 units is appended on the top of the dropout layer, and ReLU is applied as an activation function in that layer. Moreover, with a rate of 0.25, the final dropout layer is applied to discard 25% of neurons. As we have both binary and ternary outcomes, we have taken different approaches in the output layer. For binary outcome, one unit is used to identify COVID-19 and other pulmonary cases. Here, sigmoid as an activation function and binary cross-entropy as loss function are utilized. One unit is used for a ternary outcome to identify COVID-19, pulmonary disease, and healthy

Table 1 Summary of proposed model (sequence of layers and number of parameters)

Name of layer	Output shape	No. of parameters
conv2d_1 (Conv2D)	(None, 62, 62, 32)	96
conv2d_2 (Conv2D)	(None, 60, 60, 64)	18,496
max_pooling2d_1	(None, 30, 30, 64)	0
Dropout_1	(None, 30, 30, 64)	0
max_pooling2d_2	(None, 15, 15, 64)	0
Dropout_2	(None, 15, 15, 64)	0
flatten_1 (Flatten)	(None, 14,400)	0
dense_1 (Dense)	(None, 128)	1,843,328
dropout_3 (Dropout)	(None, 128)	0
dense_2 (Dense)	(None, 256)	33,024
dropout_4 (Dropout)	(None, 256)	0
Output layer		
(Binary class) dense_3 (Dense)	(None, 1)	257
(Ternary class) dense_3 (Dense)	(None, 3)	771
		Total parameters = 1,896,001 (**binary class**) = 1,896,515 (**ternary class**)

cases. Here, softmax as an activation function and categorical cross-entropy as loss function is utilized. Proposed CNN model contains 883,393 parameters, which is presented in Table 1.

As an optimization algorithm, Adam is considered for both binary and ternary outcomes.

3.3 *Performance Evaluation Measures*

In this research, six (6) different evaluation measures are considered, such as accuracy, AUC, sensitivity, specificity, precision, and recall. The confusion matrix (CM) manifests the overall performance of any prediction model. The confusion matrix is introduced in Table 2. By utilizing this confusion matrix, those six measures can be calculated.

Table 2 Confusion matrix

Original value	Predicted value	
	Positive	Negative
Positive	True positive (TP)	False negative (FN)
Negative	False positive (FP)	True negative (TN)

4 Results and Discussion

The entire experiment is executed on Google Colaboratory [27], which is a cloud-based service for developing applications based on the python programming language. To construct the proposed CNN model, Keras, a neural network library package on the virtual Tensor Processing Unit (TPU) provided by the Google Colab is used.

In this research, we used data from 3 classes to develop a generalized prediction model. We divided our entire experiment into two parts; the first one is ternary classification where we used three class data, namely healthy, other pulmonary, and COVID-19, and the second one is the binary classification where we combined healthy and other pulmonary images as non-covid. The entire operation was performed up to 50 numbers of epochs. We randomly split the experimental dataset into three parts: train, validation, and test data where 80% for training plus validation (90% training, 10% validation) and 20% for testing. We created the model based on training and validation data and then evaluated the generated model based on testing data.

In order to perceive the performance of our proposed model, we have plotted model accuracy and loss curve in Fig. 3 for ternary class and in Fig. 4 for binary class (Table 3).

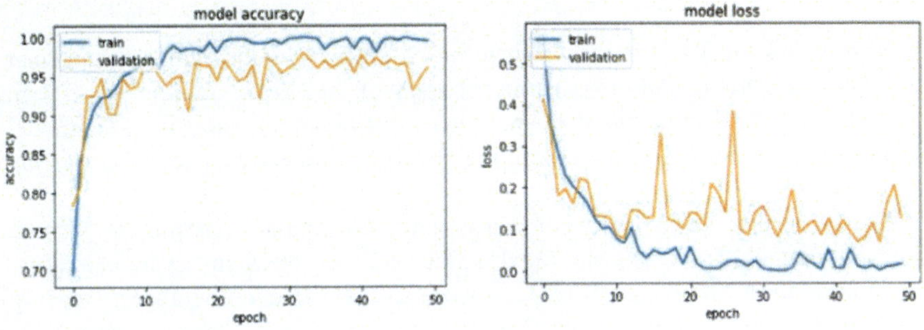

Fig. 3 Model accuracy and loss curve of ternary classification

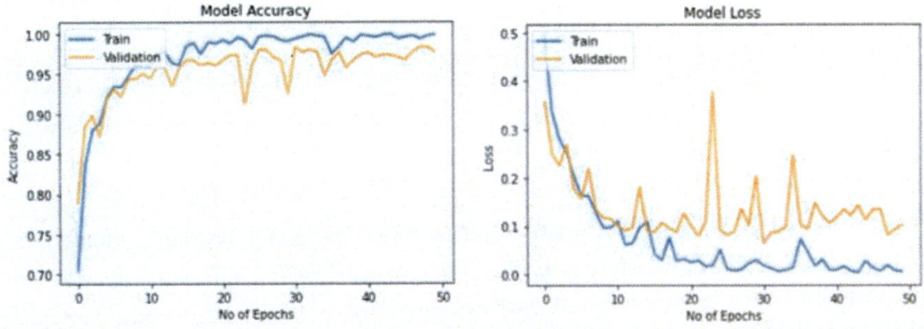

Fig. 4 Model accuracy and loss curve of binary classification

Table 3 Accuracy of the proposed model

Classification type	Training (%)	Validation (%)	Testing (%)
Ternary	99.22	96.05	96.00
Binary	99.85	97.70	96.00

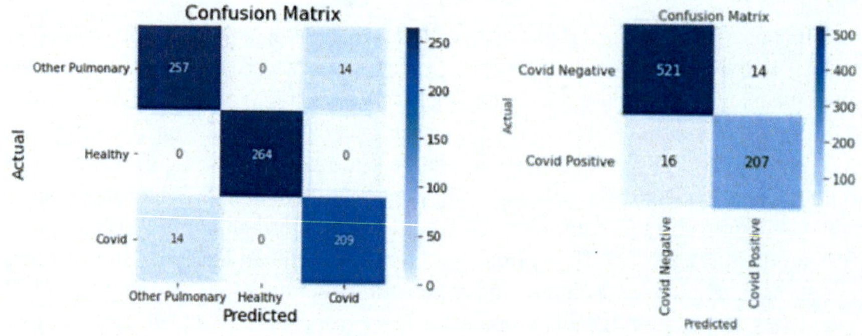

Fig. 5 Confusion matrix

While inspecting the performance in terms of ternary classification, we found that our proposed CNN model obtained an accuracy of 99.22% in training, 96.05% in validation, and 96% in testing data and in terms of binary classification, the model acquired an accuracy of 99.85% in training, 97.70% in validation, and 96% in testing data.

We have also included the CM of ternary and binary classification in Fig. 5 to see the exact number of correctly classified samples of every class. In ternary classification, none of the healthy cases were classified as COVID-19 cases or other pulmonary cases. Some (14) COVID-19 cases were falsely identified as other pulmonary cases, but these false cases can be later correctly spotted by healthcare practitioners while treating the patients. The other evaluation measures like precision, recall, and f1 score are specified classwise in Table 4.

ROC curve of binary and ternary classification is delineated in Fig. 6.

5 Conclusions

RT-PCR has been used as yet to detect the coronavirus, which regularly engenders false-negative results. Therefore, chest CT images can play an active role to combat this problem. Here, a model is developed employing CNN on chest CT images to detect and classify coronavirus cases from healthy and other pulmonary diseases. Scrutinizing the test results, it is comprehensible that CNN architecture of deep

Table 4 Classwise performance

Type of classification		Precision (%)	Recall (%)	F1 score (%)
Binary	Non-covid	97	97	97
	Covid	94	93	93
	Average	96	96	96
Ternary	Other pulmonary	95	95	95
	Healthy	100	100	100
	Covid	94	94	94
	Average	96	96	96

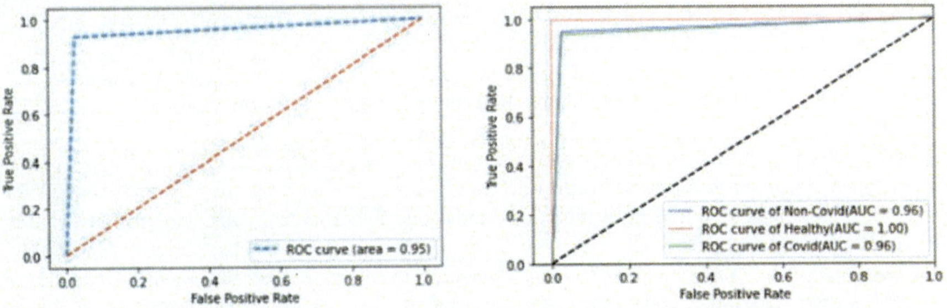

a. Binary Class

b. Ternary Class

Fig. 6 ROC curve of binary and ternary classification

learning has a notable role in the automatic extraction of key information from chest CT images akin to the diagnosis of coronavirus cases. However, our Proposed CNN model has produced superior accuracy, accounted for 98.78% and 98.25% for binary and ternary classification respectively. This model can be applied as a substitutive tool, even an assistive tool along with RT-PCR, in distant areas where there has been lacking test kits and a shortage, or no experienced physicians at all. Therefore, proposed CNN model can be cost-effective for screening COVID-19 cases and giving test results in a minute so that the affected patients can be isolated more quickly, and community spread can be prevented. Since the large dataset of COVID-19 is not easily accessible, therefore, we have used a new publicly available dataset. If the size of the dataset was even bigger, then we could create a more accurate and precise model to classify COVID-19 cases. This model can be preserved in the cloud so that the disease can be detected even more quickly, and the patient can be isolated immediately. In the future, we intend to accumulate chest CT images of real COVID-19 patients from local hospitals in Bangladesh and assess them with our developed model.

References

1. Wu Z, McGoogan JM (2020) Characteristics of and important lessons from the coronavirus disease 2019 (COVID-19) outbreak in China: summary of a report of 72 314 cases from the Chinese center for disease control and prevention. JAMA 323(13):1239–1242. https://doi.org/10.1001/jama.2020.2648
2. COVID-19 CORONAVIRUS PANDEMIC (2021) https://www.worldometers.info/coronavirus/. Accessed 28 Mar 2021
3. Bleve G, Rizzotti L, Dellaglio F, Torriani S (2003) Development of reverse transcription (RT)-PCR and real-time RT-PCR assays for rapid detection and quantification of viable yeasts and molds contaminating yogurts and pasteurized food products. Appl Environ Microbiol 69(7):4116–4122. https://doi.org/10.1128/aem.69.7.4116-4122.2003
4. Pathak Y, Shukla PK, Tiwari A, Stalin S, Singh S, Shukla PK (2020) Deep transfer learning based classification model for COVID-19 disease. IRBM. https://doi.org/10.1016/j.irbm.2020.05.003
5. Zu ZY, Jiang MD, Xu PP, Chen W, Ni QQ, Lu GM, Zhang LJ (2020) Coronavirus disease 2019 (COVID-19): a perspective from China. Radiology 200490. https://doi.org/10.1148/radiol.2020200490
6. Bernheim A, Mei X, Huang M, Yang Y, Fayad Z, Zhang N, Diao K, Lin B, Zhu X, Li K, Li S, Shan H, Jacobi A, Chung M (2020) Chest CT findings in coronavirus disease-19 (COVID-19): relationship to duration of infection. Radiology 295:200463. https://doi.org/10.1148/radiol.2020200463
7. Long C, Xu H, Shen Q et al (2020) Diagnosis of the coronavirus disease (COVID-19): RT-PCR or CT? Eur J Radiol 126:108961. https://doi.org/10.1016/j.ejrad.2020.108961
8. Dong D, Tang Z, Wang S et al (2020) The role of imaging in the detection and management of COVID-19: a review [published online ahead of print, 2020 Apr 27]. IEEE Rev Biomed Eng10.1109/RBME.2020.2990959
9. Shi F, Wang J, Shi J, Wu Z, Wang Q, Tang Z, He K, Shi Y (2020) Review of artificial intelligence techniques in imaging data acquisition, segmentation and diagnosis for COVID-19. IEEE Rev Biomed Eng 1-1. https://doi.org/10.1109/RBME.2020.2987975
10. Litjens G, Kooi T, Bejnordi BE, Setio A, Ciompi F, Ghafoorian M, van der Laak J, van Ginneken B, Sánchez CI (2017) A survey on deep learning in medical image analysis. Med Image Anal 42:60–88. https://doi.org/10.1016/j.media.2017.07.005
11. Ozturk T, Talo M, Yildirim EA, Baloglu UB, Yildirim O, Rajendra Acharya U (2020) Automated detection of COVID-19 cases using deep neural networks with X-ray images. Comput Biol Med 121:103792. https://doi.org/10.1016/j.compbiomed.2020.103792
12. Apostolopoulos ID, Mpesiana TA (2020) Covid-19: automatic detection from X-ray images utilizing transfer learning with convolutional neural networks. Phys Eng Sci Med. https://doi.org/10.1007/s13246-020-00865-4
13. Li L, Qin L, Xu Z, Yin Y, Wang X, Kong B, Bai J, Lu Y, Fang Z, Song Q, Cao K, Liu D, Wang G, Xu Q, Fang X, Zhang S, Xia J, Xia J (2020) Artificial intelligence distinguishes COVID-19 from community. Radiology 200905. https://doi.org/10.1148/radiol.2020200905
14. Bai HX, Wang R, Xiong Z, Hsieh B, Chang K, Halsey K, Tran TML, Choi JW, Wang D-C, Shi L-B, Mei J, Jiang X-L, Pan I, Zeng Q-H, Hu P-F, Li Y-H, Fu F-X, Huang RY, Sebro R, Yu Q-Z, Atalay MK, Liao W-H (2020) AI augmentation of radiologist performance in distinguishing COVID-19 from pneumonia of other etiology on chest CT. Radiology 201491. https://doi.org/10.1148/radiol.2020201491
15. Ali AA, Alireza RK, Rajendra Acharya U, Nazanin K, Afshin M (2020) Application of deep learning technique to manage COVID-19 in routine clinical practice using CT images: results of 10 convolutional neural networks. Comput Biol Med 121:103795. https://doi.org/10.1016/j.compbiomed.2020.103795
16. Kang H, Xia L, Yan F et al (2020) Diagnosis of coronavirus disease 2019 (COVID-19) with structured latent multi-view representation learning [published online ahead of print, 2020 May 5]. IEEE Trans Med Imaging. https://doi.org/10.1109/TMI.2020.2992546

17. Xu X, Jiang X, Ma C, Du P, Li X, Lv S, Yu L, Chen Y, Su J, Lang G et al (2020) Deep learning system to screen coronavirus disease 2019 pneumonia. arXiv preprint arXiv: 2002.09334
18. Butt C, Gill J, Chun D, Babu BA (2020) Deep learning system to screen coronavirus disease 2019 pneumonia. Appl Intell 1–7. https://doi.org/10.1007/s10489-020-01714-3
19. Tang Z, Zhao W, Xie X, Zhong Z, Shi F, Liu J, Shen D (2020) Severity assessment of coronavirus disease 2019 (COVID-19) using quantitative features from chest CT images. arXiv: 2003.11988
20. Shi F, Xia L, Shan F, Wu D, Wei Y, Yuan H, Jiang H, Gao Y, Sui H, Shen D (2020) Large-scale screening of COVID-19 from community acquired pneumonia using infection size-aware classification. arXiv: 2003.09860
21. Alom MdZ, Rahman MMS, Nasrin M, Taha T, Asari V (2020) COVID_MTNet: COVID-19 detection with multi-task deep learning approaches. arXiv: 2004.03747
22. Özkaya U, Öztürk Ş, Barstuğan M (2020) Coronavirus (COVID-19) classification using deep features fusion and ranking technique. arXiv: 2004.03698
23. Jin C, Chen W, Cao Y, Xu Z, Zhang X, Deng L, Zheng C, Zhou J, Shi H, Feng J (2020) Development and evaluation of an AI system for COVID-19 diagnosis. medRxiv 2020.03.20.20039834; https://doi.org/10.1101/2020.03.20.20039834
24. Ying S, Zheng S, Li L, Zhang X, Zhang X, Huang Z, Chen J, Zhao H, Jie Y, Wang R, Chong y, Shen J, Zha Y, Yang Y (2020) Deep learning enables accurate diagnosis of novel coronavirus (COVID-19) with CT images. medRxiv2020.02.23.20026930; https://doi.org/10.1101/2020.02.23.20026930
25. Eduardo S, Plamen A. SARS-COV-2 Ct-scan dataset. Kaggle. https://doi.org/10.34740/KAGGLE/DSV/1100240
26. China National Center for Bioinformation: National Genomics Data Center. https://bigd.big.ac.cn/. Accessed 28 Mar 2021
27. Google Colaboratory. colab.research.google.com. Accessed 28 Mar 2021

Real-time Face Recognition System for Remote Employee Tracking

Mohammad Sabik Irbaz, MD Abdullah Al Nasim, and Refat E Ferdous

Abstract During the COVID-19 pandemic, most of the human-to-human interactions have been stopped. To mitigate the spread of deadly coronavirus, many offices took the initiative so that the employees can work from home. But, tracking the employees and finding out if they are really performing what they were supposed to turn out to be a serious challenge for all the companies and organizations who are facilitating "work from home." To deal with the challenge effectively, we came up with a solution to track the employees with face recognition. We have been testing this system experimentally for our office. To train the face recognition module, we used FaceNet with KNN using the Labeled Faces in the Wild (LFW) dataset and achieved 97.8% accuracy. We integrated the trained model into our central system, where the employees log their time. In this paper, we discuss in brief the system we have been experimenting with and the pros and cons of the system.

Keywords Face recognition · Face detection · Computer vision

1 Introduction

Face recognition is an innovative identifying system that works on recognizing the face of a potential individual after comparing it with a source image stored in a database by a vision system. This technology extracts features from an input image and recognizes them by detecting the specific and particular information about an individual's face. We can see many applications of this, for example, access and security, criminal identification, online payments, and design of a smart home. In the face recognition and detection process, no human cooperation is needed, which makes it the best method for person identification [1]. There are various biometric

M. S. Irbaz (✉) · M. A. Al Nasim · R. E. Ferdous
Machine Learning Team, Pioneer Alpha Ltd., Dhaka, Bangladesh
e-mail: sabikirbaz@iut-dhaka.edu

M. A. Al Nasim
e-mail: nasim.abdullah@ieee.org

M. S. Arefin et al. (eds.), *Proceedings of the International Conference on Big Data, IoT, and Machine Learning*, Lecture Notes on Data Engineering and Communications Technologies 95, https://doi.org/10.1007/978-981-16-6636-0_13

authentication systems available, face recognition is one of them. In biometric-based strategies, one's physiological and behavioral attributes have been analyzed with a specified end goal to determine their uniformity [2]. Hence, compared to other biometric-based systems, face recognition has been considered as one of the most significant biometric authentication techniques among the most constitutive applications.

Starting from 2020, during the COVID-19 pandemic, many offices took the initiative so that the employees can work from home. But, tracking the employees and finding out if they are really there to work had been a serious challenge for the administration. Existing time tracker solutions can be easily fooled, which we have already seen in our company. To deal with the challenge and build a system that cannot be fooled, we came up with a solution to track the employees with face recognition. We have been testing this system internally. To train the face recognition module, we used FaceNet with KNN using the Labeled Faces in the Wild (LFW) dataset [3].

We organized our paper as follows: In Sect. 2, we presented the existing literature methods that helped us to reach our goal. In Sect. 3, we gave a description of the dataset we used in our model. Section 4 provides a summary of our methodology. Experimental evaluations have been discussed in Sect. 5. Finally, in Sect. 6, we give a brief discussion on the future work and conclusion.

2 Related Works

One of the earliest works on face recognition can be found in 1991 [4]. This model proposes a person's identification from faces which integrates the role of the context during learning as well as during recognition.The aim is to specify the functioning of two mechanisms by applying the encoding specificity principle, which assumes that traces in memory are contextualized. The structural encoding component provides a perceptual description of the presented face, which can match a previously stored representation. By activating FRU, it can then access person identity representations held in person identity nodes (PINS). The FaceNet network comprises the minimum number of layers necessary to model which lead to familiarity estimation and identification. The first layer is the input layer, which is divided into two parts: one representing face input and the other context input, each containing 25 cells. The second layer is a hidden layer with 80 cells which are divided into 3 parts. This third layer is composed of only 20 cells wholly connected to the face-context association. The network's outputs are the emergent memory representations of the system and can be seen as equivalent to ecphoric information. This system learns according to the algorithm of "gradient backpropagation" discovered by Rumelhart, Hinton, and Williams (1986a). For the recognition test, firstly, the faces were presented in their encoding context, and secondly, 15 old contexts which had never been associated with the four test faces were presented to FaceNet, once with each of these faces.

In a review till 2006 on face recognition [5], we can find various databases for face recognition. This is the start of face recognition revolution. Some of them are AT&T

(ORL) Database, FERET Database, AR Face Database, CMU-PIE Database, Yale Face Database, UMIST Face Database, MUCT Face Database, Face94 Database, Indian Database, and Grimace Database.

We find more interesting works in 2011 by Deniz et al. [6]. As HOG operates on local cells, it is invariant to geometric transformations and photometric transformations, except for object orientation. This experiment has used single-angle orientation to allow more differentiation between patterns. Overlapping in HOG significantly improves the performance of detection and identification which would have been otherwise quite difficult in presence of poor lighting conditions. To remove redundancy in the data and avoid over-fitting, they proposed to use dimensionality reduction in the HOG representation. To provide robustness to facial feature detection, they propose uniform sampling of the HOG features.

Fast forward to very recent works, we can find Prasad et al. [7] using deep learning-based representations. They input raw data in convolve filters in different levels that automatically detect underlying high levels from labeled or unlabeled data. Face verification approaches are classified into three sets. First one is used to extract face feature vectors and process them by classifiers. Second one directly enhances the proof loss for matching and non-matching pairs. Third one combined identification and verification constraints to improve the deep face model. For identifying faces, two approaches are explained here. First one is effective convolutional neural network designs for biometric face recognition. Second one is explanation of a face plan which is based on two models (VGGFace and lightened CNN). This study has discovered that deep learning-based face recognition is more robust to misalignment facial images.

3 Literature Review

3.1 FaceNet

FaceNet model was developed by Google researchers Schroff et al., which can reduce the difficulties in face detection and verification process and achieve the desired result [8]. The FaceNet algorithm works by taking an input of an image of a person's face and converting the high-dimensional vectors of that image into a 128 low-dimensional Euclidean space. This conversion process is also known as embedding. The FaceNet model uses the deep convolutional networks, which makes the most effective use of its embedding, compared to using intermediate bottleneck layers as a test of previous deep learning approaches [9]. This method is known as one-shot learning. In this way, triplet loss has been trained over the produced FaceNet model, which can achieve a result of uniformity and distinction over the given collections of images. After collecting the result of similarities between the faces, we can consider the FaceNet embedding as feature vectors. After creating the vector space, operations like face recognition, classification, verification, and clustering could be implemented more spontaneously. Additionally, FaceNet is able to train any difficult learning system of any single model that illustrates an entire goal system by collecting all of the factors at the same time, and this is the most significant part of the FaceNet model.

3.2 Triplet Loss

In deep neural networks, the triplet-based loss function used to learn the mapping is an adaption of Kilian Weinberger's large margin nearest neighbor (LMNN) classifier [10]. Triplet loss is one of the best ways to learn good 128-dimensional embedding for each individual face [11]. As it is an equivalent of the loss function, a comparison between the baseline input with positive input and negative input has been observed here. That is why, to compare to other loss functions, triplet loss is more compatible with face verification. The triplet loss function has been categorized into three main parts. They are:

- Anchor
- Positive
- Negative

Here, anchor is the starting input point, which is used for comparison. Positive denotes similar identity like the anchor, and negative denotes the different identity from the anchor. The distance between the anchor and a chosen image will be minimized if that image is positive, which denotes that the compared two images have a similar identity. In contrast, the distance will be maximized for different images.

3.3 Triplet Selection

Triplet selection is a core part of improving the performance of triplet loss. So, it is essential to select all hard triplets. Training with randomly hard triplets almost does not converge, whereas training with the hardest triplets often leads to a bad local solution. There are two types of triplet loss methods. They are online and offline triplet mining methods. Offline triplet mining worked with a full pass on the training set to generate triplets. Since it needs to update regularly, this method is not efficient. So, we used online triplet mining.

First of all, we picked all the necessary triplets. Depending on the loss, we classify the triplets in three categories: easy, semi-hard, and hard. As easy triplets are with loss 0 or very small loss, so we intended not to take them. Next, the semi-hard triplets are also far away from the anchor than the ideal model. In a hard batch, we select a set of hard anchor-positive pairs and then select the hardest negatives within the mini-batch. Here small, mini-batches are used since they improve the process of converging.

3.4 MTCNN

MTCNN or multi-cascade convolutional neural network is an improved neural network algorithm in the detection of faces and facial landmarks on images. The purpose of the proposed MTCNN is to form an avalanched structure and use it as material for multi-task knowledge to forecast the location of the face, and it is marked in a coarse-to-fine method. And, in its application, MTCNN is able to detect real time with fairly high accuracy [13]. The MTCNN architecture is made up of three convolutional networks, which are connected in a cascade. The first one is, Proposal Network(P-Net), which main task is giving a boundary box for each face bounds. In this process, a large number of face detection and false detection have been taken place. The second one is, Refine Network(R-Net), which is slightly similar to P-Net. But, R-Net includes more appropriate bounding boxes compared to P-Net. Thus, making a refinement of the result by eliminating most of the false detection and aggregate bounding boxes [12]. The last one is, Output Network (O-Net), which produces an output different from P-Net and R-Net. There are three types of output that O-Net produces. The output of the first layer is used for measuring face probabilities in the box. The second layer is used to give the boundary coordinates in the box. And, the last layer is used for the coordinates of the five landmarks of the faces [9] (Fig. 1).

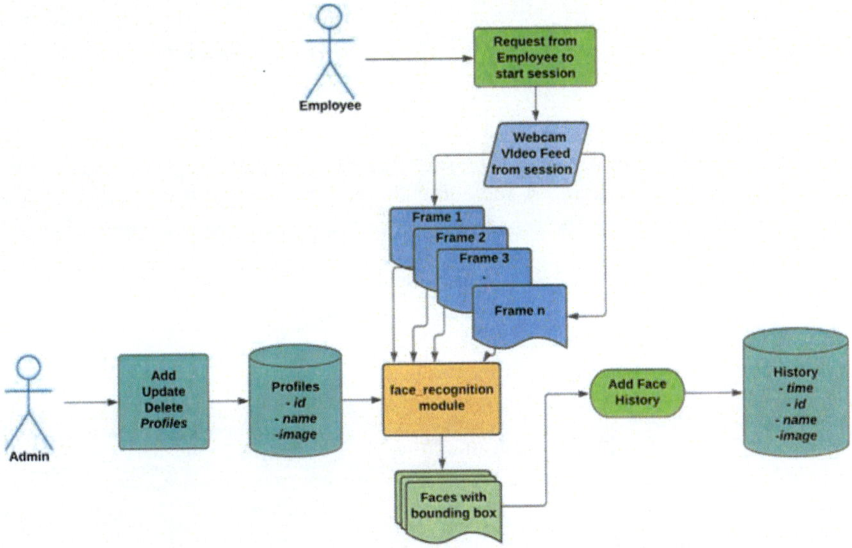

Fig. 1 Full pipeline overview of our methodology

4 Dataset

We used Labeled Faces in the wild (LFW) [3] here to uphold the effectiveness of the face recognition process. It has been said that the LFW dataset is a standard for the face verification process which is mainly developed to perusing the unconstrained problem of face recognition. This dataset consists of around 13K images collected from the web. Also, each image has been labeled with the name of that specific person. There are four different sets of LFW images, including the original and three different types of "aligned" images. Among them, deep funneled images produce superior results for most face verification algorithms over the original and funneled images. Hence, the dataset we used here is the deep funneled version.

5 Methodology

To track the presence of the employees, we initially put all the information and images of the employees in our central database. Then, we used those images during the face recognition period. Later, we took the history and kept that in a separate relational database. So, in short, our system has three stages.

5.1 Adding, Updating, and Deleting Employee Details

An admin can add new users when a new employee joins the company, update user details if anything changes about the employee, and delete employee details when an employee leaves the company. While doing so, the face recognition system does not need to change anything. The recognition system can adapt with the user changes dynamically.

5.2 Face Recognition

We used the module in Fig. 2. We trained the MTCNN with triplet loss to train the module to detect face encodings from faces. After that, we compared the employee images using the KNN classifier. This returned us with the class, i.e., name of the person and bounding box.

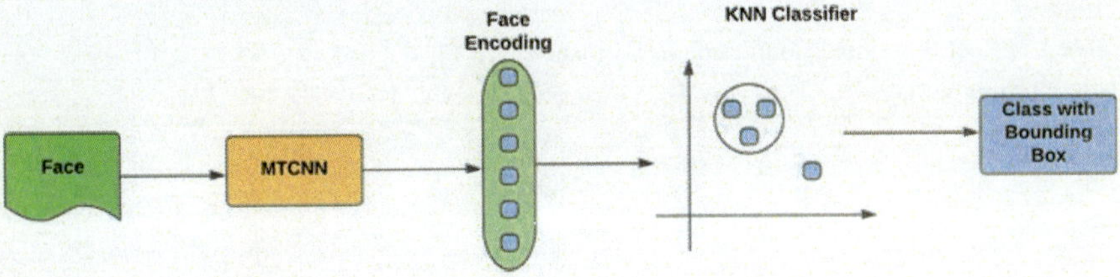

Fig. 2 Face recognition module

5.3 *Tracking and Storing Face Recognition History*

Every time, any employee starts their work session. We take the video feed from the webcam randomly from different portions of the session. If an employee who started his session is not found in video feeds consecutively, then an automatic alert goes to the admin and the employee. All the face recognition history is stored with employee details

6 Experiment Analysis

Experimental Setup

We did all our experiments using Google Colaboratory which is a hosted Jupyter notebook service. We used it because Google Colaboratory provides free GPU for 12 h a day. In our experiments, we used Numpy, Pandas, etc. for data processing and Tensor for training and testing. PyTorch [12] is an open-source machine learning framework. We chose 10% randomly as validation set from the LFW dataset. We trained all the models for ten epochs with Adam optimizer [13]. After taking a pair of two face images, we measured a squared L2 distance threshold, which is mainly used for classifying the similar and different images. The optimal threshold value we got here is 1.24. That means, this threshold value can accurately classify all pairs of images.

Result Analysis

Here, 128-dimensional float vector was used throughout the training season. This 128-byte dimensional vector tightly illustrates each face, which is most suitable for recognition and clustering. Here, we took input images of size 220 * 220 pixels and trained our model through this. But still, it works good on 80 * 80 images, which is admissible. Table 1 shows the size of training images and their corresponding accuracies.

Table 1 Effect of training samples on performance

No. of training images	Accuracy
70	90.5
700	92.3
7000	95.6
Full dataset	**97.8**

An important thing we must have to mention here that from Table 1, we can see that the outcome of trained our model with a large amount of data is immensely leading the accuracy to increase on our test sets. Here, Zeiler and Fergus [14] architecture along with 1 * 1 convolutions and inception-based models were applied. The inception-based model reduces the model size by a strikingly large amount.

Finally, face recognition task was completed by our pre-trained model based on FaceNet model. A number of facial encodings were generated by our trained model and by using Euclidean distance, which can show a comparison of facial coordinates for different samples. This results in a distance, which directly assembles to the measure of resemblance. This acquainted our model whether the face of input image matches with anyone or not. And, that is how we trained our model. For LFW dataset, we achieved a classification accuracy of 97.8% . So, we can say that, in face recognition method, FaceNet with KNN shows up the accurate results with confident.

7 Qualitative Evaluation

In Fig. 1, we get a full overview of our working prototype. Our prototype takes the webcam video feed given by users which passes through the Flask [15] backend to the face recognition module. Later, we get the face encoding and bounding box from the module. We store detected faces in the database as history. We used this system experimentally in our company on remote employees. In short, they also acted as evaluators. Our backend can effectively extract faces and recognize them in a dynamic way. The evaluators were mainly interested in how well the software can perform rather than the user interface. The software was very much self-explanatory, and most of our evaluators were expert enough. We asked them to look through the eyes of general people. They did not face any confusion while evaluating. Some evaluators complained that if they were looking somewhere else during random time video feed extraction, their faces were not detected. This is a limit of the dataset also which we plan to extend in the future. However, in most cases, even with blurry cameras, our system could identify the faces. We did not find any case where our model predicted the wrong evaluator. We found out that even when our evaluators added 10 years old images, our system could find them.

8 Applications

Even though, we used this whole pipeline for remote employee tracking. This can be used in various other cases:

- **Access and Security**:
  (a) Instead of using passcode, mobile phones, laptops, and other consumer electronics will be accessed via the owner's facial feature. Apple, Samsung, Xiaomi already installed FaceTech in their phones.
  (b) In the near future, consumers will get into their cars, houses, and other physical locations simply by looking at them.
  (c) Innovative facial security could be helpful for any company or organization where sensitive data need to keep tight control on who enters their facilities.

- **Payments**:
  (a) In 2016, Mastercard launched a new selfie pay app called "Mastercard Identity" check. Customers open the app to confirm a payment using the camera.
  (b) With FaceTech, customers would not even need their cards.
  (c) In the future, we can do the same for online payments.
  (d) Automatic face recognition to prohibit deduplication of identity to authentication of mobile payment. This is mainly used for face spoof attacks also known as biometric sensor presentation attacks, where a photograpgh or video of an authorized person's face could be used to gain access.

- **Criminal Identification**: FaceTech can be used to keep unauthorized people out of facilities.

- **Smart Home**: The design of smart homes or cities has become one of the things that many researchers have focused on. Especially, people with special needs or patients.

- **Video Surveillance**:
  (a) Surveillance used for protection, intelligent gathering, searching for drug offenders, CCTV control, power grid surveillance.
  (b) CCTV cameras can be used to monitor any well-known criminals, and authorities are notified if one is located. But, this is quite challenging for light illumination, pose variation, and facial expression.

- **Video Indexing**: Labeling faces in video [1].

9 Conclusion

Nowadays, face recognition is an unbounded exploration and development subject for research. We can use face recognition technology for remote employee tracking. Some extreme secure applications like person authentication or controlling entrance at certain areas are carried out using this technology. In our research, we try to represent a practical approach for recognition systems using the FaceNet model. Unlike the other existing methods, for example, principal component analysis (PCA) or support vector machine (SVM), our model does not require any extra operations like classifying and grouping different images or creating a decision surface. Instead, our model shows improved performance on the recognition process as we used an end-to-end learning approach. Our method is more capable of improving the percentage of collecting valuable information and produces more distinct feature information, which basically increases the face recognition rate.

In our future works, we will try to improve our model by diminishing existing limitations and try to gain more efficiency. Again, a face spoofing attack is nowadays a big issue in security and authentication systems. So, we will make an attempt to develop an anti-spoofing method, which will help to improve the security of a biometric system.

References

1. Marcialis GL, Roli F (2003) Fusion of face recognition algorithms for video-based surveillance systems. In: Multisensor surveillance systems. Springer, pp 235–249
2. Kak SF, Mustafa FM, Valente P (2018) A review of person recognition based on face model. Eurasian J Sci Eng 4(1):157–168
3. Learned-Miller E, Huang GB, RoyChowdhury A, Li H, Hua G (2016) Labeled faces in the wild: a survey. In: Advances in face detection and facial image analysis. Springer, pp 189–248
4. Schreiber AC, Rousset S, Tiberghlen G (1991) Facenet: a connectionist model of face identification in context. European J Cogn Psychol 3(1):177–198
5. Tolba A, El-Baz A, El-Harby A (2006) Face recognition: a literature review. Int J Signal Process 2(2):88–103
6. Déniz O, Bueno G, Salido J, De la Torre F (2011) Face recognition using histograms of oriented gradients. Pattern recogn Lett 32(12):1598–1603
7. Prasad PS, Pathak R, Gunjan VK, Rao HR (2020) Deep learning based representation for face recognition. In: ICCCE 2019. Springer, pp 419–424
8. Jose E, Greeshma M, Haridas MT, Supriya M (2019) Face recognition based surveillance system using facenet and MTCNN on Jetson tx2. In: 2019 5th international conference on advanced computing and communication systems (ICACCS). IEEE, pp 608–613
9. William I, Rachmawanto EH, Santoso HA, Sari CA et al (2019) Face recognition using facenet (survey, performance test, and comparison). In: 2019 fourth international conference on informatics and computing (ICIC). IEEE, pp 1–6
10. Schroff F, Kalenichenko D, Philbin J (2015) Facenet: a unified embedding for face recognition and clustering. In: Proceedings of the IEEE conference on computer vision and pattern recognition, pp 815–823
11. Weinberger KQ, Saul LK (2009) Distance metric learning for large margin nearest neighbor classification. J Mach Learn Res 10(2)

12. Paszke A, Gross S, Chintala S, Chanan G, Yang E, DeVito Z, Lin Z, Desmaison A, Antiga L, Lerer A (2017) Automatic differentiation in Pytorch
13. Kingma DP, Ba J (2014) Adam: a method for stochastic optimization. arXiv:1412.6980
14. Zeiler MD, Fergus R (2014) Visualizing and understanding convolutional networks. In: European conference on computer vision. Springer, pp 818–833
15. Grinberg M (2018) Flask web development: developing web applications with python. O'Reilly Media Inc.

Data Science and Big Data

Large Scale Image Registration Utilizing Data-Tunneling in the MapReduce Cluster

Amit Kumar Mondal, Banani Roy, Chanchal K. Roy, and Kevin A. Schneider

Abstract Applications of image registration tasks are computation-intensive, memory-intensive, and communication-intensive. Robust efforts are required on error recovery and re-usability of both the data and the operations, along with performance optimization. Considering these, we explore various programming models aiming to minimize the *folding* operations (such as *join* and *reduce*) which are the primary candidates of data shuffling, concurrency bugs and expensive communication in a distributed cluster. Particularly, we analyze modular MapReduce execution of an image registration pipeline (IRP) with the external and internal data (data-tunneling) flow mechanism and compare them with the compact model. Experimental analyzes with the ComputeCanada cluster and a crop field data-sets containing 1000 images show that these design options are valuable for large-scale IRPs executed with a MapReduce cluster. Additionally, we present an effectiveness measurement metric to analyze the impact of a design model for the Big IRP, accumulating the error-recovery and re-usability metrics along with the data size and execution time. Our explored design models and their performance analysis can serve as a benchmark for the researchers and application developers who deploy large-scale image registration and other image processing tasks.

Keywords MapReduce · Image registration · Modularity · Design gain

A. Kumar Mondal (✉) · B. Roy · C. K. Roy · K. A. Schneider
University of Saskatchewan, Saskatoon, Canada
e-mail: amit.mondal@usask.ca

B. Roy
e-mail: banani.roy@usask.ca

C. K. Roy
e-mail: chanchal.roy@usask.ca

K. A. Schneider
e-mail: kevin.schneider@usask.ca

M. S. Arefin et al. (eds.), *Proceedings of the International Conference on Big Data, IoT, and Machine Learning*, Lecture Notes on Data Engineering and Communications Technologies 95, https://doi.org/10.1007/978-981-16-6636-0_14

1 Introduction

Storing, managing, and processing a large collection of images are a Big Data problem. As a result, many of the issues for image processing application design with Big Data are emerging, starting from simple image-enhancing [10] to contents analysis [13]. Image registration is widely used [4] in remote sensing, agriculture, medical imaging, etc. The application of it is mainly employed for multiview analysis—larger representation of acquired images from different viewpoints, multitemporal analysis—find or evaluate changes of scenes gathered from different times and conditions, multimodal analysis—integrate different information obtained from different sensors (sources), and scene to model registration. However, registration of a moderate collection of images requires hours and days even with the high-performance computing [17].

It is well established among the researchers that image registration with a multicore CPU is better than a single high-performance GPU [22]. A few of the techniques have already been proposed for the large-scale image registration with the multi-CPU cluster utilizing the distributed frameworks [25, 26] mainly focusing on execution time and data size. Nevertheless, there are still opportunities to increase the performance by tuning the design model with this framework. Moreover, an image registration task is often applied as a pipeline (IRP) that co-locates with other applications (such as panoramic view generation and object detection) within a system that shares the same image collection to a certain extent. This ecosystem requires a robust focus on quality-efficiency trade-offs, error recovery, data management, and re-usability of both the image processing steps (such as grayscale conversion and feature extraction) and the produced data entities of those steps. Therefore, an efficient design is essential in the deep layers (distributed storage, IO mechanisms, partitioning, parallelism, modularity, data-flow, etc.) of implementation to address these concerns.

Nowadays, MapReduce ecosystems such as Hadoop [11] and Spark [23] provide a flexible environment hiding the low-level complexities of the cluster processing by some restricted functions (i.e., *map, reduce, join*, etc.) for implementing high-performance computing jobs even with the commodity hardware. Although the restricted interface is provided by the MapReduce frameworks, in reality, granularity can be achieved in all the layers employing intuitive design mechanisms. When designing a task with this ecosystem, one of the key things is how to handle the complexity [14] of folding operations: *reduce*, *join*, *union*, *collect*, etc. From a given set of inputs, these operations propagate an enormous amount of external and internal data-flow among the jobs and stages in the life-cycle of a task, the main source of overhead. We call the internal data-flow from one job to another that requires no concrete memory shuffling among the cluster nodes as *data-tunneling*. Modularizing image registration pipelines (IRPs) could reduce the execution performance significantly due to the above-mentioned facts (and being data and computation-intensive). However, the performance of an IRP can be optimized by minimizing the MapReduce folding operations, which is a challenging task. Yet, an effectiveness measure is essential to limit the arbitrary modularization (since it reduces the performance) in the cluster processing.

In this paper, we explore various design models of an IRP with the MapReduce framework concerning quality-efficiency trade-offs, error recovery, data management, and re-usability by utilizing various data flow mechanisms. Among the explored models, the data-tunneling technique inducing the internal data overhead is more promising as it minimizes the *folding* operations, which are communicatively expensive for a distributed cluster. We also demonstrate a way of evaluating the design gain where error-recovery and re-usability metrics are accumulated along with the data size and execution time for Big Data applications. Our experimentation with the ComputeCanada[1] cluster and a crop field data-sets containing 1000 images (having several millions of feature and descriptor points) reveal that modularization increases the execution time of the IRP significantly (around two minutes for the best configuration) but reflects the better design gain. In between the two data-flow design models, the data-tunneling model increases both the execution performance (by 8% with five worker nodes) and design gain significantly.

In the next subsequent sections, we describe the related work, background of image registration with Big Data, our proposed models, and the experimental results.

2 Related Work

Although image registration started decades ago [2], researchers have been working to reduce processing time and errors [16, 21] up-to-the recent years. In our work, we also focused on reducing the execution time for a large collection of images. High-performance image registration with multi-CPU, multi-core, and GPU is an active research area [22]. A handful of techniques are found in the survey paper by Shams et al. [22]. Likewise, other existing literature recommended horizontal parallelism with either multi-CPU or multi-GPU clusters. Therefore, we also focused on this direction, aiming for a more efficient design to create Big Data veracity and value in the image registration domain.

There are various applications of IRP co-locating and associating with other analysis tasks sharing the same image collections utilizing various cluster processing frameworks. White and Davis [25] employed the distributed computing frameworks (OpenMP and CUDA) for a real-world application of feature-based (SIFT and ORB) image registration in Geocomputing. Experimenting with eight high-resolution images, they have shown that processing with CPU-core (total 48 cores) based cluster outperforms the GPU (four) based computation. Although the design concern is not analyzed in their work, we followed a similar algorithm for image registration in our experiment. In another study, Yang et al. [26] describe how they implemented a large-scale image registration workflow with CometCloud framework utilizing a cluster where they mainly focused on resource allocation based on the number of image pairs. In an early study, Plishker et al. [17] presented the implementation of distributed processing with heterogeneous hardware clusters. However,

[1] https://www.computecanada.ca/.

we present various modular design models and their impact on the trade-offs of efficiency, error-recovery, and re-usability, which are absent in the previous studies.

Trade-offs analysis is important [1] for evaluating and benchmarking the design models of Big image registration. However, existing works focus mainly on implementation performance with the cluster resource allocation [18, 26] and hardware acceleration. To fill the gap, we focused on both the algorithmic complexity and design gain with the distributed MapReduce framework along with the resource allocation.

3 Background

Image Registration: In image registration, a single or collection of images is aligned based on a reference image. Among many available algorithms [4], feature-based (i.e., SIFT [15], SURF, etc.) technique is widely used [12]. In the feature-based method, [12], at first, feature points and their descriptors are calculated from the pixels of the reference image and target images using either SIFT or SURF (due to scale and rotation invariant of the images, which is essential for remotely captured images) algorithm. They are utilized to find matched feature points between two images by measuring their distance metrics. KNN [27] and Brute-force [20] scheme are widely used for point matching. To eradicate the false points, RANSAC [6] algorithm is applied. These points are filtered using a distance threshold, and the final list of feature points is used as control points for registering the image pair. In many techniques, a homograph matrix [9] is used for this purpose. Finally, transformation functions are applied using the control points on the target image to construct the aligned image with the reference image. Alignment through perspective warping [7] at the final stage is popular in the registration task.

Challenges and Paradigms of IRP in Real-world Application: In implementing, a single-handed image registration algorithm, design concern is trivial as the most compact program is employed. However, when it is used within a real-world application [25] with hundreds and thousands of images co-locating and sharing data with other tasks (such as mosaic generation, plot segmentation and region clustering on the merged image), the design concerns are robust. In a complete workflow, the contents (including the metadata) of all the images must be relevant and consistent with one another; otherwise, the full workflow may fail after running for a significant amount of time. An IRP should have the capability to integrate plugins due to the degree of freedom to leverage various standard techniques in the intermediate steps. The source of the images might have different formats from different sensors prompting a new data structure in the distributed environment. Moreover, it is complex and computation-intensive [21]; even processing moderate size of data requires hours with high-performance computing [25]. Furthermore, the execution time may vary in different operations within a complete registration task. Researchers [22] found that the final step of transformation (warping) requires less time with GPU (for geometric calculation), but feature calculation and matching points extraction

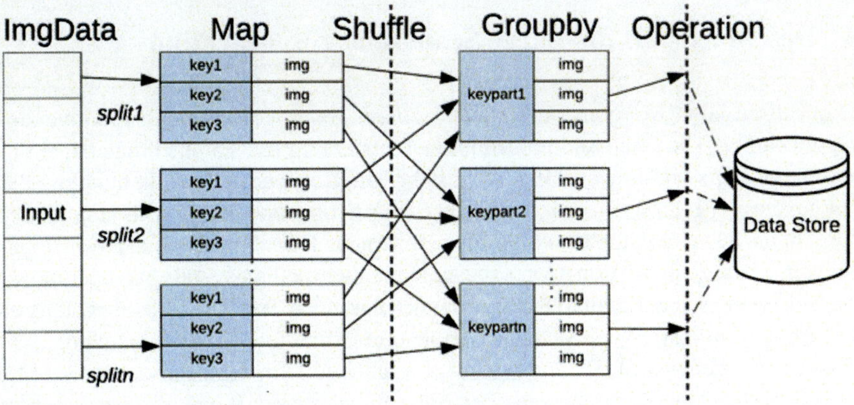

Fig. 1 Image registration with MaprRduce; keypart means pattern matching on part of the image name.

need more time with GPU than CPU. Above all, the optimization still depends on both the context of how an IRP is designed and the hardware selection.

Design Concept as a Big Data Application: Typical steps of implementing a data-analysis task in a MapReduce cluster are storing data into distributed storage, loading data into execution cluster, processing data through the *map* and *folding* operations, and finally save the result to the storage again. While the concept of modularity for localized processing following data processing steps is straightforward, in the MapReduce framework, operations are ultimately squeezed to *map, reduce, filter, groupby*, etc., which are distributed jobs holding many sub-tasks; more jobs mean more data-flow and execution time. In Big Data literature [5, 19], we can find many design solutions used in machine-learning, text-data, graph-data and GIS data processing. However, there are some critical distinctions in IRPs. During data analysis, both approximate (based on prediction and sample selection) and exact model generation are acceptable [5, 19], but image processing tasks require employing operations on all the image samples (such as plot mosaic generation). There is also a significant difference regarding accessing data between image processing tasks and the above-mentioned data analysis tasks. For efficient processing with the MapReduce cluster, the large collection of images are required to put into distributed storage such as HDFS as text, binary or object data, and before processing, they need to convert back to image objects introducing extra overhead. Furthermore, in the last stage, raw images and converted images are required (i.e., plot registration) along with the extracted features for former ones while features, relatively lower data size, are sufficient for the latter kind of tasks for most of the cases (i.e., text classification). This imposes a more significant overhead for image processing tasks as the MapReduce cluster suffers network-induced non-determinism, bandwidth overload, and data shuffling. Also, the impact of error recovery [8] for deploying such an application is an increasing research area due to the hardship in getting the best outcome of a complex application.

4 Design Models for Large Scale Image Registration

In our discussion, we focus on the feature-based image registration technique discussed earlier. The discussed procedure in Sect. 3 of an IRP can be implemented in two major steps (as shown in Fig. 1) with the MapReduce framework: bundling the raw images with *groupBy* procedure and then perform operations with *map* procedure. In the last phase, the result is collected with the *folding* procedure. However, in a pipeline based application, to get and store the intermediate results and operations, the task needs to be modularized at a standard granular level. The abstract steps of the registration task: (i) gray image, (ii) feature extraction, (ii) area matching, (iv) homography generation [9], and (v) image alignment with perspective warping [7]. In each of the steps, both data and operation can be re-used. It will also help to track and recover errors which is a characteristic of the veracity and value of Big Data [24]. Various errors are common, while deploying an application in the cloud cluster and have grave consequences, while processing a time-consuming application. Error recovery methodology that can efficiently restart partially completed application deployments is really important for cloud-cluster application deployment [8]. One of the best techniques of recovering from failure is splitting an application into multiple jobs, but that can significantly increase execution time. To address this concern, we introduce error recovery and re-usability metrics to analyze trade-offs later section in the paper.

Let's consider the re-usability and error recovery in the following ways:

$$\text{probability of error recovery } P_E = \frac{E_R}{E},$$
$$\text{probability of reuse } P_R = \frac{R_D + R_M}{D_E + \#S} \quad (1)$$

Here, E_R=number of (#) errors that are possible to recover, E = #major layers of error, R_D and R_M are #produced results and Modules that are reused, and D_E = #produced entities in the algorithmic steps (S). In the ideal case, value of P_E and P_R is 1. In an IRP for a real-world application, the possible major operations are gray conversion, scaling, feature calculation, matching points calculation, homograph, warping. These seven operations may generate seven different results which can be reused with the operations. For example, for crop field images, extracted features can be reused for mosaic image generation, matching and clustering the similar and dissimilar regions of the fields. Usually, four data entities: converted or enhanced image, features, filtered points/homograph and aligned image are reused. Again, each of the steps is complex and susceptible to error. Therefore, if we consider P_E and P_R, modularizing a task in these steps would be worthwhile. Although more modular splits are simple with localized processing, it is different when working with a cloud cluster due to the mentioned challenges in Sect. 3. Traditionally, a task is implemented with the minimum number of MapReduce jobs. However, if the seven operations are feasible to implement with three independent jobs (a.k.a. three modules), then three result-entities can be reused. So, $P_E = (3)/(7) = 0.42$ and $P_R =$

$(3+3)/(7+4) = 0.54$. Because of a massive amount of data, in the practical case, we might have to compromise the value of P_E and P_R with the trade-offs of execution time and capability of handling data size. In the next sections, we discuss these in detail with three design models.

4.1 *Implementation with Minimum Overhead with No Modularity*

In our description, we will discuss the implementation with Spark RDD (resilient distributed dataset) that provides one of the most efficient in-memory distributed computations. Compact implementation of the registration task is shown in Fig. 1. Performance validity with the Spark [23] cluster of this compact model is shown in Figs. 3 and 4. It requires two jobs along with the loading and mapping operations. One job first bundles the images using the *groupby* operation based on the pattern in the image name (which is defined during the capturing time with a drone). Another job employs all the algorithmic steps. After loading and mapping, the data structure in the mapped memory looks like this:

$$[[\text{name01}, \text{img1}], [\text{name02}, \text{img2}][\text{name11}, \text{img3}][\text{name12}, \text{img4}]] \quad (2)$$

Here, name0, name1 are groups based on how they are captured by a drone. We have to make a bundle for each group. After the bundling step, the data structure would be [[name0, [img1, img2]]; [name1, [img3, img4]]]. After that the rest of the operations are performed with the map procedure on each of the bundles parallelly with the worker nodes. No internode communication is needed for the Map tasks. With this model, there is no extra concrete data-flow among the nodes, and the least execution time is expected.

Algorithm 1: Modular image registration with external data-flow.

Data: I is the images, B-bundles, G-gray converted images, F-features, H-homograph.
Result: Warped images (W) with key-value mapping for Rdd $rdd[key, W]$

begin
- $rdd[name, I] \Leftarrow$ loadAndMap(source, partitions)
- $rdd_R[key, B] \Leftarrow$ groupBy(pattern($rdd[name, I]$))
- $rdd[key, G] \Leftarrow$ mapConverion($rdd_R[key, B]$))
- $rdd[key, F] \Leftarrow$ mapFeatures($rdd[key, G]$))
- $rdd[key, H] \Leftarrow$ mapHomograpgy($rdd[key, F]$))
- $rdd[key, BH] \Leftarrow$UnionNReduce($rdd_R[key, B], rdd[key, H]$)
- $rdd[key, W] \Leftarrow$ mapWarping($rdd[key, BH]$))

end

Algorithm 2: Modular image registration with Data-tunneling model.

Data: BG-new datatype of bundled + converted images, BF-new datatype of bundled images + extracted features.

Result: Warped images with key-value mapping for Rdd $rdd[key, W]$

1 **begin**

2 $rdd[name, I] \Leftarrow$ loadAndMap(source, partitions)

3 $rdd[key, B] \Leftarrow$ groupBy(pattern($rdd[name, I]$))

4 $rdd[key, \mathbf{BG}] \Leftarrow$ mapConverion($rdd[key, B]$))

5 $rdd[key, \mathbf{BF}] \Leftarrow$ mapFeatures($rdd[key, \mathbf{BG}]$))

6 $rdd[key, BH] \Leftarrow$ mapHomography($rdd[key, \mathbf{BF}]$))

7 $rdd[key, W] \Leftarrow$ mapWarping($rdd[key, BH]$))

8 **end**

4.2 Modularity with Merging Model

A modular design algorithm similar to the five abstract steps is shown in Algorithm 1. Considering re-usability, in this algorithm, we split the operation after bundling into four independent MapReduce modules/jobs. Each of these modules is paralleled with *map* procedure. Matching area points and homograph [9] generation from them are executed in one job (line 6). However, the raw images are required in the last step for warping [7] and alignment, which is easy to implement through array indexing in localized processing. But, in cluster processing, one solution is to perform merging raw images and generated Homograph (line 7) employing external data-flow. Well, the data structure produced after line 3 is equivalent to Eq. (4). Homography generation up to line 6 is straightforwardly producing the following data structure:

$$[[\text{name0}, [H1, H2]]; \quad [\text{name1}, [H3, H4]]] \tag{3}$$

The order of H is important here as they are the corresponding homography of each of the images within the bundle. Now, in line 8, this produced bundle of homography H is employed with the bundle of raw/gray images B (maintaining perfect order is a must need here) from either line 3 or 4. As such, the complexity of sending B into warping operation is huge, which would not be problematic if a few images are sent through the broadcasting operation. The solution is to merge B and H into same key as $name$. It is done by first doing union operation and then reducing the keys. Thus, after line 7, the produced data structure:

$$[[\text{name0}, [\text{img1}, \text{img2}, H1, H2]]; \quad [\text{name1}, [H3, H4, img3, img4]]] \tag{4}$$

One crucial thing to note in Eqs. (2) and (4) is that the order of images (img_N) and Homographs (H_N) are random, and track them is not easy (this phenomenon is not detected until we run the task with a larger dataset and it required a great effort). To solve this challenge, we have to implement some extra processing causing overhead (before warping operation we have to separate the images and Homographs locally

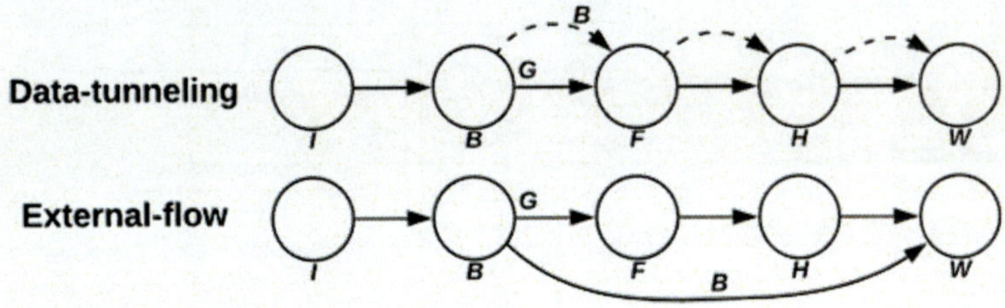

Fig. 2 Abstract view of different data-flow options for modular image registration task. Dotted line indicates that the data persists in nodes.

within each map operation). However, communication from Map tasks to Reduce tasks in line 7 is likely to be across the interconnect of the cluster nodes.

4.3 *Modularity with Data-Tunneling*

We have seen some challenges and overhead in the previous section for merging B and H in the warping phase causing external data-flow among the connected nodes. However, as can be seen, *folding* operations are not required in the other processing steps; we can persist the raw images into the respective nodes after the bundling process up to the warping stage from the conversion step, and data is moved locally from one operation to another. We call this technique data-tunneling as practically no extra data-movement among the nodes (which is shown in Fig. 2). This technique is presented in the Algorithm 2 where BG and BF are two new distributed data types that are retained in the RDD memory subsequently. In this way, we do not need the *join* and *reduce* operation as required in the previous algorithm; ultimately, data-shuffling is minimized. All is required is to track the order of intake and outgoing data-entities by synchronizing the key-value pairs with an intermediate layer of a module/job.

Although the computation has greater simplicity with the data-tunneling approach, there might be an implicit memory overhead. However, we cannot measure the memory overhead explicitly compared to the merging model as the reduce operation also create numerous temporary files. We will discuss the evaluation of these design approaches in the experimental section.

5 Experimental Result

Our experimental dataset consists of hundreds of crops field images captured by drone; each image size (1280×960) is 1.6 MB, and each group has five images (bundle). We implemented the models on the underlying algorithm of [4] with Python,

Fig. 3 Execution time of the models with different datasets (4 slower workers, 4×2 cores and 4×6 GB memory).

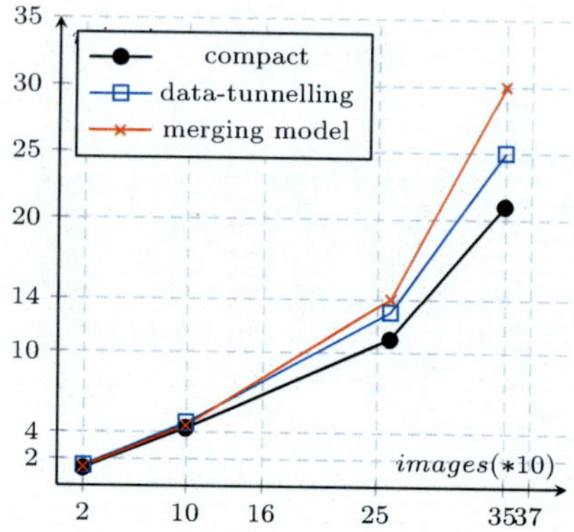

Fig. 4 Execution time with 500 images (with grouping and IO). ComputeCanada–5 workers, 5×8 cores and 5×25 GB memory.

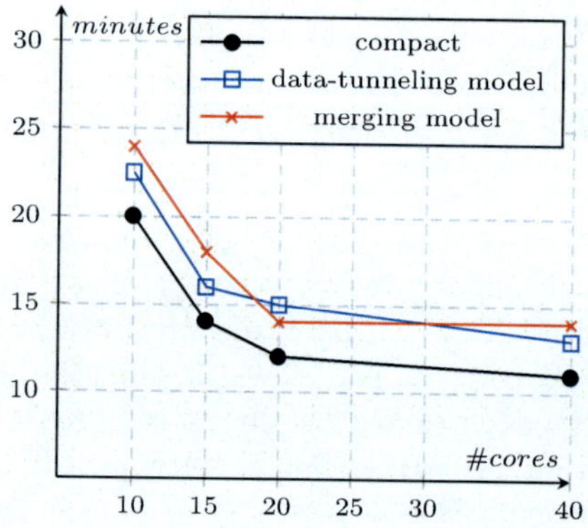

Spark and OpenCV [3]. For point matching, we used Brute-force [20] algorithm. The dataset is loaded from HDFS (Hadoop Distributed File System) as binary files, and registered images are saved in both text formats and object file formats. SIFT [15] algorithm generates up to 19,411 features and 19,411 descriptor points for a single image, and 1000 images totalling around 15 million of them. The performance evaluation of the proposed design models are shown in Figs. 3 and 4. It is undeniable that the compact version shows the best performance regarding execution time, as are seen in Figs. 3 and 4. For a small data-set of up to 100 images and lower configuration, the merging model in Algorithm 1 is slightly better, with 0.2 min less execution time (same for four workers of a total of 20 cores). However, with the increment of the size of the dataset, its performance starts to decrease significantly compared to other models.

For the 260 images, the data-tunneling model is 2 min faster than the merging model; this difference is 4 min for 350 images. Execution time for 500 images with the higher configuration cluster is shown in Fig. 4; the compact, data-tunneling, and merging models require 11, 13 and 14 min, respectively, with five nodes (each has eight cores). With a machine of the same speed as the master node, 16 cores and 70 GB RAM, non-parallel version takes around 64 min. Previous work showed [26]

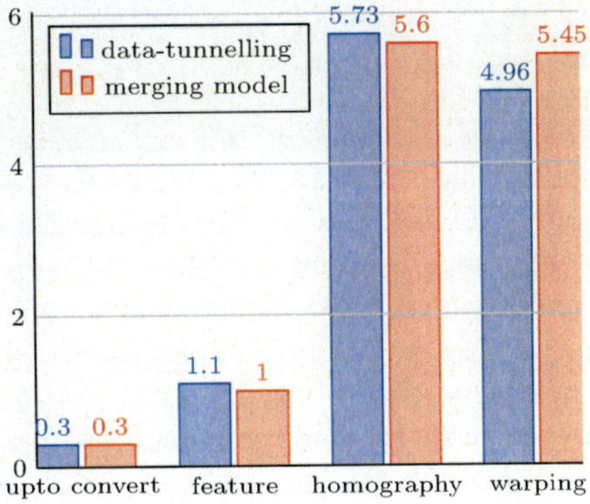

Fig. 5 Execution time of each module for 260 images (with 5 × 24 cores workers). Without count but with IO, merge-model, data-tunneling and compact finished in 5.1, 4.6, and 4 min.

that the usual version with the CometCloud framework took around 12 min with 20 nodes (each has eight cores) for 460 image pairs (although the image size and IO inclusion are not mentioned). However, with the better cluster, in the best case, the merging model is one minute slower than the data-tunneling model. Therefore, the performance increase for the data-tunneling is 8 and 20% for the higher and lower configuration cluster, respectively. This is so because the raw images are moved locally within the corresponding node during the life-cycle of a task in this model, which means there are minimum memory shuffling operations, whereas *reduce* and *groupby* operations require many memory shuffling (also creates temporary files in each node) tasks. However, the traditional non-parallel IRP has no significant impact on the execution time for the modular version.

However, the execution time of the individual modules (along with the count operation) for both the models is shown in Fig. 5. Please note that we employ the count operation to get the exact execution time fusing the result data for the individual modules. The feature extraction and homograph modules require slightly more time for the data-tunneling model (0.13 min for the latter). In contrast, in the last stage, the warping module requires significantly less time (0.5 min) for the data-tunneling model as no merging task is occurring. Overall, the execution time is significantly reduced in the next runs if, for example, extracted features are saved. Let us consider a factor α of extra time required for saving the result in each of the modules. For the next five runs (either image registration or image clustering), the execution time will be reduced by $5 * 1.1/\alpha = 5.5/\alpha$ minutes. Likewise, if the byproduct of the homography module is stored in the next two runs (for registration and mosaic/point cloud generation), $2 * 5.73/\alpha = 11.46/\alpha$ minutes will be reduced. However, if each of the modules works with the previously stored result, during the loading and mapping operation, the content of B in lines 4, 5 and 6 in Algorithm 2 are set to *Empty* or *None*. One important thing to note here that the warping module in the merging model is more independent when the operation is separately reused. Above all, in summary, the data-tunneling model is the best design model until the total data size fits in the main memory.

6 Trade-Off Analysis

Communication Cost Analysis: For MapReduce programs, complexity analysis is not straightforward like the traditional program. Leskovec et al. [14] introduced the complexity theory of the MapReduce algorithm. According to their theory, communication among the interconnected nodes is the greatest cost of such an algorithm and indicates its efficiency. Communication cost is the input size into the map and folding operation. Let, $N$is the # of images, B is the # of bundles, σ = communication cost. Then, σ for folding operation of N is $\mathcal{O}(n)$ (since it is unlikely that this Reduce task is executed at the same compute node in the subsequent phases) and map operation for B is b. We measure the communication cost of the three models excluding the IO operation as follows:

Communication cost of the compact version is, $\sigma_C = \mathcal{O}(n) + b$, and the data-flow model is, $\sigma_{df} = \mathcal{O}(n) + 4b$, $4b$ for four map operations. The cost of the merging model is, $\sigma_M = \mathcal{O}(n) + \mathcal{O}(b_1 + b_2) + 4b$. In line 7 of the Algorithm 1, Cost of Join and reduce operation is $\mathcal{O}(b_1 + b_2)$. Therefore, the communication cost of the compact and data-flow version is not much different. Whereas this difference is significant for the merging model. This divergence in the complexity is approximately reflected in the execution time in Figs. 3 and 4. Another complexity factor is the replication rate, $r = \frac{\#\text{produced key-value pair}}{\#\text{input}}$. r for each of the mapping operation is 1 in all three algorithms, for bundling operation it is n/b, and for line 7 in Algorithm 1 is $(b_1 + b_2)/b = 2$, b_1 and b_2 are same here. As a result, the replication rate r for compact, data-tunneling, and merging algorithm are $(\frac{n}{b} + 1) \sim (\frac{n}{b} + 4) \sim (\frac{n}{b} + 5)$, respectively, ($n$ is large compared to 4 and 5), and the differences among them reflect our experimental result. Thus, the communication cost and replication rate analysis predict the time-efficiency of the image registration algorithms with MapReduce cluster despite data-tunneling technique has internal data overhead.

Design Gain analysis: An existing study [26] analyzed the execution performance of the registration task considering the number of EC2 nodes allocation, task scheduling and the associated hardware price. However, our context is designing registration tasks co-located with other tasks sharing the data-set in a system. For an IRP with large-scale data, design factors are primarily associated with error-recovery, re-usability, input data size, and execution time. Thus, the following equations can be an indicator of design gain considering the computation resources as constant:

$$G_D = F_{D_i} - F_{D_{i-1}}, \quad F_D = \frac{N_J(P_E + P_R)(1 + \Delta D)}{1 + \Delta T} \tag{5}$$

Here, G_D=design gain which positive value means better design, F_{D_i} = is the current design factor, N_J = number of jobs, ΔD = data size increase and ΔT = execution time increase in a certain unit compared to previous model; P_E and P_R were discussed earlier in Sect. 4. Following the Algorithms 1 and 2, the registration task can be implemented with four independent jobs/modules; here $P_E = 4/7$ and if three of the result data are re-used then $P_R = (4 + 3)/(7 + 5)$. Let's consider the highest load (500 images) with the best configuration (all the cores of five nodes). For the compact model, $F_{D_0} = 2(0.15 + 0.25)(1 + 0)/(1 + 0) = 0.8$; data-tunneling model, $F_{D_1} = 4(0.57 + 0.58)(1 + 0)/(1 + 2) = 4.6/3 = 1.53$; for merging model, $F_{D_2} = 4.6/(1 + 3) = 1.15$. As a result, $G_D = 0.73$, $F_{D_1} > F_{D_0}$ and $F_{D_2} > F_{D_0}$; hence data-tunneling technique is the best design choice here. For this cluster configuration, $FD1 = 1.53 > FD2 = 1.15$. However, in this configuration, if the ΔT for a design model is > 5 then the modular design is inefficient. We run the models in the 40 cores configuration with 1000 images (having several millions of features and descriptor) and the execution time are 25, 30, and 31 min, respectively, for the compact, data-tunneling, and merging model; hence $F_{D_0} > F_{D_2}$. This analysis indicates that execution efficiency needs to be compromised with P_E, P_R and vice-versa.

7 Conclusion

In this paper, we proposed three design models for large-scale image registration with MapReduce frameworks reflecting the quality-efficiency trade-offs, error recovery, and re-usability. In our study, we showed that the modular design of large-scale image registration extensively depends on the data-flow in the MapReduce cluster. We also presented the complexity analysis and design gain of the proposed models. The experimental outcome and trade-offs analysis reveal that among the models, despite internal data overhead, the data-tunneling model is the best design choice concerning the properties such as communication cost and design gain, which are also associated with Big Data Vs. This design can also be used in other image processing tasks as a reference model. Overall, our proposed design models and performance analysis will serve as a benchmark for the researchers and application developers who deploy the large-scale image registration tasks.

Acknowledgements We thank Dr. Kevin Stanley for sharing the data to experiment. This research is supported by the Natural Sciences and Engineering Research Council of Canada (NSERC), and by two CFREF grants coordinated by the Global Institute for Food Security (GIFS) and the Global Institute for WaterSecurity (GIWS).

References

1. Baeza-Yates R, Liaghat Z (2017) Quality-efficiency trade-offs in machine learning for text processing. In: 2017 IEEE international conference on big data (big data), pp 897–904
2. Barnea DI, Silverman HF (1972) A class of algorithms for fast digital image registration. IEEE Trans Comput 100(2):179–186
3. Bradski G (2000) The OpenCV library. Software tools for the professional programmer
4. Brown LG (1992) A survey of image registration techniques. ACM Comput Surveys (CSUR) 24(4):325–376

5. Chen T, Guestrin C (2016) Xgboost: a scalable tree boosting system. In: Proceedings of the 22nd ACM SIGKDD international conference on knowledge discovery and data mining, pp 785–794
6. Fischler MA, Bolles RC (1981) Random sample consensus: a paradigm for model fitting with applications to image analysis and automated cartography. Commun ACM 24(6):381–395
7. Gallo O, Pulli K, Hu J (2014) Methods and apparatus for registering and warping image stacks. US Patent 8,923,652
8. Giannakopoulos I, Konstantinou I, Tsoumakos D, Koziris N (2016) Recovering from cloud application deployment failures through re-execution. In: International workshop of algorithmic aspects of cloud computing. Springer, pp 117–130
9. Gledhill D, Tian GY, Taylor D, Clarke D (2003) Panoramic imaging—a review. Comput Graph 27(3):435–445
10. Gu K, Tao D, Qiao JF, Lin W (2017) Learning a no-reference quality assessment model of enhanced images with big data. IEEE Trans Neural Netw Learn Syst 29(4):1301–1313
11. Hadoop: http://hadoop.apache.org/. June 2017
12. Hasan M, Jia X, Robles-Kelly A, Zhou J, Pickering MR (2010) Multi-spectral remote sensing image registration via spatial relationship analysis on sift keypoints. In: 2010 IEEE international geoscience and remote sensing symposium, pp 1011–1014
13. Huang L, Xu W, Liu S, Pandey V, Juri NR (2017) Enabling versatile analysis of large scale traffic video data with deep learning and HiveQL. In: 2017 IEEE international conference on big data, pp 1153–1162
14. Leskovec J, Rajaraman A, Ullman JD (2014) Mining of massive datasets. Cambridge University Press
15. Lowe DG (2004) Distinctive image features from scale-invariant keypoints. Int J Comput Vision 60(2):91–110
16. Pandey P, Guy P, Hodgson AJ, Abugharbieh R (2018) Fast and automatic bone segmentation and registration of 3d ultrasound to CT for the full pelvic anatomy: a comparative study. Int J Comput Assist Radiol Surg:1–10
17. Plishker W, Dandekar O, Bhattacharyya S, Shekhar R (2007) Towards a heterogeneous medical image registration acceleration platform. In: 2007 IEEE biomedical circuits and systems conference, pp 231–234
18. Plishker W, Dandekar O, Bhattacharyya SS, Shekhar R (2008) Towards systematic exploration of tradeoffs for medical image registration on heterogeneous platforms. In: 2008 IEEE biomedical circuits and systems conference, pp 53–56
19. Ramírez-Gallego S, Fernández A, García S, Chen M, Herrera F (2018) Big data: tutorial and guidelines on information and process fusion for analytics algorithms with MapReduce. Inf Fusion 42:51–61
20. Ranade S, Rosenfeld A (1980) Point pattern matching by relaxation. Pattern Recogn 12(4):269–275
21. Ruppert GC, Chiachia G, Bergo FP, Favretto FO, Yasuda CL, Rocha A, Falcão AX (2017) Medical image registration based on watershed transform from greyscale marker and multi-scale parameter search. Comput Methods Biomech Biomed Eng Imaging Visual 5(2):138–156
22. Shams R, Sadeghi P, Kennedy RA, Hartley RI (2010) A survey of medical image registration on multicore and the GPU. IEEE Signal Process Mag 27(2):50–60
23. Spark: http://spark.apache.org/. June 2017
24. Storey VC, Song IY (2017) Big data technologies and management: What conceptual modeling can do. Data Knowl Eng 108:50–67
25. White DA, Davis CR (2017) A fully automated high-performance image registration workflow to support precision geolocation for imagery collected by airborne and spaceborne sensors. Adv Geocomput:383–394
26. Yang L, Kim H, Parashar M, Foran DJ (2011) High throughput landmark based image registration using cloud computing. MICCAI2011-HP/DCI, pp 38–47
27. Zhang K, Li X, Zhang J (2014) A robust point-matching algorithm for remote sensing image registration. IEEE Geosci Remote Sens Lett 11(2):469–473

Incorporation of Kernel Support Vector Machine for Effective Prediction of Lysine Formylation from Class Imbalance Samples

Md. Sohrawordi and Md. Ali Hossain

Abstract A post-translational modification (PTM) named lysine formylation discovered recently is a reversible and dynamic biological process primarily found on histone proteins of the organism that plays strong roles on modulation of chromatin conformations and the process of gene activation. A large number of traditional laboratory-based experimental methods and computational methods are currently available for identifying the formylated lysine residues. But the experimental methods are more costly and time consuming than computational methods. In order to predict formylated lysine sites, the existing computational methods are not satisfactory to select reliable non-formylated sites for balancing the training samples. A useful bioinformatics model named PLF_RNS is developed in this study by using various sequence-based features with support vector machine algorithm and F-score feature selection method. For this purpose, the verified formylated lysine samples are labeled as positive and reliable negative samples that are filtered from remaining samples using evolutionary information based on BLOSUM62 matrix are labeled as negative samples. The experimental result shows that PLF_RNS has acquired an average accuracy of 95.09% on tenfold cross validation, which is better compared to other currently available models. Therefore, it may be helpful for a better understanding of those types of molecular mechanisms and the development of drugs for related diseases.

Keywords Post-translational modification · Formylated lysine · Biological mechanism · Feature descriptor · Feature selection · Support vector machine

1 Introduction

A chemical change occurred on a protein after its formation is referred to as post translation modification (PTM), which plays a vital role in the regulatory mechanism and pathological cellular physiology [1]. Lysine formylation is a type of PTMs that is

Md. Sohrawordi (✉) · Md. Ali Hossain
Rajshahi University of Engineering and Technology, Rajshahi, Bangladesh

M. S. Arefin et al. (eds.), *Proceedings of the International Conference on Big Data, IoT, and Machine Learning*, Lecture Notes on Data Engineering and Communications Technologies 95, https://doi.org/10.1007/978-981-16-6636-0_15

generated with the help of isopeptide bonds from the attachment of lysine residues and formylation group (–CHO), which has a great role in the cellular regulating process including protein-protein interactions, DNA repair, DNA transcription, and DNA binding [2–4]. Besides, the oxidative damage of DNA is also responsible for lysine formylation which causes a large number of dangerous diseases such as tumor and cancer [2, 5]. Therefore, it is extremely important to decipher the biological function of lysine formylation in cellular regulation and disclose the reasons for associated diseases and develop the drugs by accurately recognizing the lysine formylation in proteins.

As formylation is a newly discovered PTM, the identification of formylation sites from proteins is actually hard and challenging work [3, 6]. The experimental techniques are more laborious and time consuming than the computational approaches. Therefore, to develop a computational predictor for accurately identifying the formylated lysine sites is essential in terms of economic and time-saving benefits. To fulfill this requirement, currently available some mathematical predictors named LFPred [2], CKSAAP_FormSite [3], Formator [4] and LyFor [7] to identify the lysine formylation sites have already been developed using various feature construction processes and classifier algorithms. The predictive model named LFPred [2] was proposed and developed by Ning et al. [2] with KNN algorithm and binary profile features (BPF), amino acid index (AAI), and amino acid composition (AAC) sequence-based features after selecting more reliable negative samples. Only one sequence-based feature descriptor, CKSAAP, was applied for transforming each sample into a numerical vector which was used to train CKSAAP_FormSite [3] by biased support vector machine algorithm. Besides, Formator [4] was trained by SVM with K-nearest neighbor (KNN), AAindex bi-profile bayes (BPB), and CDT sequence-based features. On the other hand, LyFor [7] was trained with the features processed by principal component analysis (PCA) techniques and SVM algorithm, which acquired better performance than other existing systems. When we analyzed those currently existing models, a few numbers of drawbacks were seen in the selection of negative samples. Except LFPred [2] and Formator [4], the authors of KSAAP_FormSite [3] and LyFor [7] used the peptide fragments that are not experimentally verified as negative samples. As those negative samples are not experimentally annotated as non-formylated, those might be formylated peptide fragments. Although a reliable negative sample selection method was introduced in LFPred [2] and Formator [4], they did not advance the estimation capability of their model at a satisfactory label.

In this study, we suggest a powerful predictive model known as PLF_RNS that was trained with a support vector machine (SVM) classification algorithm by resolving the above-mentioned problems. Because of the unbalanced benchmark dataset, the synthetic minority oversampling (SMOTE) [8] for positive samples and a proposed method with evolutionary information for the selection of the negative samples were implemented to protect the classifier from the effect of the unbalanced data. Each peptide fragment was encoded as numerical feature by applying dipeptide composition (DC), binary encoding (BE), and amino acid composition (AAC), and the optimal number of features selected by F-score feature selection methods were taken for

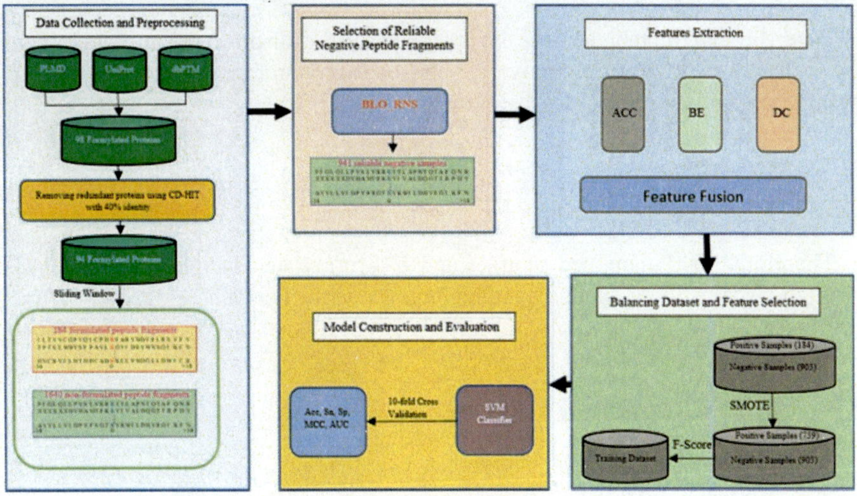

Fig. 1 Overall framework of PLF_RNS model

training the model. The overall process of developing the proposed model is shown in Fig. 1. Finally, PLF_RNS model proposed by us provided well performance than other currently available models on the task of the formylated lysine sites prediction in tenfold cross validation.

2 Methods and Materials

2.1 Benchmark Dataset

For the purpose of making a benchmark dataset, we downloaded experimentally verified 98 protein sequences that contain formylated lysine sites from Protein Lysine Modifications Database (PLMD) [9], UniProt [10] and dbPTM [11] publicly well-known databases and got 94 protein sequences by removing those protein sequences whose similarities were higher than 40% using CD-HIT program [12]. Then by using a sliding window method of window size $2n + 1$ residues, the peptide fragments containing lysine residue (K) at the center were extracted. The protein segment containing lysine with less than $2n + 1$ residues were filled with 'X' to ensure the same length of each fragment. In this study, each protein segment had an optimal length of N ($N = (2n + 1) = 29$ where n is 14) residues and represented as follows

$$A_{-n}, A_{-(n-1)}, \ldots, A_{-2}, A_{-1}, K, A_1, A_2, \ldots, A_{+(n-1)}, A_{+n} \tag{1}$$

Here, A_{-n} and A_{+n} represent the nth upstream and the nth downstream residue from the central lysine residue respectively. Then, each protein fragment are classified as positive and undefined samples as follows

$$P(k) \in \begin{cases} \text{PS if it's center } k \text{ is formylated} \\ \text{US otherwise} \end{cases}$$

Therefore, 1640 undefined samples and 184 formylated samples were obtained by applying the sliding window method to generate the benchmark dataset.

2.2 Feature Extraction

Amino Acid Composition (ACC). AAC is a simple, popular, and well-known process to extract features from protein sequences. The probability of each amino acid occurring on a protein sequence fragment is given by ACC [13, 14]. As a dummy amino acid "X" was used to fill the empty position, we got a 21-dimentional feature vector in this study. In short, the process of this encoding for a peptide fragment of length L can be given as follows

$$P_k = \frac{f_k}{L} \tag{2}$$

$$V = [P_1, P_2, P_3, \ldots, P_{20}, P_{21}] \tag{3}$$

Here, f_k and P_k represent the frequency and the probability of an amino acid k respectively. On the other hand, L represent the length of peptide fragment and V represent the feature vector of 21-dimention for a given protein segment.

Binary Encoding (BE). The positional and compositional properties of amino acids are obtained by binary encoding technique from a protein sequence [15, 16]. According to BE, the sequence "ACDEFGHIKLMNPQRSTVWYX" represents the order of all amino acids and amino acid was represented by a 21-dimentional feature vector. For example, two 21-dimentional feature vectors "100,000,000,000,000,000,000" and "010,000,000,000,000,000,000" were obtained for amino acid A and C respectively. But, especially "000,000,000,000,000,000,000" was taken for representing the dummy amino acid "X". Therefore, a feature vector of 609-dimention ($29 \times 21 = 609$) was generated for every peptide fragment of length 29 of our benchmark dataset.

Dipeptide Composition (DC). Another famous and widely used feature construction method named DC provides not only the compositional information but also the nearby properties of amino acids in a sequence [17]. As one dummy and 20 standard amino acids were used in our study, we got total 441 possible dipeptides. Then, according to the dipeptide composition described in [18, 19], all amino acid pairs in a peptide segment were gained using Eq. 4.

$$\mathrm{DPC}_i = \frac{n_i}{N} \tag{4}$$

where i represent one out of 441 dipeptides, n_i represent the total number of dipeptide (i), N represent the sum of all possible dipeptides and DPC_i represent the dipeptide frequency of dipeptide (i) appearing in a given protein.

2.3 *Imbalanced Dataset Management*

The number of formylated and undefined peptide segments was 184 and 1640 respectively in our dataset. So, the ratio of the number of them approximately was 1:9 that led to the benchmark dataset being highly imbalanced. The model trained by this imbalanced dataset considered the minority class as noisy instances and was strongly biased to the non-formylated (majority class) [20]. Therefore, a method, BLO_RNS, proposed by us was applied to select reliable non-formylated samples from undefined samples and SMOTE [21] resampling method implemented in various studies [16, 22–25] was used to increase the number of positive samples for avoiding this problem. The steps of the proposed technique for the selection of reliable negative samples are shown in Fig. 2 and the algorithm for the selection of reliable negative samples can be given as follows.

I. Measure the similarity score between an undefined sample (US) and each formylated sample (PS) using BLOSUM62 matrix.
II. Take the mean similarity score (ss) and assign it to that undefined sample (USS).

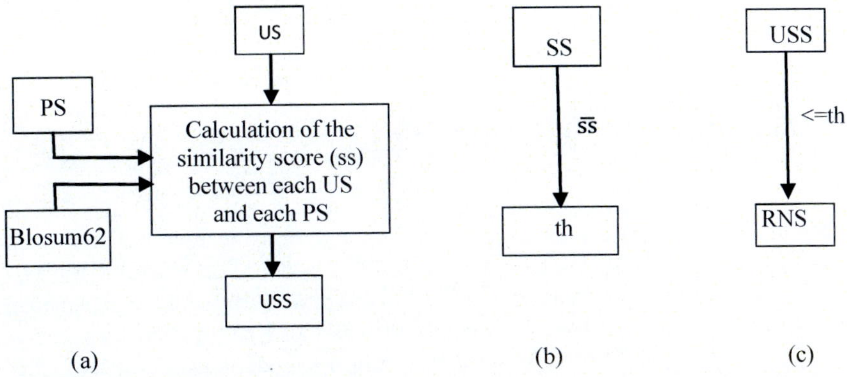

Fig. 2 Steps for selection reliable negative samples

III. Repeat steps I and II for each undefined sample.
IV. Take the mean value of the similarity score of all USS samples as the threshold (th).
V. Finally, select all undefined samples as reliable negative samples (RNS), which similarity score is bellow or equal to threshold.

In this study, 941 reliable negative samples were obtained by the proposed method. Then, the ratio between formylated and non-formylated sites became to 9:11 after increasing the total amount of formylated samples by SMOTE. Therefore, 941 negative and 770 positive samples were used for model training purposes.

2.4 Feature Selection

In our current work, a 1071-dimensional feature vector was constructed by fusing all feature descriptors, which may affect the performance of our classifier. Hence, the optimal features were chosen by a feature selection method called F-score, which had a larger discrimination degree between the formylated and non-formylated sites. A feature with a larger F-score value has larger discrimination degree to classify the samples accurately. Therefore, F-score value of the ith feature can be obtained using Eq. 5.

$$F_i = \frac{\left(\overline{x}_i^{(+)} - \overline{x}_i\right)^2 + \left(\overline{x}_i^{(-)} - \overline{x}_i\right)^2}{\frac{1}{n_+-1}\sum_{k=1}^{n+}\left(x_{k,i}^{(+)} - \overline{x}_i^{(+)}\right)^2 + \frac{1}{n_--1}\sum_{k=1}^{n-}\left(x_{k,i}^{(-)} - \overline{x}_i^{(-)}\right)^2} \tag{5}$$

Here, $n+$ and $n-$ represent the number of positive and negative samples respectively, $x_{k,i}^{(+)}$ represents the ith feature of the k the positive samples and $x_{k,j}^{(-)}$ represents the ith feature of the k the negative samples. Finally, we got an optimal number of 30 features using F-score, which is used to train our model.

2.5 Support Vector Machine (SVM)

SVM is a popular and well-known supervised machine learning algorithm proposed by Vapnik and Cortes [26] that is enable for solving classification and regression problems [27, 28]. It provides an optimal hyperplane between the classes that maximally separates the training datasets. As it is capable of working in higher dimensions with higher accuracy, it is already implemented in various types of PTM sites prediction [1–4]. In this study, SVM was used to train the model because of its higher performance according to the shown result of various classification algorithms in Table 3. For a given protein sample P_t in this study, the identification of formylation sites was performed by SVM using the following discriminant function

$$f\left(x^t\right) = \sum_{i=1}^{n} \alpha_i y_i K\left(x^i, x^t\right) + b \tag{6}$$

where, $yi \in \{-1, +1\}$ is the class label of a training sample x_i and $K\left(x^i, x^t\right)$ is RBF kernel function and it can be expressed as the following equation.

$$K\left(x^i, x^t\right) = \exp\left(-\gamma \left\| |x^i - x^t \right\|^2\right) \tag{7}$$

$$\gamma = \frac{1}{2\sigma^2} \tag{8}$$

where, α_i is the Lagrange multipliers and σ is the width of the function.

2.6 *Cross Validation and Performance Evaluation*

A model validation technique called cross validation is mostly implemented to measure the classification capability of a system by using new data that is not used in training it. For measure, the ability of formylated lysine site prediction of PLF_RNS on tenfold cross validation, five well-known parameters such as area under ROC curve (AUC), sensitivity (Sn), specificity (Sp), accuracy (Acc), and Matthew's correlation coefficient (MCC) was used and those measurements are represented as

$$\text{Sn} = \frac{\text{TP}}{\text{TP} + \text{FN}} \tag{9}$$

$$\text{Sp} = \frac{\text{TN}}{\text{TN} + \text{FP}} \tag{10}$$

$$\text{Acc} = \frac{\text{TP} + \text{TN}}{\text{TP} + \text{FP} + \text{TN} + \text{FN}} \tag{11}$$

$$\text{MCC} = \frac{\text{TP} * \text{TN} - \text{FP} * \text{FN}}{\sqrt{(\text{TP} + \text{FN}) * (\text{TP} + \text{FP}) * (\text{TN} + \text{FN}) * (\text{TN} + \text{FN})}} \tag{12}$$

Here, the number of false positive, false negative, true positive, and true negative are denoted as FP, FN, TP, and TN respectively. Besides, we also draw a ROC curve that expresses the measurement of predictive performance of PLF_RNS.

Table 1 The performance comparisons between each feature extraction method

Samples	Acc (%)	Sn (%)	Sp (%)	MCC (%)	AUC (%)
AAC	91.82	86.81	94.15	84.39	98.27
BE	84.98	79.22	89.69	69.62	88.26
DC	85.27	81.18	88.09	70.19	93.96
ACC + BE + DC	**95.09**	**95.06**	**95.11**	**90.09**	**99.70**

The highest value in a column is marked by bold

3 Results and Discussion

3.1 *Effect of Feature Extraction Methods on Prediction Results*

The performance of a trained model mainly depends on the numerical features which are how efficiently extracted from protein sequences. So, the selection of feature extraction methods has a crucial role in the construction of the best prediction model that more accurately classifies non-formylation and formylation sites. In this work, DC, BE, and AAC were singly used to train three prediction models to measure the prediction capability of each feature encoding method. Besides, another model with support vector machine algorithm was also constructed by fusing all types of features. Then, the predictive average performance of each model for distinguishing the formylation and non-formylation sites was calculated using a tenfold cross validation process and obtained prediction results are summarized in Table 1.

From Table 1, we have seen that when a model was trained by fusing all types of features, it obtained the highest average value of Acc, Sn, Sp, MCC, and AUC with 95.09%, 95.06%, 95.11%, 90.09%, and 99.70% than those of other methods. Hence, in this study, a multi-feature fusion process was applied to extract features from protein sequences for formylation and non-formylation site perdition.

3.2 *Influence of BLO_RNS and SMOTE on Prediction Results*

As our benchmark dataset was highly imbalanced, the model might become strongly biased to the majority class. In this study, a method, BLO_RNS, proposed by us was implemented for the subset selection of non-formylated samples and the SMOTE resampling method was used to increase the number of positive samples for avoiding this problem. After using SMOTE and BLO_RNS, 650 positive samples and 741 reliable negative samples were got for training our proposed system. To investigate the effect of BLO_RNS and SMOTE, the dataset was processed separately and applied

Table 2 The effect of balancing the dataset on the performance of PLF_RNS

Samples	Acc (%)	Sn (%)	Sp (%)	MCC (%)	AUC (%)
ALL	89.98	1.08	100.00	0.07	98.01
ALL (SMOTE)	91.27	87.85	94.08	82.39	98.45
ALL (BLO_RNS + SMOTE)	**95.09**	**95.06**	**95.11**	**90.09**	**99.70**

The highest value in a column is marked by bold

to train the model. The results of using BLO_RNS and SMOTE for balancing the dataset are shown in Table 2.

From the result shown in Table 2, we have seen that when the dataset processed by both BLO_RNS and SMOTE at the same time was used to train PLF_RNS, the average accuracy Acc was 95.09%, which was higher than that of others and the values of sensitivity, specificity, MCC, and AUC are also satisfactory. Therefore, it has been proved that after using BLO_RNS and SMOTE, the prediction capability of PLF_RNS is improved for distinguishing the formylation and non-formylation sites.

3.3 The Effect of the Classifier Algorithm on the Prediction Results

For the construction of a powerful predictive model that more accurately identifies the formylation sites, the selection of appropriate classification algorithms is a crucial part of building a model. In this study, the dataset of features processed by BLO_RNS and SMOTE and feature selection is used to train the most common and popular five classifier algorithms including Support Vector Machine (SVM), *K*-nearest Neighbor (KNN), Random Forest (RF), Logistic Regression (LR), and Naïve Bayes (NB). The performance of each classifier is measured by using a tenfold cross validation process. The prediction results and ROC curves for the five classification algorithms are displayed in Table 3 and Fig. 3 respectively.

In Table 3, we have seen that the prediction capability of NB, LR, RF, KNN, and SVM in terms of accuracy were 88.07%, 90.12%, 92.28%, 90.18%, and 95.09%

Table 3 Performance comparison of various classification algorithms on training dataset

Algorithm	Acc (%)	Sn (%)	Sp (%)	MCC (%)	AUC (%)
Naive Bayes	88.07	96.74	81.63	78.71	90.51
Logistic regression	90.12	92.46	88.20	80.32	96.26
Random forest	92.28	89.74	94.36	84.40	98.50
K-nearest neighbor	90.18	86.80	94.46	80.50	98.11
Support vector machine	95.09	95.06	95.11	90.09	99.70

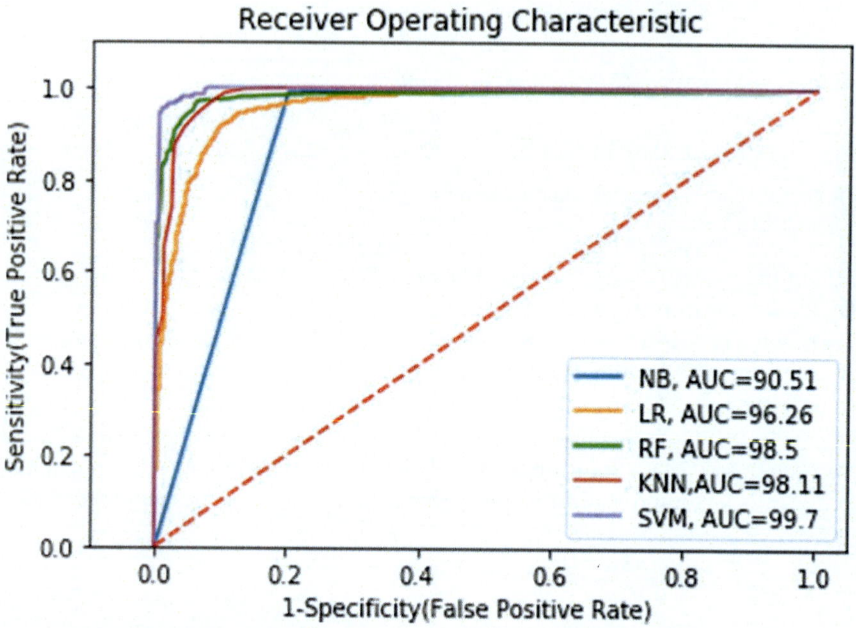

Fig. 3 ROC curves of the various classification algorithm on same training dataset

respectively. So, the prediction capability of SVM in terms of accuracy was 7.02%, 4.97%, 2.81%, and 4.91% larger than other four classification algorithms including NB, LR, RF, KNN respectively.

From Fig. 3, we have also seen that the area under the ROC curves (AUC) for all five classifiers including NB, LR, RF, KNN, and SVM was 90.51%, 96.26%, 98.5%, 98.11%, and 99.70%. So, the AUC value of SVM was also 9.19%, 3.44%, 1.20%, and 1.59% larger than the other four classification algorithms respectively. Therefore, SVM was taken as best classification algorithm to train our model in this paper.

3.4 Comparison with Other Methods

For measuring the validity degree of PLF_RNS in prediction of the formylation and non-formylation sites, prediction capability was compared with currently available models. All models were trained with our benchmark dataset and the obtained results are shown in Table 4.

The ROC curves of all models are also displayed in Fig. 4. In our current work to measure the predictive capability of the PLF_RNS model and all existing models, we applied the same training dataset with a tenfold cross validation method.

Table 4 Performance comparison between PLF_RNS and other models on training dataset

Model	Acc (%)	Sn (%)	Sp (%)	MCC (%)	AUC (%)
LyFor	90.02	88.80	90.36	75.34	91.08
Formator	84.59	86.80	82.47	69.29	91.08
CKSAAP_FormSite	76.39	71.66	77.25	38.48	89.37
LFPred	80.99	82.60	79.11	61.75	91.12
PLF_RNS	**95.09**	**95.06**	**95.11**	**90.09**	**99.70**

The highest value in a column is marked by bold

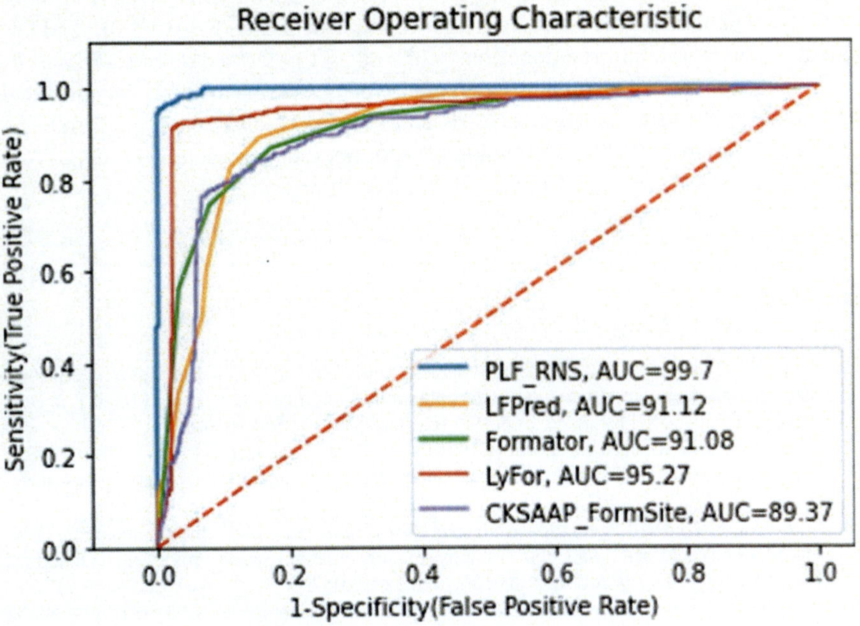

Fig. 4 ROC curves of PLF_RNS and other existing models on tenfold cross validation

After observing Table 4 and Fig. 4, we have known that the average Acc, Sn, Sp, MCC, and AUC values of PLF_RNS were 95.09%, 95.06%, 95.11%, 90.09%, and 99.70%, respectively, which are larger compared to the existing models. Therefore, the above results show that the PLF_RNS acquired better prediction performance than other existing models in the prediction of formylation sites.

4 Conclusion

Formylation regulates the process of gene activation and the modulation of chromatin conformations in the cell body. As traditional experimental processes are expensive and time consuming, computational models with machine learning become very popular for better understanding the mechanism and functional roles of formylation modification on organisms. In this paper, a computational model named PLF_RNS is developed for protein formylation sites prediction. The formylated proteins are collected and used to generate the peptide samples. Then, reliable non-formylated samples are selected by using the proposed method named BLO_RNS from non-formylated peptide segments. After that, all samples are used by ACC, BE, and DC to extract the feature information. Finally, the dataset processed by SMOTE and *F*-score is applied for model training. Compared to other existing methods, our proposed method acquires better performance. Hence, because of outstanding performance, PLF_RNS may be a very useful biological tool not only for accurately identifying the formylated sites but also for other PTMs site prediction.

References

1. Yu B, Yu Z, Chen C et al (2020) DNNAce: prediction of prokaryote lysine acetylation sites through deep neural networks with multi-information fusion. Chemom Intell Lab Syst 200(5):103999–104014. https://doi.org/10.1016/j.chemolab.2020.103999
2. Ning Q, Ma Z, Zhao X (2019) dForml(KNN)-PseAAC: detecting formylation sites from protein sequences using K-nearest neighbor algorithm via Chou's 5-step rule and pseudo components. J Theor Bio 470(7):43–49. https://doi.org/10.1016/j.jtbi.2019.03.011
3. Ju Z, Wang S (2020) Prediction of lysine formylation sites using the composition of k-spaced amino acid pairs via Chou's 5-steps rule and general pseudo components. Genomics 112(1):859–866. https://doi.org/10.1016/j.ygeno.2019.05.027
4. Jia C, Zhang M, Fan C et al (2019) Formator: predicting lysine formylation sites based on the most distant undersampling and safe-level synthetic minority oversampling. IEEE/ACM Trans Computat Biol Bioinf. https://doi.org/10.1109/tcbb.2019.2957758
5. Jiang T, Zhou X, Taghizadeh K et al (2006) N-formylation of lysine in histone proteins as a secondary modification arising from oxidative DNA damage. Proc Nat Acad Sci 104(1):60–65. https://doi.org/10.1073/pnas.0606775103
6. Machida Y, Chiba T, Takayanagi A et al (2005) Common anti-apoptotic roles of parkin and α-synuclein in human dopaminergic cells. Biochem Biophys Res Commun 332(1):233–240. https://doi.org/10.1016/j.bbrc.2005.04.124
7. Sohrawordi M, Hasan M (2020) LyFor: prediction of lysine formylation sites from sequence based features using support vector machine. 2020 IEEE Region 10 Symp (TENSYMP), 250–253. https://doi.org/10.1109/tensymp50017.2020.9230689
8. Blagus R, Lusa L (2013) SMOTE for high-dimensional class-imbalanced data. BMC Bioinf. https://doi.org/10.1186/1471-2105-14-106
9. Xu H, Zhou J, Lin S et al (2017) PLMD: an updated data resource of protein lysine modifications. J Genet Genomics 44(5):243–250. https://doi.org/10.1016/j.jgg.2017.03.007
10. Bairoch A, Apweiler R, Wu CH et al (2009) The universal protein resource (UniProt) in 2010. Nucleic Acids Res 38(1):D138–D142. https://doi.org/10.1093/nar/gkp846

11. Huang K, Lee T, Kao H et al (2018) dbPTM in 2019: exploring disease association and cross-talk of post-translational modifications. Nucleic Acids Res 47(D1):D298–D308. https://doi.org/10.1093/nar/gky1074
12. Fu L, Niu B, Zhu Z et al (2012) CD-HIT: accelerated for clustering the next-generation sequencing data. Bioinformatics 28(23):3150–3152. https://doi.org/10.1093/bioinformatics/bts565
13. Zhang L, Dong B, Teng Z et al (2020) Identification of human enzymes using amino acid composition and the composition of k-spaced amino acid pairs. BioMed Res Int 1–11. https://doi.org/10.1155/2020/9235920
14. Li S, Yu K, Wu G et al (2021) Pcysmod: prediction of multiple cysteine modifications based on deep learning framework. Front Cell Dev Biol. https://doi.org/10.3389/fcell.2021.617366
15. Ning Q, Zhao X, Bao L et al (2018) Detecting succinylation sites from protein sequences using ensemble support vector machine. BMC Bioinf 19(1):237–235. https://doi.org/10.1186/s12859-018-2249-4
16. Liu Y, Yu Z, Chen C et al (2020) Prediction of protein crotonylation sites through LightGBM classifier based on SMOTE and elastic net. Anal Biochem 609:113903–113910. https://doi.org/10.1016/j.ab.2020.113903
17. Gupta S, Mittal P, Madhu M, Sharma VK (2017) IL17eScan: a tool for the identification of peptides inducing IL-17 response. Front Immunol. https://doi.org/10.3389/fimmu.2017.01430
18. Liu M-L, Su W, Wang J-S et al (2020) Predicting preference of transcription factors for methylated DNA using sequence information. Mol Therapy Nucleic Acids. https://doi.org/10.1016/j.omtn.2020.07.035
19. Atanaki F, Behrouzi S, Ariaeenejad S et al (2020) BIPEP: sequence-based prediction of biofilm inhibitory peptides using a combination of NMR and physicochemical descriptors. ACS Omega 5:7290–7297. https://doi.org/10.1021/acsomega.9b04119
20. Yahav S, Bhole G (2020) Learning from imbalanced data in classification. Int J Recent Technol Eng 8:1907–1916. https://doi.org/10.35940/ijrte.e628 6.018520
21. Chawla N, Bowyer K, Hall L, Kegelmeyer W (2002) SMOTE: synthetic minority over-sampling technique. J Artif Intell Res 16:321–357. https://doi.org/10.1613/jair.953
22. Wang M, Cui X, Yu B et al (2020) SulSite-GTB: identification of protein S-sulfenylation sites by fusing multiple feature information and gradient tree boosting. Neural Comput Appl 32:13843–13862. https://doi.org/10.1007/s00521-020-04792-z
23. Kumari C, Abulaish M, Subbarao N (2020) Using SMOTE to deal with class-imbalance problem in bioactivity data to predict mTOR inhibitors. SN Comput Sci 1. https://doi.org/10.1007/s42979-020-00156-5
24. Wu L, Gao C, Xiang P et al (2020) CT-imaging based analysis of invasive lung adenocarcinoma presenting as ground glass nodules using peri- and intra-nodular radiomic features. Front Oncol 10. https://doi.org/10.3389/fonc.2020.00838
25. Mishra S, Mallick PK, Jena L, Chae G-S (2020) Optimization of skewed data using sampling-based preprocessing approach. Front Public Health 8. https://doi.org/10.3389/fpubh.2020.00274
26. Cortes C, Vapnik V (1995) Support-vector networks. Mach Learn 20:273–297. https://doi.org/10.1007/bf00994018
27. Cervantes J, Garcia-Lamont F, Rodriguez-Mazahua L, Lopez A (2020) A comprehensive survey on support vector machine classification: applications, challenges and trends. Neurocomputing 408:189–215. https://doi.org/10.1016/j.neucom.2019.10.118
28. Atasever S, Aydin Z, Erbay H, Sabzekar M (2019) Sample reduction strategies for protein secondary structure prediction. Appl Sci 9:4429. https://doi.org/10.3390/app9204429

Extreme Gradient Boost with CNN: A Deep Learning-Based Approach for Predicting Protein Subcellular Localization

Md. Ismail and Md. Nazrul Islam Mondal

Abstract Optimal protein subcellular localization provides physiological context for a protein's activity. Traditionally, laboratory approaches are used for this purpose. However, these methods can be time consuming and tedious. Detecting the optimal position of proteins using machine learning techniques is a challenging task because of the varying length of sequential data. This study proposes a machine learning model that leverages Convolutional neural networks (CNNs) with extreme gradient boosting (XGBoost) technique. The research contributes to take a deep learning approach for the classification of ten types of protein locations using the benchmark *DeepLoc* data set. Our study comes out with better accuracy and $F1$ score of 79.30% and 73.2%, respectively, compared to some other state-of-the-art works.

Keywords Subcellular localization · Sparse data · XGBoost · CNN

1 Introduction

Proteins are responsible for different types of mechanisms in the body. A protein can act as an antibody, an enzyme, a messenger, storage, etc. [1]. The location of a protein defines its mechanism. Hence, predicting the subcellular location is crucial to understand a protein's functionality and the relationship between the protein and its location [2, 3]. Information about the location of a protein is used in medicine design [3]. So misleading information about it can lead to evolving new diseases during the development of a remedy [2, 4].

Proteins are found in different locations of a cell [4]. Some proteins have multiple locations. Thus, the problem of predicting the protein subcellular localization can be defined as a multi-class multi-label problem [4]. However, to reduce complexity, a small number of proteins with multiple locations are removed from the data set [2].

Md. Ismail (✉) · Md. N. Islam Mondal
Department of Computer Science and Engineering, Rajshahi University of Engineering and Technology, Rajshahi 6204, Bangladesh

M. S. Arefin et al. (eds.), *Proceedings of the International Conference on Big Data, IoT, and Machine Learning*, Lecture Notes on Data Engineering and Communications Technologies 95, https://doi.org/10.1007/978-981-16-6636-0_16

Therefore, we consider the protein subcellular localization problem as a multi-class classification problem with ten different classes.

The computational approaches of predicting protein subcellular localization can be divided into two groups—one is traditional machine learning-based methods and another one is deep learning-based methods [3, 4]. SVM, K-nearest neighbor, random forest, etc., algorithms have been used in several studies based on traditional machine learning approaches [2–8]. These methods need manual feature extraction where the features can be represented in sequence-based or annotation-based representation. Sequence-based feature representation can lose the effect of sequence order where annotation-based feature representation may contain information about protein functionalities. Though both representation techniques can be used together, manual feature extraction may lose some crucial information resulting in poor performance of a model [3].

In recent years, deep learning-based methodologies are being broadly used in the prediction of protein subcellular localization. As the features are extracted by a learning mechanism like Convolutional neural networks (CNNs), more relevant information can be captured by the models and improve the performance significantly [2, 3, 9]. CNNs have been proven to be more effective to extract sequence motif information. On the other hand, an LSTM network can be effective while working long sequential data [2, 9]. To adopt the strength of both CNN and LSTM networks, an approach has been proposed by [2].

As shown in a previous study, Bidirectional Long Short Term Memory (BiLSTM) is not well suited to handle sparse data [10]. Hence, in this study, we propose a CNN and Extreme Gradient Boost (XGBoost) [11] based approach to identify the subcellular location of proteins using sparse data.

2 Data Set

DeepLoc data set was used for this study. This data set contains ten different protein localizations: nucleus, cytoplasm, extracellular, mitochondrion, cell membrane, endoplasmic reticulum, chloroplast, Golgi apparatus, lysosome/vacule, and peroxisome. The data used in this data set were extracted from the UniProt database [12] and filtered using certain criteria: eukaryotic, not fragments, encoded in the nucleus, longer than 40 amino acids, and experimentally annotated [2]. The data set distribution is given in Table 1.

3 Proposed Approach

The primary goal of this study is to develop a system that can predict the subcellular location of a protein. Figure 1 explicates a schematic process of the overall approach of this study. The approach is comprised of three major parts: data preprocessing, structuring the model, and performance analysis of the predicted outcomes.

Table 1 DeepLoc data set distribution

Location	Number of proteins
Nucleus	4043
Cytoplasm	2688
Extracellular	1973
Mitochondrion	1510
Cell membrane	1340
Endolpasmic reticulam	862
Plastid	757
Golgi apparatus	356
Lysosome/vacule	321
Peroxisome	154
Total	14004

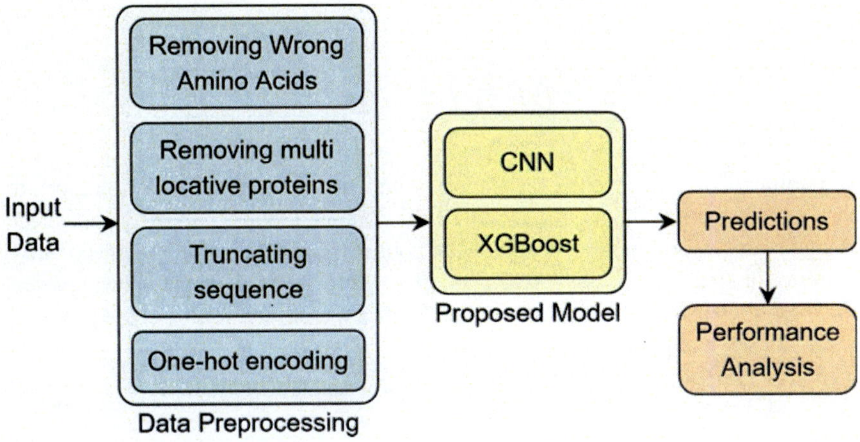

Fig. 1 Overall workflow

3.1 Data Prepossessing

Protein sequences are polymers of amino acids connected by a peptide bond. Generally, there are twenty (20) types of amino acid. In the data set, three wrong amino acids (B, U, and X) were observed. These can mislead the classifier. Thus, the wrong amino acids were removed from the data set by not considering for this experiment. Some proteins were found in multiple locations, and they are called N-locative proteins [4]. As discussed earlier, these N-locative proteins were removed from the data set by not considering for the experiment. As the model needs fixed-length sequences, all protein sequences were converted to 1000 length sequences by padding or truncating. Finally, the sequences were converted to one-hot encoded vectors.

3.2 *Structuring the Model*

The proposed model leverages CNN and XGBoost techniques. CNN extracts the higher-level features and the XGBoost acts as the predictor.

3.2.1 CNN

Convolutional neural network (CNN) is a class of deep neural networks that perform a mathematical linear operation—convolution or cross-correlation operation at least one layer [13]. CNN is consists of three different types of layers—(i) convolutional layer, (ii) pooling layer, and (iii) fully connected layer. In CNN, a kernel window of size $k \times k$ is selected. This convolution operation gives us a matrix that works as a feature vector. The convolution is performed between the input and the kernel vector using Eq. 1, where f and g are two functions.

$$(f * g)(t) \triangleq \int_{-\infty}^{\infty} f(\tau)g(t-\tau)\tau' \tag{1}$$

The pooling layer reduces the overlapping of information at the convolution layer. It can also downsample the input and reduce the computational complexity and select the significant information or feature. Max pooling selects maximum value covered by a filter of size $(F \times F)$, whereas average pooling selects average value covered by a filter of size $(F \times F)$ from the output of the convolution layer [13, 14].

3.2.2 XGBoost

XGBoost stands for extreme gradient boost, which is a decision tree-based ensemble machine learning algorithm that provides efficient distributed gradient boosting. It adopts parallel tree boosting approaches. In XGBoost, two special regularization techniques are used to minimize the loss efficiently. One of these is $L1$ or Lasso regularization technique which not only reduces weight values but also removes some weights from the weight vector. The cost function is optimized as Eq. 2.

$$\text{cost function} = \text{Loss} + \frac{\lambda}{2m}\sum ||W|| \tag{2}$$

$L2$ or Ridge regularization reduces the weights and handles the overfitting problem. It optimizes the cost function as shown in Eq. 3.

$$\text{cost function} = \text{Loss} + \frac{\lambda}{2m}\sum ||W||^2 \tag{3}$$

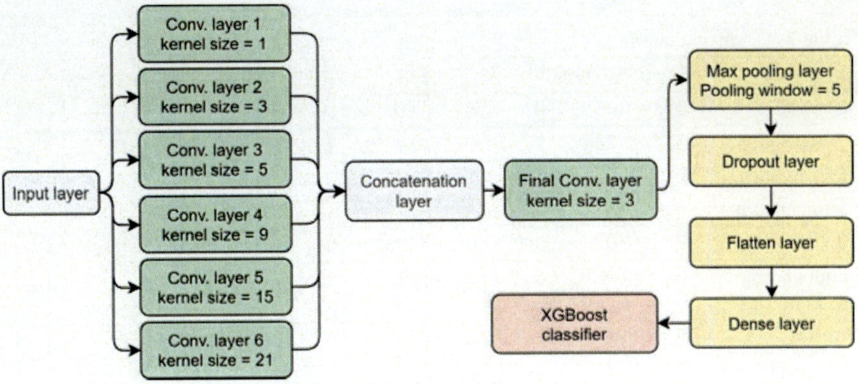

Fig. 2 Structure of the proposed CNN-XGBoost model

In both of the equations, λ and m are the tuning parameters. Finally, the XGBoost creates a decision tree-like iterative dichotomiser (ID3), classification and regression tree (CART), chi-squared automatic interaction detector (CHID), etc. In the decision tree, the attribute list is partitioned by using information gain, gain ratio, and Gini index [11]. XGBoost is able to handle missing data, zero values, and highly sparse with the help of loss function [15].

3.2.3 Proposed Model—CNN with XGBoost

The basic building blocks of the model are CNN and XGBoost as discussed before. As depicted in Fig. 2, the model starts with an input layer, which takes the input vectors. The outputs of the input layer are passed to 6 parallel 1D convolutional layers. As the base model, the kernel sizes of the layers are 1, 3, 5, 9, 15, and 21. The outputs of these layers are concatenated and passed to a final convolutional layer having a kernel size of 3. Next, a pooling layer was added to get the maximum values of each pooling window of size 5. To avoid overfitting, a dropout layer followed by a flatten layer and dense layer was used. Dropout probability was set to 0.25. Finally, the output from this dense layer is passed to an XGBoost classifier.

4 Experiments

To start the experiments, the data set was split into training data set and testing data set. The training data set was further split into train data set and validation data set. Finally, 64% of total data was used for training, 16% of total data was used for validation, and the remaining 20% of total data was used for testing.

Table 2 Confusion matrix

Location	Predicted protein								
Nucleus	123	42	0	0	0	0	2	0	0	1
Cytoplasm	35	230	1	3	4	1	2	0	0	7
Extracellular	0	2	144	1	14	2	0	1	5	0
Mitochondrion	0	6	0	84	1	1	8	0	1	1
Cell membrane	0	0	5	0	238	0	0	2	0	3
Endoplasmic reticulam	0	0	4	0	7	26	0	1	2	0
Plastid	0	6	1	8	0	0	75	0	0	0
Golgi apparatus	0	0	1	0	2	2	0	19	4	2
Lysosome	1	0	9	3	1	4	1	0	13	1
Peroxisome	3	9	0	2	0	1	1	0	1	15

During the training phase of a model, hyperparameters are important factors. Determining the optimal hyperparameters allows a model to get the optimal performance. During the experiments, we tried different values for several hyperparameters like learning rate and batch size. It was observed that the model gives the best performance with a learning rate, $\eta = 0.0025$ and batch size of 128. To determine these values, a manual hyperparameter tuning approach was taken.

To evaluate the model's performance, the test scores were considered. Table 2 represents the confusion matrix of the model using the DeepLoc data set. From the table, it can be seen that the model has performed the best for the cell membrane proteins. It has classified the cell membrane proteins with 95.97% accuracy. On the other hand, the model has the worst classification accuracy on the lysosome proteins with only 39.39% accuracy. It also archived 85.71, 83.33, 82.35, 81.27, 73.21, 65.00, 63.33, and 46.87% accuracy scores for extracellular, plastid, mitochondrion, cytoplasm, nucleus, endoplasmic reticulum, Golgi apparatus, and peroxisome proteins accordingly. However, Table 1 indicated that DeepLoc is an unbalanced data set. So, accuracy scores might not explain the performance of the proposed model. Thus, precision, recall, and $F1$ scores are given in Table 3. To reflect the performance of the proposed XGBoost based model, a comparison has been shown between the base model (BiLSTM based model) and the proposed model (XGBoost based model). Table 3 indicates location-wise precision, recall, and $F1$ scores. It can be observed that for some locations including nucleus, extracellular, and plastid, the BiLSTM based model performs better in terms of these metrics. However, for other locations, the proposed CNN and XGBoost model performs better. It also can be observed, for a low number of samples, the CNN + XGBoost model outperforms the CNN + BiLSTM model significantly. Like for "Golgi apparatus," the DeepLoc data set has only 356 samples. Our model achieves 96.39, 97.78, and 97.22% better precision, recall, and $F1$ scores, respectively, than the BiLSTM based model. The XGBoost model also performs better on average scores. Visual comparison on average scores between these models is shown in Fig. 3.

Table 3 Performance of proposed the *CNN* + *XGBoost* model and the *CNN* + *BiLSTM* model

Location	Precision		Recall		*F*1 score	
	BiLSTM	XGBoost	BiLSTM	XGBoost	BiLSTM	XGBoost
Nucleus	0.85	0.76	0.84	0.73	0.84	0.75
Cytoplasm	0.66	**0.78**	0.71	**0.81**	0.68	**0.80**
Extracellular	0.91	0.87	0.93	0.85	0.92	0.86
Mitochondrion	0.85	0.83	0.82	**0.82**	0.83	**0.83**
Cell membrane	0.77	**0.89**	0.74	**0.96**	0.75	**0.92**
ER	0.66	**0.70**	0.69	0.65	0.67	**0.68**
Plastid	0.86	0.84	0.92	0.83	0.89	0.84
Golgi apparatus	0.03	**0.83**	0.014	**0.63**	0.02	**0.72**
Lysosome	0.16	**0.50**	0.17	**0.39**	0.17	**0.44**
Peroxisome	0.33	**0.50**	0.27	**0.47**	0.29	**0.48**

Bold indicates the best performance or result score between BiLSTM and XGBoost based model

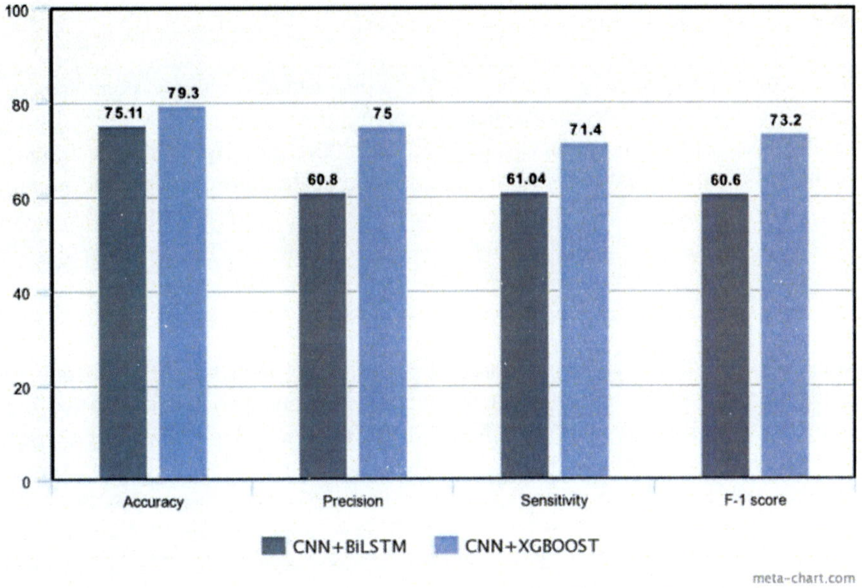

Fig. 3 Average performance of the proposed *CNN* + *XGBoost* model and the *CNN* + *BiLSTM* model.

Figure 3 indicates overall accuracy, precision, recall, and *F*1 scores for the CNN and BiLSTM model to be 75.11%, 60.8%, 61.04%, and 60.6%, respectively. Again for the proposed CNN and XGBoost model, the scores are 79.3%, 75.00%, 71.4%, and 73.2%, respectively, for test data. Overall accuracy for train data is 80.2%. Clearly, the proposed model outperforms the BiLSTM based model on the average accuracy, precision, recall, and *F*1 scores by 4.19, 14.2, 10.36, and 12.6% accordingly.

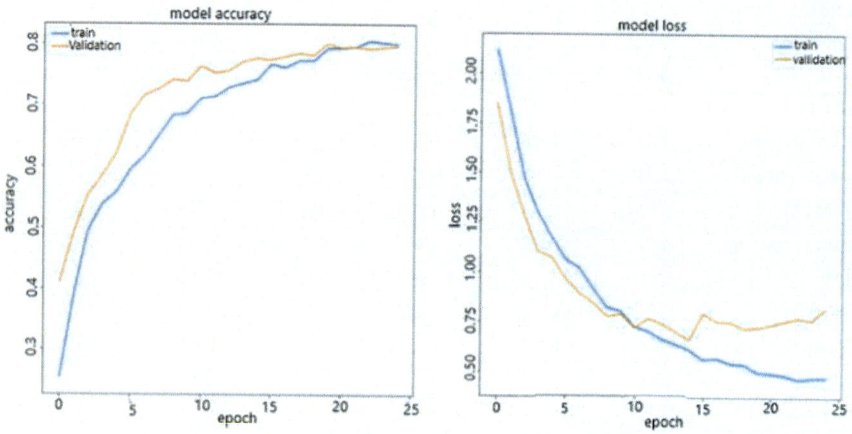

(a) Accuracy curves (b) Loss curves

Fig. 4 Training–validation accuracy and loss curves

Accuracy and loss curves can be helpful to understand if a model converges or not, how quickly it converges, overfitting, etc. Figure 4 shows both the training and validation accuracy and loss curves. To avoid overfitting, an early stopping technique was used. Figure 4a, b indicates that it took 25 epochs to the convergence of the model.

5 Conclusion

In this paper, a deep learning approach is presented to predict protein subcellular localization. We have compared our results with the recent research work on this topic. The overall accuracy of the CNN and BiLSTM model as shown in previous work [2] is 75.11%, whereas our proposed approach is a CNN and XGBoost based model, which achieves an overall accuracy of 79.3%. Also, the CNN+XGBoost approach outperforms the existing CNN+BiLSTM approach by 14.2, 10.4, and 12.6% in terms of precision, recall, and $F1$ scores, respectively. We have a plan to continue research for multi-class multi-label with multiple features in other data set [3] as only multi-class is considered in this paper.

References

1. Khan Academy (2015). Introduction to proteins and amino acids
2. Armenteros JJA, Sønderby CK, Kaae Sønderby S, Nielsen H, Winther O (2017) Deeploc: prediction of protein subcellular localization using deep learning. Bioinformatics 33(21):3387–3395

3. Wei L, Ding Y, Ran S, Tang J, Zou Q (2018) Prediction of human protein subcellular localization using deep learning. J Parall Distrib Comput 117:212–217
4. Pang L, Wang J, Zhao L, Wang C, Zhan H (2019) A novel protein subcellular localization method with CNN-XGBoost model for Alzheimer's disease. Frontiers Genet 9:751
5. Höglund A, Dönnes P, Blum T, Adolph H-W, Kohlbacher O (2006) Multiloc: prediction of protein subcellular localization using n-terminal targeting sequences, sequence motifs and amino acid composition. Bioinformatics 22(10):1158–1165
6. Blum T, Briesemeister S, Kohlbacher Oliver (2009) Multiloc2: integrating phylogeny and gene ontology terms improves subcellular protein localization prediction. BMC Bioinform 10(1):274
7. Shatkay H, Höglund A, Brady S, Blum T, Dönnes P, Kohlbacher O (2007) Sherloc: high-accuracy prediction of protein subcellular localization by integrating text and protein sequence data. Bioinformatics 23(11):1410–1417
8. Zhou H, Yang Y, Shen H-B (2017) Hum-mPloc 3.0: prediction enhancement of human protein subcellular localization through modeling the hidden correlations of gene ontology and functional domain features. Bioinformatics 33(6):843–853
9. Kaae Sønderby S, Kaae Sønderby C, Nielsen H, Winther O (2015) Convolutional lstm networks for subcellular localization of proteins. In International conference on algorithms for computational biology. Springer, pp 68–80
10. Liu S, Mocanu DC, Pechenizkiy M (2019) Intrinsically sparse long short-term memory networks. arXiv:1901.09208
11. Chen T, Guestrin C (2016) XGBoost: a scalable tree boosting system. In: Proceedings of the 22nd ACM SIGKDD international conference on knowledge discovery and data mining, pp 785–794
12. Apweiler R, Bairoch A, Wu CH, Barker WC, Boeckmann B, Ferro S, Gasteiger E, Huang H, Lopez R, Magrane M et al (2004) UniProt: the universal protein knowledgebase. Nucleic Acids Res 32(suppl_1):D115–D119
13. Albawi S, Mohammed TA, Al-Zawi S (2017) Understanding of a convolutional neural network. In 2017 international conference on engineering and technology (ICET). IEEE, pp 1–6
14. O'Shea K, Nash R (2015) An introduction to convolutional neural networks. arXiv:1511.08458
15. Brownlee J (2020) Data preparation for gradient boosting with XGBoost in python

Enhancing the Performance of 3D Rotation Perturbation in Privacy Preserving Data Mining Using Correlation Based Feature Selection

Mahit Kumar Paul and Md. Rabiul Islam

Abstract A large amount of valuable data is being produced every day with the development of technologies. To retrieve knowledge and information from these data, mining and analysis are mandatory. But, the data may contain sensitive information of the individuals like medical diagnostic reports which they do not want to expose. Privacy preserving data mining, i.e., PPDM can help in this issue keeping the sensitive information private as well as preserving the data utility. Rotation-based perturbation technique contributes to satisfying both aspects of PPDM, i.e., individuals' privacy and data utility besides other PPDM techniques. In this work, we proposed a way for generating the triplet (set of three features) for 3D rotation perturbation technique using correlation among the features. This triplet generation is a fundamental step in 3D rotation perturbation technique. The analysis of information entropy, privacy protection and utility analysis elucidates that correlation-based triplet generation provides better data privacy and utility than existing triplet generation for 3D rotation perturbation technique.

Keywords Privacy · Data utility · Correlation · Perturbation · Rotation

1 Introduction

Extraction of knowledge and necessary information from data is the fundamental task in data mining, and the extracted knowledge is useful in decision-making activities [1]. But the data available for data mining tasks may contain sensitive information of the individuals, and they do not want to expose it for analysis purposes. For example, analyzing the students' semester-wise performance, we can have insight about the students' weaknesses, scope of developments, drop-out reasons and so on [2]. But, the students' detailed reports of some educational institutes are confidential and cannot be made available for data mining tasks. On the other hand, we need to analyze

M. K. Paul (✉) · Md. R. Islam
Department of Computer Science & Engineering,
Rajshahi University of Engineering & Technology, Rajshahi 6204, Bangladesh

M. S. Arefin et al. (eds.), *Proceedings of the International Conference on Big Data, IoT, and Machine Learning*, Lecture Notes on Data Engineering and Communications Technologies 95, https://doi.org/10.1007/978-981-16-6636-0_17

the data to take decisions. That is why it is required to perturb the original data in some ways for analyzing. In this case, the main concerned issue is the privacy of the individuals' as well as the data utility because perturbed data may loss its underlying distribution while preserving privacy. Privacy preserving data mining (PPDM) can deal with this issue. PPDM aims to keep the data utility while maintaining the privacy of the individuals [3, 4].

PPDM targets to mitigate the risk of information leakage of the individuals and organizations. That is why, the data is released in such a way so that the intruder cannot estimate the original information, and the data utility has to be kept at the same time. Data perturbation approaches release the aggregate information regarding the data set. This aggregate information is useful to find out the knowledge through data mining algorithms. Also, this aggregate information introduces uncertainty of the individual values and thus reduces the chance of revealing private information [5].

Rotation-based perturbation approach preserves the geometric properties such as Euclidean distance and inner product of the data set [6]. Available rotation-based perturbation techniques can be of two types: two-dimensional rotation transformation (2DRT) [7] and three-dimensional rotation transformation (3DRT) [8]. In case of 2DRT, the direction of rotation is constantly orthogonal to the z-axis, i.e., the xy-plane. On the other hand, the direction of rotation can be chosen independently for 3DRT. It can be x-axis, y-axis or z-axis depending upon the inherent planes. In this work, we considered the 3DRT perturbation technique. In [8], the authors used a straightforward approach for generating the triplet for 3DRT. They simply selected three consecutive features for generating a triplet without considering the correlation among the features. We utilized the correlation among the features of the data set to generate the triplet for 3DRT in this paper.

The residual of this paper is organized in several sections. In Sect. 2, perturbation work done by many researchers in the field of PPDM is discussed. Throughout Sect. 3, the workflow of our work is described precisely. Different metrics used for privacy and data utility measurement are introduced in Sect. 4. The experimental throughput of our work is analyzed in Sect. 5. Finally, in Sect. 6, the conclusion of our work is given.

2 Related Work

Researchers have developed many methods which attempt to deal with the trade-off between preservation of privacy and maintenance of data utility in PPDM. Olivera and Zaiane proposed two-dimensional rotation-based transformation technique which is not dependent on any clustering algorithm [7]. Data features are rotated pairwise depending upon feature concerned threshold values. In [9], the authors familiarized a set of geometric data perturbation techniques such as translation, scaling and rotation-based data perturbation as well as hybridization of data perturbation. These methods only alter the sensitive features of the data set. A three-dimensional rotation perturbation technique is introduced in [8] which makes triplets of features and rotates the triplets at a time. The rotation direction can be any of the three x, y or z axes. For stream data mining, two distinguished data perturbation techniques are proposed in [10] for privacy preservation. Random projection and random translation along with two different forms of additive noise are used to develop the two perturbation techniques in [10]. Chamikara et al. proposed a non-invertible and extendable perturbation algorithm called PABIDOT for the preservation of privacy of big data. PABIDOT consists of multidimensional geometric transformations, reflection, translation and rotation succeeded by randomized expansion and random tuple shuffling [4]. For the privacy preservation of stream data and big data, P^2R_oCAl (privacy preserving rotation-based condensation algorithm) is proposed in [11]. P^2R_oCAl combines the proficiency of condensation and accuracy of rotation to deal with both of the aspects of PPDM. A new privacy preserving technique called secure and efficient data perturbation algorithm utilizing local differential privacy (SEAL) is discussed in [12]. SEAL utilizes Chebyshev interpolation and Laplacian noise. Combining these two, SEAL gives a good harmony between privacy and utility of PPDM.

3 Methodology

In this paper, a method to enhance the performance of three-dimensional rotation perturbation technique is proposed using the correlation among the features of the data set. The detailed workflow of our work is provided in Fig. 1. The input of our procedure is the normalized data set D^N. For normalization, z-score is used because it provides better performance among the other normalization techniques for the full feature set [13]. In three-dimensional rotation, the axis of rotation can be of x, y or z-axis. For three-dimensional rotation perturbation technique, double rotation matrices are used, where the data are rotated along the axis pairs xy, yz or xz and the rotation matrices are formulated such as in [8]. After selecting the axis pair, the step of triplet generation is implemented. In the existing three-dimensional perturbation technique, consecutive three features are used to generate triplets regardless of the correlation among the features. In our procedure, the correlation among the features is considered and given emphasized in the generation of triplets. The mostly correlated three features are put together to generate the first triplet, next mostly correlated three features are put together to generate the second triplet and so on. This can be explained using the correlation matrix of the BODS data set (Table 2) provided in Table 1. The correlation-based generated triplets are (Attr2, Attr3, Attr4), (Attr6, Attr7, Attr8), (Attr5, Attr9, Attr10) and (Attr2, Attr3, Attr1). In the last triplet, Attr2 and Attr3 are reused to generate triplet with Attr1. After rotation, these reused triplets are discarded. This way of triplet generation outperforms than the existing consecutive triplet generation schema which is illustrated in the performance analysis section. After triplet generation, each of the triplets is rotated for several angles θ in the range $0.1 \leq \theta < 360$. In the next step, the variances between the original and rotated features are determined for each triplet.

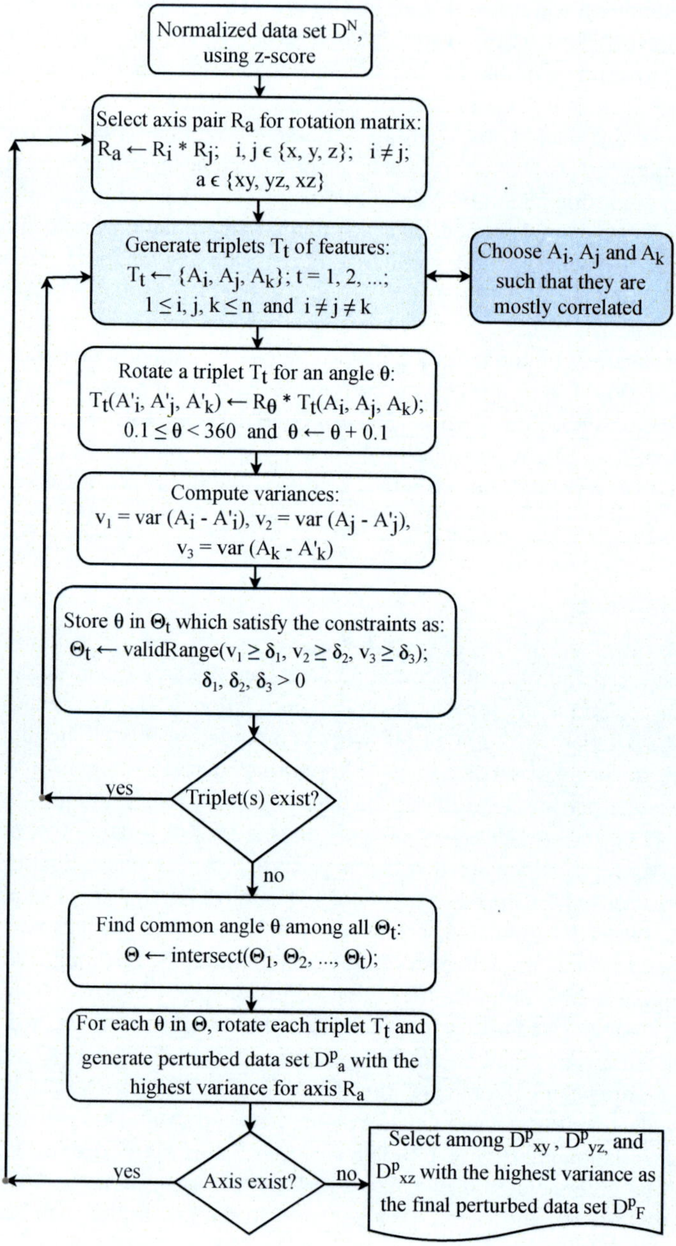

Fig. 1 Proposed workflow

Table 1 Correlation matrix for the data set BODS

	Attr1	Attr2	Attr3	Attr4	Attr5	Attr6	Attr7	Attr8	Attr9	Attr10
Attr1	1	−0.06	−0.04	−0.04	−0.07	−0.05	−0.1	−0.06	−0.05	−0.04
Attr2	−0.06	1	0.64	0.65	0.49	0.52	0.59	0.55	0.53	0.35
Attr3	−0.04	0.64	1	0.91	0.71	0.75	0.69	0.76	0.72	0.46
Attr4	−0.04	0.65	0.91	1	0.69	0.72	0.71	0.74	0.72	0.44
Attr5	−0.07	0.49	0.71	0.69	1	0.59	0.67	0.67	0.6	0.42
Attr6	−0.05	0.52	0.75	0.72	0.59	1	0.59	0.62	0.63	0.48
Attr7	−0.1	0.59	0.69	0.71	0.67	0.59	1	0.68	0.58	0.34
Attr8	−0.06	0.55	0.76	0.74	0.67	0.62	0.68	1	0.67	0.35
Attr9	−0.05	0.53	0.72	0.72	0.6	0.63	0.58	0.67	1	0.43
Attr10	−0.04	0.35	0.46	0.44	0.42	0.48	0.34	0.35	0.43	1

Then these triplet variances are compared with previously defined thresholds δ_1, δ_2 and δ_3. These thresholds help to generate a random range of rotation angles Θ_t for a specific triplet T_t. Thus, the ranges $\Theta_1, \Theta_2, \ldots, \Theta_t$ for each of the triplets $T_1, T_2, \ldots, T_t$ are generated. Then the common angle range Θ among all Θ_t is computed by intersection. For each of the angles θ in Θ, all of the triplets are rotated, and the perturbed data set D_a^P with the highest variance for the axis pair R_a is generated. In this way, the perturbed data sets D_{xy}^P, D_{yz}^P and D_{xz}^P for the axis pairs xy, yz and xz are generated. Finally, the perturbed data set with the highest variance among D_{xy}^P, D_{yz}^P and D_{xz}^P is selected and output as the final perturbed data set D_F^P.

4 Evaluation Metrics

Performance measurement of a perturbation technique is very crucial in PPDM. To asses the performance of 3D rotation perturbation technique, different privacy and data utility metrics are used in this work as follows.

4.1 Increase in Information Entropy

The information entropy of a data set can be measured by using Shannon's entropy formulation given in Eq. 1 [4].

$$H(X) = -\sum_{i=1}^{n} P(x_i) \log_2 P(x_i) \tag{1}$$

where $H(X)$ is the entropy of X which is a discrete random variable with $X =< x_1, x_2, \ldots, x_n >$, $\sum$ indicates the summation over X, and $P(x_i)$ is the probability of the occurrence of x_i. The values of the average increase in information entropy are calculated by using Eq. 2 [4].

$$\text{AIE} = \frac{\sum_{i=1}^{n}(H(x_i') - H(x_i))}{n} \tag{2}$$

where AIE stands for an average increase in information entropy and $H(x_i')$ and $H(x_i)$ are entropy for each feature of perturbed data set and original data set, respectively. When the values of AIE are positive, it indicates that the features of the perturbed data set contain more impurity as compared to the original data set and hence preserve more privacy.

4.2 *Privacy Protection of Data*

The privacy protection of data is measured using six different metrics in this paper. They are privacy [8], value difference (VD) [14], rank differences such as RP, RK, CP and CK [14]. Generally, privacy of a perturbation technique is defined as the variance between the original and perturbed data values [8]. The value difference (VD) between the original and perturbed data values is defined as Frobenius norm [14]. The rank-based privacy metric RP is used to measure the average changes of rank of all the data features [14]. RK [14] can represent the percentage of data values that preserves their individual ranks of magnitude in every feature afterward the perturbation. The metric CP is used to denote the change of rank of the average value of the features [14]. Resembling to RK, CK is used to measure the percentage of the features that preserve their ranks of the average value after the perturbation [14].

4.3 *Utility of Data*

To measure the utility maintenance of 3D rotation perturbation technique, three metrics are used in this paper—accuracy, $F1$-score and area under the ROC curve, i.e., AUC. Accuracy gives an insight of accurate decisions made by the data mining algorithms. $F1$-score, alternatively known as F-measure or F-score, resembles a balance between precision and recall provided by a data mining algorithm. AUC values resemble the betterment of a data object being classified between two separated groups. In ROC curve, true positive rate $\text{TPR} = \text{TP}/(\text{TP} + \text{FN})$ is plotted as a function of false positive rate $\text{FPR} = \text{FP}/(\text{FP} + \text{TN})$.

5 Performance Analysis

5.1 *Data Set, Classifiers and Settings*

In this paper, we used five data sets with varying number of features and instances to evaluate the performance. The details of the used data set are given in Table 2. To measure the classification performance, we used C4.5 which classifies instances by generating trees. Weka is used to implement C4.5 with the default parameters. MATLAB R2020a (v9.8.0.1323502) is used as the implementation platform for three-dimensional rotation perturbation technique. We worked on a computer having the configuration as processor: Intel(R) Core(TM) i5-8250U CPU @ 1.60 GHz 1.80 GHz; RAM: 8.00 GB; system type: 64-bit operating system, x64-based processor.

5.2 *Analysis of Privacy*

In this paper, three-dimensional rotation perturbation technique with correlation-based feature selection is denoted as 3DRT-CF and with non-correlated, i.e., consecutive feature selection-based approach is denoted as 3DRT-NCF. In the bar graph of Fig. 2, the average increase in information entropy (AIE) values returned by 3DRT-CF and 3DRT-NCF are shown. We see that all of the AIE values are positive which indicates both 3DRT-CF and 3DRT-NCF preserve more privacy than the original data set. Alongside looking at the numeric value comparison from the bar graph in Fig. 2, it is observed that 3DRT-CF preserves more privacy than 3DRT-NCF.

Table 2 Data set used for analysis tasks

S. No.	Original data set name	Abbreviation	#Instances	#Features	Feature types
1	Electricity	ELDS	45312	8	Integer, real
2	Breast Cancer Wisconsin (original)	BODS	699	10	Integer
3	Diabetic Retinopathy Debrecen Data Set	DRDS	1151	19	Integer, real
4	Breast Cancer Wisconsin (diagnostic)	BDDS	569	31	Real
5	SPECTF Heart	SHDS	267	44	Integer

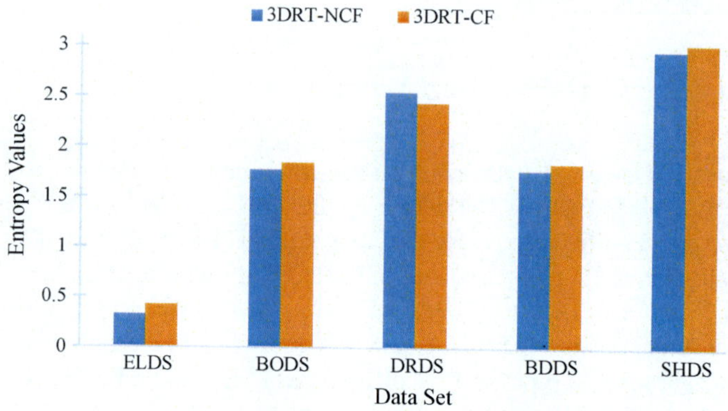

Fig. 2 Average increase in information entropy

Table 3 Privacy protection by 3DRT-NCF and 3DRT-CF

Data set	Methods	Privacy	VD	RP	RK	CP	CK
ELDS	3DRT-NCF	1275.264	1.238	17304.53	0	3.75	0
	3DRT-CF	1314.695	1.255	19337.64	0	3.5	0
BODS	3DRT-NCF	1.282	1	247.534	0.002	4	0.1
	3DRT-CF	1.361	1	244.167	0.001	3	0.1
DRDS	3DRT-NCF	255.711	1.003	437.345	0.001	5.474	0.053
	3DRT-CF	258.816	1.013	479.992	0.001	6.421	0
BDDS	3DRT-NCF	9436.237	1	234.932	0.002	11.613	0.065
	3DRT-CF	9667.235	1	255.409	0.001	13.032	0
SHDS	3DRT-NCF	1.149	1.002	114.518	0.002	14.955	0
	3DRT-CF	1.153	1.002	115.356	0.003	16.364	0

In order to show the efficacy of 3DRT-CF over 3DRT-NCF, more six privacy preservation metrics are implemented and the corresponding experimental values are provided in Table 3. The larger values of privacy, VD, RP and CP, and the smaller values of RK and CK indicate higher privacy preservation [14]. The values of privacy resemble that 3DRT-CF outperforms 3DRT-NCF for all of the data set. From the values of VD and RK, it is observed that 3DRT-CF performs better or equal with 3DRT-NCF. 3DRT-CF performs 80% better than 3DRT-NCF concerning RP values. Also, considering CP and CK values, 3DRT-CF outperforms 3DRT-NCF except for few cases.

5.3 Analysis of Utility

Alongside the preservation of privacy, the other primary property of a perturbation technique in PPDM is to keep the utility of the perturbed data set close to the original data set. To evaluate this property, we measured the accuracy, $F1$-score and AUC values corresponding to 3DRT-CF and 3DRT-NCF returned by C4.5. Figures 3, 4 and 5 represent the curves for accuracy, $F1$-score and AUC values, respectively. From Figs. 3, 4 and 5, we see that the curves for 3DRT-CF perturbed data set are more close to the the curve for original data set, and at some points the two curves lie on each other. On the other hand, the curve for 3DRT-NCF perturbed data set is far apart from the curve for original data set. Furthermore, the curves for accuracy, $F1$-score and AUC values have almost the same shape which indicate the results

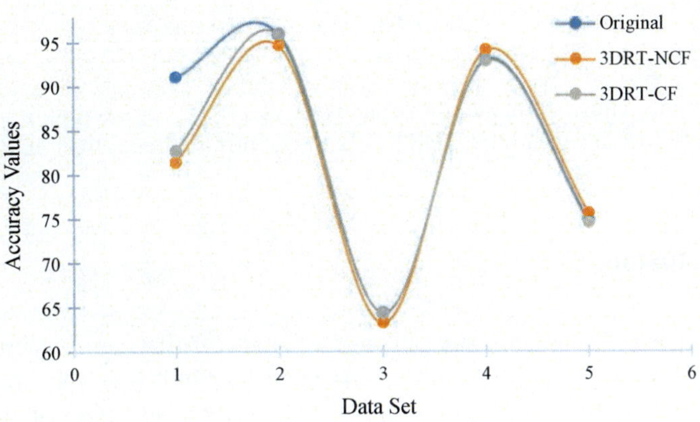

Fig. 3 Accuracy comparison

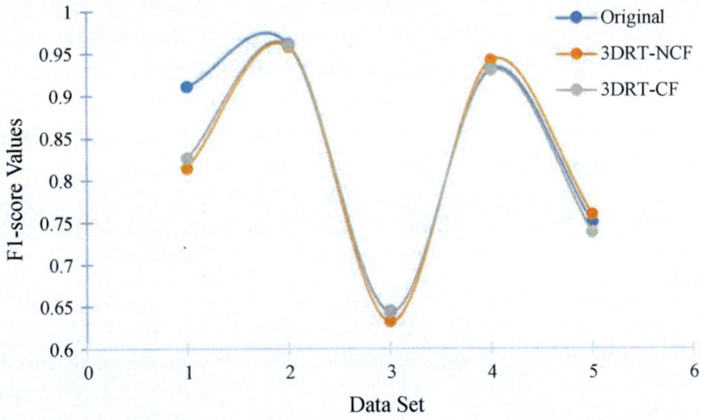

Fig. 4 $F1$-score comparison

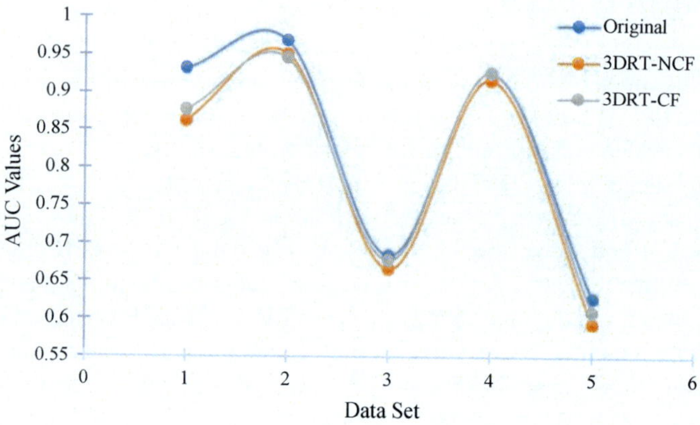

Fig. 5 AUC value comparison

are consistent. Thus, from the above analysis of privacy preservation and utility maintenance, it can be said that 3DRT-CF can perform far better than 3DRT-NCF.

6 Conclusion

In this paper, we used the correlation among the features of a data set to generate triplet (set of three features) for three-dimensional rotation perturbation technique instead of consecutive selection of features for triplet generation. Both aspects of PPDM, i.e., privacy and utility of the data, are considered while analyzing the performance. Five data sets having variations in the number of features and instances are used for experimental purposes. From the analysis of privacy and utility, it is observed that for the proposed triplet generation approach, three-dimensional rotation perturbation technique performs better than the existing triplet generation approach. Therefore, it would be a good choice to consider the correlation among the features while generating triplets. However, analyzing the effects of classifier ensembles on correlation-based triplet generation can be an extension of our work.

References

1. Salloum SA, Alshurideh M, Elnagar A, Shaalan K (2020) Mining in educational data: review and future directions. In: Advances in intelligent systems and computing. Springer International Publishing, Berlin, pp 92–102
2. Peña-Ayala A (2014) Educational data mining: a survey and a data mining-based analysis of recent works. Expert Syst Appl 41:1432–1462
3. Afrin A, Paul MK, Sattar AHMS (2019) Privacy preserving data mining using non-negative matrix factorization and singular value decomposition. In: 2019 4th International conference on electrical information and communication technology (EICT). IEEE, pp 1–6

4. Chamikara MAP, Bertok P, Liu D, Camtepe S, Khalil I (2020) Efficient privacy preservation of big data for accurate data mining. Inf Sci 527:420–443
5. Agrawal R, Srikant R (2000) Privacy-preserving data mining. In: Proceedings of the 2000 ACM SIGMOD international conference on management of data—SIGMOD'00. ACM Press, pp 439–450
6. Chen K, Liu L (2010) Geometric data perturbation for privacy preserving outsourced data mining. Knowl Inf Syst 29:657–695
7. Oliveira S, Zaiane O (2004) Data perturbation by rotation for privacy-preserving clustering. University of Alberta Libraries
8. Upadhyay S, Sharma C, Sharma P, Bharadwaj P, Seeja KR (2018) Privacy preserving data mining with 3-D rotation transformation. J King Saud Univ Comput Inf Sci 30:524–530
9. Stanley RMO, Osmar RZ (2010) Privacy preserving clustering by data transformation. J Inf Data Manage 1:37
10. Denham B, Pears R, Naeem MA (2020) Enhancing random projection with independent and cumulative additive noise for privacy-preserving data stream mining. Expert Syst Appl 152:113380
11. Chamikara MAP, Bertok P, Liu D, Camtepe S, Khalil I (2018) Efficient data perturbation for privacy preserving and accurate data stream mining. Pervasive Mob Comput 48:1–19
12. Chamikara MAP, Bertok P, Liu D, Camtepe S, Khalil I (2019) An efficient and scalable privacy preserving algorithm for big data and data streams. Comput Secur 87:101570
13. Singh D, Singh B (2020) Investigating the impact of data normalization on classification performance. Appl Soft Comput 97:105524
14. Xu S, Zhang J, Han D, Wang J (2006) Singular value decomposition based data distortion strategy for privacy protection. Knowl Inf Syst 10:383–397

Developing a Text Mining Framework to Analyze Cricket Match Commentary to Select Best Players

Ratul Roy, Md. Rashadur Rahman, M. Shamim Kaiser, and Mohammad Shamsul Arefin

Abstract Both the unpredictable nature of the game and the wide range of performance among players pose difficulties in selecting a cricket squad. When selecting players, it is important to consider their previous track record on the field. However, comparing performance indicators from previous games with those that are about to take place is far from a realistic approach. It was human opinion—live commentary (i.e., expert opinion)—that enabled us to make this a reality. Commentary is the best source of actual thoughts from veteran players at the time of any event because it is provided in real time. As a result, any knowledge gained from the commentary is beneficial to any player's overall performance metric. During our research, we established a framework for collecting actual commentary and analyzing it to determine performance indicators. For demonstrating the successful proposals from our framework, we have conducted many variants of testing. In our trial review, we discovered that our system could collect commentary and recommend the most likely best players for any forthcoming match with high efficiency.

Keywords Data mining · Commentary analysis · Performance · Team selection · Text analysis

1 Introduction

Cricket was brought to North America through the English colonies in the early seventeenth century and reached other areas of the world in the eighteenth century. AQ1

R. Roy · Md. R. Rahman · M. S. Arefin (✉)
Department of Computer Science Engineering, Chittagong University of Engineering and Technology, Chattogram 4349, Bangladesh
e-mail: sarefin@cuet.ac.bd

M. Shamim Kaiser
Wazed Miah Science Research Centre (WMSRC), Institute of Information Technology, and Applied Intelligence and Informatics (AII), Jahangirnagar University, Savar, Dhaka 1342, Bangladesh

M. S. Arefin et al. (eds.), *Proceedings of the International Conference on Big Data, IoT, and Machine Learning*, Lecture Notes on Data Engineering and Communications Technologies 95, https://doi.org/10.1007/978-981-16-6636-0_18

Over the years, cricket has become one of South East Asia's most popular games. The supporters are insane about each game and put pressure on the team squad selector.

The major aim of the commentators is to highlight the weakness and strength of each action of the players and to present a realistic image of what is being performed on the playground. Commentators are usually star players who are considered experts in the game. These views are most valuable with analysis of the match.

In this work, we have proposed a framework that takes account of a commentary that says something significant about any player and analyzes his/her performance value based on the final score that is generated. We are relying on expert opinions—sports commentary. This can be justified because a commentator is usually a veteran sportsman giving us his/her two cents on everything happening on the field. So, through his lens of experience, we get an insightful analysis of the match, venue, teams and most importantly—the players. So if we can build a framework that can analyze these opinions, we would get a sense of a players' current ability on the field. A steady perforce streak invariably indicates the player's ability to hold against his opponents.

The remaining sections are divided as follows: Sect. 2 contains related work; Sect. 3 contains methodology; Sect. 4 contains the results of the experiments, and Sect. 5 contains a conclusion of the study.

2 Related Work

There are works on how we can analyze a sentence with a parsimonious analyzer, named VADER [6]. It uses rule-based sentiment analysis. In this model, researchers combined qualitative and quantitative methods to produce a gold standard sentiment lexicon, which was later empirically validated against especially receptive micro-blog-type contexts. VADER combines important lexical features obtained from five generalized rules that embody grammatical and syntactical conventions of human speech. It also retains the advantages of conventional sentiment lexicons such as LIWC [12, 13]. San Vicente and Saralegi [14] explored three strategies to build polarity lexicons: interpreting existing lexicons from other languages, explaining sentiments from lexical knowledge bases and extracting polarity lexicons from corpora.

There are works on feedback analysis using secure frameworks [5]. It describes how a sentence can be awarded a valence-based rating and without disclosing the source and rule-based sentiment analysis procedures in detail.

There are quite a few rules by which we can distinguish different cricket events from commentary. Arefin et al. [1] have demonstrated how we can link event mentions to match reports.

Zhang et al. [17] have done extensive work in analyzing commentary. Their work is based on football commentary, and their target is not player performance but generating a comprehensive match summary that can be published as a news article.

They have utilized learning to rank (LTR)-based supervised sentence extraction that has the ability to leverage task-dependent features [15].

Arif et al. [2] have proposed a bowler's gaming prowess can be determined from related commentary. Bowler's strengths and limitations that can be determined from textual data are also applicable when deciding his chance in the team for upcoming matches.

In sports commentary, intrinsic data of game aspects like scores, performance metrics, a batsman's strike rate, a bowler's given run and wickets per innings are present in [8]. Overall valuable game data can be found in commentary text. They additionally give a double banalization strategy effortlessly portioning text in the key subtitles.

The work done in [9] considered net contribution against the individual impact on the match because the latter is incomparable for different categories of players. His method enables comparison of different categories of players, i.e., batsmen and bowlers to be compared objectively. Johnston et al. [7] suggested a performance evaluation framework using Clarke's dynamic programming model in one-day cricket. Clarke [3] expanded on the author's earlier work is that the Duckworth/Lewis model [4] used to evaluate the impact a player makes in contrast with that expected for the innings.

Uma Maheswari and Rajaram [16] established a computationally sound framework to compress existing data that is easier to mine. This new principal component analysis-based association rules mining produces frequent patterns that compress the database and increases the efficiency of statistical analysis on cricket data.

Passi and Pandey [11] propose a machine learning-based framework that estimates the performance of a player. Multiclass SVM, Naive Bayes, decision tree and the random forest are used in this study. They have leveraged supervised machine learning algorithms to generalize a performance prediction for any player in an arbitrary match, whereas Muthuswamy and Lam [10] used neural networks to estimate a bowler's wicket-taking ability, but their approach was narrowed down to eight Indian bowlers.

3 Methodology

In our framework, there are four different modules. The crawling module crawls commentary, and our storage module stores them. The filtering module processes commentary before ranking. The ranking module ranks players based on a cumulative score obtained from the ranking module units. The overall system architecture is depicted in Fig. 1.

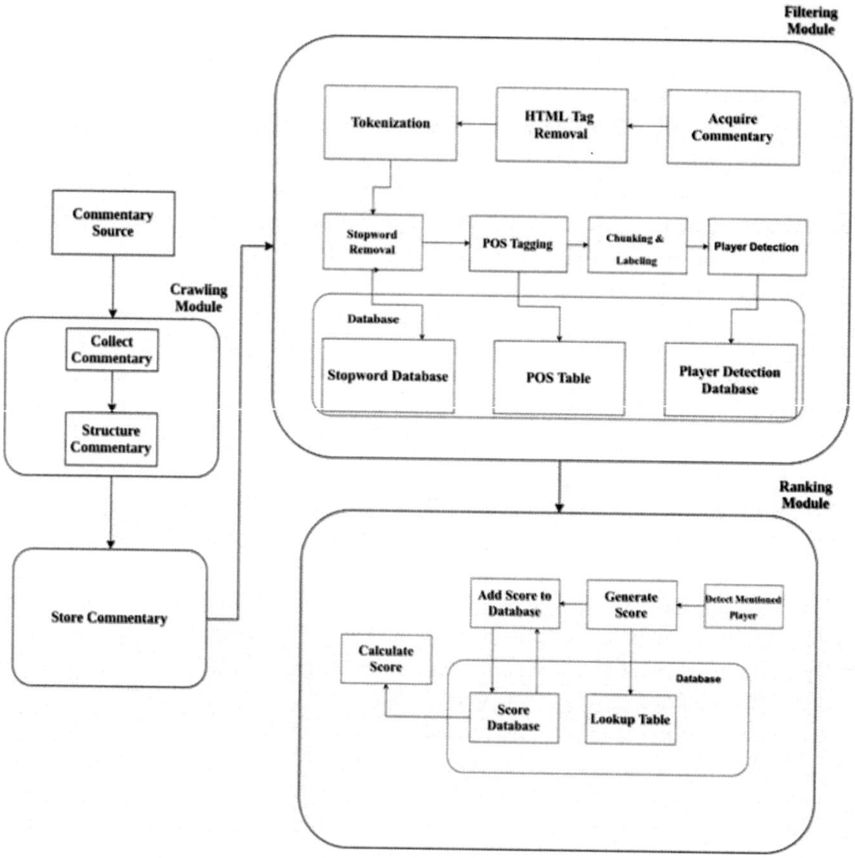

Fig. 1 System architecture

The crawling module collects commentary from the ESPNCricinfo Web site. We have crawled the Web site for commentaries (see Algorithm 1). The Web site has live commentary available match by match, ball by ball. We collect them all and store categorically innings by innings.

HTML Tags removal Crawled commentary is usually riddled with markup keywords. These are removed with the help of the beautiful soup library, after which we get written commentary text.

Tokenization At first, we split a commentary into a a number of sentences and then into words. We split up a complete sentence into small tokens of single or more words for next section.

Algorithm 1: Crawl commentary from source

```
Data: Commentaries associated with matches and innings
initialization
for each unique team id do
    Generate API string for a match
    for each innings in a match do
        Generate API for the innings
        Get all commentary for that innings of that match
        Store commentary data in JSON file
    end
end
```

Parts of Speech (POS) Tagging Then we tag each word of a sentence into which parts of speech they are. POS tagging helps us define which words we would use to identify players. Because a commentator may or may not use a player's full name.

Detecting a Person We detect a person in a sentence by their POS tag. Usually, POS-NNP (Proper Noun) points to a person. When a person is detected in a sentence, we can analysis it as a usable commentary. Figure 1 represents the filtering module.

The ranking module takes the filtered commentary and ranks players based on them. The players and their names and variations of names are stored in the database beforehand. This module finds the commentaries in which the players are located, analyzes the commentary and gives each of the commentaries a valence-based rating from the VADER module.

VADER is a combined module of qualitative and quantitative methods to produce a gold standard sentiment lexicon, which was later empirically validated against especially receptive micro-blog-type contexts. VADER combines important lexical features obtained from five generalized rules that embody grammatical and syntactical conventions of human speech. It also retains the advantages of conventional sentiment lexicons such as LIWC [13] and [12] . San Vicente and Saralegi [14] explored three strategies to build polarity lexicons: interpreting existing lexicons from other languages, explaining sentiments from lexical knowledge bases and extracting polarity lexicons from corpora.

Preprocessing: In preprocessing step, input English text is tokenized. At first, words, emoticons and all capitalized words are tokenized. Later, words and punctuation marks are tokenized. Some punctuation marks affect the valence of words, which is kept with the word.

Boosting: Once tokenization is complete all the tokens are checked for valence boosting purposes. VADER uses a bi-gram and trigram model for boosting. If a boosting word such as "extremely, very, great" is found, then the valence of the word is boosted. Then all capital words are considered, and the valence is further boosted. The text is also checked for idioms and phrases, if found then the valence is boosted again. Later, it is looking for the "but" word, if found then the sentence is divided into two parts, and valence is calculated on two different parts. Next, overall valence is calculated for such a sentence.

Valence Calculation: In this step, the valence score of a sentence is measured, which is between −4 and +4. This calculated value is further normalized to range between −1 and +1. In this way, every sentence is assigned with its respective polarity.

Lexicon modification for VADER A lexicon is the vocabulary of a man, dialect or department of expertise that stocks the lexemes in that linguistics. Polarity lexicons are those that have a listing of words with an initial level of polarities. AQ2

Here, we have used the lexicon as our lookup table. A summarized view of data flow of the ranking module is shown in Fig. 1.

Sentiment ratings from ten independent human raters (all pre-screened, trained and quality checked for optimal inter-rater reliability). Over 9000 token features were rated on a scale from"(−4) Extremely Negative" to "(4) Extremely Positive," with allowance for "(0) Neutral (or Neither, N/A)".

For subjective cricket usage, we have added cricket-ish words like "out" or "wicket" in the lexicon to generate a more accurate result.

Algorithm 2: Rank the players

```
Data: Commentary, Player list
initialization
for each sentence do
    Detect players for each sentence
    if Player is found then
        Analyze the sentence and get a score
        Put the score in the database
    end
end
Calculate corresponding value of each player from database
```

In Fig. 2 diagram, we can see flow of the whole framework across all modules.

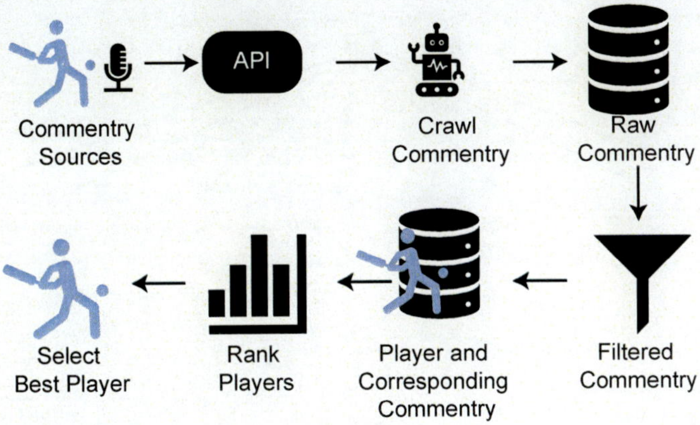

Fig. 2 Flow diagram of commentary analysis framework

4 Experiments

Commentaries are live discussions of what is happening on the field. So, primary source of any commentary would be a live broadcast. But that is outside of the scope of our work. We would like to limit our efforts to analyzing the commentary not extracting it, so a static source is what we have opted for here.

4.1 Crawling Commentary

Usually, commentaries should/would come with well-documented APIs. There are a few services like sportsmonk. But not all of them are ideal. To keep things less complicated, we will crawl the commentary that exists on the Cricinfo Web site.

After that, we detect whether a person is in a commentary. We do it using named entity tagging and chunking. We detect a person in a sentence by their POS. Usually, POS-NNP (Proper Noun) points to a person. When a person is detected, we choose it as a usable commentary. For example, the sentence "Soumya and Tamim are headed out to the middle as are the South African team." yields the result in Fig. 3.

4.2 Ranking Players

The ranking module takes the filtered commentary and ranks players based on them. The players and their names and variations of names are stored in the database beforehand. This module finds the commentaries in which the players are located,

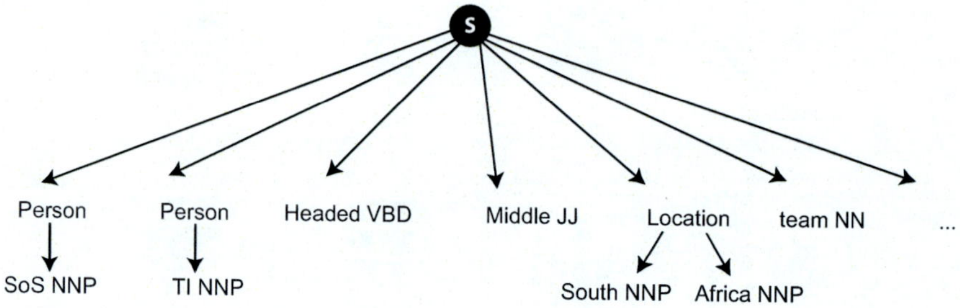

Fig. 3 Example of person detection

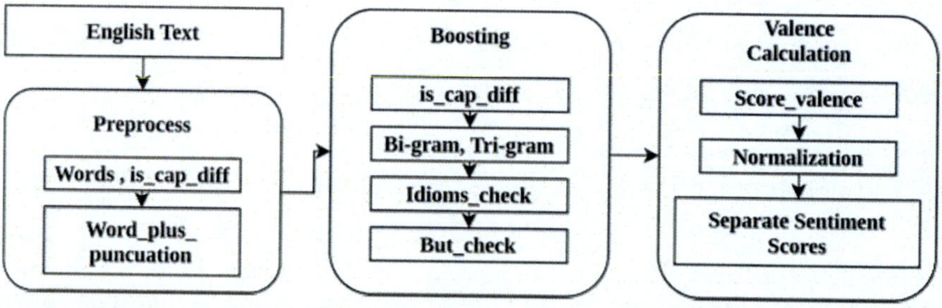

Fig. 4 Workflow of VADER module

analyzes the commentary and gives each of the commentaries a valence-based rating from the VADER module (see Fig. 4).

4.2.1 Lexicon Modification

We have already discussed how we can build polarity lexicons. Here, we have used the lexicon as our lookup table. For specific cricket usages, we have added cricket words like "out" or "wicket" in the lexicon to generate a more appropriate result. Here are a few examples of commentary scoring (Fig. 6).

4.3 Score Generation

We have processed total *15968* commentaries to examine our proposal. After analyzing the complete data set, it is visible that only a few of the commentary portions contain inherently usable commentary.

The external module, VADER sentiment analyzer, analyzes each sentence. The modified VADER is given sports-related words and their respective scores as mentioned in [6]. These words give us the ability to analyze sports-related comments and give them a valence score.

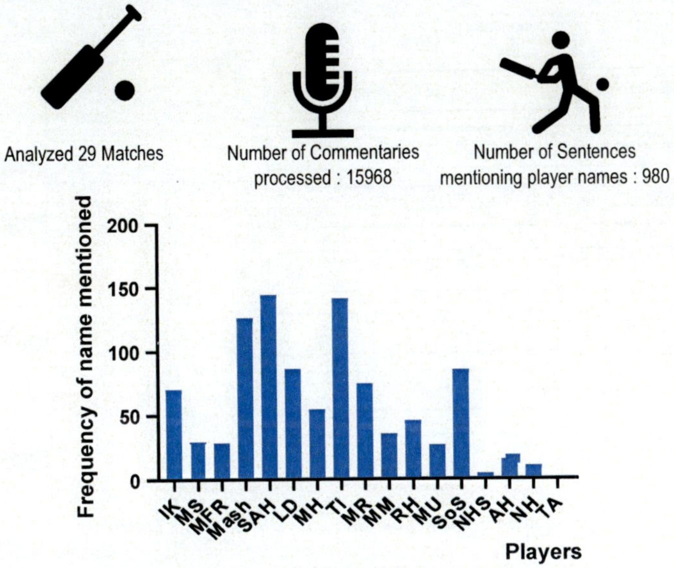

Fig. 5 Frequency of the name mentioned by different players during analysis

Then, we take the sentences that have a player's name and calculate a score for each player. We have separated those and put it through the filtering module. Then the players are ranked. For example, this is a summary of the analysis for 29 match—number of **Commentaries** processed: 15968; number of **Sentences** mentioning player names: 980; players and their names mentioned in sentences: IK-70, MS-30, MFR-28, Mash-126, SAH-144, LD-86, MH-54, TI-141, MR-74, MM-35, RH-45, MU-26, SoS-85, NHS-4, NH-10 and TA-7 times (Fig. 5).

Figure 6 shows the commentary, player and their score after analyzing ten-match commentary analysis.

4.4 Result Comparison

From the comparison of results, we can see a visibly mention worthy outcome. Table 1 is a comparison table for the actual match squad selected for the Bangladesh versus England match. The squad here is suggested only taking into account five matches. The result seems pretty good while we consider only five matches. But it still lacks quality confirmation.

Table 2 is a comparison table for the actual match squad selected for the Bangladesh versus England match. We have arrived at this suggestion for a squad after adjacent ten matches.

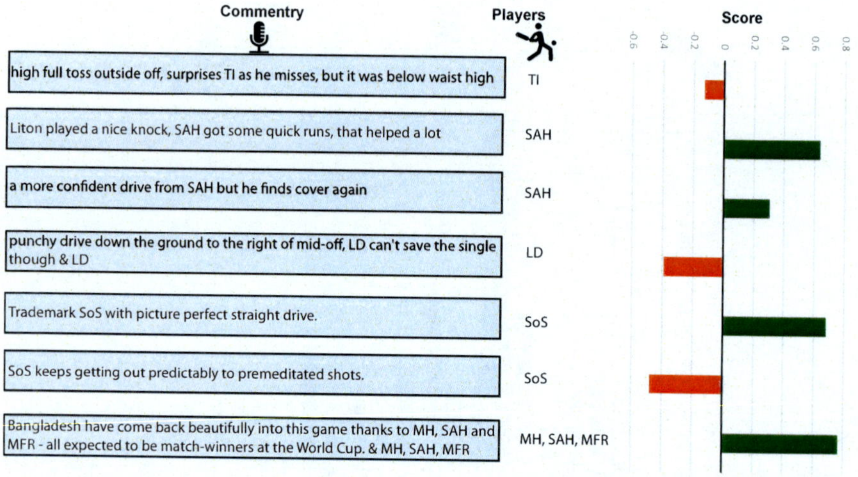

Fig. 6 Frequency of the name mentioned by different players during analysis

Table 1 Squad comparison from five-match analysis

Predicted squad	Actual squad
Mash	Mash
MH	MH
NHS	**SoS**
MS	MS
MU	MU
MM	MM
TI	TI
MFR	MFR
LD	LD
MR	MR
SAH	SAH

Table 3 is a comparison table for the actual match squad selected for the Bangladesh versus England match. The squad here is the result of analyzing adjacent 20 matches.

Table 4 is a comparison table for the actual match squad selected for the Bangladesh versus England match. The squad here is suggested by considering adjacent 29 matches.

As shown in Fig. 7, we have used simple percentage accuracy calculations. From this chart, it is evident that when we use more match data the predicted squad becomes much closer to the actual squad. We have already showed which player mismatches with the final squad, from this comparative chart, the difference and effect of given data volume are much clearer.

Table 2 Squad comparison from ten-match analysis

Predicted squad	Actual squad
Mash	Mash
MH	MH
SoS	SoS
MS	MS
MU	MU
MM	MM
TI	TI
RH	**MFR**
LD	LD
Najmul Hossain Shanto	**MR**
SAH	SAH

Table 3 Squad comparison from 20-match analysis

Predicted squad	Actual squad
Mash	Mash
MH	MH
SoS	SoS
MS	MS
MU	MU
MM	MM
TI	TI
RH	**MFR**
LD	LD
IK	**MR**
SAH	SAH

Table 4 Squad comparison from 29-match analysis

Predicted squad	Actual squad
Mash	Mash
MH	MH
SoS	SoS
MS	MS
IK	**MU**
MM	MM
TI	TI
MFR	MFR
LD	LD
MR	MR
SAH	SAH

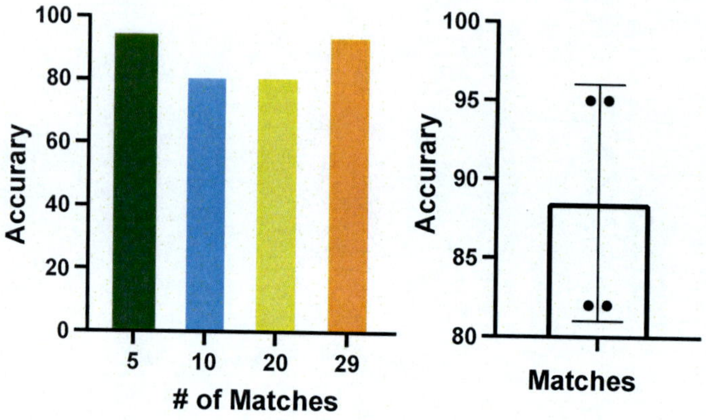

Fig. 7 Accuracy comparison (in percentages) and the variance

5 Conclusion

We focused our efforts on commentary analysis and the strongest team selection. For this purpose, we have gathered raw commentary from many web sources, filtered it and then evaluated players based on the discussion that has been written about them. This provided a small window into how the players are currently performing and how they are most likely to perform in the future. Then, for the following match, the eleven players who had the greatest impact on the game are suggested. Due to the fact that we had previously worked with a few matches, we had the opportunity to test our suggestions against the picked squad. It is simply a matter of adding a new match ID number to the existing list in order to create new matches. Then, when the code is executed, the new remark will be automatically added to the account while the player rating is being calculated. Only textual content that has been made accessible as commentary has been taken into consideration. A combination of the textual reaction of an experienced commentator with additional hard fact values such as scores, the scenario, the number of overs remaining and so on could be used to improve this theory even further in the future. Incorporating commentary from other sources will also help to improve it.

References

1. Arefin MS, Mukta RBM, Morimoto Y (2014) Agent-based privacy aware feedback system. In: International conference on advanced data mining and applications. Springer, Berlin, pp 725–738
2. Arif S, Umair M, Naqvi SMK, Ikram A, Ikram A (2018) Detection of bowler's strong and weak area in cricket through commentary. In: Proceedings of the 2nd international conference on future networks and distributed systems, pp 1–14

3. Clarke SR (1988) Dynamic programming in one-day cricket-optimal scoring rates. J Oper Res Soc 39(4):331–337
4. Duckworth FC, Lewis AJ (1998) A fair method for resetting the target in interrupted one-day cricket matches. J Oper Res Soc 49(3):220–227
5. Gupta M (2015) Cricket linking: linking event mentions from cricket match reports to ball entities in commentaries. In: Proceedings of the 38th international ACM SIGIR conference on research and development in information retrieval, pp 1033–1034
6. Hutto C, Gilbert E (2014) Vader: a parsimonious rule-based model for sentiment analysis of social media text. In: Proceedings of the international AAAI conference on web and social media, vol 8
7. Johnston MI, Clarke SR, Noble DH et al (1993) Assessing player performance in one-day cricket using dynamic programming. World Scientific
8. Jung C, Lee SY, Kim J (2008) Robust detection of key captions for sports video understanding. In: 2008 15th IEEE International conference on image processing. IEEE, pp 2520–2523
9. Lewis A (2005) Towards fairer measures of player performance in one-day cricket. J Oper Res Soc 56(7):804–815
10. Muthuswamy S, Lam SS (2008) Bowler performance prediction for one-day international cricket using neural networks. In: IIE Annual conference. Proceedings. Institute of Industrial and Systems Engineers (IISE), p 1391
11. Passi K, Pandey N (2018) Increased prediction accuracy in the game of cricket using machine learning. arXiv preprint arXiv:1804.04226
12. Pennebaker JW, Booth RJ, Francis ME (2007) Linguistic inquiry and word count: LIWC [Computer software]. Austin, TX: liwc.net 135
13. Pennebaker JW, Francis ME, Booth RJ (2001) Linguistic inquiry and word count: LIWC 2001. Mahwah: Lawrence Erlbaum Associates 71
14. San Vicente I, Saralegi X (2016) Polarity lexicon building: to what extent is the manual effort worth? In: Proceedings of the tenth international conference on language resources and evaluation (LREC'16), pp 938–942
15. Shen C, Li T (2011) Learning to rank for query-focused multi-document summarization. In: 2011 IEEE 11th International conference on data mining. IEEE, pp 626–634
16. Uma Maheswari P, Rajaram M (2009) A novel approach for mining association rules on sports data using principal component analysis: for cricket match perspective. In: 2009 IEEE International advance computing conference. IEEE, pp 1074–1080
17. Zhang J, Yao JG, Wan X (2016) Towards constructing sports news from live text commentary. In: Proceedings of the 54th annual meeting of the association for computational linguistics. Volume 1: Long papers, pp 1361–1371

Indexed Top-k Dominating Queries on Highly Incomplete Data

H. M. Abdul Fattah, K. M. Azharul Hasan, and Tatsuo Tsuji

Abstract Top-k dominating (TKD) query returns the top-k data items in a dataset that dominate other objects. This is an important decision-making tool for any business organizations because it gives data analysts an insightful way to find dominating objects. It incorporates the benefits of skyline and top-k queries and is used in a variety of decision-making applications. Due to system malfunction, privacy protection, data loss, and other factors, incomplete data occurs in a wide range of actual applications. We design an algorithm for answering the top-k dominating queries applying the idea of data bucketing. Each of the buckets is implemented with a B+ tree. A key is generated for each of the incomplete data record and inserted into the B+ tree. Using the key, we reduce the computation cost for comparison applying the B+ index structure. Since the B+ tree is well accepted for efficient indexing for commercial databases, we get the facility of high retrieval performance. The proposed approach clearly outperforms in terms of query performance compared to many other existing approaches for incomplete data.

Keywords Top-k query · Top-k dominating query · Skyline query · Incomplete data · B+ tree · Dominance relationship

---

H. M. Abdul Fattah (✉) · K. M. Azharul Hasan
Department of Computer Science and Engineering, Khulna University of Engineering and Technology, Khulna 9203, Bangladesh
e-mail: hussainfattah@cse.kuet.ac.bd

K. M. Azharul Hasan
e-mail: az@cse.kuet.ac.bd

T. Tsuji
Faculty of System Design Engineering, University of Fukui, Fukui, Japan
e-mail: tsuji@pear.fuis.fukui-u.ac.jp

M. S. Arefin et al. (eds.), *Proceedings of the International Conference on Big Data, IoT, and Machine Learning*, Lecture Notes on Data Engineering and Communications Technologies 95, https://doi.org/10.1007/978-981-16-6636-0_19

1 Introduction

Top-k dominating queries rank records by the number of records in a dataset S that are dominated by any record o and return k records that dominate the most records in the set S. It selects the query's dominance using a ranking function. An evident advantage of dominating query is that it controls the selectivity of the query output with the parameter k and a ranking function [1, 2]. It combines the facilities of skyline query and top-k query [3]. Due to system malfunction, privacy protection, data loss, data unavailability and many other factors, incomplete data is very common in actual applications such as sensor networks, decision making and location-based services [4]. In an incomplete dataset, many records have attribute values missing in some dimensions and cannot be predicted. Researchers have recently become very interested in querying such incomplete data that has missing values [5]. On incomplete records, Top-k dominating (TKD) queries have some advantages, i.e., its performance is governed by a parameter k, making it insensitive to the size of incomplete datasets in various dimensions. Let o and o' be two records having n attributes. If all of the following conditions are true, the record o dominates the record o', denoted as $o \prec o'$ [1]:

- For each attribute $i(0 \le i \le (n-1))$, $o[i]$ is no smaller than $o'[i]$ or there is a missing attribute.
- When both records have values in attribute $j(0 \le j \le (n-1))$, i.e., attributes values of $o[j]$ and $o'[j]$ are not missing, and then $o[j]$ is greater than $o'[j]$ for at least one attribute.

For example, consider the movie review dataset S in Table 1 where $S = \{m1, m2, m3, m4\}$. Movie $m2$ dominates movie $m3$. Since $m2$ is greater than $m3$ for the common attribute values $a2$ and $a3$, i.e., $m2[2] > m3[2]$ and $m2[3] > m3[3]$ and $m2$ are no less than $m3$ in any of the dimensions. The score of $m2$ is 2 since it dominates $\{m1, m3\}$. Similarly, score of $m1$, $m3$ and $m4$ are all 0. Since $m2$ has the best score, it is the the output for top-1 dominating query (T1D).

For large and incomplete dataset, bucketing is an efficient technique to answer the TKD queries [1]. In this paper, we apply efficient bucketing by applying B+ tree index structure. If there are d distinct buckets, then d B+ trees are introduced called bucket trees. Using the bucket trees, a dominating B+ tree is constructed to find the

Table 1 Sample of movie review dataset

Movie ID	Ratings				
	a1	a2	a3	a4	a5
$m1$	–	–	3	4	2
$m2$	5	3	4	–	–
$m3$	–	2	1	5	3
$m4$	3	1	5	3	4

TKD queries. We create keys for each of the buckets from the incomplete dataset, and this key is inserted into the B+ tree. The proposed computational model is not only suitable for top-k dominating queries but also other database applications such as skyline computations [6], database systems [7] and recommender system [4].

2 Related Works

There are many reasons for generating large number of incomplete data that is why query processing over incomplete data has received much attention from the database community. The brute force method [5] that uses extensive comparisons is used to find the TKD from incomplete dataset. But due to large number of comparisons, the method is not efficient when the size of the dataset and target data is large. The skyline-based and upper bound-based techniques are used to reduce the candidate set in a TKD [1]. The approach uses the upper bound score pruning technique to reduce the target data. As a variant of skyline queries [7], the top-k dominating queries applying the nearest neighbor search on the standard complete dataset using R-tree indexing are proposed. To improve performance [5], two approaches for dealing with the TKD problem based on the aR-tree are suggested. A systematic investigation of TKD queries on incomplete data is described in [1] which includes missing dimensional values in the dataset. To improve query performance, they use some techniques including partial score pruning, bitmap pruning and upper bound score pruning. For processing k-skyband queries on incomplete data [8], two efficient algorithms for processing k-skyband (kSB) query, namely VP algorithm and kISB algorithm, are presented. Their kISB algorithm exploits the intrinsic characteristic of the k-skyband query on incomplete data and outperforms baseline algorithm and VP algorithm. The use of virtual points in VP algorithm incurs extra costs in the query processing. With the difficulty of dealing with missing information in datasets [9], introduce a method to compute the skyline using crowd enabled databases. A decentralized algorithm is proposed to address the problem of distributed top-k dominating query processing in [10] using space filling curves for complete dataset. Ding et al. [11] also implemented top-k dominating query processing on incomplete data in distributed environments. In [12], a top-k dominating query model is presented over uncertain data where pruning techniques are proposed by utilizing the spatial indexing and statistic information. A balanced dominating top-k query semantic and algorithms to identify the top-k answers are proposed in [13]. A dynamic dominate model to get the largest number of uncertain objects is proposed in [14] from uncertain objects to reduce the search space to answer the queries by a pruning approach. The dynamic environment is also handled in [15] for incomplete data by event-based method to answer top-k dominating query. A parallel approach for processing TKD queries on incomplete data using MapReduce is proposed in [16]. A probabilistic model for query processing is presented in [17] and formulated several types of ranking queries on the model based on partial orders. In this paper, we introduce B+ tree-based indexing to compute the TKD. We generate keys and used the keys in the B+ tree as container of the generated keys. We reduce the computation cost for extensive comparison [1, 5] in our proposed method. Therefore, it shows good retrieval performance.

3 Top-k Dominating Query Computation Model for Incomplete Dataset

The database community has put forward a lot of effort in response to incomplete data. For example, in a real movie review dataset, it is common that some user's ratings are missing, since users prefer to only rate movies that they are familiar with. Therefore, each film rating is a n-tuple with some tuple values are missing. Table 1 shows some tuples having empty values. The fact might be that the reviewer a2 watches the movies m2, m3 and m4 but not the movie m1. Therefore, a2 only rates movies m2, m3 and m4. The missing values are denoted by the symbol '-'. As a result, the collection of film ratings dataset is an incomplete dataset. This incompleteness increases with increase of number of movies and users for practical situation. Hence, TKD on incomplete data is an important research problem for database researchers [18].

The overall functional block diagram of our proposed model is illustrated in Fig. 1. We apply the bucketing of data based on the symbol '–', and then we generate key for each of the records in the bucket. Using the key, we construct B+ trees for each of the buckets, and taking the top items from the bucket tree, we construct one more B+ tree that holds the dominating objects. From the dominating objects, we compute the scores by brute force comparison [1] with each other. Therefore, the candidate records for comparison reduce significantly, and TKD processing becomes fast in our algorithm.

3.1 Data Bucketing

The process of data bucketing is illustrated in Fig. 2. The objects are first grouped into buckets based on their bucket_id. To get bucket_id, missing dimensional rating values are represented as 0, and present dimensional rating values are represented as 1. When the first object B3 in Fig. 2 is evaluated, a bucket is generated corresponding to the bucket_id, i.e., $id_{B3} = 0011$, and B3 is the first object enrolled. For the sample dataset in Fig. 2, four unique buckets are formed, each containing five items.

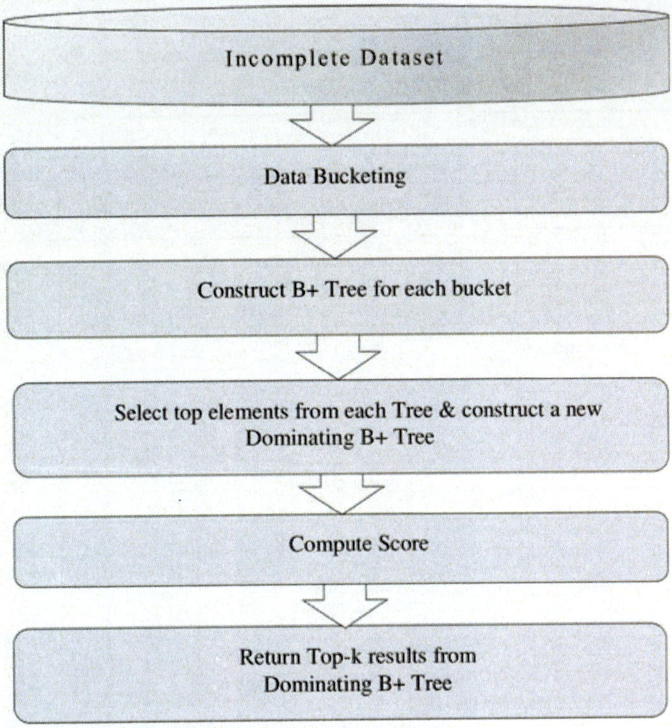

Fig. 1 Overall functional block diagram of proposed model

3.2 *Bucketing Using B+ Tree*

Definition 1 *(Bucket Tree)*: Bucket tree (BT) is a structure of B+ tree where each of the distinct keys is stored in sorted order of the leaf node of the tree. Each distinct keys of the BTs is composite keys of the form $< v, id >$.

B+ tree is a B tree extension that allows faster addition, deletion and search operations. Records can only be stored on leaf nodes in a B+ tree, while key values can only be stored on internal nodes. To make search queries more effective, the leaf nodes of a B+ tree are linked together in the form of singly linked lists called sequence set, and the non-leaf nodes, called index set, acts as an index to reach to the goal. One more important property of B+ tree is that the keys are sorted in sequence set. In comparison with B tree, the tree's height remains balanced and is lower as far as searching is concerned. The data stored in a B+ tree can be accessed both sequentially and directly. If the height of the tree is h, it needs to search the height only once then sequential search is performed to find the target record since the data is stored on the leaf nodes in sorted order sequentially. Therefore, queries are faster. Due to these advantages, we chose B+ tree in our model. A sample B+ tree model is shown in Fig. 3.

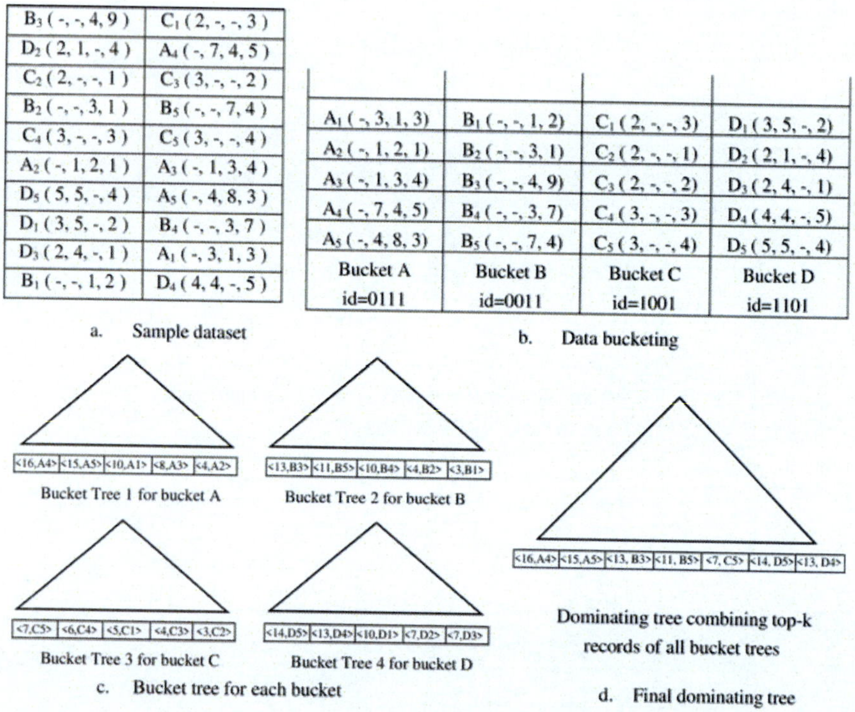

Fig. 2 Overall functional block diagram of proposed model

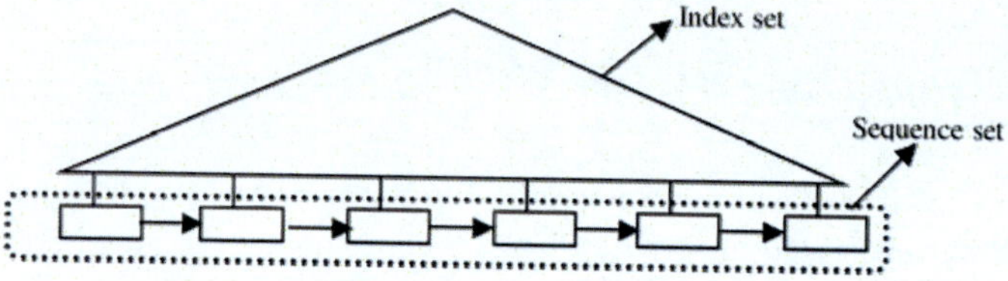

Fig. 3 Logical separation of B+ tree in index set and sequence set

Using the partitioning data (Fig. 2b), keys are generated for bucketed records, and these keys are unique and inserted to a BT.

3.3 Key Generation

We generate a composite key of the form $< v, id >$ where v is the score and id is the identification number of the data record. We use a b bit operating system of which higher b_1 bits are used for v and lower b_2 bits are used for id where $b = b_1 + b_2$.

Therefore, the keys are stored in the tree as an order of v. For example, for two keys $k1$ and $k2$, $k1 > k2$ if $v1 > v2$. If the id values are unique, then every key is unique.

For each record in a bucket, the v part of the key is the sum of the data present in the record, for example, review score of users of the movie dataset. For large value of v, it is more likely to be dominating. For example, in bucket B of Fig. 2b, B1 has the score 3 (1+2) and B3 has the score 13 (4+9). Therefore, B3 is more likely to be dominating over B1. From this bucket trees, we retrieve more likely dominating objects.

3.4 Dominance Computation

Definition 2 *(Dominating Tree):* Taking the top-k elements of each bucket tree (BT), we create a new B+ tree called dominating tree (DT) where we insert the top-k elements of each bucket trees. Since the keys are unique, it will be sorted in the order of v (i.e., score) in the DT. These top elements of each BT are the candidates for top-k dominating queries.

To trim the number of records, the idea is to generate a new B+ tree called dominating tree by combining top records of each bucket. Since top elements of BT hold the records that are more likely to dominate other records, we select top elements from each BT and construct the DT. For a dataset of d distinct buckets, in our model there are d B+ trees and one B+ tree for dominating tree. Therefore, our model contains $d + 1$ B+ trees. Hence, the DT contains $m = d \times k$ key values. Now we calculate the dominance score by comparing the m objects only by a naive or brute force method such as [1]. If the original dataset contains N records, we reduce the search items from N to m where $m << N$. This reduces the unnecessary records to compute dominance that eventually makes our method faster.

Example 1 We use publicly available MovieLens dataset [19, 20]. The dataset contains unique ID for each movie and ratings of the users in the range of [1, 5]. We use movie_id as id and sum of the ratings provided by the users as v (See definition 1). The maximum movie_id is of 6 digit integer number. Therefore, the largest 6 digit decimal number can be represented using 20 bits. From Fig. 2, let us consider record B3 (-, -, 4, 9) having the unique movie_id 200003. We first make the sum of ratings, which is equal to 13 for B3. For a 64-bit operating system, the higher 44 bits are used to represent total sum of ratings (v of the composite key), and lower 20 bits are used to represent the movie ID (id of the composite key). This constructs the composite key of a record of the incomplete dataset. Therefore, the keys are sorted according to the rating values in the index set (see Fig. 3) . For each bucket, we construct a BT (see definition 1) in descending order of keys in the index set. A key having greater value indicates greater sum of ratings. Greater sum of ratings indicates the record is more likely to dominate other records.

4 Performance Analysis

4.1 Dataset

For performance evaluation, we have used the MovieLens dataset publicly available from a movie recommender framework (https://www.imdb.com/) [19, 20]. MovieLens is a collection of films with audience ratings, where each film is depicted to an audience rating in the range of [1, 5]. A higher rating usually implies a higher level of appreciation. The dataset contains 9950 movie records (i.e., rows) and 757 reviewers (i.e., columns). The dataset has a 95% missing values, meaning that only 5% of the ratings are available. Therefore, we say highly incomplete data.

4.2 Results

In this section, we evaluate the performance of our proposed algorithms for TKD queries over incomplete data and verify the effectiveness of our model. All the algorithms are written in Python, and all tested on a machine with an Intel Core i5 Duo 2.6GHz processor and 4GB of RAM having Microsoft Windows 10 Professional Edition. This section examines the success of the algorithms on real data and illustrates the significance of the top-k dominating records. The algorithms that we considered for comparison are shown in Table 2.

In our experiments, we look at a number of variables, including k, data size N, missing rate ρ and attribute cardinality c. Table 3 summarizes the settings for all of these parameters.

In addition to ESB and our proposed B+ tree-based (BTB) algorithm, the intuitive Naive approach (exhaustive pairwise comparisons) is also implemented as the baseline for TKD queries on incomplete data. Figure 4 shows the TKD performance for various values of k for varying number of data size, N. We experimented with different dataset sizes to see how well our model performed. We observed that our proposed BTB algorithm outperforms the EPC [1] and ESB [6] model. From Fig. 4, we see that the proposed model shows improved results for BPC and ESB. The reasons are as follows:

Table 2 Description of the algorithms to which compared

Name	Description
EPC [1]	Exhaustive pairwise computation
ESB [6]	Extended skyband-based algorithm
BTB (proposed method)	B+ tree-based algorithm

Table 3 Parameters for the prototype system

Parameter	Description	Values
k	Number of results for TKD	1, 4, 8, 16, 32, 64
N	Number of records in each dataset	200, 650, 885, 3000, 7500 K
ρ	The percentage of missing rate of attribute	95% (approximately)
c	Number of attribute for each dataset	100, 100, 440, 440, 757

- Applying the B+ tree index structure gives the indexing facility that facilitates for faster data retrieval. The proposed scheme is two-level TKD system. First it takes top-k from each bucket with BT, and then it generates DT which combines top-k from each buckets. The search space is reduced significantly in DT. Since the keys in DT are compared pairwise instead of large N.
- The comparison is done with integer keys. Because of this key generation, the comparison time is quite short.
- The generated keys take small space with respect to original records. Hence, it is easy to manage.

5 Conclusion

Considering the broad variety of uses for top-k dominating (TKD) queries, as well as the prevalence of incomplete data, in this paper, we look at the problem of running a TKD query on incomplete data with missing dimensional values. First we applied the Naïve approach (exhaustive pairwise comparison) which is inefficient. To effectively solve this, we used the ESB algorithm, which prunes the search space using a local skyband technique. Despite the fact that the pairwise comparison model, the skyline-based algorithm and other models are applicable to the problem, their performance is low. We present our proposed BTB algorithm, which uses a B+ tree data structure to further minimize the cost of score computation. Using our algorithm, we significantly reduced the search space by predicting top-k dominating query result. One important direction of the scheme is to apply parallel computation for multiprocessor environment. This is because, the buckets are independent to each other and can be applied to parallel environment easily for handling large dataset.

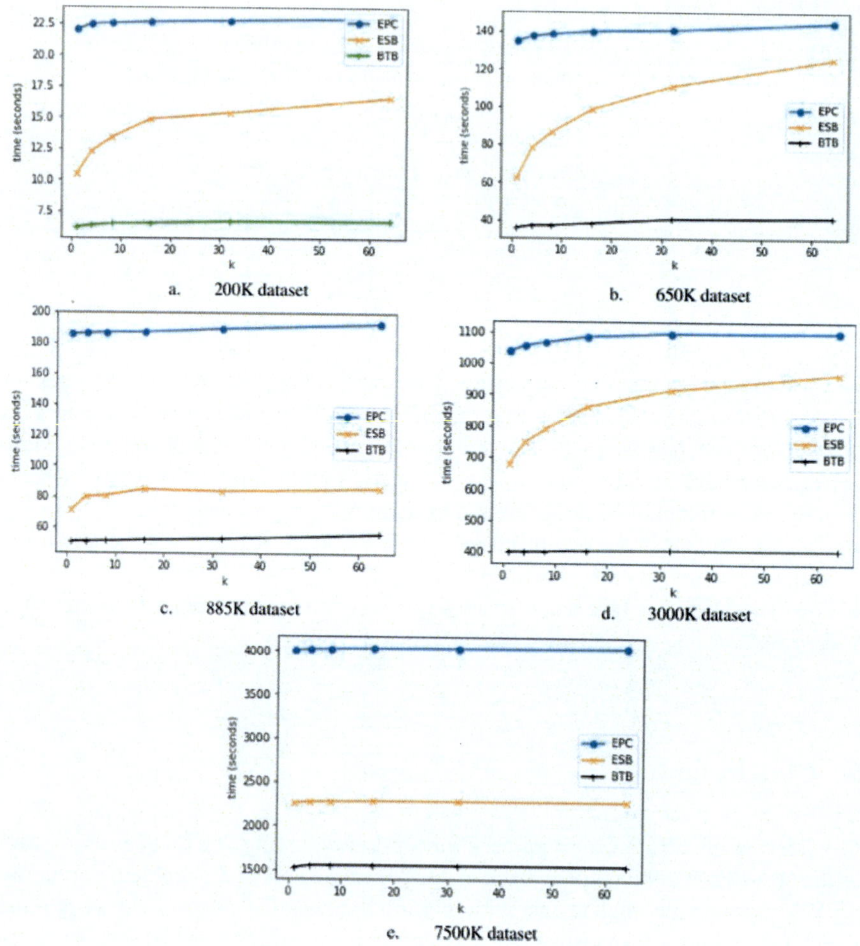

Fig. 4 TKD query cost on incomplete data for various values of *k*

References

1. Miao X, Gao Y, Zheng B, Chen G, Cui H (2015) Top-k dominating queries on incomplete data. IEEE Trans Knowl Data Eng 28:252–266
2. Han X, Li J, Gao H (2017) Efficient top-k dominating computation on massive data. IEEE Trans Knowl Data Eng 29(6):1199–1211
3. Zhang K, Gao H, Han X, Cai Z, Li J (2017) Probabilistic skyline on incomplete data. In: Proceedings of the 2017 ACM on conference on information and knowledge management, pp 427–436 (2017)
4. Soundararajan R, Kumar SR, Gayathri N, Al-Turjman F (2021) Skyline query optimization for preferable product selection and recommendation system. Wirl Per Commun 117(4):3091–3108
5. Yiu ML, Mamoulis N (2007) Efficient processing of Top-k dominating queries on multi-dimensional data. VLDB 7:483–494

6. Khalefa ME, Mokbel MF, Levandoski JJ (2008) Skyline query processing for incomplete data. In: 2008 IEEE 24th international conference on data engineering, pp 556–565
7. Papadias D, Tao Y, Fu G, Seeger B (2005) Progressive skyline computation in database systems. ACM Trans Database Syst (TODS) 30(1):41–82
8. Gao Y, Miao X, Cui H, Chen G, Li Q (2014) Processing k-skyband, constrained skyline, and group-by skyline queries on incomplete data. Exp Syst Appl 41(10):4959–4974
9. Lofi C, Maarry KE, Balke W-T (2013) Skyline queries in crowd-enabled databases. In: Proceedings of the 16th international conference on extending database technology, pp 465–476
10. Amagata D, Hara T, Onizuka M (2018) Space filling approach for distributed processing of top-k dominating queries. IEEE Trans Knowl Data Eng 30(6):1150–1163
11. Ding X, Yan C, Zhao Y, Yang Z (2018) Efficient processing of Top-K dominating queries on incomplete data using MapReduce. In: International conference on cloud computing and security, pp 478–489
12. Zhan L, Zhang Y, Zhang W, Lin X (2014) Identifying top k dominating objects over uncertain data. In: International conference on database systems for advanced applications, pp 388–405
13. Li W, Li P (2020) Balanced dominating Top-k queries over uncertain data. In: Proceedings of the 2020 5th international conference on cloud computing and internet of things, pp 69–76
14. Lian X, Chen L (2013) Probabilistic top-k dominating queries in uncertain databases. Inform Sci 226:23–46
15. Santoso BJ, Permadi VA, Ahmad T, Ijtihadie RM, Sektiaji B (2018) Continuous Top-k dominating query of incomplete data over data streams. In: 2018 International conference on sustainable information engineering and technology (SIET), pp 21–26
16. Ding X, Yan C, Zhao Y (2018) Parallel processing of Top-k dominating queries on incomplete data. In: 2018 IEEE 4th international conference on computer and communications (ICCC), pp 1785–1791
17. Soliman MA, Ilyas IF, Ben-David S (2010) Supporting ranking queries on uncertain and incomplete data. VLDB J 19(4):477–501
18. Miao X, Gao Y, Guo S, Liu W (2018) Incomplete data management: a survey. Front Comput Sci 12(1):4–25
19. MovieLens 20M Dataset. https://www.kaggle.com/grouplens/movielens-20m-dataset
20. MovieLens Homepage. https://grouplens.org/datasets/movielens

Development of an Efficient ETL Technique for Data Warehouses

Md Badiuzzaman Biplob and Md. Mokammel Haque

Abstract In the area of knowledge science, data warehouse plays an important role in data mining, data analytics, and decision making. Extraction Transformation and Load (ETL) methodology are utilized widely in a developed data warehouse. In today's competitive business world, mergers and acquisitions unit techniques are quite common. It desires extraction, transformation, and loading of a huge amount of structured data movement. This paper is associated with the improvement of Dynamic ETL (D-ETL) by adding noise-free filtering and missing data handling methods. This existing approach is modified to use the standard technique of extraction. ETL methodology is the progressive extraction procedure among the entire extraction. In this paper, we propose a new Efficient ETL technique which is an updated version of the D-ETL adding an attribute selection and noise reduction technique.

Keywords E-ETL · Noise-free filtering · Missing data handling · Improved D-ETL · IMICE

Abbreviations

ETL	Extraction transformation, and load;
D-ETL	Dynamic extraction, transformation and load;
MICE	Multiple imputations by chained equations;
IMICE	Improved multiple imputations by chained equations;
AI	Artificial intelligence;
EHR	Electronic health records;
DW	Data warehouse;
CSV	Comma-separated values;

Md Badiuzzaman Biplob (✉) · Md. Mokammel Haque
Department of Computer Science and Engineering, Chittagong University of Engineering and Technology, Chattogram, Bangladesh

Md. Mokammel Haque
e-mail: mokammel@cuet.ac.bd

M. S. Arefin et al. (eds.), *Proceedings of the International Conference on Big Data, IoT, and Machine Learning*, Lecture Notes on Data Engineering and Communications Technologies 95, https://doi.org/10.1007/978-981-16-6636-0_20

OMOP	Observational medical outcome partnership;
SQL	Structured query language;
INFFC	Iterative noise filter based on the fusion of classification

1 Introduction

Data mining [1–3] is the observation by evaluating large datasets to retrieve essential data. This data [4] can later be used to develop many informative systems or retrieve hidden data by analyzing patterns of huge datasets to predict vital outcomes. An informed system unit for measuring refined code that uses Artificial Intelligence (AI) along with a data repository to help in real-world selections. The involvement of up-to-date systems and technology inside the health business is widespread. Although this affiliation has been a gift for many years, new ideas, and experiments are administered daily. The scope for improvements in this sector remains immense. The knowledgeable systems facilitate the automation of the healthcare industry [5, 6], whereas by using completely different processing methodologies, researchers try to predict or collect valuable data. Within the health business, the impact of such systems is extremely important because they agitate people's varied lives. A mistaken decision or prediction could cause a non-public or perhaps the whole country varied awful conditions.

Expert systems or data processing algorithms typically do not make mistakes unless the info it uses itself is imperfect. One of the main tasks of the whole process is therefore to refine the information that the system will be trained. This part of the mining process is called data pre-processing, which is recognized by the most important and difficult elements among researchers in information science. Refinement is required due to the continuously imperfect measurement of real-world information squares. In many cases, duplication of information is incredibly common errors, leading to incorrect forecasts. The Data warehouse [7, 8] is used to store enormous information amounts. This information can be used to call and applied mathematical analysis. The data warehouse construction method must be extremely efficient and reliable. Applied mathematical strategies are primary alternatives for tasks such as forecasts or classifications. When AI evolves, people are very vulnerable to relying on AI for these tasks. The goals of the paper [9] are based on machine learning and AI to classify information, as these strategies will improve over time and therefore the resulting square measurement typically exceeds the standard approach of applied mathematics. The paper [9] compares a number of these machine learning methods based mainly on strategies and ancient strategies, but each of these strategies works. We tend to be ready to check that square strategies measure higher suitable attributes and that we were jointly fortunate to propose a technique that had a much better model than the previous ETL. The present ETL techniques [7, 10, 11] are not much effective for Data Warehouses. They [7–9] use all the features to execute any task. But, all the features are not needed to execute and also not much important equally.

Proper attribute selection and noise reduction techniques are missing in the existing model. So, we have proposed and developed an Efficient ETL Technique for Data Warehouses, where we have added attribute selection and noise reduction techniques.

2 Related Work

2.1 Background of ETL

For the data warehousing structure, ETL [12–14] is an important component. The method comprises of the extraction of information from various information sources, the transformation of extracted data reliable with trade necessities, and loading of that information into the data warehouse [15]. Figure 1 shows the ETL work processes in an ETL approach.

Electronic health records (EHRs) include complex clinical info on non-standard codes and structures that contain proprietary formats [16, 17]. The EHR information needs to be restructured and reworked to traditional formats and usual terminologies in multi-location clinical analysis networks and optically combined with entirely different data sources. Dynamic ETL [9] has been enforced to do ETL [9] activities that load data from an assortment of sources with altogether different data pattern structures into the common OMOP. D-ETL supports a flexible and clear method for transferring health information and products into a target info model. There is no attribute selection technology available in D-ETL, and there is no correct noise

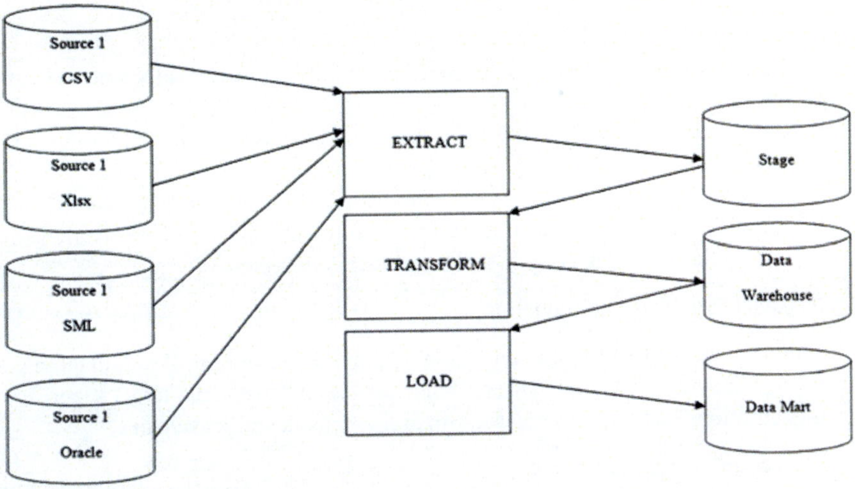

Fig. 1 ETL work processes [15]

reduction procedure. The D-ETL measures unit requires direct access to backend data, which requires extensive SQL skills.

2.2 D-ETL

Current approaches to data transformation unit of measurement are usually not versatile [9]. The ETL procedure to help data harmonization in some cases incorporates two sequential stages (schema mapping; information programming).

In the D-ETL paper [9], they have shown design with implementation a very specific rule-based D-ETL, and their contribution to this work D-ETL supports a flexible and clear methodology. EHR info wants to cooperate in clinical analysis networks on multiple sites [9]. To reduce the knowledge contribution barriers an expert tool is needed. From heterogeneous sources first, load completely different data schema structures into the OMOP common data model [9, 18]. Automatic question generation to synchronize the datasets offered.

2.3 Workflow of D-ETL

Figure 2 shows the work processes of the D-ETL way which deal with mix two source datasets.

D-ETL Approach:

The approach of D-ETL is considered on four main elements.

Extensive ETL determinations.

- D-ETL rules have been created in plain text format and the rules unit is also human-intelligible.
- From ETL rules full SQL statements have created and also remodel, emulated, and merchandise the information into destination tables.
- ETL designers can easily access the automatically generated SQL statements; check the principles associated with them, following an associated degree of varying legality, and detecting and deleting errors in methodology.

ETL determinations and modeling:

- Contains data regarding the provision and target schemas. Word mappings between data elements and values at intervals the provision and target schemas, and definitions and conventions for info at intervals the target schema.

Data extraction and validation:

- Necessary data elements which are from the provision process unit of measurement turned out to a quick memory and these are reworked and processed into the destination information.

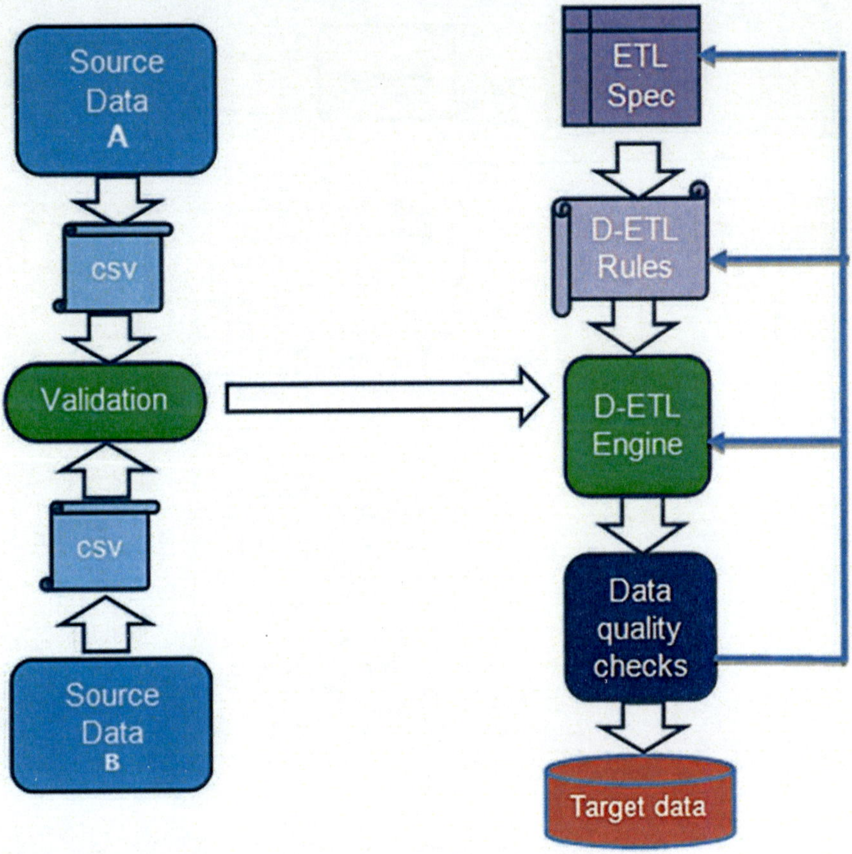

Fig. 2 Work processes of the D-ETL, which deal with mix two datasets [9]

- The D-ETL model utilizes CSV text records for information interchange as a result of its extensive use and characteristic.
- Then, the extracted data pass through by a data checking method.

Figure 3 shows the iterative noise filter based on the fusion of classification. Noise information reduction process.

- Preliminary filtering: Contains provision data and target schemes. Word mapping between info elements and values at intervals of provision and target schemes, and info definitions and conventions at intervals of target schemes.
- Noise-free filtering: A new filtering (see Fig. 3) made from the part clean information from the previous step is applied to all the coaching examples in the current iteration, which is followed by two sets of examples: a clean and strident set. This filtering is predicted to be much correct than the previous one since the noise filters measure squarely designed cleaner information.

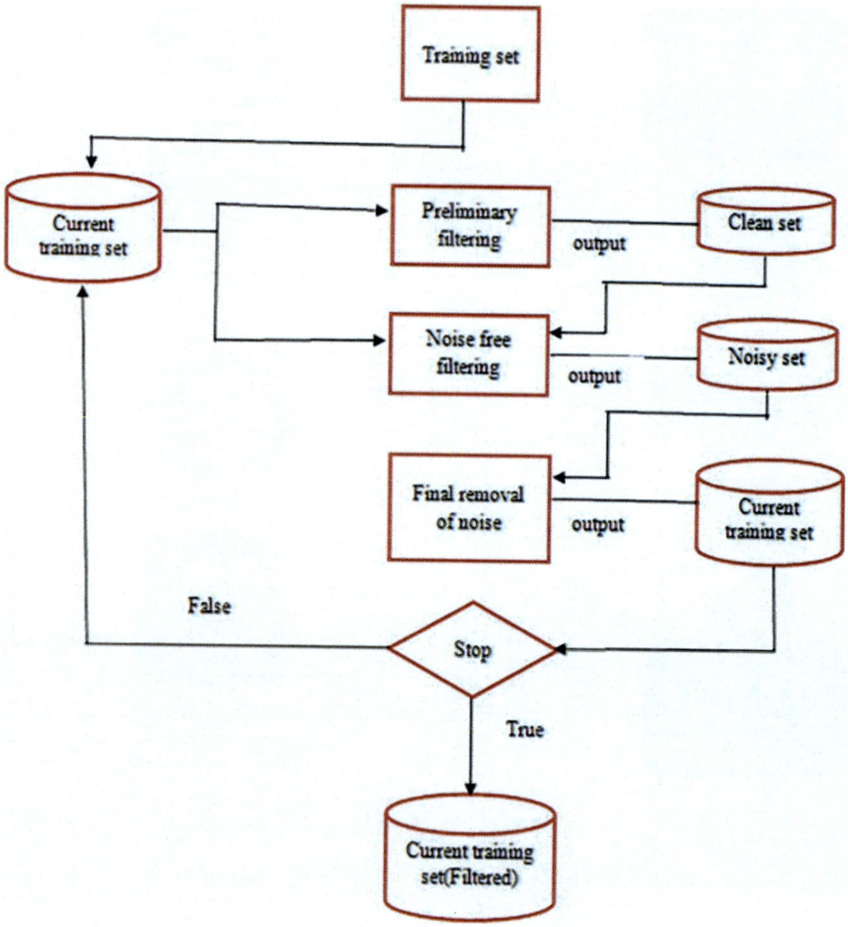

Fig. 3 Iterative noise filter based on the fusion of classification [19]

- Final removal of noise: A noise score [19] is calculated from every probably strident example of the strident set obtained in the previous step to determine which of them is finally eliminated.

Dedicated to comparing the time quality of INFFC [19] with two of these different ensemble-based filters. They introduce all three steps in each iteration to remove strident examples with the strongest liability, that is, those square measure strident examples with good confidence. In this way, examples that are striking or not, square measure left in the coaching for the post-coaching process. Making sure that only the examples that measure square to be noise square measure removed means that noise-free examples that could hurt the educational method are less likely to be found.

3 Proposed E-ETL

This work proposed an improvement of ETL, which we have called efficient extraction, transformation, and loading (E-ETL) (see Fig. 4). In E-ETL, the full processes have been completed in six steps. In the first step, we have collected the data from several available resources. The second step is feature selection where we have designed a feature selection technique [1], which is called "efficient feature selection".

Figure 4 illustrates efficient extraction, transformation, and loading (E-ETL) technique for data warehouses. This is our proposed model where we have shown all the steps clearly.

The 3rd step is the phonetic algorithm where we have used the namevalue algorithm. The 4th step is for noise reduction, where we have used IMICE (Improved multiple imputations by chained equations) technique which is the updated version of MICE [20]. The 5th step is the extraction, transformation, and loading (ETL) process. At the last step, we can get our targeted data after completing the full ETL

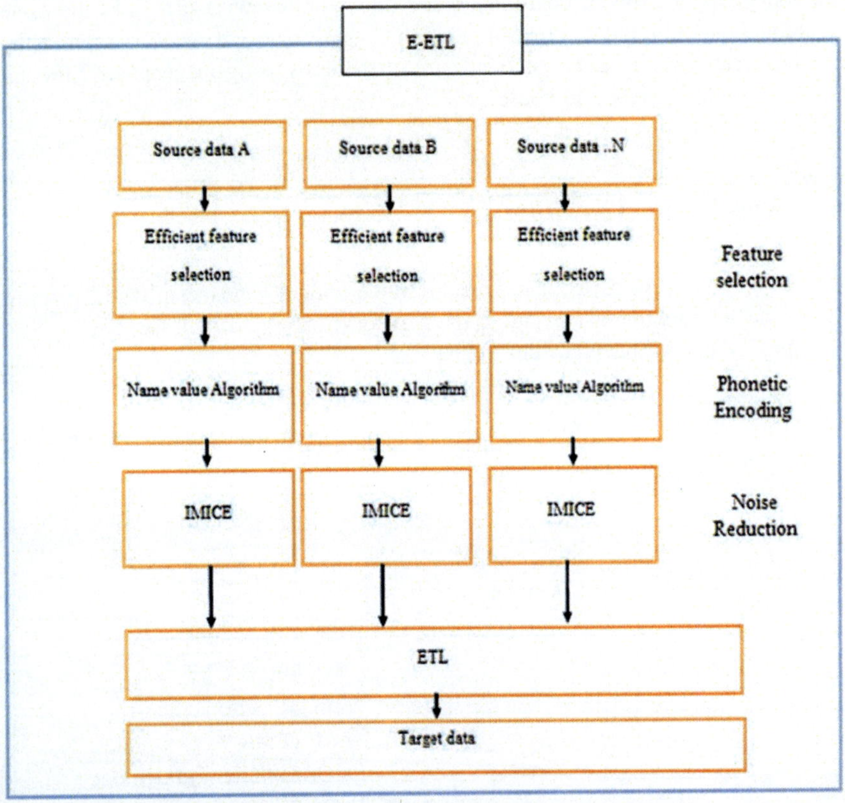

Fig. 4 Proposed efficient extraction, transformation, and loading (E-ETL)

process. The present feature selection techniques are not good in ETL that's why we have improved the feature selection technique by adding phonetic encoding and noise reduction process for better data which reduces complexity and duplexity in the data warehouse.

4 Results and Discussion

In the result section we have discussed three phonetic encoding algorithms (namevalue, soundex, and metaphone algorithm) and also found the best algorithm. Then we have improved that algorithm. We have also improved the existing noise reduction techniques. The existing system is MICE and our improved system is improved multiple imputations by chained equations (IMICE). Although multivariate imputation by chained equations (MICE) has developed as a systematic way of dealing with missing data [21, 22], significant difficulties and limits of the approach must be acknowledged. While MICE has a number of advantages over other missing data strategies in terms of flexibility, it has one major drawback: it lacks the same theoretical underpinning as other imputation techniques. So, it needs to improve the existing system. We have also get our desire attributes by using our proposed Efficient ETL technique for data warehouses.

4.1 *Namevalue Algorithm*

We have compared the performance of namevalue, soundex and metaphone algorithm and we have found that namevalue is the best for masking "patient name".

Improved namevalue algorithm steps:

1. Delete salutation
2. Delete a/A, e/E, i/I, o/O, u/U unless beginning of the name or after white space
3. Convert g/G/j/J/z/Z to j
4. Convert k/K/q/Q to k

Table 1 Ambiguity in patients input name

Actual name	Inputted name
ABU NASER	MR. ABU NASER
	MD. ABU NASER
	MR. MD. ABU NASER
	ALHAZ MD. ABU NASER
	MOHAMMAD ABU NASER
	MUHAMMAD ABU NASER
	ABU NASER

	NAMEVALUE	SOUNDEX	METAPHONE
Total run time (10 runs)	0.9076 Seconds	0.774 Seconds	1.1965 Seconds
Used memory	8397272	7858832	7886896
Accuracy (Naïve Bayes)	90%	58.9%	77.8%

Fig. 5 Performance comparison namevalue, soundex, and metaphone algorithm for patient name

5. Map all characters according to code table

The performance comparison of namevalue, soundex, and metaphone algorithm for patient name.

Figure 5 illustrates the three phonetic encoding algorithms (namevalue, soundex, and metaphone algorithm) to find out which one is best for the patient name attribute. Here we have used naïve bayes classification algorithm which is extremely fast, and it also works with categorical and numerical variables to compare with other algorithms.

4.2 *Noise Reduction in IMICE*

After the feature selection and phonetic encoding techniques, we can get our desire attributes from the total 16 attributes in our dataset [1, 23]. Now we have needed to apply the improved multiple imputations by chained equations (IMICE) technique which is the updated version of multiple imputations by chained equations (MICE) [20] for noise reduction on those attributes.

Tables 2, 3, 4 and 5 showing the output of noise reduction in IMICE for gender, age, department, and sample attributes which we get after feature selection. Here, we use a total of 40% data from the dataset because the lack of high configuration computer. When the 40% data from a dataset and missing data is 10% that time the accuracy is high compared with when the missing data is 20% and 30%. We have also shown the accuracy, precision, recall, and F1 score for all of these attributes.

Table 2 Output table of gender

Data from main dataset (*d*) %	Make missing data from *d* in %	Mean square error	Square root mean square error	Accuracy	Precision	Recall	F1 score
40	10	0.04704	0.21689	0.95295	0.95318	0.95295	0.95291
40	20	0.09954	0.31550	0.90045	0.90316	0.90045	0.90000
40	30	0.14643	0.38266	0.85356	0.85458	0.85356	0.85305

Table 3 Output table of age

Data from main dataset (*d*) %	Make missing data from *d* in %	Mean square error	Square root mean square error	Accuracy	Precision	Recall	F1 score
40	10	47.08177	6.861616	0.90199	0.91880	0.90199	0.90676
40	20	91.93061	9.588045	0.80556	0.84269	0.80556	0.81623
40	30	124.9475	11.17799	0.71186	0.79320	0.71186	0.73389

Table 4 Output table of department

Data from main dataset (*d*) %	Make missing data from *d* in %	Mean square error	Square root mean square error	Accuracy	Precision	Recall	F1 score
40	10	0.18082	0.42523	0.93864	0.943059	0.938649	0.93918
40	20	0.38829	0.62313	0.87352	0.885273	0.873528	0.87488
40	30	0.60516	0.77792	0.82040	0.836192	0.820409	0.82152

5 Conclusions and Future Work

In the present data warehousing research, ETL operations are a major issue. We have proposed and implemented a very critical challenge in current data warehousing research in this paper. The present ETL techniques are not much effective for data warehouses. All the attributes are not needed to execute and also not much important equally for any task. Proper attribute selection and noise reduction techniques [24] are missing in the existing model. So, we have proposed and developed an Efficient ETL technique for data warehouses, where we have added attribute selection and noise reduction techniques.

Security [25] to the data is that the key challenge of a locality of concern in today's world. Current approaches for the modeling of ETL do not address the protection issues inside the ETL modeling. To improve the security issue and protect the data, we will include a security component in our research in the future. A typical provider of

Table 5 Output table of sample

Data from main dataset (*d*) %	Make missing data from *d* in %	Mean square error	Square root mean square error	Accuracy	Precision	Recall	F1 score
40%	10%	0.57481	0.75816	0.95561	0.95595	0.95561	0.9554111035247863
40%	20%	1.08166	1.04002	0.91587	0.91638	0.91587	0.9156965076921142
40%	30%	1.72244	1.31241	0.87364	0.87509	0.8736441264471267	0.8720239416596348

problems in ETL in associate in nursing extremely large varies of dependencies among ETL jobs. Most use of parallelism: to load data into two datasets one can run the plenty in parallel. This analysis work proves and shows the event in attribute selection and noise reduction in the ETL technique.

References

1. Badiuzzaman Biplob M, Khan S, Sheraji G, Shuvo J (2020) Hybrid feature selection algorithm to support health data warehousing. Learn Analytics Intell Syst 103–112
2. Gini R, Schuemie M, Brown J, Ryan P, Vacchi E, Coppola M, Cazzola W, Coloma P, Berni R, Diallo G, Oliveira J, Avillach P, Trifirò G, Rijnbeek P, Bellentani M, Van Der Lei J, Klazinga N, Sturkenboom M (2016) Data extraction and management in networks of observational health care databases for scientific research: a comparison among EU-ADR, OMOP, mini-sentinel and MATRICE strategies. eGEMs (Generating Evidence & Methods to improve patient outcomes) 4:2
3. Jayaram B (2019) Mining social media data using R and WEKA tools. Int J Psychosoc Rehabil 23:243–253
4. Schilling L, Kwan B, Drolshagen C, Hosokawa P, Brandt E, Pace W, Uhrich C, Kamerick M, Bunting A, Payne P, Stephens W, George J, Vance M, Giacomini K, Braddy J, Green M, Kahn M (2013) Scalable architecture for federated translational inquiries network (SAFTINet) technology infrastructure for a distributed data network. eGEMs (Generating Evidence & Methods to improve patient outcomes) 1:11
5. Khan S, Hoque A (2015) Towards development of national health data warehouse for knowledge discovery. Adv Intell Syst Comput 385:413–421
6. Pawlak Z (1982) Rough sets. Int J Comput Inf Sci 11:341–356
7. El-Sappagh S, Hendawi A, El Bastawissy A (2011) A proposed model for data warehouse ETL processes. J King Saud Univ Comput Inf Sci 23:91–104
8. Santos V, Belo O (2013) Modeling ETL data quality enforcement tasks using relational algebra operators. Procedia Technol 9:442–450
9. Ong T, Kahn M, Kwan B, Yamashita T, Brandt E, Hosokawa P, Uhrich C, Schilling L (2017) Dynamic-ETL: a hybrid approach for health data extraction, transformation and loading. BMC Med Inform Decision Making 17
10. Apolloni J, Leguizamón G, Alba E (2016) Two hybrid wrapper-filter feature selection algorithms applied to high-dimensional microarray experiments. Appl Soft Comput 38:922–932
11. Cateni S, Colla V, Vannucci M (2014) A method for resampling imbalanced datasets in binary classification tasks for real-world problems. Neurocomputing 135:32–41
12. Wijaya R, Pudjoatmodjo B (2016) Penerapan extraction-transformation-loading (ETL) dalam data warehouse (Studi Kasus: Departemen Pertanian). Jurnal Nasional Pendidikan Teknik Informatika (JANAPATI) 5:61
13. Biswas N, Chattapadhyay S, Mahapatra G, Chatterjee S, Mondal K (2019) A new approach for conceptual extraction-transformation-loading process modeling. Int J Ambient Comput Intell 10:30–45
14. Sreemathy J, Joseph VI, Nisha S, Prabha IC, Priya RMG (2020) Data integration in ETL using TALEND. In: 2020 6th international conference on advanced computing and communication systems (ICACCS), pp 1444–1448
15. Badiuzzaman Biplob M, Sheraji G, Khan S (2018) Comparison of different extraction transformation and loading tools for data warehousing. In: 2018 international conference on innovations in science, engineering and technology (ICISET), pp 262–267
16. Sox H (2009) Comparative effectiveness research: a report from the institute of medicine. Ann Intern Med 151:203

17. Danaei G, Rodríguez L, Cantero O, Logan R, Hernán M (2011) Observational data for comparative effectiveness research: an emulation of randomised trials of statins and primary prevention of coronary heart disease. Stat Methods Med Res 22:70–96
18. Sills M, Kwan B, Yawn B, Sauer B, Fairclough D, Federico M, Juarez-Colunga E, Schilling L (2013) Medical home characteristics and asthma control: a prospective, observational cohort study protocol. eGEMs (Generating Evidence & Methods to improve patient outcomes) 1:3
19. Sáez J, Galar M, Luengo J, Herrera F (2016) INFFC: an iterative class noise filter based on the fusion of classifiers with noise sensitivity control. Inf Fusion 27:19–32
20. Azur M, Stuart E, Frangakis C, Leaf P (2011) Multiple imputation by chained equations: what is it and how does it work? Int J Methods Psychiatr Res 20:40–49
21. Peugh J, Enders C (2004) Missing data in educational research: a review of reporting practices and suggestions for improvement. Rev Educ Res 74:525–556
22. Ferguson J, Hannigan A, Stack A (2018) A new computationally efficient algorithm for record linkage with field dependency and missing data imputation. Int J Med Inform 109:70–75
23. Biplob M. Feature selection and data visualization with encoding categorical values and handling missing values in Python. https://www.linkedin.com/pulse/feature-selection-data-visualization-encoding-values-handling-biplob/. Accessed 2 July 2021
24. Xiong H, Pandey G, Steinbach M, Kumar V (2006) Enhancing data analysis with noise removal. IEEE Trans Knowl Data Eng 18:304–319
25. Chen Y, Horng G, Lin Y, Chen K (2013) Privacy preserving index for encrypted electronic medical records. J Med Syst 37

Informatics for Emerging Applications

Downlink Performance Enhancement of High-Velocity Users in 5G Networks by Configuring Antenna System

Mariea Sharaf Anzum, Moontasir Rafique, Md. Asif Ishrak Sarder, Fehima Tajrian, and Abdullah Bin Shams

Abstract A limitation of bandwidth in the wireless network and the exponential rise in the high data rate requirement prompted the development of Massive Multiple-Input-Multiple-Output (MIMO) technique in 5G. Using this method, the ever-rising data rate can be met with the increment of the number of antennas. This comes at the price of energy consumption of higher amounts, complex network setups, and maintenance. Moreover, a high-velocity user experiences unpredictable fluctuations in the channel condition that deteriorates the downlink performance. Therefore, a proper number of antenna selections is of paramount importance. This issue has been addressed using different categories of algorithms but only for static users. In this study, we proffer to implement antenna diversity in closed loop spatial multiplexing MIMO transmission scheme by operating more number of reception antennas than the number of transmission antennas for ameliorating the downlink performance of high-velocity users in case of single user MIMO technology. In general, our results can be interpreted for large scale antenna systems like Massive MIMO even though a 4×4 MIMO system has been executed to carry out this study here. Additionally, it shows great prospects for solving practical-life problems like low data rate and call drops during handover to be experienced by cellular users traveling through high-speed transportation systems like Dhaka Metro Rail. The cell edge users are anticipated to get benefits from this method in case of SU-MIMO technology. The proposed

M. S. Anzum (✉) · M. Rafique · Md. Asif Ishrak Sarder · F. Tajrian
Islamic University of Technology, Gazipur 1704, Bangladesh
e-mail: marieasharaf@iut-dhaka.edu

M. Rafique
e-mail: moontasir@iut-dhaka.edu

Md. Asif Ishrak Sarder
e-mail: asifishrak@iut-dhaka.edu

F. Tajrian
e-mail: fehimatajrian@iut-dhaka.edu

A. B. Shams
University of Toronto, Toronto, ON M5S 3G8, Canada

M. S. Arefin et al. (eds.), *Proceedings of the International Conference on Big Data, IoT, and Machine Learning*, Lecture Notes on Data Engineering and Communications Technologies 95, https://doi.org/10.1007/978-981-16-6636-0_21

method is expected to be easily implemented in the existing network structures with nominal difficulties.

Keywords Massive MIMO · SU-MIMO · 5G · Antenna configuration · Resource scheduling · User mobility

1 Introduction

The ever-increasing cellular devices and exponential rise of wireless connection, demand for lower latency, higher spectral efficiency, and ultra-high data speed. 5G technology is expected to meet these requirements. Thus, massive multiple-input-multiple-output (MIMO), an extension of MIMO technique becomes an essential requirement of this new standard technology. In the massive MIMO system, base stations use a huge number of antenna arrays to connect with users. These huge number of base station antennas can enable energy to concentrate in a small region bringing better improvements by several folds in transmission gain and user throughput than a MIMO system. By integrating the single user MIMO (SU-MIMO) technique with the spatial multiplexing techniques, multiple number of data streams can be transmitted over various antennas to single user equipment (UE). This results in a throughput gain with improved spectral efficiency.

However, practical implementation of a high number of antennas requires a larger amount of resources, high power supply, highly complex system, and larger-scale antenna channel estimates. The increasing number of users is adding to these limitations due to loss in desired QoS (quality of service) resulting from spatial interference. For improving the QoS for each user, particle swarm optimization can be used to select the minimum number of transmitter antenna elements [1]. SUS algorithm and JASUS with a pre-coding scheme could be combined to select a limited number of antennas for specific users for lessening the complexity of the whole system delivering maximum average sum rate [2]. Then again, the large number of base station antennas contribute to the degradation of energy efficiency. This problem can be solved using the antenna selection method based on channel state information [3]. Additionally, a deep learning strategy has been proposed to optimize antenna selection patterns for massive MIMO channel extrapolation [4]. The aforementioned research works proffered optimum antenna selection patterns or algorithms for mitigating some of the limitations of Massive MIMO mentioned earlier but did not consider practical scenarios like UE mobility which was investigated in recent research works. The high-velocity users experience a degradation of performance of the schedulers, spatial multiplexing techniques, and network capacity because mobility of UE causes Doppler shift which results in rapid variations in channel quality [5, 6]. Due to the frequent and rapid variations of the channel quality, the optimum antenna combination also changes over the multiple transmission time intervals (TTIs). Therefore, the optimum antenna combinations may not be feasible for practical implementation without considering mobility. Moreover, deep learning algorithms can cause high latency in real-time applications which contradicts one of the goals of 5G technology.

In this paper, we have proposed to implement antenna diversity in spatial multiplexing technique by keeping the number of receiver antennas more than the number of transmitter antennas. This diversity of antennas have enabled overall throughput for high-velocity users to improve. This simple strategy will work under any scheduler and transmission schemes considering users with low, medium, and high velocity. To imitate a system with a high number of transmission and reception antennas and to circumvent the simulation complexity, a 4×4 MIMO system has been implemented to conduct our study. In general, our results can be interpreted for large scale antenna systems like Massive MIMO. It is believed that the proposed strategy will contribute to solving real-life problems like call drops, low data speed to be faced by cellular users traveling through high-speed transportation systems like Dhaka Metro Rail.

The remaining parts of the proffered paper are assembled in the following sequence. Section 2 contains the system model where spatial multiplexing, antenna diversity, resource scheduling schemes, network model, and various performance parameters are discussed. After that, simulation model is discussed in Sect. 3 before discussing the simulation results in Sect. 4. In the end, conclusions are presented in Sect. 5.

2 System Model

2.1 Spatial Multiplexing

Spatial multiplexing is one of the transmission modes used in a communication system. Generally, the data which needs to be sent to any UE are divided into several streams and sent over multiple channels using same frequency band.

Through spatial multiplexing, the ability to use several channels simultaneously for transmitting data helps to increase system capacity. Based on theoretical deduction, the capacity of any channel increases with the increment of number of data streams keeping a linear relationship according to the following equation [7]:

$$C = MB\log_2\left(1 + \frac{S}{N}\right) \tag{1}$$

Here, capacity is represented by C for bandwidth B whereas number of data streams is M. And $\frac{S}{N}$ is signal to noise ratio.

There are mainly two categories of spatial multiplexing-close loop spatial multiplexing (CLSM) and open loop spatial multiplexing (OLSM). For the proffered strategy, CLSM is adopted as the spatial multiplexing technique.

Closed loop spatial multiplexing (CLSM): The maximum number of data streams M in T transmitter and R receiver network ($T \times R$) that can be transmitted over the

Fig. 1 Network architecture

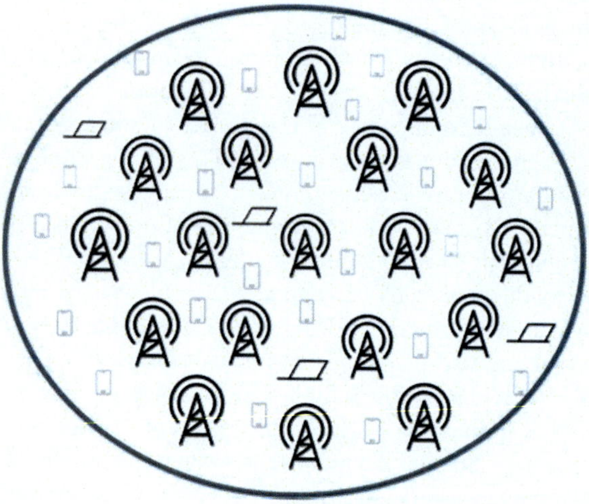

network follows the given condition in case of CLSM [5].

$$M \leq \min(T, R) \tag{2}$$

Under CLSM, UE reported CQI gives indication about channel condition which is utilized to select the suitable MCS (Modulation and Coding Scheme) by base station for users. Then rank indicator is utilized to select the number of layers under the selected MCS and channel. In this paper, MCS level can be selected as 16-QAM for the simulation. And then PMI feedback helps UE to adjust quickly to the frequently changing condition of the channel.

2.2 Network Model

A typical network comprised of 19 macro cell base stations is used here. Base stations are arranged in a hexagonal manner creating a two tier network. The distance between any two base stations is 500 m. Every base station antenna is taken as a tri-sector antenna to achieve 360-degree coverage. An architecture of the adopted model has been depicted in Fig. 1. To design these antennas, the Kathrein 742215 antenna model is utilized here [8].

2.3 Resource Scheduling

Resource scheduling method is basically utilized to allocate resources among the users in any cell in a specific way considering some criteria like system efficiency, users demands, etc. Round Robin (RR) and Proportional fair (PF) have been adopted for our simulation results.

Proportional Fair and Round Robin

The main consideration of RR is to ensure optimum fairness among all users in a cell adapting the cyclic distribution of resources. On the other hand, proportional fair primarily considers to boost the throughput of users keeping fairness as the secondary consideration. Each user equipment is taken as the priority coefficient from the priority function given in the following equation [6].

$$P = \frac{T^{\alpha}}{R^{\beta}} \tag{3}$$

Here, T is feasible throughput and R is average data rate. Using parameters, α and β fairness can be tuned. In case of PF, $\alpha \approx 1$ and $\beta \approx 1$ are used whereas $\alpha \approx 0$ and $\beta \approx 1$ are used in case of RR technique [5]. But in order to enhance user throughput and fairness of resource allocation, an improved PF scheduling algorithm has been proffered where average user throughput $T_k(t)$ can be calculated using the following equation [9].

$$T_k = \left(1 - \frac{1}{t_c}\right) T_k(t) + \frac{1}{t} \sum_{s=1}^{s} R_{s,k}(t) \tag{4}$$

Here, the throughput of the user k at sub band S is represented by $R_{s,k}(t)$, throughput averaging time window is defined by t_c. The value of t_c can be selected to carry out an optimum trade-off between fairness and system capacity. On the other hand, RR distributes resources among users without considering channel conditions in a cyclic manner. As a result, it can ensure fairness amidst users but degrades user throughput.

2.4 Key Performance Parameters

Average UE Throughput: The position of users in the whole network is random. So, the distance between them and the base stations is not fixed which results in variations in path loss. Consequently, SINR and throughput of a UE have a huge range of different values. Moreover, it is well known that throughput of a $T \times R$ system is proportional to min(T, R) where T is transmitter antenna number and R is receiver antenna number. Then again, average throughput of UE (T_{AVG}) can be expressed like the following equation [10].

$$T_{\text{AVG}} = \frac{\sum_{k=1}^{n} T_k}{n} \tag{5}$$

Here, total throughput of kth user is represented by T_k and the total number of users is n.

Cell Edge Throughput: Users near the edge of a serving cell feel signals from the neighboring cells as interferences for them. Adding to that, the strength of the signal degrades owing to the distance from the base station causing lower speed of data. Cell edge throughput is considered as the five percent of the throughput ECDF of UE. For avoiding call drop, continuous coverage and minimum data rate are required during handover.

Spectral Efficiency: Spectral efficiency can be stated as the speed of data over a specific bandwidth. This parameter can be represented by the following equation [7].

$$s = \frac{\sum_{k=1}^{n} T_k}{\text{BW}} \tag{6}$$

Here, system bandwidth is BW and total throughput for kth user is T_k. Thus, when total throughput over a specific bandwidth increases, spectral efficiency increases.

Fairness Index: Fairness index is the parameter that is utilized to dictate how the resources may be divided among the UEs. A well-known method, Jain's Fairness Index is adopted to examine the fairness in terms of resources among the UEs. Jain's Fairness Index can be calculated for n users from the following equation [11].

$$J(T) = \frac{\left[\sum_{k=1}^{n} T_k\right]^2}{n\left[\sum_{k=1}^{n} T_k^2\right]} \tag{7}$$

Here, average throughput is represented by T_k for kth user. $J(T) = 1$ if all the UEs can achieve the same share of resources distributed among them.

3 Simulation Model

At first, performance parameters are numerically investigated under proportional fair resource scheduling technique for 2×2, 2×3, 2×4, 4×2, 4×3, 4×4 MIMO schemes. After that, the investigation was done for the MIMO schemes for round robin resource scheduling technique at second phase. In both of the phases, impact of UE mobility spanning over a range of 0–120 kmph velocity was observed and then results were plotted. There were 10 UEs per sector with a total of 570 UEs arbitrarily taken within the geometrical area of the network during the simulation. To integrate mobility of UEs in the simulations, random walk model has been selected. Using random walk model, a user can be assumed to take a random step away from

Table 1 Parameters for the simulation of the network

Simulation parameters	
Channel model	WINNER+
Frequency	2.45 GHz
Bandwidth	20 MHz
No. of transmitter/receiver	4
Simulation time	50TTI (PF), 50TTI (RR)
BS height	20 m
Transmission mode	CLSM (2 × 2, 2 × 3, 2 × 4, 4 × 2, 4 × 3, 4 × 4)
BS power	45 dB
Receiver height	1.5 m
Antenna azimuth offset	30 degree
BS transmitter power	45 dBm
Antenna gain	15 dBi

the previous position in each period. For link prediction for UEs, Mutual Information Effective SINR Mapping (MIESM) is chosen here owing to its better accuracy than other ESM algorithms especially for higher modulation schemes [12]. Simulations for round robin have been done for 50 TTIs. Additionally, 50 TTIs have been also considered for proportional fair. The Vienna LTE-Advanced simulator has been utilized here [13]. The macroscopic path loss model for the macro-cells considering an urban environment can be represented by the following equation [7].

$$L = 40\left(1 - 4 \times 10^{-3} h_{Bs}\right) \log_{10}(R) - 18 \log_{10}(h_{Bs}) + 21 \log_{10}(f_c) + 80 \text{ dB} \quad (8)$$

Here, R is taken as the separation between the UE and the base station in kilometers, carrier frequency in MHz is represented by f_c and height of the antenna is h_{Bs} in meters. Rest of the parameters are displayed in Table 1.

4 Results and Discussions

Impact of mobility on average UE throughput under PF and RR with different MIMO schemes can be observed from Fig. 2a, b respectively. Due to mobility, variations in channel conditions cause SINR to have low values. Thus, average throughput degrades at increasing velocity under both schedulers. From Fig. 2a, it can be observed that PF achieves better average UE throughput for the whole velocity range of 0–120 kmph under 2 × 4, 2 × 3 and, 2 × 2 MIMO schemes compared to 4 × 2, 4 × 3 and, 4 × 4 MIMO schemes. On the other hand, RR shows best performance for 4 × 4 MIMO at low velocity but seems to have degrading performance compared to 2 × 4 and 2 × 3 MIMO at high velocity as displayed in Fig. 2b. Because more

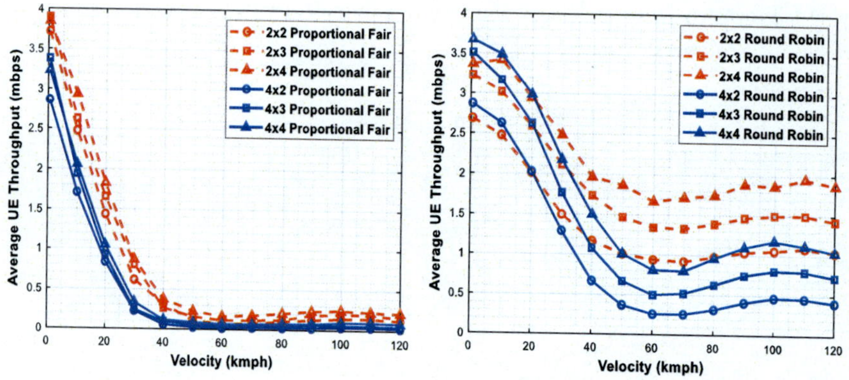

Fig. 2 Impact of mobility on average UE throughput under **a** PF and **b** RR scheme

number of receiving antennas compared to number of transmitting antennas enable a receiver to have the ability to receive signals with better SINR. Consequently, a receiver can have enhanced reception quality, better link performance along better throughput.

PF shows good cell edge throughput at low velocity but ends up going to zero cell edge throughput at increasing velocity under all MIMO schemes as shown in Fig. 3a. The same does not happen in case of RR as it does not consider channel conditions. From Fig. 3b, it can be observed that RR shows a slow decrement and ends up with very low but necessary to pursue handover from one cell to another at the cell edge. Moreover, it can be noticed that cell edge throughput shows slow a decline under 2 × 4 and 2 × 3 MIMO and holds up better performance than other MIMO schemes as velocity increases up to 120 kmph. This occurrence indicates that increasing the number of receiver antennas than that of transmitter antennas is an efficient way to reduce call drops and link failures due to better cell edge throughput ensuring a ubiquitous network.

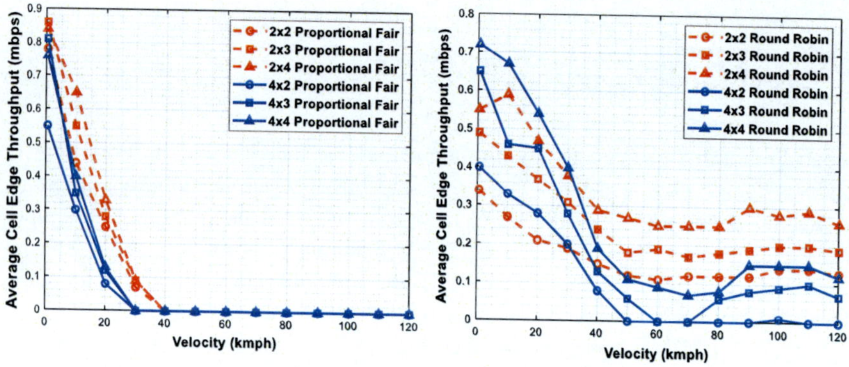

Fig. 3 Impact of mobility on cell edge throughput under **a** PF and **b** RR scheme

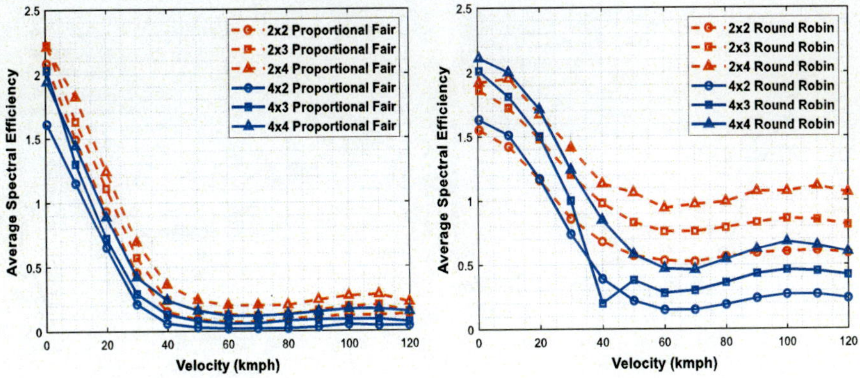

Fig. 4 Impact of mobility on average spectral efficiency under **a** PF and **b** RR scheme

Then the effect of mobility on spectral efficiency can be observed from Fig. 4a, b under both scheduling techniques along with different antenna configurations. Spectral efficiency seems to decline a lot as velocity of a user increases under PF scheduling technique. Here, 2×4 MIMO seems to show better performance than any other schemes under PF over the whole velocity range of 0–120 kmph. From before, it was noted that average throughput was also better for antenna configuration where transmitting antennas are less than receiving antennas. Consequently, spectral efficiency shows the same characteristic for PF. Then coming to RR, it can be observed that 4×2 and 4×3 MIMO scheme seem to show better performance at low velocity due to good SINR and throughput but seems to show degrading performance than 2×4 and 2×3 at high velocity. This incidence happens because receiver gets to avoid the signals with low SINR value and select the strong signal with robust link performance when receiving antennas are more in number. As a result, spectral efficiency is better whenever receiver antennas are more than transmitter antennas as high UE throughput and cell edge throughput can be achieved efficiently utilizing that specific bandwidth.

Finally, from Fig. 5a, b, performance of different antenna configurations can be analyzed in terms of fairness index under both schedulers. Under RR, fairness index does not drop that much comparing to PF as RR does not take channel conditions into consideration. Moreover, both schedulers seem to achieve better fairness index whenever transmitting antennas are less than receiving antennas over the investigated range of velocity due to better average throughput and cell edge throughput. Additionally, more receiving antennas can help achieve better SINR which ultimately results in UE sending better CQI value to the base station which allocates the resources accordingly.

Table 2 mentions some of the previous works where impact of mobility has been considered to analyze performance of downlink of cellular networks. Previous works mentioned in the table were proposed for LTE and LTE-Advanced networks whereas

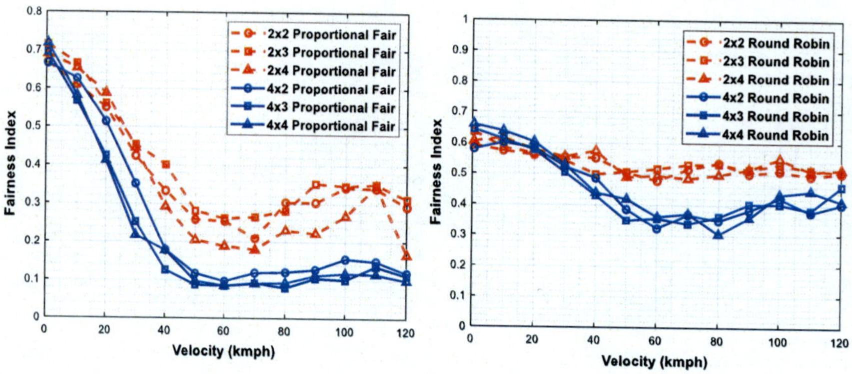

Fig. 5 Impact of mobility on fairness index under **a** PF and **b** RR scheme

Table 2 Comparison with previous research works

Research works	Transmission mode	Number of transmitting and receiving antennas	Diversity in spatial multiplexing
[5]	CLSM and OLSM	2 × 2	No
[6]	Transmit diversity	2 × 2	No
[7]	CLSM	4 × 4	No
This paper	**CLSM**	**2 × 2, 2 × 3, 2 × 4, 4 × 2, 4 × 3, 4 × 4**	**Yes**

this work which is presented in bold font is proposed for upgrading downlink performance of 5G network. None of these works investigated different antenna configurations in high velocity for single user Massive MIMO system. It is best to our knowledge that antenna diversity in SU-MIMO technology in Massive MIMO setup has not ever been investigated with respect to velocities ranging from 0 to 120 km/h under different resource scheduling schemes. Additionally, evaluating all performance parameters brought up above, it can be suggested that implementing antenna configuration where receiving antennas are more than transmitting antennas ensures enhanced downlink performance for a high-velocity user reducing multipath fading and channel fluctuations. It is also believed that introducing the concept of antenna diversity in SU-MIMO technology will pave the way for a cheaper and simple solution for high-velocity users giving better possibility of reception quality at the receiver end through both of the resource scheduling schemes considered in this study.

5 Conclusion

In this paper, we proffer to introduce antenna diversity in spatial multiplexing MIMO transmission scheme by operating more number of reception antennas than the number of transmission antennas to improve the downlink performance of high-velocity users. To circumvent the simulation complexity, a 4 × 4 MIMO system has been implemented to conduct the study. The simulation results show that this technique can work independent of the type of applied resource scheduler and the same is expected under any transmission scheme. A linear increase in data rate with higher reception antennas is also observed. The proposed method can be easily implemented in the existing network architectures with minimal difficulties. Also, it has the potential for solving real-life problems like call drops and low data rates to be experienced by cellular users traveling through high-speed transportation systems like Dhaka Metro Rail.

References

1. Park D (2017) Transmit antenna selection in massive MIMO systems. In: International conference on information and communication technology convergence. ICT convergence technologies lead fourth industrial revolution, ICTC2017.2017-Decem, pp 542–544. https://doi.org/10.1109/ICTC.2017.8191036
2. Sheikh TA, Bora J, Hussain MA (2018) Combined user and antenna selection in massive MIMO using precoding technique. Int J Sens Wirel Commun Control 9:214–223. https://doi.org/10.2174/2210327908666181112144939
3. Jin G, Zhao C, Fan Z, Jin J (2019) Antenna selection in TDD massive MIMO systems. Mob Netw Appl. https://doi.org/10.1007/s11036-019-01297-5
4. Yang Y, Zhang S, Gao FF, Xu C, Ma J, Dobre OA (2020) Deep learning based antenna selection for channel extrapolation in FDD massive MIMO. In: 12th international conference wireless commununication signal process. WCSP 2020, pp 182–187. https://doi.org/10.1109/WCSP49889.2020.9299795
5. Shams AB, Abied SR, Asaduzzaman M, Hossain MF (2017) Mobility effect on the downlink performance of spatial multiplexing techniques under different scheduling algorithms in heterogeneous network. In: ECCE 2017—international conference on electrical, computer and communication engineering, pp 905–910. https://doi.org/10.1109/ECACE.2017.7913032
6. Shams AB, Abied SR, Hoque MA (2017) Impact of user mobility on the performance of downlink resource scheduling in Heterogeneous LTE cellular networks. In: 2016 3rd international conference on electrical engineering and information communication technology iCEEiCT 2016. https://doi.org/10.1109/CEEICT.2016.7873091
7. Shams AB, Meghla MR, Asaduzzaman M, Hossain MF (2019) Performance of coordinated scheduling in downlink LTE-a under user mobility. In: 4th international conference on electrical engineering and information & communication technology iCEEiCT 2018, pp 215–220. https://doi.org/10.1109/CEEICT.2018.8628126
8. Gunnarsson F, Johansson MN, Furuskär A, Lundevall M, Simonsson A, Tidestav C, Blomgren M (2008) Downtilted base station antennas—a simulation model proposal and impact on HSPA and LTE performance. In: IEEE vehicular technology conference, pp 1–5. https://doi.org/10.1109/VETECF.2008.49
9. Hojeij MR, Abdel Nour C, Farah J, Douillard C (2018) Weighted proportional fair scheduling for downlink nonorthogonal multiple access. Wirel Commun Mob Comput. https://doi.org/10.1155/2018/5642765
10. Noliya A, Kumar S (2020) Performance analysis of resource scheduling techniques in homogeneous and heterogeneous small cell LTE-A networks. Springer, US. https://doi.org/10.1007/s11277-020-07156-x

11. Jain R, Chiu D, Hawe W (1998) A quantitative measure of fairness and discrimination for resource allocation in shared computer systems. http://arxiv.org/abs/cs/9809099
12. Aguilar FL, Cidre GR, López JML, Paris JR (2010) Mutual information effective SNR mapping algorithm for fast link adaptation model in 802.16e. Lect Notes Inst Comput Sci Soc Telecommun Eng 45 LNICST, 356–367. https://doi.org/10.1007/978-3-642-16644-0_31
13. Rupp M, Schwarz S, Taranetz M. Signals and communication technology the vienna LTE-advanced simulators up and downlink, link and system level simulation

Artificial Bee Colony and Genetic Algorithm for Optimization of Non-smooth Economic Load Dispatch with Transmission Loss

Mohammad Hanif and Nur Mohammad

Abstract Optimization plays a crucial role in economic load dispatch (ELD) due to the rising competition in the electricity market. In dealing with complex and challenging optimization problems, swarm intelligence has already demonstrated its skill. Hence, in this study, the optimization of non-smooth economic load dispatch (ELD) has been implemented by employing two popular metaheuristic algorithms, namely Artificial Bee Colony (ABC) and Genetic Algorithm (GA). However, before implementing a metaheuristic algorithm, it is crucial to understand how it performs in contrast to others. Owing to this, the comparative study between these two algorithms (ABC and GA) in terms of convergence characteristics, statistical performance, and computation time in ELD is outlined. The results reveal that ABC outperforms GA in the ELD problem when it comes to minimizing fuel costs, despite the fact that ABC's computational time is slightly longer than that of GA.

Keywords Optimization · Metaheuristic algorithm · Constraints · Statistical comparison · Computational time

1 Introduction

Optimization is the process of determining the best value for the variables in order to maximize or minimize the objective function while satisfying all constraints. For a constrained optimization, to find the optimum value of variable x, it can be mathematically expressed as [1]:

$$\text{minimize } f(x), \; x = (x_1, x_2, \ldots, x_n) \in R_n \tag{1}$$

M. Hanif (✉) · N. Mohammad
Department of Electrical & Electronic Engineering,
Chittagong University of Engineering & Technology, Chittagong 4349, Bangladesh

N. Mohammad
e-mail: nur.mohammad@cuet.ac.bd

M. S. Arefin et al. (eds.), *Proceedings of the International Conference on Big Data, IoT, and Machine Learning*, Lecture Notes on Data Engineering and Communications Technologies 95, https://doi.org/10.1007/978-981-16-6636-0_22

$$\text{Subjected to: } \begin{cases} g(x) = 0 \\ h(x) \le 0 \end{cases} \tag{2}$$

Here, the objective function is $f(x)$, and the optimization problem is a minimization problem. This optimization is subjected to an equality constraint $g(x)$ and an inequality constraint $h(x)$.

In power optimization operations, economical load dispatch (ELD) is a key strategy for planning and controlling power generation. The primary purposes of this ELD are to schedule and manage the power generators in such a way that power generation costs are kept as low as possible while upholding equality and inequality constraints of the optimization [2, 3]. ELD optimization can be divided into two categories. One includes the valve point effect (non-smooth ELD), while the other does not (smooth ELD). Due to the non-linear and non-convex nature of the cost functions, it is difficult to solve the non-smooth ELD optimization problem [4]. Every generator has a power generation limit, and optimization techniques in ELD attempt to utilize every generator in a balanced way within that limit in order to fulfill the power demand. In addition, while balancing power generation and demand, transmission loss should be taken into account. As the distance between the generators and the load increases, the transmission loss grows as well. A typical power plant with N generators and a transmission loss network are depicted in Fig. 1a. In addition, Fig. 1b shows a pictorial representation of the relationship between fuel cost and power output, where $P_{\min}$ denotes a generator's minimal power generation and $P_{\max}$ denotes the generator's maximum power generation. The fuel cost curve becomes non-smooth and fluctuating whenever the valve point effect is considered.

To tackle constrained optimization problems, many deterministic and stochastic algorithms are developed. The applicability of the deterministic approaches is limited due to their inability in handling the dynamic problems that arise in day-to-day lives. This is because deterministic techniques rely on certain assumptions, such as the continuity of the objective functions. Stochastic algorithms, on the other hand, are not

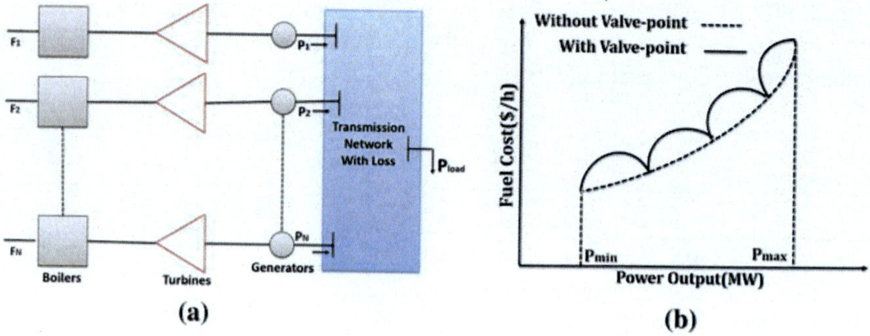

Fig. 1 **a** Power plant with transmission loss [9]. **b** Relations between fuel cost and generated power

dependent on these types of assumptions. For these reasons, stochastic algorithms can easily take the role of conventional PLC or microcontroller-based automation systems [5]. What's more, these algorithms can adapt to the unpredictable and dynamic nature of real-world problems. To put it another way, these algorithms have the power to self-organize. As a result, stochastic algorithms have emerged as a valuable tool for solving complex optimization problems [6].

Artificial Bee Colony (ABC), a stochastic algorithm, mimics the behavior of the swarm of honey-bee [7]. For solving optimization problems, Karaboga and Basturk [7] coined this algorithm. On the other hand, the Genetic Algorithm (GA) uses the behavior of the genes. This algorithm was first proposed by Holland [8]. The GA algorithm has widely been implemented in a variety of optimization problems, demonstrating its ability to solve complicated and dynamic challenges. When it comes to ELD optimization, both ABC and GA can capable of maintaining demanding outcomes. However, prior to employing an algorithm in a real-world situation, engineers require comparative knowledge about the performances of these algorithms. Because of this, the implementation of ELD, as well as the comparative studies, for the selected two algorithms (ABC and GA) are investigated in this research. Although ABC and GA-based ELD were previously implemented, the transmission loss, the ramp-rate constraints, and the performance comparison between these algorithms were rarely taken into account. Owing to this, the authors are motivated to undertake the ELD optimization, along with the comparative investigations of ABC and GA performances in ELD, considering transmission loss, and ramp-rate constraints. The analysis of convergence characteristics, statistical comparison, and computational time will be a guideline for future designers to select a proper algorithm for optimizing ELD in real-world situations.

This paper is organized as follows: Sect. 2 describes previous works, while ABC and GA are briefly introduced in Sect. 3. Section 4 outlines the problem formulation for implementing ELD using ABC and GA. Finally, the paper ends with the simulation work and result analysis followed by conclusion and references.

2 Related Works

In ELD, several techniques have already been applied to reduce the fuel cost of power generation. The conventional methods of ELD are Linear Programming (LP), Lagrange multiplier, Newton Raphson methods, and Dynamic Programming (DP) [9, 10]. These methods, however, have a number of limitations and drawbacks, especially when dealing with complex optimization issues. Furthermore, owing to the valve point effect, the ELD exhibits non-smooth properties. In that case, neither the Lagrange Multiplier nor LP approaches can handle ELD optimization[11]. The DP technique, on the other hand, has a problem with dimensionality. This is because when precision is crucial and the number of generators is increased, the execution time and storage requirements significantly rise [3]. What's more, due to ELD's non-linear behavior, conventional methods are slow to convergence and tend to get trapped in local minima instead of global optima [4].

Swarm intelligence, on the other hand, has the potential to offer useful and meaningful results in ELD. Many researchers have previously proposed several swarm intelligence-based ELDs. In ELD optimization, GA can discover global optima even when the problem is non-smooth, non-linear, or non-convex [12]. Abido [13] proposed a radical GA-based strategy for ELD that incorporates environmental goals into account. Considering the transmission loss and ramp-rate constraint, a GA-based deregulated power system was investigated by Hosseini and Kheradmandi [14]. However, the penalty factor was not taken into account by the authors in [14]. For short-term scheduling, GA-based systems, consisting of diesel generators, solar photovoltaics, batteries, and wind power were reported by Hong and Li [15]. In the past, several other metaheuristic approaches, such as Simulated Annealing [16], Particle Swarm Optimization [17], Grey Wolf Optimization [18], and so on, were implemented in ELD. Even yet, in this study, ABC-based and GA-based non-smooth ELDs with transmission loss and ramp-rate limits are investigated for a 10-unit power plant. In addition, the convergence properties and the computational time of ABC and GA are compared in order to evaluate their performances in the ELD.

3 Brief Details of ABC and GA

3.1 Artificial Bee Colony Algorithm

ABC algorithm has three types of bees [19]. They are: (1) Employed bee, (2) Onlooker bee, and (3) Scout. Employed bees make up half of the population, while onlooker bees account for the other half [19, 20]. Bees, waiting in the dance area, take the decision to select a food source are known as onlookers. On the other hand, the employed bees travel to the food source previously visited by it. Scout bees are bees that are constantly searching for random food sources. For every food source, there is only one employed bee. While a food source is depleted, the employed bee, associated with that food source, becomes a scout bee. In ABC, the placements of food sources indicate potential solutions, whereas the fitness is determined by nectars [7]. The searching of the artificial bee is as follows:

- Employed bees are able to locate food sources within the vicinity of their memory.
- Employed bees transmit their knowledge to observers, who subsequently choose a food source [20].
- The employed bees whose food sources are abandoned become scout, and subsequently, search for a random food source [7].

Hence, ABC has three control parameters: (1) Number of food sources (SN), (2) Number of limits, and (3) Maximum number of cycle [7]. Onlooker bees select a food source depending on the foods' probability, which is calculated by the following equation:

$$p_i = \frac{\text{fit}_i}{\sum_{n=1}^{\text{SN}} \text{fit}_n} \tag{3}$$

Here, p_i = Probability of ith food source, fit_i = Fitness of ith food source, and SN = Number of food sources.

In order to produce the new viable food source from an old one, the ABC employs the following expression:

$$v_{ij} = x_{ij} + \emptyset_{ij}\left(x_{ij} - x_{kj}\right) \tag{4}$$

Here, $k = \{1, 2, 3, \ldots \text{SN}\}, j = \{1, 2, 3, \ldots D\}$ (selected random index), and $\emptyset_{ij} =$ is the random value between $[-1, 1]$ that regulates the development of new neighbor food source.

While k is selected at random, i is distinct from k. When the difference between x_{ij} and x_{kj} are reduced, the perturbation of x_{ij} decreases; and the solution approaches towards the optimum value. If a food source does not improve after a predetermined number of cycles, it is abandoned and superseded by a new food source. This predetermined number of cycles, also called limit, is a crucial metric in the ABC algorithm [3]. If x_i be the abandoned food source, then scout bee discovers a fresh food source by employing the following expression:

$$x_i^j = x_{\min}^j + \text{rand}(0, 1)\left(x_{\max}^j - x_{\min}^j\right) \quad \text{where } j = \{1, 2, 3 \ldots D\} \tag{5}$$

The new food source's fitness is compared to that of the old one and greedy selection is utilized in the selection process. As a result, a new food source is chosen if it processes fitness better than the old ones. The memory, on the other hand, preserves its former food source. Following the fulfillment of the termination condition, the best food source is selected as the best solution.

3.2 *Genetic Algorithm*

Genetic Algorithm (GA) mimics the mechanism of the natural selection process. To create new powerful offspring or generations in accordance with the theory of Darwin, this algorithm executes the genetic operators, named selection, crossover, and mutation. The individual generation is known as a chromosome. This chromosome indicates the candidate solution of the optimization problem. The mechanism of GA is as follows:

- At first, the random population is created, and the fitness of each population is evaluated.
- Based on the fitness score, chromosomes are selected for crossover.
- Mutation, which is the flipping of the genes (bits), is operated after crossover. This mutation is beneficial to diversity and prevents early convergence.
- Subsequently, the chromosomes with the fittest ratings are elected for the next iteration.
- On termination, the chromosome which possesses the fittest score is identified as an optimal solution.

Since GA is an iterative algorithm, selecting the generation size of GA is important. Large generation size necessitates high computational time. The small generation size, on the contrary, causes it to converge prematurely. In GA, there are few types of selecting methods. Among them, roulette wheel selection is the most popular. The crossover is also primarily three types. They are: (1) Single-point, (2) Double-point, and (3) Uniform.

4 Problem Formulation

To solve the ELD problem using ABC and GA, at first, the objective function and the constraints are required to formulate. The objective function and constraints of the ELD problem are discussed in the following sections.

4.1 Objective Function

The goal of ELD is to keep the power plant's total fuel cost as low as possible [21]. As a result, the objective function is the sum of all the generators' fuel costs. Without taking into account the valve point impact, the fuel cost equation of ith generators can be written as:

$$F_i(P_i) = a_i P_i^2 + b_i P_i + c_i \ (\$/\text{h}); \quad \forall i = 1, 2, 3, \ldots, N. \tag{6}$$

However, if the valve point impact is taken into account, the fuel cost function of a non-smooth ELD will be:

$$F_i(P_i) = a_i P_i^2 + b_i P_i + c_i + |e_i \sin(f_i(P_{\min} - P_i))| \\ (\$/\text{h}); \ \forall i = 1, 2, 3, \ldots, N. \tag{7}$$

$$\text{FT}_{\text{ELD}} = \sum_{i=1}^{n} F_i(P_i) \tag{8}$$

Here, a_i, b_i, c_i, e_i and f_i are fuel cost coefficients that depend on generators' properties. The power generated by ith unit is P_i, while $F_i(P_i)$ is the fuel cost to generate P_i power. Finally, to determine total fuel cost (FT_{ELD}), each generator's quadratic fuel cost function must be added.

4.2 Constraints

The constraints in ELD are divided into two categories [22]. These are: (1) Equality constraints, and (2) Inequality constraints.

Equality Constraint. The equality constraint is the basic load flow equations of active and reactive power. In general, only the real power is taken into account in the optimization of ELD. In every period, the generators require to produce power that is equal to the forecasted power demand and transmission loss. As a result, it is possible to write:

$$\sum_{i=1}^{n} P_i - P_{\text{loss}} - P_{\text{D}} = 0 \tag{9}$$

Here, $\sum_{i=1}^{n} P_i$ is the total power generation, P_{D} is power demand, and P_{loss} is transmission loss.

Inequality Constraint. Several inequality constraints are considered for the optimization of ELD. In this paper, the power generation constraints of generators, and ramp-rate constraints are mainly accounted as inequality constraints.

Capacity Constraints of Generators. The power generated by every generator is limited by capacity constraints for the smooth operation of the boiler, feedback pump, and other machinery [23]. The lower value of each generator's power generation serves as the lower bound of capacity constraints, while the greatest amount a generator can produce serves as the upper bound of capacity constraints. Each generator's output power is limited by these upper and lower limits [24].

$$P_i^{\min} \leq P_i \leq P_i^{\max} \quad \forall i = 1, 2, 3, \ldots, N. \tag{10}$$

Ramp-Rate Constraints. This constraint is the operational limit for generators, which keep them inside two specific operation zones [18]. A generator's power output in a given time must not exceed a specified amount compared to the power generated by that unit in the preceding period. It is known as the upper ramp-rate limit (UR_i). Likewise, a generator's lowest power output at any given time should not go below a specified threshold compared to that unit's previous power output. It is called lower ramp-rate limit (DR_i). Mathematically, it can be expressed as:

$$P_i - P_i^{t-1} \leq \mathrm{UR}_i \tag{11}$$

$$P_i^{t-1} - P_i \leq \mathrm{DR}_i \tag{12}$$

It is assumed that the ELD's power output can be adjusted instantly. In practice, however, the ramp-rate limits restrict the generator's power output [2]. This is because the instantaneous fluctuation in the power output may destabilize the power plant [23]. Therefore, the modified power output can be written using ramp-rate limits as follows:

$$\max(P_i^{\min}, P_i^{t-1} - \mathrm{DR}_i) \leq P_i \leq \min(P_i^{\max}, P_i^{t-1} + \mathrm{UR}_i) \tag{13}$$

4.3 *Transmission Loss Formulation*

To achieve a proper balance of power generation and demand, the transmission loss in the ELD should be taken into account. If the transmission line loss is substantial, a power generator with a low incremental cost may have a high operational cost. In addition, it is important to consider transmission loss when the distance between the power unit and the load is large [9]. In general, two different types of transmission loss formulas are used. The penalty factor approach is one, and the *B* coefficient method is another [24]. The *B* matrix formula is the most extensively used formula for transmission loss. Kron's formula is another name for it. The transmission loss calculated by this *B* matrix formula is as follows:

$$\boldsymbol{P}_{\text{loss}} = \boldsymbol{P}^{\mathrm{T}}[B]\boldsymbol{P} + B_0^{\mathrm{T}}\boldsymbol{P} + B_{00} \tag{14}$$

where $\boldsymbol{P}$ = All generators' output in vector (MW)

$[B]$ = B-matrix same dimension as **P**

B_0^T = Transpose of vector as the same length of **P**

B_{00} = Constant.

It is also possible to write (14) as follows:

$$P_{\text{loss}} = \sum_{i=1}^{N}\sum_{j=1}^{N} P_i B_{ij} P_j + \sum_{i=1}^{N} B_{i0} P_i + B_{00} \tag{15}$$

5 Simulation

The experimental results of ABC and GA in the implementation of ELD are discussed in the following section. Moreover, the performances of these two algorithms are compared. For simulation, the parameters of the algorithms and generators are initially chosen. After that, ABC and GA algorithms are used to implement the ELD. The simulation works are carried out in MATLAB software.

5.1 Parameter Selection

Table 1 provides the information about the parameters of the ABC and GA, while Table 2 lists the lower and upper generation limits, as well as cost coefficients and ramp-rate limits of 10 generators in a power plant. The initial power generation of each generator, **B**-matrix, $\boldsymbol{B}_0$, and $\boldsymbol{B}_{00}$ are also declared prior to simulation.

Table 1 Selected parameters of ABC and GA for ELD implementation

ABC		GA	
Items	Value	Items	Value
Population	100	Population	100
Iteration	100	Iteration	100
Limit	300	Crossover	Uniform
No. of generator	10	Mutation	0.1
Acceleration coefficient	1	Selection	Roulette wheel

Table 2 Selected input for 10 generators

Generator number	Generation limit		Generation cost coefficients					Ramp-rate limit	
	$P_{\min}$ (MW)	$P_{\max}$ (MW)	a_i (\$/MW^{2}h)	b_i (\$/MWh)	c_i (\$/h)	e_i (\$/h)	f_i (rad/MW)	DR (MW/h)	UR (MW/h)
P_1	150	470	0.1524	38.5397	786.7988	450	0.041	80	80
P_2	135	470	0.1058	46.1591	4513.251	600	0.036	80	80
P_3	73	340	0.0280	40.3965	1049.998	320	0.028	80	80
P_4	60	300	0.0354	38.3055	1243.531	260	0.052	50	50
P_5	73	243	0.0211	36.3278	1658.570	280	0.063	50	50
P_6	57	160	0.0179	38.2704	1356.659	310	0.048	50	50
P_7	20	130	0.0121	36.5104	1450.705	300	0.086	30	30
P_8	47	120	0.0121	36.5104	1450.705	340	0.082	30	30
P_9	20	80	0.1090	39.5804	1455.606	270	0.098	30	30
P_{10}	10	55	0.1295	40.5407	1469.403	380	0.094	30	30

$B = 10^{-4}\times$	[0.49	0.14	0.15	0.15	0.16	0.17	0.17	0.18	0.19	0.20
	0.14	0.45	0.16	0.16	0.17	0.15	0.15	0.16	0.18	0.18
	0.15	0.16	0.39	0.10	0.12	0.12	0.14	0.14	0.16	0.16
	0.15	0.16	0.10	0.40	0.14	0.10	0.11	0.12	0.14	0.15
	0.16	0.17	0.12	0.14	0.35	0.11	0.13	0.13	0.15	0.16
	0.17	0.15	0.12	0.10	0.11	0.36	0.12	0.12	0.14	0.15
	0.17	0.15	0.14	0.11	0.13	0.12	0.38	0.16	0.16	0.18
	0.18	0.16	0.14	0.12	0.13	0.12	0.16	0.40	0.15	0.16
	0.19	0.18	0.16	0.14	0.15	0.14	0.16	0.15	0.42	0.19
	0.20	0.18	0.16	0.15	0.16	0.15	0.18	0.16	0.19	0.44]

Initial Power generation of 10 units, $\boldsymbol{P}_0 = [180\ 215\ 310\ 125\ 300\ 170\ 300\ 120\ 330\ 340]$.

$\boldsymbol{B}_0 = 10^{-3}\times [-0.3908\ -0.1279\ 0.7047\ 0.0591\ 0.2161\ -0.6635\ -0.3908\ -0.1279\ 0.7047\ 0.0591]$.

$\boldsymbol{B}_{00} = 0.056$.

5.2 *Implementation of ELD*

To implement the ELD, the authors consider four separate MATLAB scripts. The first script is for the objective function, the second and third scripts are for the ABC and GA source code consecutively, and the fourth script is for the ELD implementation. The fourth script is also known as the main script. The objective function, shown in (8), and the constraints, discussed in Sect. 4.2, are coded in a single MATLAB script (objective function script). In this objective function script, the selected inputs of all the 10 generators and the transmission loss coefficients (**B**-matrix, $\boldsymbol{B}_0$, $\boldsymbol{B}_{00}$) are also provided. After selecting the parameters of the algorithms, the source codes for ABC and GA are completed in two separate scripts. Finally, the implementation of the ELD problem (satisfying the constraints) is performed by calling (linking as functions) the objective function code, the ABC source code, and the GA source code in the main script (fourth script) of MATLAB. In this case, the power demand is considered 2000 MW, which is declared in the first script with the objective function. The flow charts of Fig. 2a, b briefly present the ABC-based and GA-based ELD optimization methods.

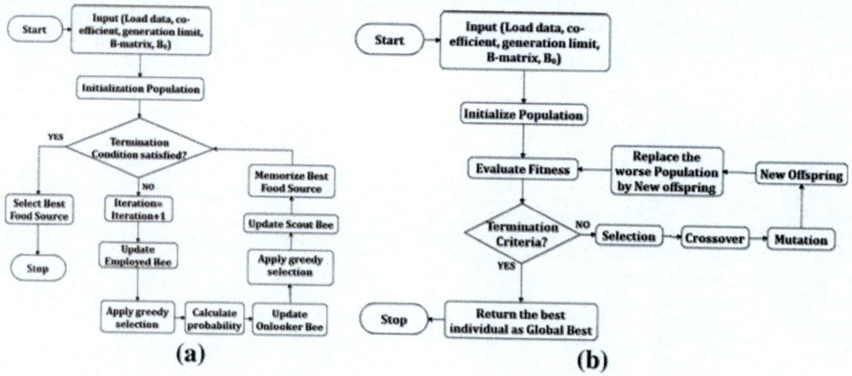

Fig. 2 Flow chart for ELD optimization. **a** ABC-based, **b** GA-based

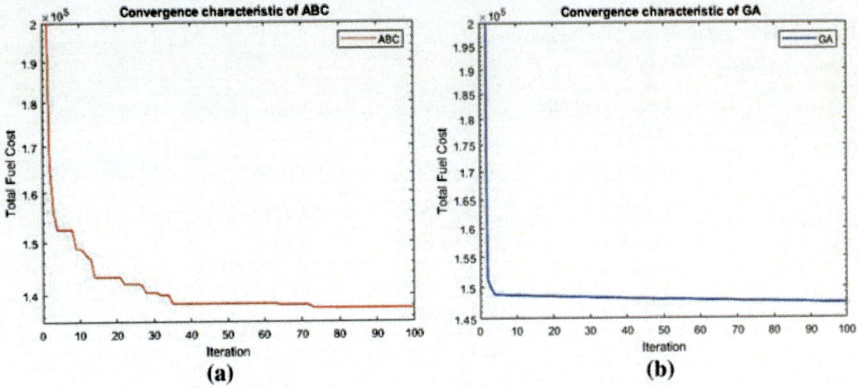

Fig. 3 Convergence characteristics of **a** ABC-based ELD, **b** GA-based ELD

6 Result and Discussion

After completing the MATLAB code, the output power, fuel cost, and transmission loss data for ABC and GA are collected. This section summarizes the simulation results and examines them in terms of convergence and computational time aspects.

6.1 Output Data

The convergence behaviors of ABC and GA are shown in Fig. 3. Moreover, Table 3 presents the optimum power output and fuel cost of each generator, as well as the total power loss of the plant. The ABC algorithm has a substantially lower ideal fuel cost than the GA approach. Using the ABC algorithm, the total generated power is 2078.14 MW, and the power loss due to transmission is 78.14 MW. Hence, the power demand of 2000 MW is maintained. On the other hand, GA finds 2079.96 MW as optimal power, where the transmission loss is 79.96 MW. However, GA cannot able to utilize the most suitable generators to fulfill the power demand, as ABC does. As a consequence, the total fuel cost is high for GA. The total fuel cost to generate this power is 137,203.52 $/h for ABC, while it is 145,933.16 $/h for GA.

6.2 Performance Comparison

To compare the performance of ABC and GA, the comparative analysis is performed both in cases of convergence and computational time properties. To analyze the convergence behavior, 100 trial simulations are conducted; on the other hand, 10 trial simulations are performed for computational time comparison. The typical

Table 3 Generated power and fuel cost for ABC and GA

Generator No.	ABC		GA	
	Power generate (MW)	Fuel cost ($/h)	Power generate (MW)	Fuel cost ($/h)
P_1	254.3231	20,853.6661	303.0601	26,467.3592
P_2	397.0142	39,519.9448	453.9285	47,796.9551
P_3	340.0000	18,319.0196	339.8754	18,311.2045
P_4	299.9279	15,940.2886	250.4625	13,178.21146
P_5	243.0000	12,000.8193	234.7712	11,544.5457
P_6	160.0000	8239.8853	155.3413	8043.5662
P_7	129.2188	6380.1628	127.5626	6357.0141
P_8	119.7397	6102.4134	99.9152	5536.0269
P_9	80.0000	5425.5725	68.3882	4942.1102
P_{10}	54.9191	4421.7392	46.6517	3756.1678
Total	2078.1431	137,203.5121	2079.9569	145,933.1615
Total power loss (MW)	78.1431		79.9569	

convergence characteristics of ABC and GA-based ELD for the selected parameters are displayed in Fig. 3. The algorithms start exploring the global cost from a very large value. The minimum total fuel costs are discovered as the number of iterations increases. Whenever the algorithms get the global best (optimum value), the convergence curves become horizontal straight lines (i.e., converge to the global optima). From Fig. 3a, the global best obtained by ABC is less than 1.4×10^5 ($/h), whereas the global optima found by GA is over 1.45×10^5 ($/h), which is depicted in Fig. 3b.

Convergence Characteristic Comparison. The convergence curves of ABC converge to approximately identical total fuel costs in 100 trial simulations, although it takes a high time to converge in global best than GA. In practically every scenario, the ABC algorithm can deliver the best results. On the other hand, there are some pre-mature convergences for GA, which lead to incorrect results. The 100 trials of the comparison reveal that in almost 20 cases, GA provides very poor results. While 100 trials are performed, the best total fuel cost obtained by GA is 145,933.16 ($/h), although the mean total fuel cost is 154,784.69 ($/h) for GA. Figure 4 shows the typical convergence behavior of ABC and GA simultaneously, while Fig. 5 presents the result for 100 trial simulations, where the superior performance of ABC is substantiated. The optimal value of ABC-based ELD is always less than that of GA which is explicitly proved from Fig. 5a. For statistical analysis, the boxplot of Fig. 5b has been plotted based on 100 independent trials, where the better performance of ABC is also demonstrated. The best cost, worst cost, mean cost, and standard deviation of these 100 trials are illustrated in Table 4. For ABC, the standard deviation is only 186.52, while it is 3994.58 for GA-based ELD problem.

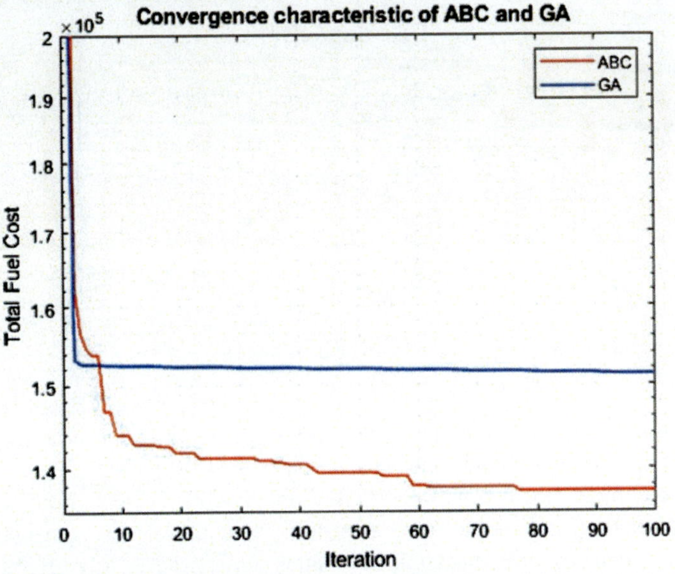

Fig. 4 Comparison of a typical convergence of ABC and GA

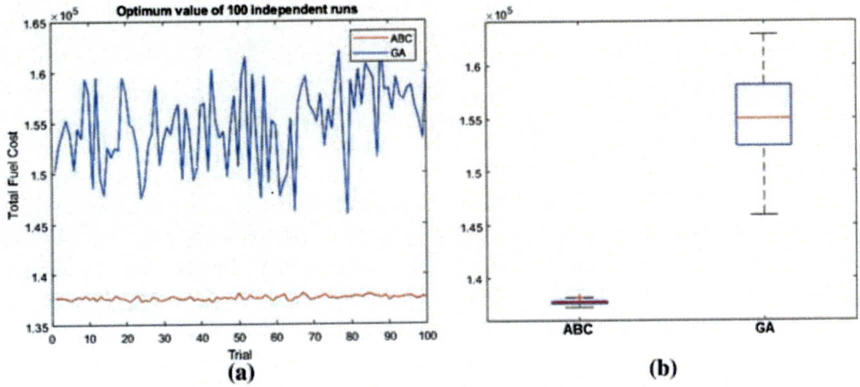

Fig. 5 Comparison **a** optimum value of 100 simulation trials, **b** boxplot of 100 optimal values of ABC and GA

Table 4 Statistical comparison between ABC and GA in ELD optimization

Metaheuristics	Best cost ($/h)	Worst cost ($/h)	Mean cost ($/h)	Standard deviation
ABC	**137,203.5121**	**138,137.2517**	**137,646.3124**	**186.5151**
GA	145,933.1615	162,848.0557	154,784.6869	3994.5774

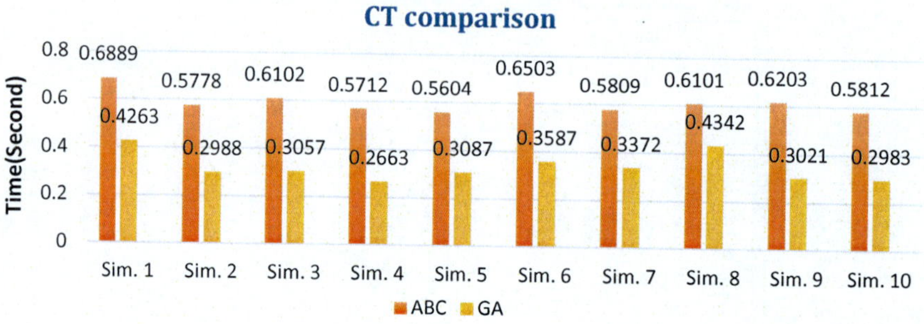

Fig. 6 Computational time comparison for 10 random simulations

Computational Time Comparison. The average computational time for ABC is about 0.60 s, while this value is 0.30 s on average for GA. Figure 6 shows a comparison of the computational time of the ABC and GA in the ELD problem for 10 random simulations, where the ABC algorithm has demonstrated somewhat worse performance than GA. From the comparison, it is explicit that ABC is slightly slower than GA in solving the ELD problem. However, this computational time will not be a barrier in real-life ELD optimization.

7 Conclusion

In the optimization of non-smooth ELD having the transmission loss, it is evident that ABC outperforms the GA in terms of minimizing the fuel cost, even though some computational time is needed to sacrifice. For more thorough examinations, it is important to consider 20, 40, 80, or 140 units power plants. Furthermore, different types of fuels must be taken into account, as fuel properties impact a generator's fuel cost coefficients. Before reaching a concrete decision for ABC, it is essential to compare the result of this algorithm with other recently proposed metaheuristic algorithms. The comparative analysis with these algorithms will provide better reliability for the ABC algorithm than GA. To utilize the GA, it is necessary to hybridize this algorithm with other metaheuristic algorithms.

References

1. Karaboga D, Basturk B (2007) Artificial bee colony (ABC) optimization algorithm for solving constrained optimization problems. In: 12th international fuzzy systems association world congress. Springer, Berlin, Heidelberg, pp 789–798
2. Mahor A, Prasad V, Rangnekar S (2009) Economic dispatch using particle swarm optimization: a review. Renew Sustain Energy Rev 13:2134–2141

3. Singh OV, Singh M (2020) A comparative analysis on economic load dispatch problem using soft computing techniques. Int J Softw Sci Comput Intell 12:50–73
4. He X, Rao Y, Huang J (2016) A novel algorithm for economic load dispatch of power systems. Neurocomputing 171:1454–1461
5. Hanif M, Mohammad N, Harun B (2019) An effective combination of microcontroller and PLC for home automation system. In: 2019 1st international conference on advances in science, engineering and robotics technology (ICASERT). IEEE, pp 1–6
6. Sinha N, Chakrabarti R, Chattopadhyay PK (2003) Evolutionary programming techniques for economic load dispatch. IEEE Trans Evol Comput 7:83–94
7. Karaboga D, Basturk B (2007) A powerful and efficient algorithm for numerical function optimization: artificial bee colony (ABC) algorithm. J Glob Optim 39:459–471
8. Kumar M, Husain M, Upreti N, Gupta D (2010) Genetic algorithm: review and application. Int J Inf Technol Knowl Manag 2:451–454
9. Wood AJ, Wollenberg B (1996) Power generation operation and control, 2nd edn. Wiley
10. Islam MA, Hasan N, Mohammad N (2020) Power system optimization model using economic load dispatch. In: 2020 IEEE region 10 symposium (TENSYMP). IEEE, pp 469–472
11. Ravikumar Pandi V, Panigrahi BK (2011) Dynamic economic load dispatch using hybrid swarm intelligence based harmony search algorithm. Expert Syst Appl 38:8509–8514
12. Warsono W, Ozveren CS, King DJ, Bradley D (2008) A review of the use of genetic algorithms in economic load dispatch. In: 43rd international universities power engineering conference. IEEE, pp 1–5
13. Abido MA (2003) A novel multiobjective evolutionary algorithm for environmental/economic power dispatch. Electr Power Syst Res 65:71–81
14. Hosseini SH, Kheradmandi M (2004) Dynamic economic dispatch in restructured power systems considering transmission costs using genetic algorithm. In: Canadian conference on electrical and computer engineering 2004 (IEEE Cat. No. 04CH37513). IEEE, pp 1625–1628
15. Hong YY, Li CT (2006) Short-term real-power scheduling considering fuzzy factors in an autonomous system using genetic algorithms. IEE Proc Gener Transm Distrib 153:684–692
16. Roa-Sepulveda CA, Pavez-Lazo BJ (2003) A solution to the optimal power flow using simulated annealing. Int J Electr Power Energy Syst 25:47–57
17. Safari A, Shayeghi H (2011) Iteration particle swarm optimization procedure for economic load dispatch with generator constraints. Expert Syst Appl 38:6043–6048
18. Pradhan M, Roy PK, Pal T (2016) Grey wolf optimization applied to economic load dispatch problems. Int J Electr Power Energy Syst 83:325–334
19. Karaboga D, Ozturk C (2011) A novel clustering approach: artificial bee colony (ABC) algorithm. Appl Soft Comput 11:652–657
20. Karaboga D, Basturk B (2008) On the performance of artificial bee colony (ABC) algorithm. Appl Soft Comput 8:687–697
21. Sadiq M, Mohammad N, Nadeem A (2019) Optimized energy generation model and pricing strategy to solve economic load dispatch. In: 2019 IEEE international conference on power, electrical, and electronics and industrial applications (PEEIACON). IEEE, pp 74–78
22. Banerjee S, Maity D, Chanda CK (2015) Teaching learning based optimization for economic load dispatch problem considering valve point loading effect. Int J Electr Power Energy Syst 73:456–464
23. Zhang R, Zhou J, Mo L, Ouyang S, Liao X (2013) Economic environmental dispatch using an enhanced multi-objective cultural algorithm. Electr Power Syst Res 99:18–29
24. Rao DLVN (2014) PSO technique for solving the economic dispatch problem considering the generator constraints. Int J Adv Res Electr Electron Instrum Eng 3:10439–10454

Forecasting Closing Price of Stock Market Using LSTM Network: An Analysis from the Perspective of Dhaka Stock Exchange

Md. Mohsin Kabir, Aklima Akter Lima, M. F. Mridha, Md. Abdul Hamid, and Muhammad Mostafa Monowar

Abstract As an essential ingredient of the financial market, the stock exchange has been involved by several researchers. Specific commercial prognostication is of magnificent possible engagement to both private and organizational investors. How to perceive the stock market trend and forecast the stock price is a dilemma many researchers investigate. In earlier studies, the prognostication techniques essentially concentrate on statistical principles and conventional neural network architectures that are comparatively familiar in contemporary times. In this study, Long Short Term Memory (LSTM) based novel architecture of forecasting stock closing price is investigated under the perspective of Dhaka stock exchange data. As the LSTM architecture has a long-term memory function, it performs the correct commercial time-series prognostication. The performance of the proposed architecture is measured using MAE, RMSE, and MAPE evaluation matrixes. Our experiments show how MAE, RMSE, and MAPE scores significantly decrease after using the updated gate to train an LSTM network.

Keywords Stock market · Long short term memory (LSTM) · Recurrent neural network (RNN) · Deep learning

---

Md. Mohsin Kabir (✉) · A. A. Lima · M. F. Mridha
Bangladesh University of Business and Technology, Dhaka, Bangladesh

M. F. Mridha
e-mail: firoz@bubt.edu.bd

Md. Abdul Hamid · M. M. Monowar
King Abdulaziz University, Jeddah, Saudi Arabia
e-mail: mabdulhamid1@kau.edu.sa

M. M. Monowar
e-mail: mmonowar@kau.edu.sa

M. S. Arefin et al. (eds.), *Proceedings of the International Conference on Big Data, IoT, and Machine Learning*, Lecture Notes on Data Engineering and Communications Technologies 95, https://doi.org/10.1007/978-981-16-6636-0_23

1 Introduction

Stock price indexes are indeed fundamental in global financial markets since they are the primary criteria for measuring the role of investments and the stock market. Among the most challenging problems for the AI community has been predicting stock prices. Stock market analysts concentrate on designing methods to predict stock prices to maximize income effectively. The booming stock market prediction concept produces the desired results with the least necessary inputs and the simplest stock market model. Many researchers have become interested and are working on predicting the stock price. An accurate prediction of a future price can result in higher profit growth for investors with stock investments. Widely known algorithms, such as Support Vector Machine (SVM), Reinforcement Learning, Multilayer Perceptron (MLP), Recurrent Neural Network (RNN), Genetic Algorithm (GA), Markov Chain Model (MCM), Deep Neural Network (DNN), and Long Short Term Memory (LSTM), were shown to be efficient in predicting stock price. Besides, these models increase the benefit of stock price purchases while keeping risk low [1].

Artificial Neural Networks (ANN) are a standard option for stock price prediction, which have been shown to perform well [2, 3]. Deep learning (DL) has recently emerged as an improved approach to traditional neural networks and applied in financial markets to predict stock prices [4, 5]. As we have seen, a comparison analysis of various DL models of stock market prediction has been conducted [6]. Experiments have been carried out to evaluate its growth and efficiency as a trading mechanism in the stock index futures markets. Nevertheless, there is still a scarcity of research on deep learning as a component of a trading framework [7]. LSTM is a well-known deep learning model for classifying, processing, and generating predictions depending on time series data in the stock price prediction approach. Memorization of early phases in the LSTM can be accomplished using gates with a long memory line integrated. The primary purpose of this article is to forecast future closing prices and create a booming stock market prediction framework with a simple LSTM model. We collected and gathered seven big banks and organizations closing stock prices from the Dhaka Stock Exchange (DSE) data archive for the stock price prediction. We obtained the claimed accomplishments in our work by utilizing those data:

- We investigated thoroughly to assemble some essential frameworks on future closing price prediction and analyzed them.
- We collected Dhaka Stock Exchange (DSE) data which consists of the seven big banks and organizations' data on the closing price of stocks and made a dataset.
- We trained our dataset with the LSTM framework computed MAE, RMSE, MAPE values to inspect the error between predicted and tested values and update the framework by diminishing the errors, and it performed satisfactory results.

The sequence of this empirical article is prepared as follows: The premature research is explained in Sect. 2. Section 3 describes the methodology of the study.

Section 4 presents the dataset, evaluation, reports on the experiments as well. Lastly, Sect. 5 draws the article to a conclusion.

2 Related Work

Several artificial intelligence and machine learning methods have been practiced to forecast the stock market over the past decade. Hiransha et al. [8] proposed a day-by-day closing price forecast on the NSE and NYSE stock exchanges utilizing four forms of deep learning, namely multilayer perceptron, RNN, LSTM, including Convolutional Neural Network (CNN). The authors trained the system solely on a single NSE business and expected five NSE and NYSE companies. Shen et al. [9] suggested a new prediction that uses SVM to anticipate the next-day price trend. The same algorithm is also used in conjunction with various regression algorithms to map the specific increment and used to compare the output of the proposed prediction algorithm to other benchmarks. Mehtab presented a hybrid method based on NIFTY50 indexes of the stock closing price. Various classification techniques and regression models were used to predict price, developed an LSTM to predict closing prices, and compared accuracy with other machine learning algorithms. A cross-validation method based on self-organizing fuzzy neural networks was used for sentimental analysis on social media data, reflecting the public sentiment of stock prices [10].

Choudhry and Garg [11] suggested a hybrid method for stock price prediction based on genetic algorithm and SVM. The authors used a selection of 35 these technical indicators collected from the projected stock as candidates for input features used by financial experts [12]. The result shows the hybrid GA-SVM system outperforms the SVM system. A stock market system predicts the index price of the Singapore stock market and uses the FTSE Straits Time Index (STI) using a feed-forward deep neural network. DNN forecasted stock prices for the following t days using historical data prices and a trading framework to make selling and buying decisions. RMSE and MAPE are used to assess DNN results [13]. Zhang and Zhang [14] suggested a Markov methodology for stock market trends and expected outcomes representing probability with a particular state of stock prices over time instead of being in an absolute state. They investigated the use of the Markov chain in the stock market and found it to be successful. Gao et al. [15] proposed Deep Belief Networks (DBNs), a form of deep learning algorithm model, are implemented as a novel approach to predicting the closing price of the stock market, together with stock technical indicators (STIs) and two-dimensional principal component analysis.

Patel and Marwala [16] designed an application that uses artificial neural networks such as Multilayer Perceptron and Radial Basis Function neural network architectures to help investors make financial decisions. Long short term memory is used in a novel stock closing price predicting system that outperforms and combines Empirical Wavelet Transform (EWT)-based preprocessing and outlier robust extreme learning machine (ORELM)-based post-processing. In addition, the dropout strategy and the Particle Swarm Optimization (PSO) algorithm are used to optimize LSTM [17] jointly. This study focused on neural and neuro-fuzzy techniques developed and implemented in stock market forecasting [18]. Input data, forecasting methodology, performance assessment, and performance indicators utilized are all classified. It has been discovered that neural networks and fuzzy neuro models were appropriate for stock market forecasting. Studies showed that soft computing methods outperform in the vast majority of instances. According to the results of this study, neuro-computational systems are valuable tools for predicting stock market changes in developing economies. For predicting KSE closing price, they used MLP neural networks and generalized regression neural networks. Their research shows that the quasi-Newton training algorithm generates a lesser prediction error than other algorithms [19].

Furthermore, over the last few decades, a significant number of deep learning-based stock price prediction approaches on its closing price have been suggested. Our present paper examines the closing prices used in Dhaka Stock Exchange, but it can be applied to any dataset from everywhere.

3 Methodology

The long short term memory deep learning architecture is applied to the mentioned dataset for forecasting stock prices. This section explains the data preprocessing techniques, LSTM networks, and the proposed architecture.

3.1 Data Preprocessing

The data preprocessing technique implies two-stage of data processing: data cleaning and normalization. Data cleaning refers to remove irrelevant entries and filling the missing values in the dataset. The missing values are replaced with the median of the relevant columns values. However, as the dataset contains stock data from diverse organizations, the data must be beneath a standard scale. Normalizing the data sets decreases the inconstancy and noise within the data. So we applied the following data normalization equation to the dataset.

$$X_{\mathrm{norm}} = \frac{X - X_{\mathrm{min}}}{X_{\mathrm{max}} - X_{\mathrm{min}}} \tag{1}$$

where, X_{norm} refers to the normalized entry, X means the real data, X_{min} defines the smallest value in the dataset and X_{max} defines the largest value in the dataset.

3.2 LSTM Network

LSTM [20] is a particular representation of RNN. LSTM networks are experts in acquiring long-term dependences. In the LSTM structure, the regular hidden layers are substituted with LSTM cells. Specific networks are intended to avoid long-term dependence difficulty but learn knowledge for an extended period. The cells of LSTM networks are formed with numerous gates that can regulate the input stream. Figure 1 presents a picturesque depiction of the LSTM cell.

An LSTM cell typically contains four gates: input gate, cell state, forget gate, and output gate. Besides, the LSTM cell has a sigmoid layer, tanh layer, also pointwise multiplication operation. The explanation of these gates are given below:

- Input gate: It contains the input data. x_t refers to the input vector and i_t refers to the input gate vector. i_t can be calculated as follows:

$$i_t = \sigma\left(W_{\mathrm{i}} \cdot \left[h_{t-1}, x_t\right] + b_{\mathrm{i}}\right) \tag{2}$$

- Cell State: Springs through the whole network and can append or extract information with the help of gates. c_t defines the cell state vector and can be calculated as follows:

$$c_t = \tanh\left(W_{\mathrm{c}} \cdot \left[h_{t-1}, x_t\right] + b_{\mathrm{c}}\right) \tag{3}$$

- Forget gate layer: Determines the portion of the knowledge to be conceded. f_t means the forget gate vector, and the following equation calculates f_t:

$$f_t = \sigma\left(W_{\mathrm{f}} \cdot \left[h_{t-1}, x_t\right] + b_{\mathrm{f}}\right) \tag{4}$$

- Output gate: It determines the output produced by the LSTM architecture. o_t refers to the output gate vector, and it can be calculated as follows:

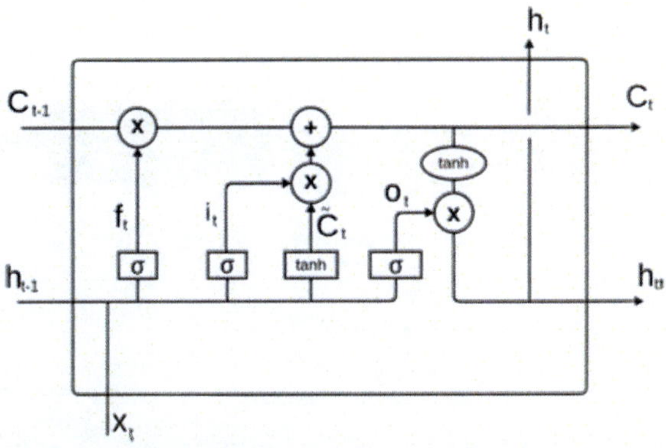

Fig. 1 Long short term memory network. Here C_{t-1} defines old cells state, C_t means current cell state, h_{t-1} refers to the output of the prior cell, h_t defines the output of the current cell, i_t means input gate layer, f_t refers to forget gate layer, and O_t represents output sigmoid gate layer

$$o_t = \sigma\left(W_o \cdot \left[h_{t-1}, x_t\right] + b_o\right) \tag{5}$$

Finally, h_t refers to the output vector, and the following equation calculates h_t:

$$h_t = o_t * \tanh(c_t) \tag{6}$$

here, W and b are the parameter matrix and vectors.

3.3 System Architecture

The proposed stock price forecasting system is evaluated based on the seven big banks and organizations stock observation dataset collected from the Dhaka stock exchange data archive. First, the dataset is cleaned and normalized by mentioned data preprocessing techniques. Then the dataset split into train and test set according to 80–20% formation. Then the LSTM network is applied to the dataset to train the model. After completing the training, the future time steps are forecasted. The future time steps have been predicted two times. Firstly, after completing initial training and secondly, after updating the network. The forecasted values are compared with the 20% test data separated previously in each forecast. Also, the root MAE, RMSE, MAPE values have been measured in both training times. The reason to calculate the RMSE is to find out how much accurate prediction this method can provide. In the end, completing those procedures, we have got a final result with reliable forecasting (Fig. 2).

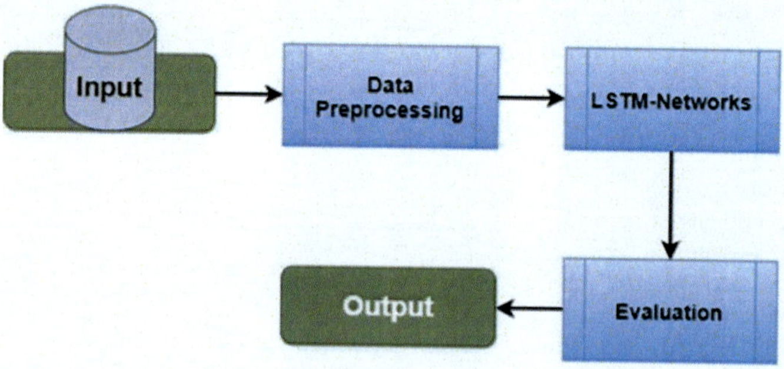

Fig. 2 Flow diagram of the proposed stock market forecasting architecture

4 Evaluation

The proposed LSTM network-based architecture is evaluated and experimented in this section. This section also briefly explains the dataset, evaluation metrics, system setup, and experiments and comparisons.

4.1 Dataset

The dataset that has been exploited in our study was obtained from Dhaka Stock Exchange's data archive. We have collected 327 observations for each of the seven big banks and organizations. The banks and organizations are ACI Formulations (ACIF), Brac Bank (BRAC), Beximco Pharmaceuticals (BXPH), Bata Shoe Company Bangladesh Ltd (BATA), City Bank Ltd (CTBK), Grameenphone Ltd (GRAE), Islami Bank Bangladesh Ltd (ISLB). We have collected the lower, upper and last prices for each of the observations. The data collection timeline was 5 January 2020 to 04 May 2021. To forecast the closing prices, we have implemented the proposed model by exploiting these data. Data have been split into train and test after stacking the data to the system. The first 80% of the data have used for training the network, and the last 20% for testing the forecasted data. After splitting into train and test, we have standardized the train data and split it into two parts for training the network.

4.2 Evaluation Metric

The mean absolute error (MAE), root mean square error (RMSE), and mean absolute percentage error (MAPE) evaluation matrixes are applied to measure the efficiency of the proposed architecture.

The most straightforward matrix of forecast accuracy is called mean absolute error (MAE). The following formula calculates it:

$$\text{MAE} = \frac{1}{N}\left(\sum_{t=1}^{N}\left|y(t) - \hat{y}(t)\right|\right) \tag{7}$$

RMSE is a conventional technique to estimate the deviation of a method in forecasting quantitative data. The following equation is used to calculate RMSE:

$$\text{RMSE} = \sqrt{\sum_{i=1}^{n}\left(\hat{y}_i - y_i\right)^2} \tag{8}$$

The mean absolute percentage error (MAPE) is specially used for forecasting problems. The following equation measure MAPE:

$$\text{MAPE} = \frac{1}{N}\left(\sum_{t=1}^{N}\left|\left(y(t) - \hat{y}(t)\right)/y(t)\right|\right) \tag{9}$$

where $(\hat{y}_1, \hat{y}_2, \ldots, \hat{y}_n)$ are predicted values, $(y_1, y_2, \ldots, y_n)$ are actual values, and n is the number of observations.

4.3 Experimental Setup

To implement the LSTM model and evaluate the dataset, we have used Python, Pandas, Numpy, Keras, and TensorFlow. These technologies implemented the model and analyzed the results. The implementation and evaluation process is conducted in the Google Colab environment.

4.4 Experiments and Comparisons

First, the dataset is split into train and test sets, and the LSTM network is applied to the train set. After complete the training of the network, we have forecasted values with the trained data. Here, we predict the costs of various time stages in the forthcoming. For the prediction, an update state method has been exploited to forecast time stages one after another and renovate the network state at every forecast. Table 1 illustrates the MAE, RMSE, MAPE values of the proposed model on initial training and updated training on the proposed dataset.

Table 1 The table presents the RMSE (in %), MAE, MAPE (in %) values of initial training (In.) and updated training (Up.) on the created dataset

Banks	RMSE (%)		MAE		MAPE (%)	
	In.	Up.	In.	Up.	In.	Up.
ACIF	27.53	3.71	0.89	0.011	22.25	2.583
BRAC	13.72	1.68	0.77	0.017	14.17	1.478
BXPH	15.57	1.87	0.88	0.009	13.62	1.089
BATA	13.64	1.43	0.96	0.082	12.28	1.025
CTBK	11.23	1.23	0.68	0.008	11.76	1.023
GRAE	7.13	0.89	0.74	0.014	6.74	0.778
ISLB	26.52	2.87	0.96	0.026	22.45	2.324

Table 2 The table presents the RMSE, MAPE (in %) values of the different algorithms on three other datasets

Models	Dataset I		Dataset II		Dataset III	
	RMSE	MAPE (%)	RMSE	MAPE (%)	RMSE	MAPE (%)
Sun's [21]	0.243	1.35	0.146	1.027	0.148	0.608
Roondiwala's [22]	0.727	4.34	1.567	14.757	0.999	4.129
Basak's [23]	0.068	0.33	0.077	0.545	0.108	0.447
Liu's [17]	0.008	0.041	0.018	0.155	0.029	0.107
Proposed model	0.078	0.039	0.016	0.149	0.031	0.113

Table 1 illustrated that the RMSE, MAE, MAPE values for each organization are much higher than updated training. The table demonstrates that after the updated training of the LSTM networks, the RMSE, MAE, MAPE values reduce significantly, greatly increasing the proposed architecture's prediction accuracy.

Furthermore, to prove the significance of the proposed model, the model is applied to three different datasets and benchmarked with three popular stock price forecasting methods Sun's model [21], Roondiwala's model [22], Basak's model [23]. Dataset #I is elected from the Standard and Poor's 500 Index; dataset #II comes from China Min-sheng Bank (CMSB), and dataset #III comes from Dow Jones Industrial Average [17].

Table 2 exhibits that the proposed model gives the best RMSE and MAPE scores for Dataset I and Dataset II but gives slightly higher for dataset III than other popular models. For Dataset III, Liu and Long [17] model provides the highest result. So, we can conclude that the proposed LSTM-based model gives significant performance to stock price prediction problems.

5 Conclusion

This study involved the construction of a method that could predict the closing price of the stock market. The Dhaka stock exchange archive data of the seven banks and organizations' are used to train an LSTM-based model to forecast the closing price accurately. This article investigates the applicability of long short term memory in the stock market and delivers comparatively excellent outcomes. Besides, the study identifies a few research work directions regarding stock price prediction. First, exploring related deep learning architectures to find more effective results, such as the Deep Boltzmann machine and Deep Q-networks. Political and economic factors can be added to the dataset to get a more reliable forecast. Lastly, including market-specific region information inside the method might assist in producing more reliable predictions.

Acknowledgments We thankfully acknowledge the assistance of the Advanced Machine Learning lab for their resource sharing and support.

References

1. Huang W, Nakamori Y, Wang SY (2005) Forecasting stock market movement direction with support vector machine. Comput Oper Res 32(10):2513–2522
2. Vui CS, Soon GK, On CK, Alfred R, Anthony P (2013) A review of stock market prediction with artificial neural network (ANN). In: 2013 IEEE international conference on control system, computing and engineering, Nov 2013. IEEE, pp 477–482
3. Yetis Y, Kaplan H, Jamshidi M (2014) Stock market prediction by using artificial neural network. In: 2014 world automation congress (WAC), Aug 2014. IEEE, pp 718–722
4. Akita R, Yoshihara A, Matsubara T, Uehara K (2016) Deep learning for stock prediction using numerical and textual information. In: 2016 IEEE/ACIS 15th international conference on computer and information science (ICIS), June 2016. IEEE, pp 1–6
5. Day MY, Lee CC (2016) Deep learning for financial sentiment analysis on finance news providers. In: 2016 IEEE/ACM international conference on advances in social networks analysis and mining (ASONAM), Aug 2016. IEEE, pp 1127–1134
6. Selvin S, Vinayakumar R, Gopalakrishnan EA, Menon VK, Soman KP (2017) Stock price prediction using LSTM, RNN and CNN-sliding window model. In: 2017 international conference on advances in computing, communications and informatics (ICACCI), Sept 2017. IEEE, pp 1643–1647
7. Deng Y, Bao F, Kong Y, Ren Z, Dai Q (2016) Deep direct reinforcement learning for financial signal representation and trading. IEEE Trans Neural Netw Learn Syst 28(3):653–664
8. Hiransha M, Gopalakrishnan EA, Menon VK, Soman KP (2018) NSE stock market prediction using deep-learning models. Procedia Comput Sci 132:1351–1362
9. Shen S, Jiang H, Zhang T (2012) Stock market forecasting using machine learning algorithms. Department of Electrical Engineering, Stanford University, Stanford, CA, pp 1–5
10. Mehtab S, Sen J (2019) A robust predictive model for stock price prediction using deep learning and natural language processing. Available at SSRN 3502624
11. Choudhry R, Garg K (2008) A hybrid machine learning system for stock market forecasting. World Acad Sci Eng Technol 39(3):315–318
12. Kaufman PJ (2011) Alpha trading: profitable strategies that remove directional risk, vol 455. Wiley
13. Yong BX, Rahim MRA, Abdullah AS (2017) A stock market trading system using deep neural network. In: Asian simulation conference, Aug 2017. Springer, Singapore, pp 356–364
14. Zhang D, Zhang X (2009) Study on forecasting the stock market trend based on stochastic analysis method. Int J Bus Manag 4(6):163–170
15. Gao T, Li X, Chai Y, Tang Y (2016) Deep learning with stock indicators and two-dimensional principal component analysis for closing price prediction system. In: 2016 7th IEEE international conference on software engineering and service science (ICSESS), Aug 2016. IEEE, pp 166–169
16. Patel PB, Marwala T (2006) Forecasting closing price indices using neural networks. In: 2006 IEEE international conference on systems, man and cybernetics, Oct 2006, vol 3. IEEE, pp 2351–2356
17. Liu H, Long Z (2020) An improved deep learning model for predicting stock market price time series. Digit Signal Process 102:102741
18. Atsalakis GS, Valavanis KP (2009) Surveying stock market forecasting techniques—part II: soft computing methods. Expert Syst Appl 36(3):5932–5941

19. Mostafa MM (2010) Forecasting stock exchange movements using neural networks: empirical evidence from Kuwait. Expert Syst Appl 37(9):6302–6309
20. Guresen E, Kayakutlu G, Daim TU (2011) Using artificial neural network models in stock market index prediction. Expert Syst Appl 38(8):10389–10397
21. Sun Y, Gao Y (2015) An improved hybrid algorithm based on PSO and BP for stock price forecasting. Open Cybern Syst J 9(1)
22. Roondiwala M, Patel H, Varma S (2017) Predicting stock prices using LSTM. Int J Sci Res (IJSR) 6(4):1754–1756
23. Basak S, Kar S, Saha S, Khaidem L, Dey SR (2019) Predicting the direction of stock market prices using tree-based classifiers. N Am J Econ Finance 47:552–567

A Dimensionality Reduction Based Efficient Multiple Voice Disease Recognition Scheme Using Mel-Frequency Cepstral Coefficients and K-Nearest Neighbors Algorithm

Shovan Bhowmik, Mahedi Hasan, and Muhammad Ataul Hakim

Abstract Disease diagnosis in medical healthcare leveraging machine learning is an area of significant interest in the whole world. Audio data analysis has facilitated the route of identifying voice disorders in a non-invasive manner. As vocal cord malady can immerse anytime from various bad habits (loud sound, smoking, extra force on vocal cords, etc.) and neurological imbalance, early recognition of voice disorders can save people from causing long-term damage. In this research, the three most injurious voice diseases have been recognized exploiting the LDA based MFCC feature matrix and KNN algorithm. Our empirical study has successfully predicted four voice class labels using this scheme with the foremost 96.49% accuracy and surpassed five other mining algorithms including Artificial Neural Network. Moreover, this model can generate output within 22.37 ms which is also the lowest execution time compared to the traditional clustering methods. Our proposed model can be easily implemented to design end-to-end services for efficient voice disease classification.

Keywords Voice pathology · Linear discriminant analysis · MFCC · K-nearest neighbors · Artificial neural network · Voting classifier

1 Introduction

With the advent of powerful computational devices and the conception of Machine Learning (ML), researchers have shifted their experiments towards non-invasive disease identification in medical science which can help people to identify malady in a fast, low-cost, and trustworthy way. These non-invasive technology dependent modules can assist people not only in early diagnosis of disease but also aid quick recovery from ailments. Voice pathology detection has been a promising

S. Bhowmik (✉)
Bangladesh Army International University of Science and Technology, Cumilla Cantonment, Bangladesh

M. Hasan · M. A. Hakim
Khulna University of Engineering and Technology, Khulna, Bangladesh

M. S. Arefin et al. (eds.), *Proceedings of the International Conference on Big Data, IoT, and Machine Learning*, Lecture Notes on Data Engineering and Communications Technologies 95, https://doi.org/10.1007/978-981-16-6636-0_24

area where extracting different acoustic features, for example, Shimmer (%), Jitter (%), Fundamental Frequency (F0), Mel-Frequency Cepstral Coefficients (MFCC), Wavelet Packet Decomposition (WPD), etc. along with ML and Deep Learning (DL) algorithms can produce disorder recognition framework [1].

Generally, voice pathologies are created due to the existence of tissue contagion, breathing disturbance, neurological imbalance, muscular substitution, vocal fold contraction, vocal surface vexation, tissue cell alteration, and other factors [2]. There are around 71+ voice diseases associated with voice disorder which include, Dysphonia, Laryngitis, Reinke's Edema, Vocal Fold Nodules and Polyps, Vocal Cord Paralysis, etc. [3, 4]. Dysphonia is a voice disease that occurs because of the creation of nodules in the vocal cords, swelling in the larynx or sudden traumatic event in the cords. Around 10% people of in the world encounter this type of disease [5]. Another commonly occurring voice disease is Laryngitis which is caused by the swelling of vocal folds. This disease can become acute sometimes if viruses attack the vocal folds [4]. Reinke's Edema can happen by taking overstress or because of immoral habits like smoking, loud shouts, etc. Other voice abnormalities can appear for the aforementioned causes as well. Almost all the voice diseases make the sound scratchy and gruff. As voice is produced from neurological signals, these problems can also hamper brain cells [6]. Thus, careless attitudes obviate severe situations which can be sometimes not curable by surgeries and might lead to gruesome cancer.

Early detection of voice pathology can reduce the risk of grievous circumstances. So far, most of the works related to voice disease recognition using ML and DL are based on the binary classification that predicts a voice sample whether it is healthy or pathological [7, 8]. Among traditional ML algorithms, Support Vector Machine (SVM), Gaussian Mixture Model (GMM), Decision Tree (DT), K-Nearest Neighbors (KNN), etc. had been widely used for voice disease detection. Several DL algorithms, e.g., Artificial Neural Network (ANN), Convolutional Neural Network (CNN), Long Short-Term Memory (LSTM)—CNN Hybrid, Bidirectional LSTM, had been considered for this task in previous works too. Previously, multiple public voice datasets were incorporated for this job. For instance, MEEI, SVD, and VOICED datasets have been constantly employed for classifying healthy and diseased voices [9]. Unfortunately, the majority of the research works focused on only two class labels and the accuracy was ranged approximately between 70 and 94%.

In this work, we have proposed a Linear Discriminant Analysis (LDA) based multiple voice disease distinguishing framework by exploiting MFCC features and the K-NN algorithm. We have introduced three crucial voice disorder classifications, e.g., Dysphonia, Laryngitis, and Renkei's Edema along with healthy voices by training our model with the SVD dataset. Moreover, our study has shown that LDA based K-NN algorithm can reach up to 97% accuracy in multiple voice pathology categorization by taking only 13 MFCC attributes. Additionally, four cross-validation techniques have been included in this work for evaluating model performance as our dataset was imbalanced. To benchmark our diagnosis scheme, we have employed ML based supervised polynomial kernel-based SVM, probabilistic GMM, ensemble-based Random Forest (RF), a voting classifier including SVM, KNN, and RF, and DL based 3-Layer ANN. Our experiment has found that the accuracy achieved by applying KNN with Shuffle Split and Stratified Shuffle Split cross-validation techniques can outperform the ANN model. Another highlighted characteristic of this experiment is the intensive comparison among models on execution time.

2 Literature Review

ML and DL models have been devised for voice pathology detection in many previous studies. MFCC and Acoustic Features (AF) had been extracted for applying XGBoost, Isolation Forest, and Dense Net to distinguish diseased voices from normal ones in [10]. Four well established public voice databases namely AVPD, MEEI, PDA, and SVD was availed with audio files consisting of /a/e/o/ vowels. The highest 0.733 $F1$-Score and 0.759 precision were achieved from this experiment where dimensionality reduction was not applied for training purposes. Dysphonia was detected by applying SVM, KNN, and RF in [11]. In this paper, Shimmer, Jitter, MFCC, etc. features were considered for three vowel pronunciation recordings from the SVD database. The highest 91.3% classification accuracy was attained for the RF model. Principal Component Analysis (PCA) was applied here for reducing dimensionality which results in a low accuracy ranging from 65 to 80% on average to classify two labels. A comparative analysis was illustrated among Bagging, Boosting and LibD3c ensemble learning models for multiple audio features like Pitch, Intensity, Global Features, MFCC, WPD, CA, etc. in [1]. Although around 96% accuracy was gained in this work, the authors did not validate their model using cross-validation. Moreover, specific voice diseases were not classified in this work. However, KNN and LDA based voice ailment recognition was done in [12] which achieved 93% accuracy with a private dataset.

Voice malady identification was established in [13] using Deep Neural Network architecture. A hybrid combination of the CNN-LSTM-Dense layer was proposed in this paper. The validation report demonstrated only 71.36% accuracy in diagnosing pathology in this study. In [14], a CNN framework was devised for the binary classification of voice samples using the SVD database. The authors of this paper achieved 95.41% accuracy in this case with a balanced dataset. However, Cyst, Polyp, Paralysis, and healthy voice samples were clustered using VGG16 and CaffeNet models for MEEI and SVD databases in [8]. This work accomplished at most 94.5% accuracy by analysis of Fourier Transform of voice spectrums.

The majority of the above-mentioned studies gave priority to only two classes and only several vowel utterances were considered for drawing out voice attributes. Cross-validation was rarely taken into account. Another limitation was the absence of calculating processing time for model execution. Although the speech recognition area has few pieces of research on the distinction between LDA and PCA performance regarding diabetes prediction [15], speech recognition lacks any solid study on it. This research work has shown state-of-art performance in classifying four voice classes utilizing full sentence utterance recordings of the SVD database.

3 The Proposed Multiple Voice Disease Recognition Scheme

In our study, we have emphasized recognizing three types of voice diseases, particularly, Dysphonia, Laryngitis and Reinke's Edema. Several ML and DL algorithms have been selected to fit the MFCC features after reducing the dimensions of the voice attributes using LDA for all the voice samples to finally predict class labels. The overall working principle has been illustrated in Fig. 1.

3.1 Dataset Description

We have selected 367 voice samples from a well-known and commonly used audio database namely the "Saarbruecken Voice Database" (SVD) [16]. Whereas most of the works highlighted in the previous section selected few vowels for their research activities, we have taken a full sentence audio clip of ".wav" format to check model performance on lengthy sequence data. Each of the audio clips picked for this work utters *"Guten Morgen, wie geht es Ihnen?"* and all of the voices are sampled at 50 kHz in this database. Additionally, we have taken only those voices where each voice clip consists of only a single class label like Dysphonia or Laryngitis or Renkei's Edema or Healthy. The total audio dataset statistics are shown in Table 1.

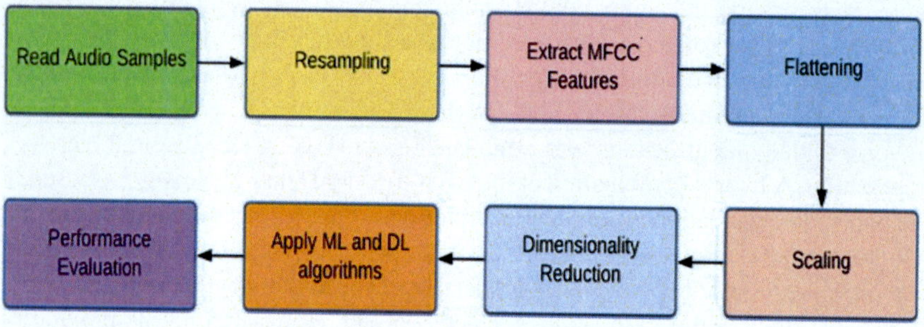

Fig. 1 The overall workflow of the multiple voice disease detection model

Table 1 Dataset information

Category	Gender		No. of .wav file
	Male	Female	
Dysphonia	10	42	52
Laryngitis	76	47	123
Renkei's Edema	30	22	52
Healthy	70	70	140
Total no. of voice samples			367

3.2 Classification Process

This section presents the step-by-step classification process of voice disease detection using ML and DL algorithms.

Audio Signal Reading and Resampling

The voice samples have been downloaded from the SVD database and read using "Librosa" [17] which is a speech analysis bundle. The usage of this package has facilitated the conversion of all the signals into a "mono" channel by normalizing the whole dataset and resampled all the speeches into 22 kHz. Among all four class labels, the waveform representation for the "Laryngitis" class is displayed in Fig. 2.

MFCC Feature Extraction

Mel-Frequency Cepstral Coefficients (MFCC) are computed based on the sample rate and bit value of all the input audio signals to determine the spectral information of the voice [18]. Commonly, 12–20 MFCC values are measured for voice characteristic analysis [19]. In our research, we have calculated 13 MFCC features for each voice signal using Eq. 1.

$$\mathrm{MFCC}_m = \sum_{k=1}^{M}\left(\log(E_k) \times \cos\left[m(k-0.5)\frac{\pi}{M}\right]\right) \quad (1)$$

In (1), M = Total no. of bands in the signal, k = ith band number, E_k = Energy value of the kth band frequency and m = No. of frequency.

The highest number of frames for a single input speech is 213 for our dataset. 13 MFCC feature values have been obtained for all the frames of every individual recording.

Flattening and Scaling

After retrieving the MFCC features for every frame of all 367 voice samples, we have flattened all the MFCC attributes in a single vector. For the highest length audio

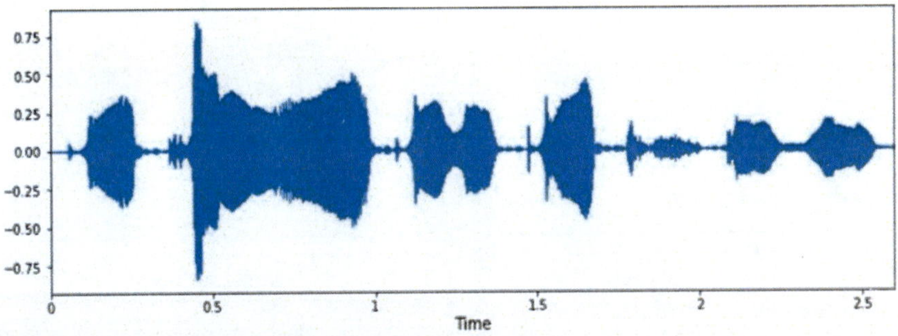

Fig. 2 The waveform of Laryngitis category

file, the (213 × 13) matrix has been converted into a (1 × 2769) vector. As the span for all the signals is not the same, we have padded the less no. of frame signals with "0" based on the largest speech vector. Consequently, we have produced a (367 × 2769) MFCC feature matrix after flattening all the sound waves.

As long as the feature values correspond to high numeric numbers, it is required to normalize those numbers into a smaller range for building a cost-effective model. Thus, we have used "Standard Scaler" for scaling the large floating numerals into small decimals.

Dimensionality Reduction

Dimensionality reduction can make the training faster with good accuracy. Moreover, it can visualize the dataset efficiently to resolve inspection [20]. Although PCA works remarkably for category type features and clustering algorithms have gained good performance occupying PCA to label several classes [21], LDA utilization has provided better results for sequential data modeling in our study. When we have visualized our dataset using Scatter Plot and taking the two most explained values of both PCA and LDA, the data points have been placed almost in a similar range for the PCA survey. On the other hand, LDA consideration has distinguished all four labeled data constructively. As a result, we have fit all the data using LDA to reduce the dimensionality before applying ML and DL algorithms. The employment of LDA has converted the (367 × 2769) feature matrix to (367 × 3) MFCC attribute matrix. The characteristic matrix has been reduced to (367 × 367), while we have applied PCA. The Scatter Plot diagram for our dataset is illustrated in Fig. 3.

Applying ML and DL Algorithms

After dimensionality reduction with the help of LDA, we have fit the decreased audio feature matrix to some clustering algorithms such as SVM, GMM, KNN, RF, and a voting model comprising SVM, KNN, and RF. We have chosen one supervised

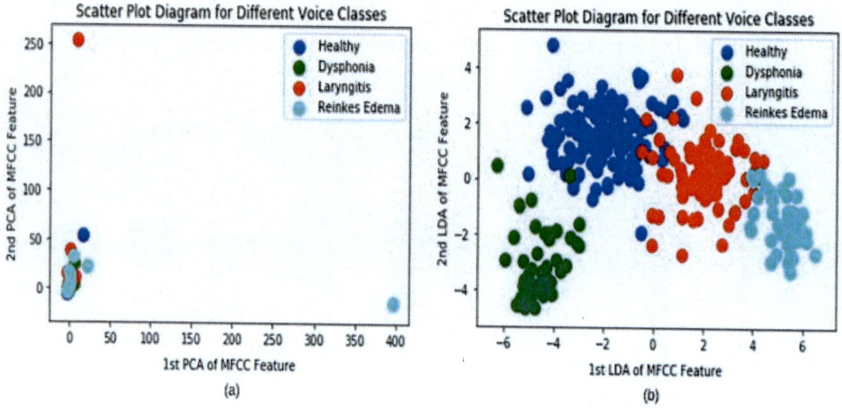

Fig. 3 Scatter plot diagram for our dataset for most the two important values of **a** PCA, **b** LDA

(SVM), one probabilistic (GMM), one unsupervised (KNN), and one ensemble-based (RF) ML algorithm. Polynomial Kernel of degree '3' has been selected when SVM is considered. We have tried varying parameters for GMM, but none has achieved satisfactory performance. But for comparison, we have taken 'Spherical' as the covariance type with 300 as the maximum number of iterations. The default parameters have been chosen for KNN and RF. In the voting classifier, we have selected the same parameters considered in the training of the dataset for each discrete algorithm. For the ANN architecture, we have added three dense layers where the first two layers consist of 256 neurons and both are followed by a dropout layer. "Relu" has been used as the activation function and "Adam" optimizer with the learning rate of 0.0001 has been selected for the ANN model. The batch size for the ANN model has been 10 and 100 epochs have been picked for this three-layer neural network model. The model fitting on the reduced feature matrix is visualized in Fig. 4.

Performance Evaluation

Because of the imbalanced dataset, we have measured accuracy, $F1$-Score and AUC area for the whole dataset using a cross-validation approach. Four popular cross-validation techniques have been employed namely "K-Fold", "Stratified K-Fold", "Shuffle Split" and "Stratified Shuffle Split". We have calculated the accuracy of the recognition framework for all the mentioned techniques and computed the other two performance evaluation metrics only for the best cross-validation models. In our study, we have placed the voice features into 10-folds for the efficacy assessment.

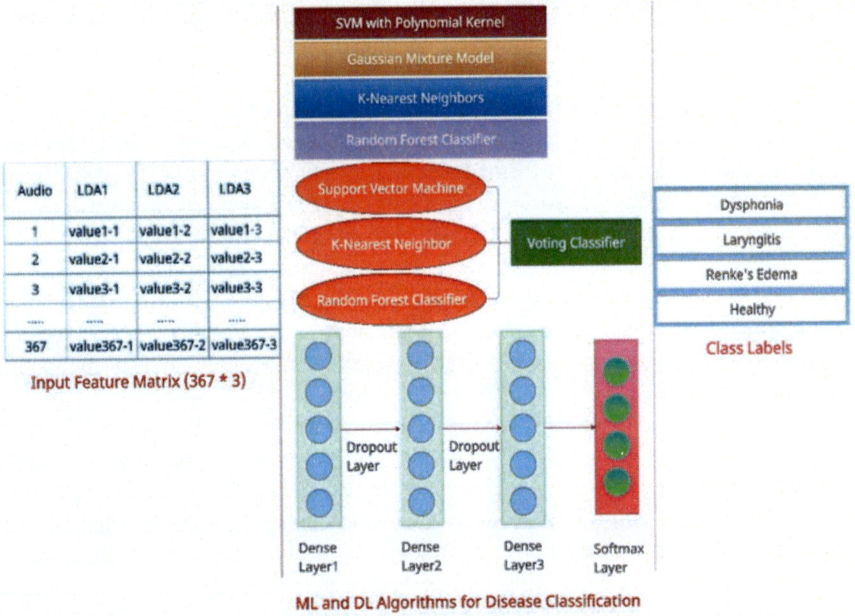

Fig. 4 Applying ML and DL algorithms to determine four class labels

Moreover, we have also determined the total time needed for executing all four cross-validation models by each algorithm.

4 Experimental Results

This research has dwelled on the performance of LDA based dimension minimization for sound waves. The finding of this work implies significant results when KNN has been applied to detect different voice diseases. Other ML and DL algorithms have also performed saliently on LDA based features.

This work has mainly aimed to find the performance of ML and DL algorithms accompanying dimensionality reduction techniques to detect multiple audio classes. Therefore, we have taken only the voice samples which are related to a single category. Due to the lack of a single labeled dataset available in the public domain, we have to rely on a small bunch of audio samples. Moreover, according to the "Hopkins Medicine Organization" [3], people suffer more from Laryngitis, Dysphonia, and Renkei's Edema that occur because of vocal cord trouble.

In this proposed model, the dimensionality of the MFCC feature matrix has been brought into (367×367) for PCA based classification and (367×3) for LDA based prediction. PCA works on the correlation between the features by placing the data points in the subspace which has the maximum variance [22]. This has resulted in an unsupervised separation of features in the orthogonal axes and a covariance matrix has been calculated to find the eigenvectors which are multiplied with the features to reduce the dimension. Consequently, we have found 367 principal components (PC) for individual audio. These PCs have explained the majority of the characters in each audio sample. In contrast, LDA always tries to sum up the vector components. Because of the known class labels, 1-D mean vectors are generated for each class from where scatter matrices are formed. Each scatter matrix computed from the mean vectors is added together to create a within-class matrix. Between-class matrices are also devised from the whole dataset by subtracting the overall mean from each feature value of the dataset. Finally, the within-class and between-class metrics are multiplied to get the reduced feature matrix [22, 23]. By applying these steps, LDA has identified three essential features for each sample that can explain the distinct characters of each audio file. As a result, we have achieved a smaller matrix compared to PCA based reduction.

Table 2 shows the mean accuracy for the several cross-validation methods while taking MFCC voice features without decreasing dimensions, applying PCA, and considering LDA respectively.

From Table 2, it can be easily observable that exploiting LDA has helped far outweigh the accuracy of the detection models using PCA and without dimensionality reduction schemes. We have achieved the best accuracy of 96.49% when KNN has been applied to LDA based reduced attribute matrix and when shuffle split has been considered as the cross-validation approach. For the same validation technique, 95.68% accuracy has been achieved for the voting classifier as well. Moreover, the

Table 2 Performance analysis of several ML and DL algorithms for four cross-validation techniques

Dimensionality of the voice features	ML and DL algorithms	Cross-validation methodology accuracy (%)			
		K-fold	Stratified K-fold	Shuffle split	Stratified shuffle split
No dimensionality reduction (367 × 2769)	SVM	39.46 (±15)	38.19 (±2)	35.41 (±12)	37.84 (±1)
	GMM	47.22 (±13)	55.56 (±19)	**56.77 (±2)**	35.14 (±18)
	KNN	31.40 (±18)	39.45 (±11)	41.08 (±15)	38.92 (±16)
	Random forest	39.38 (±19)	48.60 (±13)	46.49 (±18)	46.22 (±16)
	Voting classifier	39.20 (±10)	41.41 (±8)	42.16 (±20)	39.46 (±4)
	ANN	44.50 (±12)	48.27 (±9)	48.92 (±14)	47.84 (±12)
Dimensionality reduction using PCA (367 × 367)	SVM	45.36 (±14)	43.10 (±6)	44.33 (±11)	44.60 (±10)
	GMM	48.77 (±4)	33.33 (±17)	37.84 (±13)	32.43 (±10)
	KNN	41.40 (±19)	39.45 (±11)	39.46 (±7)	40.54 (±5)
	Random forest	38.76 (±8)	34.89 (±9)	41.08 (±10)	38.64 (±14)
	Voting classifier	45.73 (±12)	43.22 (±5)	47.84 (±14)	44.89 (±11)
	ANN	**59.06 (±16)**	44.96 (±6)	45.95 (±14)	43.78 (±13)
Dimensionality reduction using LDA (367 × 3)	SVM	91.56 (±8)	93.86 (±6)	94.87 (±3)	94.87 (±3)
	GMM	50.32 (±19)	69.44 (±10)	74.38 (±3)	62.97 (±10)
	KNN	93.73 (±5)	95.14 (±3)	**96.49 (±3)**	95.14 (±4)
	Random forest	91.83 (±8)	93.28 (±6)	92.16 (±5)	95.41 (±4)
	Voting classifier	91.82 (±8)	94.64 (±5)	**95.68 (±4)**	95.41 (±3)
	ANN	92.37 (±5)	94.25 (±5)	93.78 (±6)	**96.22 (±3)**

Stratified Shuffle Split method has shown more than 96% accuracy for the same feature vector. Other algorithms have also obtained good accuracy of more than 91% except for the GMM model which has conveyed a lower performance in multiple voice disease recognition processes. However, our findings exhibit poor output for the other two criteria. For instance, direct feed of MFCC values in ML and DL algorithms has provided accuracy ranging from 35 to 57%. GMM has achieved the largest accuracy of only 56.77% which is not a satisfactory value. Furthermore, PCA based diagnosis models have performed poorly too. In this case, 59.06% accuracy has been reached by ANN when the K-Fold method has been realized. Because of splitting our dataset into 10 folds, the accuracy has varied obviously, let alone the extent of which is mentioned in the table.

Since LDA is a supervised algorithm that incorporates Bayesian theory to discriminate multiple classes by maintaining linear distance between-class labels, it can outperform unsupervised PCA algorithm in the case of multi-label clustering with a smaller dataset [24]. This probabilistic modeling might have helped to classify three diseases by separating categories depending on the highest probability of the feature values with an upper success rate than PCA.

As LDA depended clustering framework has outperformed the other two criteria by a massive margin, we have observed the $F1$ Score, AUC area, and execution time for the best cross-validation results for each model. Table 3 provides the same.

Our findings from Table 3 imply that LDA based approach has carried out salient achievement for $F1$-Score, AUC, and Execution time accordingly. Excluding the probabilistic GMM model, all the other models have gained more than 95% $F1$-Score value which indicates significant precision and recall values for this purpose. Among all the models, KNN provides the foremost outcome with extraordinary $F1$-Score and AUC value along with only 22.37 ms execution time. ANN has been praiseworthy too, but it is costly with an enormous execution time. Even though the mean accuracy for multiple cross-validation techniques is almost similar for SVM, VC, KNN, and ANN, the AUC result is much more perfect for KNN with just over 97%. This suggests the balance between the prediction of true positive rate (TPR) and false positive rate (FPR) is more satisfactory. Moreover, the time for model training is also lower than even SVM for the KNN algorithm. As a result, we have placed KNN on the priority list in the multiple voice disease recognition scheme.

Figure 5 portrays the average accuracy of all the cross-validation methods accomplishing the recognition task.

From Fig. 5, it can be interpreted that, KNN functions better in measuring the mean accuracy among all the algorithms. As execution time of a model is also an important parameter as computationally expensive models often face resource constraints [25], the execution time comparison in model fitting for the KNN algorithm is illustrated in Fig. 6.

Table 3 $F1$-score, AUC area, and execution time of the ML and DL algorithms

Name of the algorithm	$F1$-score (%)	AUC	Execution time (ms)
SVM	94.32	0.9506	23.99
GMM	63.28	0.6639	69.96
KNN	**95.83**	**0.9709**	**22.37**
Random forest	95.13	0.9572	141.17
Voting classifier	95.05	0.9583	199.14
ANN	95.54	0.9669	103,135.49

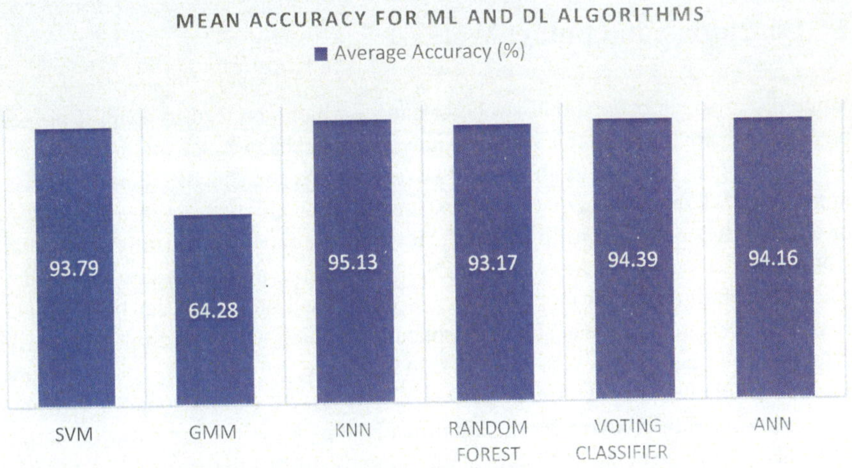

Fig. 5 Mean accuracy of the cross-validation approaches for LDA based disease recognition

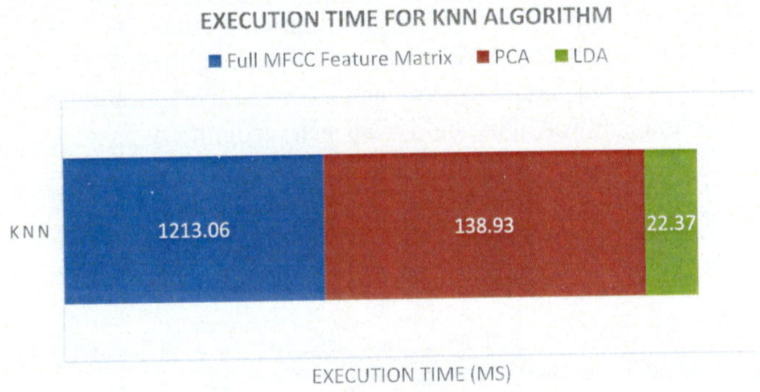

Fig. 6 Execution time comparison among three criteria for KNN algorithm

When LDA features are selected, KNN takes only around 23 s to complete the prediction as stated in Fig. 6. The other two criteria take a long time than LDA. The less execution time for LDA based feature extraction can be realized in a way that the feature matrix for LDA is (367 × 4) in size whereas PCA based attribute matrix is off (367 × 367) dimensions and which is extended massively for no reduction scheme with (367 × 2769) area. Thus, LDA has provided more correct predictions with a shorter amount of time for this sequential data.

Finally, it can be interpreted from the above result analysis is that LDA based feature minimization has been useful for detecting four labeled voice categories and the ML based KNN algorithm has even surpassed accuracy, $F1$-Score and AUC area of even DL based ANN architecture.

5 Conclusion and Future Works

Voice diseases can cause brain stroke or cancer which can lead to death if proper precaution is not taken. In this work, we have proposed LDA based multi-label voice disease recognition model with ML and DL algorithms and shown a remarkable rise in accuracy while using LDA as a dimensionality reduction system instead of PCA or without dimension exclusion strategy. Besides, we have wonderfully found that KNN can even outweigh ANN when LDA is considered predicting disorder within a less amount of time. Our multiple voice disease detection models can be deployed in various web and mobile healthcare applications which will be fruitful in the early detection and treatment of voice diseases. In the future, we will try to compare more ML and DL algorithms like Decision Tree, CNN, LSTM, Bidirectional LSTM, etc. for vigorously validating our work. Additionally, the dataset we have utilized in this study has a shortage in the number of audio samples. To ensure performance stability and for achieving more unbiased results, we will incorporate more labeled speech signals. Moreover, we have taken only 13 MFCC features for each voice sample which will be extended with more MFCC features and other voice characteristics mentioned in various studies previously. Furthermore, there is a plan to evaluate the model's robustness by testing audio recordings of general people in real-time. We will essentially apply our model to more audio related works for discovering more robust methods and analysis in the field of speech recognition.

References

1. Mythili J, Vijaya MS (2018) Pathology voice detection and classification using ensemble learning. Int J Eng Sci Invent (IJESI) 7(8):1–8
2. Titze IR, Martin DW (1998) Principles of voice production. Prentice-Hall, Englewood Cliffs
3. Voice disorders. https://www.hopkinsmedicine.org/health/conditions-and-diseases/voice-disorders. Accessed 14 May 2021
4. Internal medicine. https://www.medstarsouthernmaryland.org/our-services/internal-medicine/conditions/ear-nose-and-throat-conditions/voice-and-swallowing-disorders/. Accessed 14 May 2021
5. Martins RHG, do Amaral HA, Tavares ELM, Martins MG, Gonçalves TM, Dias NH (2016) Voice disorders: etiology and diagnosis. J Voice 30(6):761-e1
6. Uma Rani K, Holi MS (2016) A hybrid model for neurological disordered voice classification using time and frequency domain features. Artif Intell Res 5(1):87–94
7. Verde L, De Pietro G, Sannino G (2018) Voice disorder identification by using machine learning techniques. IEEE Access 6:16246–16255
8. Alhussein M, Muhammad G (2018) Voice pathology detection using deep learning on mobile healthcare framework. IEEE Access 6:41034–41041
9. Verde L, De Pietro G, Alrashoud M, Ghoneim A, Al-Mutib KN, Sannino G (2019) Leveraging artificial intelligence to improve voice disorder identification through the use of a reliable mobile app. IEEE Access 7:124048–124054
10. Harar P, Galaz Z, Alonso-Hernandez JB, Mekyska J, Burget R, Smekal Z (2020) Towards robust voice pathology detection. Neural Comput Appl 32:15747–15757
11. Dankovičová Z, Sovák D, Drotár P, Vokorokos L (2018) Machine learning approach to dysphonia detection. Appl Sci 8(10):1927

12. Boyanov B, Hadjitodorov S (1997) Acoustic analysis of pathological voices. A voice analysis system for the screening of laryngeal diseases. IEEE Eng Med Biol Mag 16(4):74–82
13. Harar P, Alonso-Hernandezy JB, Mekyska J, Galaz Z, Burget R, Smekal Z (2017) Voice pathology detection using deep learning: a preliminary study. In: 2017 international conference and workshop on bioinspired intelligence (IWOBI). IEEE, Funchal, pp 1–4
14. Mohammed MA, Abdulkareem KH, Mostafa SA, Ghani MKA, Maashi MS, Garcia-Zapirain B, Alhakami H, Al-Dhief FT (2020) Voice pathology detection and classification using convolutional neural network model. Appl Sci 10(11):3723
15. Choubey DK, Kumar M, Shukla V, Tripathi S, Dhandhania VK (2020) Comparative analysis of classification methods with PCA and LDA for diabetes. Curr Diabetes Rev 16(8):833–850
16. Pützer M, Koreman J (1997) A German database of patterns of pathological vocal fold vibration. Phonus 3:143–153
17. McFee B, Raffel C, Liang D, Ellis DP, McVicar M, Battenberg E, Nieto O (2015) Librosa: audio and music signal analysis in python. In: Proceedings of the 14th python in science conference, vol 8, pp 18–25
18. Gupta S, Jaafar J, Ahmad WW, Bansal A (2013) Feature extraction using MFCC. Signal Image Process Int J (SIPIJ) 4(4):101–108
19. Poorjam, Hossein A. Why we take only 12–13 MFCC coefficients in feature extraction? https://rb.gy/2mimzc. Accessed 31 May 2018
20. Arjmandi MK, Pooyan M, Mohammadnejad H, Vali M (2010) Voice disorders identification based on different feature reduction methodologies and support vector machine. In: 2010 18th Iranian conference on electrical engineering. IEEE, Isfahan, pp 45–49
21. Bhowmik S, Reno S, Sultana S, Ahmed M (2021) Clusterization of different vulnerable countries for immigrants due to covid-19 using mean probabilistic likelihood score and unsupervised mining algorithms. In: 2021 international conference on information and communication technology for sustainable development (ICICT4SD). IEEE, Dhaka, pp 285–290
22. Ottensen C. Comparison between PCA and LDA. https://dataespresso.com/en/2020/12/25/comparison-between-pca-and-lda/. Accessed 25 Dec 2020
23. Sarkar P. What is LDA: linear discriminant analysis for machine learning. https://www.knowledgehut.com/blog/data-science/linear-discriminant-analysis-for-machine-learning. Accessed 30 Sept 2019
24. Mehta A. Everything you need to know about linear discriminant analysis. https://www.digitalvidya.com/blog/linear-discriminant-analysis/. Accessed 04 Jan 2020
25. Priya R, de Souza BF, Rossi AL, de Carvalho AC (2013) Predicting execution time of machine learning tasks for scheduling. Int J Hybrid Intell Syst 10(1):23–32

Dynamic Topology Reconstruction on Next Generation WLAN Using Spatial Reuse Gain by DBSCAN Clustering Algorithm

Maleeha Sheikh, Syeda Myesha Mashuda, Redwan Abedin, and Md. Obaidur Rahman

Abstract The succeeding peers high density based WLAN set-up is seemingly appropriate for expansion drift in manufacturing wireless sensor-based networks. A spatial group-based multi-user full-duplex orthogonal frequency division multiple access (OFDMA) (GFDO) multiple access control (MAC) protocol was projected previously for the static topology with enhanced spatial reuse-gain since power-control technology decreases the signal coverage which also reduce cluster size, resulting in elimination of the overlapping area, minimize the occurring collisions. According to our research, we approached to build a dynamic topology with varying densities using the adaptive DBSCAN clustering technique that can maintain both power and increase signal range, while eliminating the overlapping areas as the nodes can now be mobile and densities of the network is balanced throughout the topology. The adaptive DBSCAN demonstrates work efficiency for groups with changed densities that can work significantly well for classifying clusters with varying-densities that achieve a successful efficiency rate for dataset where clusters have the constant density of data points, also significant attributes for this algorithm are noise cancelation.

Keywords WLAN · DBSCAN · GFDO

1 Introduction

Today's generation for Internet of Things (IoT), wireless local area network (WLAN) is endlessly utilized for its pay of minimal effort and easy to understand organization, also has become an exploration area of interest in industry and the scholarly world

M. Sheikh · S. M. Mashuda · R. Abedin
Department of Information and Communication Technology, Bangladesh University of Professionals (BUP), Mirpur Cantonment, Dhaka 1216, Bangladesh

Md. O. Rahman (✉)
Department of Computer Science and Engineering, Dhaka University of Engineering & Technology (DUET), Gazipur, Bangladesh
e-mail: orahman@duet.ac.bd

M. S. Arefin et al. (eds.), *Proceedings of the International Conference on Big Data, IoT, and Machine Learning*, Lecture Notes on Data Engineering and Communications Technologies 95, https://doi.org/10.1007/978-981-16-6636-0_25

[1, 2]. During the previous decade, with the fast improvement of portable Internet, numerous new applications and necessities have arisen and individuals' interest for remote traffic has expanded quickly at a build yearly development pace of 47% from 2016 to 2021 [3], where the prerequisites for transmission deferral and jitters are more severe. IEEE 802.11b keeps on OFDMA innovation and MU-MIMO innovation presented by IEEE 802.11ax, and takes multi-AP agreeable and multi-band operation (MB-Opr) as its fundamental advancements [4].

Spatial group-based Multi-User Full-Duplex OFDMA MAC-Protocol for the Next-Generation WLAN was implemented before which was limited to some extent providing a static topology leading to disadvantages of reduced signal coverage area for minimizing collisions that can improve spatial reuse gain.

The problem as stated is that as for reduced signal coverage area, it is quite problematic when it comes to areas that are not visited or areas that are not covered if it is wider. Another disadvantage is that if area is reduced, for covering greater areas, more access points or antenna towers are needed for maintaining small clusters and covering the inbound which increases cost. It also increases density that leads to slower rate for transfer of signals or BSRs (Buffer State Reports) or inter-node interference within neighboring clusters or other zones.

The objectives of this paper are pointed as: (i) Enhance and implement to a dynamic topology, (ii) Wider signal area covered, (iii) Reduce delay for signal transmission, (iv) Network is distributed in accordance to density, (v) Formulation of Inter-Cluster and Inter-Zone transmission.

DBSCAN is considered as perhaps the most impressive and most referred to density based grouping calculations [5] that can relate to huge precision clusters of arbitrary shape and size in enormous databases distorted with noise. One of the primary benefits of the DBSCAN calculation is that fate of the quantity of bunches (or clusters) isn't needed on datasets [6]. As the DBSCAN calculation is fit for dealing with the noises accurately and adequately [7], it is more appropriate to discover a gathering encompassed by noise just as various different gatherings. Once more, it is viewed as uncaring toward commotion and exceptions. In any case [8], two basic boundaries are the fundamental necessity for applying the DBSCAN calculation: (i) Eps and (ii) MinPts. Then, by the study of Adaptive DBSCAN (ADBSCAN) [9], its calculation is proposed to decide a proper Eps and MinPts esteems with the goal that the calculation can distinguish all the bunches in the datasets. These contributions are made through the whole research and can ensure through this method can effectively reconstruct an maintain a topology that would be dynamic and adapt to shifting of nodes and density that can cluster varying areas.

The space groups will be considered as each cluster with group heads as centroids and group members as the nodes. Each neighboring node will be visited following the parameters, MinPts and Epsilon, covering wider areas and continuous cluster formation alongside throughout the network for every visits. Information is passed using Access Points (AP) to Aps of other areas following cloud upload criteria.

The other parts of the paper are outlined into several sub-contents as prearranged beneath: Section 2 deliberates the literature overview of the content that dictates the contribution of the research and implementations based on the new algorithm and efficiency of the upgrade. Section 3 defines the methodology preparations of the whole paper, the methods and techniques used for the implementation of the experimental model. Section 4 calibrates on the performance analysis and observation that validates the improvement achieved in this experimental paper. Section 5 is the conclusion that implicates the whole point of view of the paper and directs further work based on future aspects following the research.

2 Related Work

Basicallym, cluster-based MAC protocols are categorized into two categories: one is symmetric and another is asymmetric MAC protocols. In case of symmetric protocols, all the STAs and AP have full duplex proficiency with full duplex MAC protocol. Frequency Domain Coordination [10] proposed next generation WLAN by full duplex MAC protocol using FDC where STAs must be full duplex capable. Full Duplex Carrier Sense Multiple Access/ Collision Detect [11] proposed CSMA/ CD where sub channels using MAC frame preface detection technology where all nodes require full duplex proficiency. In contrast with asymmetric protocols, the STAs are half duplex device and AP has full duplex proficiency with full duplex MAC protocol.A single user Fuplex [12] proposed next generation WLAN where AP has full duplex competence but STAs does not have. Multi-user FuPlex [13] where the collection of BSR information leads to high overhead on AP pure scheduling. Power control MU-FuPlex [14] based on AP pure scheduling where probability of forming FD link is expressed. FD-OMAX protocol [15] works on trigger free MU-FuPlex which shows a scenario of high-density deployment and improves system throughput. EnFD-OMAX protocol which establishes full duplex link by improving the success probability and the spectrum efficiency (Table 1).

A spatial group-based multi-user full-duplex OFDMA MAC protocol (GFDO) [16] is a recent MAC protocol which is a combination of power regulator and spatial alliance technology. This paper illuminate low in effectiveness and obstructions dis-

Table 1 Comparing MAC protocols [16]

Protocol	Category	Approach	Characteristic
FDC [10]	Symmetric	Centralized	Scheduling BSR assortment
FD-CSMA/CD [11]	Symmetric	Distributed	Random argument in sub-channel
FuPlex [12]	Asymmetric	Distributed	Simple and reliable full duplex MAC protocol
MU- FuPlex [13]	Asymmetric	Centralized	MU OFDMA, Scheduling based channel access
PCMU- FuPlex [14]	Asymmetric	Centralized	Probability of establishing FD link, MU OFDMA Power Control
FD-OMAX [15]	Asymmetric	Distributed	MU OFDMA Inter-node interference assortment
EnFD-OMAX [15]	Asymmetric	Distributed	Refining the success possibility and the spectrum proficiency

semination issue in WLAN. In GFDO MAC protocol, they follow two algorithms. First one is two level BSR information collection-mechanism and second one is cascading method. In first algorithm flow, AP sends trigger frame to GM and GH. GM and GH respectively send RTS and CTS frame to AP. In cascade method flow, AP sends trigger frame to GH and GM. Within each other they did data transmission through GFDT and ACK transmission happens in both direction.

3 Implementation Details of the Proposed Work

To come in help of improving the status, we thought of laying out a conceptual improvement to a modified version of this system based on: Topology, Inter-Cluster Communication and Clustering Algorithm.

3.1 Cluster Formation Using DBSCAN Algorithm

We first focused on finding and analyzing Cluster algorithms with Varying Densities. The main points were to find an optimal type of Algorithm which will pin out the density of nodes and separate noise. The primary discussion starts with the Algorithm, where Cluster is formed with noise detection.

DBSCAN: A simple data clustering Algorithm, which does not require specifying beforehand, regarding the number of clusters we are looking for and we can work on Random Datasets. It does not try to cluster every single point and identifies the more separated ones as noise. It focuses on two definite parameters [17];
Eps: The distance that specifies the adjacent neighborhoods.
MinPts: The least number of information focuses to characterize a cluster (Fig. 1).

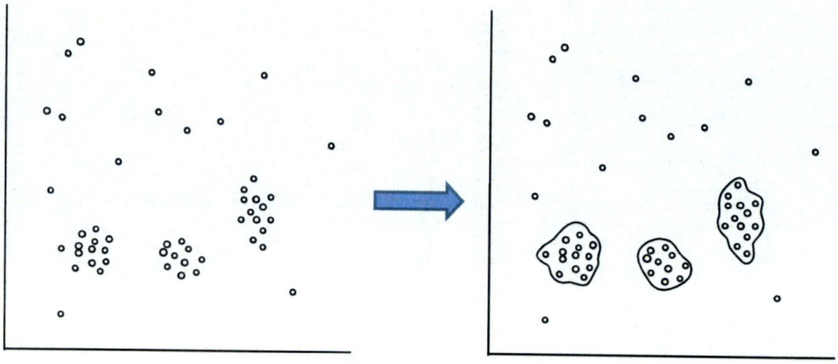

Fig. 1 Cluster formation by DBSCAN algorithm

Algorithm 1 Cluster formation through DBSCAN algorithm

Input: Dataset M, Eps and MinPts
Output: Find noise and total number of Clusters with proper graphs
Assume: N = Any positive integer [> 0]
Update: Eps and MinPts [Parameters to build the Clusters]
1: **Initialization:**
2: i = Total marked set of points
3: if index (i) = −1
4: Define as Noise
5: else [i = 1, 2, ..., N]
6: Define in a respective Cluster
7: All Clusters are plotted with noise [Plot of Clusters with Noise]
8: End if

Specialized Adaptive DBSCAN (ADBSCAN): It is an improvement to the DBSCAN Algorithm were nodes with varied distances are also marked into a Cluster. Through a small parameter change and specified variation in our DBSCAN Algorithm, we implemented our conceptual Specialized Adaptive DBSCAN, which is later taken in consideration for Zone-wise Separation or BSS deployment [9].

3.2 *Analyzing GFDO Network Model*

The network model (system model) which we are analyzing had its own benefits of solving the interference diffusion problem as well as the issue of efficiency access being low in highly deployed area in respect to WLAN. It had adopted two precise Algorithms (for BSR and Scheduling) maintaining both node-to-AP and AP-to-node connection for data transmission.

Two-Level Mechanism of Information-Collection (BSR): It includes the BSR-information collection of SG and CMs in (a specific) SG. By the help of trigger frames, the two distinct levels perform;

First-Level: In first level, CH is used in collecting the BSR-information of the CMs in SG and record the necessary information of interference of other CHs.

Second-Level: In second level, it adopts P-probability (Persistence Probability) OFDMA random-access method in reporting AP with two-kinds of information: the first-level information and the interference-information between the SGs.

CFDT with Cascading Method: The full-duplex transmission [18] initiates, while scheduling the CFDT in a cascading method, with respect to the BSR-information. It is focused into action when demand-collected for uplink by a cluster is greater than the number of data-points in the system. These two algorithms were put to action with accordance to the Protocol.

3.3 Inter-Cluster Transmission without Accessing AP

The proposed Inter-Cluster initiates without AP, it starts from the Cluster Head, CH. The Cluster Member, CM sends information to CH requesting for communicating to another Cluster. The CH, while collecting interference information checks other CHs for Inter-Cluster communication based on the request of CM. If found it requests the respective CH for transmission. Once the CH accepts, it designates the specified CM to get ready for Communication. Request to Send (RTS) and Clear to Send (CTS) frames are also used in confirming the status before transmission.
Again, if all the CH (in that BSS or Zone) does not match with the requested location, it then gets ready for inter-zone Communication. An algorithm has been analyzed by the stated interference-information collection without accessing the AP for transmission (Fig. 2).

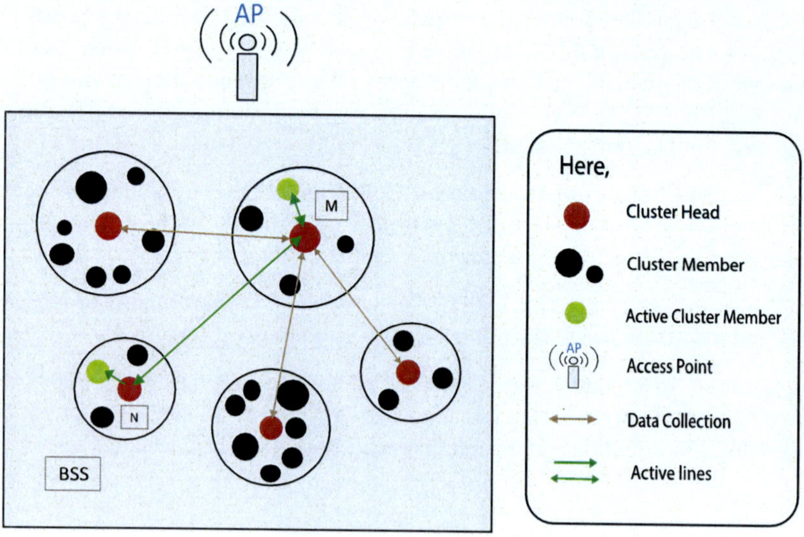

Fig. 2 Inter-cluster communication between two SGs (clusters); M and N

Algorithm 2 Inter-Cluster Transmission without the use of AP

Input: Information of N_2, Total cluster C
Output: Find N_2, get ready to Transmit Inter-Cluster, collect interference Information
Assume: Inter-cluster communication between N_1 and N_2
Update: Interference Information [Parameters to build the Clusters]
Initialization:
loop = 1 to go through all CH
Interference Information = I_{ii}
while CH is collecting Interference Information, I_{ii} [Based on the reference [1] used]
  if loop < (C – 1)
    if CH N_n == CH N_2
      N_2 is found, ready for transmission with I_{ii}
    else [i = 1, 2, ..., N]
      Updates only I_{ii}
    exits connection to respective CH
    loop + +
  else
    exit loop and initiate inter-zone
    End if [Plot of Clusters along with Noise]
  End if
Collected all I_{ii} and ready to start transmission
End while

3.4 Cloud-Based Model in Implementing Dynamic Topology

The proposed inter-zone initiates once the CH passes all its information from CM to AP. An algorithm given earlier also portrays small view of AP-to-AP path formation through Cloud.

Now, the RTS and CTS frames are put to action in clearing and preparing to send. The AP then updates its information in the Cloud store and requests to find the Designated AP of the specified Cluster (CM and CH). The Cloud responds and tries to match receiving AP location. Once found, the AP is accessed with the transmitting AP's Information along with its location. The AP forwards it to the required Cluster with CTS and RTS frames. When the path is clear, it requests the Cluster. As the specified node receives the information,it passes to its CH and CM accordingly. The CH then compares the interference information to that of its other CHs in the same zone (Fig. 3).

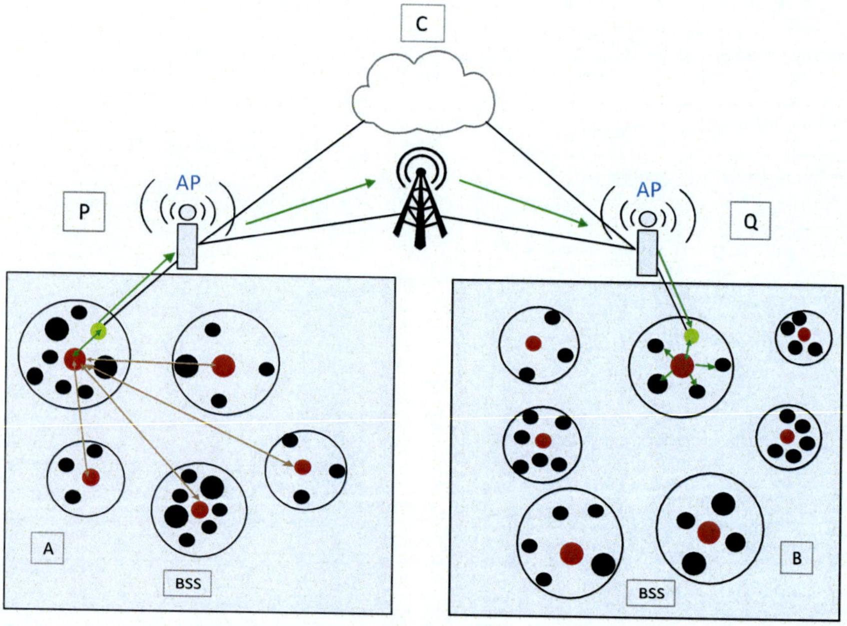

Fig. 3 Inter-zone communication between two BSSs (zones); A and B

Algorithm 3 AP_T Accessing AP_R by the Help of Cloud System

Input: Request to cloud with designated AP information
Output: Find AP (requested) of Receiving end and Store (requesting) AP's data
Assume: AP (requesting) as P, AP (requested) as Q [with specified Address]
Update: Cloud-Storage Information [Updating Information]

```
Initialization:
Store P's information (updated) in cloud
for all unvisited APs in Cloud
  while P requesting to connect to Q in Cloud from list of APs
     The Cloud processes the Request, starts searching and matching the AP
     if AP == Q
       Define Q found and stop search
     else
       Keep on searching and mark AP visited        [Continue AP == Q]
     end if
  end while
AP found will then initiate AP-to-AP connection, completing inter-zone Path of Transmission
end for
```

The details of proposed work concludes with initiating the discussion of the optimal results that were able to be drawn out in our analysis.

Table 2 Conditional statements for Eps and MinPts

Parameter	Value	Statement
MinPts	Increased	Total number of cluster decrease and more noise
	Decreased	Total number of clusters increase and less noise
Eps	Increased	Total number of cluster decrease and noise decreased
	Decreased	Total number of clusters increase and noise increased

4 Performance Analysis and Observation

4.1 *Data and Parameter Analysis*

Basing on our density-based study of data points (or nodes), we are focused on forming cluster with respect to varying high-density based deployment of data points. We recognize Noise with '−1' and definite member of Clusters with any number greater than '0.' Parameter analysis gives out a table which was concluded through our output and simulated Cluster results. Table 2 gives us a short idea regarding the parameters and how these influence in construction and recognition of Clusters.

4.2 *Dynamic Validation with Noise Optimization*

In WLAN, identification of varying number of clusters instead of fixed number of clusters is considered as a dynamic system. As we analyze the outputs (see Fig. 4), we can see variation in how they include different number of nodes with respect to their Minpts and Eps values. Clusters are varying with change of parameters and noise is also being varied with each change. The noise or outlier points were designated differently; we can control and optimize it. We can view the table of parameter-analysis, the cluster header which secures a definite path of communication, Specialized Adaptive DBSCAN that increase cluster number with a decrease in Noise are certain ways we found outcomes in noise optimization.

With Dynamic validation, the latency and throughput of system were theoretically analyzed. The varying number of clusters, decrease in the amount of access in APs and increased number of cluster through Specialized Adaptive DBSCAN concluded our statement in the decrease of latency and increase of throughput of the system.

4.3 *Result Optimization and Observation*

Optimal output and observation were mainly focused on the general output of DBSCAN algorithm. A small range was considered with respect to satisfying our

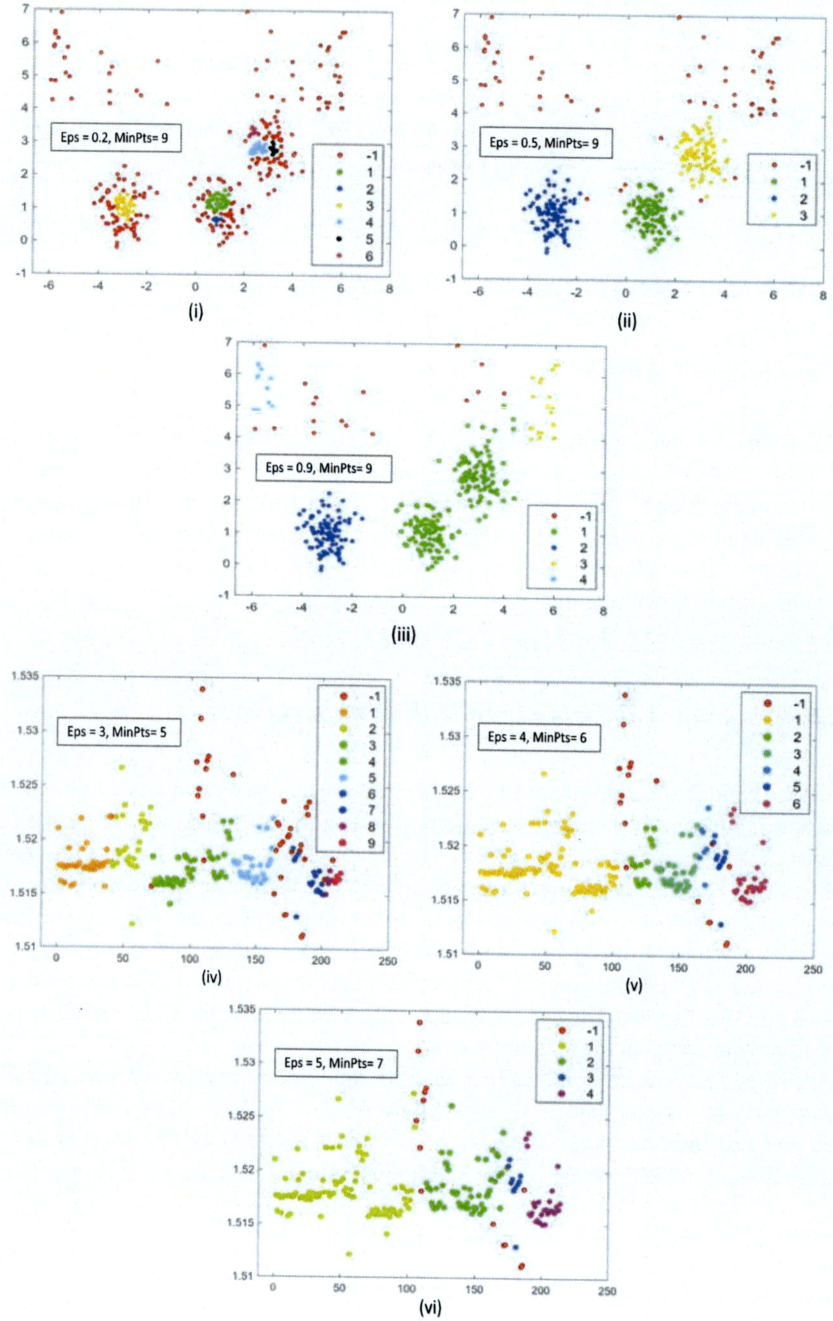

Fig. 4 Inter-zone communication having dynamic validation with noise optimization outlook, constant and varying parameters: (i)–(iii) MinPts is constant here with 9; (iv)–(vi) both Eps and MinPts are increased with uniform value

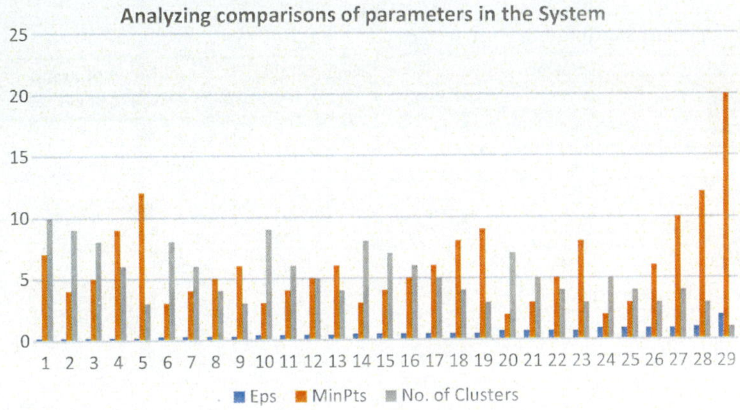

Fig. 5 Analyzing Parameters in the System

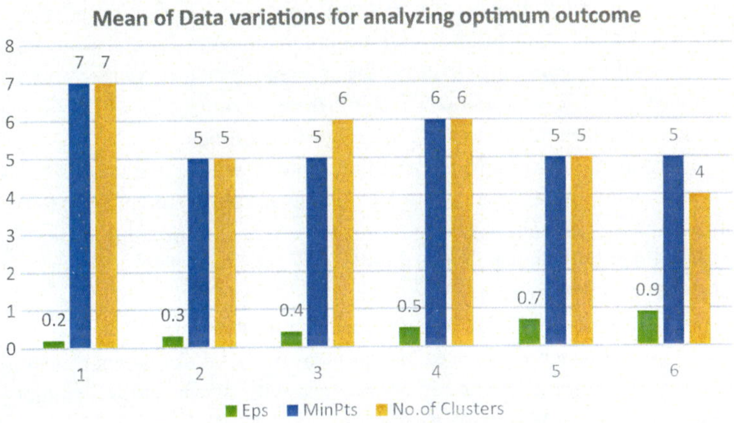

Fig. 6 Data variation arrangement to analyze optimum outcome

mentioned theoretical analysis, through which we marked our optimized state of system.

With the graphs (see Figs. 5 and 6) and data (see Table 3), we observed that points: [0.5, 6], [4, 6], [3, 0.3] and [3, 0.7] gives out the most suitable optimized outcomes with accordance to our previously discussed points to establish this topology. The outcomes not only show the dynamic nature but also induces a balance between the range of variations for a better throughput and noise optimization.

Throughout the whole system analysis, we observed several outcomes based on varying datsets, which when brought together gave us certain graphical representations assisting us to circle out the more balanced sets of Parameters to prepare the system to be more feasible and flexible for further study.

Table 3 Optimal value for setting parameters

Formation	Eps	Mean MinPts	Mean of number of clusters
Spherical	0.4	5	5
	0.5	6	5
Arbitrary	4	6	6
	4	6	5
Specialization	0.3	3	8
with adaptive nature	0.7	3	5

5 Conclusion and Future Work

The problem as stated is that as for reduced signal coverage area, all nodes are not covered or visited, also more access points or antenna towers are needed for maintaining small clusters and covering the inbound which increases cost. It also increases density that leads to slower rate for transfer of signals or BSRs or inter-node interference within neighboring clusters or other zones.

Our main proposal is to build a dynamic topology with varying densities using the DBSCAN (including its specialized adaptive form) clustering technique that can maintain both power and increase signal range, while eliminating the overlapping areas as the nodes will then become mobile and densities of the network will be balanced throughout the topology.

According to our contribution, we enhanced and implemented the dynamic topology so that it covers wider signal area; this reduces the delay for signal transmission. The network is made to distributed in accordance with density with formulation of inter-cluster and inter-zone transmission.

The recreation results show that the effectiveness of our proposed convention is efficient and more optimized than the past protocol. It has been concluded, after comparing the results with the problem stated in accordance with our studied paper, in a proper theoretical aspect including varied data set, our system decreased latency to some extent and signal throughput was enhanced, increased performance that optimizes the cluster balance.

Future scopes of our research work are: Core utilization of Adaptive DBSCAN and will implicate the algorithm along with certain upgrades for better efficiency with respect to our optimal outcome. The existing algorithms with pace to our system will be compared and analyzed with respect to the efficiency of similar existing works near future. The model is to be more refurbished in future by combining Adaptive DBSCAN and Mean-Shift cluster technique leading to a robust network.

References

1. Pirayesh H, Sangdeh PK, Zeng H (2019) EE-IoT: an energy-efficient IoT communication scheme for WLANs. In: IEEE INFOCOM 2019—IEEE conference on computer communications. IEEE
2. Teng R, Yano K, Kumagai T (2018) Scalable distributed-sensing scheme with prioritized reporting for multi-band WLANs. In: 2018 IEEE 4th World Forum on Internet of Things (WF-IoT), pp 580–585. IEEE
3. Index CVN (2017) Global mobile data traffic forecast update, 2016–2021 whitepaper. Cisco
4. López-Pérez D, Garcia-Rodriguez A, Galati-Giordano L, Kasslin M (2019) Doppler: IEEE 802. 11 be extremely high throughput: the next generation of wi-fi technology beyond 802. 11 ax. IEEE Commun Mag 57:113–119
5. Ester M (1996) A density-based algorithm for discovering clusters in large spatial databases with noise. inKdd. 96:226–231
6. Arya R, Sikka G (2014) An optimized approach for density based spatial clustering application with noise. In: ICT and critical infrastructure: proceedings of the 48th annual convention of computer society of India—Vol I, pp 695–702. Springer International Publishing, Cham (2014)
7. Borah B, Bhattacharyya DK (2004) An improved sampling-based DBSCAN for large spatial databases. In: Proceedings of international conference on intelligent sensing and information processing, pp 92–96. IEEE
8. Campello RJGB, Moulavi D, Sander J (2013) Density-based clustering based on hierarchical density estimates. Adv Knowl Discover Data Min. Springer, Berlin, pp 160–172
9. Khan MMR, Siddique MAB, Arif RB, Oishe MR (2018) ADBSCAN: adaptive density-based spatial clustering of applications with noise for identifying clusters with varying densities. In: 2018 4th international conference on electrical engineering and information & communication technology (iCEEiCT), pp 107–111. IEEE
10. Ahn H, Lee J, Kim C, Suh Y-J (2019) Frequency domain coordination MAC protocol for full-duplex wireless networks. IEEE Commun Lett 23:1–1
11. Wang X, Tang A, Huang P (2015) Full duplex random access for multi-user OFDMA communication systems. Ad Hoc Netw 24:200–213
12. Qu Q, Li B, Yang M, Yan Z, Zuo X (2015) FuPlex: a full duplex MAC for the next generation WLAN. In: Proceedings of the 11th EAI international conference on heterogeneous networking for quality, reliability, security and robustness, pp 239–245. IEEE
13. Qu Q, Li B, Yang M, Yan Z, Zuo X (2017) MU-FuPlex: a multiuser full-duplex MAC protocol for the next generation wireless networks. In: 2017 IEEE wireless communications and networking conference (WCNC), pp 1–6. IEEE
14. Qu Q, Li B, Yang M, Yan Z (2018) Power control based multiuser full-duplex MAC protocol for the next generation wireless networks. Mob Netw Appl 23:1008–1019
15. Peng M, Li B, Yan Z, Yang M (2020) A trigger-free multi-user full duplex user-pairing optimizing MAC protocol. Lecture notes of the Institute for Computer Sciences. Social informatics and telecommunications engineering. Springer International Publishing, Cham, pp 598–610
16. Peng M, Li B, Yan Z, Yang M (2020) A spatial group-based multi-user full-duplex OFDMA MAC protocol for the next-generation WLAN. Sensors (Basel) 20:3826
17. Ram A, Jalal S, Jalal AS, Kumar M (2010) A density based algorithm for discovering density varied clusters in large spatial databases. Int J Comput Appl 3:1–4
18. Yang M, Li B, Yan Z, Yan Y (2019) AP coordination and full-duplex enabled multi-band operation for the next generation WLAN: IEEE 802.11be (EHT). In: 2019 11th international conference on wireless communications and signal processing (WCSP), pp 1–7. IEEE

Aggressive Fault Tolerance in Cloud Computing Using Smart Decision Agent

Md. Mostafijur Rahman and Mohammad Abdur Rouf

Abstract Application of cloud computing is increasing gradually. It is a useful model for a collection of configurable computing resources such as data-centers, servers, data storage and application services in real-time. Due to the emergence of cloud computing, providing reliable service becomes vital issue. Transient faults may affect temporary unavailability of services and timeout to get response. These types of faults can be catastrophic in cloud applications such as, scientific research, financial and safety critical applications. To reduce the effect of such errors, a fault tolerant mechanism is required. We propose an aggressive fault tolerant (AFT) technique to detect and recover from faults in cloud environment. Aggressive fault detection and recovery module detects faults and recovers from these faults using a smart decision agent. A smart decision agent takes decision on different types of hardware, software and communication faults. It reduces complexity and improves performance of fault tolerant schemes compared with other existing techniques such as checkpointing, resubmission and replication techniques. The proposed scheme achieves 98.7% error coverage while it is 1.5 times faster than checkpointing, 2.0 times faster than resubmission and 2.5 times faster than replication technique.

Keywords Cloud computing · Transient faults · Fault tolerance · Fault detection · Fault recovery · Availability · Fault injection

1 Introduction

Cloud computing system is a new paradigm of transparent distributed system. It handles the resources on a larger scale with cost-effective and location independent manner. Since the use of cloud computing is increasing in broad spectrum of applications, fault free services are required [1, 2]. A cloud is more effective and reliable when it is more fault tolerant and more scalable. Fault tolerance considers effective

Md. M. Rahman · M. A. Rouf (✉)
Department of Computer Science and Engineering, Dhaka University of Engineering and Technology, Gazipur, Bangladesh
e-mail: marouf.cse@duet.ac.bd

M. S. Arefin et al. (eds.), *Proceedings of the International Conference on Big Data, IoT, and Machine Learning*, Lecture Notes on Data Engineering and Communications Technologies 95, https://doi.org/10.1007/978-981-16-6636-0_26

steps to recover failure. It ensures system reliability by improving the fault detection and recovery mechanism [3, 4]. Fault tolerance is a scheme that can operate a system even in faulty environment. Proactive fault tolerant technique recovers from faults by predicting before these errors affect the system [5]. It proactively replaces the mistrust components. An application has a feedback-loop mechanism which always monitors and resolves faults which is called preemptive migration [6, 7]. Backward recovery is a rollback technique that starts backward processing from a prior state. It needs extra time for rolling it back [8]. Forward error recovery is a scheme that can proceed forward even a fault is occurred. The fault is detected later by duplex system and recovered by re-execution or detected and recovered by triple modular redundancy (TMR) [9, 10].

Reactive fault tolerance requires error recovery after faults are detected [8, 11]. Checkpoint is a snapshot of the full state of the process. It runs the failed system from the recently checked point rather than from initial state [12–14]. The identified failed work is re-executed using the same resources in real time, and it is called the task re-execution. If the cloudlet is failed or canceled, then it will be resubmitted. Failed tasks are re-executed by replicas with different resources [9]. This technique has a primary virtual machine, and other one is replica (or secondary) virtual machine. When a cloudlet is failed to execute on primary virtual machine, the replica re-executes the cloudlet from the initial state. It needs more overhead than hundred percent [11, 15].

We propose aggressive fault tolerance (AFT) technique consists of smart combination of checkpointing, resubmission, and replication methods. The aggressive fault detection (AFD) module monitors the message using heartbeat mechanism. The aggressive fault recovery (AFR) module recovers the faults using a smart decision agent [16]. The performance of proposed AFT is 1.5 times faster than checkpointing, 2.0 times faster than resubmission technique and 2.5 times faster than replication technique.

The contributions of this paper are:

- We propose an aggressive faults detection and recovery module by extending the datacenter broker policy.
- The proposed scheme detects and recovers the transient, permanent, omission and timeout faults.
- The proposed scheme integrates checkpointing, replication and resubmission methods and a smart decision agent which can choose the detection and recovery technique based on fault types.
- We have verified the proposed model by using fault injection mechanism.

The paper is presented as follows. In Sect. 2, background work is presented. In Sect. 3 various existing works are discussed. Section 4 presents proposed aggressive fault tolerance (AFT) technique, aggressive fault detection and recovery module are described. Experimental setup, results and analysis are given in Sect. 5. Section 6 concludes the paper.

2 Background Works

Different types of faults may occur in the cloud environment, such as transient and intermittent faults may occur in processing elements (PEs) of virtual machines (VMs). These types of faults may also occur in hosts and datacenters. Different types of faults usually are occurred in processing elements and memory modules as discussed in [9]. Unavailability of host, datacenter and VM may be occurred due to disk full or disk error. On the other hand memory faults and other faults are discussed in [17, 18]. The software faults are software state transition faults, early and late timing faults, timing overhead, protocol incompatibilities, data fault, logical fault, numerical exception, operating system faults, link timeout fault, user defined exception and unhandled exception faults [19, 20]. The communication faults are sending and receiving omission faults, early or late timing faults, packet corruption and packet loss faults [21]. The transient faults may cause single bit, multiple bits and burst-bits errors. The permanent faults are physical damage of host, PEs and memory [22]. The main components of cloud architecture are shown in Fig. 1. Firstly, Cloud Information Service (CIS) in which datacenter is registered, and the information of the cloud components are stored in a table of CIS. Secondly, the datacenter broker acts as a coordinator between software-as-as-services (SaaS) and cloud providers. The main responsibility of broker collects the available resources and provides quality of service to clients of cloud system. Thirdly, CIS sends the acknowledgment to the broker about available resources of cloud.

Fourthly, broker connects to datacenter. A datacenter is a collection of virtualized hosts, virtual machines, processing elements, virtual networks and virtual storage for cloud users. A datacenter consists of operating system, virtual machine monitor (VMM), host list, memory, bandwidth and storage. A host consists of multiple virtual machines. The VMs are allocated in a host with the best-fit mechanism. The parameters of the host are processing capacity usually measured in million instructions per second (MIPS), memory size is in megabyte (MB), storage size is in terabyte (TB) and communication bandwidth is in megabyte per second (Mbps). Fifthly, cloudlet is

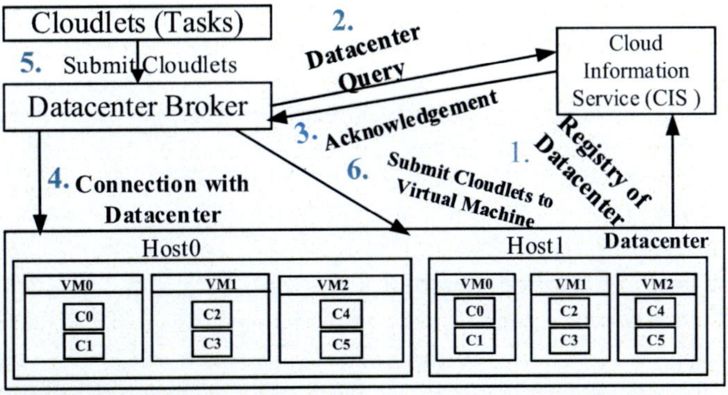

Fig. 1 Architecture of cloud computing

an application which consists of million instructions (it is also known as task such as social networking, content delivery and business application, etc.). These cloudlets are executed by processing element (PE). The parameters of a cloudlet are cloudlet Id, user Id, length (in million instructions), number of PEs, input and output size (in MB). Sixthly, the cloudlets are submitted to virtual machine and are executed by PEs.

3 Related Works

Many researchers are working on fault detection and recovery mechanism. A comprehensive overview of a fault tolerance in cloud computing is given in [1]. It emphasizes different significant concepts, architectural details and techniques. Mittal and Agarwal [2] proposed heartbeat algorithm to detect the hardware and software faults. Ledmi et al. [3] discussed the optimizing fault tolerance in the distributed system. Gokhroo et al. [4] proposed fault detection and mitigation using two fault detection time algorithms (fault detection time-algorithm 1 and time-algorithm 2). These two algorithms identify the faults and correct them. Amoon [5] used selection of fault tolerant algorithm to detect and prevent faults for responding customer requests. They observe the overhead of replication and checkpointing technique for increasing number of customers. Liu et al. [6] illustrated the proactive fault tolerance approach against five related approaches in terms of the overall overheads (such as transmission, network resource consumption and total execution time). Amin et al. [8] explain the heartbeat algorithm to detect the faults. This algorithm monitors the correctness of resources by synchronous time manner. Rouf and Kim [9] discussed the state-of-the-art techniques to combat with soft errors using different techniques broadly categorized in three types: (i) software based schemes, (ii) hardware based schemes, (iii) hardware and software based co-design schemes. Abdelfattah et al. [11] proposed a technique which execute the failed tasks by the best reliable node. Reject message is sent back if it cannot be recovered.

Azaiez et al. [13] propose a hybrid fault tolerant model that consists of checkpointing and replication techniques. Mohammed et al. [14] proposed an integrated virtualized failover strategy that managed the faults reactively. The faults are detected and recovered using the checkpointing technique. However, the overhead of checkpointing can degrade the performance of a system. Jaswal and Malhotra [16] proposed model for fault tolerance in cloud environment. Trust model in cloud computing is the mostly on-demand mechanism that helps in building secured communication in cloud environment. Siddiqui et al. [18] have proposed a single-bit error detection scheme based on hardware fault tolerance for huge data in cloud computing. This scheme uses concurrent error detection (CED) mechanism that is able to detect hardware faults.

Chinnaiah and Niranjan [20] proposed algorithm that achieved reliability for depth critical configuration of a software system. They explained different critical configurations of system and observed the effects of fault tolerance. Buyya et al. [23]

proposed a scheme which can work whenever the demand of cloud users are variable on scalable and virtualized entities. They explain the relationship among entities and events. They illustrate the performance between the federated and without federated network. Nivitha and Pabitha [24] developed a dynamic fault monitoring algorithm for virtual machine. Jhawar et al. [25] implement a fault tolerant management system that consists of a replication manager, a fault detection and recovery manager. They use the gossip and heartbeat algorithm to detect the faults. Rajesh and Devi [26] propose a technique that improves the reliability. It has a forward and backward recovery mechanism and it can calculate the reliability of node and takes decision based on reliability. Bosilca et al. [27] propose a fault injection module that can inject fault in cloud environment. Zhang et al. [28] propose a heuristic ant colony algorithm that VMs are placed by the ant colony algorithms. Wang [29] proposed ARCMeas conforms system architecture, component system's reliability and sensitivity to ensure the effectiveness of recognizing the critical component systems to be settled a fault tolerance mechanism. Neto et al. [30] work on resilient environment using intelligent agents and transient instances. A smart agent-based architecture provides an efficient approach to detects and protects the faults in cloud computing. Al Obaidy et al. [31] proposed an agent-based fault tolerant system to behave the powerful, reliable, predictable, scalable more flexible, autonomous and capable of intelligent behavior. Dähling et al. [32] proposed a cloneMAP algorithm to achieve scalability and fault-tolerance. The cloneMAP eliminates the single points of failure in cloud environment. This scheme enables horizontal scalability and fault tolerant in the distributed data storage. Deshkar [33] proposes an intelligent (smart) and software agents that are used in fuzzy inference system based upon fuzzy set theory for creating a decision system.

3.1 Motivation

Since the use of cloud computing is increasing in different critical application, different fault tolerant techniques are needed. Replication and resubmission technique work from initial state if a fault is occurred. If the faults of VM, host, and PE are occurred, then replicas are used to execute the failed cloudlets from initial state. So the overhead of replication is more than hundred percent. The checkpointing technique moves to the last checkpoint and compares with failed part of the task [14]. The creation of checkpoint increases overhead. The checkpointing, replication and resubmission techniques are backward recovery and solve the problems using roll back technique [11]. It also increases workload and execution time of cloud infrastructure. A maximum efficient smart decision is required to minimize the overhead of fault detection and recovery.

4 Proposed Aggressive Fault Tolerant Technique in Cloud

Cloud environment consists of some entities such as datacenter, datacenter broker, host, virtual machine, cloud information service (CIS) and cloudlet as depicted in Fig. 2. Virtual machines are allocated in host using the allocation policy. The processing capacity of a host is default setup for allocated virtual machines. In virtual machines VM_0, VM_1, VM_2 … VM_V and replica virtual machine $VM_{0'}$, $VM_{1'}$, $VM_{2'}$, …, VM_V have processing elements PE0, PE1, …, PEp. The cloudlets are scheduled by the scheduler and are processed by the PEs. Every entity communicates among each other via event handler. Firstly, the events are stored in future queue during the event execution. Future queue is a queue to store the events to be executed in future. The events of future queue are moved to deferred queue whenever the events cannot be executed due to some faults. The events are triggered using first in first out (FIFO) manner until deferred queue is empty. This type of a fault injection module injects the faults and changes the triggering status, such as cloudlet failure, host failure, PE failure and virtual machine failure. The aggressive fault tolerant (AFT) technique can detect and correct transient, permanent, omission and time out faults in the cloud environment. In the AFT architecture, cloud users U0, U1, …, Un are managed by cloud manager. They submit the cloudlets to datacenter broker. The entities of datacenter are registered in the cloud information service (CIS). When virtual machines or hosts are created, the CIS table of resources are updated. The broker knows the available information of datacenter from the CIS.

The broker is connected to datacenter. A cloudlet Ci is submitted to the available virtual machine (VMj) by broker. The processing capacity of virtual machine is

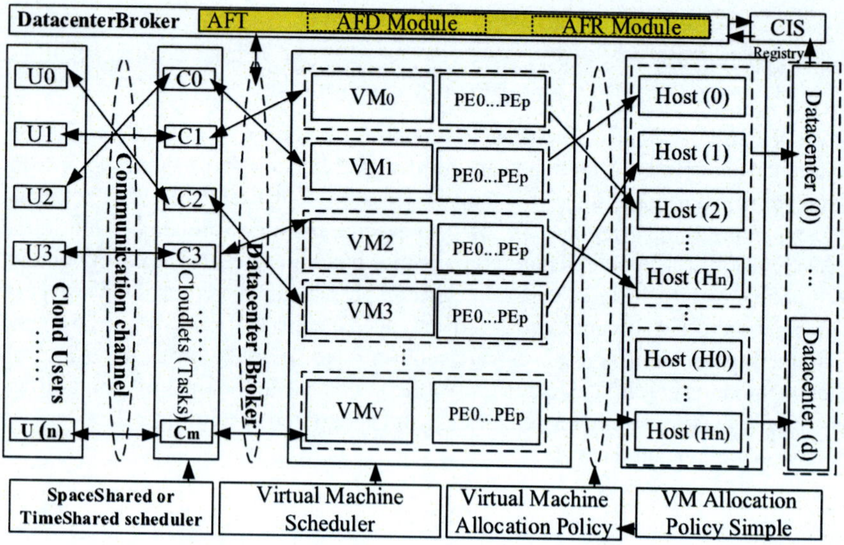

Fig. 2 Proposed architecture of aggressive fault tolerance (AFT) in cloud computing

represented by million instructions per second (MIPS). A certain length of cloudlet is executed by processing elements (PEs). If any fault is occurred during the cloudlet processing, then the AFT technique can detect and recover the faults. AFT has two modules *(i) Aggressive faults detection (AFD) module, (ii) Aggressive faults recovery (AFR) module.* The AFD module detects the fault using heartbeat protocol and AFT recovers the faults using a smart decision technique. Heartbeat protocol is the most popular to detect the faults. In a heartbeat protocol, every process Pr sends "Alive now" message to other process Qp after a static interval time. Accept the process Pr is suspect list if it cannot be reached to Qp within the time threshold which is called timeout. This protocol determines suspect process Pr. It increases independently the accuracy of AFD module within the timeout.

4.1 Aggressive Fault Detection (AFD) Module

The AFD module monitors the events triggering and detects the faults if it is occurred. Every tasks are processed by processing elements (PEs). The heartbeat algorithm detects the faults during message passing. When the status of virtual machine is down, a fault is notified to broker. Whenever, the VM is over utilized, and an unavailability fault is notified to broker. The PEs and memory module are failed due to transient faults. If the number of available PEs are less than required number of PEs, then the status of PEs is failed based on the rules of quality of service. The virtual machine is considered as failed if all PEs are failed. A host is damaged if all virtual machines are down. Whenever a host is down, the broker finds out another new available host. If the availability acknowledgment is not received before the threshold time, then a timeout fault is occurred.

4.2 Aggressive Fault Recovery (AFR) Module

The AFR module recovers dynamically after a fault is occurred, and it is recovered after fault detection. The quality of service (service level agreement) ensures the commitment between the service provider and end-users. If the available level is less than the agreed level, then availability and reliability level are decreased. AFR module makes a replica of cloudlet, replica of virtual machine and replica of host. If a part of system is failed, then the replica ensures recovery from fault. The replica is defined as an available entity which executes the failed cloudlets. Whenever virtual machine is over utilized by overflow of memory or bandwidth, the tasks migrate to the replica virtual machine. Mean time to failure (MTTF) is defined as the average time of a non-repairable entity works before it fails. The mean time to repair (MTTR) is the time to repair the failed entities. The mean time between failures (MTBF) is calculated by the arithmetic average time between failures of a system as given in Eq. 1 [24]. The availability ensures the measurement of available resources for usability. The availability is calculated using Eq. 2.

$$\text{MTBF} = \text{MTTF} + \text{MTTR} \tag{1}$$

$$\text{Availability} = \frac{\text{MTTF}}{\text{MTBF}} = \frac{\text{MTTF}}{\text{MTTF} + \text{MTTR}} \tag{2}$$

We can calculate the Service Level Agreement (SLA) of available resources using Eq. 2. If the MTBF is zero, the highest availability of a node is 100%. The highest availability of virtual machine re-executes the failed cloudlets. The data structure of AFT is given in Fig. 3. It has the following classes:

(a) AFD class (b) AFR class (c) Cloudlets (d) Host (e) Virtual machine	(f) CIS (g) CPUBaseFaults (h) Datacenter broker (i) Smart decision agent

The AFD detects the faults of cloudlet, host, virtual machine and PEs as described in Algorithm 1. A smart decision agent takes the right decision according to fault status. If the status of cloudlet might be canceled or failed, memory or bandwidth is over utilized, virtual machine is either down or over utilized, hardware might be damaged, and the acknowledgment might be timeout, then smart decision agent decides according to fault status.

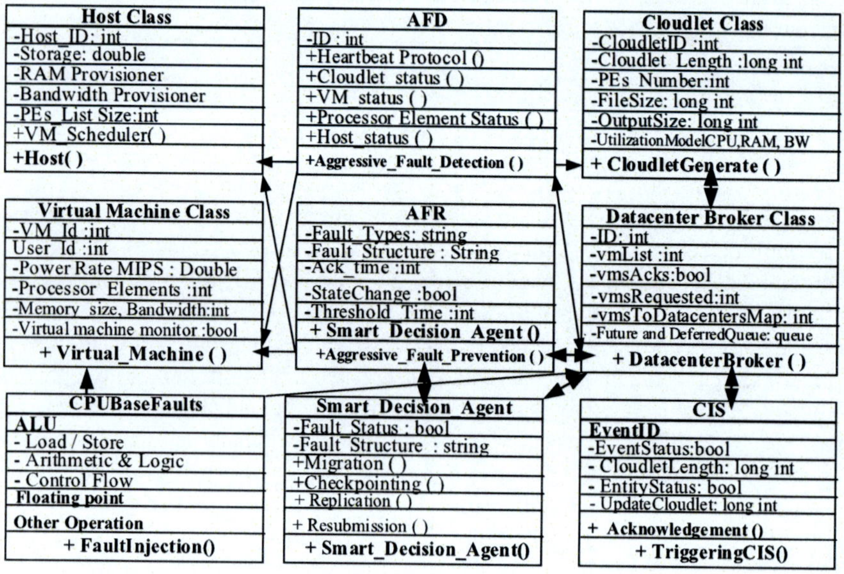

Fig. 3 Class diagram of aggressive fault tolerance (AFT)

Algorithm 1: Aggressive Fault Tolerance (AFT): Fault Detection and Recovery

```
AFT (cloudletList, VMs, hosts, PEs)
// Cloudlet length, input and output size, status.
// Smart Decision Agent monitor the status of cloudlets,
VMs, Hosts and PEs
Input: cloudletList, VMs, hosts, PEs.
Output: Response time, FaultDetected, status.
 Foreach (each cloudlet in cloudletList)
     Process cloudlet Ci on VMj of PEk
   If cloudletCi statusequal-toexecutionthen

   {

    RemainedLenCi = cloudletLengthCi - executedCi;

    Do

    {

         Process cloudlet Ci.
         RemainedLengthCi = RemainedLenCi - executedCi;
         Checkpointing(P, ti+1) checkpointing pivot(P);
          // ti+1 dynamic time interval
         Process cloudlet Ci and continue;

         IF instructions== fail  then
            Smart_Decision_Agent(status);
            Rollback ( ) last saved point;
            FaultDetected++;
         End IF

    } While (RemainedLengthCi !=0);

   }
   Else if(cloudletequal-tocancel OR fail )
    {
          Resubmission( cloudlet );
          FaultDetected++;
   }
   Else if (failure equal-toPEORVMORHost)
   {
           Replication (replica);
           Migrated( );
   FaultDetected ++;
   }

   Else
     Traversehigher VM and migrated the tasks.
   Smart_Decision_Agent(percepts status)
   Actuators condition-actions.

 End Foreach              // End of cloudlet List
```

Each cloudlet length is executed on PEs of virtual machine from the cloudlet list as shown in Algorithm 1. The processing capacity of virtual machine processes the cloudlet of a certain length and updates the remained cloudlet length. If the cloudlet status is canceled or failed, then this task is resubmitted. Whenever the status of PE or VM or host is failed, the tasks are migrated to replica VM. The instructions are executed until it finishes the remaining length of cloudlet. RemainedLengthCi is updated of the cloudlet length and faults occurred when checkpoint ensures last saved pivot point. Otherwise, failed cloudlet is executed by available (or healthy) virtual machine. The Smart_Decision_Agent (status) method percepts the status, and actuator takes the smart decision.

5 Experimental Results

We use the CloudSim simulator (CloudSim Plus) [23] to implement AFT technique in cloud environment. The parameter of the simulator is given in Table 1. The cloudlet length varies from 1 to 10 K. We use the integrated development environment Eclipse IDE version: 4.16.0 and Java Runtime Environment version 1.8.0. The configuration of datacenter uses X86 architecture.

We have compared the execution performance of AFT with other existing techniques (such as checkpointing, resubmission and replication) as given in Fig. 4. The *X*-axis shows the length of each cloudlet in million instructions (MI), and the *Y*-axis shows the execution time in second. For every cloudlet batch, the replication performs worst by taking maximum time, followed by the checkpointing and resubmission methods. We can see that the performance of the proposed scheme increases with increasing cloudlet size. For example, AFT technique reduces 26 s compared to the replication technique for the cloudlet of size 1 K million instructions. On the other hand, for the cloudlet of size 10 K million instructions, the proposed AFT technique reduces 202 s compared to replication technique. On average the AFT is 1.5 times faster than checkpointing, 2.0 times faster than resubmission and 2.5 times faster than replication.

Table 1 The parameter of experimental setup for host and virtual machine

Parameters	Host	Virtual machine
Processor	Intel(R) Core i7	Intel(R) Core i7
Processing elements	8	2
Processing capacity (MIPS)	1000–6000	1000
Memory (GB)	10	2
Storage	2 TB	10 GB
Bandwidth (Mbps)	8024	1000
VM manager	Xen	Xen
Operating system	Windows 10	Windows 10

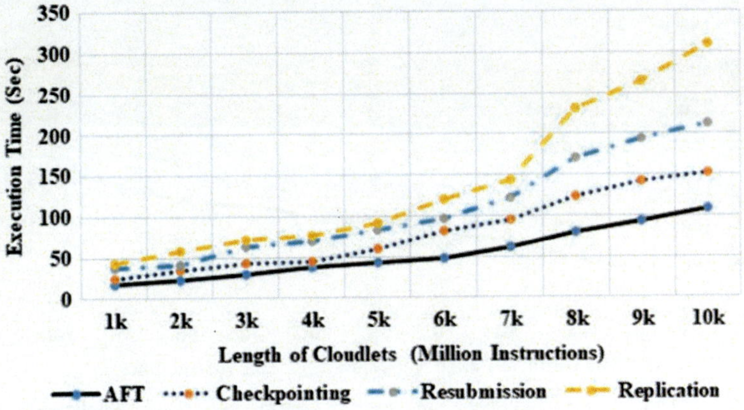

Fig. 4 Comparison of execution time of different length (MI) of cloudlets

In the same way the proposed approach saves a huge amount of time compared with the checkpointing, the resubmission and the replication technique when the size of data set are high. These time savings are very important in cloud computing fault management, because cloud handles a huge amount of data.

The bar graph shown in Fig. 5 illustrates the number of faults recovered according to different length of cloudlets from 1 to 10 k. The *X*-axis shows length of cloudlets, and the *Y*-axis shows the number of faults occurred and recovered. The number of faults recovered by AFT is greater than the other techniques.

The *X*-axis shows the size of each batch of failed cloudlets, and the *Y*-axis shows the recovery execution time required to handle the batch as shown in Fig. 6. Comparing with replication technique the time from 105 s to only 50 s which is an improvement of 55 s for a batch-size of 1 K data. For a batch-size of 10 K data,

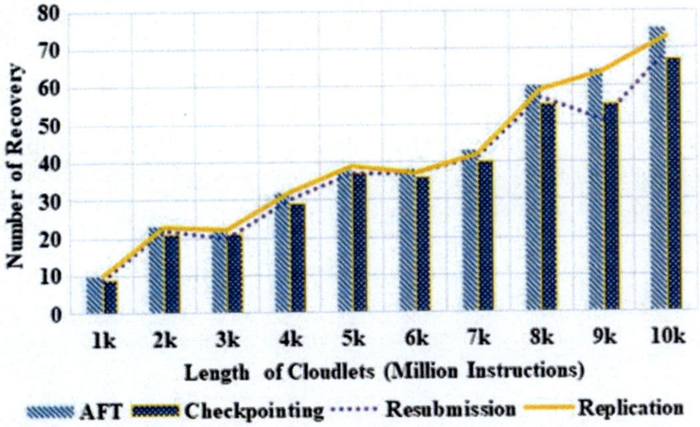

Fig. 5 Faults recovered for different length of cloudlets

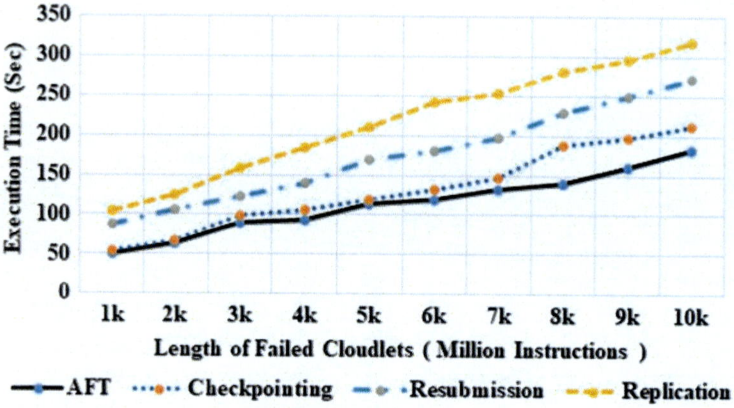

Fig. 6 Execution time of failed cloudlets length

the replication takes 318 s, and our approach reduces it to 183 s. That is a saving of 135 s in performance. Our proposed AFT technique is 1.14 times faster than checkpointing, 1.55 times faster than resubmission and 1.92 times faster than replication technique.

5.1 *Fault Injection and Coverage Analysis*

The different types of faults are injected randomly and their effects are observed in Table 2. We have used the Poisson's distribution for different types of injected faults based on fixed time interval and different locations [24, 27, 34].

In the AFT module, fault injector class is extended the SimEntity class that injects faults randomly during the occurrence of an event among the entities. Fault event class has extended the SimEvent class that has tag type of host failure, cloudlet failure and VM failure. The fault handler of datacenter class has extended the datacenter class that handled the VM migration. It updates the cloudlet execution and status according to the fault event types. The main purpose of this module finds out the boundary of application [13, 35].

The overhead of fault injection (F_{oft}) is given in Algorithm 2.

Table 2 Coverage of injected faults

Instructions type	Total injected faults	Activated faults	Faults recovered	Coverage (%)
ALU	200	44	43	97.8
Floating point	98	23	23	100
Others	57	9	9	100
Total	355	76	75	98.7

C_L Cloudlet_length,
Rate Million instruction per second (MIPS),
F_{fi} Total_Faults_Injected.

$$\text{Overhead } (F_{oft}) = (F_{fi} \times \text{Rate})/C_L \quad (3)$$

Algorithm 2: Host Fault Injection and Coverage Analysis

```
HostFaultsInjection (Datacenter, Statis_Distribution)
CL   := Cloudlet_length.
Rate:= Million instruction per second (MIPS).
Ffi := Total_Faults_Injected.
SD   := Statis_Distribution.
FaultsMeanValue = Statis_Distribution.mean;
SD = Statis_Distribution.Sample ( );
    If SD > FaultsMeanValue then
        Injected_Faults ( )
    Else
       Continue work
   End IF
Coverage: = (Ffi * Rate) / CL
Return Coverage
```

The cloudlet length consists of set of instructions. The set of instructions include ALU operations, floating point operations and others as given in Table 2. The fault injector injects fault and changes the state different types of instructions. Among 200 faults are injected in ALU operations from which 44 errors are activated. From the activated errors, there are 43 errors are recovered so that the coverage is 97.8%. For the floating point operations, total injected faults are 98 and 23 faults are activated. There are 23 faults recovered. For the other operations, total injected faults are 57, activated faults are 9 and all the faults are recovered so that coverage is 100%. For total 355 injected faults, total 76 faults are activated. From these 76 faults, there are 75 faults detected and recovered so that coverage is 98.7%.

6 Conclusion and Future Work

To achieve higher availability of cloud services, the fault tolerant mechanism is required. We propose an aggressive fault tolerant (AFT) technique that combines checkpointing, resubmission and replication with smart decision agent, which can efficiently detect and recover different types of faults in cloud environment. Our proposed technique can detect around 98.7%. The experimental result shows that the proposed approach is 1.5 times faster than checkpointing, 2.0 times faster than resubmission and 2.5 times faster than replication scheme. In future, we shall design a real cloud environment in laboratory that can implement the proposed AFT in physical cloud environment.

References

1. Rouf MA, Shahariar Parvez AHM, Robiul Alam Robel M, Podder P, Bharati S (2020) Effect of fault tolerance in the field of cloud computing. Lect Notes Netw Syst 98:297–305
2. Mittal D, Agarwal N (2015) A review paper on fault tolerance in cloud computing. In: 2015 international conference on computing for sustainable global development, INDIACom 2015, pp 31–34
3. Ledmi A, Bendjenna H, Hemam SM (2018) Fault tolerance in distributed systems: a survey. In: International conference on pattern analysis and intelligent systems (PAIS), pp 1–5
4. Gokhroo MK, Govil MC, Pilli ES (2017) Detecting and mitigating faults in cloud computing environment. In: 3rd IEEE international conference on computational intelligence & communication technology (CICT)
5. Amoon M (2016) Adaptive framework for reliable cloud computing environment. IEEE Access 4(c):9469–9478. https://doi.org/10.1109/ACCESS.2016.2623633
6. Liu J, Wang S, Zhou A, Kumar SAP, Yang F, Buyya R (2018) Using proactive fault-tolerance approach to enhance cloud service reliability. IEEE Trans Cloud Comput 6(4):1191–1202, Oct.-Dec. https://doi.org/10.1109/TCC.2016.2567392
7. Talwani S, Singla J (2019) Comparison of various fault tolerance techniques for scientific workflows in cloud computing. In: 2019 International conference on machine learning, big data, cloud and parallel computing (COMITCon), 2019, pp 454–459. https://doi.org/10.1109/COMITCon.2019.8862211
8. Amin Z, Singh H, Sethi N (2015) Review on fault tolerance techniques in cloud computing. Int J Comput Appl 116(18):11–17
9. Rouf MA, Kim S (2019) A review on fault tolerant techniques and issues in recent generation processors. DUET J 5(ii):1–6. Available at: https://www.duet.ac.bd/wp-content/uploads/2020/10/5.pdf
10. Dos Santos VA, Manacero A, Lobato RS, Spolon R, Cavenaghi MA (2020) A systematic review of fault tolerance solutions for communication errors in open source cloud computing. In: 2020 15th Iberian conference on information systems and technologies (CISTI), vol 2020, pp 1–6. https://doi.org/10.23919/CISTI49556.2020.9140933
11. Abdelfattah E, Elkawkagy M, El-Sisi A (2018) A reactive fault tolerance approach for cloud computing. In: 2017 13th international computer engineering conference (ICENCO), vol 2017, pp 190–194. https://doi.org/10.1109/ICENCO.2017.8289786
12. Villamayor J, Rexachs D, Luque E (2017) A fault tolerance manager with distributed coordinated checkpoints for automatic recovery. In: 2017 International conference on high performance computing & simulation (HPCS) 2017, pp 452–459. https://doi.org/10.1109/HPCS.2017.73
13. Azaiez M, Chainbi W, Ghedira K (2019) Hybrid fault tolerance model for cloud dependability. In: 2019 IEEE 21st international conference on high performance computing and communications; IEEE 17th international conference on smart city; IEEE 5th international conference on data science and systems (HPCC/SmartCity/DSS 2019), pp 2436–2444. https://doi.org/10.1109/HPCC/SmartCity/DSS.2019.00340
14. Mohammed B, Kiran M, Awan IU, Maiyama KM (2016) An integrated virtualized strategy for fault tolerance in cloud computing environment. In: 2016 Intl IEEE international conference on ubiquitous intelligence and computing, advanced and trusted computing, scalable computing and communications, cloud and big data computing, internet of people, and smart

world congress (UIC/ATC/ScalCom/CBDCom/IoP/SmartWorld), pp 542–549. https://doi.org/10.1109/UIC-ATC-ScalCom-CBDCom-IoPSmartWorld.2016.0094
15. Swetha S, Kumar DSV (2018) Fault detection and prediction in cloud computing. Int J Trend Sci Res Dev (ijtsrd) 2(6):878–880. ISSN: 2456-6470
16. Jaswal S, Malhotra M (2019) Trust and fault tolerance models in cloud computing: a review. Int J Trend Sci Res Develop (ijtsrd) 8(11):1273–1285, ISBN: 2277-8616
17. Qu C, Calheiros RN, Buyya R (2018) Auto-scaling web applications in clouds: a taxonomy and survey. ACM Comput Surv 51(4):1–33
18. Siddiqui ZA, Lee JA, Park U (2018) SEDC-based hardware-level fault tolerance and fault secure checker design for big data and cloud computing. Sci Program 2018:16 Article ID 7306837
19. Santiago Pinto VHC, Souza SRS, Souza PSL (2019) A preliminary fault taxonomy for multi-tenant SaaS systems. In: 2019 19th IEEE/ACM international symposium on cluster, cloud and grid computing (CCGrid) 2019, no 1, pp 178–187. https://doi.org/10.1109/CCGRID.2019.00032
20. Chinnaiah MR, Niranjan N (2018) Fault tolerant software systems using software configurations for cloud computing. J Cloud Comput 7(1)
21. Mcmanus JP, Day TG (2019) The effects of latency, bandwidth, and packet loss on cloud-based gaming services. Interact Qualif Proj (All Years), p 58
22. Giannakopoulos I, Konstantinou I, Tsoumakos D, Koziris N (2017) AURA: recovering from transient failures in cloud deployments. In: Proceedings of 2017 17th IEEE/ACM international symposium on cluster, cloud and grid computing, CCGRID 2017, pp 762–765
23. Buyya R, Ranjan R, Calheiros RN (2009) Modeling and simulation of scalable cloud computing environments and the cloudsim toolkit: challenges and opportunities. In: Proceedings of 2009 international conference on high performance computing & simulation, HPCS 2009, pp 1–11
24. Nivitha K, Pabitha P (2020) Fault diagnosis for uncertain cloud environment through fault injection mechanism. In: 2020 4th international conference on intelligent computing, information and control systems (ICICCS) 2020, pp 129–134. https://doi.org/10.1109/ICICCS48265.2020.9121168
25. Jhawar R, Piuri V, Santambrogio M (2013) Fault tolerance management in cloud computing: a system-level perspective. IEEE Syst J 7(2):288–297, June 2013. https://doi.org/10.1109/JSYST.2012.2221934
26. Rajesh S, Devi RK (2014) Improving fault tolerance in virtual machine based cloud infrastructure. Int J Innov Res Sci Eng Technol 3(3):2163–2168
27. Bosilca A, Nita MC, Pop F, Cristea V (2014) Cloud simulation under fault constraints. In: 2014 IEEE 10th international conference on intelligent computer communication and processing, ICCP 2014, pp 341–348. https://doi.org/10.1109/ICCP.2014.6937019
28. Zhang W, Chen X, Jiang J (2021) A multi-objective optimization method of initial virtual machine fault-tolerant placement for star topological data centers of cloud systems. Tsinghua Sci Technol 26(1):95–111
29. Wang L (2019) Architecture-based reliability-sensitive criticality measure for fault-tolerance cloud applications. IEEE Trans Parallel Distrib Syst 30(11):2408–2421
30. De Araujo Neto JP, Pianto DM, Ghedini Ralha C (2018) A resilient agent-based architecture for efficient usage of transient servers in cloud computing. In: 2018 IEEE International conference on cloud computing technology and science, CloudCom, vol 2018, pp 218–22. https://doi.org/10.1109/CloudCom2018.2018.00050
31. Al Obaidy AT, Al Doori MMS (2014) The future for adaptive software development in cloud computing environment using multi agent system. Eng Tech J 3(1):25–36
32. Dähling S, Razik L, Monti A (2021) Enabling scalable and fault-tolerant multi-agent systems by utilizing cloud-native computing. Auton Agent Multi Agent Syst 35(1):1–27
33. Deshkar M (2021) The intelligent agent-based information security model for cloud. Int J Adv Res Ideas Innov Technol 7(3):38–45
34. Malik MK (2020) Host fault injection using various distribution functions. Int J Comput Sci Mob Comput 9(12):1–10

35. Feinbube L, Pirl L, Tröger P, Polze A (2017) Software fault injection campaign generation for cloud infrastructures. In: 2017 IEEE international conference on software quality, reliability and security companion, QRS-C 2017, pp 622–623. https://doi.org/10.1109/QRS-C.2017.119

Classification of Functional Grasps Using Hybrid CNN/LSTM Network

C. Millar, N. Siddique, and E. Kerr

Abstract Gestures made by a human can be classified using Electromyography (EMG) signals collected from the forearm; even with low-frequency devices. Numerous steps are required from data collection and pre-processing through to final classification. Traditionally, an important part of EMG signal classification is extracting features from the raw signal to reduce dimensionality. It is predominantly carried out manually before the signals are input into a neural network. In this research, we successfully used a CNN to extract the features automatically, and an LSTM layer was utilised to classify the gestures. This network architecture removes a step in the gesture classification process. Using the raw signals input into a CNN/LSTM hybrid increased classification when compared with an LSTM network that required features to be manually extracted from the raw signals.

Keywords LSTM · CNN · Gesture classification · sEMG

1 Introduction

Over the past half century, robots have shown that they can perform very well in repetitive environments where repetitive actions are performed time after time and there is no variation in the environment the robot is performing in, i.e. car assembly/manufacturing lines. However, the world does not consist of strict environments that never change. Humans interact with unstructured environments that are filled with objects of all shapes and sizes. To bring this kind of adaptive grasping to the robotics field is a huge challenge [1]. Anthropomorphic robotic grasping is a

C. Millar (✉) · N. Siddique · E. Kerr
Ulster University, Magee Campus, Derry, NI, UK
e-mail: Millar-c21@ulster.ac.uk

N. Siddique
e-mail: Nh.siddique@ulster.ac.uk

E. Kerr
e-mail: Ep.kerr@ulster.ac.uk

M. S. Arefin et al. (eds.), *Proceedings of the International Conference on Big Data, IoT, and Machine Learning*, Lecture Notes on Data Engineering and Communications Technologies 95, https://doi.org/10.1007/978-981-16-6636-0_27

highly complex process to perform successfully, and it requires data-driven development that uses data collected in real-world experiments to improve the performance of such a system [2–4]. A lot of research has been carried out in the area of data-driven grasping that have employed varied technologies for recording data produced by the human demonstrator, i.e. data gloves [5] or vision-based motion capture [6]. Beyond these methods, it is possible to analyse and classify biological signals produced in the muscles of the human body when performing physical movements of the hand and fingers.

Electromyography (EMG) is the process of detecting the electrical activity produced in contracting skeletal muscles during a movement of the body. More specifically in terms of grasping the muscles of the forearm control the extension and flexion of the fingers when shifting the fingers into position to form grasp poses associated with different objects or when performing tasks. A popular method is known as surface EMG (sEMG). This is a non-invasive method that requires placing electrodes on the surface of the skin above the muscles that are controlling the movement. It is the goal of the researcher to detect the biological signals and utilise them as an input to an algorithm that can accurately classify the movement being performed to allow for human–computer interaction (HCI) or potentially control an anthropomorphic robotic hand.

Classification of sEMG signals can prove to be a difficult task. This is due to various different factors that affect the ability of some classifiers to accurately classify the movements being performed. The signals themselves are inherently weak but also the signal patterns are complex. Everything involved in the process from the data collection to the machine learning algorithm and its parameters contribute to the final outcomes. For example, different subjects have different physiological structures and therefore produce different signals despite completing the same movements [7] which can add further complexity to classifying signals. The device used to detect the sEMG signal is also one of the key contributing factors. Most devices used in the literature are medical grade equipment that have large sampling rates (in some cases up to 10 kHz) that allow a researcher to collect more data per sample. This is particularly important when compared with more recently developed commercial devices, i.e. Myo Gesture Control Armband[1] that have lower sample rates and therefore less data per sample.

The combination of deep learning networks and sEMG signal processing has highlighted the potential for creating a control mechanism that can be reliably used as an input for the control of anthropomorphic robotic hands or highly dexterous prosthetic limbs. The eventual application of this research will be to demonstrate to and teach a robotic system how to use functional grasps to pick up everyday objects. This paper investigates the classification of sEMG signals generated during the execution of functional grasps that are associated with activities of daily living using commercial wearable sensor in conjunction with a selection of deep learning techniques.

[1] www.bynorth.com.

The rest of the paper is structured as follows: Sect. 2 contains the literature review of the related work in this field of interest. Section 3 describes the device used in this research along with the methodologies applied during the experimental process as well as descriptions of the deep learning networks and their architectures that were used. Section 4 summarises the experiment protocol that were followed. Section 5 reports the results of the experiments conducted. Section 6 contains the conclusion.

2 Related Work

The use of sEMG signals has been used in many different fields of research. Most commonly, it has been used to diagnose muscular problems [8] and as a control mechanism for prosthetic limbs [9–12]. A further use of sEMG is also for gesture classification. This application of sEMG signals focuses on classification of movements of the wrist, hand and fingers as well as complete gestures or functional grasps. Various research has been conducted that demonstrates the efficacy of using biological signals, i.e. sEMG as an input for controlling systems or devices such as intuitive musical devices [13] or robotic hands [14–16]. Using sEMG as the input to such systems instead of other traditional methods, i.e. data gloves, cameras/motion capture offers other advantages in terms of application. Data gloves have some drawbacks when it comes to accuracy, e.g. the gloves are not tailored for different sized hands and detecting the finger movements can be difficult depending on which sensors the gloves are equipped with. Motion capture can have difficulties with occlusion of the markers or duplication. However, commercially available EMG systems are wireless, typically with an onboard power supply offering the user an intuitive portable solution. Wearable sensors like the aforementioned Myo Gesture control armband have helped broaden the research area due to the fact that the prior more commonly used systems were medical grade devices that were expensive and can require an element of expertise to calibrate and set up.

Using sEMG devices in conjunction with contemporary machine learning networks has demonstrated the overall efficacy of sEMG gesture classification [17–26]. Various forms of recurrent neural networks (RNN) have been shown to classify these biological signals with relatively high accuracies. The authors in [17–19] used the Myo armband to collect the sEMG signals and a Convolutional Neural Network (CNN) to classify a distinct small set of single-finger movements, i.e. flexion movements of the thumb, index, ring and little fingers. The authors in [17] classified the raw signal after it was passed through various filters, whereas [18] converted the signals into images to take advantage of the CNN's ability to recognise patterns within the images. Chen et al. [19] transformed the raw sEMG signals using a continuous wavelet transform (CWT) and compared various classifiers and other feature extraction methods showing that CWT and their own CNN network named 'EMGNET' performed the best [19]. In previous work conducted by the authors, they demonstrated how Long Short Term Memory (LSTM) networks can be used to classify movements of all the fingers and thumb [25] and functional grasps [26] using sEMG signals. Further research has been introduced that classify basic gestures of the hand and wrist, i.e. open/close hand or flexion/extension of the wrist using CNN, LSTM and hybrid CNN/LSTM architectures [20–24]. In [20], it was shown that the hybrid LSTM-CNN (LCNN) network classified the signals with the highest accuracy when compared with LSTM. Moreover, [21] demonstrated that an attention-based CNN/LSTM hybrid was effective in classifying sEMG signals from some publicly available datasets. The authors converted the sEMG signals into various different types of image representations and used them as inputs into the hybrid network. The authors in [24] used an attention-based CNN to classify 53 movements of the hand from the NinaPro database [27].

The authors in [23] introduced a novel CNN architecture that combined 'slow fusion' and 'inception' models that then combine the outputs of the these 2 layers, and they propagate through the rest of the 3D CNN. They successfully demonstrated that this hybrid architecture performed better than using any of the individual models singly. Another use of sEMG signals with a CNN was exhibited in [7], where the authors classified 30 Korean sign language gestures with high degree of accuracy. They used the CNN's ability to extract features from the raw signal and presented a comparison between CNN architecture with a dropout layer and a CNN without dropout layer. Whilst having a dropout layer did show to have some effect on the classification it varied between subjects and on average decreased classification by around 1%.

In the literature reviewed above, a series of papers that carried out research into various CNN/LSTM hybrid architectures when investigating classification of sEMG signals was presented. These papers focused on sEMG signals that were generated when performing basic hand gestures and wrist movements, i.e. open/close hand or extension/flexion of the wrist [20–23].

The authors of this paper have applied a novel variation of these architectures to sEMG data gathered from human demonstrations of functional grasps. These are grasps that would typically be performed during activities associated with daily living, i.e. power grasp, precision sphere. The benefit of the hybrid network proposed is that it removes the need for a manual feature extraction process which can be time consuming. The application of these networks and future goal of this research is to use the trained models as a control mechanism for a robotic grasping system.

3 Methodology

This section describes the methods and technologies applied in this research along with the architectures of the neural network classifiers used to accurately predict the gesture using the raw sEMG signal detected using the Myo armband.

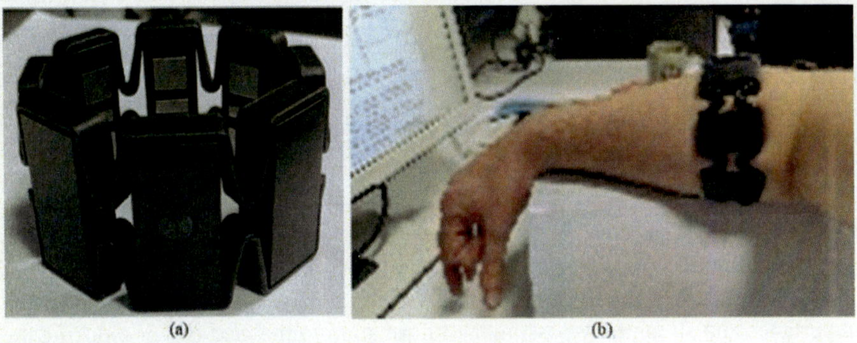

Fig. 1 **a** Myo gesture control armband, **b** subject wearing the Myo

3.1 sEMG Detection

The data used in this work has been collected using the Myo Gesture Control Armband, as shown in Fig. 1. This device is a commercially available wearable solution that allows a user to interact with various software using just the motions of the forearm, hand and wrist. The device itself consists of 8 dry electrodes that fit uniformly around the forearm of the user. The electrodes detect the inherently weak EMG signal produced by the skeletal muscles of the forearm and represents the signals as an arbitrary normalised range of values from −128 and 128. The electrodes sample signals at 200 Hz, 200 samples per second. In some instances, this has been highlighted as a drawback for the device as it produces less data than some of the more expensive medical grade equipment that is available. The medical grade devices can have sample rates as high as 10 kHz which provide a lot more data per second than the Myo is capable of. Despite the differences in the sample rates, it has been shown that despite a lower sample rate, the use of the Myo is still viable with the lower sample rate equating to less the 5% difference in the final accuracy. Additionally, the device also has advantages over medical grade equipment in that it is an intuitive device that offers portability to the user and requires no expertise to use.

3.2 Network Architecture

The classification of sequential data is the cornerstone of the sEMG gesture recognition models. RNN's have been popular amongst researchers studying various types of sequential data, i.e. handwriting classification, speech recognition and other time-sequence data [28–32]. A commonly used variation of an RNN used for these types of applications is an LSTM network. LSTM networks were designed to overcome the exploding/vanishing gradient problem that was associated with networks that use a gradient based learning function or backpropagation [33]. LSTM networks have previously presented as a viable option for classification of sEMG signals generated through various types of finger movements [25] and functional grasps [26]. LSTM networks are often preferred for these types of scenarios as they specialise in learning the dependencies between the different time steps of the sequential data. These networks can be used to classify signals and also as a method of predicting future time steps in the sequence. The network enforces constant error flow between the internal states of the special units. The special units contain multiplicative input gates, forget gates and output gates that control the flow of data depending on the importance and strength by passing the information to the next cell or blocking the information. This process is controlled by the modification of the networks weights via its learning process [32]. The LSTM architecture used in the experiments is shown in Fig. 2. This architecture has proven to perform accurate classification of both finger movements and functional grasps in [25, 26]. A shallow 6-layer network has been used comprising of a sequential input layer where the inputs are normalised using *z*-score normalisation. These normalised sequences are fed into a bi-directional LSTM layer (bi-LSTM) that contains the LSTM networks hidden units which enable the network to learn the bi-directional dependencies of the time series data. Following similar architectures, a dropout layer was added. Dropout layers have been shown to reduce over-fitting [34]. The dropout layer randomly deactivates hidden units with a

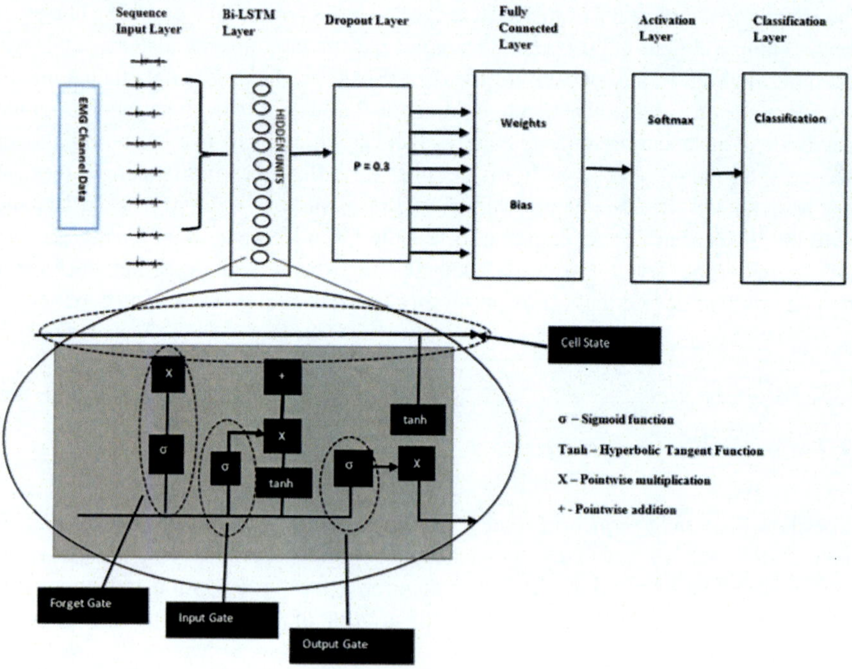

Fig. 2 LSTM network architecture and close up view of LSTM hidden cell

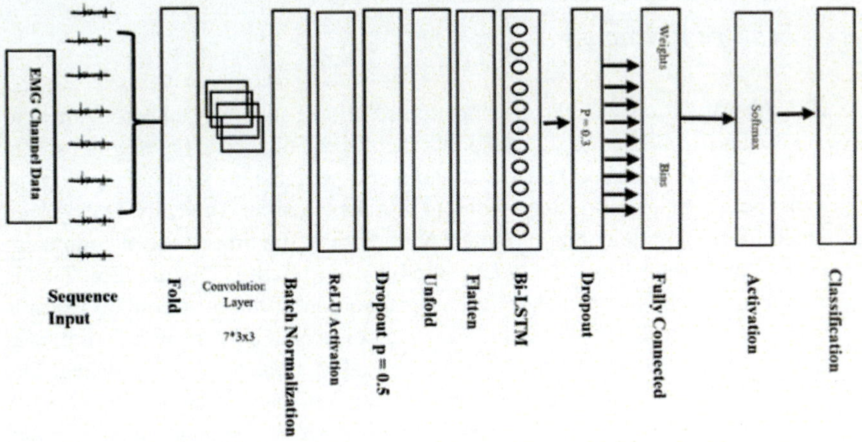

Fig. 3 CNN/LSTM network architecture

probability of p, in this network $p = 0.3$ [25, 26]. The fully connected layer applies the weights and biases of the network to the data through multiplication of the weight matrix and the bias vectors. A softmax activation function calculates the probability distribution in a multiclass classification problem. The final layer in the architecture is the classification layer that computes the cross-entropy loss from classification problems with mutually exclusive classes.

Further to the LSTM network, this research evaluates a hybrid CNN-LSTM structure as these have demonstrated to produce accurate classification of sEMG signals. Traditionally, CNN has been used in image classification problems [35]. The CNN specialises in detecting patterns found within high-dimensional data sequences or images. CNN's contain convolutional layers that use filters of a predefined size to traverse, or convolve, over the input data. The filter calculates the dot product of the data contained in the filter, creates a feature map and passes it to the next layer.

Figure 3 illustrates the architecture of the hybrid network used in this research. The purpose of combining these network architectures is to take advantage of the strengths provided by both of these types of networks. The z-score normalised sequential input data is fed into a folding layer that converts the sequences to an array of sequences. A convolution layer follows the folding layer. An empirically determined 3×3 filter has been used in this architecture. The data in the feature map are normalised in the batch normalisation layer and are processed in a rectified linear unit (ReLU) layer that removes any values less than 0. In this final CNN section of the architecture, a dropout layer is added where $p = 0.5$. Once the convolution process has been conducted, the feature maps are unfolded and flattened. The unfolding layer restores the original sequence structure and the flattened layer converts the data into a 1-dimensional array. This 1-dimensional array is fed into the bi-LSTM layer of the network where the long-term dependencies between the features are processed. The final 3 layers of the network follow the same arrangement as the LSTM network.

3.3 Feature Extraction

As with other traditional signal classification solutions, an important procedure is to extract features from the raw signal so that the dimensionality of the original signal can be reduced. When using LSTM networks or other machine learning algorithms, this must be completed prior to being input into the classifier. Typically, features are extracted from the time domain, frequency domain or the time–frequency domain. Time domain features being the most commonly used in the literature [36]. This is due to the efficiency associated with the computation of the time domain features but also due to their demonstrated improvement of overall classification in previous works [12, 25, 26, 35–38]. In this research, an LSTM architecture is evaluated. The LSTM was tested using time domain feature data sequences as input to the LSTM network. As aforementioned, this process requires greater computational cost and needs careful consideration depending on the requirements of the model that is being produced. The authors in [12] specified an acceptable processing time of 300 ms when performing classification of sEMG signals with the view of controlling a prosthetic hand in real-time. Further delay than 300 ms would be too slow and will affect the performance of the prosthetic in terms of responsiveness. This processing window would be required to be similar for controlling robotic hands. If humans are to control robotic hands in a responsive manner where the mimicry is achieved with almost instantaneous recreation of the human action then the processing time of the input signal will need to be less than the 300 ms window that was suggested [12]. Further experiments were completed in this research that take advantage of the CNN's ability to extract feature information from the original raw signal. This ability of the CNN makes it a very attractive option for the researcher as searching for the optimum manually selected features can be time consuming and also vary across research due to the differences in detection devices being used, the movements being performed and the classifiers that will classify the sequences.

The features selected in this research are commonly found throughout the literature associated with EMG signal classification as they are computationally efficient and have contributed to more accurate models being developed [25, 26, 35]. Table 1 contains the 11 features, and their associated equations used in this research.

4 Data Acquisition and Pre-processing

This section outlines the protocols followed during the data collection phase of the experiment and the steps to process the raw signal data. The authors also incorporated a publicly available database that was first made available in [27].

Table 1 Features extracted from original signal

Feature	Expression	Feature	Expression	Feature	Expression
Mean absolute value (MAV)	$\frac{1}{N}\sum_{k=1}^{N}\lvert x_k\rvert$	Willison amplitude (WAMP)	$\sum_{k=1}^{N-1} f(\lvert X_k - x_{k+1}\rvert)$	Slope sign change (SSC)	$\sum_{k=2}^{N-1}\left[f((x_k - x_{k-1})(x_k - x_{k+1}))\right]$
Waveform length (WL)	$\sum_{k=1}^{N}\lvert x_{k+1} - x_k\rvert$	Root mean square (RMS)	$\sqrt{\frac{1}{N}\sum_{k=1}^{N} x_k^2}$	Zero crossings (ZC)	$ZC = f\left(\sum_{k=1}^{N-1} f([x_k x_{k+1}][x_k - x_{k+1}]) - Z_1 - Z_2\right)$
Variance (VAR)	$\frac{1}{N-1}\sum_{k=1}^{N} x_k^2$	Standard deviation (STD)	$\sqrt{\frac{1}{N-1}\sum_{k=1}^{N}(x_k - \mu)^2}$	Kurtosis (KURT)	$\frac{\frac{1}{N}\sum_{k=1}^{N}(x_k-\mu)^4}{\left(\frac{1}{N}\sum_{k=1}^{N}(x_k-\mu)^2\right)^2}$
Autoregressive modelling (AR)	$X_k = \sum_{i=1}^{\rho} a_i x_{k-i} + e_k$	Mean absolute deviation (MAD)	$\frac{1}{N}\sum_{k=1}^{N}\lvert x_k - \mu\rvert$		

N number of samples; k k-th sample; x_k current sample; μ feature mean of all windows; N_w wth window of sample N; p order; e_k white noise error; a_i AR coefficients; $f(x)$ takes 1 if $x \geq 0$ and otherwise 0. Z_1, Z_2 Threshold

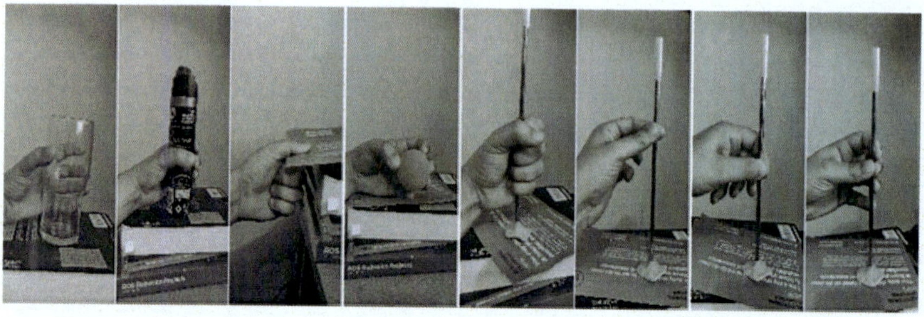

Fig. 4 Functional grasps performed: top row (from left to right). 1. Power grasp, 2. Medium wrap, 3. Lateral grasp, 4. Precision sphere, 5. Small diameter grasp, 6. Tip pinch, 7. Writing tripod, 8. Prismatic quad grasp

4.1 Data Collection

This paper focuses on functional grasps that are associated with activities of daily living. A series of such grasps were performed by a single subject following the same protocol outlined in [25]. The Myo gesture control armband was placed around the subjects left forearm just below the elbow. The subject was then tasked with performing a select set of functional grasps, shown in Fig. 4, on various objects to produce realistic signals that are impacted by the object's geometry instead of the user pretending to grasp an object. Each grasp was carried out by the subject a total of 200 times. This was to make sure that the networks had a large amount of data to train and attempt to avoid over-fitting of the model. This database is referred to as Grasp Database 1 in the rest of the paper.

4.2 NinaPro Database 5

CNN's require large amounts of data to train effectively. Therefore, a publicly available benchmark database, NinaPro database [39], was utilised and incorporated into the dataset. The NinaPro database contains sEMG signals from multiple right-handed subjects performing a wide range of dexterous hand and finger movements including functional grasps. The NinaPro database has multiple databases within that were recorded from both able bodied and amputee subjects using a range of medical grade EMG systems as well as a database recorded using a dual Myo configuration. Data was used from 9 of the available subjects and the sEMG signals from the corresponding movements, shown in Fig. 4, were the only movements considered in this research.

4.3 Data Pre-processing

All the subject data was concatenated into a singular sequence array, known as a cell array in Matlab. This required taking each subject from the NinaPro database and separating each of the movements and creating sequence samples of the movement classes from Fig. 4. The master set of data that contained the data recorded for this research combined with the 9 subjects from the NinaPro database was split into training (70%), validation (15%) and testing sets (15%). No other processing was applied to the signals being used for the hybrid CNN/LSTM network. However, for the LSTM network that was also trained a set of time domain features were used as aforementioned in Table 1, found in Sect. 3.3.

5 Experiments and Results

This section outlines the experiments carried out, and a summary of the results for each experiment is provided. Each of the networks were trained 50 times using multiple hyper-parameters to find the best combination to produce the highest classification accuracies. Various parameters were tested throughout the experimentation process which are found in Table 2. In this research the authors have evaluated the best performing network as the network with the highest accuracy attained using the validation set during training. Table 3 contains the parameters of the best network for each experiment. The hyper-parameters were adjusted following the Bayesian optimization strategy which automates the procedure of tuning the parameters [40]. The aim of this strategy is to maximise the generalisation of the algorithm.

5.1 Experiment 1—LSTM Classification of Raw Signals

The initial network trained was an LSTM network. This was carried out to give a baseline accuracy for comparison with the hybrid CNN/LSTM network. This initial network was trained using the raw EMG signals with no feature extraction carried out on the signal. 50 networks were trained, using both the Grasp Database 1 and the Ninapro database, using a set of 1828 training samples (70%) and a set of 229 validation samples (15%). During the training process the validation samples are used to tune the learning process to optimize the generalization of the network and reduce overfitting. The average classification accuracy on the training data was 71.87% (9.27%±) across the 50 networks. The optimazation process was focused on maximising validation accuracy. The average validation accuracy was 57.70% (3.80%±) and the average test accuracy was 74.07% (8.33%±). In this experiment, trial 29 produced the highest training accuracy of 95.18% and a validation accuracy of 62.45%. When tested with an unseen test set trial 29 acheived an accuracy of

Table 2 Parameters and the range values used for bayesian optimization of the networks

Parameter	Exp 1	Exp 2	Exp 3	Exp 4	Exp 5
No. of Filters	N/A	N/A	[4 9]	[4 9]	[1 10]
Mini Batch Size	[80 120]	[60 160]	[80 120]	[80 120]	[60 160]
Max Epochs	[15 30]	[10 25]	[12 25]	[12 25]	[5 25]
Initial Learn Rate	[0.001 0.003]	[0.001 1]	[0.001 0.003]	[0.001 0.003]	[0.001 1]
L2 Regularisation	[1e−21 1]	[1e−21 1]	[1e−21 1]	[1e−21 1]	[1e−21 1]
No. of Hidden Units	s50 200]	N/A	N/A	N/A	N/A

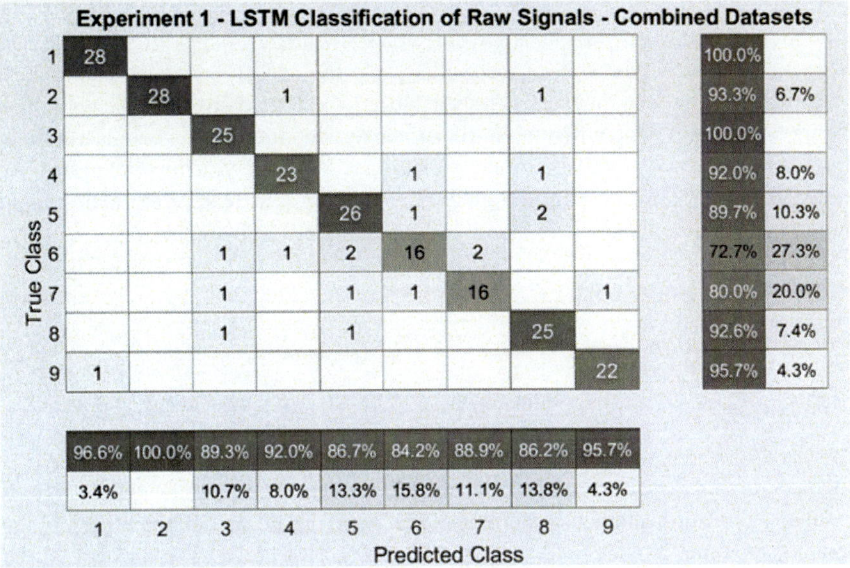

Fig. 5 Experiment 1 test data confusion matrix for trial 29

Table 3 Parameters of the best performing networks

Parameter	Exp 1	Exp 2	Exp 3	Exp 4	Exp 5
No. of Filters	N/A	N/A	8	9	7
Mini Batch Size	107	69	90	84	64
Max Epochs	23	25	25	23	25
Initial Learning rate	0.0025	0.8350	0.0028	0.0021	0.0032
L2 Regularisation	N/A	0.0366	0.0171	0.0306	0.0110
No. of Hidden Units	158	60	60	60	60

91.27%. Figure 5 shows the confusion matrix for the classification of the test set on trial 29.

5.2 *Experiment 2—LSTM Classification of Processed Signals (Time Domain Fea-tures)*

The data for this network contained the time domain features that were extracted from the raw data signal over a 200 ms sliding window with overlapping increments of 25 ms [25, 26]. The parameters of the best performing network of the 50 that were trained are found in Table 3. The average training accuracy 93.34 (4.18%±) and validation accuracy was 75.49% (2.85%±) with 5 of the trained networks achieved a maximum validation accuracy of 79.04%. An average classification accuracy of 96.87% (2.94 %±) was achieved across all 50 networks using the unseen test data. The highest classification accuracy of unseen test set achieved was 99.13% by multiple networks. Figure 6 illustrates the confusion matrix for the test set of trial 17. There is a distinct improvement demonstrated by this network when compared with the previous experiment that classified raw sEMG signal data showing that feature extraction is an important step when using LSTM network to classify sEMG signals.

5.3 Experiment 3—CNN/LSTM Hybrid Using Grasp Database 1

The CNN/LSTM network was trained using the single-subject database recorded for the purpose of this research. The network was trained using the raw sEMG signals as inputs. Table 2 displays the parameters that were optimised and the ranges of the parameters. Table 3 contains the parameters of the best performing network of the 50 that were trained. On average the training accuracy reached 97.79% (8.25%±). The average classification accuracy of the unseen test data was 98.50% (7.00%±). The validation accuracy peaked at 97.22% for trial 20 and was tested using unseen data where it achieved 100% accuracy. Figure 6 shows the confusion matrix for the test set of trial 20.

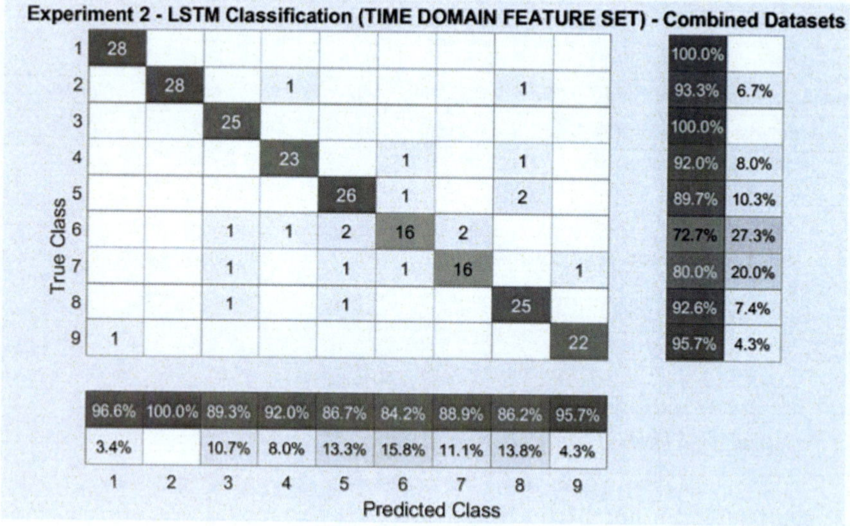

Fig. 6 Experiment 2 test data confusion matrix, trial 17

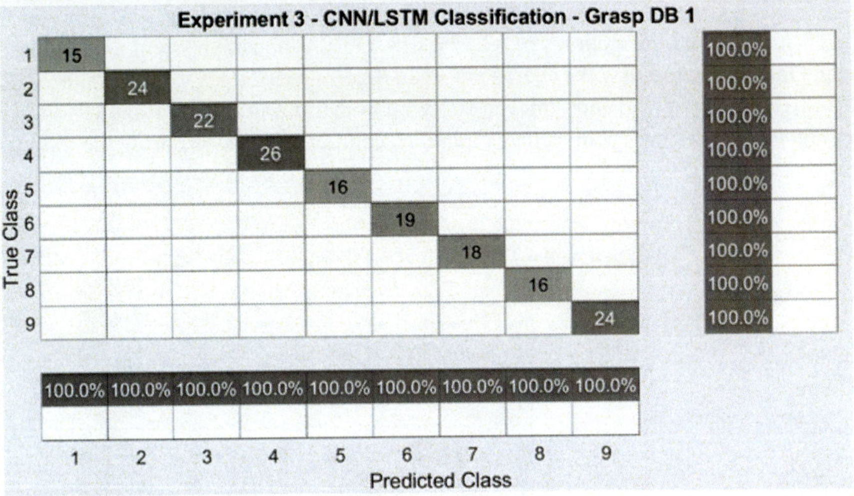

Fig. 7 Experiment 2 confusion matrix test data, trial 20

5.4 Experiment 4—CNN/LSTM Hybrid Using NinaPro Database 5

A separate network was trained using only the data from the NinaPro DB5. The CNN/LSTM parameter ranges are displayed in Table 2, and the parameters of the best performing network are contained in Table 3. The Ninapro dataset is made up of samples from multiple different subjects whose signals will vary due to the various physiological differences between the subjects. The average training accuracy was 99.30% (3.59%±) and the average validation accuracy was 50.25% (5.22%±). However, the unseen test set still achieved an average accuracy of 99.76% (1.44%±) with trial 10 achieving a training and test accuracy of 100% along with a 59.18% validation accuracy. Figure 8 illustrates the confusion matrix for the test set during the training process.

5.5 Experiment 5—CNN/LSTM Hybrid Classification—Combined Dataset

The CNN/LSTM network was trained to classify the sequences produced by the performance of the 9 functional movements. This network was trained using the raw sEMG signal data only of the combined datasets. The parameters of the best performing network are found in Table 3. It should be noted that an additional parameter for the number of filters used in the convolution layer were searched in this experiment. The network achieved an average training accuracy of 80.14% (30.82%±) and an average validation accuracy of 63.57% (23.52±). When the networks were tested using the unseen test data the average reached 82.10% (30.21%±). The best network, trial 16, performed well and achieved 100% for both training and test accuracy, shown in Fig. 9. This network achieved a validation accuracy of 83.41% during the training process.

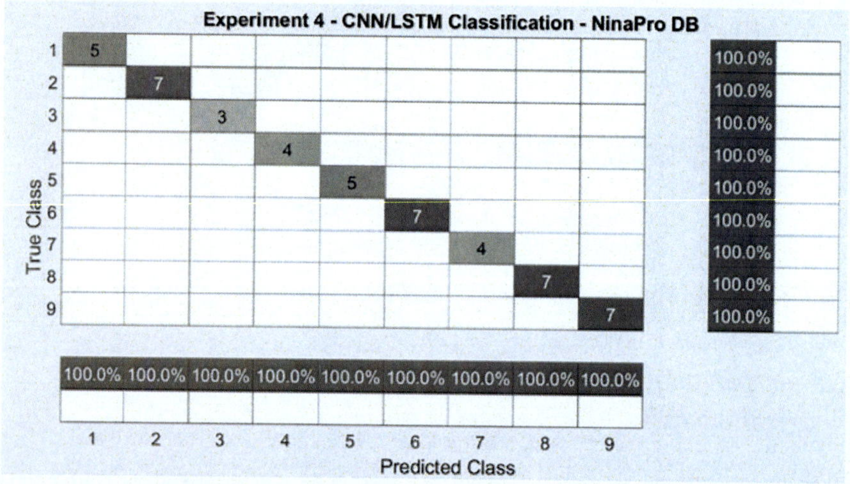

Fig. 8 Experiment 4 test data confusion matrix, trial 10

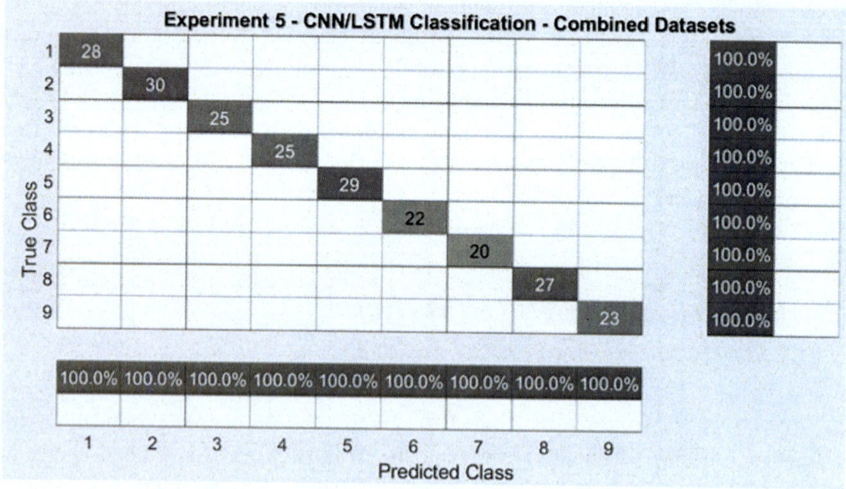

Fig. 9 Experiment 5 test data confusion matrix, trial 16

6 Conclusion

This paper presents the results of a hybrid CNN/LSTM classifier that can classify raw sEMG signals generated when performing functional grasps. A LSTM network was developed as a benchmark classifier and produced the lowest average classification of 74.07% with the best network achieving 91.27% for test data. The novel hybrid CNN/LSTM architecture reached 82.1% on average with the best network reaching 100% classification for all the datasets used in this research. The main advantage of using the hybrid CNN/LSTM network architecture is that it removes the need to manually process the raw signal to extract time domain features. This task is now automatically completed by the network using the CNN section of the architecture to calculate feature maps and then passes these feature maps as inputs in the LSTM section of the network.

The future application of these models will be in conjunction with an anthropomorphic robotic hand. The models created in this research will be applied in an offline capacity in order to develop a robotic grasping system that can perform dexterous functional grasps associated with activies of daily living in a domestic setting.

References

1. Hodson R (2018) A gripping problem. Nature 557(7704):S23–S25
2. Bohg J, Morales A, Asfour T, Kragic D (2014) Data-driven grasp synthesis—a survey. IEEE Trans Rob 30(2):289–309. https://doi.org/10.1109/TRO.2013.2289018
3. Spiers AJ, Liarokapis MV, Calli B, Dollar AM (2016) Single-grasp object classification and feature extraction with simple robot hands and tactile sensors. IEEE Trans Haptics 9(2):207–220. https://doi.org/10.1109/TOH.2016.2521378
4. Goldfeder C, Allen PK (2011) Data-driven grasping. Auton Robot 31(1):1–20. https://doi.org/10.1007/s10514-011-9228-1
5. Lin Y, Sun Y (2015) Robot grasp planning based on demonstrated grasp strategies. Int J Robot Res 34(1):26–42. https://doi.org/10.1177/0278364914555544
6. Gustus A, Stillfried G, Visser J, Jörntell H, Van der Smagt P (2012) Human hand modelling: kinematics, dynamics, applications. Biol Cybern 106(11–12):741–755. https://doi.org/10.1007/s00422-012-0532-4
7. Shin S, Baek Y, Lee J, Eun Y, Son SH (2018) Korean sign language recognition using EMG and IMU sensors based on group-dependent NN models. In: 2017 IEEE symposium series on computational intelligence, SSCI 2017—proceedings, Jan 2018, pp 1–7. https://doi.org/10.1109/SSCI.2017.8280908
8. Nor T, Tengku S, Abdullah AR, Shair EF, Saad NM (2019) Classification of EMG signal for health screening task for musculoskeletal classification of EMG signal for health screening task for musculoskeletal disorder, Apr 2019
9. Harada A, Nakakuki T, Hikita M, Ishii C (2010) Robot finger design for myoelectric prosthetic hand and recognition of finger motions via surface EMG. In: 2010 IEEE international conference on automation and logistics, ICAL 2010. IEEE, pp 273–278. https://doi.org/10.1109/ICAL.2010.5585294
10. Krasoulis A, Kyranou I, Erden MS, Nazarpour K, Vijayakumar S (2017) Improved prosthetic hand control with concurrent use of myoelectric and inertial measurements. J NeuroEng Rehabil 14(1):1–14. https://doi.org/10.1186/s12984-017-0284-4

11. Brunelli D, Tadesse AM, Vodermayer B, Nowak M, Castellini C (2015) Low-cost wearable multichannel surface EMG acquisition for prosthetic hand control. In: Proceedings—2015 6th IEEE international workshop on advances in sensors and interfaces, IWASI 2015, Mar 2016, pp 94–99. https://doi.org/10.1109/IWASI.2015.7184964
12. Hudgins B, Parker P, Scott RN (1993) A new strategy for multifunction myoelectric control. IEEE Trans Biomed Eng 40(1):82–94. https://doi.org/10.1109/10.204774
13. Nymoen K, Romarheim M, Alexander H, Jensenius R (2015) MuMYO—evaluating and exploring the MYO armband for musical interaction. In: New interfaces for musical expression (NIME'15), paper 179
14. Meattini R, Benatti S, Scarcia U, De Gregorio D, Benini L, Melchiorri C (2018) An sEMG-based human-robot interface for robotic hands using machine learning and synergies. IEEE Trans Compon Packag Manuf Technol 8(7):1149–1158. https://doi.org/10.1109/TCPMT.2018.2799987
15. Meattini R, Benatti S, Scarcia U, Benini L, Melchiorri C (2015) Experimental evaluation of a sEMG-based human-robot interface for human-like grasping tasks. In: 2015 IEEE international conference on robotics and biomimetics, IEEE-ROBIO 2015, Dec 2015, pp 1030–1035. https://doi.org/10.1109/ROBIO.2015.7418907
16. Lee HJ, Kim SJ, Kim K, Park MS, Kim SK, Park JH, Oh SR (2011) Online remote control of a robotic hand configurations using sEMG signals on a forearm. In: 2011 IEEE international conference on robotics and biomimetics, ROBIO 2011, pp 2243–2244. https://doi.org/10.1109/ROBIO.2011.6181628
17. Srinivasan VB, Islam M, Zhang W, Ren H (2018) Finger movement classification from myoelectric signals using convolutional neural networks. In: 2018 IEEE international conference on robotics and biomimetics (ROBIO), pp 1070–1075. https://doi.org/10.1109/ROBIO.2018.8664807
18. Stephenson RM, Chai R, Eager D (2018) Isometric finger pose recognition with sparse channel spatiotemporal EMG imaging. In: Proceedings of the annual international conference of the IEEE engineering in medicine and biology society, EMBS, 2018, July 2018, pp 5232–5235. https://doi.org/10.1109/EMBC.2018.8513445
19. Chen L, Fu J, Wu Y, Li H, Zheng B (2020) Hand gesture recognition using compact CNN via surface electromyography signals. Sensors 20(3):672. https://doi.org/10.3390/s20030672
20. Wu Y, Zheng B, Zhao Y (2019) Dynamic gesture recognition based on LSTM-CNN. In: Proceedings 2018 Chinese automation congress, CAC 2018, pp 2446–2450. https://doi.org/10.1109/CAC.2018.8623035
21. Hu Y, Wong Y, Wei W, Du Y, Kankanhalli M, Geng W (2018) A novel attention-based hybrid CNN-RNN architecture for sEMG-based gesture recognition. PLoS ONE 13(10):e0206049. https://doi.org/10.1371/journal.pone.0206049
22. Simão M, Neto P, Gibaru O (2019) EMG-based online classification of gestures with recurrent neural networks. Pattern Recogn Lett 128:45–51. https://doi.org/10.1016/j.patrec.2019.07.021
23. Tutak Erozen A (2020) A new CNN approach for hand gesture classification using sEMG data. J Innov Sci Eng (JISE) 4(1):44–55. https://doi.org/10.38088/jise.730957
24. Josephs D, Drake C, Heroy A, Santerre J (2020) sEMG gesture recognition with a simple model of attention. In: JMLR: workshop and conference proceedings, vol 136, pp 126–138
25. Millar C, Prof. Siddique N, Dr. Kerr E (2020) LSTM classification of sEMG signals for individual finger movements using low cost wearable sensor. In: The international symposium on community-centric systems. https://doi.org/10.1109/CcS49175.2020.9231515
26. Millar C, Siddique N, Kerr E (2021) LSTM classification of functional grasps using sEMG data from low-cost wearable sensor. In: 2021 7th international conference on control, automation and robotics (ICCAR). IEEE, pp 213–222. https://doi.org/10.1109/ICCAR52225.2021.9463477
27. Atzori M, Gijsberts A, Kuzborskij I, Heynen S, Mittz Hagger A-G, Deriaz O, Castellini C, Müller H, Caputo B (2013) A benchmark database for myoelectric movement classification. Trans Neural Syst Rehabil Eng
28. Graves A, Jaitly N, Mohamed AR (2013) Hybrid speech recognition with deep bidirectional LSTM. In: 2013 IEEE workshop on automatic speech recognition and understanding, ASRU 2013—proceedings, pp 273–278. https://doi.org/10.1109/ASRU.2013.6707742

29. Graves A, Mohamed A, Hinton G (2013) Speech recognition with deep recurrent neural networks. In: IEEE international conference on acoustics, speech and signal processing, pp 6645–6649
30. Graves A, Liwicki M, Fern S, Bertolami R, Bunke H (2008) A novel connectionist system for unconstrained handwriting recognition. IEEE Trans Pattern Anal Mach Intell 31(5):855–868. https://doi.org/10.1109/TPAMI.2008.137
31. Malhorta P, Vig L, Shroff G, Agarwal P (2015) Long short term memory networks for anomaly detection in time series. In: ESANN 2015 proceedings, European symposium on artificial neural networks, computational intelligence and machine learning, Ciaco, pp 89–94
32. Petenehazi G (2019) Recurrent neural networks for time series forecasting, pp 1–22
33. Hochreiter S, Schmidhuber J (1997) Long shortterm memory. Neural Comput 9(8):1735–1780
34. Tatarian K, Couceiro MS, Ribeiro EP, Faria DR (2019) Stepping-stones to transhumanism: an EMG-controlled low-cost prosthetic hand for academia, Nov 2019, pp 807–812. https://doi.org/10.1109/is.2018.8710489
35. Pak M, Kim S (2018) A review of deep learning in image recognition. In: Proceedings of the 2017 4th international conference on computer applications and information processing technology, CAIPT 2017, Jan 2018, pp 1–3. https://doi.org/10.1109/CAIPT.2017.8320684
36. Phinyomark A, Phukpattaranont P, Limsakul C (2012) Feature reduction and selection for EMG signal classification. Expert Syst Appl 39(8):7420–7431. https://doi.org/10.1016/j.eswa.2012.01.102
37. Phinyomark A, Khushaba RN, Scheme E (2018) Feature extraction and selection for myoelectric control based on wearable EMG sensors. Sensors (Switzerland) 18(5):1–17. https://doi.org/10.3390/s18051615
38. Wang G, Wang Z, Chen W, Zhuang J (2006) Classification of surface EMG signals using optimal wavelet packet method based on Davies-Bouldin criterion. Med Biol Eng Comput 44(10):865–872. https://doi.org/10.1007/s11517-006-0100-y
39. Atzori M, Gijsberts A, Castellini C, Caputo B, Hager AGM, Elsig S, Giatsidis G, Bassetto F, Müller H (2014) Electromyography data for non-invasive naturally-controlled robotic hand prostheses. Sci Data 1:1–13. https://doi.org/10.1038/sdata.2014.53
40. Snoek J, Larochelle H, Adams RP (2012) Practical Bayesian optimization of machine learning algorithms. Relig Arts 17(1–2):57–73. https://doi.org/10.1163/15685292-12341254

Internet of Things for Smart Applications

Fire Safety and Supervision System: Fire Hazard Monitoring Based on IoT

Ahsan Habib, Srejon Sharma, Mohammad Riduanur Rahman, Md. Neamul Haque, and Mohammad Ariful Islam Bhuyan

Abstract Fire safety is the most significant thing in this era and the people of the twenty-first century. The fire occurred everywhere, basically homes, industries, buildings and lost more property and lives. So we developed an Internet of things (IoT)-based fire regulate and response system which takes the protection before fire occurs and takes action on the fire situation. This requirement has been achieved in this proposed system by employing fire controlling and fire detection using Arduino Uno and Raspberry Pi b+. This system consists of perspective sensors like gas or smoke and temperature sensors. All sensor data has been sent to a server for prediction and checked data validation using a support vector machine (SVM) algorithm. Invalid data refers to a system that creates an error and immediately notify the authority. Using a trained dataset, machine learning method takes proper action automatically which reduces complexity and delivers the best accuracy than existing systems.

Keywords IoT · Fire hazards monitoring · Mobile messages · Gas cylinder regulator off · Fire extinguisher · Exhaust fan · Raspberry Pi b+ · Sensor data validation

1 Introduction

Fire prevention can save lives and reduce the damage caused by fire. People cannot always stay at home or at an institution. Many applications include a gas sensor, a flame sensor, a camera module, water sprinkler and a fire fighting robot. That is costly and that is used individually, which is maintained by big authorities. So for this reason, we build a technology which can protect before of fire occurrence and fire occurs situation. After all, the existing system is not cost efficient, installment is not easy and performance is not fast, that has been fault tolerance.

IoT is an essential part of smart homes and smart buildings. These structures are built up with devices which are connected and controlled by autonomous systems.

A. Habib (✉) · S. Sharma · M. R. Rahman · Md. N. Haque · M. A. I. Bhuyan
Premier University, Chattogram, Bangladesh

M. S. Arefin et al. (eds.), *Proceedings of the International Conference on Big Data, IoT, and Machine Learning*, Lecture Notes on Data Engineering and Communications Technologies 95, https://doi.org/10.1007/978-981-16-6636-0_28

Besides, one can perform various tasks within the building remotely with the help of these connected devices. Now the question is how does the system works to detect fire hazards. At first, this system observes any suspicious changes in the environment, such as different types of gas leakage, temperature crossing the threshold value or the presence of smoke. If the gas sensor senses gas leakage, it is on the exhaust fan, turns off the gas cylinder regulator with a servo motor, sends data in the cloud, that means a server, and sends a message to the owner's mobile.

The most important role of the IoT in fire safety is detecting data validation. The sensor sends the data to a server and checked whether the data is valid or not by using machine learning. Invalid data means the system has a faulty connection and notify the user or authority system that the system has faced a problem or error.

The benefits of this proposed system are the most effective means of fire control, data validation, and user notification. When properly installed, this proposed system can be a highly effective safeguard against the loss of life and property.

2 Related Works

In this section, we try to know about some proposed system with references.

This is an Internet of things (IoT) based on a proposed system, where a webcamera monitors the area and detects fire using image processing [1]. If the system detects fire, an e-mail has been sent to the security and nearest fire department with an attachment photograph. But in this system if dust falls on the camera, it cannot detect fire properly, and the mail communication process is also not reliable.

A device efficient for fire detection and warns users about the fire. Whenever a fire is detected by the system, it will start the smoke alarm and extinguish the fire [2]. There is a chance that the fire transmitter sensed false fire activity, but it will simulate the smoke alarm and extinguisher without confirmation of fire.

The other model highlights the capable feature of wireless sensor networks (WSN) as a probable solution to the challenge of identifying fires quickly [3]. The proposed scheme relies on wireless sensor networks to help in earlier detection of any fire threat. But this system does not mention how users will be notified if a fire incident occurs.

3 Proposed System

The system detects gas or smoke and temperature using perspective sensors. If gas or smoke sensor data is increased above the threshold value, the system will send messages to the user mobile, turn off the gas cylinder regulator by using a servo motor, and turn on the exhaust fan. If a flame or temperature sensor detects unexpected fire, the fire extinguisher will turn on by using a solenoid valve, the fire alarm sounds and sends messages to the owner's mobile and fire department. All sensor data has

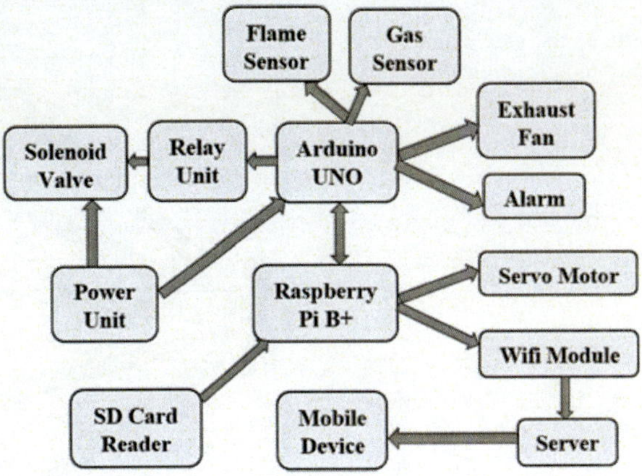

Fig. 1 Structural design of system

been sent to a server for prediction and checked data validation after a period of time using machine learning using support vector machines (SVM). Invalid data means the system has a fault that means the connection is lost and notify the authority.

3.1 *Structural Design of System*

The structural design of the system is based on Arduino Uno and Raspberry Pi 3b+. Flame and gas sensors are connected in Arduino Uno for taking analog signals as an input before fire occurrence. The exhaust fan and alarm are connected as an output signal to take action at the time of the fire. The relay unit likes a switch which uses a solenoid valve which joint in Arduino Uno and a solenoid valve to give the analog signal for doing turn on the fire extinguisher if a fire is detected. Raspberry Pi 3b+ is joined not only to the Arduino Uno but also to the connected Servo motor. Raspberry Pi 3b+ other works transmit a sensor value in the Web server to give the message in the user device and it is to take system connection losses or faulty. The power unit gives proper current voltage in solenoid valves, Arduino Uno, and Raspberry Pi (Fig. 1).

3.2 *Use Case Diagram*

The use case diagram represents the user interface. Basically the use case diagram has shown what will be able to see a user such as—the user will see valid or invalid data, view detection data from the detector. A user can know system errors as well as receive notification from the Web server. The use case diagram for mobile application improvement is proven in Fig. 2.

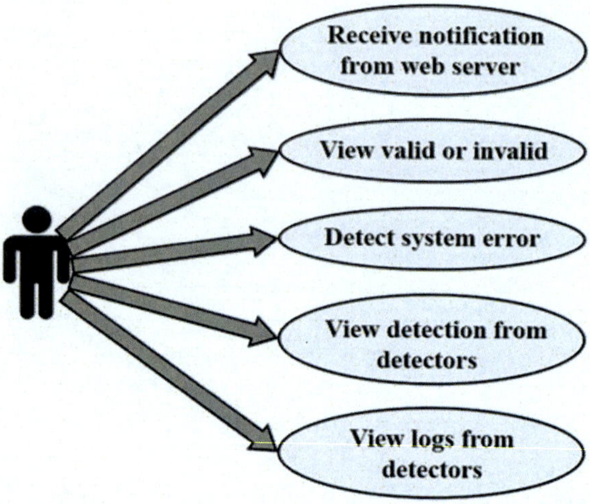

Fig. 2 Use case diagram of user mobile application

4 System Implementation

4.1 Fire Hazard Detection Using SVM

The support vector machine (SVM) algorithm is an expert to categorize the data points like temperature, smoke, and flame into a positive class as fire outbreak with threshold value 1 or negative class as no fire outbreak with threshold value 0. The input data as training data are formulated as given in Eq. 1;

$$(p_1, q_1) \ldots (p_n, q_n) \tag{1}$$

Here:
p is the set of components
q is the threshold value
Now here is $p_i = p_i^1, p_i^2, \ldots p_i^d$
Where: p_i^j is a real value
Then another is $q_i = 0$ and 1
Where: 0 refers to fire outbreak and 1 refers to no fire outbreak.
The Radial Basis Function kernel is employed to sketch the non-separable training data from input space to feature space in addition to finding an optimized hyperplane that accurately secludes the data. The RBF kernel is existent in Eqs. 2 and 3, respectively.

$$K(\vec{p}_l, \vec{p}_j) = \varnothing(\vec{p}_l)^T * \varnothing(\vec{p}_l) \tag{2}$$

$$K(\vec{p}_l, \vec{p}_j) = exp(-\gamma * ||p_i - p_j||^2) \quad (3)$$

Where: γ is $1/(2\sigma^2) > 0$
$\vec{p}_i$ is the support vector points
$\vec{p}_j$ is the feature vector points in the transformed space
and $K(\vec{p}_i, \vec{p}_j)$ is the kernel function.
The kernel function computes the point of the sketch data points in the transformed feature space. The optimal hyperplane that secludes between the two classes, like to fire outbreak and no fire outbreak, is found using Eq. 4;

$$w^T p + r = \sum_{i=1}^{l} \alpha_i q_i \varnothing(\vec{p}_l)^T \varnothing(\vec{p}_l) + r \quad (4)$$

The alignment frontiers are found using;

$$w\varnothing(p) + r = 1 \quad (5)$$

This points are defined as fire outbreak.

$$w\varnothing(q) + r = 0 \quad (6)$$

This points are defined as no fire outbreak.
The optimal weight vector (w) is given by:

$$\vec{w} = \sum_{i=1}^{l} \alpha_l q_i \varnothing(\vec{p}_l) \quad (7)$$

Here employed the dual training of SVM algorithm presented as a maximization problem over alpha has been given in Eq. 8;

$$max \sum_{i=1}^{l} \alpha_i - 1/2 \sum_{i=1}^{l} \sum_{j=1}^{l} \alpha_i \alpha_j q_i q_j \varnothing(\vec{p}_l)^T \varnothing(\vec{p}_l) \quad (8)$$

Subject to:
$0 < \alpha < C$ and $\sum_{i=1}^{l} \alpha_i \, q_i = 0$
Here: alpha is represent the weight vector; q: is formulated the label vector and C: represents the detach
The decision function d used in making prediction is given as:

$$d(\vec{p}) = sgn(\vec{w}^T \vec{p} + r) => sgn\left(\sum_{i=1}^{l} \alpha_i q_i \varnothing(\vec{p}_l)^T \varnothing(\vec{p}_l) + r\right) \quad (9)$$

Table 1 SVM Model training set and accuracy

Data	Training set and accuracy	
Sets	Training data	Testing data
Normal data	1000	100
Fire data	1000	100
Warning data	300	30
Accuracy	99%	99%

Where: $g(\overrightarrow{p})$: is the predicted label, sgn: is the sign of $(\overrightarrow{w}^T\overrightarrow{p} + r)$ (i.e., 1 or 0), α_i: is the value of weight vector.
The ordinary steps taken to execute the classification and prediction of fire outbreak are given in below:
Process 1: Task to input training data
Process 2: Isolate data into $\overrightarrow{p}_i$ as component set and $\overrightarrow{q}_i$ as labels
Process 3: By using RBF kernel sketch data from input to component space
Process 4: Search an optimal hyperplane using the formula (6)
Process 5: Finally, find the classification frontiers as support vectors.

4.2 SVM Model Training Set and Accuracy

The system decides whether it is a sensor fault or a spark ignition around the sensor. As it returns to normal data after a short time, this is not considered a fire situation and only a warning message is sent as confirmation. The possibility of a final fire is calculated from the fire situation and the normal situation, which is identical to the SVM algorithm. Finally, accuracy is up to 99% successfully (Table 1).

4.3 Fire Extinguisher Working Area Measurement

Calculating the quantity of class, an extinguisher is needed, which roughly equates to five 13 A extinguishers covering 1000 m^2. Floor area (m^2) = 1000 and calculation = 1000 * 0.06/13 = 4.613.
Result = A total of 5 * 13 A extinguishers is required.

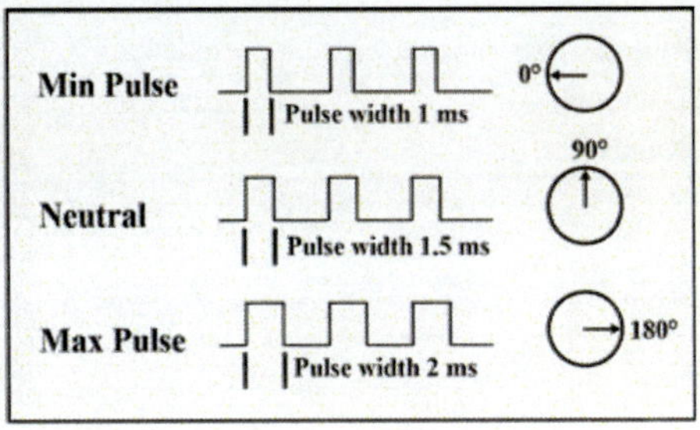

Fig. 3 Servo motor activity

4.4 Gas Cylinder Regulator Controlling by Servo Motor

Servo motors are controlled by sending a variable width electrical pulse or pulse width modulation (PWM) using controlwise. There is a minimum pulse, a maximum pulse, and a repetition rate. A motor can typically rotate 90° in either direction during 180° movements (Fig. 3).

5 Result and Performance Analysis

5.1 Evaluation Methods

In addition to comparison, the success of using data augmentation methods is to prosper the performance of the procedure, the role of some hyperparameters and how to growth improve rates against some other SVM models. We have been used by evaluation methods which are defined as follows:

$$\text{DR} = (Pp/Qp) * 100\% \tag{10}$$

$$\text{FAR} = (Np/Qn) * 100\% \tag{11}$$

$$\text{AR} = (Pp + Nn)/(Qn + Qp) * 100\% \tag{12}$$

Where: DR, FAR, and AR are detection rate, false alarm rate, and accuracy rate, respectively. Moreover, Qp is explained as positive data which is the total number of smoke data in our test data and Qn is explained as the number of negative examples. Pp and Np are the number of correctly classified positive examples and a number of falsely classified negative examples respectively.

Table 2 Experiments with data augmentation

Data sets	Data accuracy methods		
Number	AR%	DR%	FAR%
Set 1	94.85	93.12	0.63
Set 2	95.87	93.41	0.60
Set 3	95.21	93.91	0.58
Set 4	**98.15**	**97.98**	**0.23**

Table 3 Performance appraisal of subsystem modules

Subsystem	Subsystem performance	
Modules	Work in (%)	Remarks
Microcontroller unit	100	Works in all the time
Sensor unit	94	Sometimes lead to unreliable
SVM method	98.15	Provide high accuracy levels
Fire notification	100	Shows fire notifications in real time

5.2 *Experiments with Data Augmentation*

The testing results listed in the Table 2 show that when we train our network on the larger datasets, we can achieve better accuracy and detection rates as well as a lower error rate. For an example, when we train the network on the Set 2, AR increases from 94.85 to 95.87%, and DR increases from 93.12 to 93.41%, while there is the minor drop in FAR from 0.63 to 0.60%. Experimental results of SVM trained on the Set 3, which consists of data generated using the sensors are better than the network has trained on the original dataset, the Set 1.

5.3 *Performance Evaluation*

Based on the evaluation and performance tests, all modules of the subsystem of the advanced hardware and software systems were tested, and the results are recorded and listed in Table 3.

5.4 *Gas and Flame Sensor Activity*

Figure 4 represents the date in the x-axis, which takes the data continuously and the gas level in the y-axis which contains the threshold value. If the gas level is up, its threshold value contains 0 or 1. In this performance, it is 1.

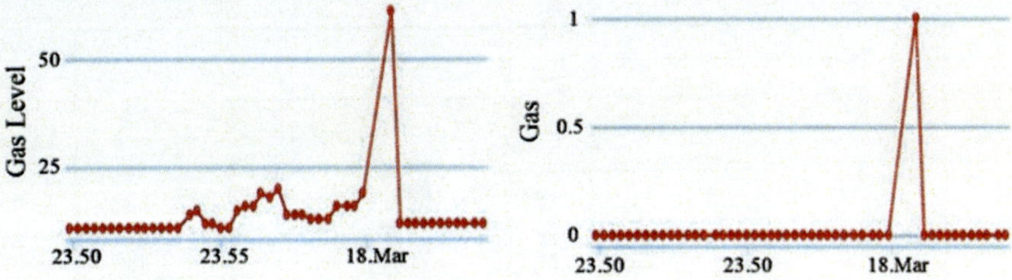

Fig. 4 Gas sensor data performance

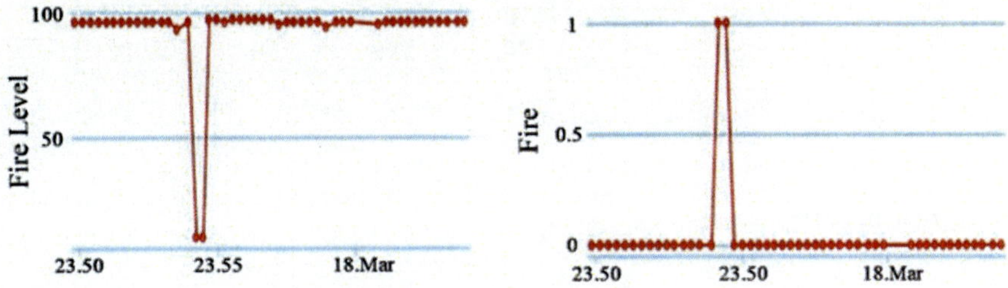

Fig. 5 Flame sensor data performance

The date in x-axis, which takes the data continuously and flame level in y-axis which contains threshold value is represented in Fig. 5. If the flame level is down, its threshold value contains 0 or 1. In this performance it is 1.

6 Discussion

Other models have been used for fire hazard detection, usually based on KNN, LDR, LR, NB, and CART, but the accuracy of this model prediction and training dataset is less than 98.15%. When classifying and predicting datasets based on fire outbreaks, the results show that support vector machine-based fire detection can solve problems related to fire outbreaks by regularly monitoring environmental changes that lead to fires such as temperature, smoke, and flames. The model can detect fire accidents by knowing that it may cause a fire. The system uses the SVM model in the event of a fire, with an accuracy rate of 98.15% and a minimum error rate of 1.85%. We built prototype sensor systems to work in various environments. Based on the dataset, we use supervised machine learning processes in different environments to observe undetected events. Achieve the expected pattern and predict the severity of the risk. The characteristics of the difference model are shown below, and SVM proposes the best solution (Fig. 6).

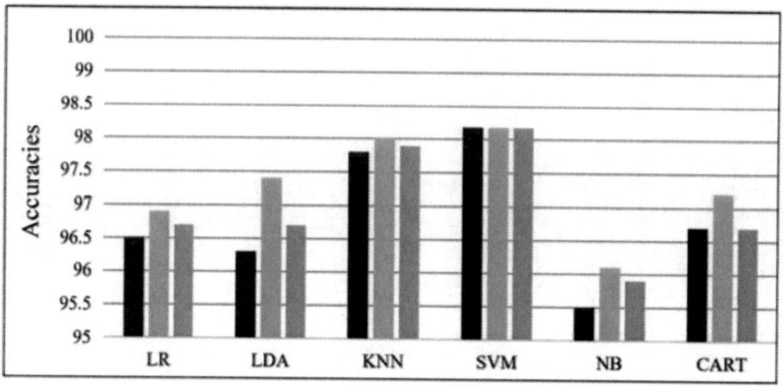

Fig. 6 Difference model performance analysis

7 Conclusion

This project is not only a successful representation of early fire detection but also provides better features and performance. We have implemented supervised machine learning processes on sensor data validation, because an invalid sensor data can cause panic and create unnecessary problems in an office or at home, etc. If this system is installed and implemented successfully, it can save many lives and resources.
In the later part, we will collect more additional data and integrate other machine learning algorithms to promote model accuracy and reduce false positives. We also have the trend toward real-time analytic with cloud services.

References

1. Divya TK (2019) Dharshini: IoT Enabled forest fire detection and early warning system. In: 2019 International conference on systems computation automation and networking, 2019. Accessed 12 July 2020
2. Prabha (2019) An Iot based efficient fire supervision monitoring and alerting system. In: 2019 third international conference on I-SMAC (IoT in Social, Mobile, Analytics and Cloud), pp 414–419. Accessed 07 July 2020
3. Angeline AS, Abishek (2019) Fire alarm system using IOT. Int J Innov Technol Explor Eng (IJITEE) 8(6S3):110—112. ISSN: 2278-3075. Accessed 02 July 2020
4. Uduak EU, Nyoho (2019) Support vector machine-based fire outbreak detection system. Int J Soft Comput Artif Intell Appl (IJSCAI) 8(2):2019. Accessed 06 June 2020
5. Rashid MR, Azad (2018) An automated fire suppression mechanism controlled using an Arduino, pp 49–54. Accessed 24 June 2020
6. Kumar RP, Smys (2018) A novel report on architecture, protocols and applications in Internet of Things (IoT). In: 2018 2nd international conference on inventive systems and control (ICISC), pp 1156–1161. IEEE, 2018. Accessed 28 June 2020
7. Ravi AV (2018) MQTT implementation of IoT based fire alarm network. In: 2018 international conference on communication computing and Internet of Things (IC3IoT), pp 143—146. Accessed 30 June 2020

8. Namozov AA, Cho YI (2018) An efficient deep learning algorithm for fire and smoke detection with limited data. In: 2018 Advances in electrical and computer engineering. ISSN: 1582–7445. Accessed 28 July 2020
9. Sathyakala VK, Aishwarya (2018) Computer vision based fire detection with a video alert system. In: 2018 international conference on communication and signal processing (ICCSP), pp 725–727. Accessed 15 July 2020
10. Ahmed TR, Kamrul MSA, Saad (2017) An IoT based fire alarming and authentication system for workhouse using raspberry Pi 3. In: 2017 international conference on electrical, computer and communication engineering (ECCE). Accessed 16 June 2020

Development of an Optimal Design and Subsequent Fabrication of an Electricity-Generating Ground Platform from Footstep

Sudipta Mondal, Md. Tazul Islam, Arnab Das, and Mayeen Uddin Khandaker

Abstract This study presents a solution to the modern power crisis with the non-renewable power source by optimally using electricity-generating ground platform which utilizes the energy from human footsteps. An optimal design is proposed using both rack, and pinion mechanism and piezoelectric transducer mechanism which is fabricated and analyzed both theoretically and practically. The results are promising; a single device can produce up to 33 V. These ground platforms can solve the problems regarding electricity generation; load shedding, environment pollution, lack of non-renewable power sources, etc. if they are mass-produced and deployed in busy places and roads.

Keywords Generator · Piezoelectric transducer · Footstep · Rack and pinion

1 Introduction

Energy is needed in every form of our life. One form of this energy electricity has become a basic need for us nowadays. Some people say that this is the era of electricity. In a common way, we have been producing this form of energy using fossil fuels which have a limited source, and these sources have been consumed to a great extent. Now the time has come to look for alternative means to harvest

S. Mondal · Md. Tazul Islam · A. Das
Department of Mechanical Engineering, Chittagong University of Engineering and Technology, Chittagong 4349, Bangladesh
e-mail: u1503077@student.cuet.ac.bd

Md. Tazul Islam
e-mail: tazul2003@cuet.ac.bd

A. Das
e-mail: u1703003@student.cuet.ac.bd

M. U. Khandaker (✉)
Centre for Applied Physics and Radiation Technologies, School of Engineering and Technology, Sunway University, Bandar Sunway, 47500 Selangor, Malaysia
e-mail: mayeenk@sunway.edu.my

M. S. Arefin et al. (eds.), *Proceedings of the International Conference on Big Data, IoT, and Machine Learning*, Lecture Notes on Data Engineering and Communications Technologies 95, https://doi.org/10.1007/978-981-16-6636-0_29

electricity from other sources for this huge population of this world. We may call them renewable or regenerative energy sources. We can also use the implementation of the techniques for optimum utilization of conventional sources of conservation of energy. A recent power generating technique has caught the eye of many researchers, that is, electricity-generating ground platform.

The operating principle of generators that use electromagnetic flux change to induce electricity was discovered in the years 1831–1832 by Michael Faraday known as Faraday's law. Using this theory, later many concepts are invented to use natural kinetic energy for electrical energy such as wind turbines, wave electricity generators, etc. And ground electricity generator is one of them. Researchers of Hull University started working at the very beginning to transfer the locomotive energy of a person to electric energy. Mohan et al. [1] showed three different methods of footstep power generation namely piezoelectric method, rack and pinion method, and fuel piston method and after a comparative study, rack and pinion system is found to be more efficient with a moderate cost of operation and maintenance. Later Lowattanamart et al. [2] used rack and pinion system with a rectifier circuit and generated about 5 V–500 mA regulated power supply. Joydev et al. [3] used magnetic coil to produce electricity and got about 80 V–40 mA from their first invention. Later inventions provide about 95 V–50 mA from one coil which can be used to light LED and run DC motor including charging of batteries. The revolution per minute (RPM) value of the dynamo or motor used for this type of device has a great impact on the performance. Mahmood [4] used a pneumatic fluid pressure based system where water is used for the fluid- flown through nozzle and rotating turbines for electricity. Pascual [5] fabricated a model from stainless steel, recycled tires, and aluminium; making the device nature friendly and green technology.

There are also a lot of researches on the piezoelectric method where piezoelectric transducers are used to convert pressure energy into electric energy. The invented and experimented footstep power generating devices based on piezoelectric method are shown in Table 1.

These devices rather have rack and pinion system or have piezoelectric method according to the literature review. But this study combines these two methods of power generation into one particular device with a novel design proposed that can optimally produce electricity from footsteps. The main problem behind using these footstep power generators is, they are too much costly to be used for same amount of power generation like coal fired electricity generators. But the proposed design with both rack and pinion system and piezoelectric transducer system is cheaper compared to other available devices of this kind. An electric circuit to store the generated electricity is also presented in the present study. Mass people can produce this proposed device locally at a low cost apply these devices in their homes or workplaces to generate electricity to be used to recover the loss of load shedding in 3rd world countries like Bangladesh, Pakistan, etc.

Table 1 List of devices based on Piezoelectric method

Company-Technology	Generated energy	Price/Egyptian pounds	References
Waynergy Floor	Waynergy Floor	4000	[6]
Sustainable Energy Floor (SEF)	Up to 30 W of continuous output. Typical power output for continuous stepping by a person lies between 1 and 10 W nominal output per module (average 7 W)	15,000	[7]
Pavegen tiles	5 Watts continuous power from footsteps	35,000	[8]
Sound Power	0.1 W per 2 steps	N/A	[9]
Drum Harvesters-Piezo buzzer Piezoelectric Ceramics	Around 2.463 mW	Estimated 500/tile	[10]

2 Materials and Method

The electricity-generating device from footsteps was fabricated under two key objectives: user-friendly and cost effective. The materials that are used for the fabrication of this device were selected as well as actuated with these goals in mind. A brief information of the components with some of their key features is presented in Table 2.

Table 2 Components list

Component name	Key features	Advantages
Dynamo	• Max voltage output: 12 V • Max current output: 1–2 A • Max torque: 3 kg • Max rpm: 300 rpm	• Sufficient torque for this study • Doesn't have excessive torque • Optimal Size
Piezoelectric transducer	• Voltage range: 10-33 V • Current range: 1e−7–10e−7A • Diameter: 35 mm	• Easy connection • Cheap price • Easily available
Helical spring	• Number of turns: 20 • Inner diameter: 3.6 cm • Outer diameter: 4.3 cm	• Can resist weight up to 200 kg without any deflection of the structure
Rectifier	• 1N5408	• Easily available

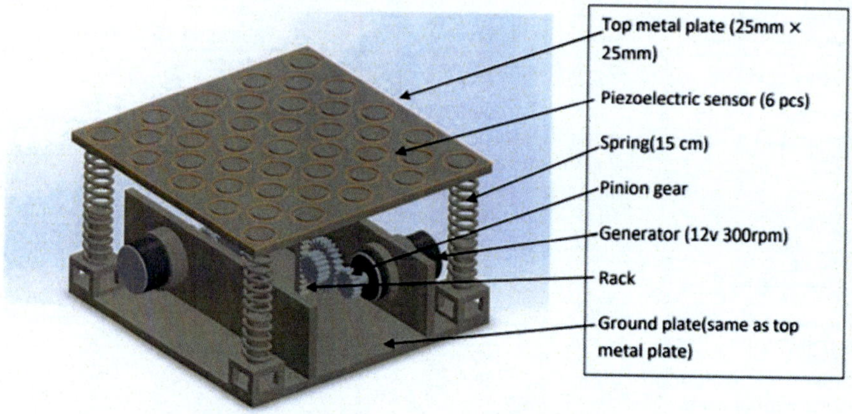

Fig. 1 3D isometric design of the fabricated electricity-generating device

2.1 Design and Modeling

The output performance of any system is determined by a variety of factors, including its structural reliability, resistance to environmental change, etc. These factors mainly depend on the design and architecture of that device. The Electricity-Generating Ground Platform device is designed in CAD software and an isometric view of the 3D design with specific dimension parameters is shown in Fig. 1.

The gear mechanism used in this device is reck and pinion gear mechanism. Here in this study, two types of pinion gears and 4 springs with different specifications are used which are shown in Table 3.

2.2 Theoretical Aspect

The electricity-generating device was designed based on some numerical assessment. The reck and two types of pinions are used to convert the vertical force into rotational force. The setup is shown in Fig. 2.

Here Module of Gear A = Module of Gear B = Module of Gear C = Module of Rack.

Teeth of gear A, TA = Teeth of gear C, T_C.

In this study, amount of Teeth of gear A and C is 16, and the number of teeth of gear B, T_B is 32.

As the teeth number is same and both are of same shaft, the speed and Torque of gear A and gear B is same but the speed of gear C is different.

The relation between the speed and torque of the gears is expressed as,

Table 3 Specifications of the components

Components	Specifications	Value
Gear (pinion) (small)	Number of teeth (N)	16
	Outside diameter (D_o)	25.4 mm
	Pitch circle diameter (D)	22.58 mm
	Module (m)	1.41
	Diametric pitch (P)	0.7
	Addendum (a)	1.41 mm
	Dedendum	1.785 mm
	Pressure angle (Φ)	20°
Gear (pinion) (Big)	Number of teeth (N)	32
	Outside diameter (D_o)	48 mm
	Pitch circle diameter (D)	45.17 mm
	Module (m)	1.41
	Diametric pitch (P)	0.7
	Addendum (a)	1.415 mm
	Dedendum	1.785 mm
	Pressure angle (Φ)	20°
Rack specifications	Module	1.41
	Pitch circle radius (r)	1.27 cm
	Length of rack	100 cm
	Length (l)	15 cm
Spring specifications	Total length	15 cm
	Number of turns	20
	Inner diameter	3.6 cm
	Outer diameter	4.3 cm

$$\frac{N}{N_c} = \frac{T_C}{T_B} \tag{1}$$

Here, N represents the speed of gear A and gear B, N_C is the speed of gear C. T_C and T_B are the Torque at gear C and gear B.

The torque generated can be found from the equation,

$$T = r \times F \tag{2}$$

Here T is the generated torque, r represents the radius of the gears and F is the force exerted.

The mechanical power generated can be calculated from,

$$P = T \times \omega \tag{3}$$

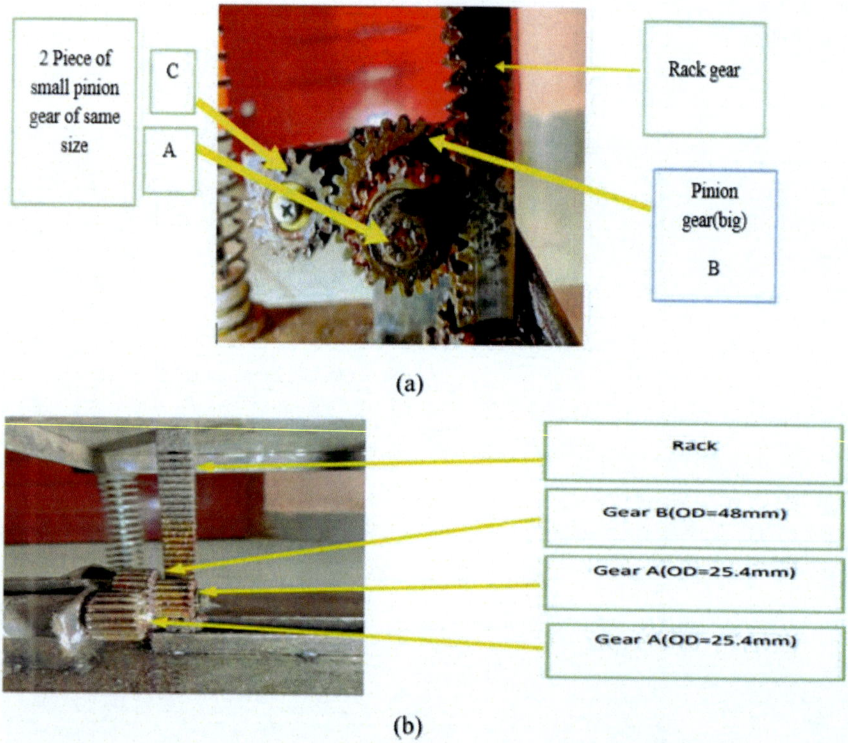

Fig. 2 Gear mechanism of the electricity-generating device, **a** front view, **b** side view

where, P is the mechanical power generated, T is torque generated and ω is angular velocity of the gear

$$\omega = \frac{2\pi N}{60} \tag{4}$$

Here N is the rotation per minute speed of the gears.

As there are 2 sets of gear of same size each set produce same amount of power.

Total power generated from the gear = 2× power generated at each set of gear.

2.3 Power Collection System:

In this study, two types of power sources are used-12 V capacitive motor or dynamo and piezoelectric transducers. A 9 V rechargeable Li-Po Battery is used in this study to store the generated electricity. An ammeter is used to measure the current (mAh) level of the battery. Full wave bridge rectifier is used to make the charging single

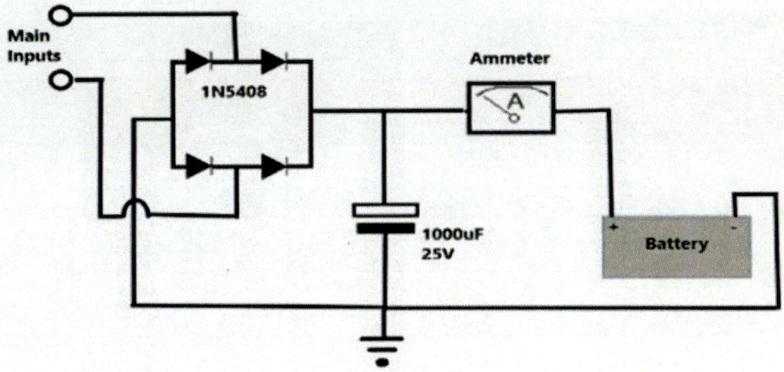

Fig. 3 Schematic diagram of the battery charging circuit

way and to prevent discharging of the battery. The schematic of the designed circuit is shown in Fig. 3.

Here dynamo and piezoelectric transducers should be connected to the main input connectors of this circuit shown in Fig. 3.

This is a very simple charging circuit that charges batteries very slowly and optimally. As neither the current nor the voltage generated from the proposed device of the present study is constant, as they mostly vary on the weight exerted on the device; It could harm the battery life. To prevent such events, this simple circuit is used so that the battery can charge slowly and the excessive currents or voltage can not harm the battery health or battery life.

3 Fabrication Process

The proposed device in this study was fabricated manually in the workshop. The structure had to undergo different machining processes including welding, as the structure is totally made of stainless steel and mild steel which are easily available. The gears are also made manually using lathe machining process. The piezoelectric transducers are placed on the top plate of this device. The placement of the dynamo and the piezoelectric sensors are shown in Fig. 4.

The power collecting circuits and the battery are placed underneath the device.

4 Working Principle of the Device

The whole working process is expressed in a flowchart in Fig. 5.

Fig. 4 **a** Front view of the electricity-generating device, **b** isometric view showing the array of the piezoelectric transducers

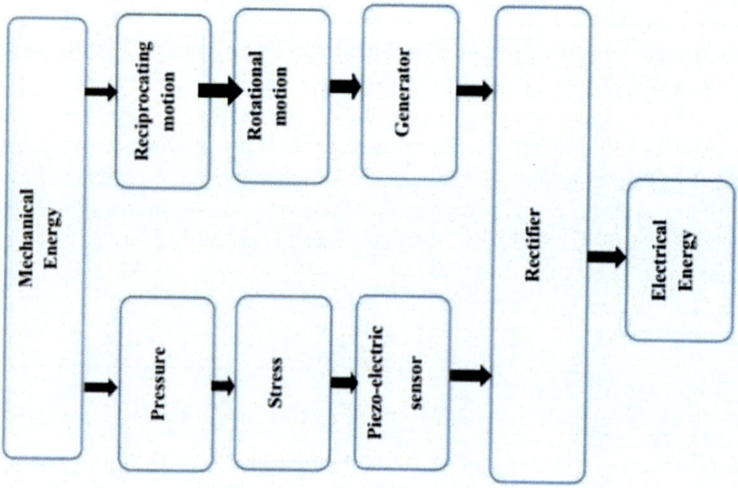

Fig. 5 Flowchart of the working process of the proposed device

The proposed device of this study changes mechanical energy into electric energy. This mechanical energy comes directly from the weight exerted by the human footsteps and also the vehicles on the road which is generally get wasted otherwise. The weight pressure moves the reck which causes the bigger gear rotate at a certain speed for both push and release process of the footstep. The bigger gears then transfer the rotational energy to the smaller gears increasing the rotation per minute or speed of the gear. This gear is directly connected to the shaft of the dynamo from which the electric energy is generated by the change of the magnetic fluxes in the electric coil. The piezoelectric transducers produce electric voltage when pressure is exerted on them as they are placed on the upper plate of the device. These electric energies are then collected and reserved in a battery through an electric circuit.

5 Results and Discussion

Different loads have been applied on the setup and output data have been noted. The setup obtains kinetic energy from human body mass in the form of gear rotation. This gear rotation of gear helps the dynamo to produce electrical voltage. The setup also converts pressure energy from human mass into electrical voltage signals with the help of piezoelectric sensors. About 100 trials were recorded and among them, 5 trials are shown in Table 4 to show the relation between the weight and generated power.

In This device, in order to increase the rotation speed of the rotor, gear reduction mechanism is used which will increase speed of the gear with smaller diameter about two times the rotation caused by the rack.

The visual representation of the voltage and electricity generation is presented in Fig. 6.

Here Fig. 6a shows the change of produced voltage with respect to mass that has been used as input. Here we can see the voltage from piezoelectric sensors is higher than the dynamo. But the Fig. 6b, c shows fully opposite results. The electricity produced from dynamo is much higher than the current flow from piezoelectric sensor. But this phenomenon is according to the nature of piezoelectric transducer. Here the current flow from dynamo ranges from 0.0863 to 0.289 A but the current flow from piezo is at micro ampere level. Current flow from piezo ranges from 0.2 to 0.8 μA depending on the input mass. The main reason behind these opposite characteristics is that dynamo has a large coil and magnet which can generate a good amount of current flow balancing with the generated voltage, whereas, the piezoelectric transducers working principle is based on pressure based electron transfer through chemicals which can create a large amount of voltage difference but can not produce enough electricity and as a result dynamo generated power is higher than the piezo generated power.

Downward gravitational force for different mass creates torque on the gear which results in angular rotation of the gear. From this angular rotation and torque the power at the gear has been calculated. Gear reduction has been used to get higher angular speed at the dynamo. But loss in gear reduction has been neglected in the calculation.

Table 4 Summary of output: mechanical power, voltage, and electricity with respect to weight

Trial No.	Mass (kg)	output at gear (W)	Voltage from 2 dynamo (V)	Current Flow from dynamo (A)	Voltage from Piezo-sensor (V)	Current flow at piezo-sensor (A)
1	57	20.8	2.5	0.0863	13	0.2×10^{-6}
2	68	27.69	3.9	0.134	18	0.3×10^{-6}
3	73	35.20	4.8	0.1654	22	0.5×10^{-6}
4	80	50.04	6.1	0.2105	25	0.7×10^{-6}
5	86	62.77	8.4	0.289	33	0.8×10^{-6}

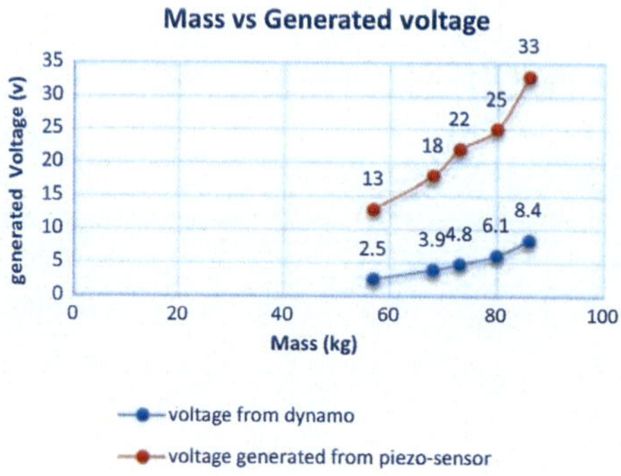

(a)

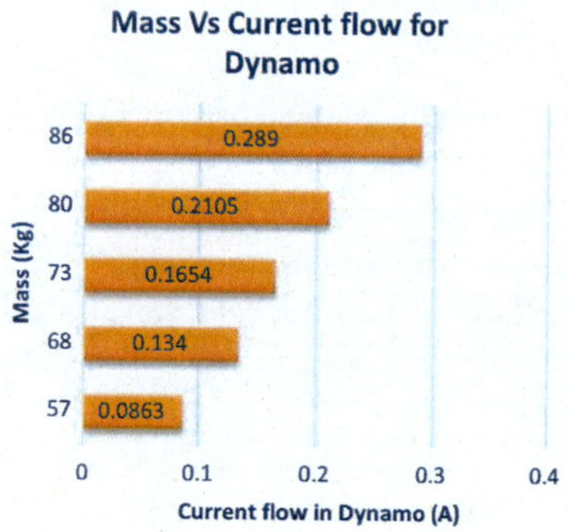

(b)

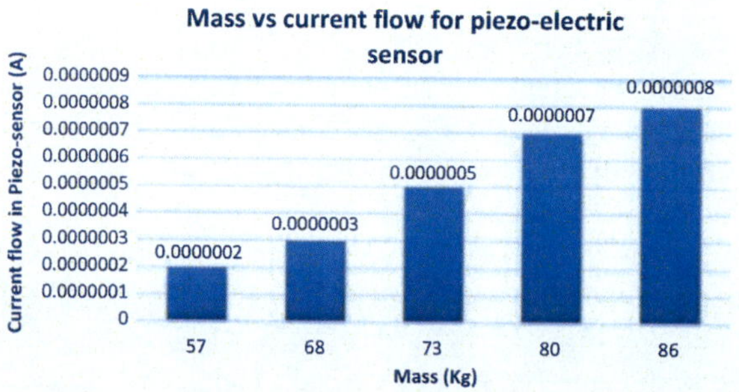

(c)

Fig. 6 **a** Mass versus generated voltage for dynamo and piezo. **b** Mass versus current flow at dynamo, **c** Mass versus current flow for piezoelectric transducer

This gives an approximate vision of the nature of the output. The downward force creates rpm ranging from 28 to 56 rpm to gear 'A' meshing with the rack. There might be slight difference in those values that have been measured from calculation with the help of the slow-motion videos from mobile. Here the gear A and Gear B is at the same shaft o the torque has been to be the same. Again the power loss has been neglected in calculation hence the power produced at gear A is considered to be the same as the power in Gear 'C' which is connected to the dynamo. These power generated at the gear increases with the increasing weight. Here power output is higher than the power from the dynamo as there is a slip between the coupler and the motor. Solving this problem can increase the output voltage. The total mechanical power generated from the gear mechanism is presented in Fig. 7.

Output from various input mass has been shown. Due to gear reduction, the rotational speed of dynamo and voltage generated from dynamo has increased. In previous studies absence of gear reduction caused low voltage production which is solved in this study. Electrical power generation based comparison is shown in Table 5 based on the previous studies.

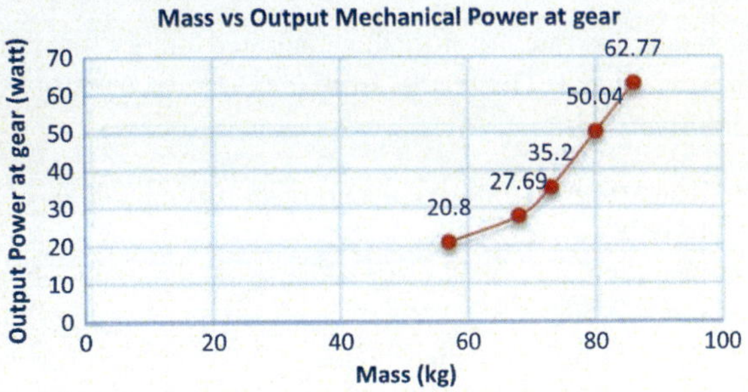

Fig. 7 Mass versus output power at gear

Table 5 Power generation comparison

Company—technology	Generated electric energy	References
Sustainable energy floor (SEF)	average 7 W	[7]
Pavegen tiles	5 W	[8]
Sound Power	0.05 W	[9]
Drum Harvesters - Piezo buzzer Piezoelectric Ceramics	Around 2.463 mW	[10]
Developed device	Avg. 1.049 W	

Also in case of previous studies, an integrated circuitry system to store the generated power was not shown or clearly described. But in this study authors designed and fabricated an electric circuitry system to store the power.

Usage of piezoelectric sensors also adds some voltage production though the increased power is less in comparison. Here from dynamo both downward motion and upward motion of the platform give voltage output with the help of a rectifier.

But the piezoelectric sensors only use the downward pressure of the footstep. The goal of this project is concerned with capturing energy that is lost in our daily life without using any fossil fuel. Though the generated power is not that much but massive implement can lead to considerable power output in regular days. Again overcoming limitations of this setup from further research can increase its efficiency. This may contribute to the portion of renewable energy that is a very much important phenomenon in upcoming days.

6 Conclusion

This study is based on the concept of harvesting electricity from energy that is not converted into other forms in regular life with no utilization. Nowadays looking for alternative sources for energy is a new challenge. This present study deals with this fact and so the electricity-generating ground platform has been built using dynamo and piezoelectric transducer. The concept of the device was designed first and then fabricated and tested with different weights. A circuitry system has been developed and demonstrated to store the generated energy form the fabricated device which is a novel concept. The results from the tests were analyzed and the concluding remarks are given as;

- The voltage from dynamo ranges from 2.5 to 8.4 V with current flow of 0.0863 to 0.289 A.
- The range is from 13 to 33 V in case of piezo with current flow ranging 0.2 to 0.8 μA.
- The efficiency of the device is 2.67% based on the average mechanical power generated and average electric power generated.
- Average power generation is 1.049 W per step.

Optimization of the gear reduction mechanism is recommended as future recommendation to increase the output power. Also number of piezoelectric transducers can also be increased along with the number of dynamo for better performance. Systems to avoid slip of gear and dynamo shaft can increase the efficiency of the device.

Acknowledgments Authors pay gratitude to Sunway University for providing the registration fee for the conference proceedings.

References

1. Mohan S, Nitesh Ganar SK (2018) Power generation using vehicle suspension. Int J Sci Res Dev
2. Lowattanamart W et al (2020) Feasibility on development of kinetic-energy harvesting floors. IOP Conf Ser: Earth Environ Sci. https://doi.org/10.1088/1755-1315/463/1/012107
3. Ghosh J et al (2013) Electrical power generation using foot step for urban area energy applications. In: Proceedings of the 2013 international conference on advances in computing, communications and informatics, ICACCI 2013. https://doi.org/10.1109/ICACCI.2013.6637377
4. Mahmud I (2018) Electrical power generation using footsteps. Eur Sci J, ESJ. https://doi.org/10.19044/esj.2018.v14n21p318
5. Pascual EL (2020) Electricity generation using spring-powered floor pad. Eng Technol J (2020). https://doi.org/10.47191/etj/v5i12.01
6. da Rodrigues JACC (2011) Waynergy: the way for energy harvesting: business model design
7. Energy floors—human-powered floor tiles—Rhine capital partners, http://www.rhinecapital.com/energy-floors-human-powered-floor-tiles/. Last accessed 2021/05/15
8. Pavegen, https://pavegen.com/. Last accessed 2021/05/15
9. エネルギーハーベスティング-振動力発電 | 株式会社音力発電, http://www.soundpower.co.jp/work/vibration.html#ttl_N7. Last accessed 2021/05/15
10. Mishra R et al (2015) Vibration energy harvesting using drum harvesters. Int J Appl Eng Res

An Automated Planning Approach for Scheduling Air Conditioning Operation Using PDDL+

Amina Shaikh Miah, Fazlul Hasan Siddiqui, and Md. Waliur Rahman Miah

Abstract Scheduling and planning of a hybrid (mixed discrete—continuous) system for a real-world problem are very challenging. Fortunately, PDDL+ shows a way to model such systems. An air conditioning system is an example of a hybrid system which provides thermal comfort to the occupants of any residence at an expense of energy. Scheduling the air conditioner with a wise plan can reduce the energy consumption without losing thermal comfort. In this work, we create schedules for air conditioning operations and predict the amount of energy required to execute the schedules for maintaining the thermally comfortable zone temperature. For this purpose, we design two hybrid domain models, namely linear durative model and process-event model. For optimized operations of an air conditioner, we implement those models with automated planning using PDDL+ and show the comparisons between two domains. For simulation, a state-of-the-art hybrid planner SMTPlan+ is used.

Keywords Automated planning · Hybrid planning · Energy optimization · PDDL+

1 Introduction

The air conditioning (AC) system is one of the essential appliance that provides thermally comfortable environment by cooling or heating the zone air. People tend to either turn on or turn off the AC system periodically or leave it turned on for the whole duration. They usually select a setpoint temperature for the AC thermostat with little or no knowledge of thermal comfort and energy optimization. The running period of the AC depends on human interventions as well. These arise several issues. First,

A. S. Miah (✉) · F. H. Siddiqui · Md. W. R. Miah
Dhaka University of Engineering and Technology, Gazipur, Bangladesh
e-mail: siddiqui@duet.ac.bd

Md. W. R. Miah
e-mail: walimiah@duet.ac.bd

M. S. Arefin et al. (eds.), *Proceedings of the International Conference on Big Data, IoT, and Machine Learning*, Lecture Notes on Data Engineering and Communications Technologies 95, https://doi.org/10.1007/978-981-16-6636-0_30

often the AC system runs at wrong temperature setpoint. Second, the AC system does not turn off automatically until the user turns it off or alters the thermostat temperature setpoint. Third, the AC system stays on longer than necessary, which in turn wastes energy. Energy optimization is an interesting contemporary research field. However, researchers in this field mostly focused on optimizing the energy consumption for larger systems, such as heating, ventilation and air condition (HVAC) system [20], while small AC systems received less attention. Different approaches are applied in different works for optimizing a HVAC system [5, 7, 10–17, 21, 24]. Those works are carried out with different approaches such as neural networks, evolutionary algorithms, genetic algorithm, particle swarm optimization (PSO) algorithm, linear programming, and few other techniques.

The automated hybrid planning is an attractive discipline to AI researchers for optimizing any feature such as time, energy, and actions/operations [2, 8]. In this field the optimization is achieved by clever scheduling. Recently, the automated temporal planning has showed promising results in optimizing energy consumption in fields like urban traffic management [23], personalized medication [1] and petroleum refinery operation [6].

Though promising, to the best of our knowledge, the automated hybrid planning field still remains unexplored for optimizing the energy consumption of an AC system. This research gap motivates us in pursuing the current work. Under this circumstance, we pose ourselves few research questions. First, can the human interactions for operating an AC system be reduced by automatic planning? Second, can the thermal comfort of a zone be maintained automatically with scheduling, while optimizing energy consumption for the AC system? Third, can the automated hybrid planning approach be applied for scheduling the operations of the AC system? Throughout this research, we will gradually find out the answers of these questions.

In this work, we present a novel approach for energy optimization by automatically scheduling the operations of air conditioning system. We model the system by automated hybrid planning using planning domain definition language, PDDL+ [6]. Our main contribution in this paper is a novel automated planning model that can be applied to an air conditioning system for scheduling its operation. The idea is operating the AC system by following a plan with time and duration generated from our proposed AC scheduling Domain Model. In this regard, our model automatically generates a plan of the AC system operations along with a predicted amount of required energy. When the plan is executed, the AC system consumes optimal amount of energy and the zone temperature becomes thermally comfortable. The generated plan confirms that the AC system do not run for longer time than necessary. In this way, human intervention is reduced to periodically turning on or off the AC system for maintaining the thermal comfort over a dedicated time period.

The remaining part of this paper is organized as follows: Sect. 2 provides background literature related to automated hybrid planning and thermodynamics of an air conditioning system. A formal problem is defined for an AC system in Sect. 3. Related discrete and continuous variables are also explained in that section. Section

4 designs the AC scheduling domain model with automated hybrid planning. Our experiments and results are presented in Sect. 5. Finally, we conclude and provide some future directions in Sect. 6.

2 Background

The automated hybrid planning allows to model a mixed discrete-continuous hybrid systems in the real world. It considers the changing of both discrete and continuous variables of any hybrid system with respect to time [6, 8]. Petroleum refinery production problem and Mars planetary lander activities are two examples of hybrid planning [6]. In petroleum refinery production problem, raw materials (for example crude oil and natural gas condensate) go through a number of processing units for refinement. It involves in changing the control settings which is discrete. The volumes and rates of flow, material properties and the time-dependent properties along with filling and emptying different tanks are continuous in nature. In the case of mars lander, it is designed to operate with limited energy and time. Both the problems have discrete-continuous effects which are the results of discrete-continuous causes presented by real world. Generator problem, where the fuel consumption needs to be optimized, is another benchmark hybrid planning problem [2]. The generator runs and refuels continuously when necessary, until all the available tanks are not empty. Two most recent examples of hybrid planning are (1) an activity scheduling and planning model of personalized medication doses [1], and (2) urban traffic models to reduce the traffic congestion at the junctions [23].

The thermodynamics of an air conditioning (AC) system own mixed discrete-continuous properties. The major explicit variable parameters of an AC system are temperatures, consumable energy, and the operational states of the AC system. These parameters have both discrete and continuous effects which are caused by real world changes. For example, the air conditioner changes its status from OFF to ON in order to turn on the cooling system, which is discrete. Again, when the cooling system is ON, the room temperature drops gradually. Hence, the AC system spends energy in a continuous manner. These discrete and continuous properties make the AC system an excellent candidate for hybrid planning domain.

An automated hybrid planning domain generates hybrid plans. A hybrid plan is a sequence of timed actions that change the environment from initial state to goal state. The actions are modeled by considering the mixed discrete-continuous parameters. An automated planning domain consists of predicates (facts), numeric predicates (numeric function) and actions (durative or instantaneous). The body of a durative action is divided into three parts—duration, precondition and effect. An action becomes active when the preconditions become true. The effect is the result of the action. The preconditions of temporal action are three types which must be true at different stages of the action. The stages are (1) at the beginning of the action, (2) at the end of the action and (3) for whole period of time. The effect of temporal action takes place at two stages, (1) at the start of the action and (2)at the end of the action. In the case of hybrid planning, the domain consists of two more constructs, *process* and *event* [6, 8]. A process models the flow of continuous parameters, where an event models the change of discrete parameters. An *action* or an *event* triggers a *process* to start by making the precondition(s) true. Then, the *process* keeps changing the numeric predicates until the precondition(s) holds truth. Finally, the *process* and its effects stop when intended goals are achieved.

The planning domain definition language (PDDL+) is dedicated for implementing hybrid domain. It comprises of continuous *process*, exogenous *event*, PDDL2.1 features, and basic PDDL features. A hybrid planner takes the domain model and problem instance created using PDDL+ as input and produces the plan. In this work, we used a modern hybrid planner namely SMTPlan+ [2]. The SMTPlan+ posses all the features of PDDL2.1 along with PDDL+, which is required to model the durative linear and non-linear changes of numeric variables of the system. Also, it shows extraordinary good results compared to another hybrid planner such as UPMurphi [3] for the benchmark problems [2].

3 AC Problem Definition

Assume, the air conditioning system is turned on when the zone temperature (T_{zone}) is higher (during summer) than the thermostat setpoint (T_{set}). So, the cooling process begins and the zone temperature starts to drop. In order to cooling, the AC system requires energy (electricity) as fuel. While the cooling continues, energy consumption continues, as well. So, there are two variable parameters which changes continuously over time. These are the zone temperature (T_{zone}) and the total energy ($\text{energy}_{\text{total}}$).

The zone temperature (T_{zone}) changes in two ways. First, while the AC is turned off, the zone temperature increases by a changing rate (t_{rate}). Second, when the cooling system is turned on the zone temperature decreases by a cooling rate ($t_{\text{cooling\_rate}}$). The cooling rate is modeled as negative of the changing rate (t_{rate}) because it decreases the value. The total energy ($\text{energy}_{\text{total}}$) is changed by energy consumption rate (e_{rate}). These thermodynamics of air conditioning system are defined in Eqs. (1), (2) and (3) as below:

$$\text{Temperature Change Rate, } t_{\text{rate}} = \frac{\mathrm{d}T_{\text{zone}}}{\mathrm{d}t} \tag{1}$$

$$\text{Temperature Cooling Rate, } t_{\text{cooling\_rate}} = -\frac{\mathrm{d}T_{\text{zone}}}{\mathrm{d}t} \tag{2}$$

$$\text{Energy Consumption Rate, } e_{\text{rate}} = \frac{\mathrm{d}\,\text{Energy}_{\text{total}}}{\mathrm{d}t} \tag{3}$$

The invariant parameter of air condition system is the zone thermostat setpoint (T_{set}) temperature. This setpoint temperature has to be thermally comfortable.

Accepted metric of thermal comfort in existing literature are percentage of mean vote (PMV) and predicted percentage of dissatisfaction (PPD) [4]. The PMV-PPD index indicates the temperatures where majority of the people in a zone can be thermally satisfied. This index is calculated based on six parameters: air temperature, mean radiant temperature (MRT), humidity, air speed, clothing, and metabolic rate [4, 19]. Equation (4) presents the ASHRAE Standard 55-2004 [18] approved PMV-PPD index.

$$-0.5 < \text{PMV} < 0.5 \quad \text{and} \quad \text{PPD} < 10\% \tag{4}$$

We have used a python package $pythermalcomfort$ [22] to calculate the PMV and PPD, which indicates the thermally comfortable zone thermostat setpoint (T_{set}) temperature for our system. In this way, we confirm the zone thermal comfort for occupants.

Using above discrete and continuous parameters we define the following constraints for modeling the AC system.

- The AC must stay OFF, as long as the T_{zone} remains under thermostat setpoint (T_{set}). In this state the zone temperature T_{zone} increases by the rate t_{rate} .
- The AC must turn ON to reduce the T_{zone} temperature before T_{zone} exceeds the setpoint T_{set}. The zone temperature decreases by cooling rate $T_{\text{cooling\_rate}}$.
- While T_{zone} drops, a safety boundary temperature $T_{\text{lower\_limit}}$ need to be employed to prevent T_{zone} from dropping to thermally uncomfortable temperature range.
- The AC can be turned ON and OFF as many time as required to maintain the thermal comfort over the period of time.
- The AC can be turned ON and OFF at any time point. And remain ON or OFF as long as necessary.

In the next section, we discuss modeling of above constraints using PDDL+ and PDDL2.1.

4 AC Scheduling Domain Model

In this research, we develop two distinct domain models for the air conditioning system. One of them is a model for linear domain and the other one is a model with pure hybrid event-process feature. The linear domain is modeled using temporal planning features only. The event-process domain is modeled using hybrid and temporal planning features. There are more than one way to design and develop a problem in automated planning field. The outcome of differently designed domains varies in several ways, such as number of actions in a plan, time duration of the actions, time span of the plan, CPU runtime. The outcomes of any particular problem can be more desirable with a domain that is designed and developed in a specific way. Following subsections elaborate our models.

```
(:durative-action air_conditioning
    :parameters (?z - zone ?c - chiller)
    :duration (= ?duration 120)
    :condition (and
        (at start (not (system-on ?z)))
        (over all (for ?c ?z))
        (over all (<= (zone-temperature ?z) (set-temperature ?z)))
    )
    :effect (and
        (at start (system-on ?z))
        (at end (air_conditioned ?z ?c))
        (increase (zone-temperature ?z)
                (* #t (temperature_change_rate ?z)))
        (at end (not (system-on ?z)))
    )
)
```

Fig. 1 PDDL model for (Air_Conditioning) action in linear domain

4.1 Air Condition Domain Model: Linear

The air condition domain is designed with two components—an AC system attached to the zone and the zone itself. The linear domain comprises of two durative actions: `Air_conditioning` (Fig. 1) and `Cooling` (Fig. 2). These two actions are responsible for maintaining zone temperature under thermostat set temperature over the period of time.

The action `Air_conditioning` is the main action that handles the zone temperature. The `(zone-temperature ?z)` and `(set- temperature ?z)` are the numeric predicates that capture the values of zone temperature and thermostat set temperature, respectively.

The predicate `(for ?c ?z)` confirms that the AC system `?c` is dedicated for the zone `?z` and it will remain that way over the period of time. The predicate `(system_on ?z)` makes sure that the AC system does not turned ON accidentally. It permits the AC to turn ON or OFF when required. The changes in zone temperature is handled as the effect of the action. The statement `(increase (zone-temperature ?z) (* #t (temperature_change_rate ?z))` indicates that at each `#t` timestamp zone temperature increases at `(temperature_change_rate ?z)`.

The `Cooling` action mimics the operation of cooling system of AC. The duration of this action is not fixed, which means it is left for planner to decide. The condition is that `cooling` must remain active until `(zone-temperature ?z)` reaches the safety temperature `(temperature_limit ?z)`. The predicate `(chiller_on`

```
(:durative-action cooling
    :parameters (?z - zone ?c - chiller)
    :duration (>= ?duration 0)
    :condition (and
                    (at start (not (chiller_on ?c)))
                    (over all (system-on ?z))
                    (over all (>= (zone-temperature ?z)
                                  (temperature_limit ?z)))
                  )
    :effect (and (at start (chiller_on ?c))
                 (decrease (zone-temperature ?z)
                                  (* #t (cooling_rate ?z)))
                 (increase (total-energy) (* #t (energy_rate ?z)))
                 (at end (not (chiller_on ?c)))
                 )
)
```

Fig. 2 PDDL model for (Cooling) action in linear domain

?z) being true indicates AC system is ON. As effects, the zone temperature drops at the rate of (cooling_rate ?z). Consequently, total energy consumption (total_energy) increases at rate (energy_rate ?z).

4.2 *Air Condition Domain Model: Process-Event*

The second domain is a process-event model. In this model, we replace the durative action (cooling) of previous (linear) model with an action-process-event trio combination. The main durative action (air_conditioning) remains as it is. In this domain, action (cooling) in Fig. 3 activates the process (cooling_begins), and event (stop_cooling) deactivates it.

As the time passes, the zone temperature increases as well. The planner decides appropriate time to initiate (cooling) action which in turn activates the (cooling_begins) process, so that (zone-temperature ?z) never cross the thermally comfortable set temperature (set-temperature ?z). The process (cooling_begins) activates as soon as the predicate (cooling ?chil ?z) becomes true. The process (cooling_begins) is active meaning that the AC system is turned ON in real world. While process (cooling_begins) is in progress, the (zone-temperature ?z) starts cooling. To stop that process (turn OFF the AC system in real world), the predicate (cooling ?chil ?z) needs to become false. This is done by the event (stop_cooling). An events trig-

```
(:action cooling
    :parameters (?z - zone ?c - chiller)
    :precondition (and (not (cooling ?c ?z))
                       (system-on ?z))
    :effect (and (cooling ?c ?z)))


(:process cooling_begins
    :parameters (?z - zone ?c - chiller)
    :precondition (and (cooling ?c ?z))
    :effect (and
        (decrease (zone-temperature ?z)
                  (* #t (cooling_rate ?z)))
        (increase (total-energy)(* #t (energy_rate ?z)))))


(:event stop_cooling
    :parameters (?z - zone ?c - chiller)
    :precondition (and (cooling ?c ?z)
                       (<= (zone-temperature ?z)
                           (temperature_limit ?z)))
    :effect (and (not (cooling ?c ?z))))
```

Fig. 3 PDDL model for action, event and process used to model the process–event domain

gers when certain condition is met. The event `(stop_cooling)` triggers as soon as `(zone-temperature ?z)` hits the comfort limit `(temperature_limit ?z)`. As the effect `(:event stop_cooling)` set the predicate `(cooling ?chil ?z)` false, which stops the process `(cooling_begins)`. Figure 3 shows the process and event for the cooling operation. In the resultant plan, only action(s) and durative action(s) can be visible, hence it does not show any trace of process and event. However, by validating the plan with validation tool, one can see the background activities of `(:process)` and `(:event)`.

4.3 *Initial State and Goal*

A PDDL problem includes initial states, goal condition and objective functions. Initial states indicate the beginning status of predicates and numeric predicates used in the domain. For the air condition problem, the predicates `(for ?c ?z)`, `(not (chiller_on ?c))` and `(not (system-on ?z))` are initially true. These predicates make sure that the dedicated AC system `?c` of zone `?z` is not operating at this moment and is ready for future operation.

At the end of the period, durative action `air_conditioning` makes predicate `(air_conditioned ?z ?c)` true. This indicates that the zone `?z` is conditioned by AC system `?c` over the period of time, hence the goal is achieved.

4.4 Plan Metrics

The (:metric) token instructs the planner to minimize or maximize certain numeric measure to optimize the resultant plan. In this problem, we have used only one (:metric), which is (total-energy) which indicates the total amount of energy used by the AC system over the duration. Our motive is to minimize the total energy consumption.

5 Experimental Result

In this work, we develop two distinct hybrid domains for an air conditioning system. Each domain is simulated against six problems with different time periods. To generate plans of the problem, we choose SMTPlan+ hybrid planner. The output of a planner is a plan, which is then validated by the validation tool VAL [9].

Input values of parameters are presented in Table 1. We keep the parameter values invariant for all five problems. For this experiment, the set_ temperature is chosen 23°C according to PMV-PPD index (equation (4)) described in Sect. 3. The safest lower limit temperature for cooling is set 18 °C, chosen by PMV-PPD index as well. We assume that room temperature changes at 20% rate while the AC system is OFF. The AC system runs with full potential (100%), while cooling. The energy is consumed by 0.5 KW/min. Each domain is tested for six different time periods of 30, 60, 90, 120, 150 and 180 min.

In this section, we describe elaborately the results for the time period 150 min. Table 2 displays the generated plans for linear domain (right) and Process-Event domain (left). The validation tool VAL confirms that the domains are modeled correctly and can generate correct plans for the AC condition system. also the plans are successful in keeping the zone temperature within the 18 and 23 ° C, evidence is shown in Fig. 4.

In addition to that the plans generated by the two domains are almost identical except around timestamp 30–40 min. We can see that the plan generated from process-event domain suggests cooling earlier (at 32.6 min), while the plan of linear domain

Table 1 Input parameters to the problem

Parameter	Value
zone_temperature	20 °C
set_temperature	23 °C
temperature_limit	18 °C
cooling_rate	1.0 (100%)
temperature_change_rate	0.2
energy_rate	0.5 KW

Table 2 Generated plan: process-event domain (right) and linear domain (left)

	Process-event domain		Linear domain	
Timestamp	(Action)	Timestamp	(Action)	[Duration]
0.0:	(air_conditioning z1 ac)	0.0:	(air_conditioning z1 ac)	[150.0]
14.5:	(cooling z1 ac)	14.5:	(cooling z1 ac)	[6.12]
32.6:	(cooling z1 ac)	45.0:	(cooling z1 ac)	[3.0]
59.6:	(cooling z1 ac)	60.0:	(cooling z1 ac)	[6.0]
89.6:	(cooling z1 ac)	90.0:	(cooling z1 ac)	[6.0]
119.6:	(cooling z1 ac)	120.0:	(cooling z1 ac)	[6.0]

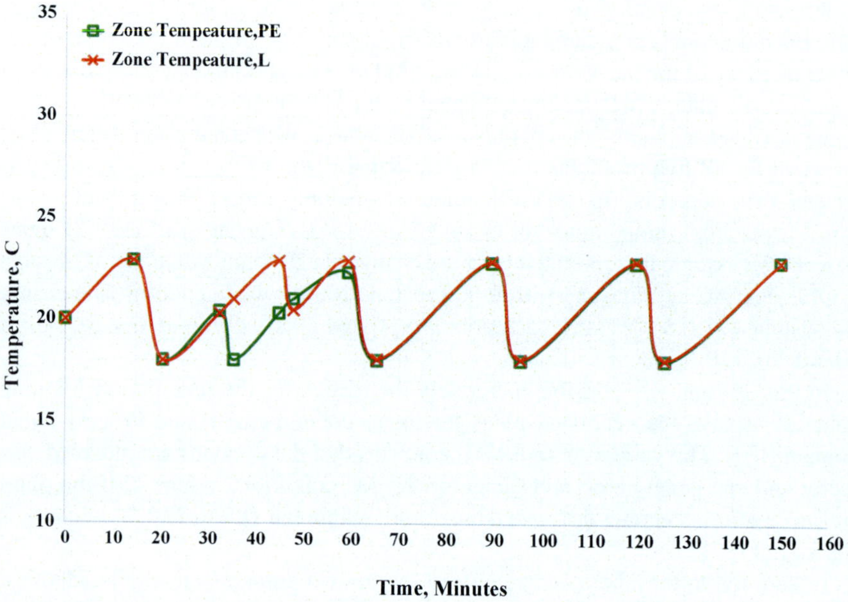

Fig. 4 Scheduled zone temperatures found by SMTPlan+ planner by Linear (L) and process–event domain (PE)

suggests cooling later (at 45.0 min). The later one is more desirable because it utilizes the cool environment inside the zone fully.

Figure 5 shows the energy consuming periods in the simulation plans of linear domain and process-event domain. Consequently, this indicates that at what time periods the AC system is become operative for each plan.

The generated plans from both domains (linear and process-event) are successfully validated with the validation tool VAL [9]. In this paper, we describe validation output report for process-event domain only. Plan validation confirms that the plans are correct. It also helps to understand the hidden activities behind the plan which

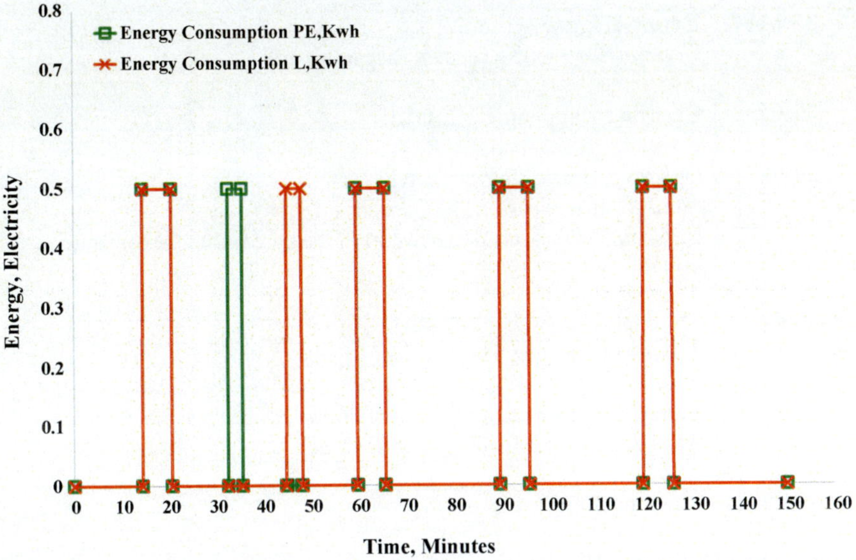

Fig. 5 Energy consumption: Linear (L) and process–event domain (PE)

is required for executing the plan successfully. For example, although the actions of two plans are similar, there is structural difference between two plans which is theoretical.

The linear domain produces plan, where each action shows the duration explicitly. The process-event domain only mentions the starting time of the operation and the operation runs by `:process` and stops by `:event` internally. However, the processing activities can be observed in plan validation report generated by VAL.

Figure 6 shows a fragment of validation report of plan generated by process-event domain. It shows that at 59.62 min zone temperature reaches to 22.8003. As the zone temperature can not go beyond the thermostat set temperature 23, the planner immediately adds predicate `(cooling ac z1)` in the planner stack, which activates the process `(cooling_begins z1 ac)`. The process keeps lowering the zone temperature until it touches the safety temperature 18. At 65.62 min, 18 is achieved, and hence, the event `(stop_cooling z1 ac)` is triggered. This event deletes the predicate `(cooling ac z1)` from planner stack, which inactivates the cooling process at once. This activation-inactivation continues whenever requires cooling during the over all time period.

Another purpose of plan validation is to predict an optimized amount of energy which is required to operate the AC system for a dedicated time period. We have measured the optimized value by minimizing `:metric (total_energy)`. The validation report confirms that if the AC system consumes electricity 0.5 kw/min, to keep zone temperature between 18 to 23 for 150 min; the system requires a total of 13.56 kw. Both linear and process-event domains exhibit the same value for `(total_energy)`.

35.6187: **Event triggered!**
Unactivated process (cooling_begins z1 ac)

59.62: Checking Happening... ...OK!

59.62: Checking Happening... ...OK!
(**zone-temperature z1**)(t) $= 0.2t + 18$
Updating (**zone-temperature z1**) (18) by 22.8003 for continuous update.

59.62: Checking Happening... ...OK!
Adding (cooling ac z1)

59.62: **Event triggered!**
Activated process (cooling_begins z1 ac)

65.6203: Checking Happening... ...OK!

65.6203: Checking Happening... ...OK!
(**total-energy**)(t) $= 0.5t + 4.56187$
(**zone-temperature z1**)(t) $= -0.8t + 22.8002$
Updating (**total-energy**) (4.56187) by 7.56203 for continuous update.
Updating (**zone-temperature z1**) (22.8003) by 18 for continuous update.

65.6203: **Event triggered!**
Triggered event (stop_cooling z1 ac)
Deleting (cooling ac z1)

65.6203: **Event triggered!**
Unactivated process (cooling_begins z1 ac)

Fig. 6 Plan validation output of a plan generated with process–event domain

As mentioned before, we have simulated two domains with five different time periods. Table 3 shows the CPU runtime (in seconds) to generate plans. In three out of five cases, Process-Event domain takes less time than linear domain. We have run our experiment on operating system Ubuntu 20.04 LTS in a Intel(R) Core(TM) i5 CPU 3.00 GHz with 4 GB of RAM.

In our experiment, we run the planner until finding the solution except for the problem instance with plan duration 180 min. For the problem instance of 180 min, the planner is unable to find any plan. Although we have allowed planner to run for 10 CPU minutes, after that we kill the process. There can be several reasons for not being able to find a plan. One of them is the problem becomes complex as the time duration gets longer. The planner needs to handle more ground predicates

Table 3 CPU runtime (in seconds) to generate the plans

Plan duration (min)	Linear domain	Process-event domain
30	0.061	0.031
60	0.089	0.129
90	0.645	0.792
120	6.792	4.829
150	99.886	43.887
180	–	–

and numeric predicates. Hence, the problem becomes unsolvable for the planner. Moreover, for applying the plan to an AC system, solution must be available instantly. For problem instance with plan duration 150 min, it took 99.9 and 43.9 s for linear domain and process-event domain, respectively. So, we can say that our domains are capable of generating solutions for 150 min. However, parameters vary in real world, for example, temperature changes in every minute. So, generating plans for long time such as two and half hours is quite ambitious and risky. Furthermore, problems with plan duration below 120 min (included) took acceptable amount of time to find the solutions for both domains.

6 Conclusion and Future Direction

In this paper, we have created two domains for air conditioning system using automated hybrid planning. The domains are modeled with PDDL+ and PDDL 2.1. The state-of-the-art planner SMTPlan+ is used to generate plans. Our domains produce schedules (plans) for operations of air conditioning system over a period of time. The resultant plans are able to maintain thermally comfortable temperature successfully by following the actions in it. The plans are also presented with optimized energy consumption. This automation reduces human interactions in operating an AC system.

We have devised six problems for testing our designed domains. Each problem takes a set of input values with different time durations. We choose the thermostat temperature and safety limit temperature according to PMV-PPD index. Our approach successfully solved five out of six problems, although the problems with smaller time duration show better performance in terms of CPU runtime of the planner. By validating the plans with VAL tool, we confirmed the correctness of the plan. For our designed domains a planner takes only a few seconds to generate a plan.

In the current work, limited number of parameters such as zone temperature, thermal comfort, and energy consumption is used for modeling. In future, this can be extended by adding more parameters (for example, humidity, air quality) which would improve and enrich the domains. This work uses only one planner SMTPlan+, in future other planners can be used. An interesting future direction would be implementing our work in a climate control system in a built environment. This would achieve satisfactory thermal comfort in that environment with improved efficiency, and optimal energy consumption. However, a successful implementation in such environment requires further research.

References

1. Alaboud FK, Coles A (2019) Personalized medication and activity planning in pddl+. In: Proceedings of the international conference on automated planning and scheduling, vol 29, pp 492–500
2. Cashmore M, Fox M, Long D, Magazzeni D (2016) A compilation of the full pddl+ language into smt. In: Proceedings of the international conference on automated planning and scheduling, vol 26
3. Della Penna G, Intrigila B, Magazzeni D, Mercorio F (2015) Upmurphi released: Pddl+ planning for hybrid systems. In: Proceedings of 2nd ICAPS workshop on model checking and automated planning, pp 35–39. Citeseer
4. Fanger PO (1973) Assessment of man's thermal comfort in practice. Occupat Environ Med 30(4):313–324
5. Fong KF, Hanby VI, Chow TT (2006) Hvac system optimization for energy management by evolutionary programming. Energy Build 38(3):220–231
6. Fox M, Long D (2006) Modelling mixed discrete-continuous domains for planning. J Artif Intell Res 27:235–297
7. Ghahramani A, Jazizadeh F, Becerik-Gerber B (2014) A knowledge based approach for selecting energy-aware and comfort-driven hvac temperature set points. Energy Build 85:536–548
8. Haslum P, Lipovetzky N, Magazzeni D, Muise C (2019) An introduction to the planning domain definition language. Morgan & Claypool Publishers
9. Howey R, Long D, Fox M (2004) Val: automatic plan validation, continuous effects and mixed initiative planning using pddl, pp 294–301. https://doi.org/10.1109/ICTAI.2004.120
10. Klein L, Jy Kwak, Kavulya G, Jazizadeh F, Becerik-Gerber B, Varakantham P, Tambe M (2012) Coordinating occupant behavior for building energy and comfort management using multi-agent systems. Autom Construc 22:525–536
11. Kusiak A, Li M (2010) Reheat optimization of the variable-air-volume box. Energy 35(5):1997–2005
12. Kusiak A, Li M, Tang F (2010) Modeling and optimization of hvac energy consumption. Appl Energy 87(10):3092–3102
13. Kusiak A, Tang F, Xu G (2011) Multi-objective optimization of hvac system with an evolutionary computation algorithm. Energy 36(5):2440–2449
14. Lim B, Van Den Briel M, Thiébaux S, Backhaus S, Bent R (2015) Hvac-aware occupancy scheduling. In: Proceedings of the AAAI conference on artificial intelligence, vol 29
15. Lin Y, Liu M, Yang W (2015) Energy efficiency measures for a high-tech campus in California based on total performance oriented optimization and retrofit (tpor) approach. Procedia Eng 121:75–81
16. Nassif N (2014) Modeling and optimization of hvac systems using artificial neural network and genetic algorithm. Build Simulat 7:237–245
17. Nassif N, Kajl S, Sabourin R (2005) Optimization of hvac control system strategy using two-objective genetic algorithm. HVAC&R Res 11(3):459–486
18. Olesen BW (1982) Thermal comfort. Technic Rev 2:3–37
19. Olesen BW, Brager GS (2004) A better way to predict comfort: the new ASHRAE standard 55-2004. ASHRAE standard 55-2004

20. Selamat H, Haniff MF, Sharif ZM, Attaran SM, Sakri FM, Razak MAA (2020) Review on hvac system optimization towards energy saving building operation. Int Energy J 20(3)
21. Seo J, Ooka R, Kim JT, Nam Y (2014) Optimization of the hvac system design to minimize primary energy demand. Energy Build 76:102–108
22. Tartarini F, Schiavon S (2020) Pythermalcomfort: a python package for thermal comfort research. SoftwareX 12:100578
23. Vallati M, Magazzeni D, De Schutter B, Chrpa L, McCluskey T (2016) Efficient macroscopic urban traffic models for reducing congestion: A pddl+ planning approach. In: Proceedings of the AAAI conference on artificial intelligence, vol 30
24. Wani M, Swain A, Ukil A (2019) Control strategies for energy optimization of hvac systems in small office buildings using energyplus tm. In: 2019 IEEE Innovative Smart Grid Technologies-Asia (ISGT Asia), pp 2698–2703. IEEE

Nuclear Power Plant Burst Parameters Prediction During a Loss-of-Coolant Accident Using an Artificial Neural Network

Priyanti Paul Tumpa, Md. Saiful Islam, Zazilah May, and Md. Khorshed Alam

Abstract Several researchers have concentrated on analyzing the nature of fuel claddings through performing burst experiments on computed loss-of-coolant accident scenarios and creating practical and conceptual computer programs. In comparison to experimental observation, the established burst criteria (a) assumes that the cladding tube deforms in a symmetrical manner (b) infers the characteristics of Zircaloy-4 cladding for mixed-phase of $\alpha + \beta$ step (c) ignores azimuthal temperature variations. To resolve all of the shortcomings of the burst criteria, this paper proposed an artificial neural network to forecast the burst parameters. In this research, a feedforward backpropagation algorithm with the logsig activation function is used to build this neural network model. A neural network architecture of 2-15-15-15-3, which is a model of three hidden layers containing fifteen neurons in each layer is designed. The mean deviation of burst temperature, burst stress, and burst strain gained from the burst criteria is 1.15%, 3.82%, and 39.41%, respectively, while these parameters are predicted by the proposed neural network includes mean deviations of 0.43%, 1.57%, and 3.85%, respectively. The proposed neural network has been discovered to be more efficient than existing models.

Keywords Artificial neural network · LOCA · Burst criteria · Zircaloy-4

P. P. Tumpa · Md. Saiful Islam (✉) · Z. May · Md. Khorshed Alam
Department of Electronics and Telecommunication Engineering, Chittagong University of Engineering and Technology, Chattogram 4349, Bangladesh
e-mail: saiful05eee@cuet.ac.bd

Z. May
e-mail: zazilah@utp.edu.my

Md. Khorshed Alam
e-mail: md.khorshed_g03456@utp.edu.my

Department of Electrical and Electronic Engineering, Universiti Teknologi PETRONAS, Seri Iskandar, 32610 Perak Darul Ridzuan, Malaysia

M. S. Arefin et al. (eds.), *Proceedings of the International Conference on Big Data, IoT, and Machine Learning*, Lecture Notes on Data Engineering and Communications Technologies 95, https://doi.org/10.1007/978-981-16-6636-0_31

1 Introduction

The pellets of uranium oxide are encased in a thin-walled tube known as Cladding for nuclear fuel that undergoes a fission reaction to produce heat. In light-water reactors, the generated heat is moved away by a coolant with high pressure circulated on the outside of the cladding. There is a drop-in heat transfer rate from the fuel and system pressure of cladding on the outside during loss-of-coolant accident (LOCA). If the external cooling system's pressure decreases, the cladding's internal pressure rises above the surrounding pressure, causing a rise in hoop stress, in the meantime a reduction even in the rate of heat transfer induces a fast rise in cladding burst temperature. As a consequence, fuel cladding begins to deform or balloon, which can ultimately cause it to burst. Furthermore, fuel cladding ballooning can cause a clog of the cooling system, reducing fuel cooling ability [1].

Several types of research have been carried to further understand the behavior of claddings of nuclear fuel, including burst experiments on a single tube of cladding during computed LOCA situations and the formation of conceptual cum theoretical predictive computer codes. In [2] an internal heater was used in Zircaloy-4 cladding in order to imitate fuel pellets inside a steam-heated environment. It was found that even a small temperature differential will cause Zircaloy-4 cladding to deform and that local temperature fluctuations are quite a key determinant of burst parameters. In [3] Burst tests on Zircaloy-4 cladding were performed and established that the burst stress is determined by oxygen concentration and temperature. Burst correlations depending on observational data integrating values burst stress and oxygen content, a stress-based burst criterion was suggested, implying symmetrical cladding deformation. In [4], it was examined the burst properties of cladding made of zircaloy-4 both in vacuum and vapor configurations. The cladding tube bends during ballooning but not until it bursts, and such effect is quite prominent for burst tests in steam experiments than for burst tests in vacuum in otherwise similar circumstances. In [5], to evaluate thermo-mechanical anisotropic creep deformation behavior throughout ballooning and explosion in LOCA, a multi-physics method was implemented. It incorporates gap heat transmission, material anisotropy with high temperature creep, and temperature and burnup dependent material characteristics. To anticipate the burst tendency of Indian PHWR fuel claddings under intermittent heating during an inert atmosphere, a burst criterion was devised [6]. A new hydrogen concentration based burst criteria is constructed in [7] for nuclear fuel cladding. In [8, 9], a burst criterion was established to estimate burst parameters within virtual LOCA situations for fuel cladding.

The established burst criteria in the existing studies [8–10] imply symmetrical cladding tube deformation, which contradicts laboratory observations of cladding tube bending during tube ballooning prior to burst [2]. During LOCA, cladding formed of a zircaloy-4 material change to β-phase gradually from α-phase as the ambient temperature increases. There is a combined $\alpha + \beta$-phase as a result of this gradual phase transition. These mixed $\alpha + \beta$-properties phases are presently unclear. As a result, the burst criteria interpolate cladding properties during the mixed $\alpha + \beta$

phase [8–10]. Even though deformations are particularly vulnerable to local cladding temperature, azimuthal temperature changes are not taken into consideration by these burst parameters [2]. So far to resolve all the issues of the burst criteria, it is well-founded that AI technology may be a smarter choice for predicting the fuel cladding burst factors at LOCA because it doesn't necessitate a conceptual interpretation of all of the occurrences.

An artificial neural network (ANN) is an element of a computational system that mimics how the human brain evaluates and processes data. It uses data to learn and simulate outputs for specified input data. Each layer of the neural network, to be precise the input, hidden, and output layers, is consisted of neurons. From forecasting to medical issue identification and risk assessment [11, 12], Artificial Neural Network (ANN) may be utilized in a wide range of applications. This technique is being used to simulate the Zircaloy-4 cladding burst parameters throughout LOCA in this present research. The output of the artificial neural network was compared to experimental results and even burst criteria values. The mean deviation of the predicted data is calculated concerning experimental and criterion data.

The paper is organized as follows: Sect. 2 presents the data collection for the neural network; Sect. 3 presents the design of a neural network. Evaluation of the outcomes of the proposed network is with a linear regression model and a comparative analysis between experimental, criterion, and predicted data of the proposed network are given in Sect. 4 followed by the conclusion in Sect. 5.

2 Data Collection

Prior to the implementation of an ANN model, data collection is an important step. The validity of the data used to train an ANN model defines its efficiency. Chung [4], Karb et al. [13], Chapman et al. [2], and Sawarn et al. [14] performed experimentation on a single tube of Zircaloy-4 fuel cladding in a virtual LOCA condition. Those experimental data are used to develop a neural network in this research. The data used in this study is particularly for non-irradiated Zircaloy-4 cladding burst experiments in a vapor surrounding.

In [4] burst experiments were performed in a steam atmosphere on Zircaloy-4 claddings for heat transfer rates ranging between 3 and 145 K/s and initial pressures ranging between 0.5 and 14.5 MPa. During the test, a video recorder with a decent speed was accustomed to monitor the tube's axial and diametrical shifts. Chapman et al. [2] investigated the deformation behavior of Zircaloy-4 cladding where the heat transfer rate and initial internal pressure were ranged between 4–30 K/s and 0.8–20 MPa [2]. Even minor temperature variations were found to have a significant impact on deformation. Sawarn et al. [14] carried out instantaneous heating tests on cladding made of Zircaloy-4 that were pressurized from inside with the gas argon at 3–70 bar. The heat transfer rate was adjusted between 5 and 19 K/s. Karb et al. [13] performed the same tests with heating rate along with initial pressure ranging from

Table 1 Specifics of the experimental data are required to design the Neural network

Researchers	External diameter (mm)	Internal diameter (mm)	Initial pressure (MPa)	Heating rate (K/s)
Chung et al. [4]	10.90	9.63	0.56–14.50	3–220
Chapman et al. [2]	10.92	9.65	0.8–20.35	4.8–30.6
Karb et al. [13]	10.75	9.30	2.6–9.40	7–19
Sawarn et al. [14]	15.20	14.40	0.3–7.10	5–19

7–9 K/s to 2.6–9.40 MPa respectively. Table 1 lists the details of all of the data used in the current study.

Cladding tubes were transiently heated and pressurized from inside in the experimentations during LOCA circumstances to better understand their rupture properties and to determine burst parameters. Initial internal pressure, p_o was transformed into initial hoop stress, σ_o in order to homogenize the various measurements of the cladding tube using the following equations [6]:

$$\sigma_o = \frac{p_o r}{s} \tag{1}$$

Here, s and r are the thickness and internal radius of cladding, respectively, and p_o is the internal pressure.

3 Design of Artificial Neural Network Structure

Neural Networks are complex systems consisting of artificial neurons that can take in several inputs and generate one or more outputs. There are three layers in every ANN structure: input layer, hidden layer, and output layer. The hidden layer learns from data and determines the nonlinear relationship between inputs and outputs. The predictive outcome is influenced by the overall ANN architecture.

3.1 Network Structure

In a neural network, the total amount of input-output parameters decides the input and output layers' neurons number. Indeed, the input-output mapping's complexity determines the hidden layers number and the also neurons in each of the hidden layers. In order to get the best results, the total amount of hidden layers and the size for each of the hidden layers are often determined by the process of trial and error. For the estimation of diametral creep of Zr2.5% Nb pressure tubes in the reactor, Sarkar et al. [15] designed ANN with a 9-12-1 configuration, which means it consists of nine

Table 2 Mean errors for different ANN architectures

ANN architecture	Mean error (%)
2-4-4-3	11.11
2-10-10-3	9.60
2-15-15-3	9.16
2-4-4-4-3	10.52
2-10-10-10-3	8.61
2-15-15-15-3	6.16

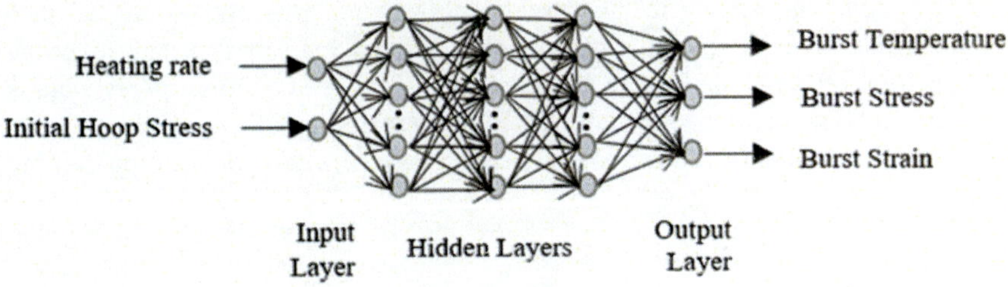

Fig. 1 The architecture of the proposed neural network

neurons for the input layer, twelve neurons in one hidden layer, and a single neuron in the output layer. When designing a neural network to analyze irradiated steels' changing temperature, the number of hidden layer neurons was changed between 2 and 15 [16]. Suman [17] modeled a neural network with a 2-4-4-3 configuration based on the theory that the number of neurons can be $n/2$, n, $2n$, and $2n + 1$ used in a hidden layer, in which n is the number of input neurons. However, if there aren't enough neurons, the outcomes may not be fully accurate.

In this study, several ANN architectures are tested to find the optimized configuration. Based on minimum mean error represented in Table 2, the proposed neural network has a 2-15-15-15-3 configuration where there are three hidden layers with fifteen neurons in each. The two-input neuron represents the input parameter: heating rate and initial pressure. Three output burst parameters are burst temperature, burst stress, burst strain. Figure 1 shows the architecture of the neural network.

3.2 *Training and Testing Data*

Since ANNs are focused on learning from data, the greater the database used to train the network, the better the prediction. In the field of material science, the information applied to train the network originates through real-world experiments. Researchers face numerous challenges in obtaining significant amounts of data for training, including resource supply, expense, and time taken to perform experiments. ANN is designed to get a reliable outcome for burst parameters by using a dataset

Table 3 Specifics of the proposed neural network

Specifics	Value
Architecture	2-15-15-15-3
Algorithm	Feedforward backpropagation
Training function	Levenberg–Marquardt (trainlm)
Activation function	Log-sigmoid (logsig)
Performance function	MSE

of 322 observations shown in this research. For neural network testing, 10 data samples from this dataset were randomly chosen in such a way that they cover the full spectrum of heat transfer rate as well as the hoop stress. From the rest data is available, 80% was chosen for training and the remaining 20% as validation data. Validation data would be utilized to evaluate network generality and to stop training once it has reached a stable state.

3.3 Neural Network Algorithm and Components

Researchers have proposed a variety of ANN algorithms for modeling a system's response, including Cascade forward, Radial basis, Feedforward Backpropagation, and others. Researchers, on the other hand, mostly use the feedforward backpropagation algorithm to predict various types of outputs as it provides more accurate results [15, 18]. Multiple activation functions, as an example, log sigmoid (logsig), linear activation function (purelin), and hyperbolic tangent sigmoid function (tansig) are also available for the ANN algorithm. According to Nalbant et al. [19], the preference of activation function is focused on the complexity of the issue.

The logsig activation function along with the feedforward backpropagation method is used in this study because this activation function has the advantage of not allowing the result to contract and expand indefinitely. The error is a distinction between the experimental outcome and the predicted outcome from the network, and this error could be a performance function of the model. The performance function selected for the proposed network is the mean squared error (MSE). The specifics of the designed network are given in Table 3.

4 Result Analysis

The established ANN model's performance is compared to non-irradiated Zircaloy-4 burst criteria developed during a loss-of-coolant accidents situation. The required training time for this ANN model is 15 s which is quite fast. This model training is done on a PC with the specifics Intel(R) Core (TM) i7-8550U CPU of 1.99 GHz frequency along with 8 GB memory. The effectiveness of the proposed network is evaluated using ten separate testing conditions known as testing data from the collected dataset that balance the overall variety of heat transfer rate and also initial hoop stress. The burst parameters are also tested and compared to the ANN model predicted outputs using the burst criteria. Burst parameters, such as burst temperature, burst stress, and burst strain, derived from the configured neural network and burst criteria [8] are compared to experimental data in Table 4.

The mean deviation of burst temperature, burst stress, and burst strain gotten from the burst criteria is 1.15%, 3.82%, and 39.41%, respectively, while the proposed neural network estimated those parameter values with mean deviations of 0.43%, 1.57%, and 3.85%, respectively. By using the artificial neural network, the accuracy of predicting burst parameters improved significantly.

All of the disadvantages can be overcome by using a neural network, which does not require any conceptual or empirical relationships. It tries to learn from data and develops complex relationships and the improved accuracy of its prediction implies it performs better than burst criteria. It could be utilized to evaluate the occurrence of rupture of cladding tubes. A ratio between experimental burst parameters and burst parameters gained from burst criteria and the ratio between experimental burst parameters and burst parameters gained from neural network has been calculated to analyze the deviations from experimental data. These ratios for the test data are plotted for each sample to determine how much they varied from the ideal ratio of 1.

Figure 2 plots the ratio of burst temperature for all of the testing records presented in Table 4. The ANN predicted value is less scattered than the criterion values and almost similar to the line of 1. Similarly, Figs. 3 and 4 plots the ratio of burst stress and burst strain for all test data respectively. In Fig. 4 burst strain for the criterion data is scattered too much than the ANN data.

In addition, linear regression is implemented to determine the coefficient of correlation (R) for each of the output parameters. Figure 5a–c represent the regression plots between ANN predicted value and experimental value for burst temperature burst stress, burst strain respectively. The neural network model's estimated burst parameter values are quite similar to the experimental values.

The value of R is 0.98047 for burst temperature, 0.9913 for burst stress, and 0.9168 for burst strain. This Neural Network is suitable for the prediction of burst parameters because R is very near to 1. The values of coefficient of correlation are compared with those of in [17] is shown in Table 5. It is quite visible that the value of R increases for burst temperature and burst stress from those of in [17]. But for burst strain, the value of R is quite less. It is due to the randomness of the training data so, there is less correlation between input data and burst strain. The values of the mean deviations for each parameter are very low compared to those of in [17].

5 Conclusion

Based on information sourced by various scholars from burst tests on Zircaloy-4 fuel cladding during virtual loss-of-coolant accident conditions, an artificial neural

Table 4 Comparison of the output of burst criteria and experimental data with predicted burst parameters from proposed ANN

Sample ID (#)	Heating rate (K/s)	Initial hoop stress (MPa)	Burst temperature (K)			Burst hoop stress (MPa)			Burst strain (%)		
			Experiment	Criteria	Proposed ANN	Experiment	Criteria	Proposed ANN	Experiment	Criteria	Proposed ANN
IPL-5	5	19.94	1166	1206	1165.54	19.53	29.96	20.37	23.09	20.36	23.04
10C	5	105.07	966	748	969.26	95.541	243.97	94.12	73	42.12	71.64
SR-19	25.9	6.88	1439	1351	1454.53	6.078	7.81	4.767	26	6.36	8.21
SR-1	24.2	161.73	961	741	974.20	144.67	249.03	144.6	16	21.58	16.39
IS-43	50	5.5	1520	1410	1476.59	5.383	5.97	5.383	32	2.64	31.31
IS-86	51	117.2	983	807	986.23	106.92	204.27	106.9	37	27.79	36.99
IPL-49	100	6.71	1448	1454	1446.95	5.99	7.4	5.99	38	4.89	37.93
IS-11	101	111.54	1053	837	1308.52	99.333	186.33	99.33	58	25.66	48.67
IS-15	130	5.58	1554	1486	1551.06	5.307	5.85	5.30	32	2.35	31.66
IPL-67	130	100.23	1068	863	1068.97	84.925	172.32	84.92	33	27.11	32.93

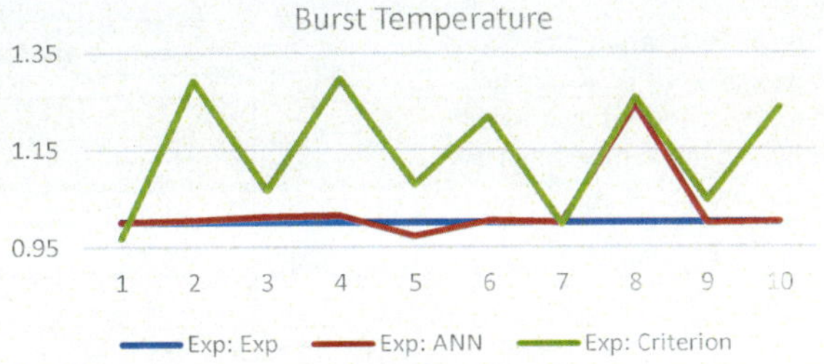

Fig. 2 Comparative analysis of deviation in burst temperature gained from burst criteria and proposed neural network contrast to the experimental burst temperature

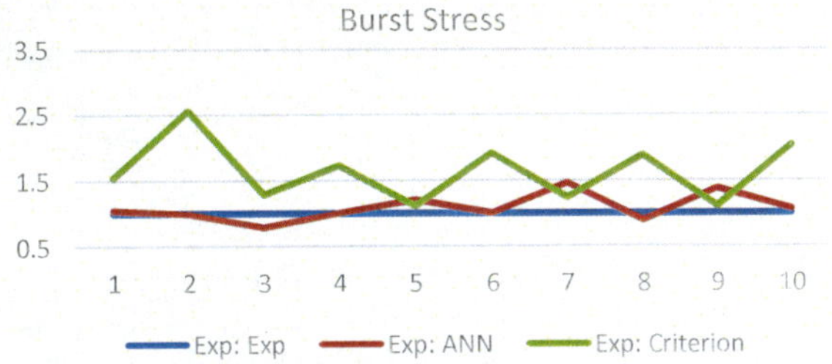

Fig. 3 Comparative analysis of deviation in burst stress gained from burst criteria and proposed neural network contrast to the experimental burst stress

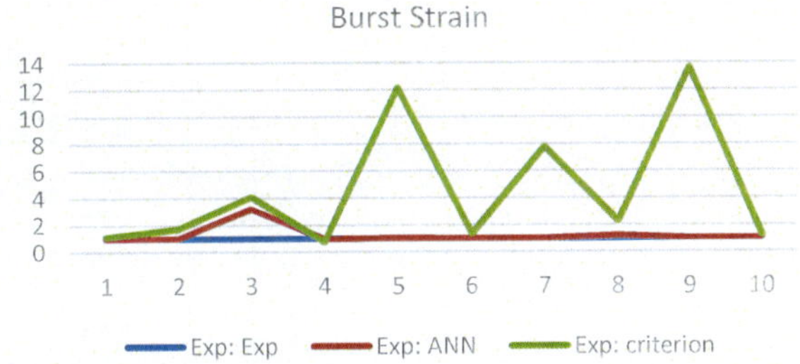

Fig. 4 Comparative analysis of deviation in burst strain gained from burst criteria and proposed neural network contrast to the experimental burst strain

network was created for estimating the Zircaloy-4 fuel cladding burst parameters. The results from the proposed ANN model for test samples are compared to previously developed burst criteria, and it has been discovered that the ANN surpasses the burst criteria. The predictions from the neural network are very similar to those of the experiments. The mean deviation of burst temperature, burst stress, and burst strain obtained from the burst criteria is 1.15%, 3.82%, and 39.41%, respectively, while the proposed neural network estimated those same parameter values with mean deviations of 0.43%, 1.57%, and 3.85%, respectively. It's also discovered that a relatively low burst temperature is related to increased initial stress for a given heat transfer rate. For a certain internal pressure, burst temperatures increase dramatically as the heating rate is increased; that means, the more the heat transfer rates increase, the higher will be the burst temperature. But for a given initial heating rate or initial pressure, burst strain changes randomly not in a specific manner.

Fig. 5 Regression plot between ANN predicted data and Experimental data for **a** burst temperature, **b** burst stress, **c** burst strain

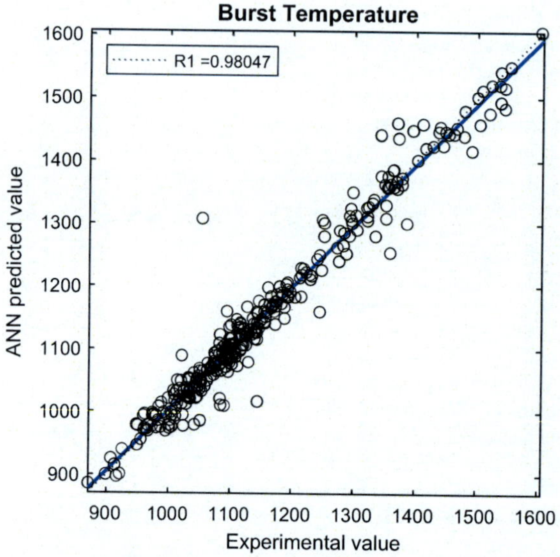

(a)

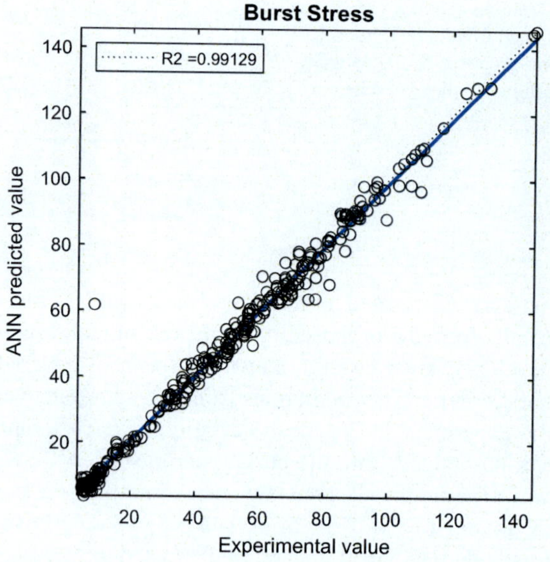

(b)

Fig. 5 (continued)

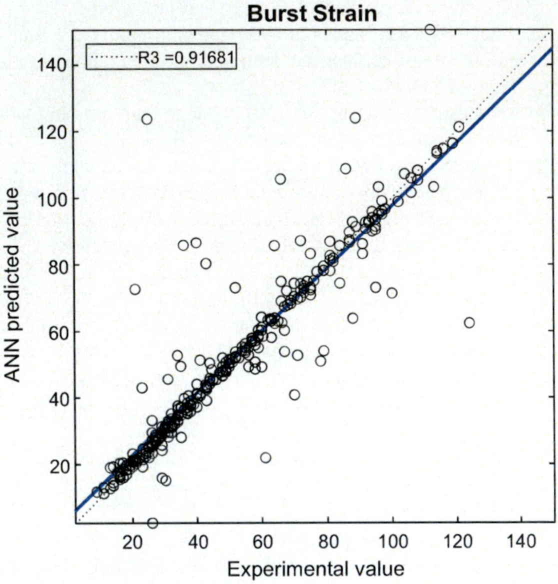

(c)

Table 5 Comparison of correlation coefficient (R) values and mean deviations

Reference	Burst parameter	R	Mean deviation (%)	Burst parameter	R	Mean deviation (%)	Burst parameter	R	Mean deviation (%)
Suman [17]	Temperature	0.966	6	Stress	0.987	2	Strain	0.973	8
Proposed method		0.981	0.43		0.991	1.57		0.916	3.85

References

1. Suman S (2020) Influence of hydrogen concentration on burst parameters of Zircaloy-4 cladding tube under simulated loss-of-coolant accident. Nuclear Eng Technol
2. Chapman RH, Crowley JL, Longest AW, Hofmann G (1979) Zirconium cladding deformation in a steam environment with transient heating, ASTM Special Technical Publication, pp 393–408
3. Erbacher FJ, Neitzel HJ, Rosinger H, Schmidt H, Wiehr K (1982) Burst criterion of zircaloy fuel claddings IN a loss-of-coolant accident. ASTM Special Technical Publication, ASTM International, pp 271–283, 100 Barr Harbor Drive, PO Box C700, West Conshohocken, PA
4. Chung HM, Kassner TF (1978) Deformation Characteristics of Zircaloy Cladding in Vacuum and steam under Transient—Heating Conditions: Summary report
5. Kim J, Yoon JW, Kim H, Lee S-U (2021) Prediction of ballooning and burst for nuclear fuel cladding with anisotropic creep modeling during Loss of Coolant Accident (LOCA). Nuclear Eng Technol
6. Suman S (2019) Burst criterion for Indian PHWR fuel cladding under simulated loss-of-coolant accident. Nucl Eng Technol 51:1525–1531

7. Suman S (2021) Impact of hydrogen on rupture behaviour of Zircaloy-4 nuclear fuel cladding during loss-of-coolant accident: a novel observation of failure at multiple locations. Nucl Eng Technol 53:474–483
8. Manngard T, Massih AR (2011) Modelling and simulation of reactor fuel cladding under loss-of-coolant accident conditions. J Nuclear Sci Technol 48:39–49
9. Suman S, Khan MK, Pathak M, Singh RN, Chakravartty JK (2016) Rupture behaviour of nuclear fuel cladding during loss-of-coolant accident. Nuclear Eng Des 307:319–327
10. Erbacher FJ, Neitzel HJ, Rosinger H, Schmidt H, Wiehr K (1982) Burst criterion of zircaloy fuel claddings in a loss-of-coolant accident, ASTM Special Technical Publication, ASTM International, pp 271–283, 100 Barr Harbor Drive, PO Box C700, West Conshohocken, PA
11. Biswas A, Islam MS (2021) Brain tumor types classification using K-means clustering and ANN approach. In: 2nd international conference on robotics, electrical and signal processing techniques (ICREST), pp 654–658
12. Sathi KA, Islam MS (2020) Hybrid feature extraction based brain tumor classification using an artificial neural network. In: IEEE 5th international conference on computing communication and automation (ICCCA), pp 155–160
13. Karb EH, Sepold L, Hofmann P, Petersen C, Schanz G, Zimmermann H (1982) Lwr fuel rod behavior during reactor tests under loss-of-coolant conditions: results of the FR2 in-pile tests. J Nuclear Mater 107:55–77
14. Sawarn TK, Banerjee S, Sheelvantra SS, Singh JL, Bhasin V (2017) Study of clad ballooning and rupture behaviour of Indian PHWR fuel pins under transient heating condition in steam environment. J Nuclear Mater 495:332–342
15. Jin M, Cao P, Short MP (2019) Predicting the onset of void swelling in irradiated metals with machine learning. J Nuclear Mater 523:189–197
16. Cottrell GA, Kemp R, Bhadeshia HKDH, Odette GR, Yamamoto T (2007) Neural network analysis of Charpy transition temperature of irradiated low activation martensitic steels. J Nuclear Mater 367:603–609
17. Suman S (2020) Deep neural network-based prediction of burst parameters for Zircaloy-4 fuel cladding during loss-of-coolant accident. Nuclear Eng Technol 52:2565–2571 (Elsevier)
18. Sarkar A, Sinha SK, Chakravartty JK, Sinha RK (2014) Artificial neural network modelling of in-reactor diametral creep of Zr2.5%Nb pressure tubes at Indian PHWRs. Ann Nuclear Energy 69:246–251
19. Nalbant M, Okkaya HG, Toktas I, Sur G (2009) The experimental investigation of the effects of uncoated, PVD- and CVD-coated cemented carbide inserts and cutting parameters on surface roughness in CNC turning and its prediction using artificial neural network. In: Robotics and computer-integrated manufacturing, vol 25. Elsevier, pp 211–223

IoT Controlled Six Degree Freedom Robotic Arm Model for Repetitive Task

Aditi Barua, Tazul Islam, Aidid Alam, and Suvrangshu Barua

Abstract The study intends to fabricate a six degree of freedom articulated robotic arm with a gripper that will be controlled by IoT. After designing the arm in Solidworks, fabrications was done using mechanical hardware like beam bracket, U and L-shaped brackets, and gripper made of aluminum alloy. The robot arm was given four revolute joints with one degree of freedom in the base, shoulder, elbow, and wrist. The end effector was given two revolute joints at the gripper, and the gripper joint to fulfill the six degrees of freedom estimation. So it can revolve in 3D plane to grip object. The study intends to experiment on the response of the arm, while it is controlled by IoT using an Android app for sending instructions to the arm. The arm was controlled successfully with higher Internet speed from different districts using IoT, and the response timing was recorded not more than 11 s. In recent times, training the robotic parts is really getting a lot of importance. The training is often done by complex calculations and time-consuming approaches. The arm built here is expected to be trained for repetitive task loops. The program setup is such that it would store the position values of the angle in the two-dimensional array and run them in loops when the proper command is given.

Keywords Robotic arm · Internet of things · IoT

A. Barua (✉) · T. Islam
Department of Mechanical Engineering, Chittagong University of Engineering and Technology, Chittagong, Bangladesh
e-mail: tazul2003@cuet.ac.bd

A. Alam
Department of Mechatronics and Industrial Engineering, Chittagong University of Engineering and Technology, Chittagong, Bangladesh

S. Barua
Department of Computer Science and Engineering, Khulna University of Engineering and Technology, Khulna, Bangladesh
e-mail: barua1407047@stud.kuet.ac.bd

M. S. Arefin et al. (eds.), *Proceedings of the International Conference on Big Data, IoT, and Machine Learning*, Lecture Notes on Data Engineering and Communications Technologies 95, https://doi.org/10.1007/978-981-16-6636-0_32

1 Introduction

Robotics is a multidisciplinary field of knowledge. Nowadays, researchers are experimenting with the arenas where robots can assist humans. Such sectors include health care, fire safety, and home maintenance. Joseph et al. [1] presents numerous examples where robots have come into the assistance of the doctors in acts as sensitive as surgery. Gray and Davis [2] shows how robots have contributed in the food industry. Moreover, with advent of Industry 4.0, robots are designed, so that they can work within the same envelop as people and achieve more precision and celerity [3]. However, the notion that robots are only suitable for life-threatening tasks that would otherwise have been quite impossible for humans still prevails. Harpel et al. [4] include an analysis where a three-joint manipulator in a hazardous environment. Anantha Raj and Srivani [5] explain how a firefighting task was accomplished by an IoT controlled robot.

Robots are not entirely self-actuated as they can only follow commands given to them by the maker with precision and accuracy. Large-scale industrial robots move at such a speed that it can be quite dangerous to be near them. So the importance of remote controlling any robotic part is indispensable.

The study also intends to build such a platform for controlling the robot arm remotely that has been built for any pick and place action. Robotic arms are suitable to perform tasks where the robot is fixed, but the end effector can reach certain distances. Instead of building a whole humanoid robot due to the cost, only the arm-shaped structure is made.

This study includes the remote operation of an articulated arm with six degrees of freedom. The base, shoulder, elbow, and the wrist will have revolute joints each, and the end effector has two revolute joints. The robotic arm will be controlled by the Internet of Things or IoT applying the master–slave concept where the master will be Node MCU and slave will be the microcontroller. Beforehand, it was planned to control the arm using only one controller hardware and that would be Node MCU alone. But the current supplied by the Node MCU was not sufficient to control more than two motors at a time. So the concept of master slaving will be applied.

2 Literature Review

The idea of automation was introduced from the beginning of the industrialization. From simple watch to today's humanoid robot, human ingenuity expressed its versatility. But the concept of robot came into reality when dangerous, heavy duty tasks were needed to be done. Though industrial robots outdo at simple tasks like hauling and welding, they are not suited for replicating such tasks which require the craftsmanship and expertise of skillful engineers. The employment of robots in the manufacturing business is still partial. Also many jobs are still dependent on the people.

This paper [6] has shown that why it is important to make the systems that help the farmers to check soil properties automatic with high reliability and cost effective as they often do not get enough training on agricultural science. Megalingam et al. [7] has introduced human dealings with the robotic arm in the concurrent situation. This paper tells about the construction and testing of a small prototype resembling a robotic controller arm using Bluetooth technology. Abhilash and Mani [8] performed study blending the IoT with the robotics. The robot is IoT controlled wheeled robot. The robot can move smoothly over a surface compared to a legged one. The wheels are basically controlled with IOT not the arm. The arm sits over the wheeled structure to perform occasional grabbing. Rajashekar et al. [9] presented a robot also based on Arduino-based controller. The robot moves in four directions with the help of servo motors. The servo motors can rotate up to 180° if given enough torque. The robot is given a arm like shape. The actuators are responsible for its movement. This paper [10] demonstrated an intelligent robotic arm (IRA) which is capable of cooking. The robotic arm identifies the kitchen accessories like utensils and spices required for the specific recipe. For object detection, they have developed a system using Open CV and Python which helps the robotic arm to recognize the correct ingredient and pick and drop it to appropriate utensil for cooking the required recipe.

3 Methodology

Experiments are conducted considering some basic aspects. First comes the assembly of hardware for performing experimental procedures. Further comes data collections and calculation. Lastly some decision regarding the observations are drawn. The robotic arm project has also considered these steps. In this project, integrating wireless communication between hardware and software has enhanced the applications of robotic operations, hence increased the chance of eradicating limitations regarding control. This study aspires to implement the concept of IoT over a six degree of freedom robotic arm. The aim is not only to control the arm from distant places using IoT but also to train the arm using Android application for any kind of pick and place operation. The arm model is designed using Solidworks. The Solidworks model is presented in Fig. 1.

In this project, Arduino IDE is used to write the command instructions into an Arduino UNO. For communication purpose, Node MCU microchip is used to accumulate data and program. As it has high processing power with inbuilt Wi-Fi or Bluetooth as well as deep sleep operating features, it is ideal for using in this IoT project. It can be easily programmed in the Arduino IDE platform. MQTT broker is optimal for overcoming constraints like low bandwidth or communicating with satellites. The devices that send data are known as publishers that establish connection with the MQTT broke. The broker then sends a package data named topic data with variety of information and shares data if there are subscribers.

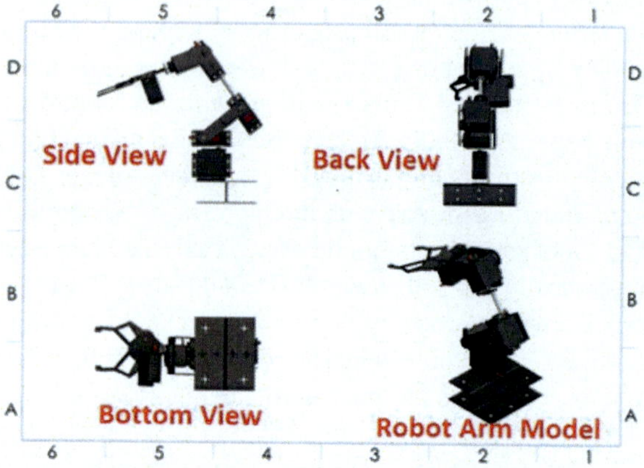

Fig. 1 3D modeling of the arm in solidworks

Figs. 2 and 3 show the complete assembly of the fabricated arm structure and the electrical setup accordingly.

All instruction for controlling the robotic arm is given by an Android application. The app has the functionality to choose modes of control as well as saving entry data in order to train the arm for repeated actions. Figure 4 shows the control interface of the application.

The concept of master slave of IoT is a applied here. Here, the Node MCU acts as a master. Master provides clock which effectively becomes the data transfer rate. Being a bidirectional bus master can write to the slave and also read from the slave. The data is shifted bit by bit. Here two bus lines, serial clock line (SCL), and serial data (SDA) are provided. The Node MCU receives instructions from the Android application. Then send it to the Arduino UNO via SDA and SCL connection set between them. The Arduino controls the motors as per the order. The application and the node is connected to the local Internet using Wi-Fi. The instructions received from MQTT server to Arduino UNO that acts as a slave. Node MCU being the master publishes the commands to Arduino UNO using I2C communication protocol. Data transmission flow should be like this: Connecting App and Node MCU to Wi-Fi ⇒ Instructions sent from Application to the MQTT Broker ⇒ Instructions received by MQTT Broker ⇒ Node MCU receives data as subscriber from MQTT Broker ⇒ Node MCU sends data to a Arduino UNO via I2C communication ⇒ Arduino UNO receives data and controls the motors.

4 Result and Discussion

The aim of the project was primarily to build and apply such a platform where the robot arm can be controlled using IOT. The observations are more visual rather being mathematical. The important measure was its accuracy and response. The results obtained from various observations are listed in the following few pages.

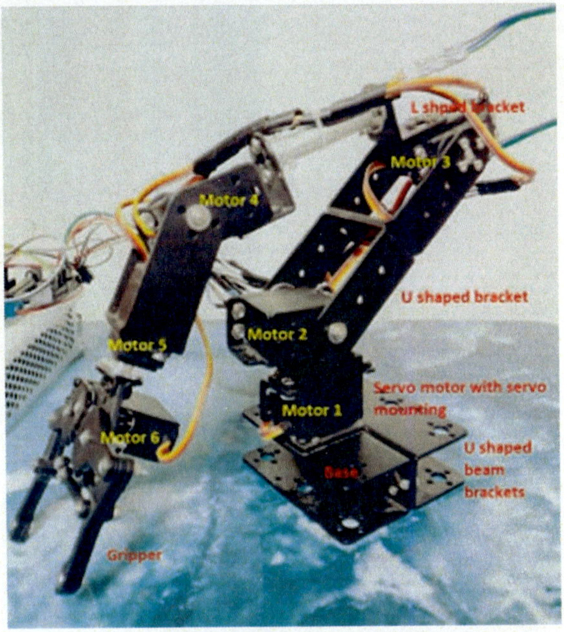

Fig. 2 Complete setup of fabricated arm model

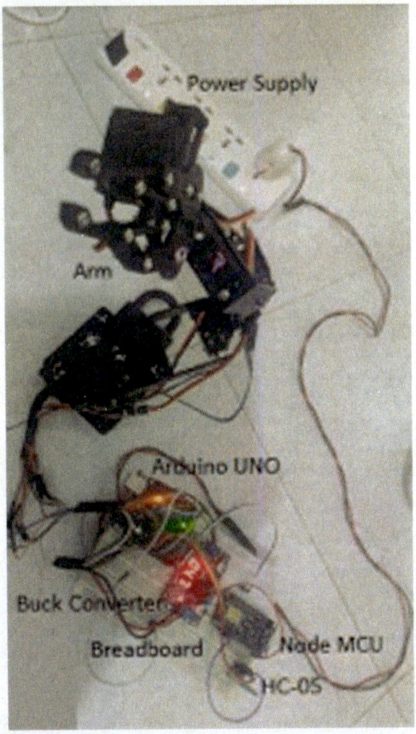

Fig. 3 Circuit setup with electrical equipment

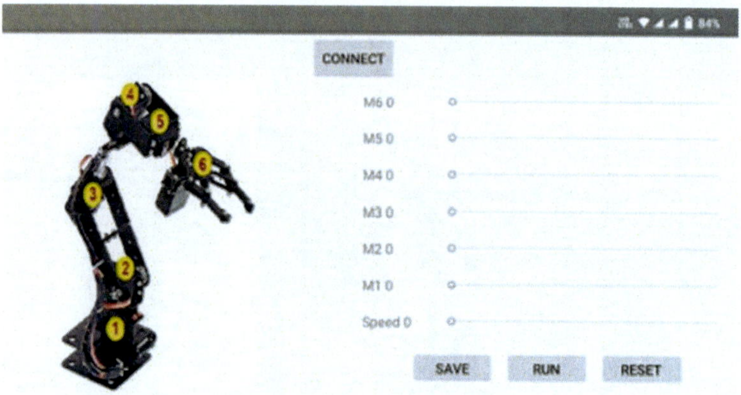

Fig. 4 Control menu of the application

4.1 Data Collection

Before performing any kind of mathematical or control actions, some visual observations were made. Table 1 shows the different criteria observed for the robotic arm.

Table 2 shows the response time for IoT control when publisher data transfer speed is 5.4 Mbps. The arm is controlled from different distances.

Table 3 shows the response time for IoT control when publisher data transfer speed is 20 Mbps. The arm is controlled from different distances.

The receiver's Internet speed was considered constant. As the data is transmitting in one way, the arm is controlled from different distances. This time the distances are not confined in one area or so. The last data is taken from a distant district.

The previous table shows the arm can be controlled with IoT. IoT enables the way to control the area from any place irrespective of geographical location.

Table 1 Criteria of the designed robot

Criteria	Type
Classification of robot	Industrial robot
Degree of freedom	6
Configuration	Articulated
Robot drive	Electric
Vision and sensing	Not used
End effector	Grapper
Robot motion	Stop to stop control
Work envelop shape	Does not have specific shape
Maximum weight to be carried	≤ kg
Maximum opening of gripper	6 cm
Weight of the complete set up	1.5 kg

Table 2 Response of robot arm during IOT control (5.4 Mbps)

Average data transmission rate of publisher and subscriber (Mbps)	Distance (m)	Input angles (°)	Time of response of each motor (s)	Total time of response (s)
5.4 mbps	1	M1 = 60	1.99	11.82
		M2 = 75	1.96	
		M3 = 80	1.95	
		M4 = 45	1.97	
		M5 = 60	1.97	
		M6 = 30	1.98	
	5	M1 = 60	1.99	11.85
		M2 = 75	1.99	
		M3 = 80	1.95	
		M4 = 45	1.97	
		M5 = 60	1.97	
		M6 = 30	1.98	
	7	M1 = 60	1.99	11.85
		M2 = 75	1.99	
		M3 = 80	1.95	
		M4 = 45	1.97	
		M5 = 60	1.97	
		M6 = 30	1.98	
	10	M1 = 60	1.99	11.84
		M2 = 75	1.97	
		M3 = 80	1.97	
		M4 = 45	1.97	
		M5 = 60	1.97	
		M6 = 30	1.98	

Table 4 shows the response of in such situations. The arm was controlled from different districts as shown in the next table.

For this type of control, the average Internet speed of the publisher was kept nearly 10 Mbps. The traffic in the MQTT server was considered fairly normal. The delay of response in within the range of 10.15–10.86. The network observed was fairly stable.

The arm contains a unique feature where it can be instructed for performing repetitive tasks. This is a feature for arms that is intended to be used in industrial setup for pick and place application. Table 5 shows the performance while the arm was performing repetitive task.

Table 3 Response of robot arm during IOT control (20 Mbps)

Average data transmission rate of publisher and subscriber (Mbps)	Distance (m)	Input angles (°)	Time of response of each motor (s)	Total time of response (s)
20 Mbps	1	M1 = 60	1.05	6.48
		M2 = 75	1.09	
		M3 = 80	1.15	
		M4 = 45	1.05	
		M5 = 60	1.07	
		M6 = 30	1.07	
	5	M1 = 60	1.06	6.53
		M2 = 75	1.11	
		M3 = 80	1.15	
		M4 = 45	1.06	
		M5 = 60	1.07	
		M6 = 30	1.98	
	10	M1 = 60	1.06	6.54
		M2 = 75	1.11	
		M3 = 80	1.16	
		M4 = 45	1.06	
		M5 = 60	1.07	
		M6 = 30	1.08	
	10	M1 = 60	1.06	6.55
		M2 = 75	1.12	
		M3 = 80	1.16	
		M4 = 45	1.06	
		M5 = 60	1.07	
		M6 = 30	1.08	

To be able to train the arm to perform repetitive task, the position data were collected in a two-dimensional 10×6 array where the column indicated number of each motor and the row represented position values of angles associated with each motor. After saving one or more positions, one after another the command runs reading data from row by row. So the command to change position from one to another is executed.

5 Discussion

In the previous section, several data were collected on the basis of performed observations and response of the arm during IoT control. The arm was controlled by two publisher of different Internet speeds, respectively. Then, the arm was tested for performing repeated tasks. For this, instructions were saved in its memory then set for running at those given commands.

Table 4 Response of robot arm during IoT control from distance places

Location of publisher and average data transmission rate of publisher (Mbps)	Location of arm	Average data transmission rate of subscriber (Mbps)	Total response delay (s)	
Chittagong	10	Dhaka	5.4	10.86
Rajshahi	10			10.15
Khulna	10			10.22
Sylhet	10			10.45

Table 5 Training the arm for repetitive action

Control type	Data transmission rate of publisher (Mbps)	Data transmission rate of subscriber (Mbps)	Range (km)	Input angle for first position (°)	Input angle for second position (°)	Number of loops (N)	Response delay (s)
IoT	10	5.4	6.3	M1 = 70	M1 = 90	5	1.5
				M2 = 75	M2 = 95		
				M3 = 80	M3 = 100		
				M4 = 45	M4 = 70		
				M5 = 100	M5 = 120		
				M6 = 30	M6 = 50		
IoT	20	5.4	256	M1 = 70	M1 = 90	5	1.3
				M2 = 75	M2 = 95		
				M3 = 80	M3 = 100		
				M4 = 45	M4 = 70		
				M5 = 100	M5 = 120		
				M6 = 30	M6 = 50		

Figure 5 shows the response of the arm during IoT control. The responses are fairly at any distance. The response time decreases as the data transmission rate of the publisher increases. For 20 Mbps, the arm responses faster than it does with 5.4 Mbps Internet speed. The arm maintained its response time well within the same range for each mbps. For 5.4 Mbps, the arm responded within 11.82–11.85 s. For 20 Mbps, the response time is fairly less. The response mainly depends on the data transmission rate of the publisher. The response is faster with less traffic in the server.

6 Conclusion and Future Work

The industrial revolution is moving toward Industry 4.0. The above study is a preparation for the upcoming revolution in the world of product manufacturing and automated production. This type studies pave the way for collaborative deployment of

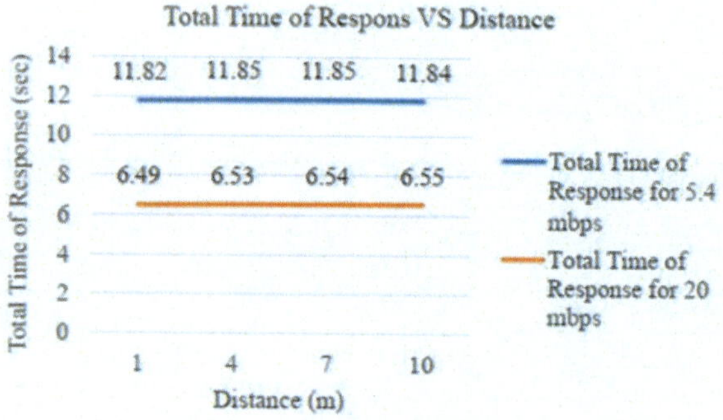

Fig. 5 Response time of the arm during IoT control

small-scale industrial robot in the vicinity of household workers. From the above study, following conclusions can be drawn:

1. The response of the arm during IoT control does not vary with distance but with Internet speed of the publisher. The response time decreases as Internet speed is increased.
2. The IoT response is faster if the MQTT broker server has less traffic.
3. The servo motors are able to rotate upto 180°. But for the grippers motor, it was limited to 30° because maximum opening of it is 6 cm which does not suffice for 180° rotation of the gripper motor.

In future practice where the robot arm will be designed for specific industrial environment, this should be kept in mind as higher torque means increased price, and hence, the project will not be cost effective. The suggested framework can be updated as following

1. Study of collaborative action, safety, and design
2. Object detection system can be introduced
3. The structure can be upgraded for different production operations by changing the end effector such as drilling and CNC operations.
4. Trials using pneumatic and hydraulic actuator for large-scale industrial job.

References

1. Joseph A, Christian B, Abiodun AA, Oyawale F (2018) A review on humanoid robotics in healthcare. MATEC Web Conf 153:1–5. https://doi.org/10.1051/matecconf/201815302004
2. Gray JO, Davis ST (2012) Robotics in the food industry: an introduction. Robot Autom Food Ind Curr Futur Technol 21–35. https://doi.org/10.1533/9780857095763.1.21
3. Ferraguti F, Pertosa A, Secchi C, Fantuzzi C, Bonfè M (2019) A methodology for comparative analysis of collaborative robots for Industry 4.0. In: Proceedings of the 2019 design, automation and test in Europe conference and exhibition, DATE 2019, pp 1070–1075. https://doi.org/10.23919/DATE.2019.8714830

4. Harpel BML, Dugan JB, Walker ID, Cavallaro JR (1997) Analysis of robots for hazardous environments. In: Proceedings of annual reliability and maintainability symposium, pp 111–116. https://doi.org/10.1109/rams.1997.571676
5. Anantha Raj P, Srivani M (2018) Internet of robotic things based autonomous fire fighting mobile robot. In: 2018 IEEE international conference on computational intelligence and computing research, ICCIC 2018, pp 1–4. https://doi.org/10.1109/ICCIC.2018.8782369
6. Sayed MA, Shams N, Zaman HU (2019) An IoT based robotic system for irrigation notifier. In: 2019 IEEE international conference on robotics, automation, artificial-intelligence and internet-of-things, RAAICON 2019, no. November, pp 77–80 (2019). https://doi.org/10.1109/RAAICON48939.2019.38
7. Megalingam RK, Boddupalli S, Apuroop KGS (2017) Robotic arm control through mimicking of miniature robotic arm. In: 2017 4th international conference on advanced computing and communication systems, ICACCS, pp 4–10. https://doi.org/10.1109/ICACCS.2017.8014622
8. Abhilash V, Mani PK (2018) IOT based wheeled robotic arm. In: Int J Eng Technol 7:16–19. https://doi.org/10.14419/ijet.v7i2.24.11990
9. Rajashekar K, Reddy H, Fathima SZ, Kauser S et al (2020) Robotic arm control using arduino. 7(6):453–455
10. Ban P, Desale S, Barge R, Chavan P (2020) Intelligent robotic arm. ITM Web Conf 32:01005. https://doi.org/10.1051/itmconf/20203201005

An ECC Based Secure Communication Protocol for Resource Constraints IoT Devices in Smart Home

Towhidul Islam, Ravina Akter Youki, Bushra Rafia Chowdhury, and A. S. M. Touhidul Hasan

Abstract The Internet of Things (IoT) is one of the emerging fields of technology, which allows the appliance of Smart Home to connect via the internet for communicating and sharing data between them. However, ensuring the security of IoT devices is challenging for their processing and power consumption. IoT devices are resource-constrained, and traditional encryption algorithms are computationally expensive for these small-size and low-powered devices. In this research, an efficient and effective lightweight cryptographic scheme is presented for securing communication between IoT Home devices based on Elliptic Curve Cryptography (ECC). It ensures complete protection against security risks such as authentication, confidentiality, integrity, and key agreement and protects against common cyber attacks. We have conducted an experimental evaluation to demonstrate that the proposed method can outperform the state-of-the-art cryptographic techniques, which significantly improves the security and reduces the storage and communication cost.

Keywords IoT · Smart Home · Confidentiality · ECC · Security · Lightweight cryptography · Authentication · Integrity

T. Islam · R. A. Youki · B. R. Chowdhury · A. S. M. T. Hasan
Department of Computer Science and Engineering, University of Asia Pacific, Dhaka, Bangladesh
e-mail: 17101135@uap-bd.edu

R. A. Youki
e-mail: 17101154@uap-bd.edu

B. R. Chowdhury
e-mail: 17101140@uap-bd.edu

A. S. M. T. Hasan (✉)
Institute of Automation Research and Engineering, Dhaka, Bangladesh
e-mail: touhid@uap-bd.edu

M. S. Arefin et al. (eds.), *Proceedings of the International Conference on Big Data, IoT, and Machine Learning*, Lecture Notes on Data Engineering and Communications Technologies 95, https://doi.org/10.1007/978-981-16-6636-0_33

1 Introduction

In the era of speedy mass globalization, living standards are improved by integrating the Internet of Things (IoT) with the home automation system called the Smart Home. Smart home is a popular application area of the IoT, and it connects smart devices over a network to share data and interconnect between the devices. Smart home is connected to various devices and has the flexibility to manage, monitor, and access. Therefore, the security and privacy of the user's data must be ensured before sharing. Several security attacks may arise in each layer of IoT architecture, such as Man-in-the-Middle (MITM), Denial of Service (DoS), node capture, fake node, Distributed Denial of Service (DDoS), replay attacks take places in the networks and the perception layer [1]. Moreover, data accessibility and authentication, data privacy, and identity security risks arise at the application layer. It is a great challenge to secure Smart Home appliances from the security perspective to prevent all the attacks and risks.

There is a need of developing new security mechanisms because of the attacks. By the lack of proper security protection, attackers can know how the messages or data are routed, enabling attackers to listen to traffic and intercept it, affecting the transmission. Attackers can prevent authorized users from accessing it, increase the power consumption, create a routing loop, and spread fake identities, affecting the security of the entire network. Moreover, fake or invalid users could have a significant impact on the availability of the entire system. Figure 1 illustrates the differences between the present and future possibilities of IoT systems. The IoT devices are connected via fog node, but IoT devices will be connected directly to each other and can share data between them soon. Thus the security of IoT layers will become a primary concern. In conclusion, the major IoT security concerns are authentication, integrity, confidentiality, availability, and privacy [2]. All of these objectives can be fulfilled with the help of cryptographic approaches.

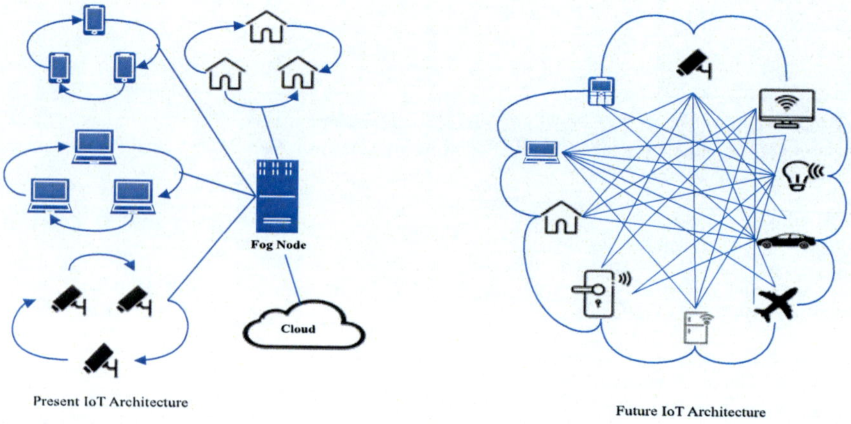

Fig. 1 IoT architecture

In this paper, we propose a security and privacy protocol for resource-constrained IoT devices of Smart Home. It ensures authentication, confidentiality, integrity, and key agreement and resists common cyber-attacks with an affordable computation and less communication and storage cost. A modified Elliptic Curve Cryptography (ECC) is applied to ensure lightweight cryptographic calculation that consumes less time in processing and data sharing. ECC is considered an efficient technique with having smaller key sizes, and for breaking the system, it takes an exponential time challenge for attackers [3].

In the proposed system, the sender and the receiver generate a pair of keys, i.e., a public key and a private key by the ECC calculation. All the parameters of the authentication and communication phases are transmitted over a secure channel by computing their hash value. The hash function is providing a digital signature. All of the received parameters are recalculated to ensure authenticity and integrity. In this way, the proposed communication protocol ensures the security of IoT Smart Home appliances.

The rest of this paper is organized as follows: Sect. 2 provides a summary of our related works. The methodology of our proposed protocol is presented in Sect. 3. In Sect. 4, the security analysis is discussed. Implementation of the proposed protocol is discussed in Sect. 5. Finally, we have discussed the conclusion and future work in Sect. 6.

2 Related Works

For the security of smart homes, there are several research works taking place. Pham et al. [4] proposed a three-phase protocol to perform authentication between device-to-device and device-to-server. But their two-phase authentication has increased the storage cost and computation cost. They have used too many parameters in communication which has increased the communication cost significantly. Authors in [5] proposed a secure IoT-enabled Smart Home system that generates a password by combining user password and fingerprint. The system can resist man-in-the-middle and online dictionary attacks. However, the use of many components has increased the system's implementation cost. Chowdhury et al. [6] proposed a low-cost system with minimum requirements for home security and home automation. However, their system has not considered an attack caused by the hacker over the internet. Singh et al. [7] proposed a home security system that is especially for the old, disabled, and children. But they have only provided hardware-based security and they ignored the security issues over the internet.

Utilizing ECC's advantage, Vasudev et al. [8] have developed a V2V authentication method, where a vehicle can request the vehicle server for real-time information. Their system has involved identification numbers, smart cards, secure hash, and nonce. However, the large number of security parameters has increased their communication cost and storage cost. Garg et al. [9] designed a mutual authentication-based key agreement protocol, where they have performed formal security analysis and proved the resistance against different attacks. Nevertheless, the schemes interaction process is relatively complicated, requiring at least two rounds of computation.

Kumar et al. [10], proposed a face recognition-based security system for IoT devices. They used Raspberry Pi 3 to increase the performance and reduce power and energy consumption. However, the implementation and storage cost are very high, and they have not considered any security measures to resist cyber attacks. Raju et al. [11], proposed a security system for the devices used in home automation by using node MCU. But their system is vulnerable to cyber attacks, as they have not considered it.

It is a big issue to ensure the security of IoT Smart Home devices as it overgrows. Our method uses mutual authentication based on ECC to satisfy security solutions such as authentication, confidentiality, integrity, and key agreement and protects against different cyber attacks. Moreover, it will also tend to have less communication and storage costs.

3 Methodology

We designed a security protocol for those IoT devices that cannot deal with high computation and large keys. The proposed protocol enables end-to-end authentication, confidentiality, data integrity, and key agreement between sender (S) and receiver (R). It can also provide complete protection against common cyber attacks. In Fig. 2, we showed the workflow diagram of our proposed protocol. It depicts that our proposed protocol is divided into three phases. In phase 1, the communicating nodes initialize the communication parameters. In phase 2 and phase 3, authentication and communication are performed, respectively, between the communicating nodes. To keep the computation within the capability of IoT devices, we used mapping as described in Table 1. The mapping table enables the system to preserve the value of the random variable A_n within the range of 0–127 and makes the selection of generator points easier. This protocol is applicable for resource constraints IoT devices that are used in home appliances. The notations that have been used throughout the methodology section are given in Table 2.

3.1 Phase 1: Initialization

Initially, both the S and the R will determine the equation of the elliptic curve E, over a field F, and selects the generator point G.
The S selects a random number A_n and mod it with 127, like $A_n = A_n \bmod 127$, to keep it within the range of our mapping table, which is 0–127. The S maps the A_n with the mapping Table 1 and get values for the A_n and G e.g. if $A_n = 736$ then $A_n = A_n \bmod 127 = 101$. The mapped values of $A_n = 101$ is $(A_n, G) = (125{,}285)$.

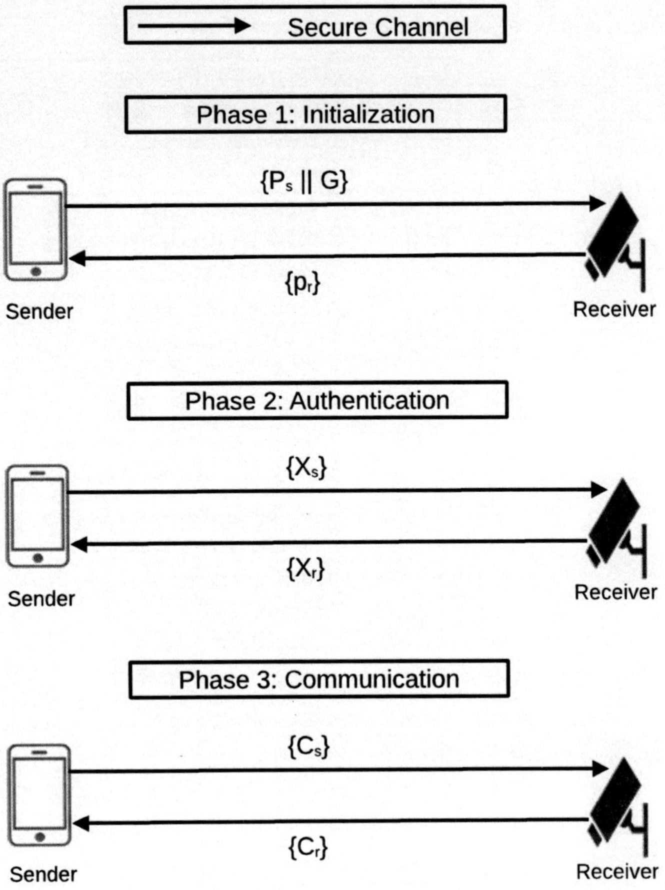

Fig. 2 Workflow diagram of the proposed protocol

Table 1 Mapping table

A_N	(A_N,G)
0	(182,221)
1	(110,310)
...	...
101	(125,285)
...	...
126	(52,147)
127	(132,338)

Table 2 Notations used in the protocol

Notations	Details
E	Equation of elliptic curve
G	Generator point of E
D_s	Private key of sender
D_r	Private key of recipient
P_s	Public key of sender
P_r	Public key of recipient
K_s	Secret key of sender
K_r	Secret key of recipient
P_m	A point on which plain text M is encoded
C_s	Cipher text generated by sender
C_r	Cipher text generated by recipient
N_s	Nonce generated by sender
N_r	Nonce generated by recipient
$h(.)$	One way hash function
$\|\|$	Concatenation operation
$\oplus$	XOR operation
/	Division sign

3.2 Phase 2: Authentication

This authentication phase helps to identify the valid nodes of the communication process and prevents the interference of the eavesdropper. A description of this process is given and shown in Fig. 3.

- The S selects a random number D_s, which is the private key of the S and calculates the public key P_s.

$$P_s = D_s \cdot G \tag{1}$$

  The S computes $P_s = h(P_s)$ and sends $\{P_s \,||\, G\}$ to the R by using a secure communication channel.
- The R selects a random number D_r which is the private key of the R. The R calculates the public key P_r and secret key $K_r = D_r \cdot P_s$.

$$P_r = D_r \cdot G \tag{2}$$

  The R computes $P_r = h(P_r)$ and sends $\{P_r\}$ to the S by using a secure communication channel.
- The S calculates the secret key $K_s = D_s \cdot P_r$. Since, the S and the R are using the same generator point G, their secret key is same, i.e., $K_s = K_r$. The S generates a

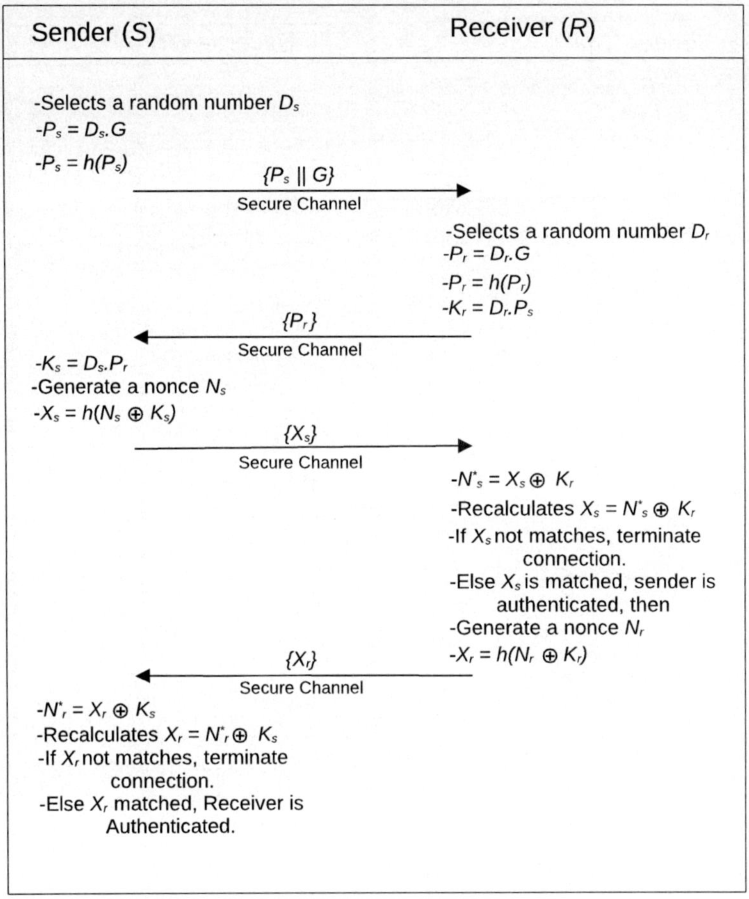

Fig. 3 Authentication between sender and receiver

nonce N_s and calculates $X_s = h(N_s \oplus K_s)$. The S sends the X_s to the R by using a secure communication channel.

- Since, $K_s = K_r$, the R calculates the $N_s^* = X_s \oplus K_r$ and recalculates the $X_s = N_s^* \oplus K_r$. If recalculated X_s is not same as the received X_s then the R terminate the connection. If recalculated X_s is same as the received X_s then the sender is authenticated and the R generates a nonce N_r, calculates $X_r = h(N_r \oplus K_r)$. The R sends the X_r to the S by using a secure communication channel.
- The S calculates the $N_r^* = X_r \oplus K_s$ and recalculates the $X_r = N_r^* \oplus K_s$. If the recalculated X_r is same as the received X_r then the R is authenticated otherwise the S terminate the connection.

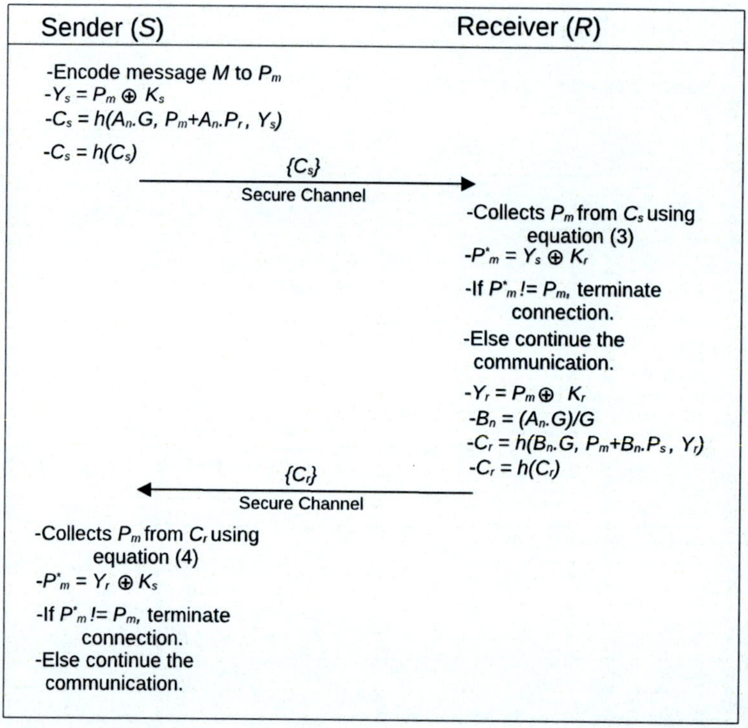

Fig. 4 Communication between sender and receiver

3.3 *Phase 3: Communication*

The communication phase describes the whole procedure of communication between sender (S) and receiver (R) and it is shown in Fig. 4.

Let's consider the message is M. At first, we will encode the message M by using a point from the elliptic curve. We can consider the point as P_m.

- The S calculates $Y_s = P_m \oplus K_s$ and creates the cipher text $C_s = \{A_n \cdot G, P_m + A_n \cdot P_r, Y_s\}$. The values of A_n and G is achieved from the phase 1. The S computes $C_s = h(C_s)$ and sends C_s to the R.
- After receiving the C_s, the R multiply it's private key D_r with the first left most value inside of the C_s and gets $A_n \cdot G \cdot D_r$. Now, the R subtract this value from the second left most value inside of the C_s.

$$P_m + A_n \cdot P_r - A_n \cdot G \cdot D_r$$

From Eq. (2) we know,

$$P_r = G \cdot D_r$$

So,

$$\begin{aligned} P_m + A_n \cdot P_r - A_n \cdot G \cdot D_r &= P_m + A_n \cdot P_r - A_n \cdot (G \cdot D_r) \\ &= P_m + A_n \cdot P_r - A_n \cdot P_r \\ &= P_m \end{aligned} \tag{3}$$

Now, the R calculates $P_m^* = Y_s \oplus K_r$. If the P_m and the P_m^* not matches then the communication is terminated. If P_m and the P_m^* matches then the communication would be continued and it also ensures the integrity of the message of the S. The R calculates $Y_r = P_m \oplus K_r$ and $B_n = (A_n \cdot G)/G$. The R creates the cipher text $C_r = \{B_n \cdot G, P_m + B_n \cdot P_s, Y_r\}$. The R computes $C_r = h(C_r)$ and sends the C_r to the S.

- After receiving the C_r, the S multiply it's private key D_s with the first left most value inside of the C_r and gets $B_n \cdot G \cdot D_s$. Now, the S subtract this value from the second left most value inside C_r.

$$P_m + B_n \cdot P_s - B_n \cdot G \cdot D_s$$

From Eq. (1) we know,

$$P_s = G \cdot D_s$$

So,

$$\begin{aligned} P_m + B_n \cdot P_s - B_n \cdot G \cdot D_s &= P_m + B_n \cdot P_s - B_n \cdot (G \cdot D_s) \\ &= P_m + B_n \cdot P_s - B_n \cdot P_s \\ &= P_m \end{aligned} \tag{4}$$

Now, the S calculates $P_m^* = Y_r \oplus K_s$. If the P_m and the P_m^* not matches then the communication is terminated. If P_m and the P_m^* matches then the communication would be continued and it also ensures the integrity of the message of the R.

4 Security Analysis

This section shows how the proposed system resists different cyber attacks and fulfills the fundamental security requirements: authentication, confidentiality, integrity, and key agreement.

4.1 Authentication and Confidentiality

The proposed scheme performs a strong authentication by computing X_s and X_r respectively at the senders and receivers end. The S sends X_s to the R and the R

sends X_r to the S. The S computes N_r^*, and the R computes N_s^* with their secret key. The S and the R recalculates the X_r and X_s respectively and verifies that the recalculated X_r and X_s is the same as the received one. The X_s and the X_r are sent over the secure communication channel. It is hard to get the data out of the secure channel. If any eavesdropper managed to get the X_s and the X_r, there is no use of it without the secret key. The secret key is computed with the private key, and without the private key, it is impossible to get the secret key. This is how our proposed protocol ensures authentication and confidentiality.

4.2 Key Agreement

Here, the S is creating a secret key K_s using its private key and the public key received from R. Similarly, the R is creating a secret key K_r using its private key and the public key received from S. We showed that both K_s and K_r are the same because they are using the same generator point G to create their public key. Thus, the key agreement is ensured.

4.3 Integrity

To ensure integrity, we have used a one-way hash function for both ciphertext and communication parameters. The one-way hash function acts as a digital signature. In the case of C_s and C_r, when a node receives them, the node recalculates the P_m and matches with the received one to make it more robust. For X_s and X_r, these parameters also being recalculated when received by a node. Thus, the recalculation process and the digital signature are providing integrity.

4.4 Resistance to Attacks

- Replay attack: In this scenario, an adversary fraudulently replies to the messages it intercepts to attain desired goals. suppose, in phase 3, an attacker has intercepted messages of the receiver and replied with the fallacious messages. Our system detects these types of deceitful messages by recalculating messages at every end. When the fraudulent reply arrives at the sender's end, the S recalculates the message with Y_r and K_S. Since the attacker does not know the secret key (K_r) the recalculated message will not match with the received one and the sender will terminate the connection.
- Impersonation attack: In this case, the attacker attempts to establish a connection by adopting the disguise of other nodes. Assume that a disguised device sends a connection request to a valid node. When X_s arrives at the valid node, it will

recalculate the X_s with its secret key (K_r) and recalculated nonce (N_s^*). Since the disguised node does not have the actual private key (D_s) it can not generate the actual secret key (K_s). Therefore, the recalculated X_s will not match with the arrived one and the valid node will abort the connection.

- Insider attack: In this attack model, the attackers are the authorized devices of the network. Because of their authorization, the attack is stronger and effective than external attacks and the attackers are hardly detected. In our proposed system, there is no scope for an insider attack. In our system, there is no administrative body to serve the communication. During a communication, each node assists itself with the required keys of the authentication and communication phases. Therefore, no other devices get the authorization to intervene in the communicating devices.
- Offline dictionary attack: In this case, the attacker tries to guess the crucial information (e.g., password, secret key, private key, etc.) by capturing the transferred messages between nodes. In our system, communicating nodes do not transfer their private key and secret key over the network. Instead, the S and the R transfer two hashed variables X_s and X_r, respectively over the secured channel to ensure the authentication. Thus, the crucial information of the system remains unguessable to the attacker.
- Man-in-the-middle attack: In this scenario, the attacker places himself in between communication and intercepts information and data by pretending as a legitimate node on both sides. Our system can successfully defend against this type of attack. Assume that the attacker gets the X_s or X_r in phase 2 by eavesdropping. But in phase 3, it can not perform communication without the actual private key and secret key. The attacker will fail to generate Y_s and K_s or Y_r and K_r. Thus, the connection will be terminated.

5 Implementation

We have implemented the authentication phase and the communication phase on a dedicated hardware device to obtain the proposed protocol's computation cost.

5.1 *Implementation on Raspberry Pi*

We have implemented the protocol on Raspberry Pi, it has an 8 GB SD card and BCM2708 System-On-Chip with an ARMv6-compatible processor. The authentication phase takes 4.202 ms, and in the communication phase, it takes 4.206 ms.

5.2 *Performance Analysis*

Here, we evaluate the performance of our proposed protocol based on the storage cost and communication cost.

5.2.1 Storage Cost

The storage cost is involved with the number of security parameters we stored in the database. Here, we do not consider the parameters that our system needs to store temporarily. According to the proposed system, the device requires to store the private key (128 bits) and the secret key (128 bits). Total memory we need = [128 + 128] = 256 bits. The storage cost of our system is 256 bits.

5.2.2 Communication Cost

In the proposed system, it transmits some parameters to ensure the security of the system. Those parameters are involved with the communication cost. The communication cost is the same for both sender and receiver. Considering the sender's end, the communication cost is $[X_s, C_s] = [128 + 128] = 256$ bits.

5.2.3 Comparison

We perform a comparison for the storage cost and communication cost with other related systems. In [4, 12–16], the authors showed their storage cost and communication cost for their ECC-based proposed system, and it is presented in Table 3. The storage and communication cost are higher than the proposed system because we have used fewer security parameters. It is observed that the proposed approach outperforms the other techniques. The storage and communication cost are reasonably low and also ensures authentication, confidentiality, key agreement, and data integrity.

Table 3 Comparison of performance with the existing methods

	Storage cost (bits)	Communication cost (bits)
Aakanksha [12]	576	1152
Jiang [13]	1218	352
Ray and Biswas [14]	1280	448
Liao and Hsiao [15]	1680	652
Pham [4]	461	2120
Abichar [16]	800	448
Proposed approach	256	256

6 Conclusion and Future Work

In this paper, we proposed a secure communication protocol for resource constraints IoT devices by leveraging the advantage of the ECC. All of the communicating nodes secure themselves by using the authentication and communication protocol of the proposed system. The system fulfills all of the essential security criteria such as authentication, confidentiality, integrity, and key agreement to enable the smart home's security. In the security analysis, we showed that our system can resist different common cyber attacks. Our storage and communication costs are relatively low compared to other methods. From the implementation of the authentication and communication protocol on the Raspberry Pi, it is evident that the proposed scheme is compatible with low-computational powered devices.

In future, we will extend the work by implementing the protocol on Arduino or more low-computational powered devices, and we will evaluate securing communication cost of the system. With the advancement of science, many smart IoT devices and sensors are incorporated into the healthcare system that collects sensitive data. We can employ our protocol to fulfill the crucial security criteria of those sensitive data.

Acknowledgements The authors thank the Department of Computer Science and Engineering of the University of Asia Pacific, Dhaka, Bangladesh, and the Institute of Automation Research and Engineering, Dhaka, Bangladesh to support this research.

References

1. Aldossari M, Sidorova A (2018) Consumer acceptance of internet of things (IoT): smart home context. J Comput Inf Syst 60:1–11. https://doi.org/10.1080/08874417.2018.1543000 Nov
2. Mohammed H, Qayyum M (2017) Internet of things: a study on security and privacy threats, Mar 2017. https://doi.org/10.1109/Anti-Cybercrime.2017.7905270
3. Daisy Bai T, Raj K, Rabara S (2017) Elliptic curve cryptography based security framework for internet of things (IoT) enabled smart card, Feb 2017, pp 43–46. https://doi.org/10.1109/WCCCT.2016.20
4. Pham C, Nguyen T, Dang T (2019) Resource-constrained IoT authentication protocol: an ECC-based hybrid scheme for device-to-server and device-to-device communications, Nov 2019, pp 446–466. ISBN: 978-3-030-35652-1. https://doi.org/10.1007/978-3-030-35653-8_30
5. Sarmah R, Bhuyan M, Bhuyan M (2019) SURE-H: a secure IoT enabled smart home system, Apr 2019, pp 59–63. https://doi.org/10.1109/WF-IoT.2019.8767229
6. Chowdhury S et al (2019) IoT based smart security and home automation system, Oct 2019, pp 1158–1161. https://doi.org/10.1109/UEMCON47517.2019.8992994
7. Singh A, Gupta D, Mittal N (2019) Enhancing home security systems using IoT, June 2019, pp 133–137. https://doi.org/10.1109/ICECA.2019.8821833
8. Vasudev H et al (2020) A lightweight mutual authentication protocol for V2V communication in internet of vehicles. IEEE Trans Veh Technol 69(6):6709–6717
9. Garg S et al (2019) Secure and lightweight authentication scheme for smart metering infrastructure in smart grid. IEEE Trans Ind Inform 1. https://doi.org/10.1109/TII.2019.2944880

10. Kumar A, Kumar P, Agrawal R (2019) A face recognition method in the IoT for security appliances in smart homes, offices and cities, Mar 2019, pp 964–968. https://doi.org/10.1109/ICCMC.2019.8819790
11. Raju KL et al (2019) Home automation and security system with node MCU using internet of things, Mar 2019, pp 1–5. https://doi.org/10.1109/ViTECoN.2019.8899540
12. Tewari A, Gupta B (2018) A mutual authentication protocol for IoT devices using elliptic curve cryptography, Jan 2018, pp 716–720. https://doi.org/10.1109/CONFLUENCE.2018.8442962
13. Jiang R et al (2013) EAP-based group authentication and key agreement protocol for machine-type communications. Int J Distrib Sens Netw. https://doi.org/10.1155/2013/304601
14. Ray S, Biswas G (2012) Establishment of ECC-based initial secrecy usable for IKE implementation, July 2012
15. Liao Y-P, Hsiao C-M (2014) A secure ECC-based RFID authentication scheme integrated with ID-verifier transfer protocol. Ad Hoc Netw 18:133–146. https://doi.org/10.1016/j.adhoc.2013.02.004 July
16. Abi-Char P, M'Hamed A, ElHassan B (2007) A fast and secure elliptic curve based authenticated key agreement protocol for low power mobile communications, Oct 2007, pp 235–240. ISBN: 978-0-7695-2878-6. https://doi.org/10.1109/NGMAST.2007.4343427

Internet of Things for Wellbeing

IoT-Based Smart Blind Stick

Asraful Islam Apu, Al-Akhir Nayan, Jannatul Ferdaous, and Muhammad Golam Kibria

Abstract Optic failure can be named as a visual shortcoming and optic misfortune. Moreover, this hindrance makes numerous challenges in their daily exercises, such as walking, socializing, reading, socializing, and driving. This research aims to implement an IoT stick that will view the image of opportunity, autonomy, and certainty. The proposed smart stick is planned with an impediment identification module, a global positioning system (GPS), pit and flight of stairs detection, water detection, and a global system for mobile communication (GSM) to perform their daily activities quickly. The impediment identification module utilizes an ultrasonic sensor alongside a water level sensor to distinguish the obstructions that insinuate recognizing the obstacles and identifying the obstructions pattern. An Arduino ATmega328 is used to advise the weakened people about the barriers and sends notifications using an earphone and a buzzer. The current location of the blind person is located using GPS and GSM modules. The stick activates an alert system in case of loss. Several test cases prove that the functionalities introduced with the stick are performing correctly. Such a stick will be a blessing for blind people having a positive impact on science and technology.

Keywords Internet of Things · Blind stick · Smart stick · Good health and well-being

A. I. Apu
University of Liberal Arts Bangladesh (ULAB), Dhaka, Bangladesh
e-mail: asraful.islam.cse@ulab.edu.bd

A.-A. Nayan · J. Ferdaous
European University of Bangladesh (EUB), Dhaka, Bangladesh

M. G. Kibria (✉)
IoT Lab, University of Liberal Arts Bangladesh (ULAB), Dhaka, Bangladesh
e-mail: golam.kibria@ulab.edu.bd

M. S. Arefin et al. (eds.), *Proceedings of the International Conference on Big Data, IoT, and Machine Learning*, Lecture Notes on Data Engineering and Communications Technologies 95, https://doi.org/10.1007/978-981-16-6636-0_34

1 Introduction

Blindness is the absence of vision. It also refers to a deficiency of optic that cannot be reformed with any other easy ways. Fragmented optic insufficiency suggests getting limited or partial optic power. Complete optic debilitation infers nothing to watch. Among the significant types of incapacity, visual hindrance is perhaps the most severe disability and influences numerous individuals around the globe. The International Arrangement of Diseases [1–3] characterizes visual deficiency as a distance vision hindrance that introduces visual keenness more regrettable than 3/60. Additionally, as indicated by the World Health Organization (WHO) [4–6], about 2.2 billion individuals are outwardly weakened worldwide. The expense of this enormous section can be huge [7–9].

The research marks out related issues and proposes a brilliant and impaired solution based on IoT that makes a weakened individual's life more straightforward and considerably more effortless. Comparing with other fundamental devices, "Smart Stick" is more intelligent and successful. As those weakened individuals need the help of an intelligent stick constantly, the device will provide them a keen and mechanical arrangement to be confident while moving. The primary aim is to offer and guarantee a strain-free living as like as normal human beings. Impaired people can be tracked via GPS and the Blynk application. If a blind person encounters any problem, he can transmit a text message to his caretaker by pressing the remote button. If he faces obstacles, then he will be guided through voice messages. All essential features will be in vain if the blind man loses the stick. As a solution, an alarm is set with the IoT-based smart blind stick associated with a remote controller. This alert helps to locate the lost stick. Many individuals who will utilize this visually impaired stick will live a joyful and productive life.

In the context of Bangladesh, it is practically difficult to fix visual deficiency, but the intelligent blind stick will offer them the chance to overhaul their lifestyle. The features will offer its users the opportunity to be liberated from most of their pressures concerning their movements.

2 Related Works

A Nava let was created by Shovlet, and a multipurpose wearing pc that detects obstacles was placed inside the house. It was based on two things; the first one was that it would sound different for different interruptions. One volume was to move forward, and another one was when it faced hindrance. Moreover, the second one was if the blind man stands in the wrong place, the intelligent stick will warn him about his position. The research work could not satisfy users for the lack of necessary features [10].

A research work described a white cane with space measurement. It was complex but time-saving technology. The stick could produce different types of vibration on various modes. Various vibration moods could identify the obstacle variations. It warned blind people for getting ready at the time of danger. However, it contained a drawback. It had data security issues. This cane had no remote detection feature [11].

Another research work described an intelligent white cane that was suitable inside the room. But it could not identify the outdoor obstructions. The features were limited [12].

Two authors proposed an innovative blind stick technique called an intelligent walking stick for the visually impaired. The group built up a stick for outwardly hindered people that helped the individual by providing an alert. But it could not send notifications like navigation or voice message alerts. The features were limited and could not fulfill user's demand [13].

Another bright stick of tiny size proposed an effective wearable device for tracking the route. Moreover, the cane could tell the user in advance about the path. If an obstacle was detected in walking the road, then it could suggest a safe shortcut route. Undoubtedly, it was a time-saving feature. And the authors planned to attach a camera to see the way. This paper introduced a plan that paved the way to monitor blind people in real-time. But the implementation was expensive, and it could not draw the attention of users [14].

Another research work mentioned a smart walking cane that provided advance notifications using infrared sensors. If obstacles were detected in front of the stick, it could warn the blind man through the vibration signal. However, the cane could only detect the front obstacles but produced no warning at the time of danger. Moreover, there contained a limitation for IR sensors. For example, it could not detect distant objects efficiently [15].

Most of the research work related to the blind stick showed unsatisfactory performances when tested. Besides, the previous results did not suggest any solution while losing the stick and network failure. Those drawbacks paved the way for us to rethink a bright blind stick that will be more accurate in detecting obstacles. Therefore, we have introduced a solution against the stick losing. Our implementation has worked out successfully, and we have described the performances elaborately in the result section.

3 Materials and Methods

3.1 Proposed System

The IoT-based smart stick for blind persons is planned with impediment discovery module, water recognition, front obstacle detection, pit and flight of stairs localization utilizing different sensors. GPS [16] and GSM [17] modules are used to make daily work easier. The GPS module can trace the impaired people when they get out of the

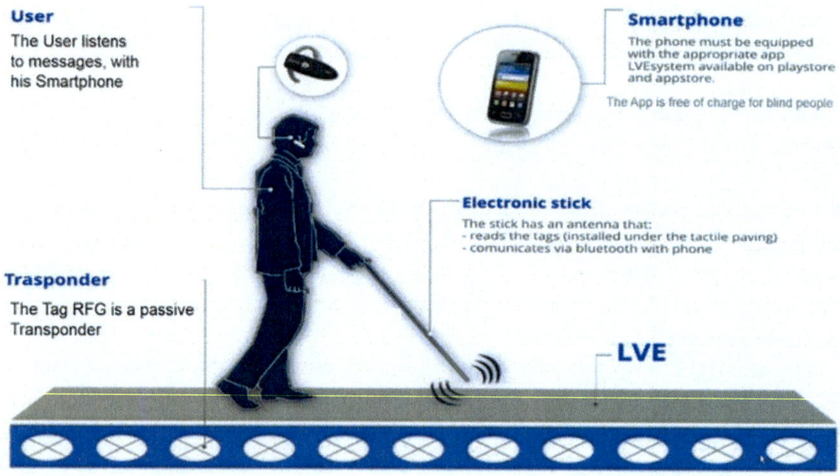

Fig. 1 Proposed system

home. A caretaker can locate him through a mobile application. It makes him tension-free to move everywhere. The GSM module sends emergency messages to the family members if the impaired person faces any problem. The obstruction identification module utilizes sonar and water level sensors to identify and recognize the type, location, and distance of obstacles from the blind person.

The system generates outputs using an airphone that provides a voice signal and a buzzer that creates sound. All specialties may fail if the stick gets lost. As a solution, an alarm system using a remote controller is introduced. The controller has two emergency buttons called "A" and "B." "A" is used to locate the stick while misplaced, and "B" is used to send a message to family members at the time of danger. The blind man can find out his stick by pressing the emergency remote controller button. All the sensors, GPS, and GSM modules are connected to an Arduino pro mini and attached with a stick. An abstract of the proposed system is shown in Fig. 1. The block diagram of the system is shown in Fig. 2, which may help to understand the stick's working procedure quickly.

3.2 *Proposed Context-Aware System Architecture*

Context-aware choice permits innovative blind stick frameworks to settle on context-aware services and depend on the user's encompassing context. Context-aware middleware is the blend of two components: the Context Management System (CMS) and the Knowledge and Notification System (ANS). The Context Control framework encourages the use of setting mindful programming and assets to distribute logical information. The knowledge and messaging frameworks empower client applications

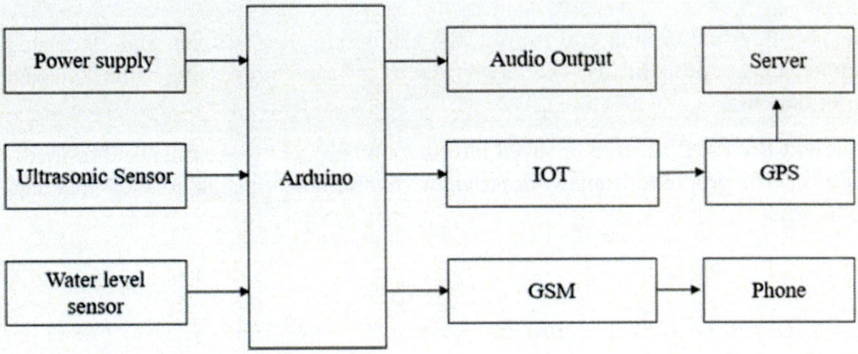

Fig. 2 Block diagram of the system

to apply guidelines that incorporate importance-based prerequisites and notice when the system has been set up [18]. The architecture plans to bring customer applications and data sources together. As a result, a middleware door is created, associating connection between ANS and CMS. The architecture is shown in Fig. 3.

According to the architecture, there contains Context source, Context supplier, Server, and Context buyer. Diverse works are done in these classifications, and for the productivity of the turn of events, the proposed framework has been determined into four parts. Their portrayals are:

Context Source: The sensors are the primary source from which the framework gets data about the specific situation and generates decisions.

Context Supplier: Microcontroller or SOC (framework on-chip) functions as setting supplier. Arduino is the gadget for interfacing or gathering data from the setting.

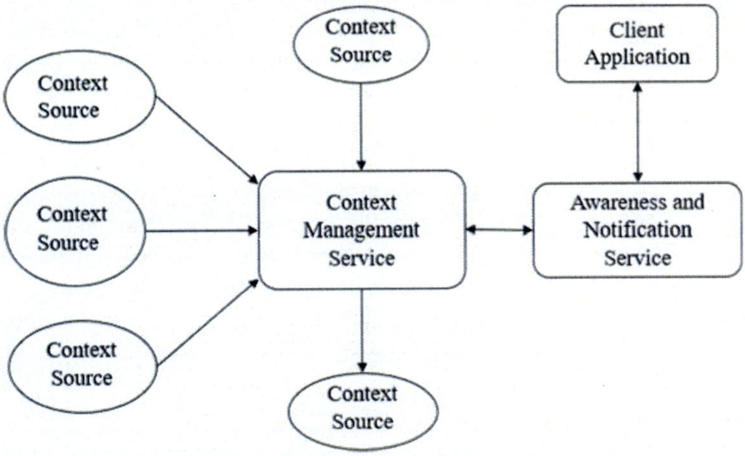

Fig. 3 Context-aware middleware architecture

Server: Information that comes from the setting supplier is saved and calculated immediately or periodically. The server is used for further processing of the collected information.

Context Buyer: Processed or saved information is served to the clients after investigation. The proposed framework provides an interface or gadget to show that data to the clients.

3.3 Required Devices and Sensors

The following devices and sensors have been used to implement the project.

- Arduino pro mini (ATmega328)
- Ultrasonic sensor
- Water level sensor
- GSM Module
- GPS Module
- RF transmitter and receiver module
- Power supply
- Blynk software

3.4 Hardware and Software Implementation

All the components, including two ultrasonic sensors [19, 20], GPS, and GSM, are related to the Arduino pro mini microcontroller. The microcontroller board model is Atmega328 [21, 22]. RF transmitters [23] provide a signal to the microcontroller through the trans receiver. The water sensor [24] provides a call to the microcontroller while moving the stick into the water. Figures 4 and 5 shows the circuit diagram and workflow diagram of the system accordingly. The microcontroller receives signals from sensors and provides outputs through a buzzer and earphone. Here, two threshold values are used (for the front, 35 cm, and deep, 20 cm). If the threshold value is more than 35 cm, it keeps moving; otherwise, it warns.

Necessary code was written installing the Arduino IDE application on a laptop computer. Different sensor's libraries were imported and launched simultaneously. Before uploading the code to the Arduino microcontroller, it was compiled and tested to solve the error. After checking the capabilities of all functions, the code was uploaded to the Arduino using a dedicated cable. To monitor the test outputs, the Arduino COM3 port was used.

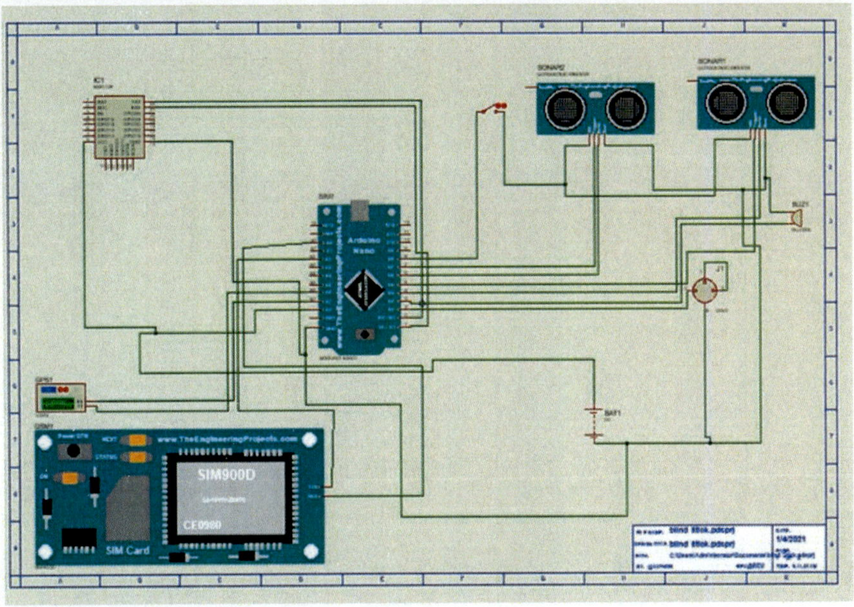

Fig. 4 Circuit diagram

3.5 *Blynk Application*

Blynk application displays the latitude and longitude values when the victim will go out of the home. Using the application, the family member can trace and locate the blind person. The GPS module will send the latitude, longitude, and the blind person's speed to the Blynk Application in a critical situation. The application will help to find out the person immediately.

4 Analysis and Result

An internet connection is essential for the stick to work smartly. A caretaker can trace the blind person via the Blynk application when the GPS module is connected to the internet.

First, the client must put a SIM card into SIM900 [25] module to use the stick. When the internet connection is established, the GPS gets connected to the satellites to find the stick. When the blind man starts his journey to his destination, the intelligent stick guides him to find the path avoiding obstacles and danger. His location is traced every second, and his family member can trace him using the Blynk application. Figure 6 shows the current location of the blind person. We ran test cases several times and got accurate results processed through the GPS module and the application.

Fig. 5 Flowchart diagram

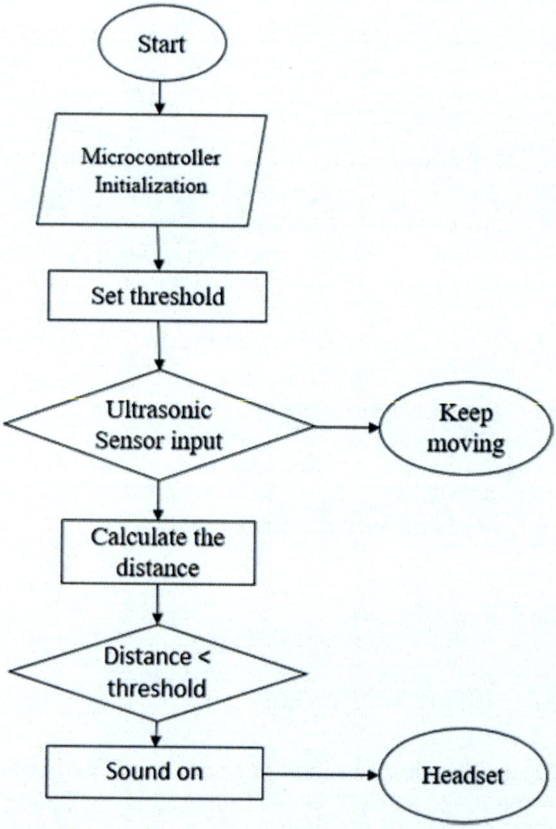

4.1 GSM and GPS Output

All the parts like GPS, GSM, and speaker are associated with each other to trade information through their conventions. These pieces are associated with Arduino pro mini ATmega328 and situated at the highest point of the stick, as appeared in Fig. 7. It is the most suitable spot for those parts. For instance, the GPS receiving wire is set at the highest point of the stick to get the signal from the satellite with no obstacle. Also, the stick is associated with a remote. Pressing the remote button, it sends text messages to his caretaker when he encounters any problem. We tested and found the messaging system functioning correctly. The testing results of sending messages have been shown in Fig. 8.

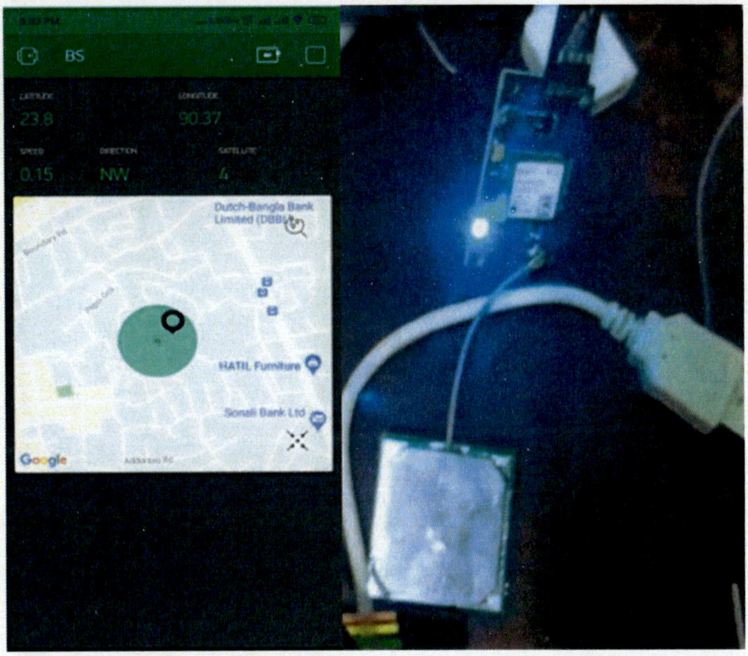

Fig. 6 Blynk application output

4.2 *Sensor Readings*

We utilized two ultrasonic sensors and a water sensor. These sensors are associated with a buzzer that produces an alarm. To investigate the system's ability, we created hindrances like a divider, a tree, or others on the way. At that point, the ultrasonic-1 recognized the forward checks by computing the distance, and the sonar sensor-2 detected the pit hole. The buzzer started to produce sound to caution the blind man, and the notice established through a headphone.

The visually impaired individual may confront impediments beneath the ground level. That is why the Ultrasonic-2 is situated at the lower piece of the stick to identify scarps like the walkway and steps. It works equivalent to the guideline portrayed beforehand. Creating a pseudo environment for testing, we found that the signal produces various sounds in various frequencies. Utilizing those sensors, the stick distinguishes the obstructions which assist the visually impaired in moving securely. In the test case, the water sensor lying at the end of the stick recognized the downpour on the road and functioned accurately to avoid unexpected incidents caused by the water. It created an alarm to express the presence of water. Table 1 shows the readings obtained from the ultrasonic sensor.

Readings from 30 to 400 cm have been shown in Table 1. Accuracy is determined in percentage. If the distance increases, then the error values increase. For example,

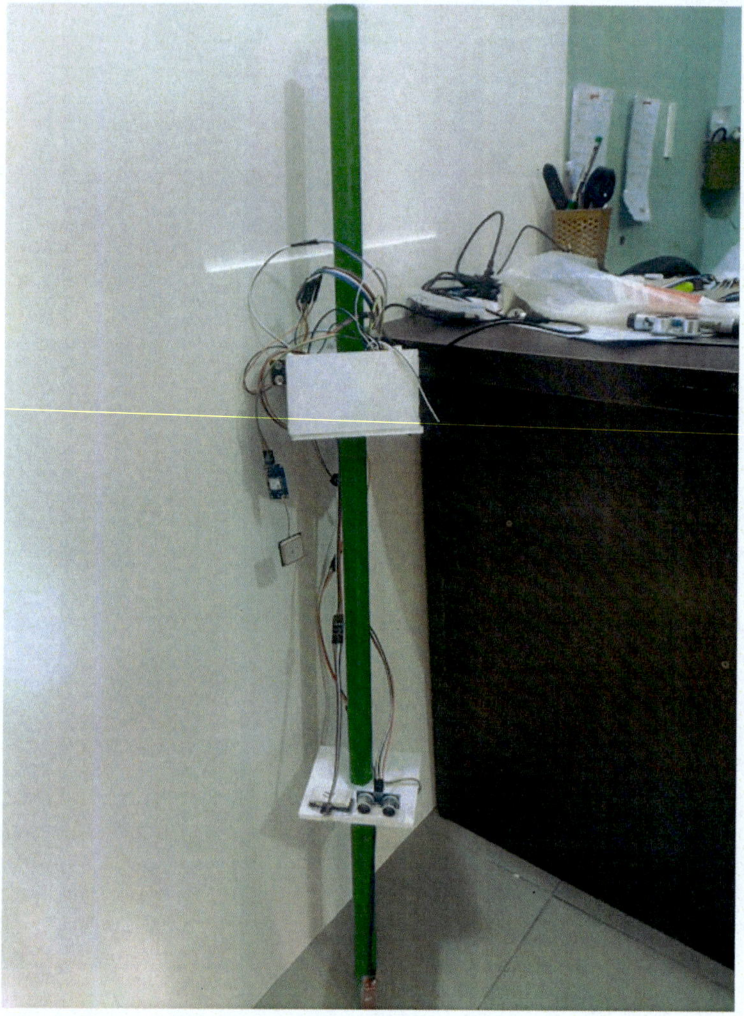

Fig. 7 Complete picture of the smart blind stick

at 30 cm, the error is 0.53, and for 400 cm, the error is 4.17. As distance increases, the accuracy becomes less gradually.

4.3 *Cost Analysis*

The cost for producing the smart blind stick has been mentioned in Table 2. Calculating the amount, we found that the production cost of the stick is cheaper.

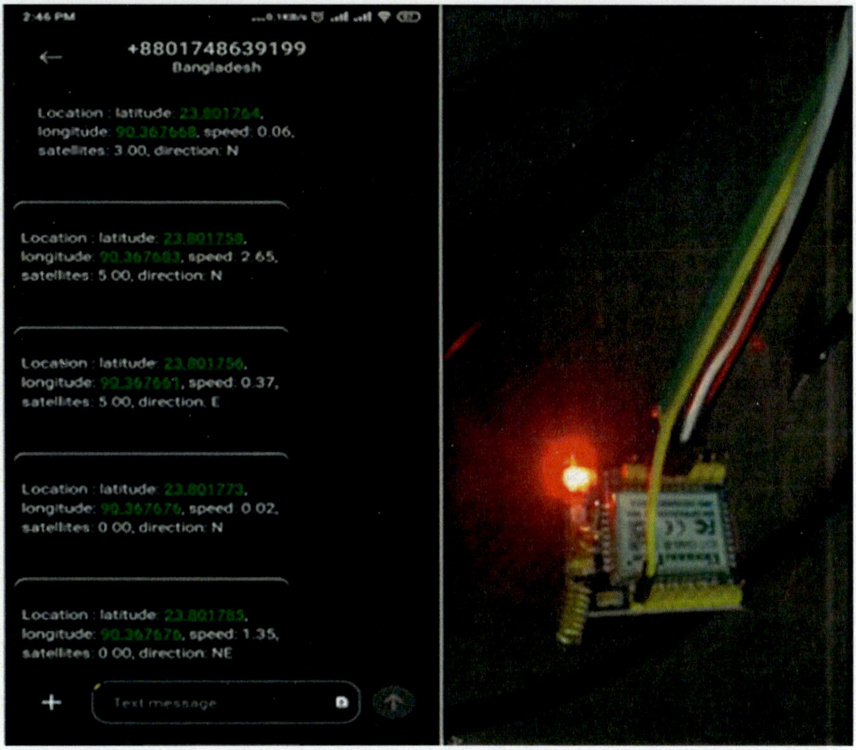

Fig. 8 Messages generated by GSM module

4.4 Comparison

The stick tumbles from the hand of optical blind people. If the blind man losses the stick, he can locate it from four meters away by pressing a button in the remote controller. It is the feature that makes the stick brighter than any other blind stick. After analyzing and comparing the production cost, we found that the smart stick is cheaper than any other blind stick. Several test cases prove the ability of the stick. The rapid production of the stick can be a blessing for the blind people of our society.

5 Conclusion

Blindness is considered a curse in society. Blind people encounter various challenges to lead their life. The challenges kill their self-respect, and the blind people start to think of themselves as a burden of socio-economic development. It destroys their interests in life. Being human, we cannot neglect them. That is why we planned to invent an intelligent blind stick that will help blind people making their daily

Table 1 Ultrasonic sensor reading

Actual Distance (cm)	1	2	3	4	5	6	Average (cm)	Accuracy (cm)	Error
30	30.02	29.99	30.08	30.01	29.02	29.87	29.84	99.47	0.53
50	50.08	49.1	49.02	48.95	49.55	49.75	49.41	98.82	1.18
100	96.2	97.97	58.3	99.1	97.17	98.31	97.84	97.84	2.16
150	147.08	144.17	148.32	145.01	146.78	146.5	146.31	97.54	2.46
250	244.89	243.44	242.21	240.45	241.66	243.98	242.77	97.11	2.89
300	284.9	287.45	286.09	298.23	290.97	292.44	288.51	96.17	3.83
400	390.34	380.43	377.38	381.9	393.9	375.87	383.3	95.81	4.17

Table 2 Cost analysis

S. No.	Description	Quantity	Amount
1	Arduino nano	1	260
2	IoT	1	400
3	GSM	1	450
4	GPS	1	1600
5	Ultrasonic sensor	2	200
6	Audio jack	1	5
7	Water level sensor	1	100
8	RF module	1	450
9	Wire		120
10	Pipe	1	75
11	Glue stick		50

life easier. With the help of a microcontroller, IoT, GPS, GSM, sensors, and other modern instruments, we have successfully implemented and tested a smart stick. The stick provides solutions related to blindness. It guides its users, creating automatic paths. Even the stick contains a unit that activates at the time of loss. The stick can automatically trace itself and sends an alert to its user to find it out. The stick transmits signals and text messages to the guardian of a user in a difficult situation. Several test cases were run that prove the ability of the stick. Such a type of stick may contribute to the welfare of humankind.

References

1. Krawczyk P, Święcicki Ł (2020) ICD-11 vs. ICD-10 - a review of updates and novelties introduced in the latest version of the WHO International Classification of Diseases. Psychiatr Pol 54:7–20
2. Rodríguez-Marín J (2004) International classification of diseases (WHO). In: Encyclopedia of applied psychology. Elsevier, pp 349–353
3. Bewley T (1979) Implementation of the 9th international classification of diseases. Psychiatr Bull R Coll Psychiatr 3:188–188
4. World Health Organization classification of visual impairment and blindness. In: Eye Care in Developing Nations, 4th edn. CRC Press (2007), pp 223–223
5. Colenbrander A (2009) The functional classification of brain damage–related vision loss. J Vis Impair Blind 103:118–123
6. Rovira K, Gapenne O (2009) Tactile classification of traditional and computerized media in three adolescents who are blind. J Vis Impair Blind 103:430–435
7. Köberlein J, Beifus K, Schaffert C, Finger RP (2013) The economic burden of visual impairment and blindness: a systematic review. BMJ Open 3:e003471
8. Chakravarthy U, Biundo E, Saka RO, Fasser C, Bourne R, Little J-A (2017) The economic impact of blindness in Europe. Ophthalmic Epidemiol 24:239–247
9. Frick KD, Gower EW, Kempen JH, Wolff JL (2007) Economic impact of visual impairment and blindness in the United States. Arch Ophthalmol 125:544–550

10. Smart cane for blind people using raspberry PI and Arduino. Regular Issue. 9, 1520–1522 (2020)
11. Billah MM, Mohd Yusof Z, Kadir K, Mohd Ali AM, Nasir H, Sunni A (2019) Experimental investigation of a novel walking stick in avoidance drop-off for visually impaired people. Cogent Eng. 6:1692468
12. Messaoudi MD, Menelas B-AJ, Mcheick H (2020) Autonomous smart white cane navigation system for indoor usage. Technologies (Basel) 8:37
13. Gend M, Pathan S, Shedge C, Potdar PDP (2021) Smart ultrasonic walking stick for visually impaired people. Int J Adv Res Sci Commun Technol 680–687
14. Nada A, Mashelly S, Fakhr MA, Seddik AF (2015) Effective fast response smart stick for blind people. In: Second international conference on advances in bio-informatics and environmental engineering—ICABEE 2015. Institute of Research Engineers and Doctors (2015)
15. Dey A, Palit K (2018) Smart ultra-sonic blind stick. Ind Sci Cruiser 32:16
16. Chambers R (2018) Module 3: encouraging assertiveness skills. In: Survival skills for GPs. CRC Press, pp 53–80
17. Vedder K (2003) The subscriber identity module: past, present and future. In: GSM and UMTS. Wiley, Chichester, pp 341–369
18. Dey A, Haque KA, Nayan AA, Kibria MG (2020) IoT based smart inhaler for context-aware service provisioning. In: 2nd international conference on advanced information and communication technology (ICAICT), Dhaka, Bangladesh, pp 410–415
19. Fung ML, A study of sensor fusion of a depth camera and ultrasonic sensor
20. New low temperature ultrasonic ranging sensor. Sens Rev 20. https://doi.org/10.1108/sr.2000.08720aad.012
21. Adler J, Aprianto E (2019) Monitoring bergesernya tanah dengan membuat simulasi data logger berbasis mikrokontroler arduino pro mini. Majalah Ilmiah UNIKOM 17:61–68
22. Kurnia Utama YA (2016) Perbandingan Kualitas Antar Sensor Suhu dengan Menggunakan Arduino Pro Mini. e-NAR. 2. https://doi.org/10.31090/narodroid.v2i2.210
23. Mobile RF transmitter and receiver design solutions. In: Mobile Handset Design. Wiley, Chichester (2013), pp 123–157
24. Sketches for Driving Water Level Sensor, HydroShare Resources
25. Ohoiwutun J (2018) Analisis dan perancangan smart dump automatic menggunakan arduino mega 2560 Rev3 dan GSM SIM900. Jelekn 4:32

Deep Learning Techniques in Cyclone Detection with Cyclone Eye Localization Based on Satellite Images

Md. Nazmul Haque, A. A. M. Ashfaqul Adel, and Kazi Saeed Alam

Abstract Contemporary progress in satellite imaging and computer vision technology has made it feasible to make automatic weather predictions accurately. To reduce environmental damages and limit the risk of human lives, we have introduced a system to predict cyclones from satellite imagery. The eye of a powerful cyclone inherits a lot of information about its behavior, indicating that detecting the cyclone's center rotating area is also an important task; we have constructed a deep learning object detection system to locate a cyclone's eye. Several models using machine learning and deep learning approaches have produced a cyclone prediction accuracy, ranging from 86 to 95% or better and cyclone eye detection accuracy over 87%. Our experiment's results have been impressive, highlighting that it can outperform conventional approaches and existing works.

Keywords Satellite imagery · Remote sensing · Machine learning · Deep learning · Cyclone detection · Cyclone eye localization

1 Introduction

A cyclone is a natural catastrophe that occurs when a low-pressure circular spinning area arises over tropical waters, causing significant damage. It can also be called a 'hurricane', 'typhoon', etc., and its center rotation can be in the clockwise or anticlockwise direction. It has caused countless damages all across the world due to its destructive characteristics like massive flooding, strong winds and heavy downpours. Cyclones such as 'Sidr', 'Aila' and 'Amphan' have destroyed numerous lives in Bangladesh. Thousands of people died in the Bay of Bengal area in 1970 as a result

Md. N. Haque (✉) · A. A. M. Ashfaqul Adel · K. S. Alam
Khulna University of Engineering & Technology, Khulna, Bangladesh
e-mail: haque1707101@stud.kuet.ac.bd

K. S. Alam
e-mail: saeed.alam@cse.kuet.ac.bd

M. S. Arefin et al. (eds.), *Proceedings of the International Conference on Big Data, IoT, and Machine Learning*, Lecture Notes on Data Engineering and Communications Technologies 95, https://doi.org/10.1007/978-981-16-6636-0_35

of the disastrous storm surge known as 'Bhola'. As a result, cyclone prediction methods are becoming increasingly important.

To provide reliable results, it is important to understand the cloud patterns [1]. Forecasters have utilized satellite data to provide early warnings for cyclone formation; many of the outcomes were determined by human judgment, which is an ineffective procedure in today's world. Researchers have used machine learning and deep learning methods to overcome some of the constraints in weather forecasting. Nonetheless, additional study is required to produce a more dependable system.

Weather forecasting issues with machine learning algorithms have been reported in studies [2, 3]. Because of data availability, deep learning methods have been brought into the remote sensing field. It has achieved incredible performance in object recognition and classification with the help of graphical processing units (GPUs). Such models need processing capacity, in which powerful GPUs deliver. Open-source deep learning software frameworks like TensorFlow and Keras have made the work simple. Research works on satellite image classification [4, 5], and deep learning algorithms in disaster detection [6] have given us the motivation to work in this field. To detect the cyclone and cyclone's eye in satellite images, we propose a deep learning-based framework with multiple models. There are three major components to our proposed methodology: (1) image preprocessing, (2) cyclone detection and (3) cyclone eye localization.

We have worked with several deep learning models and achieved accurate cyclone identification. It is also necessary to locate the center of a cyclone. To address this issue, we have implemented an object-detecting system. This stage anticipates the central rotating region with a bounding box. A version of the You Only Look Once (YOLO) model, which has been trained in Google Colab with an Nvidia Tesla GPU returned the object localization result. Experiments have been performed on a series of satellite images obtained from the web and Kaggle competition. When compared to multiple models, our deep learning-based solution produced a higher recognition rate.

2 Related Works

Several methodologies for dealing with satellite images have lately been introduced. Traditional methodologies were introduced by some researchers, while deep learning and machine learning techniques were presented more recently. We will go through some of the prior research that has been done in this field.

Matsuoka et al. detected tropical cyclones and their precursors using ensemble CNN, having data of 50,000 for cyclones and 500,000 for non-cyclones, and they attained a maximum accuracy of 91.26% [7]. Kim et al. used three distinct machine learning models (SVM, DT and RF) to predict cyclone formation, with a hit rate ranging from 94 to 96% [8]. Deep Learning models (CNN, MobileNet, NasNet, Xception) were used to classify cyclone images and a linear regression model to forecast the future direction [9] with a minimum of 90% accuracy for classification and 84% for path prediction. Panangadan et al. developed a Kalman filter-based tracker that presented the results from five satellite sources, as well as an automated graph-based approach to locate cyclone's eye from satellite infrared images [10]. Wang et al. presented a two-step deep learning object detection approach 'OSIP' and 'IP' for detecting the precise center of a cyclone [11].

Many forecasters were aided by Dvorak's invention in observing cloud characteristics changes in geostationary satellite images and estimating cyclone intensity [12, 13]. However, determining intensity in distinct cloud basins has proven to be challenging. Chen et al. proposed a modified CNN model for cyclone intensity estimation from satellite infrared and microwave data with an RMSE of 8.39 kt [14]. For cyclone track predictions, Kovordányi et al. presented a multilayered artificial neural network and obtained a 95% accuracy. However, they did not use a wide range of datasets [15]. Kussul et al. proposed an ensemble CNN architecture for the classification of land cover and crop type from Landsat-8 and Sentinel-1 satellite data and acquired 85% accuracy [16]. Deep neural network architecture such as VGG13 was applied to detect floods from the satellite images [17].

Additionally, our suggested approach can outperform many previous efforts and overcome their drawbacks. In the proposed method of Matsuoka et al., their classifiers can produce model-specific biases [7], where we have applied cross-validation techniques to prevent our models from bias and overfitting. Our method can accurately locate multiple cyclone eyes in a single image, where we have found the proposed deep learning algorithm of Shakya et al. unable to detect when images contain more than one cyclone. And their performance of classification can affect the cyclone eye localization [9]. We have performed the cyclone detection and eye localization task separately so that classification performance cannot influence the localization performance. We also do not have to rely on expert human intuition to accomplish our tasks. We have also used some hybrid architecture of the machine learning and deep learning models, although some researchers have just used traditional machine learning models, and the results of our models are notable. Kovordányi et al. used relatively very low-resolution input images and found that distinguishing cyclone from the neighboring cumulonimbus clouds is difficult [15]; in our case, we have used much better resolution input images and also applied different preprocessing techniques on our images before training the models which can overcome the problem of distinguishing cyclones from the neighbor clouds. We have used several models, the classification accuracy is higher than most of the earlier works and for eye localization, and we have used an advanced version of You Only Look Once (YOLO) that generated faster and accurate results.

3 Methodology

3.1 Dataset Description

The satellite images of the cyclone have been collected from Kaggle [18]. For the non-cyclone satellite images, some of them we have collected from the web [19], which contains cloud masks from the 2018 Level-1C Sentinel-2 archive. In addition, these images contained 11 categories of surface (dense tree cover, any snow or ice covering either land or water, farmlands or pastures, coastlines, somewhat sparsely vegetated areas, extended bodies of water, etc.) and seven categories of cloud (cumulus, cirrus, haze/fog, ice clouds, etc.). We have also gathered some images from Kaggle [20], which included Sentinel-2 satellite images of water bodies. We have created a new dataset using these images.

There are two parts to our dataset: training and testing for both detection and localization. We double-checked that no images were duplicated. For the detection purpose, the dataset contains 1460 images overall, with 837 images having cyclones and 623 images without a cyclone. The train set has 1151 images, accounting for about 80% of the whole dataset, whereas the test set has 309 images, accounting for slightly over 20% of the total data. Cyclone images have been categorized as positive, whereas non-cyclone images as negative. For localization, we have used the same 837 cyclone images. The train set contains 649 images, and the test set contains 188 images.

3.2 Data Preprocessing

Preprocessing techniques have been used to clean image data for the model's input. The images we gathered required to be preprocessed because there was the possibility of smaller cloud formations, shear and center density overcast, as well as noise [21]. It has also decreased models training time and increased execution speed. Here are some preprocessing techniques we have applied to our images as follows:

Images have been converted to grayscale. Working with grayscale images, our model will only need to keep track of one matrix per image, which reduces complexity and speeds up performance. Binary thresholding has been used to distinguish an object from the background or, in other words, it allows us to work with the areas of an image we are interested in. By applying thresholding, the cyclone's eye and the surrounding clouds became more visible for training our models. We have used a size 2 erosion, which reduced the size of objects and eliminated minor irregularities, and the output pixel's value is the least of all the pixels in the neighborhood. Our dataset includes images of different sizes. Therefore, all the images have been resized to 300×300 pixels. The INTER_AREA interpolation method has also been used for enhancing the quality of an image. Figure 1 shows an example of an original and preprocessed image for cyclone detection.

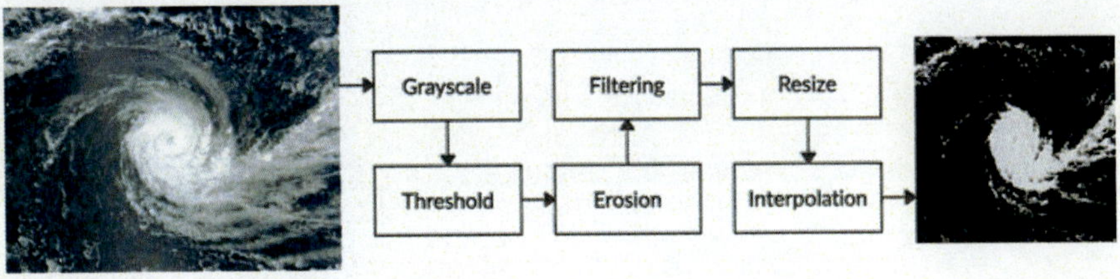

Fig. 1 Preprocessing steps and the outcome

For cyclone's eye localization, we have manually annotated the cyclone images. Using the 'LabelImg' tool, we took the images and annotated the bounding box.

3.3 Models for Detecting Cyclone

3.3.1 Convolutional Neural Network (CNN)

CNN can be thought of as artificial neural networks that can recognize patterns. In our model, we have stacked the layers and formed an architecture as shown in Fig. 2. The first layer consists of a 'convolutional' layer with 'rectified linear unit' (ReLU); we have used 16 convolutional filters of size (3 × 3), a stride of size 2. Then a 'max pooling' layer with a pool and a stride of size 2 was added. After that, a second convolutional layer with 32 filters was implemented, followed by a max pooling layer. Similarly, we have added two more convolutional and pooling layers. After these layers, the output was flattened to produce a one-dimensional vector for fully connected layers [6]. We have used two fully connected layers with the activation function ReLU. And finally, a 'Sigmoid' function was used in the last layer for final outputs. We compiled the model with 'Adam' optimizer and 'binary cross-entropy' loss.

We have also used the standard CNN architecture of VGG16 to classify images under the category of cyclone or non-cyclone. VGG16 is a convolutional neural network that is easy to understand and implement. It has 16 layers, including convolution and pooling layers that are placed on top of each other.

3.3.2 Support Vector Machines (SVM)

SVMs are primarily utilized in classification and regression tasks. It works by drawing outlines between two classes. In this proposed architecture, we have used convolutional layers to extract the features, and SVM is used for the final classification. The convolutional layer contains 16, 32, 64 and 128 filters with the size of 3 × 3 and a stride of 2, respectively. Then fully connected layers have been added with 256 and 128 units, respectively. Finally, the last output layer was an SVM classifier. In the

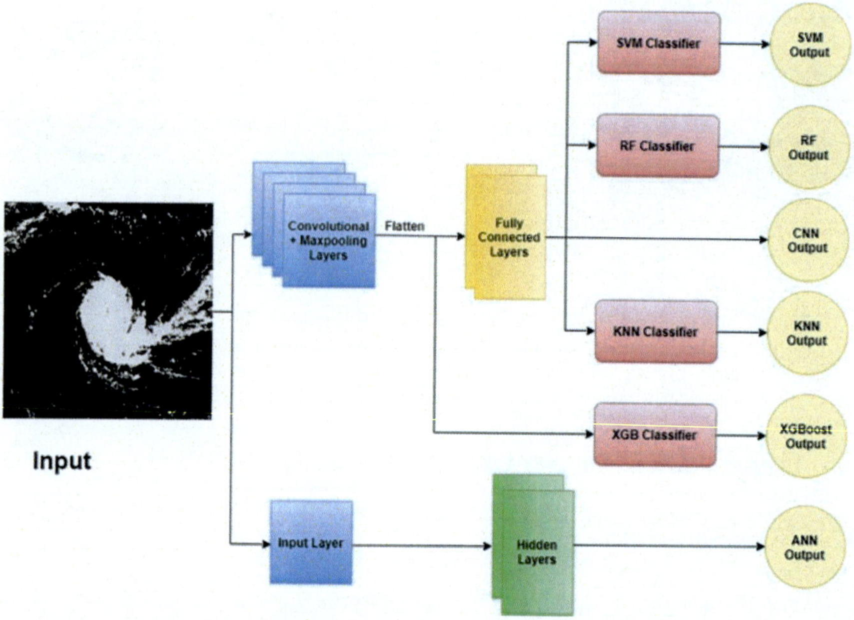

Fig. 2 Proposed architecture of different models

feature extraction layers, ReLU has been used as an activation function, and for the SVM classifier, a linear activation function with '12' regularization has been used. We have compiled the model with the 'Adam' optimizer and loss function 'Hinge'.

3.3.3 *K*-Nearest Neighbor (KNN)

KNN is based on the idea that every data point that is close to another belongs to the same class. It selects a number *k* as the closest neighbor to the data point to be identified. We have used convolutional layers for extracting features and used them as input to the KNN classifier to predict the class. This model has been recently used for image classification purposes [22]. For the implementation of the KNN classification, we utilized the Python scikit-learn toolkit.

3.3.4 Random Forest (RF)

A random forest is a collection of decision trees. This ensemble technique works well for classification and regression purposes. It uses majority voting for final classification. Because of the local minimum, vanishing gradient and overfitting difficulties, classical CNNs can face a tough time achieving the optimum generalization [23]. We have used convolutional layers as feature extractors and used them as input to random forest. By doing so, our model is more generalized and reduces the possibility of overfitting. We have used 50 trees in the random forest.

3.3.5 Gradient Boosting (XGBoost)

The boosting approach is based on the ensemble method's idea. We have used an extreme gradient boosting (XGBoost) classifier, a tree-based algorithm for classification with a custom loss function. We have built a feature extractor and used the features as the input to the XGBclassifier.

3.3.6 Artificial Neural Network (ANN)

ANN has been recently used in many satellite image classification problems [15, 24]. We have built an ANN model having two hidden layers with 128 and 64 units, respectively, and rectified linear Unit (ReLU) as an activation function. The last layer is an output layer with an activation function, 'Sigmoid'. The classifier has been compiled with the 'Adam' optimizer and the 'binary cross-entropy' loss function.

3.4 Model for Cyclone Eye Localization

3.4.1 You only Look once (YOLO)

It is an object detection algorithm that produces extremely accurate findings in a short time. We have used the YOLOv4 version for detecting a cyclone's eye or the center rotating area. Fast region-based CNN (R-CNN) can mistake the background patches as an object [25], wherein YOLOv4, the background error, can be very much reduced [26].

YOLOv4 backbone architecture is composed of three parts: CSPDarknet-53, Bag of freebies and Bag of special [27]. We have scaled the original image to (300 $\times$ 300) pixels and used it as input to our network. CSPDarknet-53 network is constructed, which is a series of convolutional and residual blocks. 'SPP' block was added over the CSPDarknet, 'PAN' used as the neck and a YOLOv3 head completes the architecture of YOLOv4. For training and validating our YOLOv4 model, we employed the Google Colab platform, which provided us with 12 GB Nvidia Tesla K18 GPU RAM. However, we have annotated our dataset in the local machine. The architecture of our YOLOv4 model is shown in Fig. 3.

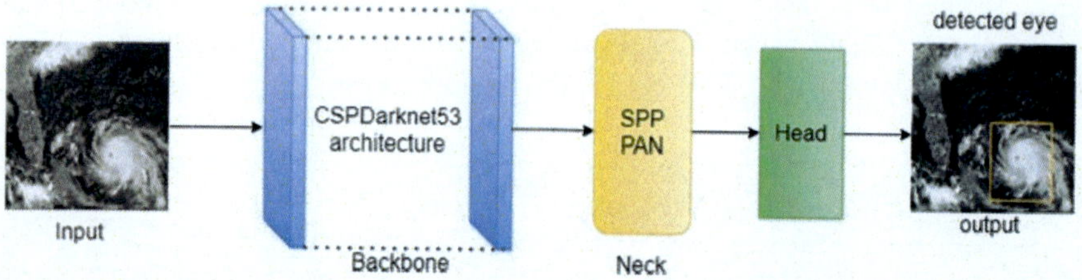

Fig. 3 YOLOv4 architecture for cyclone eye localization

4 Experimental Results

We have developed several deep learning models to determine whether a satellite image contains a cyclone or not and an object detection model to locate the cyclone's center rotating area. The detection results of the various models will be shown first, followed by the detection result of the cyclone's eye.

4.1 Model Evaluation and Validation for Cyclone Detection

We have trained our cyclone detection models for several epochs using seven distinct classifiers with 1151 images. After a few epochs of running, the decreasing behavior of training loss and enhanced accuracy indicated that models have been well trained on the dataset. The models are evaluated with 309 images, and we have displayed the results in Table 1.

From Table 1, we can see that we have reached over 95% accuracy for CNN, 94% for SVM, 93% for RF, 93% for VGG16, 91% for XGBoost, 88% for ANN and 86% with our KNN model. Using the convolutional layer as a feature extractor and classifying the class with different classifiers helped the models to improve the accuracy; the effectiveness of preprocessing can also be noticed.

Table 1 Performance summary for various classifiers

Model	F_1-Score	Accuracy (%)
CNN	0.95	95.15
SVM	0.94	94.17
KNN	0.86	86.73
RF	0.93	93.20
XGBoost	0.91	91.26
ANN	0.88	88.67
VGG16	0.93	93.20

We have calculated the accuracy by dividing the total number of estimates by the number of right estimates. We have also calculated precision (P), recall (R) and F_1-measure as evaluation metrics.

We have verified our model's performance by applying fivefold 'stratified K-fold' and 'K-fold' cross-validation techniques to prevent overfitting. We have achieved 98.42% and 98.23% accuracy with K-fold and stratified K-fold, respectively, for the CNN model. For SVM, we have reached 99.66%, 98.84% accuracy, respectively. We have also obtained improved performance for other models. Table 2 summarizes the cross-validation effects.

Following an analysis of the overall results, we have decided that the CNN model could be utilized as a benchmark model for cyclone prediction, using satellite images. In Fig. 4, a bar chart depicts the comparison among the mentioned models based on metrics evaluation.

Table 2 Accuracy based on cross-validation

Model	K-fold (%)	Stratified K-fold (%)
CNN	98.29	98.42
SVM	98.84	99.66
ANN	96.51	99.52
VGG16	93.15	94.73

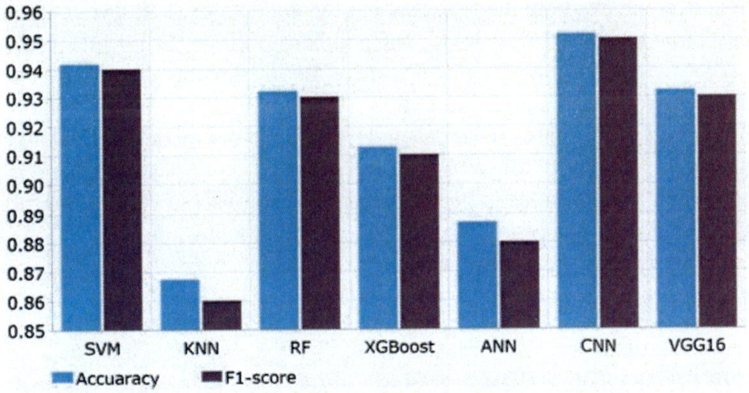

Fig. 4 Performance comparison graph for all the classifiers

Table 3 Performance evaluation metrics for YOLOv4

Model name	Recall	F_1-Score	Average IoU (%)	mAP (mean average precision) (%)
YOLOv4	0.92	0.89	65.37	87.86

4.2 *Model Evaluation and Validation for Cyclone Eye Detection*

For predicting the cyclone's center rotating area, 837 images have been manually annotated for training and testing the YOLOv4 model. The model has been trained on Google Colab for 6000 iterations. After several iterations, the training loss started to decrease. We reached a training loss of average 0.5723 with a learning rate of 0.00001.

The performance of our model has been evaluated on 188 test images. The anticipated bounding box of an image is compared to the labeled bounding box of the ground truth image. To test the model's performance, we have used precision, recall, 'mAP' and 'IoU' as evaluation metrics. Table 3 shows the validation results of cyclone eye localization.

The ratio of the intersection and union of the anticipated boundary and the ground truth (real cyclone eye boundary) is denoted by 'IoU'. A cyclone eye prediction is considered true positive (TP) if IoU ≥ 0.5 and false positive (FP) if IoU < 0.5. The mAP score indicates a model's performance; the greater the score, the better the prediction is. We have reached an mAP of 87.86% for an average IoU of 65.37%. From Table 3, we can also see the recall and F1-score of our model are 0.92 and 0.89, respectively.

After evaluating the model's performance, we can observe that it is accurate in detecting a cyclone's eye. Even if a satellite image contains multiple cyclone eyes, the model can identify them effectively.

5 Conclusion and Future Works

Considering cyclones are one of the most destructive natural calamities, an automated cyclone prediction is constantly in high demand. We have offered a solution to this major challenge using satellite images and deep learning principles. To efficiently detect a cyclone and its core cloud pattern, we have developed a two-step methodology. Although finding a benchmark dataset was a barrier to our accomplished task, our applied technique generated significant outcomes. However, we would want to broaden our future work by using additional data (wind speed, surface temperature, air pressure, etc.) for forecasting and tracking cyclones and predicting a more accurate center position. Overall, the research reveals that the suggested deep learning models could be employed for cyclone prediction with more prominent accuracy and speed than the existing works.

References

1. Bai T, Li D, Sun K, Chen Y, Li W (2016) Cloud detection for high-resolution satellite imagery using machine learning and multi-feature fusion. Remote Sens 8(9):715
2. Shi X, Chen Z, Wang H, Yeung D, Wong W, Woo W (2015) Convolutional LSTM network: a machine learning approach for precipitation nowcasting. In: NIPS
3. Grover A, Kapoor A, Horvitz E (2015) A deep hybrid model for weather forecasting. In: Proceedings of the 21th ACM SIGKDD international conference on knowledge discovery and data mining, pp 379–386
4. Pritt M, Chern G (2017) Satellite image classification with deep learning. In: IEEE Applied imagery pattern recognition workshop (AIPR), pp 1–7
5. Asokan A, Anitha J (2019) Machine learning based image processing techniques for satellite image analysis—a survey. In: 2019 International conference on machine learning, big data, cloud and parallel computing (COMITCon), pp 119–124
6. Amit SNKB, Shiraishi S, Inoshita T, Aoki Y (2016) Analysis of satellite images for disaster detection. In: 2016 IEEE International geoscience and remote sensing symposium (IGARSS)
7. Matsuoka D, Nakano M, Sugiyama D, Uchida S (2018) Deep learning approach for detecting tropical cyclones and their precursors in the simulation by a cloud-resolving global nonhydrostatic atmospheric model. Prog Earth Planet Sci 5:80
8. Kim M, Park M, Im J, Park S, Lee M (2019) Machine learning approaches for detecting tropical cyclone formation using satellite data. Remote Sens 11(10):1195
9. Shakya S, Kumar S, Goswami M (2020) Deep learning algorithm for satellite imaging based cyclone detection. IEEE J Sel Top Appl Earth Obs Remote Sens 13:827–839
10. Panangadan A, Ho S, Talukder A (2009) Cyclone tracking using multiple satellite image sources. In: Proceedings of the 17th ACM SIGSPATIAL international conference on advances in geographic information systems (GIS'09), Association for Computing Machinery, New York, NY, USA, pp 428–431
11. Wang P, Wang P, Wang C, Yuan Y, Wang D (2020) A center location algorithm for tropical cyclone in satellite infrared images. IEEE J Sel Top Appl Earth Obs Remote Sens 13:2161–2172
12. Velden C, Harper B, Wells F, Beven JL II, Zehr R, Olander T, Mayfield M, Guard C, Lander M, Edson R, Avila L, Burton A, Turk M, Kikuchi A, Christian A, Caroff P, McCrone P (2006) The Dvorak tropical cyclone intensity estimation technique: a satellite-based method that has endured for over 30 years. Bull Am Meteorol Soc 87(9):1195–1210
13. Dvorak VF (1975) Tropical cyclone intensity analysis and forecasting from satellite imagery. Mon Weather Rev 103(5):420–430
14. Chen B, Chen B, Lin H, Elsberry RL (2019) Estimating tropical cyclone intensity by satellite imagery utilizing convolutional neural networks. Weather Forecast 34(2):447–465
15. Kovordányi R, Roy C (2009) Cyclone track forecasting based on satellite images using artificial neural networks. ISPRS J Photogram Remote Sens 64(6):513–521. ISSN 0924-2716
16. Kussul N, Lavreniuk M, Skakun S, Shelestov A (2017) Deep learning classification of land cover and crop types using remote sensing data. IEEE Geosci Remote Sens Lett 14(5):778–782
17. Bischke B, Bhardwaj P, Gautam A, Helber P, Borth D, Dengel A (2017) Detection of flooding events in social multimedia and satellite imagery using deep neural networks. In: MediaEval
18. Disaster images dataset. https://www.kaggle.com/mikolajbabula/disaster-images-dataset-cnn-model. Accessed 6 Feb 2021
19. Francis A, Mrziglod J, Sidiropoulos P, Muller JP (2020) Sentinel-2 cloud mask catalogue [Data set]. In: Zenodo. http://doi.org/10.5281/zenodo.4172871

20. Water bodies images dataset. https://www.kaggle.com/franciscoescobar/satell-ite-images-of-water-bodies. Accessed 6 Feb 2021
21. Dutta I, Banerjee S, De M (2013) An algorithm for pre-processing of satellite images of cyclone clouds. Int J Comput Appl 78(15):13–17
22. Srinivas B, Rao GS (2019) A hybrid CNN-KNN model for MRI brain tumor classification. Int J Adv Sci Technol (IJAST) 127:20–25
23. Malof JM, Collins LM, Bradbury K, Newell RG (2016) A deep convolutional neural network and a random forest classifier for solar photovoltaic array detection in aerial imagery. In: 2016 IEEE International conference on renewable energy research and applications (ICRERA), pp 650–654
24. Mahmon NA, Ya'acob N (2014) A review on classification of satellite image using artificial neural network (ANN). In: 2014 IEEE 5th Control and system graduate research colloquium, pp 153–157
25. Ren S, He K, Girshick RB, Sun J (2015) Faster R-CNN: towards real-time object detection with region proposal networks. IEEE Trans Pattern Anal Mach Intell 39:1137–1149
26. Bochkovskiy A, Wang C, Liao H (2020) YOLOv4: optimal speed and accuracy of object detection. arXiv:2004.10934
27. Redmon J, Divvala S, Girshick RB, Farhadi A (2016) You only look once: unified, real-time object detection. In: 2016 IEEE Conference on computer vision and pattern recognition (CVPR), pp 779–788

Detection of Autism Spectrum Disorder by Discriminant Analysis Algorithm

Mirza Muntasir Nishat, Fahim Faisal, Tasnimul Hasan, Sarker Md. Nasrullah, Afsana Hossain Bristy, Md. Minhajul Islam Shawon, and Md. Ashraful Hoque

Abstract This paper represents an important perspective to analyze machine learning (ML) algorithms, particularly linear and quadratic discriminant analysis algorithms, in order to predict Autism Spectral Disorder (ASD) disease in a competent yet convenient way. ASD, mostly caused on account of genetic susceptibilities, is one kind of psychiatric disorder, affects significantly in social interactions, communications and behaviors. If it is detected early, a customized treatment plan will pave the way for a positive outcome by trimming the intensity of the illness. In this context, discriminant analysis algorithm has emerged as an impeccable component for its remarkable characteristics in solving classification problems. To this intent, the University of California, Irvine (UCI) data reservoir has been employed to train and test in order to construct the ML models. An extensive assessment has been executed in terms of accuracy, precision, sensitivity, Youden Index, F1 score and AUC with which it can be concluded that Quadratic Discriminant Analysis (QDA)

M. M. Nishat · F. Faisal (✉) · T. Hasan · A. H. Bristy · Md. Minhajul Islam Shawon · Md. Ashraful Hoque
Department of EEE, Islamic University of Technology, Dhaka, Bangladesh
e-mail: faisaleee@iut-dhaka.edu

M. M. Nishat
e-mail: mirzamuntasir@iut-dhaka.edu

T. Hasan
e-mail: tasnimulhasan56@iut-dhaka.edu

A. H. Bristy
e-mail: afsanahossain2@iut-dhaka.edu

Md. Minhajul Islam Shawon
e-mail: minhajulislam84@iut-dhaka.edu

Md. Ashraful Hoque
e-mail: mahoque@iut-dhaka.edu

S. Md. Nasrullah
Department of Public Health, North South University, Dhaka, Bangladesh
e-mail: sarker.nasrullah@northsouth.edu

M. S. Arefin et al. (eds.), *Proceedings of the International Conference on Big Data, IoT, and Machine Learning*, Lecture Notes on Data Engineering and Communications Technologies 95, https://doi.org/10.1007/978-981-16-6636-0_36

algorithm was the most proficient one with accuracy of 99.77% after tuning the hyperparameters.

Keywords Discriminant analysis algorithms · Accuracy · Machine learning · Autism Spectrum Disorder (ASD)

1 Introduction

The term Autism Spectrum Disorder (ASD) refers to a constellation of heterogeneous, neurodevelopmental conditions, commonly observed in the early years of life, and clinically diagnosed by the observation of atypical and underdeveloped social communication skills, with or without restrictive, repetitive patterns of activities or behavior in a child [1, 2]. Researches have revealed genetic predispositions to be responsible for this set of psychiatric disorders with more or less environmental influence [3]. Mutation of DNA, and deletion or duplication of genes involved with intellect and neuropsychiatric development of a person contribute to the multiple variants of ASD [4]. The onset of ASD may be variable and featured by an extended range of symptoms such as not meeting the eyes, keeping to himself or herself, repeating the same pattern of activity or behavior, not showing any interest in communicating with people, having unusual tone of voice, facial expression or gestures, etc. [5]. A number of other psychiatric and medical conditions are also reported frequently in patients with ASD. Common psychiatric comorbidities include attention deficit hyperactivity disorders (ADHDs), social anxiety disorders, depressive disorders, intellectual deficits, etc. [6–8]. Among the medical infirmities, gastrointestinal diseases, sleep disorders, immune dysfunctions, etc., are commonly associated with ASD [9–11].

The negative impact of autism disorders on the families can be both direct and indirect. Substantial effect in the sectors of health, housing, education and social care of the affected children can be a cause of distress. ASD also creates a financial burden that extends into their adulthood, mostly carried by the guardians [12, 13]. The annual costs of medical and non-medical care for the patients have been calculated to reach around $500 billion in the USA alone by 2025 [14]. A study has estimated the cost of care to be $3.2 million per capita in a lifetime on average [12]. These expenditures may vary depending on treatment, care and age [13]. Therefore, the rising number of autism cases in recent years is a matter of great concern. The average incidence of cases per 10,000 has risen from approximately 1.9 in 1980 to 14.8 in 2010 in Asia [15]. On the other hand, a review of studies from the regions of USA, UK, Scandinavia and Japan has estimated these numbers to be within the range of 30–60 cases per 10,000 [16]. More recent studies on 8-year old children in the United States revealed one case in every 68 children [17].

The treatment is limited, and the goal is to develop social communication skills, adjust behavior, and enhance learning and problem solving, and management of the comorbidities. Counseling the parents for a better understanding of the disorder is essential to improve the quality of care given and to make clear what to expect in future [18]. The recent advancements in medical treatment of ASD look promising, however, there is little evidence to support the benefit of most of the interventions [19]. On the other hand, researches have proved the worth of intervention during the early years of life in the intellectual development of the patients. Behavioral therapy and long-term, intensive intervention could improve the cognitive functions, adaptive capabilities and language skills of the children, making less dependent on pharmacologic treatments. That is why, early screening and diagnosis for better outcomes in treating ASD are of great importance [20].

Since ages, different algorithms have stirred the curiosity of the scientists in case of cutting-edge sectors [21, 22]. Several researchers have unfastened new windows in many sectors by accumulating almost accurate estimates with ML [23–29]. The use of machinery to acquire knowledge of healthcare techniques is critical and grows fast. In the past few years, the way device acquired expertise can be used in various industries and analysis has been widely used to be common trends. Healthcare organizations continue to take advantage of system-study methods to obtain useful statistics which can then be subsequently used to diagnose health conditions in advance as quickly as possible.

In our work, ML algorithms, particularly discriminant analysis algorithms have been instigated to modulate the conditions more precisely and timely, to have an idea of ASD for the minimal repercussions of the impact in human of different. The data containing different stages categorized for the study with overall workflow and correlation heatmap are presented in Sect. 2, while Sect. 3 consists of the study of LDA and QDA. Extensive analyses of the acquired results from the simulation with discriminant algorithms are stated in Sect. 4 where the performances of the algorithms are estimated. Finally, the overall exploration is outlined in Sect. 5 with worthwhile visions and insinuations.

2 Methodology

This research used the dataset available in the UCI Machine Learning Repository, which is drawn from the University of California Irvine Machine Learning Repository [30–32]. The first step is to preprocess the datasets and sequentially merge them. The preprocessed dataset contains 1100 instances merged from three datasets, with 104 adolescents, 704 adults, and the remainder being child instances. One hot encoding was implied to age description, which is a categorical feature because most machine learning algorithms cannot deal with categorical information directly and must be converted to a numerical format for easy access. Some instances with missing values are filled using the median method, where label encoding was used for categorical information, and the null values were filled using KNN imputer. After processing of the data, they were split into training and testing set, followed by applying cross validation and selection of hyperparameters. Hence, discriminant analysis algorithms were applied, and models were constructed. Finally, different performance parameters were observed, and best performing model was proposed.

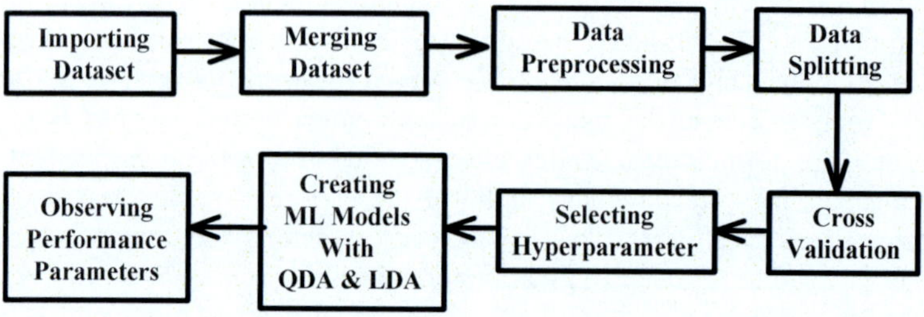

Fig. 1 Overall workflow diagram

The workflow diagram of the overall method is portrayed in Fig. 1. However, the correlation heatmap is displayed in Fig. 2.

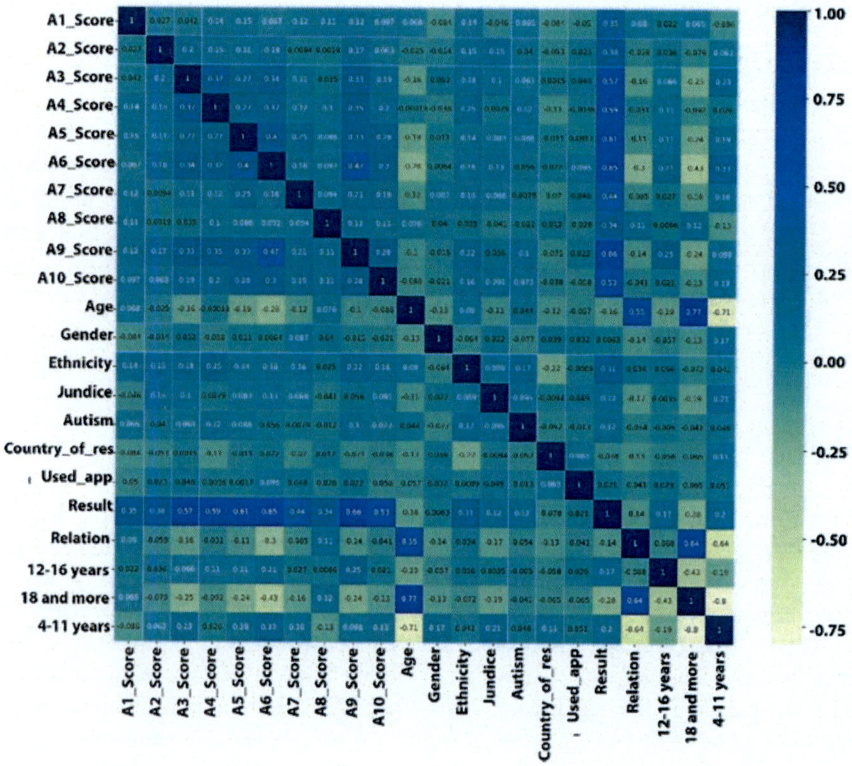

Fig. 2 Correlation heatmap

3 Study of Discriminant Analysis Algorithms

Discriminant analysis algorithms, particularly linear and quadratic are the two classic classifiers with a linear and nonlinear supervisory aspect, respectively. These algorithms are valuable for they are intrinsically multiclass, and are shown to perform well in practice with conveniently calculated solutions [33].

3.1 Linear Discriminant Analysis (LDA)

In order to solve the linear dimensionality problems in case of pattern recognition and ML, LDA is customarily brought into play. Conventionally, the LDA algorithm projects the original data matrix into a lower space if two classes with multiple features are to be separated efficiently. If only a single feature is used, it may result in unwanted overlapping. So, in order to have a correct classification, increment of the number of features is a must. LDA is a straightforward method where the generative approach for classification is applied and trained. It tends to measure and maximize the associativity of the various classes which can also be named as the interclass variance or the interclass matrix [34]. The LDA discriminant function can be written as

$$\partial_k(x) = 2\mu_k^T \sum_k^{-1} X - \mu_k^T \sum_k^{-1} \mu_k - 2\log(\pi_k) \tag{1}$$

where,

X = the set of measurements.

μ_k = mean vector.

π_k = prior probability.

$\sum_k$ = covariance matrix and.

k = class.

The goal of LDA is to evaluate the results when the criterion or dependent variable is categorical with an interval in existence of the indicator or the independent variable. This means that it is possible to create discriminating functions which are nothing but the linear combination of separate variables which distinguish perfectly between the categories of the dependent variable. The consistency of the classification is also assessed [35].

3.2 Quadratic Discriminant Analysis (QDA)

LDA can learn only linear limits, while QDA can learn quadratic limits, and is hence more mobile. QDA algorithm is a simplified LDA version provided that the measurements usually are distributed by only two groups of points. In QDA, however, it does not take into account the postulation that the covariance of each of the class is equal like LDA. Furthermore, a conic area would be the surface separating the subspaces (like a hyperbola, or parabola) [36]. The QDA discriminant function can be stated as

$$\partial_k(x) = -\frac{1}{2}(x-\mu_k)^T \sum\nolimits_k^{-1} (x-\mu_k) + \log(\pi_k) \tag{2}$$

The hypothesis that each class can be modeled by a Gaussian distribution, and the similar covariance matrix would be allotted is the basic idea for LDA. But, it is nearly incapable when the mean of the distributions are mutual, because to catch a new axis linearly separable in this case is hardly possible by LDA. Hence, nonlinear discriminant analysis algorithms are welcomed such as Quadratic Discriminant Analysis (QDA) where each class uses its individual estimate of variance or covariance.

4 Results

After going through the algorithms, the dataset was schemed to build the ML models with QDA and LDA. Here, all the simulations were performed using Python. Later on, confusion matrices of these algorithms are tabulated in Table 1 with before and after tuning the hyperparameters. Afterward, a detailed analysis is portrayed in terms of different performance parameters in Table 2 where accuracy, precision, sensitivity, Youden Index, F1 score and ROC-AUC were calculated and compared for both of the cases.

Table 1 Confusion matrix for discriminant analysis algorithms

Algorithms				Predicted	
				False	True
LDA	Before tuning	Actual	False	287	16
			True	0	137
	After tuning		False	301	2
			True	0	137
QDA	Before tuning	Actual	False	34	14
			True	7	172
	After tuning		False	302	1
			True	0	137

Table 2 Comparative analysis based on performance parameters for LDA and QDA

Performance parameters	LDA	LDA (tuned)	QDA	QDA (tuned)
Accuracy	0.9636	0.9955	0.7182	0.9977
Precision	0.8954	0.9856	0.5461	0.9928
F1_score	0.9448	0.9928	0.554	0.9964
AUC	0.9736	0.9967	0.6754	0.9983
Specificity	0.9472	0.9934	0.7888	0.9967
Youden index	0.9472	0.9934	0.3508	0.9967

In case of QDA (after tuning), improved performances are observed in all of the parameters. Although, LDA has presented more accuracy (96.36%) before tuning of the hyperparameters, QDA surpassed it after hyperparameter tuning with an accuracy of 99.77%. In case of precision, LDA (0.8954) triumphs QDA (0.5461) before tuning, while QDA outperforms LDA after tuning. For specificity, QDA (0.9967) performs better than LDA (0.9934) after tuning. Likewise, QDA also takes lead of LDA in case of other parameters (Youden Index, F1 score and sensitivity) after tuning although LDA showcased better results before tuning.

Hence, the graphical analyses have been illustrated to analyze the investigation in a quite comprehensive manner. Figure 3 depicts the comparison of Accuracy and Precision where it is evident that QDA outperforms LDA in both the cases with tuning. Next, comparison of F1 score and AUC_ROC shown in Fig. 4 can give the idea that after tuning, QDA (F1 score 0.9964 and AUC 0.9983) leads LDA (F1 score 0.9928 and AUC 0.9967) and in similar fashion, specificity and Youden index of Fig. 5 can show that tuned QDA gives the satisfactory results. This, QDA presents promising results in detecting ASD which will enable the healthcare professionals to prescribe early and timely treatments for the patients.

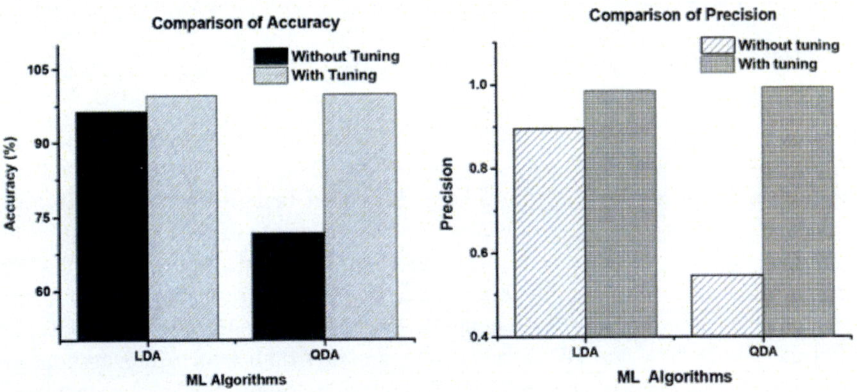

Fig. 3 Comparison of accuracy and precision

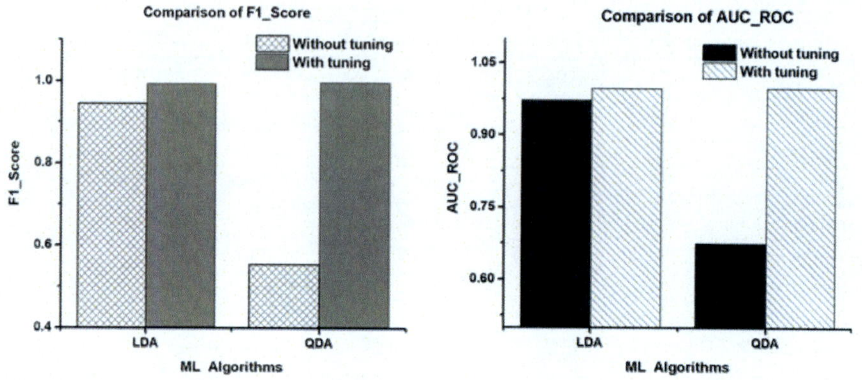

Fig. 4 Comparison of F1 score and AUC_ROC

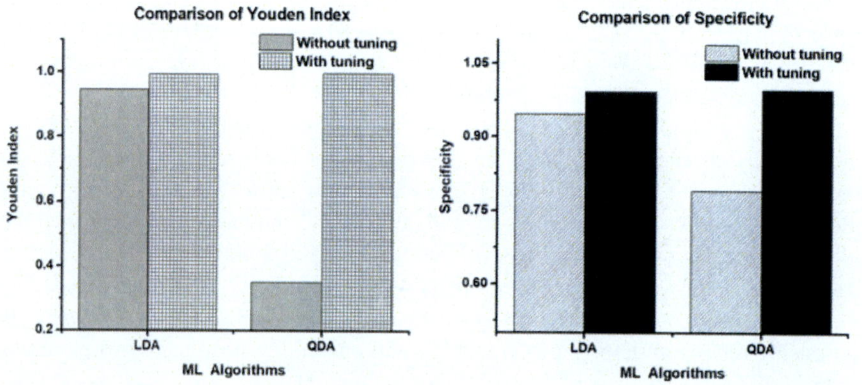

Fig. 5 Comparison of Youden index and specificity

5 Conclusion

DNA alteration of genes convoluted with intelligence and neuropsychiatric growth of a person can contribute to mild or severe ASD. This disorder can craft a noticeable impact among people at any time of life. That being the case, detection of it at the most embryonic stage is vital to limit the intensity of the disease and have fruitful medications. Subsequently, the intellectuals have started acting on employing Machine Learning (ML) algorithms due to its competent nature. In this study, discriminant analysis algorithms are investigated and well-organized analytical models are fabricated with LDA and QDA with hyperparameter tuning for better upshots. If compared, the accuracy of QDA is 71.82% which, after tuning, bolsters the maximum accuracy of 99.77% leaving behind LDA in terms of other performance metrics too. Necessarily, the intricacy of an e-healthcare system choreographed by ML can play a major role for the physicians in clinical judgment, guide the patients to potential diseases and aid the hospitals in mapping and treating them scrupulously.

References

1. Lord C, Elsabbagh M, Baird G, Veenstra-Vanderweele J (2018) Autism spectrum disorder. The Lancet 392(10146):508–520
2. American Psychiatric Association (2013) Diagnostic and statistical manual of mental disorders (DSM-5®). American Psychiatric Pub.
3. Newschaffer CJ, Croen LA, Daniels J, Giarelli E, Grether JK, Levy SE, Mandell DS, Miller LA, Pinto-Martin J, Reaven J, Reynolds AM (2007) The epidemiology of autism spectrum disorders. Annu Rev Public Health 28:235–258
4. Devlin B, Scherer SW (2012) Genetic architecture in autism spectrum disorder. Curr Opin Genet Dev 22(3):229–237
5. Ozonoff S, Heung K, Byrd R, Hansen R, Hertz-Picciotto I (2008) The onset of autism: patterns of symptom emergence in the first years of life. Autism Res 1(6):320–328
6. Matson JL, Shoemaker M (2009) Intellectual disability and its relationship to autism spectrum disorders. Res Dev Disabil 30(6):1107–1114
7. Mannion A, Leader G, Healy O (2013) An investigation of comorbid psychological disorders, sleep problems, gastrointestinal symptoms and epilepsy in children and adolescents with autism spectrum disorder. Res Autism Spectrum Disorders 7(1):35–42
8. Simonoff E, Pickles A, Charman T, Chandler S, Loucas T, Baird G (2008) Psychiatric disorders in children with autism spectrum disorders: prevalence, comorbidity, and associated factors in a population-derived sample. J Am Acad Child Adolesc Psychiatry 47(8):921–929
9. Zerbo O, Leong A, Barcellos L, Bernal P, Fireman B, Croen LA (2013) Immune mediated conditions in autism spectrum disorders. Brain Behav Immun 46:232–236
10. Mannion A, Leader G (2013) Comorbidity in autism spectrum disorder: a literature review. Res Autism Spectrum Disorders 7(12):1595–1616
11. Bauman ML (2010) Medical comorbidities in autism: challenges to diagnosis and treatment. Neurotherapeutics 7(3):320–327
12. Ganz ML (2007) The lifetime distribution of the incremental societal costs of autism. Arch Pediatr Adolesc Med 161(4):343–349
13. Buescher AV, Cidav Z, Knapp M, Mandell DS (2014) Costs of autism spectrum disorders in the U.K. and the U.S. JAMA Pediatrics 168(8):721–728
14. Leigh JP, Du J (2015) Brief report: forecasting the economic burden of Autism in 2015 and 2025 in US. J Autism Dev Disorders 45(12):4135–4139
15. Sun X, Allison C (2010) A review of the prevalence of Autism Spectrum Disorder in Asia. Res Autism Spectrum Disorders 4(2):156–167
16. Rutter M (2007) Incidence of autism spectrum disorders: changes over time and their meaning. Acta Paediatr 94(1):2–15
17. Baio J, Wiggins L, Christensen DL, Maenner MJ, Daniels J, Warren Z, Dowling NF (2018) Prevalence of autism spectrum disorder among children aged 8 years—autism and developmental disabilities monitoring network, 11 sites, United States, 2014. MMWR Surveillance Summaries 67(6):1–23
18. Mukherjee SB (2017) Autism spectrum disorders—diagnosis and management. Indian J Pediatrics 84(4):307–314
19. McPheeters ML, Warren Z, Sathe N, Bruzek JL, Krishnaswami S, Jerome RN, Veenstra-VanderWeele J (2011) A systematic review of medical treatments for children with autism spectrum disorders. Pediatrics 127(5):e1312–e1321
20. Sanchack KE, Thomas CA (2016) Autism spectrum disorder: primary care principles. Am Fam Physician 94(12):972–979

21. Faisal F, Nishat MM (2019) An investigation for enhancing registration performance with brain atlas by novel image inpainting technique using Dice and Jaccard score on multiple sclerosis (MS) tissue. Biomed Pharmacol J 12(3). https://dx.doi.org/10.13005/bpj/1754
22. Farazi MR, Faisal F, Zaman Z, Farhan S (2016) Inpainting multiple sclerosis lesions for improving registration performance with brain atlas. In: 1st international conference on medical engineering, health informatics and technology (MediTec). IEEE, pp 1–6. https://dx.doi.org/10.1109/MEDITEC.2016.7835363
23. Nishat MM, Faisal F, Mahbub MA, Mahbub MH, Islam S, Hoque MA (2021) Performance assessment of different machine learning algorithms in predicting diabetes mellitus. Biosc Biotech Res Comm 14(1):74–82
24. Nishat MM, Faisal F, Dip RR, Shikder MF, Ahsan R, Asif MAAR, Udoy MH (2020) Performance investigation of different boosting algorithms in predicting chronic kidney disease. In: 2nd international conference on sustainable technologies for Industry 4.0 (STI). IEEE, pp 1–5. https://dx.doi.org/10.1109/STI50764.2020.9350440
25. Faisal F, Nishat MM, Ashif Mahbub M, Minhajul Islam Shawon M, Mahbub-Ul-Huq Alvi M (2021) Covid-19 and its impact on school closures: a predictive analysis using machine learning algorithms. In: 2021 international conference on science and contemporary technologies (ICSCT), IEEE
26. Asif MAAR, Nishat MM, Faisal F, Dip RR, Udoy MH, Shikder MF, Ahsan R (2021) Performance evaluation and comparative analysis of different machine learning algorithms in predicting cardiovascular disease. Eng Lett 29(2):731–741
27. Asif MAAR, Nishat MM, Faisal F, Shikder MF, Udoy MH, Dip RR, Ahsan R (2020) Computer aided diagnosis of thyroid disease using machine learning algorithms. In: 11th international conference on electrical and computer engineering (ICECE). IEEE, pp 222–225. https://dx.doi.org/10.1109/ICECE51571.2020.9393054
28. Nishat MM, Dip RR, Faisal F, Nasrullah AM, Ahsan R, Shikder MF, Asif MAAR, Hoque MA (2021) A comprehensive analysis on detecting chronic kidney disease by employing machine learning algorithms. EAI Endorsed Trans Pervasive Health Technol 7(27):1–12. https://dx.doi.org/10.4108/eai.13-8-2021.170671
29. Nishat MM, Hasan T, Nasrullah SM, Faisal F, Asif MAAR, Hoque MA (2021) Detection of Parkinson's disease by Employing Boosting Algorithms. In: 2021 Joint 10th International Conference on Informatics, Electronics & Vision (ICIEV) & 5th International Conference on Imaging, Vision & Pattern Recognition (ICIVPR). IEEE, pp 1–7. https://dx.doi.org/10.1109/ICIEVicIVPR52578.2021.9564108
30. Autistic Spectrum Disorder Screening Data for Children Data Set, UCI Machine Learning Repository https://archive.ics.uci.edu/ml/datasets/Autistic+Spectrum+Disorder+Screening+Data+for+Children++
31. Autistic Spectrum Disorder Screening Data for Adolescent Data Set, UCI Machine Learning Repository https://archive.ics.uci.edu/ml/datasets/Autistic+Spectrum+Disorder+Screening+Data+for+Adolescent+++
32. Autism Screening Adult Data Set, UCI Machine Learning Repository https://archive.ics.uci.edu/ml/datasets/Autism+Screening+Adult
33. Hong H, Naghibi SA, Dashtpagerdi MM, Pourghasemi HR, Chen W (2017) A comparative assessment between linear and quadratic discriminant analyses (LDA-QDA) with frequency ratio and weights-of-evidence models for forest fire susceptibility mapping in China. Arab J Geosci 10(7):167
34. Tharwat A, Gaber T, Ibrahim A, Hassanien AE (2017) Linear discriminant analysis: a detailed tutorial. AI Commun 30(2):169–190
35. Xanthopoulos P, Pardalos PM, Trafalis TB (2013) Linear discriminant analysis. In: Robust data mining. Springer, New York, pp 27–33
36. Bhattacharyya S, Khasnobish A, Chatterjee S, Konar A, Tibarewala DN (2010) Performance analysis of LDA, QDA and KNN algorithms in left-right limb movement classification from EEG data. In: International conference on systems in medicine and biology. IEEE, pp 126–131

A BRBES to Support Diagnosis of COVID-19 Using Clinical and CT Scan Data

S. M. Shafkat Raihan, Raihan Ul Islam, Mohammad Shahadat Hossain, and Karl Andersson

Abstract In the prevailing COVID-19 pandemic, accurate diagnosis plays a vital role in preventing the mass transmission of the SARS-CoV-2 virus. Especially patients with pneumonia need correct diagnosis for proper treatment of their respiratory distress. However, the current standard diagnosis method, RT-PCR testing has a significant false negative and false positive rate. As alternatives, diagnosis methods based on artificial intelligence can be applied for faster and more accurate diagnosis. Currently, various machine learning and deep learning techniques are being researched on to develop better COVID-19 diagnosis system. However, these approaches do not consider the uncertainty in data. Deep learning approaches use backpropagation. It is an unexplainable black box approach and is prone to problems like catastrophic forgetting. This article applies a belief rule-based expert system (BRBES) for diagnosis of COVID-19 on hematological data and CT scan data of lung tissue infection of adult pneumonia patients. The system is optimized with nature-inspired optimization algorithm—BRBES-based adaptive differential evolution (BRBaDE). This model has been evaluated on a real-world dataset of COVID-19 patients published in a previous work. Also, performance of the BRBaDE has been compared with BRBES optimized with genetic algorithm and MATLAB's fmincon function where BRBaDE outperformed genetic algorithm and fmincon and showed best accuracy of 73.91%.

Keywords COVID-19 diagnosis · Hematological data · BRBES · BRBaDE

S. M. Shafkat Raihan (✉) · M. S. Hossain
Department of Computer Science and Engineering, University of Chittagong, Chittagong 4331, Bangladesh
e-mail: hossain_ms@cu.ac.bd

R. U. Islam · K. Andersson
Pervasive and Mobile Computing Laboratory, Luleå University of Technology, 93187 Skellefteå, Sweden
e-mail: raihan.ul.islam@ltu.se

K. Andersson
e-mail: karl.andersson@ltu.se

M. S. Arefin et al. (eds.), *Proceedings of the International Conference on Big Data, IoT, and Machine Learning*, Lecture Notes on Data Engineering and Communications Technologies 95, https://doi.org/10.1007/978-981-16-6636-0_37

1 Introduction

It has been more than a year since the COVID-19 pandemic began. COVID-19 infection can lead to pneumonia in patients. Certain characteristics in radiological data such as chest X-ray or CT scan might indicate the possibility that the pneumonia has been caused due to COVID-19 infection. To confirm this, RT-PCR test is conducted to diagnose COVID-19. However, RT-PCR method has been reported to have a variable false negative rate, and often it is significantly high as well [1]. Mutations and cross-reactivity can result in false negative and false positive results for RT-PCR testing [2]. Previous researches have shown that diagnosis of COVID-19 based on blood tests can be a potential faster alternative of RT-PCR [3]. Research is also being conducted on medical imaging technologies such as computed tomography (CT) and X-ray to discover a reliable alternative diagnostic system. However, X-ray-based diagnosis has been reported to have high false negative results [4]. On the other hand, CT scans have been reported to have better performance in diagnosis of COVID-19 over X-ray images [5]. Various research works have focused on the application of machine learning (ML) and deep learning (DL) algorithms on hematological data and chest CT scan data for COVID-19 diagnosis [6, 7]. However, DL and ML techniques have some limitations. DL relies on gradient-based backpropagation which induces the problem of catastrophic forgetting. This means during any learning trial, an unpredictable part of the learning memory can collapse [8]. Although the underlying inference mechanism of deep learning techniques ensure accuracy, it does not support explainability in the decision-making process [9]. Rudin [10] has noted that ML and DL models should be interpretable rather than explainable for high stake decision making. Deep learning techniques transform data manifolds assuming that learnable transfer functions exist that can facilitate the mapping [11]. However, [12] has pointed out that for data that have causal relationships, such transformations do not perform well, even for large datasets. Nguyen et al. [13] have shown that deep neural networks are prone to misclassify data with high confidence. Moreover, hematological data can have uncertainties of measurement due to the variation in reagents used in the blood analyzer machine [14]. CT scan images can have uncertainty due to misinterpretation of image and variation in image acquisition techniques [15, 16]. Besides, it might not be possible to run all the tests that assist in diagnosis. This can happen due to lack of reagents and equipment. Taking these facts into account, we have trained a belief rule-based expert system (BRBES) using the gradient-free optimization algorithm-BRB-based adaptive differential evolution (BRBaDE)—on a public clinical dataset containing hematological and CT scan parameters. BRBES can reason on data having nonlinear causal relationship [17]. It is a reasoning mechanism that can address the uncertainty in data such as ignorance, incompleteness, and imprecision of data. Also, BRBES can learn using gradient-free optimization approaches such as various evolutionary algorithms. BRBaDE [18] is one of the efficient evolutionary algorithms that provides an appropriate balance between exploration and exploitation of the search space.

The rest of the article is organized as follows: In Sect. 2 related works have been reviewed. Section 3 covers the methodology followed by Sect. 4, where the performance of the system is evaluated and discussed. Finally, Sect. 5 concludes the article and points out our future work.

2 Literature Review

In previous works, efforts were made to identify various hematological parameters related to COVID-19 [3, 19–21]. Machine learning has been applied on such hematological parameters in multiples works. Authors of [6] employed random forest (RF) and found eleven parameters that produce the best results. In [22], authors conducted a feasibility study with application of classical machine learning on routine blood tests for COVID-19 diagnosis. Similarly [23] explored the potential of machine learning on hematological dataset of Oxfordshire Research Database (IORD). An explainable ML approach using decision tree and criteria graph was used in [24] to diagnose COVID-19 from routine blood tests. Chest CT scan results can provide essential insight for COVID-19 diagnosis. Previous researches discovered various CT scan image features correlated to COVID-19 [25–29]. DL approaches were employed on chest CT scan image data of patients as well for identifying alternative diagnosis methods for COVID-19. In [30], an AI-based system was developed by training a convolutional neural network (CNN) on CT scan image data and training classical machine learning algorithms on clinical biochemical data. Authors in [7] used a voting-based approach to develop an efficientNet-based diagnosis system for chest CT image data. In [31], a DL-based computer-aided detection (CAD) system was developed to assist radiologists to diagnose COVID-19 at early stage from CT scan images. These methods rely on ML or DL algorithms whose limitations we discussed in previous section. Unlike ML and DL algorithms, BRBES can address nonlinear causal data, reason under uncertainity and can be optimized using gradient-free nature-inspired algorithms such as BRBaDE [18]. BRBES has been applied to design artificial intelligence-based diagnosis systems for various diseases and medical conditions [32–39]. In [40], BRBES optimized with a modified differential evolution (DE) algorithm has been developed to predict the severity of illness in COVID-19 patients. They used lactic dehydrogenase (LDH), lymphocytes, and high-sensitivity C-reactive protein (hs-CRP) as clinical parameters. However, they only considered hematological parameters in developing their diagnostic system. In our research, we consider four hematological parameters and one chest CT scan parameter to train our BRBES. They will be discussed in Sect. 3. BRBES can be trained using optimization algorithms such as genetic algorithm (GA), DE, and MATLAB's fmincon function. However, GA requires to encode and decode the solutions of the optimization problem to bit strings, and it does not support vectorized mutation [41]. Fmincon uses gradient-based algorithms which are prone to problems related to local optima. DE is a developed form of GA that has updating equation to allow efficient vectorized mutation [41]. Real numbers can be used to represent the solutions [41]. However, the crossover and mutation factors remain constant hindering balanced exploration and exploitation of the search space. This has been addressed in [18] where a new optimization algorithm, BRBES-based adaptive differential evolution (BRBaDE) has been used to optimize a BRBES for power usage effectiveness (PUE) prediction of data centers. In summary, BRBES can infer on hematological and chest CT scan data of COVID-19 patients that contain uncertainty and BRBaDE provides a gradient free optimization with optimal exploration and exploitation.

3 Methodology

3.1 BRBES

A belief rule-based expert system (BRBES) uses a belief rule base (BRB) as knowledge base and evidential reasoning (ER) as inference engine. Unlike assertive IF-THEN rule base, BRBES can express complex and nonlinear causal connections under uncertainty. BRB has two main components—antecedent and consequent. Each antecedent attribute has some referential values while a distribution of degrees of belief is assigned to the consequent attribute. An example of a belief rule is as follows:

IF C-Reactive Protein is Medium AND Prothrombine Time is High AND Fibrinogen is Medium AND CT percentage of lung infection is High AND Platelet Count in Low THEN COVID-19 Positivity is (Very High, 0.37), (High, 0.06), (Medium, 0.27), (Low, 0.3)

The rule is considered complete if sum of the degrees of belief (0.37 + 0.06 + 0.27+ 0.3 = 1) is one. Otherwise it is considered incomplete. This can happen because of ignorance or incompleteness in data. Since the example rule uses the conjunction operator (AND), it represents a conjunctive BRB. There are disjunctive BRBs as well that use the disjunctive operator (OR). The relationship between antecedents and consequents in a BRB can be represented with a BRB tree as shown in Fig. 1. Evidential reasoning (ER) can handle heterogeneous data and various types of uncertainties such as incompleteness, ignorance, imprecision, vagueness, and randomness. Evidential reasoning comprises of certain steps. They are input transformation, matching degree calculation, rule activation weight calculation, belief update, and rule aggregation which as shown in Fig. 2. Through input transformation, the input value of

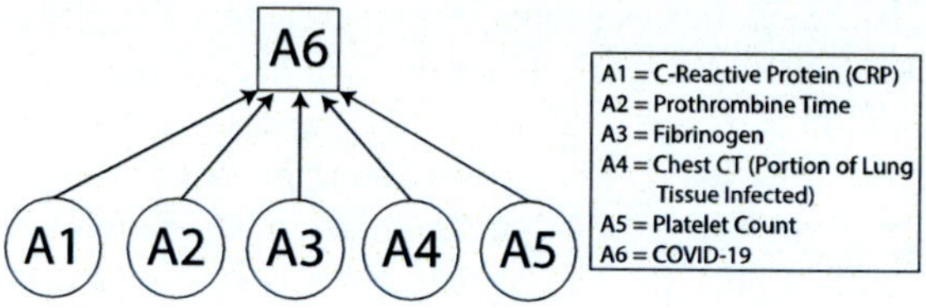

Fig. 1 A BRB tree representing the COVID-19 diagnosis framework using hematological data and CT scan lung infection data

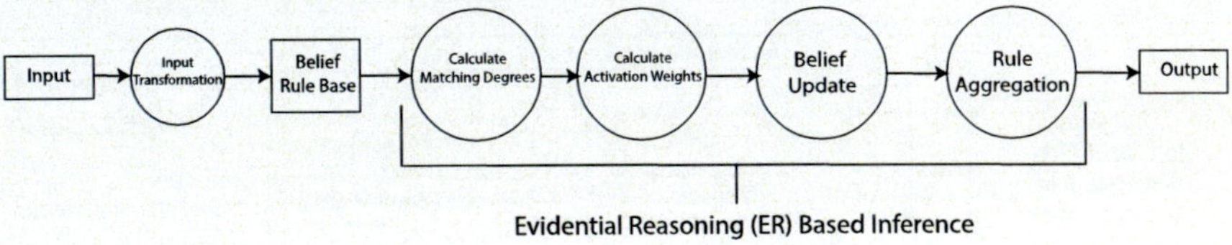

Fig. 2 Working process of a BRBES

an antecedent attribute is distributed over the referential values associated with that antecedent attribute. These transformed values are known as matching degrees. The rules are generated by using multiplication or addition operation on the individual matching degrees. The operation type depends on whether the BRB is conjunctive or disjunctive, respectively. Due to this operation, total matching degrees are obtained for each rule. Each of them represents a rule in the rule base. With matching degrees assigned, the rules then become active. Next, the activation weight of the rule is calculated using the matching degrees and relative rule weights. The activation weight of a rule will be zero if that rule is not activated. The sum of the rule activation weights of a rule base should be one. After these steps, an initial belief matrix is generated which lists degrees of belief corresponding to the consequent attribute of each rule. There may be absence of data for any antecedent attributes because of ignorance. In this case, the belief degrees associated with each rule in the rule base need to be updated to address this uncertainty due to ignorance. Lastly, all the rules are aggregated to calculate the output for the input data of the antecedent attributes using the evidential reasoning algorithm. The calculated output value against the input data will be in a fuzzy form, i.e., distributed over the referential values of the consequent. Then a crisp or numerical value is obtained from the fuzzy form. This is done using the utility score associated with each referential value of the consequent attribute. All these steps are executed by following the procedures mentioned in [32, 34].

Optimization is performed using BRBaDE. Differential evolution (DE) is a vector-based metaheuristic algorithm. It uses both exploration and exploitation in the forms of crossover (C_r) and mutation (F) operations, respectively. However, traditional differential evolution assumes the C_r and F factors to be constant over the iterations. BRBaDE takes into consideration the uncertainties in C_r and F factors as well. In BRBaDE, the C_r and F factors change in each iteration based on the change in solution vector and fitness values of the current population from the previous population. The change in solution population is denoted by PC, and change in fitness value is denoted by FC. From PC and FC, we can obtain the values d_{11}, d_{12}, d_{21} and d_{22}. PC, FC, d_{11}, d_{12}, d_{21}, and d_{22} can be calculated using Eq. 15–20 of [18]. Here, d_{11} and d_{12} are sent as antecedent attributes in one BRBES called BRBES_CR which outputs the crossover factor C_r. And d_{21} and d_{22} in another BRBES called BRBES_F, which outputs the mutation factor F. Both of these BRBES' are conjunctive and all attributes (both antecedent and consequent) have three referential values, i.e., low (L), medium (M), and high (H). The utility values assigned to antecedent and consequent

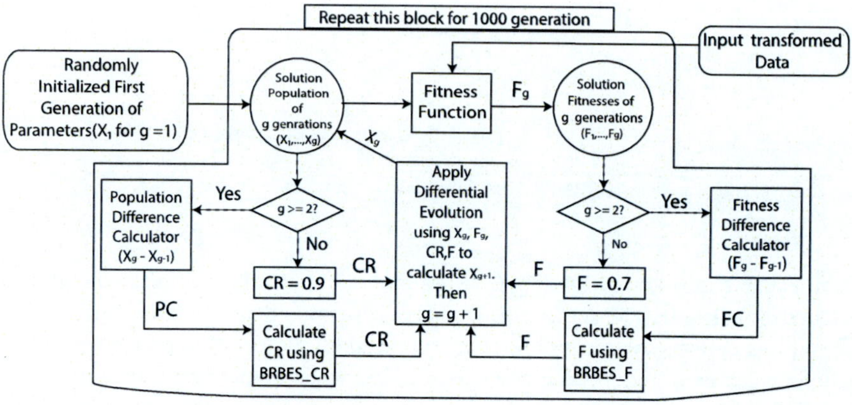

Fig. 3 Schematic diagram of BRBaDE

attributes of BRBES_CR are 0, 0.5, and 1. Utility values assigned to antecedent and consequent attributes of BRBES_F are 0, 1, and 2, since $F, d_{21}, d_{22} \in [0, 2]$.

Finally, using two separate BRBES' to calculate newer values of C_r and F provides balanced exploration and exploitation by addressing the uncertainty. In Fig. 3, the schematic diagram of BRBaDE is shown. The training is done for 1000 iterations with a tolerance of 10^{-6} with tenfold cross-validation. The objective function ξ is constructed using Eqs. 7 and 8 of [18]. The optimization of the parameters is done using the following:

$$\min_{P} \quad \xi(P)$$
$$P = P(\mu(O_j), \theta_k, \delta_k, \beta_{jk})$$

The parameters here are as follows:

1. $\mu(O_j)$ ($0 \leq \mu(O_j) \leq 1$. $j = 1, \ldots, N$ where N is the number of referential values of consequent attribute) represents the utility values of the consequent attribute's referential values. Constraint: $\mu(O_i) < \mu(O_j)$ if $i < j$.
2. θ_k ($0 \leq \theta_k \leq 1$. $k = 1, \ldots, L$ where L is the number of rules) represents the rule weights.
3. δ_k ($0 \leq \delta_k \leq 1$. $k = 1, \ldots, L$ where L is the number of rules) represents the antecedent attribute weights. It is applicable for only conjunctive BRBES.
4. β_{jk} ($0 \leq \beta_{jk} \leq 1$. $j = 1, \ldots, N$ and $k = 1, \ldots, L$ where N is the number of referential values of consequent attribute and L is the number of rules) represents the degree of belief of the jth consequent referential value of the kth rule. Constraint: $\sum_{j=1}^{n} \beta_{jk} \leq 1$.

3.2 Dataset

We have run our experiments on the dataset prepared and used in [21]. This is a real-world dataset consisting of data of 231 patients. The dataset was available in .xlsx format and collected data of ICU-transferred patients and stable patients in two separate sheets. The sheet of ICU-transferred patients contained 100 samples of which 18 were COVID-19 negative and 82 were COVID-19 positive. The sheet of stable patients contained 131 samples of which 57 were COVID-19 negative and 74 were COVID-19 positive. The two sheets had 15 common fields which are Patient Id, Gender (0: Female, 1: Male), Age (years), Age > 60 years (Yes: 1, No: 0), SARS-CoV-2 RT-PCR testing results (Positive: 1, Negative: 0), Time between the disease onset and admission to the hospital (days), CRP upon admission (mg/L), INR upon admission, PT upon admission (sec.), Fibrinogen upon admission (mg/L), Platelet count upon admission (10^9/L), Chest CT upon admission: lung tissue affected (%), CRP, 1 week after admission (mg/L), Chest CT, 1 week after admission: lung tissue affected (%) and Platelet count, 1 week after admission (10^9/L). The sheet of ICU-transferred patients contained two additional fields, namely time between admission to the hospital and transfer to ICU (days) and artificial lung ventilation in ICU (Yes: 1, No: 0). For this research, we combined the data of both sheets resulting in 231 samples. Of these fields, we selected the fields that recorded CRP, PT, fibrinogen, chest CT, and platelet count upon admission. These are the antecedent attributes used in our model. Measurement of these values taken one week later was not considered because usually RT-PCR results are available within a week. If the values recorded one week later are considered, our proposed diagnostic model will require to wait for a week as well. INR was not considered as it is value derived from PT value. The field "SARS-CoV-2 RT-PCR testing results" was taken as the consequent attribute. Some of the samples had blank field for one or two antecedent attribute. If samples with blank fields were omitted, the size of the training set would reduce in size. However, BRBES has the remarkable property of being able to address ignorance in data samples. Therefore, fields that were missing values were filled with 0, so that the complete dataset could be utilized. The final dataset contained 156 samples of COVID-19 positive patients and 75 samples of COVID-19 negative patients. The attributes selected as antecedent attributes are presented in Table 1. The RT-PCR testing result for COVID-19 (1 if positive, 0 if negative) was considered as consequent attribute. For each antecedent and consequent attribute, four referential values were considered. The utility values of each antecedent attribute were assigned based on the highest and lowest values of that attribute in the dataset. The lower bound and upper bound for the consequent attribute were 0 and 1, respectively. The crisp output is intended to be a probability value indicating the chance of being COVID-19 positive. The utility values for the antecedent attributes are presented in Table 1. We conducted tenfold stratified cross-validation on the whole dataset. On average, the training folds had a positive-to-negative class ratio of 140.4:67.5. On average, the test folds had a positive to negative class ratio of 15.6:7.5.

Table 1 Utility values of the antecedent attributes

Referential value	CRP (u.a.)	PT (u.a.)	Fibrinogen (u.a.)	CT (u.a.)	Platelet (u.a.)
Low (L)	0	0	0	0	38.0000
Medium (M)	109.3300	6.8700	3.8270	0.3300	189.6700
High (H)	218.6700	13.7300	7.6530	0.6700	303.3300
Very high (VH)	328.0000	20.6000	11.4800	1.0000	493.0000

CRP = C-reactive protein, PT = prothrombine time, CT = chest CT (percentage of lung tissue infected), platelet = platelet count, u.a. = upon admission

4 Results and Discussion

We have trained the BRBaDE for both conjunctive and disjunctive BRBES although BRBES_CR and BRBES_F were conjunctive for both. For comparison of results, we have also trained and tested by optimizing the disjunctive BRBES with genetic algorithm as well as fmincon optimization function from MATLAB Optimization toolbox.

Yang [41] states that in practice, it is more efficient and stable to use $F \in [0, 1]$ instead of $F \in [0, 2]$. Thus, we have trained the conjunctive and disjunctive BRBaDEs using $F \in [0, 1]$ as well. For this, the utility values for L, M, and H for all antecedent and consequent attributes of BRBES_F were changed to 0, 0.5, and 1, respectively. Also, Eqs. 19 and 20 of [18] were modified as follows,

$$d_{21} = d_{11}$$
$$d_{22} = d_{12}$$

We also performed a tenfold cross-validation on our dataset. The mean square error (MSE) was the objective function to be minimized. Table 2 shows the mean square error of test sets in each fold as well as the average and best values.

The model classifies samples with predicted output ≤ 0.5 as 0 (negative) and those > 0.5 as 1 (positive). Various classification metrics have been shown in Table 3 to compare the diagnosis performance of the algorithms. Accuracy shows the number of correctly classified samples. Precision represents the exactness of the classifier . Sensitivity or recall represents the completeness of the classifier . F_1 score is the harmonic mean of precision and recall. These measures are very effective when the dataset has class imbalance. The true negative (recognition) rate is represented by specificity. Receiver operator curve helps to visualize the trade-off between true positive rate and false positive rate. Similarly precision–recall curve helps visualize the trade-off between precision and recall. The area under curve (AUC) is a measure of how capable a classifier is to distinguish between classes. To visualize the performances, we have plotted ROC curve and precision recall curve of the algorithms

Table 2 Mean square error per fold for various BRBES optimization algorithms

Algorithm	Fold 1	Fold 2	Fold 3	Fold 4	Fold 5	Fold 6	Fold 7	Fold 8	Fold 9	Fold 10	Average	Best value
BRBaDE ($F \in [0, 2]$) for conj. BRBES	0.3228	0.2971	0.3117	0.3353	0.3175	0.3685	0.3729	0.4273	0.3462	0.3957	0.3495	0.2971
GA BRBES	0.2106	0.2190	0.2312	0.2269	0.2272	0.2236	0.2102	0.2154	0.2113	0.2125	0.2188	0.2102
fmincon BRBES	0.2084	0.2186	0.2466	0.2246	0.2312	0.2234	0.2080	0.2158	0.2071	0.2227	0.2206	0.2071
BRBaDE ($F \in [0, 1]$) for disj BRBES	0.2131	0.2224	0.2330	0.2255	0.2258	0.2151	0.2173	0.2090	0.2152	0.2019	0.2178	0.2019
BRBaDE ($F \in [0, 2]$) for disj. BRBES	0.2110	0.2210	0.2603	0.2249	0.2208	0.2251	0.2108	0.2216	0.2106	0.1992	0.2205	0.1992

Table 3 Best performances of the optimization algorithms

	BRBADE ($F \in [0, 2]$) for conj. BRBaDE	GA BRBES	fmincon BRBES	BRBaDE ($F \in [0, 1]$) for disj. BRBES	BRBaDE ($F \in [0, 2]$) for disj. BRBES
Accuracy	0.54167 (Fold-2)	0.69565 (Fold-1)	0.69565 (Fold-1)	0.73913 (Fold-8)	0.73913 (Fold-10)
Precision	1 (Fold-1)	0.69565 (Fold-1)	0.69565 (Fold-1)	0.72727 (Fold-8)	0.72727 (Fold-10)
Sensitivity	0.5 (Fold-2)	1 (Fold-1)	1 (Fold-1)	1 (Fold-1)	1 (Fold-1)
Specificity	1 (Fold-2)	0 (Fold-1)	0 (Fold-1)	0.14286 (Fold-8)	0.14286 (Fold-10)
F_1-score	0.59259 (Fold-1)	0.82051 (Fold-1)	0.82051 (Fold-1)	0.84211 (Fold-8)	0.84211 (Fold-10)
ROC-AUC	0.61607 (Fold-1)	0.61607 (Fold-7)	0.56667 (Fold-6)	0.63393 (Fold-10)	0.63393 (Fold-10)

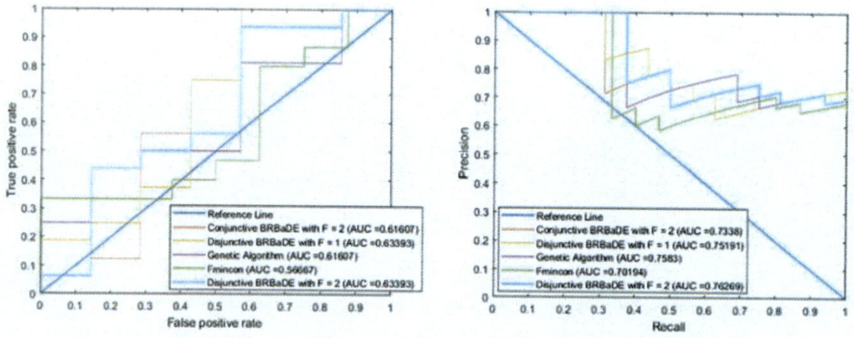

(a) ROC (b) Precision-Recall

Fig. 4 ROC and precision–recall curves

as well. They have been depicted in Fig. 4a, b, respectively. We have trained our model on a dataset of adult patients of viral pneumonia. In the COVID-19 pandemic, faster diagnosis is very important. In case of patients of pneumonia, further care is needed as the lungs get infected. Various previous studies have shown the correlation of hematological parameters with COVID-19. Other studies have identified characteristics of chest CT image data that can assist in identifying COVID-19 infection. However, to confirm the suspicion, medical professionals need to rely on RT-PCR test. It is the current standard for COVID-19 diagnosis. But RT-PCR has significant susceptibility to false positive and false negative errors. Also, since the virus is mutating, RT_PCR primers and probes needed for identifying the mutated strains need to be available at the hospitals and diagnostic institutes. This might not always be possible. If the correlation of hematological data and CT scan data to COVID-19 can be utilized to form an alternative diagnostic approach, many lives can be saved. Artificial intelligence can make that possible. And since real-word scenarios have uncertainty in them, artificial intelligence methods like BRBES will prove to be very effective in implementing efficient diagnostic systems. Another feature of BRBES is that it can operate even when some aspects of data are missing. So even if not all the tests could be run due to lack of equipment and reagents, BRBES will be able to reason based on test results that are available. Alongside, BRBES can also handle the uncertainties raised due to ambiguities and imprecision in the measurement of hematological parameters. It can also address uncertainties raised due to misinterpretation and variations in image acquisition in CT scan data.

From Table 2, we can see that both instances of disjunctive BRBES with BRBaDE optimization have performed relatively better than the rest. BRBaDE on disjunctive BRBES with $F \in [0, 2]$ has the most minimum mean square error (obtained in Fold 10). Also, the BRBaDE on disjunctive BRBES with $F \in [0, 1]$ has the lowest average mean square error. The results of MSE vary for different folds across different algorithms for various possible reasons. One of those reasons can be the presence of missing antecedent values which were set to zero. A second possible reason is that although the sum of degrees of belief can 1 or less than 1, for this article we strictly adhered to ensuring completeness of belief and hence the algorithm was coded in a way that ensures the sum of degrees of belief is one. Although this is an ideal assumption, real-world scenarios do not always have complete belief rules. Another reason can be the reduced size of the test set due to tenfold cross-validation.

From Table 3, we can observe that both instances of disjunctive BRBES have outperformed the other models in accuracy, sensitivity, F1-score, and ROC-AUC. In precision, they were outperformed only by conjunctive BRBES. It can also be observed that the two instances of disjunctive BRBaDE could not perform well in specificity. This might have happened due to the class imbalance in our dataset. The dataset has more instances of positive class than that of negative class. However, their specificity score is comparatively better than those of fmincon optimized BRBES and GA optimized BRBES. Although conjunctive BRBES is has better specificity, it has not performed well in the other metrics. Also, the rule base increases exponentially in conjunctive BRBES due to which it takes long time to train for larger iterations [18]. And if the number of variables to be optimized is large, it may run out of memory during execution as well. ROC curve plots true positive rate against false positive rate. On the other hand, a high area under the precision–recall curve represents both high recall and high precision. High precision relates to a low false positive rate and high recall relates to a low false negative rate. From Fig. 4a, b, we see that disjunctive BRBES optimized with BRBaDE has outperformed conjunctive BRBES, genetic algorithm, and fmincon. Disjunctive BRBES optimized with BRBaDE with $F \in [0, 2]$ has outperformed genetic algorithm and obtained precision–recall AUC more than 76%. It may be noticed that the overall performance of the ROC curve is quite poor. This is due to the fact that due to tenfold cross-validation on the whole dataset, the number of test values reduced. Also, as mentioned earlier, strictly maintaining completeness of belief rules and presence of missing values (which were initialized to zero) may have influenced the ROC as well.

It cannot be denied that the training procedure of BRBaDE needs further refinement for being applied to COVID-19 diagnosis. We trained on a smaller dataset to experiment on the feasibility of BRBES optimized with BRBaDE. We found that disjunctive BRBES optimized with BRBaDE performs better over disjunctive BRBES optimized with genetic algorithm and fmincon. It also performs better over conjunctive BRBES optimized with BRBaDE. To enhance the performance of disjunctive BRBES optimized with BRBaDE, it is essential to train on a larger dataset. Also, [18] has shown that BRBaDE performs very well when both structure and parameters of the BRBES are optimized. Our study has revealed that disjunctive BRBES optimized with BRBaDE possesses good potential as a metaheuristic and gradient-free optimization algorithm for COVID-19 diagnosis based on hematological and CT scan data of lung infection. Therefore, it is feasible to continue the research on disjunctive BRBES optimized with BRBaDE to enhance its performance.

5 Conclusion

In this article, we demonstrated the application of disjunctive BRBES optimized with BRBaDE on hematological and CT scan data. The objective of the model was to serve as COVID-19 diagnostic decisionmaker for adult patients with pneumonia. It showed better performance in comparison to three other metaheuristic algorithms. However, our model still requires much improvement. It is trained on a smaller dataset and can optimize parameters only. Previous studies have shown that disjunctive BRBES performance enhances when both structure and parameters are optimized jointly. This research is intended to be a feasibility study to observer how the belief rule-based expert system performs in COVID-19 diagnosis using both hematological and CT scan data. The model requires further improvement before being considered for deployment in the real field. However, to our knowledge, there has been no previous application of BRBES using both hematological and CT scan data for COVID-19 diagnosis. The findings of this research can assist other researchers who are researching on developing AI-based diagnostic tools for better diagnosis of COVID-19. In future therefore, we shall train our disjunctive BRBES optimized with BRBaDE on a larger dataset of hematological and CT scan data. Also, joint optimization of parameters and structure optimization will be experimented to obtain better results.

References

1. Kucirka LM, Lauer SA, Laeyendecker O, Boon D, Lessler J (2020) Variation in false-negative rate of reverse transcriptase polymerase chain reaction-based sars-cov-2 tests by time since exposure. Ann Internal Med 173(4):262–267
2. D'Cruz RJ, Currier AW, Sampson VB (2020) Laboratory testing methods for novel severe acute respiratory syndrome-coronavirus-2 (sars-cov-2). Front Cell Dev Biol 8

3. Ferrari D, Motta A, Strollo M, Banfi G, Locatelli M (2020) Routine blood tests as a potential diagnostic tool for covid-19. Clin Chem Lab Med (CCLM) 1(ahead-of-print)
4. Weinstock MB, Echenique A, Russell J, Leib A, Miller J, Cohen D, Waite S, Frye A, Illuzzi F (2020) Chest x-ray findings in 636 ambulatory patients with covid-19 presenting to an urgent care center: a normal chest X-ray is no guarantee. J Urgent Care Med 14(7):13–18
5. Benmalek E, Elmhamdi J, Jilbab A (2021) Comparing CT scan and chest X-ray imaging for covid-19 diagnosis. Biomed Eng Adv 100003
6. Wu J, Zhang P, Zhang L, Meng W, Li J, Tong C, Li Y, Cai J, Yang Z, Zhu J et al (2020) Rapid and accurate identification of covid-19 infection through machine learning based on clinical available blood test results. MedRxiv
7. Silva P, Luz E, Silva G, Moreira G, Silva R, Lucio D, Menotti D (2020) Covid-19 detection in CT images with deep learning: a voting-based scheme and cross-datasets analysis. Inform Med Unlocked 20:100427
8. French RM (1999) Catastrophic forgetting in connectionist networks. Trends Cogn Sci 3(4):128–135
9. Grossberg S (2020) A path toward explainable AI and autonomous adaptive intelligence: deep learning, adaptive resonance, and models of perception, emotion, and action. Front Neurorobot 14
10. Rudin C (2019) Stop explaining black box machine learning models for high stakes decisions and use interpretable models instead. Nat Mach Intell 1(5):206–215
11. Chollet F (2017) The limitations of deep learning. In: Deep learning with python
12. Mahmud M, Kaiser MS, McGinnity TM, Hussain A (2021) Deep learning in mining biological data. Cogn Comput 13(1):1–33
13. Nguyen A, Yosinski J, Clune J (2015) Deep neural networks are easily fooled: high confidence predictions for unrecognizable images. In: Proceedings of the IEEE conference on computer vision and pattern recognition, pp 427–436
14. Hassan OI, Abdelrahman A (2020) Uncertainty evaluation of blood analyser system. In: 2020 International conference on computer, control, electrical, and electronics engineering (ICCCEEE), pp 1–4
15. Fletcher JG, Leng S, Yu L, McCollough CH (2016) Dealing with uncertainty in CT images
16. Berenguer R, Pastor-Juan MDR, Canales-Vázquez J, Castro-García M, Villas MV, Mansilla Legorburo F, Sabater S (2018) Radiomics of CT features may be nonreproducible and redundant: influence of CT acquisition parameters. Radiology 288(2):407–415
17. Jamil MN, Hossain MS, Ul Islam R, Andersson K (2019) A belief rule based expert system for evaluating technological innovation capability of high-tech firms under uncertainty. In: 2019 Joint 8th international conference on informatics, electronics & vision (ICIEV) and 2019 3rd International conference on imaging, vision & pattern recognition (icIVPR). IEEE, pp 330–335
18. Islam RU, Ruci X, Hossain MS, Andersson K, Kor AL (2019) Capacity management of hyperscale data centers using predictive modelling. Energies 12(18):3438
19. Fan BE (2020) Hematologic parameters in patients with covid-19 infection: a reply. Am J Hematol
20. Formica V, Minieri M, Bernardini S, Ciotti M, D'Agostini C, Roselli M, Andreoni M, Morelli C, Parisi G, Federici M et al (2020) Complete blood count might help to identify subjects with high probability of testing positive to sars-cov-2. Clin Med 20(4):e114
21. Baranovskii DS, Klabukov ID, Krasilnikova OA, Nikogosov DA, Polekhina NV, Baranovskaia DR, Laberko LA (2020) Prolonged prothrombin time as an early prognostic indicator of severe acute respiratory distress syndrome in patients with covid-19 related pneumonia. Curr Med Res Opin 1–8
22. Brinati D, Campagner A, Ferrari D, Locatelli M, Banfi G, Cabitza F (2020) Detection of covid-19 infection from routine blood exams with machine learning: a feasibility study. J Med Syst 44(8):1–12
23. Soltan AA, Kouchaki S, Zhu T, Kiyasseh D, Taylor T, Hussain ZB, Peto T, Brent AJ, Eyre DW, Clifton DA (2021) Rapid triage for covid-19 using routine clinical data for patients attending hospital: development and prospective validation of an artificial intelligence screening test. Lancet Digital Health 3(2):e78–e87

24. Alves MA, Castro GZ, Oliveira BAS, Ferreira LA, Ramírez JA, Silva R, Guimarães FG (2021) Explaining machine learning based diagnosis of covid-19 from routine blood tests with decision trees and criteria graphs. Comput Biol Med 132:104335
25. Chung M, Bernheim A, Mei X, Zhang N, Huang M, Zeng X, Cui J, Xu W, Yang Y, Fayad ZA et al (2020) CT imaging features of 2019 novel coronavirus (2019-ncov). Radiology 295(1):202–207
26. Wu Z, Liu X, Liu J, Zhu F, Liu Y, Liu Y, Peng H (2021) Correlation between ground-glass opacity on pulmonary CT and the levels of inflammatory cytokines in patients with moderate-to-severe covid-19 pneumonia. Int J Med Sci 18(11):2394–2400
27. Mehrabi S, Safaei M, Ghandi Y, Bahrami M (2021) Chest CT features in pediatric patients with covid-19 infection: a brief review article. Int J Pediatr 9(4):13421–13427
28. Pan F, Ye T, Sun P, Gui S, Liang B, Li L, Zheng D, Wang J, Hesketh RL, Yang L et al (2020) Time course of lung changes at chest CT during recovery from coronavirus disease 2019 (covid-19). Radiology 295(3):715–721
29. Xie X, Zhong Z, Zhao W, Zheng C, Wang F, Liu J (2020) Chest CT for typical coronavirus disease 2019 (covid-19) pneumonia: relationship to negative RT-PCR testing. Radiology 296(2):E41–E45
30. Li L, Qin L, Xu Z, Yin Y, Wang X, Kong B, Bai J, Lu Y, Fang Z, Song Q et al (2020) Artificial intelligence distinguishes covid-19 from community acquired pneumonia on chest CT. Radiology
31. Ghaderzadeh M, Asadi F, Jafari R, Bashash D, Abolghasemi H, Aria M (2021) Deep convolutional neural network-based computer-aided detection system for covid-19 using multiple lung scans: design and implementation study. J Med Internet Res 23(4):e27468
32. Hossain MS, Khalid MS, Akter S, Dey S (2014) A belief rule-based expert system to diagnose influenza. In: 2014 9th International forum on strategic technology (IFOST). IEEE, pp 113–116
33. Karim R, Andersson K, Hossain MS, Uddin MJ, Meah MP (2016) A belief rule based expert system to assess clinical bronchopneumonia suspicion. In: 2016 Future technologies conference (FTC). IEEE, pp 655–660
34. Hossain MS, Ahmed F, Andersson K et al (2017) A belief rule based expert system to assess tuberculosis under uncertainty. J Med Syst 41(3):43
35. Biswas M, Chowdhury SU, Nahar N, Hossain MS, Andersson K (2019) A belief rule base expert system for staging non-small cell lung cancer under uncertainty. In: 2019 IEEE International conference on biomedical engineering, computer and information technology for health (BECITHCON). IEEE, pp 47–52
36. Hossain MS, Monrat AA, Hasan M, Karim R, Bhuiyan TA, Khalid MS (2016) A belief rule-based expert system to assess mental disorder under uncertainty. In: 2016 5th International conference on informatics, electronics and vision (ICIEV). IEEE, pp 1089–1094
37. Rahaman S, Hossain MS (2014) A belief rule based (BRB) system to assess asthma suspicion. In: 16th International conference on computer and information technology. IEEE, pp 432–437
38. Hossain MS, Hossain E, Khalid MS, Haque MA (2014) A belief rule-based (BRB) decision support system for assessing clinical asthma suspicion. In: Scandinavian conference on health informatics. Linköping University Electronic Press, pp 83–89
39. Hossain MS, Habib IB, Andersson K (2017) A belief rule based expert system to diagnose dengue fever under uncertainty. In: 2017 Computing conference. IEEE, pp 179–186
40. Ahmed F, Hossain MS, Islam RU, Andersson K (2021) An evolutionary belief rule-based clinical decision support system to predict covid-19 severity under uncertainty. Appl Sci 11(13):5810
41. Yang XS (2020) Nature-inspired optimization algorithms. Academic Press

Performance Analysis of Particle Swarm Optimization and Genetic Algorithm in Energy-Saving Elevator Group Control System

Mohammad Hanif and Nur Mohammad

Abstract Optimization of energy consumption in Elevator Group Control System (EGCS) has become a concerning issue due to the global energy crisis. To resolve this concern, only a handful amount of approaches, based on swarm intelligence, have been implemented. For this reason, this study implements and analyzes the energy-saving EGCS based on two popular metaheuristic algorithms: Particle Swarm Optimization (PSO) and Genetic Algorithm (GA). The performance analysis of these two algorithms in energy-saving EGCS reveals that both of the algorithms have some pros and cons. While PSO can optimize energy consumption much better than GA in most cases, the trapping in local minima or pre-mature convergence of PSO makes its performance worse. GA, on the other hand, is unable to reduce energy consumption in EGCS to a considerably lower level. However, the average energy consumption and standard deviation of GA are superior to that of PSO. In the case of computational time, PSO outperforms GA, which makes PSO a popular choice where faster computation is required.

Keywords Metaheuristic algorithms · Constraint · Convergence characteristic · Energy-saving · PSO · GA

1 Introduction

In high-rise buildings, EGCS generally controls three or more elevators simultaneously. The most important part of EGCS is the group control dispatching algorithm. Selecting the best elevator from a group of elevators is the basic goal of EGCS [1]. In this case, selecting the right elevator is crucial for optimizing the performance parameters of EGCS. The major performance indicators of EGCS are average waiting-time

M. Hanif (✉) · N. Mohammad
Department of Electrical & Electronic Engineering, Chittagong University of Engineering & Technology, Chittagong 4349, Bangladesh

N. Mohammad
e-mail: nur.mohammad@cuet.ac.bd

M. S. Arefin et al. (eds.), *Proceedings of the International Conference on Big Data, IoT, and Machine Learning*, Lecture Notes on Data Engineering and Communications Technologies 95, https://doi.org/10.1007/978-981-16-6636-0_38

(AWT), average journey-time (AJT), computational time (CT), and energy consumption (EC) [2–5]. Due to the contradictory relation among these performance indices, it is difficult to satisfy all the parameters at the same time [1]. When selecting parameters, the earlier studies placed a strong emphasis on minimizing AWT or AJT, which is critical for passengers' comfort. However, owing to the worldwide energy shortage, the optimization of energy consumption is being paid much attention in recent years.

The optimization of the EGCS dispatching problem is a complex and challenging task. Since passengers arrive in multi-story buildings at random, elevator scheduling is influenced by traffic patterns (i.e., passenger inflow rate). Moreover, optimization in EGCS is difficult due to the following reasons:

- The optimization in EGCS is dependent on a variety of factors, such as elevators' position and running direction, passengers' number in the elevator cars, number of stops to respond to a hall-call, and traveling distance [6].
- The performance metrics of EGCS are contradictory to one another. As a result, selecting an objective for EGCS may degrade the other indices of EGCS.
- The EGCS dispatching problem is a combinatorial problem. If an EGCS consists of $\boldsymbol{n}$ elevators and gets $\boldsymbol{p}$ hall-call at a particular time, the dispatching algorithm needs to consider $\boldsymbol{n^p}$ number of cases [6, 7].
- Because of the dynamic nature of EGCS, it is necessary to consider the possible future call. As a result, various uncertain variables must be considered prior to optimization.
- Finally, traffic patterns have an impact on EGCS scheduling. As a result, a purely mathematical method to solve this problem is not feasible [8].

In Fig. 1, the regulatory factors of EGCS to transport passengers are shown. The controlling factors are mainly divided into two parts (group control and elevator control). The group control composes of strategies and car selection. On the other

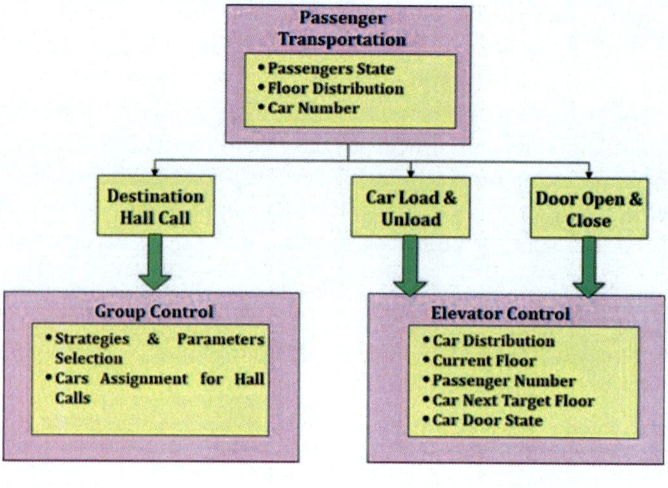

Fig. 1 Basic factors of elevator control and group control to transport passengers

hand, the elevator control depends on the elevator position, the number of passengers, and their destinations.

In the meanwhile, EGCS makes considerable use of fuzzy logic and neural networks [9]. However, due to EGCS's unpredictable and dynamic behavior, each approach has significant drawbacks [10, 11]. Moreover, because of the combinatorial nature and uncertainties, metaheuristic approaches are highly beneficial in the optimization problems of EGCS [12]. Swarm intelligence has already demonstrated its effectiveness in solving complex and dynamic problems. On top of that, this swarm intelligence-based optimization does not necessitate the use of a precise mathematical model [13]. For this reason, the authors of this study are motivated to use two popular metaheuristic algorithms (PSO and GA) to construct EGCS strategies for reducing energy usage. PSO and GA have been widely applied to solve a variety of complex, dynamic, and difficult problems, such as scheduling and optimization. Both of these algorithms are also implemented in EGCS optimization. In most cases, however, the performance parameters are AWT or AJT, neglecting the energy-saving issue. The optimization of energy consumption in the EGCS car dispatching problem is a pivotal issue, even though it has been overlooked for many years, with the exception of a few metaheuristic techniques. To address this concern, an approach to optimize the usage of energy in EGCS is implemented in this study. The authors apply PSO and GA algorithms in the EGCS dispatching problem to optimize energy consumption. In addition, the comparative analysis between these two algorithms is discussed in this research. This comparative analysis will serve as a guide for future researchers and designers in deciding which algorithm is best for the energy-saving in EGCS.

The paper is organized as follows: first, some major previous works are outlined, then the objective function and constraints of the optimization are formulated. The fourth section discusses the PSO and the GA in the EGCS problem, while the simulation methods and parameters of the optimization are described in the fifth section. Finally, the paper concludes with the analysis and comparison of the obtained result, followed by limitations and future research scopes.

2 Related Previous Works

In 1979, Otis Elevator Company produced the first elevator controller, known as Elevonic 101 [14]. The first and most commonly implemented metaheuristic algorithm for EGCS is GA. The primary concept of EGCS was mentioned by Fujino et al. [15] in 1997, where the authors implemented the EGCS employing GA. Following that in 2001 and 2003, Tyni and Ylinen [16, 17] implemented two GA approaches. Subsequently, these two authors investigated another multi-objective EGCS approach in 2006 [18]. In the elevator group, Bolat and Cortés [19] implemented GA and Tabu Search (TS) approaches in 2011. For various EGCS configurations, the authors compared these two algorithms (GA and TS). However, these approaches were solely examined by AJT, ignoring the issue of energy usage. In 2013, Bolat et al. [20] used PSO in EGCS to minimize AJT, demonstrating that PSO performed better than GA and TS in complex configurations. In a simple building, however, PSO performed poor results in that study. In 2014 and 2016, Tartan et al. [21, 22] implemented two AWT minimization approaches employing GA. Recently, in 2020, Ant Colony Optimization (ACO)-based EGCS was implemented by Le et al. [23], where AWT optimization was taken into account. In that approach, however, due to the increased operating frequency, the energy consumption performance deteriorated.

3 Problem Formulation

The formulation of the objective function of energy consumption and constraints are the most significant factors for optimizing energy consumption in EGCS. In this section, the formulations of the objective function and applied constraints for optimization have been outlined. An EGCS with four elevators is depicted in Fig. 2a, while Fig. 2b shows a counter-weight attached with a single elevator.

3.1 Objective Function Formulation

The objective function of EGCS consists of two parts. The first part is the stopping-starting energy consumption (E_a), and the other one is the uniform speed energy consumption (E_v). As a result, the energy consumption of a single elevator can be

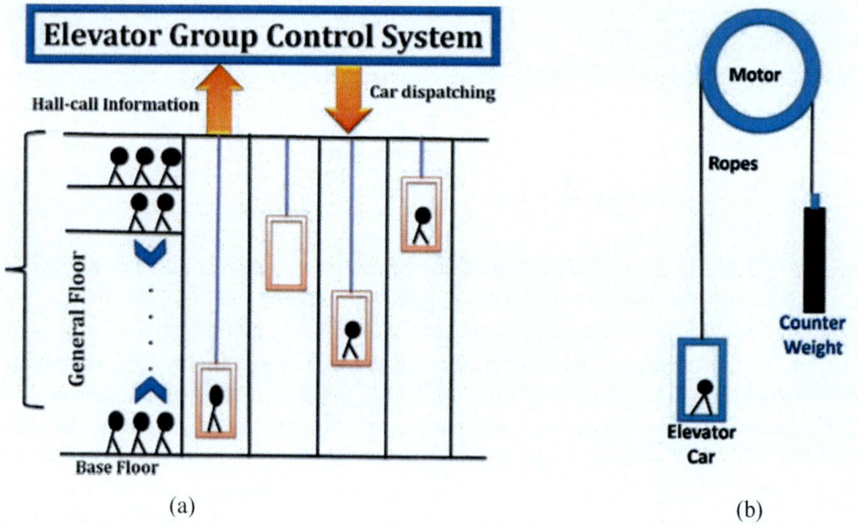

Fig. 2 **a** Elevator group system with 4 cars. **b** An elevator connected with a counter-weight dispatches to transport passengers

calculated as follows:

$$E = E_a + E_v \tag{1}$$

If the EGCS has N number of elevators, then the overall energy consumption can be estimated by the following equation:

$$E_T = \sum_{r=1}^{N} [E_a(r) + E_v(r)] \tag{2}$$

Here, $E_a(r)$: Energy consumption for acceleration-deceleration of the rth elevator.

$E_v(r)$: Energy consumption during uniform running of the rth elevator.

Acceleration-Deceleration Energy Consumption. The starting-stopping energy consumption of rth elevator depends on the number of stops of the rth elevator, when it responds to a hall-call. The following formula is used to determine this energy consumption:

$$E_a(r) = \sum_{s=1}^{q(r)} \left(P_{(r,s)} \times E_c\right) \tag{3}$$

Here, $q(r)$: Number of stops of the rth elevator.

$P_{(r,s)}$: Total number of stops of the rth elevator when responding sth hall-call.

E_c: Energy consumption of an elevator car in single stopping-starting.

Uniform Speed Energy Consumption. The uniform speed energy consumption relies on the traveling distance of elevator and the number of a passenger inside the elevator car. This energy consumption is calculated by the following equations:

$$E_v(r) = \sum_{s=1}^{q(r)} \sum_{t=1}^{p_{(r,s)}} \left[\left\{n_p(r, s, t) \times M + M_{car} - M_{cwt}\right\} g \times d(r, s, t)\right] \tag{4}$$

Here,

$n_p(r, s, t)$: Passengers' number inside the elevator.

M: Passengers' average mass.

M_{car}: Mass of a single elevator car.

M_{cwt}: Mass of single counter-weight.

$d(r, s, t)$: Distance traveled by an elevator when responding to a particular hall-call.

Final Objective Function. The overall objective function of the EGCS is the sum of the starting-stopping energy and the uniform speed energy consumption of all elevators in the group. As a result, the total energy consumption is calculated using the following equation:

$$E_T = \sum_{r=1}^{N} \left\{ \sum_{s=1}^{q(r)} (P_{(r,s)} \times E_c) + \sum_{s=1}^{q(r)} \sum_{t=1}^{p_{(r,s)}} [\{n_p(r,s,t) \times M + M_{car} - M_{cwt}\} g \times d(r,s,t)] \right. \quad (5)$$

3.2 Constraints Formulation

In the optimization of the energy consumption in EGCS, some constraints, based on the parameters of the elevator, need to consider. Since the performance parameters of the EGCS are contradictory to one another, it is crucial to optimize the energy consumption of the EGCS by keeping the AWT within a certain range. The other constraints of energy-saving EGCS are load constraint, stopping constraint, and distance constraint. A list of items that are used to formulate the constraints is presented in Table 1.

Load Constraint. The load of the elevator should be kept at a limited value. In this case, the minimum load in an elevator is zero, and the maximum load is the rated capacity of the elevator (12 persons). On the other hand, the current load is

Table 1 Selected parameters to implement EGCS

Items	Value
Distance of two adjacent floors	3 m
Time of door-opening	3 s
Time of door-closing	3 s
Velocity of elevator car	3 m/s
Acceleration time of elevator	3 s
Deceleration time of elevator	3 s
Car capacity	12 person
No. of floor	20
No. of cars in EGCS	4
Average mass of passenger	65 kg
Passenger boarding-time	1 s/person
Passenger alighting-time	1 s/person
Mass of an elevator car	800 kg
Mass of a counter-weight	850 kg

determined by the initial passenger number, the boarding passenger in the elevator, and the alighting passenger from the elevator.

$$0 \leq n_p(r, s, t) \leq 12; \tag{6}$$

Again,

$$n_p(r, s, t) = n_{pi}(r, s, t) - n_{p(\text{exit})}(r, s, t) + n_{p(\text{enter})}(r, s, t) \tag{7}$$

Here, $n_p(r, s, t)$: Current passenger number in the elevator.

$n_{pi}(r, s, t)$: Initial passenger number in the elevator before responding to a particular hall-call.

$n_{\text{p(exit)}}(r, s, t)$: Passenger exiting from the elevator on a particular hall-call.

$n_{p(\text{enter})}(r, s, t)$: Passenger entering in the elevator on a particular hall-call.

The minimum passenger number is zero, and it can never be a negative value. As a consequence, the number of exiting passengers cannot exceed the sum of the initial passenger number and the entering passenger number. Hence,

$$\begin{aligned} &\text{if } n_{pi}(r, s, t) - n_{p(\text{exit})}(r, s, t) + n_{p(\text{enter})}(r, s, t) < 0; \\ &\text{then } n_p(r, s, t) = 0. \end{aligned}$$

AWT Constraint. The AWT of the passengers must be kept within a limited value. In a single stop, the time requires by an elevator is: elevator acceleration time + elevator deceleration time + door opening time + door closing time + passenger unloading time × number of passengers unloaded + passenger loading time × number of passengers loaded $= 3 + 3 + 3 + 3 + 1 \times n_{\text{p(exit)}} + 1 \times n_{p(\text{enter})} = (12 + n_{p(\text{exit})} + n_{p(\text{enter})})$ seconds.

On the other hand, if an elevator needs to travel $d(r, s, t)$ distance to respond to a hall-call, and the uniform speed of the elevator be 3 m per second, then the AWT constraint for keeping the AWT within 5 min (300 s) is computed by the following equation:

$$\begin{aligned} &\left[12 + n_{p(\text{exit})} + n_{p(\text{enter})}\right] \times P_{(r,s)} + \frac{d(r, s, t)}{3} \leq 300; \\ &\text{Here, } P_{(r,s)} = \text{Total number of stops} \end{aligned} \tag{8}$$

Boundary Constraints. The boundary constraints of stopping are determined by considering the minimum number of stops, which is one, and the maximum number of stops, which is (number of floors -1). Hence, the stopping constraint is:

$$1 \leq P_{(r,s)} \leq 19; \tag{9}$$

The traveling distance constraints range from zero to twice the height of the building. It is due to the fact that the elevator's initial state and the hall-call may be on the same floor, resulting in the minimum possible distance. The maximum distance covered by an elevator in a round trip is twice the building's height.

$$0 \leq d(r, s, t) \leq 120; \tag{10}$$

4 GA and PSO-Based EGCS

4.1 GA-Based EGCS

The following is the optimization process in EGCS utilizing the Genetic Algorithm:

1. Generating random initial population.
2. Evaluating the fitness of each population.
3. Executing the genetic operators (Selection, Crossover, Mutation, and Deletion) to generate new offspring of each population.
4. **Selection**: Based on the fitness score, the best two populations are selected to pass their genes in the subsequent generations.
5. **Crossover**: The parents exchange their genes to produce new offspring.
6. **Mutation**: Some genes are flipped in some of the offspring to increase variety in the offspring. This is referred to as mutation.
7. **Termination**: When the maximum number of iteration completes, the GA algorithm terminates.

The flow chart of the GA-based energy-saving EGCS optimization implementation technique is shown in Fig. 3. The data from the floor is initially transferred to the solution space, and then GA is applied to the provided data. Finally, the optimal solution is passed to the elevator controller, which will dispatch optimal cars.

4.2 PSO-Based EGCS

The PSO optimizes the objective function by mimicking the behavior of the swarm. This algorithm optimizes the objective function by updating the particles' position following the local best and the global best. To update velocity, as well as position of the next iterations, the following expressions are employed:

$$V_i(t+1) = wV_i(t) + c_1r_1(P_i(t) - X_i(t)) + c_2r_2(P_g(t) - X_i(t)) \tag{11}$$

$$X_i(t+1) = X_i(t) + V_i(t+1) \tag{12}$$

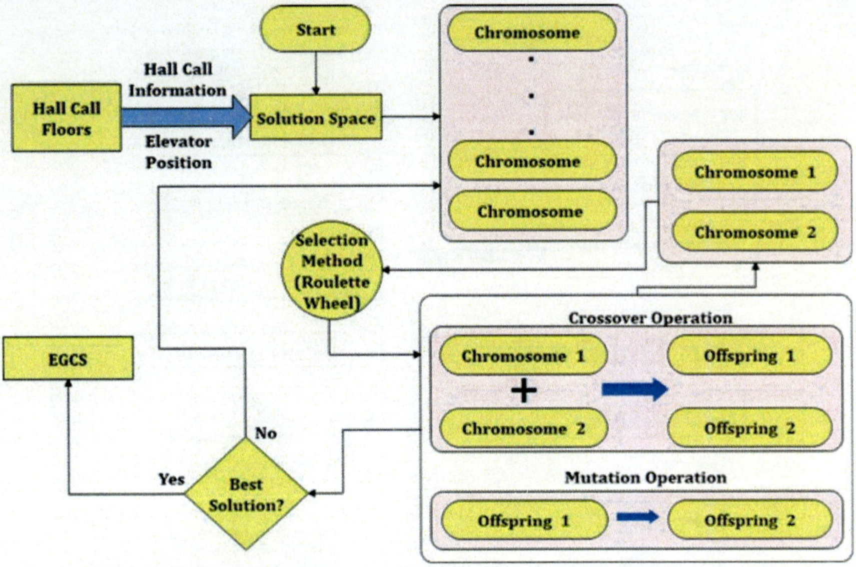

Fig. 3 Flow chart of Genetic Algorithm-based EGCS [19]

Here, $V_i(t)$ = current velocity of a particle, $V_i(t+1)$ = new velocity of a particle, $X_i(t)$ = current position of a particle, $X_i(t+1)$ = next position of a particle, c_1 and c_2 = random constants (acceleration constants), r_1 and r_2 = random values between 0 and 1, $(P_i(t) - X_i(t))$ = distance between local best and current position of a particle, $(P_g(t) - X_i(t))$ = distance between a particle's current position and global best.

Figure 4 presents the procedure to implement the PSO-based EGCS. The global best is detected based on the available information, and the elevators are accordingly scheduled to optimize the energy consumption in EGCS.

5 Simulation

5.1 Implementation

The optimization simulation is carried out by utilizing MATLAB software. In a single script, the objective function and the constraints are first converted into code. The PSO and GA algorithms are independently implemented on this objective function's script. Later, both the PSO and GA source code are simultaneously applied to the objective function to compare the performance of these two algorithms. In this study, the authors consider EGCS having four elevators in a 20-floor building. Moreover, due to the random and dynamic characteristics of EGCS, the authors use *randi*

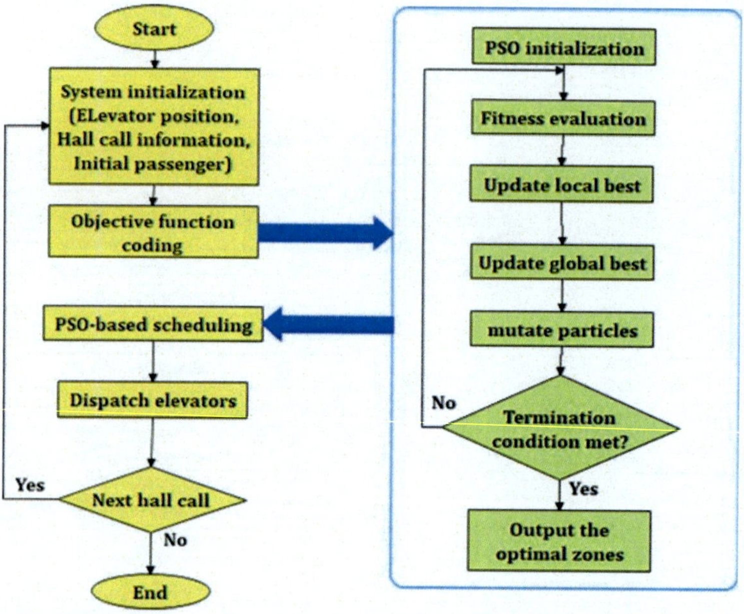

Fig. 4 Flow chart of PSO-based EGCS

function to select the number of stops of elevators, the number of initial passengers, the number of entering and exiting passengers in the elevators. This *randi* function makes the optimization more dynamic, and in every circumstance, PSO and GA are able to handle the optimization of the energy consumption in EGCS.

5.2 Parameter of GA and PSO

Parameter selection is an important task in the implementation of metaheuristic-based EGCS problems. For the simulations of GA-based and PSO-based EGCS, the selected parameters are listed in Table 2. For GA, the required population is quite large. This is due to the convergence difficulty of GA. If a small population size is used, the GA may not able to find the global best within the selected iterations.

6 Result Analysis and Comparison

In this section, the results of the simulation of optimization are discussed. The computational time of the two algorithms to calculate the optimization result is also analyzed. At first, the convergence characteristics of PSO and GA in energy-saving

Table 2 Selected parameters of GA-based and PSO-based energy-saving EGCS

GA parameter		PSO parameter	
Population	1000	Population	50
Iteration	50	Iteration	50
Crossover	1	C1	1
Mutation	0.1	C2	0.1
Selection	Roulette Wheel	W	1
Pc (Percentage of children)	0.1	Damping ratio	0.99

EGCS are separately presented. Following that in order to compare the results of PSO and GA optimization, the typical convergence characteristics of both algorithms are shown at the same time. Finally, to visualize the comparison between PSO-based and GA-based EGCS, the statistical comparison and computational time are outlined.

6.1 Convergence Characteristic

Due to the stochastic nature, the convergence characteristic of the metaheuristic algorithms differs from one another. From the random initial value, the search agents start searching for the global best. Therefore, at first, the best costs obtained by both PSO and GA are quite large. The search agents of both algorithms struggle to get lower costs as iteration increases, and if they find a lower cost, it is recorded. After reaching the global best (optimum cost), the convergence curve becomes a straight line parallel to horizontal. In most scenarios, the PSO optimizes the objective function in a better way if it is not stuck in the local minima. The typical convergence characteristics of PSO-based and GA-based energy-saving EGCS are presented in Fig. 5. In energy-saving EGCS, the comparative convergence curves of PSO and GA are also depicted in Fig. 6a, where PSO outperforms the GA. However, in some cases, PSO exhibits very poor performance when it is stuck in local minima. In that case, PSO converges pre-maturely and exhibits worse performance than GA. Figure 6b illustrates a typical pre-mature convergence of PSO. On the other hand, GA usually cannot able to optimize the energy consumption in EGCS in a better way like PSO. However, for 100 independent trials, the mean value of GA (6.4823×10^{04} Kilo-Joule) is superior to that of PSO (9.6095×10^{04} Kilo-Joule), which is the most significant advantage of GA. Moreover, the standard deviation of GA is 35,888.9, while this value is 107,643. 53 for PSO.

It is clear from the 100 trials of the two algorithms that the PSO algorithm is considerably better at finding the global minima of energy-saving EGCS than the GA algorithm. Over 100 runs, the performance of these two methods in the optimization of energy-saving EGCS is demonstrated in Fig. 7a. The boxplot of these 100 runs is shown in Fig. 7b, where a visual presentation of standard deviations of PSO and GA is presented. From 100 trials, the best cost, worst cost, mean cost, median cost, and

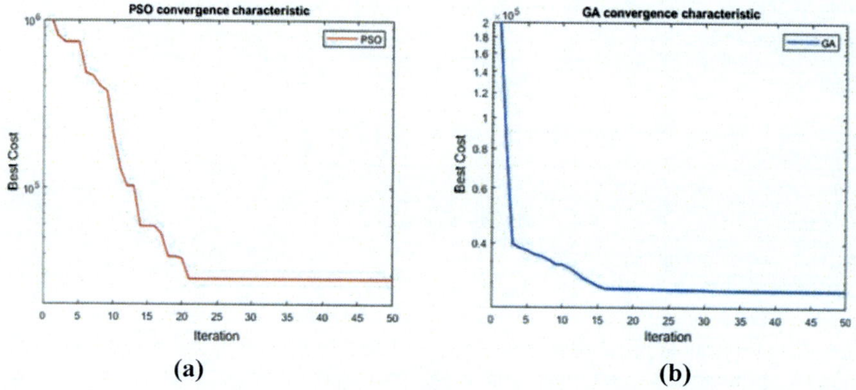

Fig. 5 Convergence characteristics of **a** PSO-based EGCS, **b** GA-based EGCS

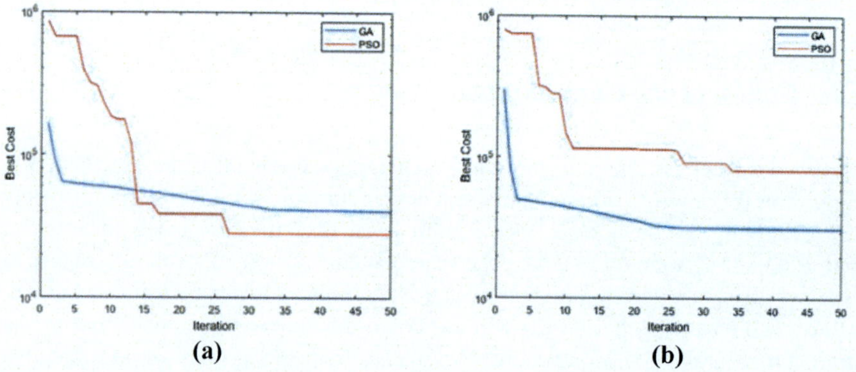

Fig. 6 **a** General convergence comparison of PSO and GA-based EGCS. **b** A typical pre-mature convergence of PSO

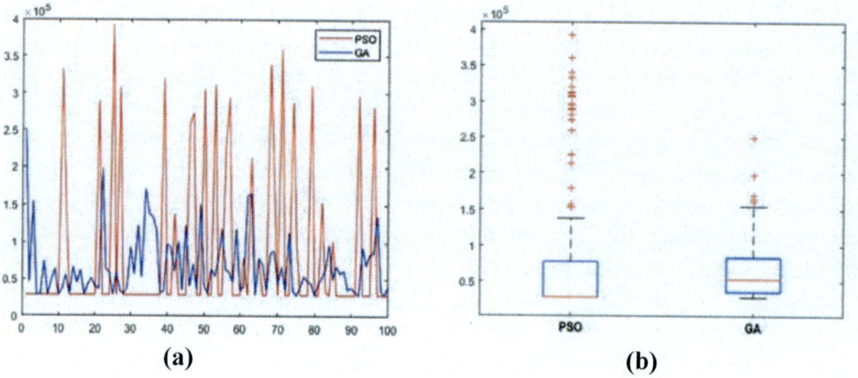

Fig. 7 **a** Plotting of 100 optimum values found from 100 independent runs of PSO and GA. **b** Boxplot of 100 runs

Table 3 Statistical comparison of PSO and GA of 100 trial simulations

Metaheuristic	Best (Kilo-Joule)	Worst (Kilo-Joule)	Mean (Kilo-Joule)	Median (Kilo-Joule)	Standard deviation
PSO	**2.84044e + 04**	3.85372e + 05	9.6095e + 04	**2.84044e + 04**	1.0764353e + 05
GA	2.86034e + 04	**1.95465e + 05**	**6.4823e + 04**	5.48051e + 04	**3.58888e + 04**

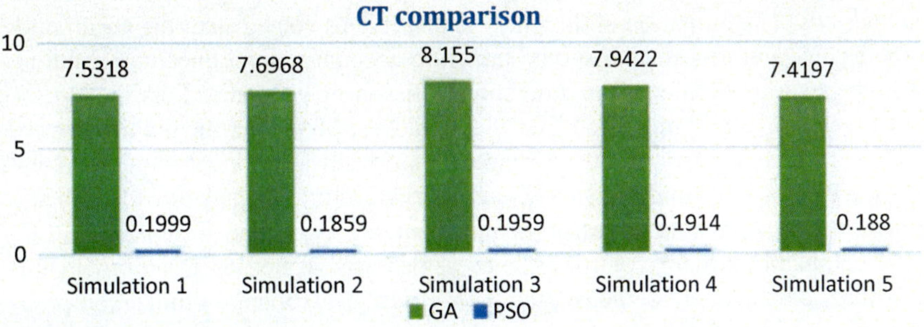

Fig. 8 Computational time comparison of GA-based and PSO-based EGCS

standard deviation of both PSO and GA are presented in Table 3, where the better performances are highlighted in bold.

6.2 *Computational Time*

PSO computes and finds the optimum result in a shorter amount of time. As a result, PSO's computational time is significantly faster than GA's. The average computational time of PSO in energy-saving EGCS is roughly 0.19 s, whereas GA requires over 7.5 s. Due to the slow convergence of GA in EGCS, a large population number requires to implement the GA approach, which increases the computational cost. Figure 8 shows the comparative result of computational time of PSO and GA for 5 random trials, demonstrating indisputably that the PSO algorithm outperforms the GA strategy in terms of computational time.

7 Conclusion

Previously, most of the GA-based and PSO-based EGCS optimizations tried to minimize the AWT or AJT of the passengers. Only a few papers considered the optimization of energy consumption with a brief detail of the procedure. In this paper, the authors attempt to handle the optimization of the energy consumption in EGCS with a detailed description of objective function and constraints. The findings reveal that GA is unable to produce optimal results in the majority of cases. This algorithm, on the other hand, has the advantage of rarely trapping in local optima. The PSO algorithm provides the best value when it does not converge pre-maturely. However, the problem is, it gives a very worse result when trapped in local minima. In the case of computational time, PSO is substantially faster than GA. As a result, PSO can compute scheduling techniques significantly faster, allowing it to better handle peak periods. The limitation of this study is that it does not consider the uncertainty in the EGCS. Future research should take into account several uncertain handling methods, such as Monte Carlo simulation, Poisson or geometric Poisson process in metaheuristic algorithm-based EGCS approaches. Moreover, the multi-objective EGCS is crucial to optimize both passenger comfort and energy consumption. In that case, the Pareto-based multi-objective metaheuristic algorithm may provide valuable results. It is also possible to utilize the metaheuristic algorithms in replacing PLC-based automatic elevator control systems [24]. Last but not least, various hybrid algorithms have already demonstrated their efficacy in solving optimization problems in a variety of domains. As a result, combining two or more algorithms to address the limitations of this study is essential, which will be the authors' future research study.

References

1. Zhang J, Zong Q, Wang F, Li J (2011) Elevator group scheduling for peak flows based on Adjustable Robust Optimization model. In: 2011 Chinese control and decision conference, (CCDC). IEEE, pp 1593–1598
2. Liu Y, Hu Z, Su Q, Huo J (2010) Energy saving of elevator group control based on optimal zoning strategy with interfloor traffic. In: 2010 3rd international conference on information management, innovation management and industrial engineering. IEEE, pp 328–331
3. Hasan MZ, Fink R, Suyambu MR, Baskaran MK (2012) Assessment and improvement of elevator controllers for energy efficiency. In: 2012 IEEE 16th international symposium on consumer electronics. IEEE, pp 1–8
4. Barney G, Al-Sharif L (2003) Elevator traffic handbook: theory and practice. Taylor & Francis, London
5. Strakosch GR (1998) The vertical transportation handbook. Wiley, New York
6. Liu J, Liu Y (2007) Ant colony algorithm and fuzzy neural network-based intelligent dispatching algorithm of an elevator group control system. In: 2007 IEEE international conference on control and automation. IEEE, pp 2306–2310
7. Kim CB, Seong KA, Lee-Kwang H, Kim JO (1998) Design and implementation of a fuzzy elevator group control system. IEEE Trans Syst Man Cybern Part ASyst Hum 28:277–287
8. Li Z (2010) Pso-based real-time scheduling for elevator group supervisory control system. Intell Autom Soft Comput 16:111–121
9. Ho M, Robertson B (1994) Elevator group supervisory control using fuzzy logic. In: 1994 proceedings of Canadian conference on electrical and computer engineering. IEEE, pp 825–828
10. Cortés P, Fernández JR, Guadix J, Munuzuri J (2012) Fuzzy logic based controller for peak traffic detection in elevator systems. J Comput Theor Nanosci 9:310–318

11. Shapiro AF (2002) The merging of neural networks, fuzzy logic, and genetic algorithms. Insur Math Econ 31:115–131
12. Nesmachnow S (2014) An overview of metaheuristics: accurate and efficient methods for optimisation. Int J Metaheuristics 3:320–347
13. Fathy A (2018) Recent meta-heuristic grasshopper optimization algorithm for optimal reconfiguration of partially shaded PV array. Sol Energy 171:638–651
14. Fernandez JR, Cortes P (2015) A survey of elevator group control systems for vertical transportation: a look at recent literature. IEEE Control Syst Mag 35:38–55
15. Chen TC, Hsu YY, Lee AC, Wang SY (2013) GA based hybrid fuzzy rule optimization approach for elevator group control system. Trans Can Soc Mech Eng 37:937–947
16. Tyni T, Ylinen J (2001) Genetic algorithms in elevator car routing problem. In: Proceedings of the genetic and evolutionary computation conference (GECCO-2001). Morgan Kaufman Publishers, San Francisco, pp 1413–1422
17. Sorsa J, Siikonen ML, Ehtamo H (2003) Optimal control of double-deck elevator group using genetic algorithm. Int Trans Oper Res 10:103–114
18. Tyni T, Ylinen J (2006) Evolutionary bi-objective optimisation in the elevator car routing problem. Eur J Oper Res 169:960–977
19. Bolat B, Cortés P (2011) Genetic and tabu search approaches for optimizing the hall call—car allocation problem in elevator group systems. Appl Soft Comput 11:1792–1800
20. Bolat B, Altun O, Cortés P (2013) A particle swarm optimization algorithm for optimal car-call allocation in elevator group control systems. Appl Soft Comput 13:2633–2642
21. Tartan EO, Erdem H, Berkol A (2014) Optimization of waiting and journey time in group elevator system using genetic algorithm. In: 2014 IEEE international symposium on innovations in intelligent systems and applications (INISTA) proceedings. IEEE, pp 361–367
22. Oner Tartan E, Ciftlikli C (2016) A genetic algorithm based elevator dispatching method for waiting time optimization. IFAC-PapersOnLine 49:424–429
23. Le Y, Shifeng Y, Huanhuan L, Zhicheng L, Xiaobing H (2020) Research on elevator group optimal dispatch based on ant colony algorithm. In: 2020 international conference on artificial intelligence and electromechanical automation (AIEA). IEEE, pp 99–102
24. Hanif M, Mohammad N, Harun B (2019) An effective combination of microcontroller and PLC for home automation system. In: 2019 1st international conference on advances in science, engineering and robotics technology (ICASERT). IEEE, pp 1–6

An Automated and Online-Based Medicine Reminder and Dispenser

Shayla Sharmin, Md. Ibrahim Khulil Ullah Ratan, and Ashraful Haque Piash

Abstract It is important to take the right medications at the right times and in the right amounts. Patients, on the other hand, often fail to take their medications at the times specified in their prescriptions, causing disease or illness to develop more slowly, especially in the elderly or those who are too preoccupied with their job. An automated and online-based medicine reminder and dispenser application is introduced in this paper. A three-part package, an LCD on top of the box, a buzzer, and a multicolored LED light were all included in this unit. This interface also reminds the consumer when it is time to take their medicine. An Android application on this device displays some of the results. The input interface and the output interface are the two components of this mobile app interface. Prescriptions are accepted or modified via the input interface. Empty lists, previous data, and whether or not to take medication are all shown on the data interface. The LED lights and buzzer will switch on when it is time to take medicine, and the LCD will show the prescription at the same time. So that the patient is conscious that drug time has arrived. The proposed method is assessed both quantitatively and subjectively. The results show that the success rate in terms of perfect functioning is 90%, and the participants scored the overall system on average 4.6 out of 5 in subjective assessment.

Keywords IoT · Arduino · Android · Medical dispenser

1 Introduction

People nowadays are susceptible to a variety of diseases, so staying healthy and fit is important and diseases can be prevent by medicines and drugs [1, 2]. Because of today's hectic lifestyles, people often fail to take their medications on time. However, patients must take the right doses at the right time to lead a healthy life, whether they are in a hospital or at home [3]. It is especially important for the elderly because they

S. Sharmin (✉) · Md. I. K. U. Ratan · A. H. Piash
Department of Computer Science and Engineering, Chittagong University of Engineering & Technology (CUET), Chattogram 4349, Bangladesh

M. S. Arefin et al. (eds.), *Proceedings of the International Conference on Big Data, IoT, and Machine Learning*, Lecture Notes on Data Engineering and Communications Technologies 95, https://doi.org/10.1007/978-981-16-6636-0_39

can mismanage and take one medication twice or fail to take the correct one [4, 5]. In certain nations, finding a caregiver for an elderly person is difficult. Adherence to medication refers to the level or degree to which a patient takes the appropriate drug according to a doctor's prescription at the appropriate time. Non-compliance has recently become a major issue, as many studies have shown that non-compliance may have a direct effect on the patient and raise healthcare costs. Medication non-adherence is a long-term, difficult, and costly problem that results in poor patient outcomes and drains healthcare resources [6]. The growing trend of prescription non-adherence can be attributed to a number of factors. Non-compliance is often due to a lack of confidence in the need for medication, a desire to prevent side effects, difficulty handling several dosages per day, or a variety of prescription regimens.

This project created an automatic dispenser that could be controlled by an Android phone, while also considering the importance of a proper intake scheduler, reminder, and monitoring system. The contribution of this work are given below:

- Developed a dispenser which has three compartments containing medicines.
- Proposed a system to remind medicine at the right time via buzzer and light.
- Proposed a model to notify when a storage is low.
- Tracked down whether users have taken their medications or not.
- Evaluated the framework with subjectively and quantitatively.

The following section outlines relevant works in Sect. 2, elaborates the proposed model in Sect. 3, and then implements and analyzes the findings in Sect. 4; before concluding the paper in Sect. 5.

2 Related Works

On the basis of the medication reminder dispenser, various works have been completed [7] recently.

Jabeena et al. [8] used hardboard to build the dispenser and also used GSM to send messages. Ulloa et al. [9] created an IoT-based smart medication dispenser, but this work does not include any warnings about low storage. In 2019, Chawariya et al. carried out a research on medication reminder systems in 2019, but their findings were not applied [10]. Park and Lim [11] and Pak and Park [12] proposed and developed a smart drug dispenser with high scalability and remote manageability. It also allows medical personnel and system administrators to deal with drug dispensers rather than end-users, which saves money and ensures safe device operation.

Ashwini et al. [13] present an Android application for patients that automatically sets reminders in the user's phone to remind them to take the proper medicines in the proper quantity at the proper time. Tiwari et al. [14] suggested robotic platforms that have the potential to extend the user interface's versatility to make personalized experiences more engaging and encouraging, as well as to proactively reach out to elderly users to assist with their healthcare delivery. A multi-robot prototype system that can deliver pills and water to a person in a real-world home environment was proposed by Emeli et al. [15]. The computer consists of a mobile robot with a tray, a stationary dispensing robot, and a smartphone kept by the user.

3 Proposed Methodology

The proposed model is described in detail, as well as the hardware which refers to the medicine dispenser and the software module thus the mobile app, in this section.

3.1 Proposed Approach

There are two sections to the proposed work. One is a hardware module, such as a dispenser, and the other is an Android app that can be mounted on any smartphone. The input in the mobile app includes the name of the drug, the amount, the time, and meal detail, among other things. These inputs will be stored in a text file in JSON format on a cloud server. The data is then processed by the Wi-Fi module and Arduino, and the countdown begins. If it is time for treatment, the prescription will appear on the LCD screen. This is where the user will find the names of all the medicines you've been given, as well as their dosages. The alarm often sounds to warn the user to take his or her medication on time. Figure 1 depicts the suggested technique.

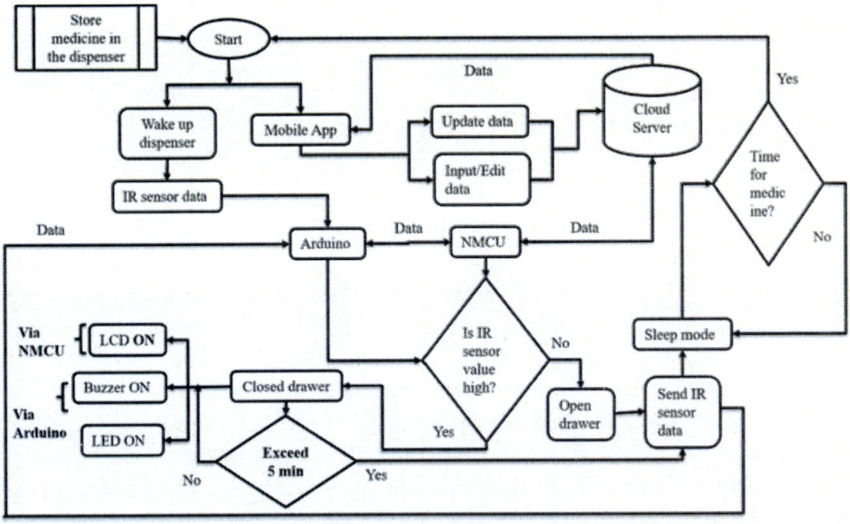

Fig. 1 Process diagram of the proposed medicine dispenser

The proposed medicine box has three drawers that divide one day into morning, noon, and night parts. Each drawer holds a different kind of medication. In addition, each drawer in the box has an LED light that indicates which medication must be taken right now. The various LED lights show the parts of the medication are taking longer. For example, if the LED light in the first drawer turns on, it means that it is time for morning medicine.

The dispenser is initially refilled with medication according to the time. In the mobile app, all of the data was collected at the start. When it was time to take medication, the Arduino-enabled LED lights and a buzzer, assisting the user in taking medicine on time. In addition, at that moment, Node MCU displays the medicine information collected from the cloud on an LCD monitor at the required time. When the user opens the drawer, the value of the IR sensors drops, indicating that the user has opened the drawer to take medication. After the running task of closing the drawer is completed, it will move on to the next task. As a result, the device is up and running, and the user is reminded at the appropriate time. If the drawer is opened within five minutes of the alarm, the system sends the message "yes" to the cloud; if the user does not open it, the system sends the message "no." These messages are stored in the cloud and shown in the mobile app, assisting the user in remembering to take medicine and alerting the caregiver that his or her patient is having difficulty taking medications. When the dispenser storage becomes zero, the Arduino activates the LED light at the top of the package, as well as the buzzer, to notify the consumer that medicine storage is limited. If the drawer is opened, the inventory status in the app will be changed.

The data from each drawer is saved in the cloud server after it is opened and closed. These data can be seen in the mobile app, and the prescription can also be changed through the app.

3.2 *Designing of the Medical Dispenser*

The medicine dispenser is permitted to store the medications and remind the patient when it is time to take them. It also alerts the customer when storage space is running low. An Arduino, a Wi-Fi module, IR sensors, a buzzer, an LCD monitor, and an LED are all included in the dispenser's configuration. The Arduino uses a cloud server to transfer data from the IR sensor to an Android app, which is then used to decide if the medicine door has been opened or not. The Node MCU connects the box to the cloud server and displays medicine data on the LCD panel.

Figure 2 shows the structure of the proposed medicine dispenser. Figure 2a shows the front view of the dispenser and an Android mobile which has the installed medicine reminder app Fig. 2b shows the top view of the dispenser showing all the circuit components.

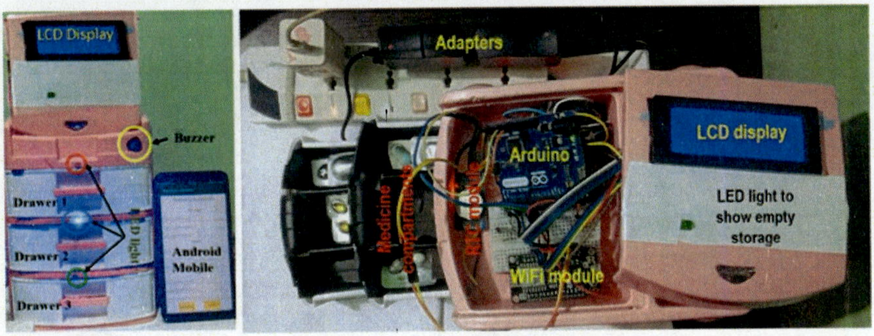

(a) (b)

Fig. 2 Proposed model **a** front view **b** top view

3.3 *Designing the Mobile App*

The mobile app has two interfaces: one is an input interface for taking detailed medication information, and the other is an output interface that displays inventory status and whether or not the drug has been taken.

In this input interface, the user is allowed to insert medicine name, time, unit along with the amount of medicines loaded. Figure 3a shows the input interface indicating the fields. Figure 3b shows the Data interface. It shows how the user can track down his/her medication.

After entering the users' details, this data interface allows them to view the inventory and prescription. If the drawer is opened, the Node MCU sends a "yes" message to the server, otherwise "no," and the app retrieves the medication consumption status. If the system receives a "yes," it will change the inventory by decreasing the amount of medication the user has taken and notifying the user that medicine has been taken, indicating "yes." If the system does not receive a "yes," the inventory will not be changed and the status will remain "no." The flowchart Fig. 4 depicts the medication monitoring process.

4 Implementation and Result

In this section the implementation of the propose method and result analysis have been described.

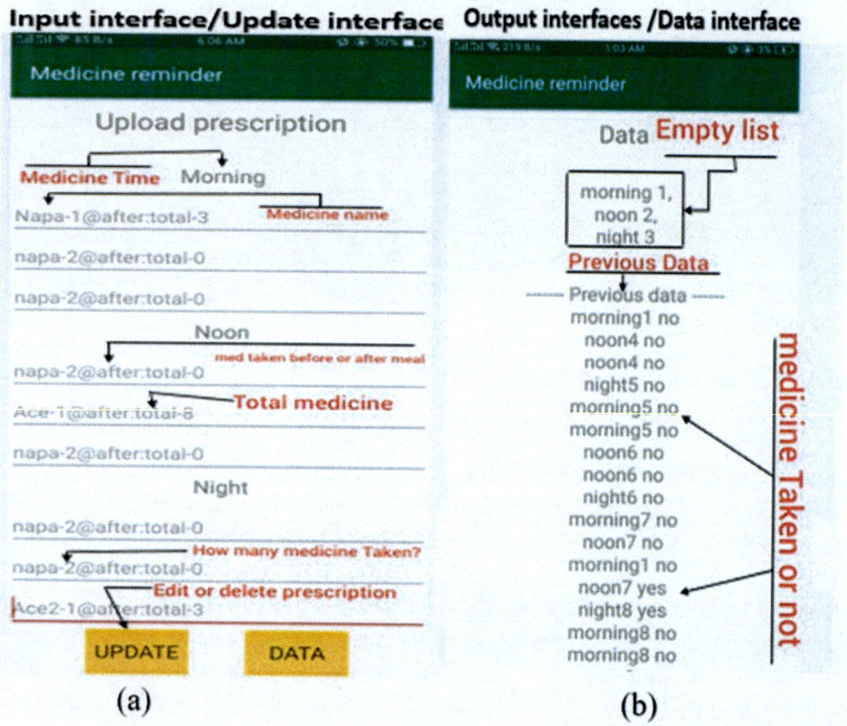

Fig. 3 Mobile app: **a** input medicine information **b** showing inventory and medicine consumption status

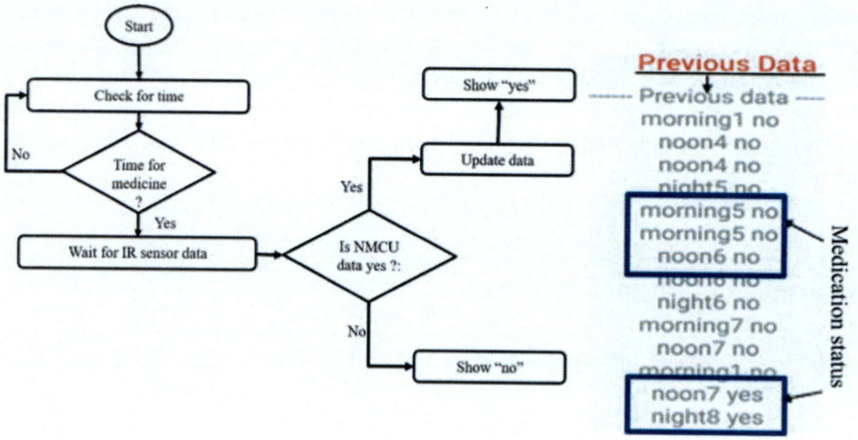

Fig. 4 Mobile app: tracking medication and updating inventory status

4.1 Implementation

To implement the proposed methodology a dispenser box has been installed at a particular place and the app is installed in at least two mobile phone. Figure 5a shows that it is time for medicine at noon as the middle drawer light is shown. Also the LCD is showing which medicine need to be taken. Figure 5 shows the alert when the dispenser is out of medicine. Figure 6a shows the initial inventory status of the dispenser, (b) shows that updated inventory status and (c) shows the message "yes" that means the user has taken the medicine. On the other hand, Fig. 7a shows the initial inventory, (b) shows that the inventory has not been updated, and (c) shows the message "no" that has been shown which refers that the user did not take the medicine.

4.2 Experimental Environment and Result Analysis

The main goal of this project is to create a medical reminder dispenser that is simple to use and keeps accurate track of medication. There is a box in this proposed architecture that holds medicines that the consumer has allocated. The patient's and caregiver's phones both have an Android app installed. For the assessment of the proposed method, both subjective and quantitative analysis were used.

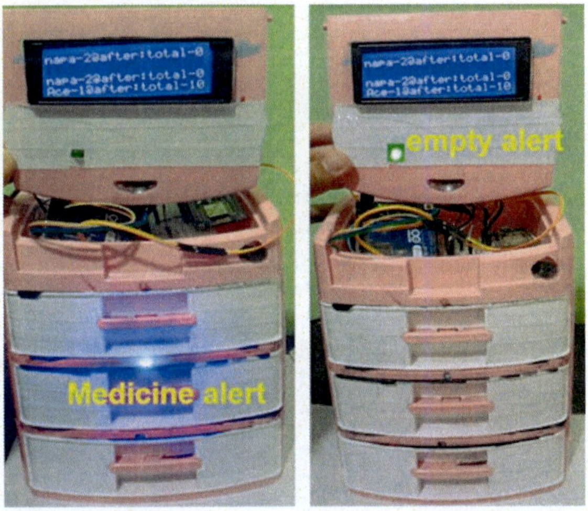

(a) (b)

Fig. 5 Medicine dispenser **a** medicine consumption alert **b** empty storage alert

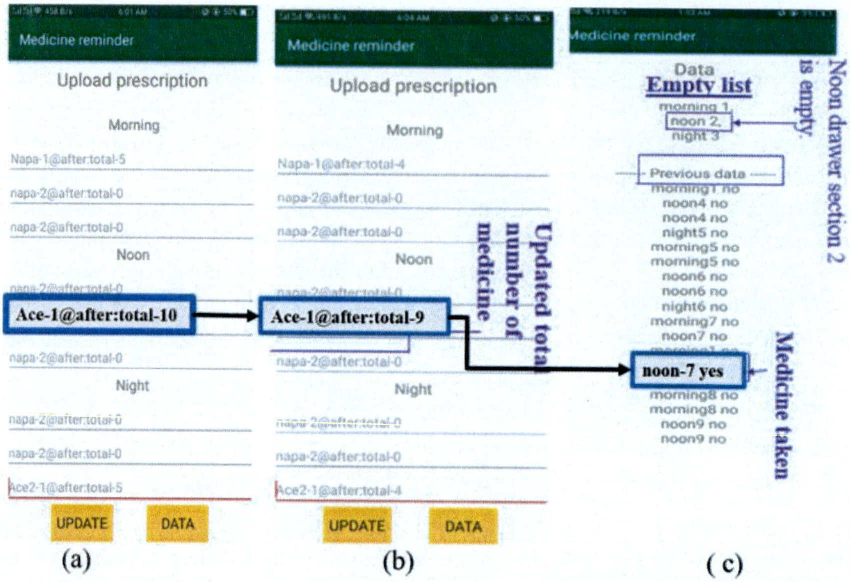

Fig. 6 Mobile app **a** initial inventor **b** updated inventor **c** medicine consumption status ("yes")

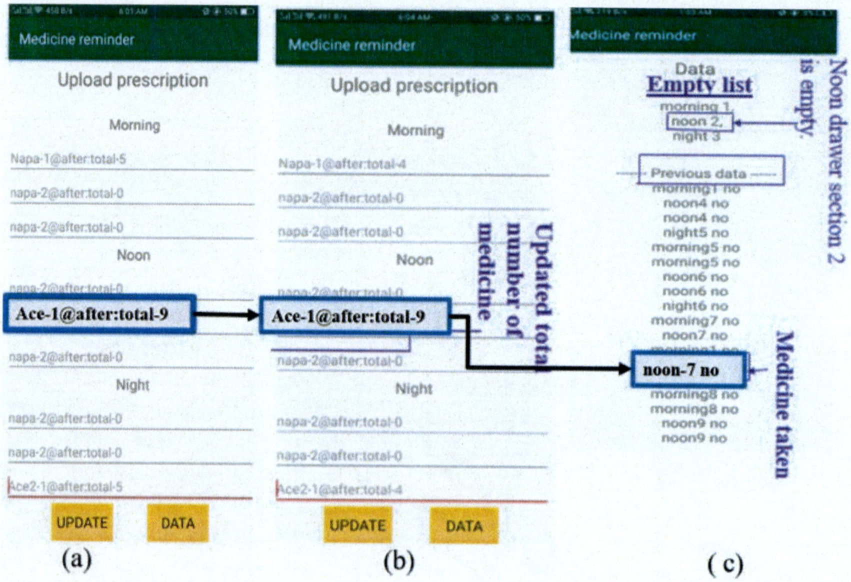

Fig. 7 Mobile app **a** initial inventor **b** unchanged inventor **c** medicine consumption status ("no")

4.2.1 Experimental Setup and Configuration

The dispenser was placed in the Tilpapara, Khilgaon, Dhaka-1219, Bangladesh, to evaluate our proposed system. We also installed the Android app in patient's and caregiver's mobile. To build this dispenser, we used Arduino leonardo, Wi-Fi module (ES8266), IR sensor, breadboard, box, buzzer, LCD display, LED, etc. The box is made of plastic with three drawers where we planned to put the medicines.

4.2.2 Participant

The proposed approach was evaluated by a total of ten patient participants. Every participant's home had the device installed for one day. The patient's phone and their caregiver's phone have both been updated with the mobile app (son, daughter, etc.). In general, two apps were installed on each unit, and the assessment included ten patients and ten caregivers.

4.2.3 Subjective Evaluation

To evaluate the system subjectively, the patients and the caregivers were given questionnaires. The participants were asked to evaluate the system by 1–5 scale Likert scale where 1 refers low and 5 represents high, and three questions that have been asked are listed below:

- Did you find that system provides accurate timing and information?
- Did you think that you without prior theoretical knowledge this system is easy to use?
- Rated overall expression on this system.

The outcome is depicted in Fig. 8. The average score for questions from patients and caregivers is 4.6, indicating that the system's time management and information delivery were satisfactory. When it came to providing theoretical expertise, the caregiver scored slightly higher (4.7) than the patients (4.3). The average score from both the patient and the caregiver was 4.6 and 4.7, respectively.

4.3 Quantitative Analysis

Patients and caregivers were allowed to use the six features of our project for quantitative analysis. The total number of attempts was twenty. Eighteen attempts were successful, and two attempts were unsuccessful because of power failure. As a result, it is concluded that project's success rate is 90%. The quantitative evaluation analysis is given in Table 1.

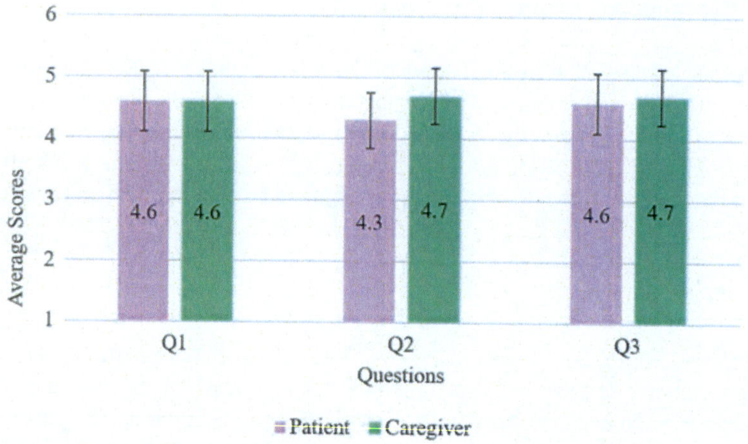

Fig. 8 Mean value of Q_1, Q_2, and Q_3 where error bar shows the standard deviation

Table 1 Quantitative analysis

# of Participants		Total attempts, T_N	Successful attempts, T_S	Unsuccessful attempts
Patient	Caregiver			
10	10	20	18	2
Success ratio = $\frac{T_S}{T_N} * 100\% = 90\%$				

4.4 Comparison with the Previous Work

We have introduced some new features which minimize the limitation of the previous works which is given in Table 2.

Table 2 Comparison with previous works

Features of previous works	Features of the proposed model
Used hardboard to build the dispenser and also used GSM to send messages [8]	Used a plastic box and Wi-Fi module which is more convenient
Did not give low storage warning [8–10]	Low storage alarm has been introduced
Developed only a Android application [13]	Developed both a dispenser and Android application
Emphasized only on elder people [14]	Kept in mind all ages

5 Conclusion

The key focus of the project is the development of a reminder system for elderly and busy people who often fail to take their medications on time. This proposed system focuses on a modern technology-based drug system. The reminding system aims to raise awareness among the elderly and workaholics by providing services that enable them to live a comfortable and stress-free life. By defining such a crucial problem, the proposed structure defines core features of the work. The device is capable of meeting requirements in any way. The proposed framework can be used everywhere. There is no need to put this in a specific location. As a consequence, the subject of setting space is unimportant. As a consequence, it is really simple to set up and use. This device's primary function is to remind them to take their drugs at the appropriate times. The participants gave the systems an average of 4.6 out of 5 for overall results, with 90 % accuracy based on providing medicine information in real time via buzzer and light. We attempted to keep the cost of this device reasonable, and the project's overall cost was 3340 BDT (39 USD).

While the proposed methodology works flawlessly, it does have some flaws, such as the possibility of it ceasing to operate due to a power outage. A Lipo battery can be used to solve this issue. The mobile app is missing details, such as a medication timer warning, which should be included. To make the framework more user friendly, it is intended to create a multi-user app and redesign the mobile interface to make it easier to understand. In addition, potential work will include automatic drawer opening and closing.

References

1. Tsai PH, Shih CS, Liu JWS (2011) Mobile reminder for flexible and safe medication schedule for home users. In: International conference on human-computer interaction, pp 107–116
2. Tsai PH, Yu CY, Wang MY, Zao JK, Yeh HC, Shih CS, Liu JWS (2010) iMAT: intelligent medication administration tools. In: The 12th IEEE international conference on e-health networking, applications and services, pp 308–315
3. Kader MA, Uddin MN, Arfi AM, Islam N, Anisuzzaman M (2018) Design & implementation of an automated reminder medicine box for old people and hospital. In: 2018 International conference on innovations in science, engineering and technology (ICISET), pp 390–394
4. Ahmad S, Hasan M, Shahabuddin M, Tabassum T, Allvi MW (2020) IoT based pill reminder and monitoring system. Int J Comput Sci Netw Secur 20:152–158
5. Kumar SB, Goh WW, Balakrishnan S (2018) Smart medicine reminder device for the elderly. In: 2018 Fourth international conference on advances in computing, communication & automation (ICACCA), pp 1–6
6. Dharmoji S, Patil A, Anigolkar A (2020) Design of smart medicine reminder (SMR) box with an android application. Int J Innov Res Sci Eng Technol 9:4207-4214
7. Suwanthara J, Noinongyao A, Vittayakorn S (2019) WiseMed: medication reminder for seniors. In: 2019 23rd International computer science and engineering conference (ICSEC), pp 409–414
8. Jabeena A, Sahu AK, Roy R, Basha N (2017) Automatic pill reminder for easy supervision. In: 2017 International conference on intelligent sustainable systems (ICISS), pp 630–637

9. Ulloa GG, Hornos MJ, Dominguez CR, Coello MF (2020) IoT-based smart medicine dispenser to control and supervise medication intake. Intell Environ 28:39–48
10. Chawariya A, Chavan P, Agnihotri A (2019) Fundamental research on medication reminder system. Int Res J Eng Technol (IRJET) 6
11. Park K, Lim S (2012) Construction of a medication reminder synchronization system based on data synchronization. Int J Bio-Sci Bio-Technol 4:1–10
12. Pak J, Park K (2012) Construction of a smart medication dispenser with high degree of scalability and remote manageability. J Biomed Biotechnol 2012:381493–381503
13. Ashwini B, Sapna K, Ishwari B, Pallavi P, Achaliya PN (2013) An Android based medication reminder system based on OCR using ANN. Int J Comput Appl 975:8887
14. Tiwari P, Warren J, Day K, MacDonald B, Jayawardena C, Kuo IH, Igic A, Datta C (2011) Feasibility study of a robotic medication assistant for the elderly. In: Proceedings of the twelfth Australasian user interface conference, vol 117, pp 57–66
15. Emeli V, Wagner AR, Kemp CC (2012) A robotic system for autonomous medication and water delivery. Georgia Institute of Technology

Pattern Recognition and Classification

Power Transformer Fault Diagnosis with Intrinsic Time-Scale Decomposition and XGBoost Classifier

Shoaib Meraj Sami and Mohammed Imamul Hassan Bhuiyan

Abstract An intrinsic time-scale decomposition (ITD)-based method for power transformer fault diagnosis is proposed. Dissolved gas analysis (DGA) parameters are ranked according to their skewness, and then ITD-based features extraction is performed. An optimal set of PRC features is determined by an XGBoost classifier. For classification purpose, an XGBoost classifier is used to the optimal PRC features set. The proposed method's performance in classification is studied using publicly available DGA data of 376 power transformers and employing an XGBoost classifier. The proposed method achieves more than 95% accuracy and high sensitivity and F_1-score, better than conventional methods and some recent machine learning-based fault diagnosis approaches. Moreover, it gives better Cohen Kappa and F_1-score as compared to the recently introduced EMD-based hierarchical technique for fault diagnosis in power transformers.

Keywords DGA · Power transformer fault · Intrinsic time-scale decomposition · XGBoost

1 Introduction

Power transformer is one of the most essential equipments for power transmission and distribution system. Monitoring the condition and fault diagnosis is very important for ensuring uninterrupted electricity supply. Due to thermal, electrical stress of insulation material and aging a variety of faults occur in power transformers. These faults have strong correlation with concentration of the dissolved gas emitted from the oil or cellulose paper. The gases include hydrogen (H_2), methane (CH_4), ethane (C_2H_6), ethylene (C_2H_4) and acetylene (C_2H_2). Dissolved gas analysis (DGA) methods such

S. M. Sami (✉) · M. I. H. Bhuiyan
Department of Electrical and Electronic Engineering, Bangladesh University of Engineering and Technology, Dhaka 1205, Bangladesh

M. I. H. Bhuiyan
e-mail: imamul@eee.buet.ac.bd

M. S. Arefin et al. (eds.), *Proceedings of the International Conference on Big Data, IoT, and Machine Learning*, Lecture Notes on Data Engineering and Communications Technologies 95, https://doi.org/10.1007/978-981-16-6636-0_40

as Duval Triangle, Rogers Ratio, IEC method and Dornenburg ratio method are widely used for the detection of power transformer faults [1–3]. A limitation of these methods is occasional poor performance and ambiguity to detect fault. To overcome those shortcomings, many machine learning-based approaches are proposed in the literature [4–7].

Recently, empirical mode decomposition (EMD) based feature extraction from ranked features is shown to provide promising results for transformer fault detection [5]. However, compared to EMD, intrinsic time-scale decomposition (ITD) has several benefits because ITD consider proper rotational property of any non-stationary signal [7]. As such it can provide more information, while being computationally efficient and more robust to noise [9, 10]. Interestingly, ITD has also been successfully to analyze a variety of non-stationary signals and related prediction with machine learning [10–13].

To the best of our knowledge, the use of ITD for the detection of transformer faults using machine learning is yet to be reported. In this paper, a number of DGA parameters used in the traditional methods are first ranked by their skewness. A subset of the DGA parameters in terms of their increasing rank are decomposed into the ITD domain. The resulting PRC components are used as features and classified by an XGBoost classifier for transformer fault detection. Note that unlike [5], the proposed approach employs a single XGBoost classifier thus, reducing complexity in the classification system and computational time. The performance of the proposed method is studied using publicly available DGA data of 376 transformers and compared with those of recently reported results.

2 The DGA Dataset

The DGA data of 376 power transformers is used. Among them, data of 239 transformers are collected from publicly available Egyptian Electricity Network samples [14]. Others are collected from different published scientific literatures [2, 14]. Six different types of faults are used. These are partial discharge (PD), low-energy discharge (D_1), high-energy discharge (D_2), and three different types of thermal faults those are T_1 (temperature less than 300 °C), T_2 (temperature between 300 to 700 °C) and T_3 (temperature greater less than 700 °C). A summary of the DGA data is provided in Table 1.

Table 1 Distribution of different fault types

Fault type	PD	D_1	D_2	T_1	T_2	T_3	Overall
Lab samples	27	42	54	70	18	28	239
Literatures sample	15	25	59	10	3	25	137
Total	42	67	113	80	21	53	376

Notice that the dataset is slightly unbalanced because about 30% of samples belongs to $_{D2}$ fault and 5.59% to T2 fault. This will be taken into account during performance analysis of the proposed method described in Sect. 4.

3 Methodology

In this section, DGA parameter generation and their ranking, ITD-based feature extraction and optimal feature selection are described. Hydrogen (H_2), methane (CH_4), ethane (C_2H_6), ethylene (C_2H_4) and acetylene (C_2H_2) are main DGA gas parameter for this purpose.

3.1 Ratio-Based DGA Parameter Generation and Ranking

In the literature, many ratio-based DGA parameters are generated from Hydrogen (H_2), methane (CH_4), ethane (C_2H_6), ethylene (C_2H_4) and acetylene (C_2H_2). We used thirty-seven DGA parameters from this work, which is collected from literature [5]. Those parameters are illustrated in Table 2. Different ratio-based parameters provide different discriminatory properties of faulty transformer.

After generating different DGA ratio-based parameters, we rank them by skewness. For this purpose, we calculate skewness of each parameter for the 376 transformers. After that the parameters are ranked from lower skewness to higher skewness. Similar work is also performed in the literature [5]. This ranking scenario is presented in Table 2. We also investigate the distribution of transformer DGA parameter among six classes of faults using Box plots (Fig. 1). One can see that with increasing skewness the distribution of DGA parameters become more indistinguishable and ambiguous.

3.2 Intrinsic Time-Scale Decomposition Based Feature Extraction and Optimal Feature Set Selection

In this section, feature extraction by using intrinsic time-scale decomposition (ITD)-based feature extraction and optimal feature set selection are discussed. ITD is an efficient tool for extracting amplitude and frequency changing pattern from any nonlinear signal. In comparison, the well-known empirical mode decomposition (EMD) has some shortcomings: (i) inaccurate outcomes when signal dynamics is considered, (ii) occasional failure to generate residual from IMF, when it has proper rotational property [8]. Thus, ITD is more effective than EMD for representing nonlinear and non-stationary signals and data decomposition [8–10].

Table 2 The DGA parameters

No.	Parameter	No.	Parameter	No.	Parameter
1	H_2/TH	14	H_2	27	C_2H_2/THD
2	CH_4/TH	15	CH_4	28	H_2/THH
3	C_2H_6/TH	16	C_2H_6	29	CH_4/THH
4	C_2H_4/TH	17	C_2H_4	30	C_2H_6/THH
5	C_2H_2/TH	18	C_2H_2	31	C_2H_4/THH
6	C_2H_2/H_2	19	TH	32	C_2H_2/THH
7	C_2H_2/CH_4	20	THD	33	H_2/TCH
8	C_2H_2/C_2H_6	21	THH	34	CH_4/TCH
9	C_2H_2/C_2H_4	22	TCH	35	C_2H_6/TCH
10	C_2H_4/H_2	23	H_2/THD	36	C_2H_4/TCH
11	C_2H_4/H_4	24	CH_4/THD	37	C_2H_2/TCH
12	C_2H_4/C_2H_6	25	C_2H_6/THD		
13	$(C_2H_4/H_2) + (C_2H_4/CH_4)$	26	C_2H_4/THD		

$TH = H_2 + CH_4 + C_2H_6 + C_2H_4 + C_2H_2$; $THD = CH_4 + C_2H_4 + C_2H_2$; $THH = H_2 + C_2H_4 + C_2H_2$; $TCH = CH_4 + C_2H_6 + C_2H_4 + C_2H_2$

Above DGA parameters ranked by skewness (lower to higher)

	DGA parameters sorted by considering by higher to lower rank
Features no. from above	28, 24, 1, 27, 31, 37, 26, 35, 36, 3, 32, 2, 34, 4, 5, 33,21,14,19,20, 13, 10, 23, 6, 22, 15, 18, 17, 7, 8, 16, 11, 9, 12, 25, 30 and 29

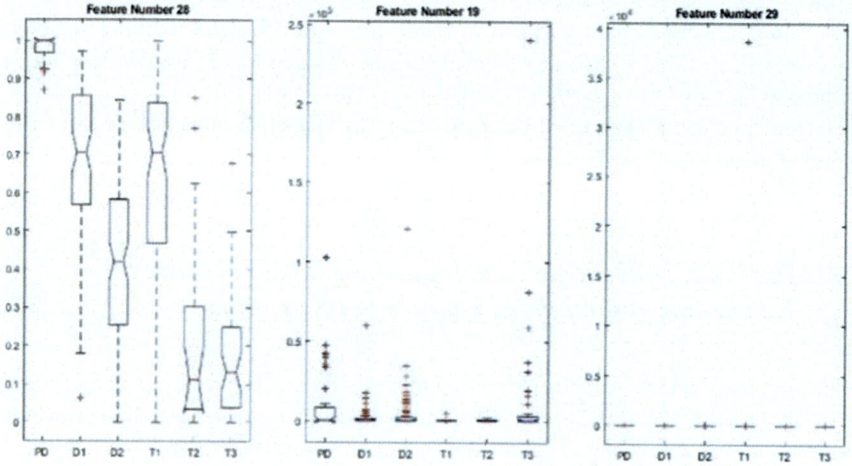

Fig. 1 Box plots of three DGA parameters (28, 19, 29 from Table 1) ranked by skewness

Intrinsic time-scale decomposition algorithm decomposes data series into integer sum of proper rotation components (PRCs) and residual signal. We decomposed each transformer's ranked features set into single-stage ITD domain and extracted the proper rotation component (PRC). For this purpose, initially, we decompose the first eighteen ranked parameters of each transformer. Those eighteen DGA parameters are ranked considering lower to higher skewness and shown in Table 2. We continue the process adding next DGA-ranked parameter and so on. Therefore, total nineteen PRC feature sets are obtained. There first set element is eighteen, and last set element is thirty-seven. For getting optimal set of features, we classify those nineteen features set by XGBoost classifier among 376 transformer's dataset, where training and testing are split in 85:15 ratio randomly. Those nineteen features set performance scenario is depicted in Fig. 2. One can see that first ranked twenty-four extracted features provide the best classification performance. Thus, PRC coefficients of ranked twenty-four DGA features (i.e., 28, 24, 1, 27, 31, 37, 26, 35, 36, 3, 32, 2, 34, 4, 5, 33, 21, 14, 19, 20, 13, 10, 23, 6) are our final feature set. The twenty-four features are then used in the classification schemes. Figure 3 shows that the PRC features provide good separation among classes. Moreover, the final twenty-four PRC features provide low p-values in general in one-way analysis of variance (ANOVA) test, average p-value being 0.0879. The thirty-seven DGA parameter is considered as thirty-seven data channels, each consisting of 376 data value. The ranking of the channels provides a pattern in which the statistically less informative channels are ordered after the more

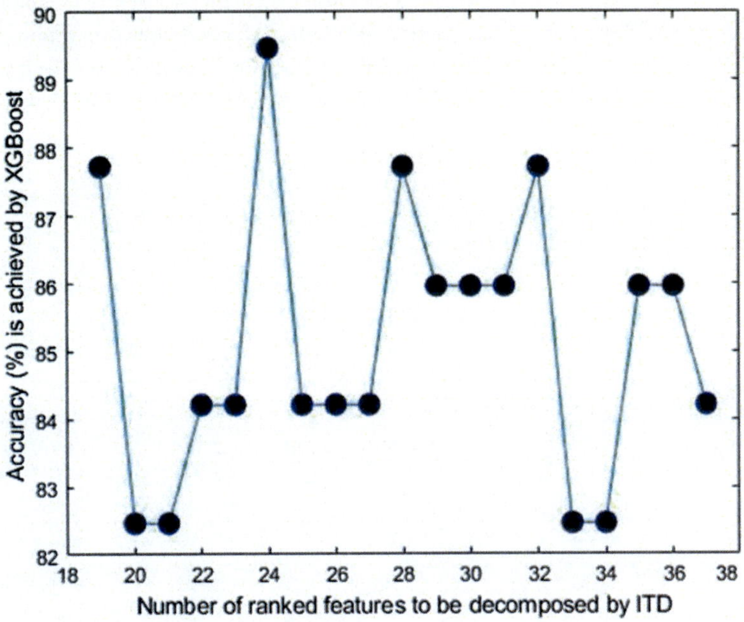

Fig. 2 Values of accuracy obtained by the XGBoost (for optimal PRC set selection)

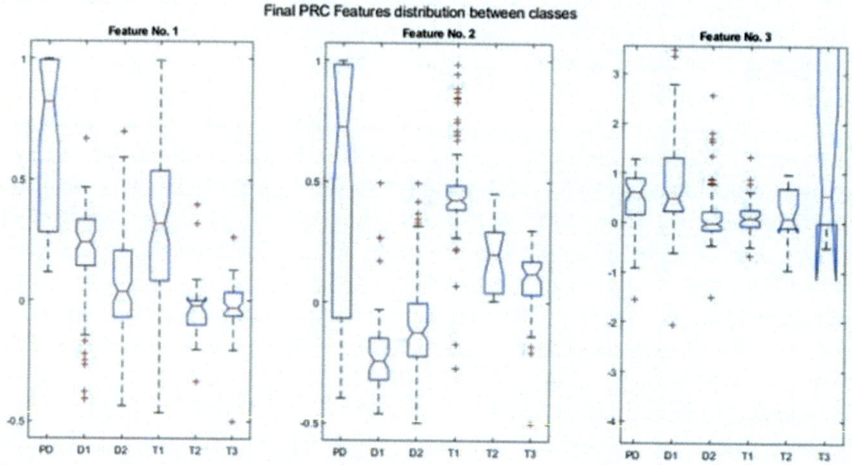

Fig. 3 Box plots of the three PRC features (coefficients) among twenty-four among the six different classes of faults

informative channels, thus providing better depiction of the fault classes. The low *p*-value of the features also indicates that. Note that ranking can also be done by Lasso, Laplacian score, Fisher score and Relief that have been used for optimal selection of DGA features [15]. The average p-value for the twenty-four PRC features are found to be 0.4789, 0.3267, 0.239 and 0.2479 Lasso, Laplacian score, Fisher score and Relief, respectively, much higher than obtained with skewness. Thus, skewness-based ranking is more discriminative than the others. In the next section, we discuss the classification procedure for fault diagnosis.

3.3 Classification Method

For classification purpose, we use single XGBoost (Extreme Gradient Boosting) classifier which is widely used machine learning-based classification schemes [7]. A single XGBoost classifier is used for detection of six class power transformer faults using the twenty-four PRC features described above. The implementation of XGBoost classifier is performed with default parameters in Python 3.7 environment. In the next section, we will discuss classification performance in our proposed method with others conventional and machine learning-based approaches.

4 Results and Discussion

In this section, the experimental setting and performance analysis of the proposed method are presented. The experiments are carried out in 4 GB RAM and 2.11 GHz Intel Core-i5 processor-based Windows-10 PC. This classification process is performed in Python 3.7 and feature extraction carried out in Matlab-2020b platform.

A total of 376 transformer DGA samples are used. Among them, randomly 333 is used for training and the rest 43 in the test set. In Table 3, the confusion matrix is presented where in bracket the number of samples in each fault class is provided. The performance of the proposed method is investigated using accuracy, sensitivity and F_1-score. The average sensitivity is 90%, and the value of F_1-score for each fault class is high, mostly near 1 or 1.

The conventional method such as IEC method and Rogers four ratio method provides 'No Fault (NF)' and 'Undefined (UD)' states. Duval triangle makes false prediction in its boundary region. To illustrate this, the classification results of six randomly selected transformers are shown in Table 4. One can see that the Duval triangle can truly predict only one transformer fault class among six, Rogers four ratio method predicts 'Undefined' states on four samples, while the IEC method predicts four 'UD' and two 'NF' falsely. On the other hand, the proposed method predicts the faults accurately.

Table 5 shows the performance comparison among various methods. In general, the proposed technique gives superior accuracy, about 32%, 9% and 4.65% higher than the conventional Duval Triangle Method, BA-PNN and EMD-based Hierarchical Ensemble Method, respectively.

The proposed method is also more efficient computationally than the EMD-based technique. For training and testing, it requires 0.108 s and 0.00201 s, respectively, whereas the EMD-based one needs 0.456 s and 0.1456 s. The lower time required and better accuracy for ITD-based method as compared the Hierarchical EMD-based technique can be attributed to; (i) the capacity of ITD of giving superior proper rotation components and hence, following the trend of a nonlinear data much better, and (ii) EMD employing spline interpolation requiring more memory and

Table 3 Confusion matrix of proposed method (among random 43 samples)

Actual	Predicted							
	PD	D_1	D_2	T_1	T_2	T_3	Sensitivity (%)	F_1-score
PD (4)	4	0	0	0	0	0	100	1
D_1 (7)	0	7	0	0	0	0	100	0.9333
D_2 (15)	0	1	14	0	0	0	93.33	0.9333
T_1 (6)	0	0	0	6	0	0	100	1
T_2 (3)	0	0	1	0	2	0	66.67	0.8
T_3 (8)	0	0	0	0	0	8	100	1

Table 4 Proposed method and conventional method performance (among different fault samples)

H_2 (ppm)	CH_4 (ppm)	C_2H_6 (ppm)	C_2H_4 (ppm)	C_2H_2 (ppm)	Actual fault	Duval method	Rogers four ratio method	ICE method	Proposed method
292	346	32	313	196	D_2	D_2	UD	UD	D_2
385	28.8	50	82.3	171	D_1	D_2	UD	UD	D_1
34	8.6	70.3	3.1	0.001	T_1	T_2	T_1	NF	T_1
157	46	76	12	0.001	T_1	T_2	T_1	NF	T_1
10	15	0.001	0.001	35	D_2	D_1	UD	UD	D_2
1651	90	33	45	2	PD	T_2	UD	UD	PD

Table 5 Performance comparison among different fault classes

Method	Fault						
	PD	D_1	D_2	T_1	T_2	T_3	Average accuracy
Duval method (%)	25	42.86	73.33	66.67	33.33	87.50	62.79
Rogers four ratio method (%)	0	0	66.67	100	33.33	25	44.19
IEC method (%)	0	28.57	53.33	66.67	33.33	62.50	46.51
Ensemble learning [7] (%)	100	57.14	93.33	100	66.67	87.50	86.05
BA-PNN [6] (%)	75	42.86	66.67	66.67	66.67	62.5	62.79
EMD-based H-XGBoost [5] (%)	100	85.71	93.33	83.33	100	87.50	90.70
Proposed method (%)	100	100	93.33	100	66.67	100	95.35

computation time. Moreover, the later Method uses three XGBoost classifiers in a hierarchical mode of classification.

As mentioned before, the dataset used in this paper is imbalanced. This can bias the obtained accuracy toward the majority classes. F_1-score and Cohen Kappa are well-known matrices for performance evaluation in classification using imbalanced dataset [16]. Also, the Synthetic Minority Oversampling Technique (SMOTE) is invariably used with imbalanced dataset for experimentations in classification [17, 18]. A fivefold cross-validation is performed with SMOTE for the proposed method as well as the EMD-based Hierarchical Technique [5]. The cross-validation techniques are employed to account for the relatively moderate size of the dataset. The proposed method yields an Cohen Kappa and F_1-Score of 0.91 and 0.92, respectively, while the EMD-based Method gives 0.89 and 0.90.

5 Conclusion

A novel transformer fault diagnosis method has been proposed in this paper. The DGA parameters of publicly available 376 transformers have been generated and then ranked by their skewness. Intrinsic time-scale decomposition (ITD) has been used to extract features from skewness-based ranked parameters. The motivation for using ITD has been its improved capacity to capture the trend in nonlinear signals at reduced computational complexity as compared the well-known EMD. An XGBoost classifier has been employed to obtain the optimal set of extracted PRC features. The classification performance of the proposed method has been studied and compared with several existing DGA-based techniques. Conventional methods give 'No fault' or 'Undefined State' in many cases or suffers from modest accuracy. The proposed method overcomes these limitations and has provided more than 95% average accuracy as well as high F_1-score and sensitivity in each fault classes, better than conventional methods and several machine learning-based power transformer fault diagnosis method. Since the DGA dataset is slightly imbalanced, the effectiveness of the proposed method has been further investigated using a fivefold cross-validation and SMOTE. The results have shown that the ITD-based proposed method provides a better performance than EMD-based hierarchical approach in terms of F_1-score and Cohen Kappa score.

References

1. Duval M, Dukarm J (2005) Improving the reliability of transformer gas-in-oil diagnosis. IEEE Electr Insul Mag 21:21–27. https://doi.org/10.1109/mei.2005.1489986
2. Duval M, de Pabla A (2001) Interpretation of gas-in-oil analysis using new IEC publication 60599 and IEC TC 10 databases. IEEE Electr Insul Mag 17:31–41. https://doi.org/10.1109/57.917529
3. Faiz J, Soleimani M (2017) Dissolved gas analysis evaluation in electric power transformers using conventional methods a review. IEEE Trans Dielectr Electr Insul 24:1239–1248. https://doi.org/10.1109/tdei.2017.005959
4. Bacha K, Souahlia S, Gossa M (2012) Power transformer fault diagnosis based on dissolved gas analysis by support vector machine. Electr Power Syst Res 83:73–79. https://doi.org/10.1016/j.epsr.2011.09.012
5. Meraj Sami S, Imamul Hassan Bhuiyan M (2021) An EMD-based method for the detection of power transformer faults with a hierarchical ensemble classifier. In: 2020 11th International conference on electrical and computer engineering (ICECE), pp 206–209. https://doi.org/10.1109/ICECE51571.2020.9393037
6. Yang X, Chen W, Li A et al (2019) BA-PNN-based methods for power transformer fault diagnosis. Adv Eng Inform 39:178–185. https://doi.org/10.1016/j.aei.2019.01.001
7. Liu Y, Li J, Li Z et al (2019) Transformer fault diagnosis model based on iterative nearest neighbor interpolation and ensemble learning. In: Proceedings of the 2019 2nd international conference on data science and information technology. https://doi.org/10.1145/3352411.3352434
8. Frei M, Osorio I (2006) Intrinsic time-scale decomposition: time–frequency–energy analysis and real-time filtering of non-stationary signals. Proc Roy Soc A Math Phys Eng Sci 463:321–342. https://doi.org/10.1098/rspa.2006.1761
9. Mohamed E, Yusoff M, Malik A et al (2018) Comparison of EEG signal decomposition methods in classification of motor-imagery BCI. Multimedia Tools Appl 77:21305–21327. https://doi.org/10.1007/s11042-017-5586-9
10. Voznesensky A, Kaplun D (2019) Adaptive signal processing algorithms based on EMD and ITD. IEEE Access 7:171313–171321. https://doi.org/10.1109/access.2019.2956077
11. Pazoki M (2018) A new DC-offset removal method for distance-relaying application using intrinsic time-scale decomposition. IEEE Trans Power Deliv 33:971–980. https://doi.org/10.1109/tpwrd.2017.2728188
12. Wang J, Zhou N, Li T, Wang Q (2016) A forecasting method for metering error of electric energy based on intrinsic time-scale decomposition and time series analysis. In: 2016 IEEE Innovative smart grid technologies—Asia (ISGT-Asia). https://doi.org/10.1109/isgt-asia.2016.7796439
13. An X, Jiang D, Chen J, Liu C (2011) Application of the intrinsic time-scale decomposition method to fault diagnosis of wind turbine bearing. J Vib Control 18:240–245. https://doi.org/10.1177/1077546311403185
14. Ghoneim S, Taha I (2016) A new approach of DGA interpretation technique for transformer fault diagnosis. Int J Electr Power Energy Syst 81:265–274. https://doi.org/10.1016/j.ijepes.2016.02.018

15. Kari T, Gao W, Zhao D et al (2018) Hybrid feature selection approach for power transformer fault diagnosis based on support vector machine and genetic algorithm. IET Gener Transm Distrib 12:5672–5680. https://doi.org/10.1049/iet-gtd.2018.5482
16. Wardhani N, Rochayani M, Iriany A et al (2019) Cross-validation metrics for evaluating classification performance on imbalanced data. In: 2019 International conference on computer, control, informatics and its applications (IC3INA). https://doi.org/10.1109/ic3ina48034.2019.8949568
17. Jiang J et al (2019) A novel multi-module neural network system for imbalanced heartbeats classification. Expert Syst Appl 1. https://doi.org/10.1016/j.eswax.2019.100003
18. Kalaivani S et al (2021) Sleep classification from wrist-worn accelerometer data using random forests. Sci Rep. https://doi.org/10.1038/s41598-020-79217-x

Human Fall Classification from Indoor Videos Using Modified Transfer Learning Model

Arifa Sultana and Kaushik Deb

Abstract With the progression of the world, older people living alone are increasing enormously because of the rising number of nuclear families. At this old age, falls are a severe issue for their injury or even death. It is high time to discover some effective ways to classify human fall action immediately for these older adults. This paper represents a video-based deep learning model that classifies indoor fall events from other household activities to overcome this problem. Initially, key frames are extracted using a frame generator. Afterward, these frames are passed to VGG16, VGG19, freezing all of the layers of VGG16 without the last four layers and similar for VGG19, freezing all of the layers of VGG16 without the last eight layers and similar operation for VGG19, and Xception for extracting spatial features as an experiment. Finally, these features are passed to the gated recurrent unit (GRU) for extracting temporal features. Experimental result shows that freezing all of the layers of VGG16 without the last four layers outperforms with an accuracy of 100% for UR fall detection dataset and 99% for multiple cameras fall dataset. This result ascertains the effectiveness of our proposed model in terms of accuracy.

Keywords Human fall · Transfer learning · Convolutional neural network · Gated recurrent unit

---

A. Sultana · K. Deb (✉)
Department of Computer Science and Engineering, Chittagong University of Engineering and Technology (CUET), Chattogram 4349, Bangladesh
e-mail: debkaushik99@cuet.ac.bd

A. Sultana
e-mail: arifa.z@eastdelta.edu.bd

M. S. Arefin et al. (eds.), *Proceedings of the International Conference on Big Data, IoT, and Machine Learning*, Lecture Notes on Data Engineering and Communications Technologies 95, https://doi.org/10.1007/978-981-16-6636-0_41

1 Introduction

Falls are commonly defined as an unaware approach on the ground, except deliberate change in position to rest in any support, wall, or other objects. Statistically stated that by 2050, adult people aged 60 or over will increase to 20 billion from 900 million [1] who are very much prone to suffer many types of health diseases. Among all the unpleasant results, fractures and other long-term disorders are the most common immediate consequences of falls, which lead to a common pathway to depression that causes diminished independence, disability, and psychological fear of falling again. However, the reduction of fall events in adult people aged 60 or over around the world can be achieved by developing automatic fall detection. As the classification of human fall action has become a hot research topic of developing intelligent detection and prevention systems, we have proposed a model freezing all of the layers of VGG16 without the last four layers to classify human fall action.
However, the key contribution of this proposed model can be summarized as follows:

- We have experimented with different transfer learning models by freezing multiple layers to achieve a higher accuracy rate along with a fewer number of parameters. This will reduce our computational time and will also help to classify human fall events in proper time.

2 Related Works

The theory on fall events can be classified between vision-based approaches and sensor-based concerns. Aspects regarding security and safety are also significant concerns. The composition of spine ratio and deflection angles is proposed by [2] that describes the varieties of human stance using skeleton extraction. But this method uses a limited data set and shows some false detection when dealing with workout motions. To resolve this case, Chen et al. [3] has focused on a technique for reorientation of abrupt fall events using the symmetry principle. That model can start only with single direction recognition, which comprehensively unable to find out fall event location accurately. Han et al. [4] has designed a model for detecting fall events with the Mobile VGG. However, the fall detection sign is not modified, and the 3D network is also not taken into account in this model. Mask R-CNN and bidirectional long- and short-term memories are applied in [5] for extracting human features from a noisy background. It is unable of separating the behavior of numerous persons residing in a room. To detect fall down events in a complex environment, an enhanced dynamic optical flow method is used by [6]. However, long-term temporal dependence is not considered here, which is a crucial point for fall classification. Human propositions are generated from human body joint position in [7] for fall event classification. However, it faces a substantial computational burden and needs parameter pruning. Islam et al. [8] has suggested an extensive analysis on fall event detection technologies using deep learning methodologies that use convolutional neural network (CNN), long short-term memory (LSTM), and other systems using auto-encoder. However, CNN is very sensitive to training data and is open to adversarial attacks, and LSTM requires high memory and a long training time. A vision centric method which is on convolutional neural network, is developed by [9] to decide if a series of frames carry a person falling where optical flow images are not used that could provide great representational power for motion.

The rest of the paper is embodied as follows. In Sect. 3.2, key frame extraction is discussed, Sect. 3.3 illustrates frame preprocessing, Sect. 3.4 describes the methodology to freeze all of the layers of VGG16 without the last four layers, Sect. 3.5 demonstrates the gated recurrent unit, Sect. 3.6 describes the function of dense layer, and Sect. 3.7 describes the classification process. Section 4 represents experimental results. At the end, Sect. 5 concludes the paper with some effective future directions.

3 Proposed Methodology

3.1 Overview

Our proposed model is designed with six basic steps, as mentioned in Fig. 1. These are video frame generation, preprocessing of extracted frames, implementing VGG16 by freezing all layers without the final four layers, passing spatial features to the gated recurrent unit (GRU), passing temporal features to dense layers, and finally classify events using the sigmoid classifier. Video frame generator extracts significant frames for faster operation. Frame preprocessing is done for data augmentation. Layer freezing in VGG16 is done to update weights according to the dataset to extract spatial features. GRU extracts temporal features using its three gates: update, reset, and finally, the current memory gate. These features are passed to the dense layer, which has multiple neurons. Finally, for binary classification of fall events, the sigmoid activation function is used.

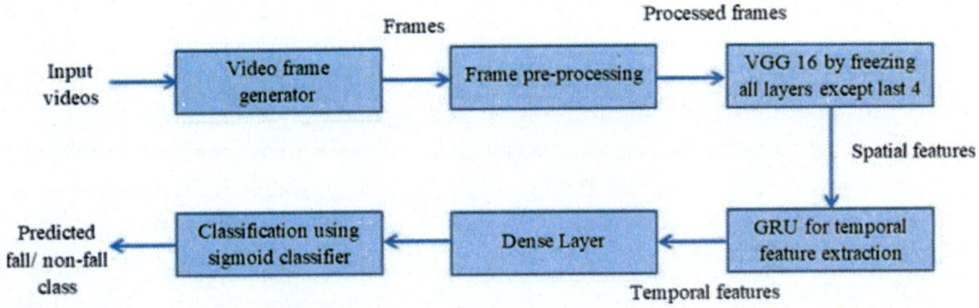

Fig. 1 Proposed model for human fall classification

3.2 Key Frame Extraction

Our dataset is constituted of approximately 262 videos of different lengths. We do not need to process each frame because of the enormous computational cost. For this reason, we have used a built-in video frame generator to extract ten significant video frames from the entire video. These key frames are extracted using frame differencing, thresholding, and dilation operation in the video frame generator.

3.3 Frame Preprocessing

Frame preprocessing is done for increasing image quality along with specific features. We have resized the frames to 150 × 150 for reducing the computational cost in the preprocessing phase. Afterward, we have performed data augmentation like zoom, rotation, width shift, horizontal flip, and height shift on the resized frames. Finally, normalization is done to convert the pixel values between 0 and 1 by scaling with 1/255.

3.4 VGG16 with Layer Freezing

We have performed several transfer learning models like VGG16, VGG19, and Xception. The transfer learning model is a kind of machine learning model designed for a task and can be reused in other related works. VGG16 is a transfer learning model which is constituted with 23 layers. There are 15 convolution layers, five max-pooling layers, and lastly three fully connected layers in the VGG16 model. A pixel-wise convolution operation is performed in the convolution layer between the image frame and the kernel to generate the convolved image. Rectified linear unit (ReLU) serves as an activation function, and it generates rectified feature map as output. This image is passed to the max-pooling layer, which produces a pooled feature map. The max-pooling layer also minimizes the dimensionality of the feature map and noisy artifacts. We have also performed a layer freezing method to hinder the weight updating during the training. Freezing the weights, this model allows updating the weights depending on the dataset. We have done several experiments freezing all layers, freezing all of the layers of VGG16 and VGG19 without the last four, and freezing all of the layers of VGG16 and VGG19 without the last eight layers. Figure 2 depicts the proposed VGG16 model after freezing all of the layers, excluding the last four layers. Here, we have used the input image of size 150 × 150, and the output of the final layer is (5,5,512). The fully connected layers of VGG16 are discarded in our model.

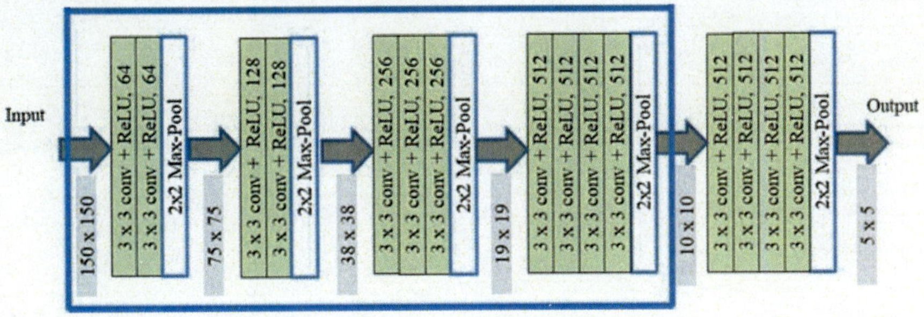

Fig. 2 VGG16 model with all frozen layers excluding last four layers

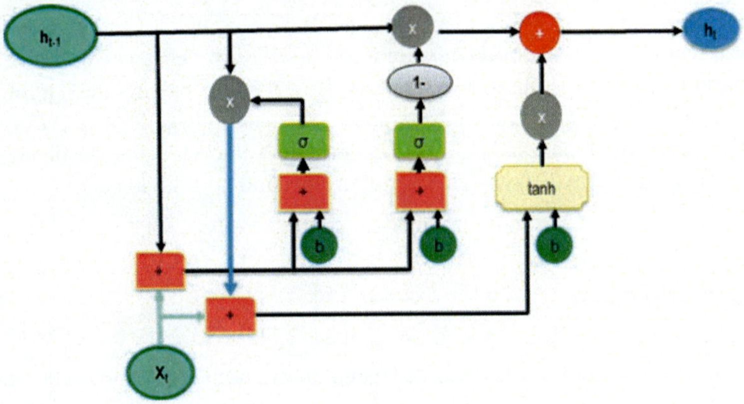

Fig. 3 Gated recurrent unit (GRU)

3.5 Gated Recurrent Unit (GRU)

To classify human fall events, we need to examine not only the spatial feature but also the temporal features. To overcome the exploding and vanishing gradient problems and to extract temporal features, spatial features from frozen VGG16 are passed to the gated recurrent unit (GRU). Figure 3 represents the cell of GRU which consists of three gates, i.e., update, reset, and lastly, the current memory gate.

Update gate marks the important features of preceding time steps and passes them to posterior time steps for further processing. The reset gate identifies the amount of insignificant information that needs to forget. Finally, the current memory gate decides the current information, which needs to take into consideration calculating with relevant information from previous states. The operations of these three gates can be calculated using the following equations acquainted from [10].

$$\text{Update gate, } Z_t = \sigma(W^{(Z)}x_t + U^{(z)}h_{(t-1)}) \tag{1}$$

$$\text{Reset gate, } r_t = \sigma(W^{(r)}x_t + U^{(r)}h_{(t-1)}) \tag{2}$$

$$\text{Current memory gate, } \hat{h}_t = \tanh(Wx_t + r_t * Uh_{(t-1)}) \quad (3)$$

$$\text{Final memory, } h_t = Z_t * h_{(t-1)} + ((1 - Z_t) * \hat{h}_t) \quad (4)$$

Here, x_t is current input, $h_{(t-1)}$ denotes hidden state of preceding time step, $W^{(z)}$ and $W^{(r)}$ represent current weight matrices, and $U^{(z)}$, $U^{(r)}$ stand for updated weight parameters.

3.6 Dense Layer

Temporal features from the GRU cell are fed to a series of dense layers. Each layer is constituted with a large number of neurons. However, for passing the output from each layer to successive layers, we have used a drop-out rate of 50% to remove the overfitting problem of the test data. There are only two neurons at the last dense layer because we need to classify two classes: fall events and non-fall events.

3.7 Classification

The output of the dense layer is predicted using the sigmoid activation function [11]. For the classification of fall actions, we have used the following equation of the sigmoid activation function.

$$\text{Sigmoid activation function, } S(x) = e^x/(e^x + 1) \quad (5)$$

Here, 'e' represents Euler's number.

4 Experimental Result

4.1 Tools and Evaluation Matrices

The configuration of our machine, which is used for conducting this experiment, is as follows: AMD Ryzen 7 2700X Octa-core processor with 3.7 GHz speed, NVIDIA GEFORCE RTX 2060 SUPER graphics with GPU memory of 8 GB, 32 GB RAM. Google colab is also used for experimental purposes. The equations of evaluation parameters that are used in our experiment from [12] are stated below.

$$\text{Accuracy} = (\text{TP} + \text{TN})/(\text{TP} + \text{TN} + \text{FP} + \text{FN}) \quad (6)$$

Table 1 Dataset details

Parameters	UR fall detection dataset	Multiple cameras fall dataset
Total number of videos	70	192
Number of fall events	30	96
Number of non-fall events	40	96
Resolution of videos	640 × 240	720 × 480
Frame rate (fps)	30	30

$$\text{Precision} = \text{TP}/(\text{TP} + \text{FP}) \tag{7}$$

$$\text{Recall} = \text{TP}/(\text{TP} + \text{FN}) \tag{8}$$

$$F1\text{-score} = (2 * \text{Precision} * \text{Recall})/(\text{Precision} + \text{Recall}) \tag{9}$$

Here, TP stands for the true positive rate, which denotes that the predicted data is a fall event which is indeed a fall action. TN denotes true negative rate, i.e., the predicted result is non-fall which is a non-fall event indeed. FP stands for false positive rate, which means that the predicted result is a fall event which is a non-fall action. FN represents false negative rate where the predicted outcome is a non-fall action, but in real-time, it is a fall event. Accuracy in Eq. (6) describes the rate of accurately classified data. The high score of precision in Eq. (7) represents that every predicted positive class belongs to the positive class in real life. Recall in Eq. (8) demonstrates the amount of correctly predicted positive data out of all positive predicted data. F1-score in (9) calculates the harmonic mean value of precision and recall.

4.2 *Dataset Description*

We have conducted our experiment with the two most prominent datasets. One is the UR fall detection dataset [13], and another dataset is the multiple cameras fall dataset [14]. The details of these datasets are illustrated in Table 1.

4.3 *Experimental Result Analysis*

In our experiment, we have used 40% data for training the proposed model, 35% for validation, and others are used for testing the model. We have conducted several experiments to choose this percentage of the dataset, and the ratio mentioned above gives better accuracy. We have also experimented with selecting an efficient number of frames. To minimize computational cost and to provide sufficient information in

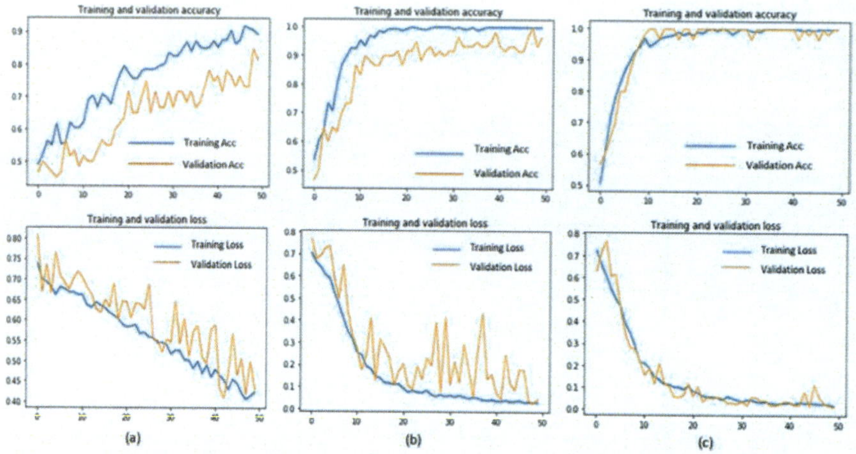

Fig. 4 Accuracy loss curve for **a** freezing all layers of VGG16, **b** freezing all layers of VGG16 excluding last eight layers, **c** freezing all layers of VGG16 excluding last four layers

the model, ten significant frames are chosen from each video. Several experiments such as VGG16, freezing all of the layers of VGG16, freezing all of the layers of VGG16 excluding the last four, and freezing all of the layers of VGG16 excluding the last eight layers have been performed. Among these, VGG16 with frozen all layers without final four gives the outstanding performance for the UR fall detection dataset, which is shown in Fig. 4. This figure illustrates that at the 50th epoch, freezing all layers without the last four layers of VGG16 achieves higher accuracy of 100% in (c). Freezing all layers without the last four layers of VGG gives better accuracy because the initial layers of VGG16 extract the shallow features. For the classification of events, we need to consider the deep features from the video frame extracted by the last layers. When we are unfreezing the last eight layers, it updates weights in the last eight layers, whereas unfreezing the last four layers only updates weights for the last four layers for all of the 40 epochs. This operation reduces our computational time by considering all the robust features from the video frames.

Confusion matrices in Fig. 5 also represent the performance of all frozen layers of VGG16 and freezing all layers of VGG16 excluding the last eight layers on test data of multiple cameras fall dataset. Figure 5a depicts that freezing all layers of VGG16 misclassifies three videos, and Fig. 5b represents that freezing all of the layers of VGG16 excluding the last eight layers misclassifies one video among 38 test data.

Figure 6 illustrates the confusion matrices of our proposed model on different datasets. The confusion matrix in Fig. 6a depicts that all the test data except one are correctly classified for multiple cameras fall dataset, and all the test data are correctly classified for the UR fall detection dataset shown in Fig. 6b.

The predicted outputs of our proposed model on different datasets are represented in Fig. 7. The first row in this figure shows the output frame sequence of a video

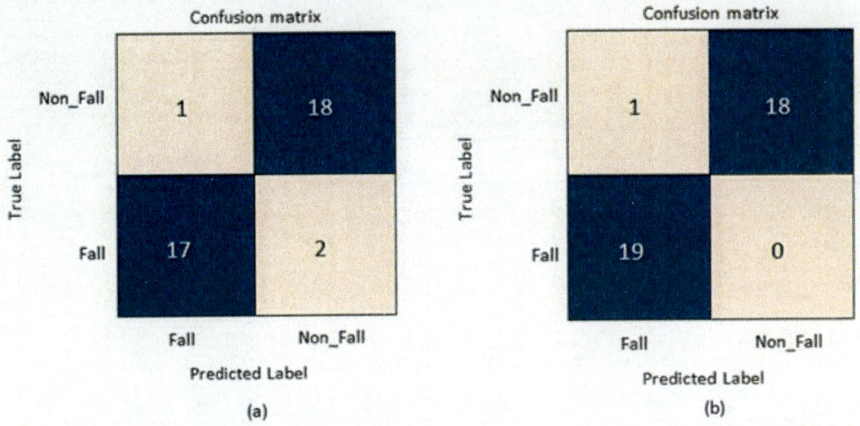

Fig. 5 Confusion matrix of **a** freezing all layers of VGG16, **b** freezing all layers of VGG16 excluding last eight layers for multiple cameras fall dataset

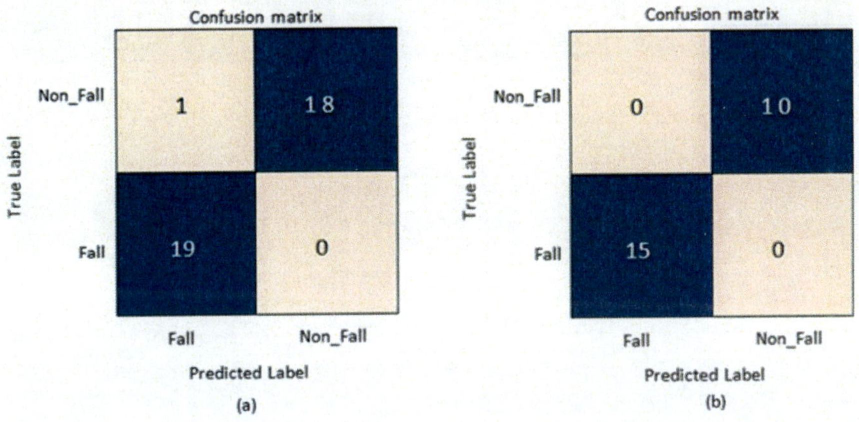

Fig. 6 Confusion matrix of proposed model for **a** multiple cameras fall dataset, **b** UR fall detection dataset

of the UR fall detection dataset, whereas the second row depicts the output frame sequence of the multiple cameras fall dataset.

An example of misclassification of non-fall action video is depicted in Fig. 8. This figure is representing the movements of non-fall action. However, our model predicts this event as fall action. The reason behind this misclassification is, among ten frames, five frames are similar to fall action. Moreover, since there is an uneven movement before lying position, which is depicted in frame number 9, our model predicts this event as fall action. An experiment is done on the outdoor human fall dataset for the generalization of our proposed model. To conduct this experiment, we have created a dataset from Youtube videos which comprises 96 videos of different lengths, which may vary from 3 to 30 s. Here, 46 videos represent fall action, and 50

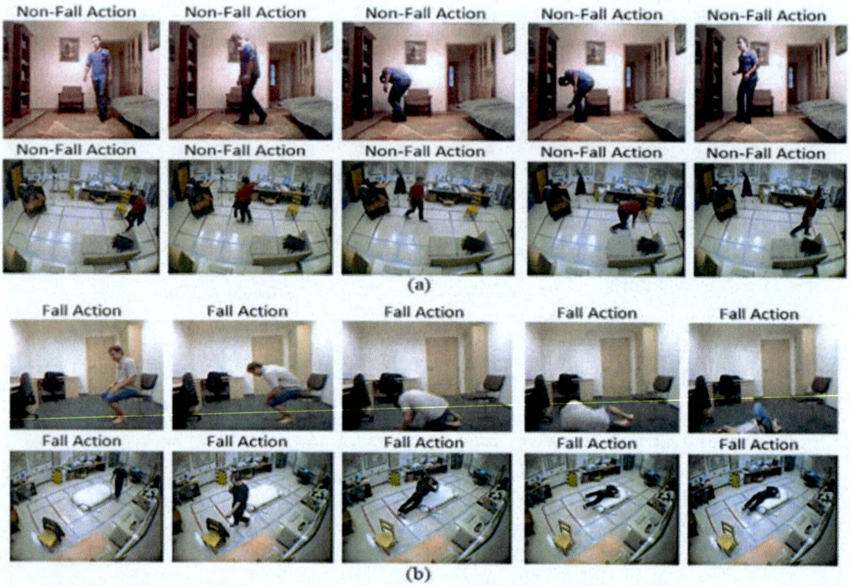

Fig. 7 Predicted output for **a** non-fall events, **b** fall events

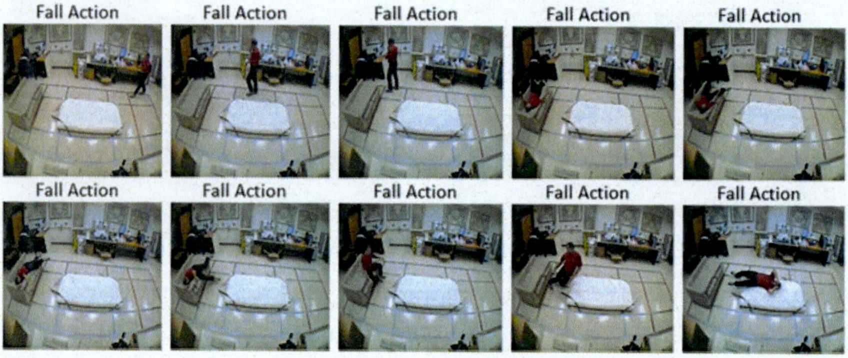

Fig. 8 Misclassified result on multiple cameras fall dataset

videos represent natural activities in various outdoor environments where the frame rate is 25 frames per second. Using this outdoor human action dataset, our proposed model achieves 100% accuracy in 40 epochs which are depicted in the accuracy loss curve in Fig. 9.

Figure 10 illustrates some correctly classified output results after implementing our proposed model on this outdoor human action dataset. Figure 10a shows the output for non-fall action, and Fig. 10b represents the output for fall down action.

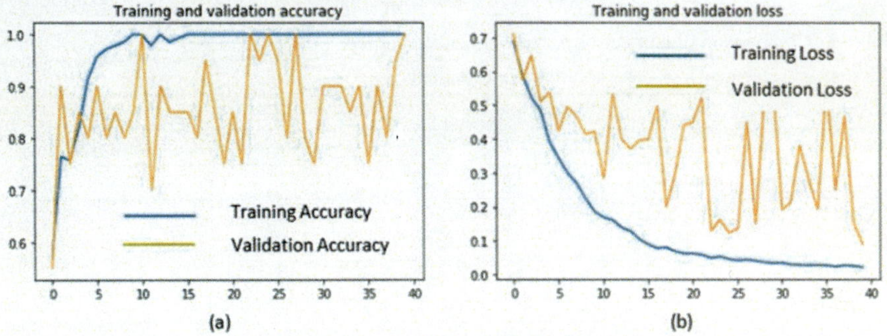

Fig. 9 Performance of our proposed model for outdoor human action dataset where **a** accuracy curve, **b** loss curve

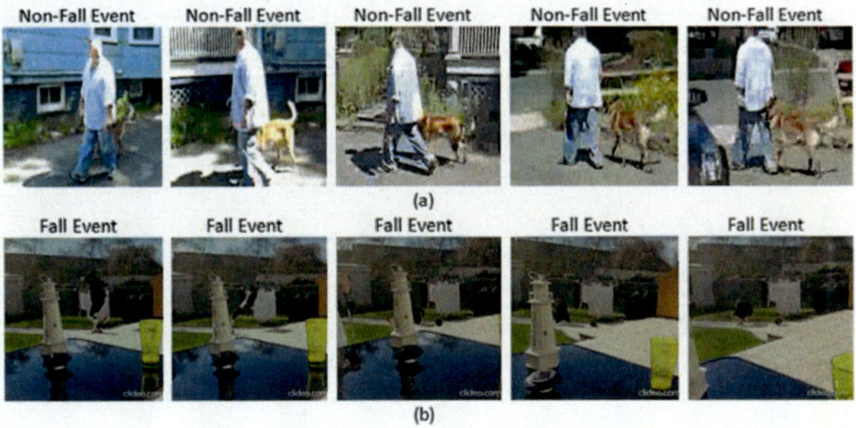

Fig. 10 Predicted output on outdoor human action dataset for **a** non-fall event, **b** fall event

4.4 *Result Evaluation and Comparison*

Freezing all layers except the last four layers, our model generates a faster response because of the fewer parameters than existing models. The comparison of the number of used parameters in our proposed model with other executed models is illustrated in Table 2.

Figure 11 graphically represents the accuracy of these different executed models, where freezing all layers except the last four layers of VGG16 outperforms others for both the UR fall detection dataset (URFD) and multiple cameras fall dataset (Multicam).

The classification report of our proposed model for the UR fall detection dataset (URFD) and multiple cameras fall dataset (Multicam) is represented in Table 3. Here, accuracy, precision, recall, F1-score, and support are calculated for evaluating the result.

Table 2 Comparison of number of parameters and accuracy

Models	Number of parameters	Mean accuracy (%)
Xception	22,910,480	98.5
VGG19	143,667,240	98
VGG19 by freezing all layers	20,024,384	92
VGG19 by freezing all layers except last 8	20,024,384	97
VGG19 by freezing all layers except last 4	20,024,384	97.5
VGG16	138,357,544	97.8
VGG16 by freezing all layers	14,714,688	93
VGG16 by freezing all layers except last 8	14,714,688	98
VGG16 by freezing all layers except last 4	14,714,688	99.5

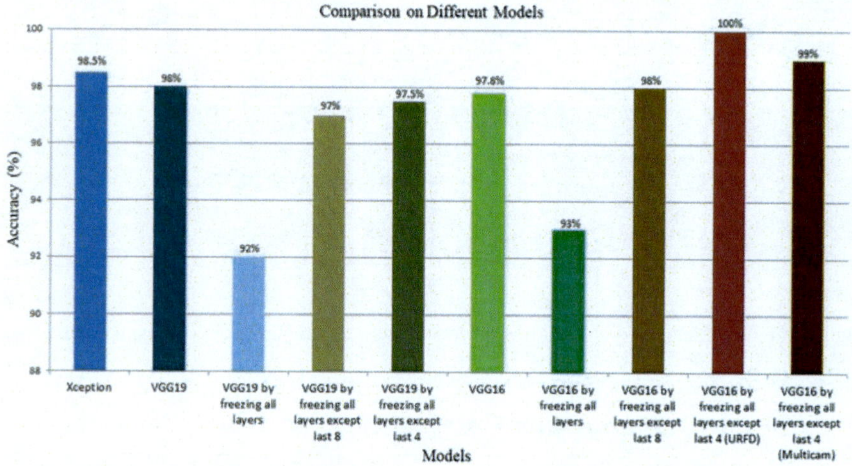

Fig. 11 Comparison on different executed models

Table 3 Classification report of our proposed model

Dataset	Classes	Precision	Recall	F1-score	Support
URFD	Fall	1.00	1.00	1.00	19
	Non-fall	1.00	1.00	1.00	19
Multicam	Fall	0.99	1.00	0.99	19
	Non-fall	1.00	1.00	1.00	19

Table 4 Performance comparison among existing research works

Models	Dataset	Accuracy (%)
Zerrouki et al. [15]	URFD	96.88
Marcos et al. [9]	URFD	95
Proposed model	URFD	100
Marcos et al. [9]	Multicam	96
Proposed model	Multicam	99

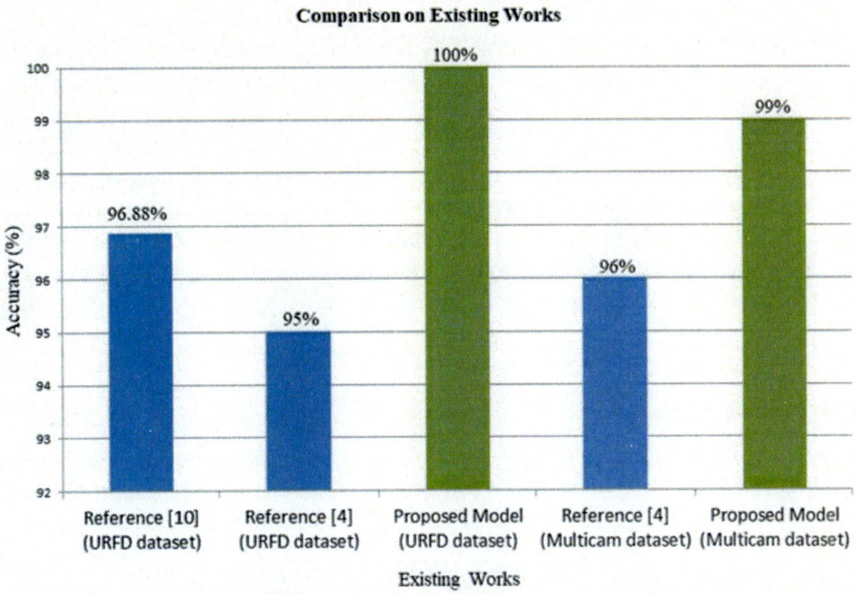

Fig. 12 Comparison of proposed model with existing works

Table 4 exhibits the comparison of our proposed model with some existing models executing on similar datasets where URFD denotes UR fall detection dataset and multicam stands for multiple cameras fall dataset. Here, Marcos et al. used the VGG16 model for feature extraction, and Zerrouki et al. used the hidden markov model to classify human fall action.

Figure 12 graphically represents the performance comparison of our proposed model with other existing research works as like as [16]. Here, blue color bars represent the performance of existing models, and green colored bars represent the accuracy of our proposed model on these datasets.

5 Conclusions and Future Works

This paper represents a deep neural network-based model to classify fall actions from household activities. As older adults are not eager to carry wearable sensors and the vision-based model will help doctors understand fall circumstances, we have proposed this model. Our key contribution is to classify fall action immediately using the proposed VGG16 model with a frozen layer along with a fewer number of parameters to reduce computational cost. Initially, from indoor videos, we have extracted the key frames using a frame generator. Afterward, these frames are preprocessed for faster execution and are passed to the VGG16 transfer learning model by freezing all of the layers, excluding the last four. Then, spatial features are fed to the gated recurrent unit for extracting temporal features. Finally, the sigmoid activation function followed by a series of dense layers classifies human fall actions from other household activities. We have also explored other pretrained models with layer freezing. However, our proposed model outperforms other state-of-the-art models. However, We can more minutely evaluate our proposed model if we enrich our dataset in the future. Furthermore, it will lessen computational cost if we extract features from the salient region rather than the whole frame. AQ1

References

1. WHO. Number of people over 60 years set to double by 2050; major societal changes required. https://www.who.int/news/item/30-09-2015-who-number-of-people-over-60-years-set-to-double-by-2050-major-societal-changes-required. Accessed 16 Apr 2021
2. Han K, Yang Q, Huang Z (2020) A two-stage fall recognition algorithm based on human posture features. Sensors 20:1–21
3. Chen W, Jiang Z, Guo H, Ni X (2020) Fall detection based on key points of human-skeleton using OpenPose. Symmetry 12:1–17
4. Han Q, Zhao H, Min W, Cui H, Zhou X, Zuo K, Liu R (2020) A two-stream approach to fall detection with mobile VGG. IEEE Access 8:17556–17566
5. Chen Y, Li W, Wang L, Hu J, Ye M (2020) Vision-based fall event detection in complex background using attention guided bi-directional LSTM. IEEE Access 8:161337–161348
6. Chhetri S, Alsadoon A, Prasad PWC, Rashid TA, Maag A (2021) Deep learning for vision-based fall detection system: enhanced optical dynamic flow. Comput Intell 37:578–595
7. Reddy GP, Geetha MK (2020) Video based fall detection using deep convolutional neural network. Eur J Mol Clin Med 7:739–748
8. Islam MM, Tayan O, Islam MR, Islam MS, Nooruddin S, Kabir MN, Islam MR (2020) Deep learning based systems developed for fall detection: a review. IEEE Access 8:166117–166137
9. Marcos A, Gorka A, Carreras I (2017) Vision-based fall detection with convolutional neural networks. Hindawi Wirel Commun Mob Comput 2017:1–17
10. Understanding GRU networks. https://towardsdatascience.com/understanding-gru-networks-2ef37df6c9be. Accessed 22 Apr 2021
11. Activation functions in neural networks. https://towardsdatascience.com/activation-functions-neural-networks-1cbd9f8d91d6. Accessed 22 Apr 2021
12. Accuracy, recall, precision, F-score & specificity, which to optimize on? https://towardsdatascience.com/accuracy-recall-precision-f-score-specificity-which-to-optimize-on-867d3f11124. Accessed 22 Apr 2021

13. UR fall detection dataset. http://fenix.univ.rzeszow.pl/~mkepski/ds/uf.html. Accessed 15 May 2021
14. Multiple cameras fall dataset. http://www.iro.umontreal.ca/~labimage/Dataset/. Accessed 15 May 2021
15. Zerrouki N, Houacine A (2017) Combined curvelet and hidden Markov models for human fall detection. Multimed Tools Appl 23:1–20
16. Adiba FI, Islam T, Kaiser MS, Mahmud M, Rahman MA (2020) Effect of corpora on classification of fake news using Naive Bayes classifier. Int J Autom Artif Intell Mach Learn 1:80–92

Road Sign Detection Using Variants of YOLO and R-CNN: An Analysis from the Perspective of Bangladesh

Aklima Akter Lima, Md. Mohsin Kabir, Sujoy Chandra Das, Md. Nahid Hasan, and M. F. Mridha

Abstract Road sign detection represents a feature that assures the safety of drivers, vehicles, and pedestrians by efficiently detecting road signs. This feature is designed to notify drivers about road signs whether he is missing the signs or not. This detecting and recognizing feature of the road signs' has improved a part of the advanced driver assistance system (ADAS). ADAS is an automated technology containing cameras and sensors intended to help the drivers with road signs, while traveling to a new road or having no knowledge about road signs. Before the work analysis, this topic has shown formidability as it has a real-time processing solution. This paper analyzed seven architectures for detecting the road signs: YOLO, YOLOv2, YOLOv3, PP-YOLO model and R-CNN, Fast R-CNN, Faster R-CNN. We have built a dataset based on Bangladesh's road sign named the "BD Road Sign 2021 (BDRS 2021)" dataset to evaluate the architectures. This dataset contains 16 categories (16 types of road-sign), and each has 168 images. Finally, we applied the seven advanced architectures to find the effective one to detect Bangladesh's road signs. This study implies that YOLOv3 and Faster R-CNN perform comparatively better for road sign detection.

Keywords Advanced driver assistance system (ADAS) · Road sign detection · You only look once (YOLO) · Region-based convolutional neural networks (R-CNN) · Deep learning

1 Introduction

Identifying road sign's position is an important area of research that continuously captures the attention of researchers in the area of Intelligent Transportation System (ITS). The road sign shows route road marking, possible hazards, and involvements

A. A. Lima · Md. M. Kabir (✉) · S. C. Das · Md. N. Hasan · M. F. Mridha
Bangladesh University of Business and Technology, Dhaka, Bangladesh

M. F. Mridha
e-mail: firoz@bubt.edu.bd

M. S. Arefin et al. (eds.), *Proceedings of the International Conference on Big Data, IoT, and Machine Learning*, Lecture Notes on Data Engineering and Communications Technologies 95, https://doi.org/10.1007/978-981-16-6636-0_42

that vehicles can encounter on the road. Adding to that it assists vehicles in route by providing valuable data and alerts. Every driver must hold their attention on the road and be aware of their surroundings while driving as road signs have variations everywhere in the world. This is not yet possible to construct the universal Traffic Sign Recognition System (TSRS) structure. When a driver is introduced to a different route, he must concentrate on the road incredibly late at night, which results in a diversion from a highway sign. Road sign detection features could be a helpful way to assist drivers and mitigate road injuries caused by the driver's lack of understanding. A system or feature should be built for Bangladeshi road signs to alert drivers to road signs without interfering with their driving concentration. As a result, we are firmly persuaded to conduct some research in this direction with the explicit aim of providing more analysis in the study of TSRS concerning the Bangladesh environment.

The driver will be directed into a favorable configuration every time with support from an Advanced Driver Assistance System (ADAS) for any encountered signs. As a result, drivers won't have to face finding out the sign's meaning, and a better TSRS structure can make it possible. TSRS structure techniques are divided into two key sections: position and identification. The design helps to support the driver in several ways to ensure their well-being, also the safety of various people and pedestrians on the path. These systems include one main goal: to identify and track road signs mostly during the driving period. With these features, the system will direct and make drivers aware of the consequences to the environment. This article focuses on creating a TSRS framework for Bangladeshi road signs that use some architecture based segment measurement and identification. An empirical analysis and its empirical setup comparing the following seven main architectures: YOLO [1], YOLOv2 [2], YOLOv3 [3], PP-YOLO [4] model and R-CNN [5], Fast R-CNN [6], Faster R-CNN for detecting signs. The overall contributions of this research are:

- We have investigated and distinguished the contemporary challenges of the Advanced Driver Assistance System.
- A recently built dataset named "BD Road sign 2021 (BDRS 2021)" is introduced that consists of 16 classes. Each class consists of 168 images.
- We applied seven baseline architectures, "YOLO, YOLOv2, YOLOv3, PP-YOLO model and R-CNN, Fast R-CNN, Faster R-CNN", to the newly created dataset and analyzed the obtained results. The analysis found Faster R-CNN and YOLOv3 more effective.

The continuation of this experimental paper is organized as follows: The previous literature is described in Sect. 2. Section 3 describes the dataset. Section 4 addresses the process, including features such as data preprocessing and design. Section 5 explains the assessment, provides a summary of the experiments as well. Finally, Sect. 6 brings the article to a close.

2 Related Work

Due to technical advancements like computing and computer vision in the modern age, a device allows quick, accurate, and automatic detection of road signs in various conditions. Many notable cutting-edge architectures have been developed in recent years. Among those listed are:

A novel approach for traffic sign detection based on deep learning architecture named capsule networks achieves excellent performance on the German road sign dataset, which is introduced in [7]. In some cases, CNN's are easily fooled by multiple adversary attacks [8], but capsule networks can overcome those attacker attacks and improve traffic sign detection accuracy. Compared to CNNs, capsule networks perform much better by correctly performing image classification and recognition tasks [9]. D. Tabernik and D. Skočaj identified and recognized a wide range of traffic sign categories appropriate for automating traffic sign inventory management. The mask R-CNN is a CNN-based approach that addresses the entire detection and recognition process with automated end-to-end learning. This method is used to detect the 200 traffic sign classes specified in the dataset. Researchers demonstrated that the deep learning-based approach could produce an outstanding performance for a wide range of traffic sign categories, along with some complex ones with high intra-class variability [10].

Wang and Guo [11] suggested the YOLO neural network model is configured using an updated CNN model focused on the YOLO model, darknet 53. By adding batch normalization and RPN networks, it can enhance network architecture for traffic sign detection. The method described in this paper will significantly improve the efficiency and detection rate of traffic signs, while also reducing the detection system's hardware specifications. The authors of [12] use the Radial Symmetry Transform to identify other geometric shapes such as octagons, squares, and triangles.

Zhang, J. suggested an end-to-end convolutional network modeled after YOLOv2. To detect minor traffic signs more effectively, they divide the input images into dense grids and generate more precise feature maps. Both experimental results based on their extended CTSD and German Traffic Sign Detection Framework (GTSDB) show that the proposed approach is faster and more stable [13]. Buyval, presented a technique for classifying and localizing road signs in 3D space using a neural network and a point cloud acquired from a laser range finder (LIDAR). A dataset was collected to achieve this goal and train the neural network (built on the Faster-R-CNN architecture). The device generates a series of images with bounding boxes and points clouds related to actual road signs [14]. The first section of a method for detecting and classifying road signs identifies the road signs on a real-time basis. The second section identifies the German traffic signs (GTSRB) dataset and produces predictions using the road signs detected during the first section. In the detection section, they used HOG and SVM to identify the road signs captured images. Later, in the classification section, a convolutional layer based on the LeNet model was used to modify [15]. A system for detecting and recognizing Bangladeshi road signs is being established. To begin, images of road signs are collected from various districts across Bangladesh to create the dataset. The photos are then numbered, and the Single Shot Multibox Detector (SSD) is then used to locate and identify road signs. A CNN-based model is being used in the classification stage [16].

Besides, a large number of deep learning-based road sign detection approaches are proposed in the last decades. Our paper mainly analyzes the road signs that are used in Bangladesh based on seven detection algorithms.

3 Dataset

We discovered that most road sign image datasets on the web are divided into four categories: regulatory signs, warning signs, information signs, and additional signs. We attempted to gather simple road signs commonly used on Bangladesh's roads from the set of every road sign. We have gathered a large number of images from the Bangladesh Road Transport Corporation (BRTC). Finally concludes, 168 images for each of the 16 types (Under four classes: regulatory signs, warning signs, information signs, and additional signs) and used 80% (2150) of the photos for training and the remaining 20% (538) for testing in our assessments (Fig. 1).

4 Methodology

To test road sign detection from image datasets "BDRS 2021," a comparative analysis of YOLO, YOLOv2, YOLOv3, PP-YOLO model, and R-CNN, Fast R-CNN, Faster R-CNN architectures is introduced and benchmarked. The framework is outlined in detail in the articles that follow.

4.1 Data Preprocessing

Data preprocessing is divided into two stages: data normalization and data augmentation. These two methods are discussed further below.

Data Normalization: Image normalization is an essential preprocessing technique. It decreases the inner-class function disparity and is regarded as intensity offsets. Since the intensity offsets are defined in the field distribution, standard deviation and Gaussian normalization can be used to normalize. Equation (1) is used to evaluate the image during normalization [17].

$$\Psi(\pi, \theta) = \frac{\xi(\pi, \theta) - \mu(\pi, \theta)}{6\sigma(\pi, \theta)} \quad (1)$$

Fig. 1 Sample images from BDRS 2021 dataset

where μ is a local mean and σ is a local standard deviation [1].

$$\mu(\pi, \theta) = \frac{1}{M^2} \sum_{k=-\alpha}^{\alpha} \sum_{n=-\alpha}^{\alpha} \xi(K + \mu, n + \theta)$$

Data Augmentation: The data augmentation techniques are applied to enlarge the dataset. We used five image appearance filters: Gaussian, disk, unsharp, average and motion, and six affine transformation matrices. This makes the dataset quantity 30 times of its actual size.

4.2 Baseline Architectures

Seven baseline architectures of deep learning-based detection algorithms are used to evaluate the dataset. This section briefly analyzes the seven algorithms. The general structure of the evaluation system presents in Fig. 2.

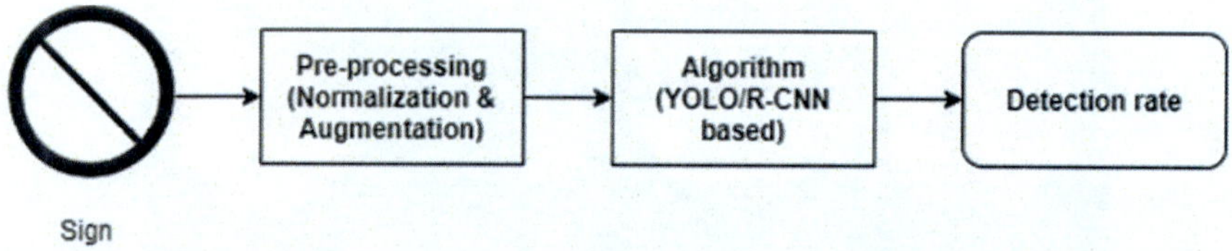

Fig. 2 The image presents the structure of the system. First, the photos were taken and passed through the preprocessing phase. Then each algorithm is applied to the processed data and results captured

YOLO: YOLO [1] is CNN-based for real-time object detection, which utilizes the entire image and splits it into regions, and estimates bounding boxes and probabilities for each image. The estimated probabilities are used to weight these bounding boxes when it reaches its high precision, while operating in real-time. The method is called "You Only Look Once" at the object in the context, which makes predictions. Then, it outputs objects predictions along with bounding boxes after non-max suppression. It generally learns applicable representations of objects, allowing it to perform better from many detection methods.

YOLOv2: YOLOv2 implements a range of enhancements to increase accuracy and batch processing. YOLOv2 [2] solves significantly higher localization error and poor recall comparison to region-based strategies by allowing batch standardization and better resolution classifiers. The Batch Normalization method is used to stabilize the input layers by modifying and measuring the activations [18]. Multi-scale instruction randomly selects a new size for every ten iterations of the system. This helps to predict well over a wide range of input measurements. The enhancement of the YOLOv2 is the really well functionality to enhance the ability to identify small items, which follow a pass-through layer method. This combines high-resolution features with low-resolution features, equivalent to ResNet identification mapping [19]. The mathematical analysis of the architecture can be found in [6].

YOLOv3: YOLOv3 [3] is published, distinguished by greater accuracy, and substitutes the softmax activation function with logistic regression and threshold. YOLOv3 is enhanced by using a multi-label classification that varies from the shared exclusive label used in the earlier versions. It utilizes a logistic classifier to measure the likelihood of the item becoming a particular mark. In classification loss, the binary cross-entropy loss by each mark is used rather than the generalized mean square error used in the earlier versions. The secondary enhancement is with a particular bounding box prediction that combines the score of one item in a bounding box anchor that overlaps the maximum likelihood object instead of others. YOLOv3 determines a bounding box anchor by each ground truth item. The third development with the use of estimation across dimensions by using the idea of feature pyramid networks. YOLOv3 forecasts boxes on three spatial dimensions, and then extracts the features from all those scales. The predicted outcome of the network is a 3D sensor that encodes the bounding box, item score, and class estimation. The fifth upgrade is the latest CNN function extractor called Darknet-53. It's a 53-layer CNN that utilizes ResNet-inspired skip connections. YOLOv3 predicts at three different scales, precisely determined by downsampling the proportions of the source images by 32, 16, and 8 pixels, respectively.

PP-YOLO: The PP-YOLO [4] (PaddlePaddle YOLO) object detection system is based on the YOLO object detection algorithm. PP-YOLO is not a novel object detection system. Instead, PP-YOLO is a revised version of YOLOv4 with faster inference and a higher mAP score. Such enhancements are made possible by utilizing a RESNET-50 backbone architecture and additional features, including larger batch size, Drop block, IOU Loss, and training models. This structure consists of 3 parts, Backbone, Detection Neck, Detection Head. DarkNet-53 with ResNet50-vd, the backbone of YOLOv3, is substituted in PP-YOLO. It replaces some of the convolutional layers in ResNet50-vd with deformable convolutional layers (DCN) in this case. Many detection models have shown the efficacy of Deformable Convolutional Networks (DCN). ResNet as a backbone network architecture itself provided an increase in effectiveness and efficiency.

R-CNN: The region-based Convolutional Network (R-CNN) [5] method achieved excellent image prediction performance through using deep ConvNet to identify input images. The R-CNN [20] process trains CNN's end-to-end to locate the region proposals through element clusters or backgrounds. R-CNN development is a multi-stage process that includes extracting features, fine-tuning a log loss infrastructure, training SVMs, and eventually constructing a bounding box. There are drawbacks like R-CNN is sluggish since it executes the forward ConvNet transfer by each object proposal without exchanging the calculation and cannot modify the co-evolutionary layers that precipitate the pooling of the structural pyramid [21].

Fast R-CNN: The Fast R-CNN [22] network uses image data and a collection of training samples as input. The network processes the entire image with many co-evolutionary and max pooling to generate a fully connected function map. A region of interest (RoI) pooling layer extracts an adjusted vector function for each model for evaluation. Fast R-CNN has many advantages, such as higher recognition performance (mAP) than R-CNN, SPPnet, separate training, multi-task failure, and training can upgrade all network layers, and no storage devices are needed for caching features. In Fast R-CNN, we reduce an optimization technique after the multi-task loss.

Faster R-CNN: The Faster R-CNN [6] model consists of two components: the Region Proposed Network (RPN) and the Fast R-CNN tracker. RPN is an entirely convoluted system used to generate regional proposals with various dimensions and rotational speeds that serve as feedback for the second method. The RPN, as well as the Fast R-CNN detector, share a specific convolutional layer. Faster R-CNN, by extension, may be composed of a single and coordinated R-CNN. Network for the identification of artifacts. The RPN is a region proposal algorithm, and the Fast R-CNN as a detection network comprises the Faster R-CNN architecture.

5 Evaluation

The supervision of the empirical analysis on estimating the recommended Road sign detection on our “BDRS 2021” dataset carried out the comparative study of seven architectures. First, we describe the data set that was used in the research. Then, we describe the experimental setup. Third, the measures used to assess method accuracy are discussed. Fourth, the comparative study of seven architectures YOLO, YOLOv2, YOLOv3, PP-YOLO model and R-CNN, Fast R-CNN, Faster R-CNN are analyzed. Finally, the results of the comparative analysis are shown.

5.1 Evaluation Metric

Precision and recall measurement metrics are used to evaluate the architecture based on the confusion matrix results. Also, mAP is used, which is primarily used for the evaluation of visual object detectors.

$$\text{Precision} = \frac{\text{TP}}{\text{TP} + \text{FP}} \tag{2}$$

$$\text{Recall} = \frac{\text{TP}}{\text{TP} + \text{FN}} \tag{3}$$

$$\text{mAP} = \frac{1}{n}\sum_{k=1}^{k=n} \text{AP}_k \tag{4}$$

where TP means true positive, FP means false positive; FN means false-negative and AP_k is the AP of class k, and n is the number of classes. AP of class k calculates by the following formula.

$$\text{AP} = \sum_{k=0}^{k=n-1} [\text{Recalls}(k) - \text{Recalls}(k+1)] * \text{Precisions}(k) \tag{5}$$

where n is the number of thresholds.

5.2 Experimental Setup

Python is used for data preprocessing, experimentation, and model evaluation in the “BDRS 2021” dataset. TensorFlow [23] and Keras [24] are used to evaluate the proposed architecture. Furthermore, NumPy [25] is used to perform mathematical operations on seven architectures that are compared in our dataset for the experiment.

5.3 *Experiments and Comparisons*

In this paper, the "BDRS 2021" dataset is used to detect Bangladesh's road signs. The experiment is done based on the mentioned seven architectures of YOLO and R-CNN. For the YOLO algorithm, we applied the Darknet implementation. The Darknet implementation of YOLO gives 0.642 mAP for the proposed dataset. As the YOLO algorithm makes many localization errors and lower recall rates, the accuracy obtained is insufficient. The YOLOv2 further reduced these problems. This architecture is developed using Darknet-19 deep architecture and increases the mAP to 0.76 for the mentioned dataset. The faster YOLO version till present days is YOLOv3. We applied the Darknet-53 as the backbone architecture of YOLOv3 and obtained massive enhancements of the result to 0.885 mAP. Afterward, the dataset is also used for the PP-YOLO variant. PP-YOLO replaced the Darknet-53 backbone with ResNet architecture and became helpful for real application scenarios. We achieved nearly the same mAP as YOLOv3 of 0.878 for PP-YOLO. Hence, the study shows that YOLOv3 with Darknet-53 backbone and ResNet backbone both give satisfactory road sign detection results.

Then, the dataset is evaluated using R-CNN, Fast R-CNN, and Faster R-CNN architecture. First, the R-CNN architecture is applied with a selective search algorithm. It takes a considerable training time, and the mAP score is only 0.683, which is much lower than any YOLO architecture. Then, the Fast R-CNN architecture is applied, which is the advanced version of R-CNN. In this time, the training time drastically reduced, and the mAP score increases to 0.795. Finally, we applied Faster R-CNN-based on the region proposal network and found satisfactory performances of 0.896 mAP. Table 1 presents the precision, recall, and mAP scores of the mentioned algorithms.

However, the study suggests YOLOv3 and Faster R-CNN for Bangladeshi Road Sign detection.

Table 1 This table presents the precision, recall, and mAP score of different sign detection architectures

Model	Precision	Recall	mAP
YOLO [1]	0.597	0.625	0.642
YOLOv2 [2]	0.732	0.751	0.760
YOLOv3 [3]	**0.882**	**0.899**	**0.885**
PP-YOLO [4]	0.878	0.877	0.878
R-CNN [5]	0.696	0.656	0.683
Fast R-CNN [22]	0.783	0.796	0.795
Faster R-CNN [6]	**0.884**	**0.901**	**0.896**

6 Conclusion

This paper represents a comparative analysis of road sign detection techniques implemented on the "BDRS 2021" dataset. It is collected from a public source and modified later according to preferences. Various architectures, specifically YOLO, YOLOv2, YOLOv3, PP-YOLO, R-CNN, Fast R-CNN were practiced for methodological modification on this dataset. Among the mentioned architectures YOLOv3 and Faster R-CNN perform better detecting the road sign more precisely on our respective datasets. The implementation of this paper shows good possibilities to recognize a road sign and reduce the risk of an accident caused by disregarding the signs by drivers. We discovered that no comparative study on this topic was conducted focusing on Bangladesh's Roads during our analysis. That is why, we think this research can enhance the factors and possibilities for the researchers, while working on this topic in future.

Acknowledgements We thankfully acknowledge the assistance of the Advanced Machine Learning lab for their resource sharing and supports.

References

1. Redmon J, Divvala S, Girshick R, Farhadi A (2016) You only look once: unified, real-time object detection. In: Proceedings of the IEEE conference on computer vision and pattern recognition, pp 779–788
2. Jo K, Im J, Kim J, Kim DS (2017) A real-time multi-class multi-object tracker using YOLOv2. In: 2017 IEEE International conference on signal and image processing applications (ICSIPA). IEEE, pp 507–511
3. Redmon J, Farhadi A (2018) Yolov3: an incremental improvement. arXiv preprint arXiv:1804.02767
4. Long X, Deng K, Wang G, Zhang Y, Dang Q, Gao Y et al (2020) PP-YOLO: an effective and efficient implementation of object detector. arXiv preprint arXiv:2007.12099
5. Benjdira B, Khursheed T, Koubaa A, Ammar A, Ouni K (2019) Car detection using unmanned aerial vehicles: comparison between faster R-CNN and Yolov3. In: 2019 1st International conference on unmanned vehicle systems-Oman (UVS). IEEE, pp 1–6
6. Ren S, He K, Girshick R, Sun J (2015) Faster R-CNN: towards real-time object detection with region proposal networks. arXiv preprint arXiv:1506.01497
7. Hinton GE, Sabour S, Frosst N (2018) Matrix capsules with EM routing. In: International conference on learning representations
8. Su J, Vargas DV, Sakurai K (2019) One pixel attack for fooling deep neural networks. IEEE Trans Evol Comput 23(5):828–841
9. Kumar AD (2018) Novel deep learning model for traffic sign detection using capsule networks. arXiv preprint arXiv:1805.04424
10. Tabernik D, Skočaj D (2019) Deep learning for large-scale traffic-sign detection and recognition. IEEE Trans Intell Transp Syst 21(4):1427–1440
11. Wang Z, Guo H (2019) Research on traffic sign detection based on convolutional neural network. In: Proceedings of the 12th international symposium on visual information communication and interaction, pp 1–5

12. Gudigar A, Jagadale BN, Mahesh PK, Raghavendra U (2012) Kernel based automatic traffic sign detection and recognition using SVM. In: International conference on eco-friendly computing and communication systems. Springer, Berlin, Heidelberg, pp 153–161
13. Zhang J, Huang M, Jin X, Li X (2017) A real-time Chinese traffic sign detection algorithm based on modified YOLOv2. Algorithms 10(4):127
14. Buyval A, Gabdullin A, Lyubimov M (2019) Road sign detection and localization based on camera and Lidar data. In: Eleventh international conference on machine vision (ICMV 2018), vol 11041. International Society for Optics and Photonics, p 1104125
15. Bouti A, Mahraz MA, Riffi J, Tairi H (2019) A robust system for road sign detection and classification using LeNet architecture based on convolutional neural network. Soft Comput 1–13
16. Ahsan SMM, Das S, Kumar S, La Tasriba Z (2019) A detailed study on Bangladeshi road sign detection and recognition. In: 2019 4th International conference on electrical information and communication technology (EICT). IEEE, pp 1–6
17. Ioffe S, Szegedy C (2015) Batch normalization: accelerating deep network training by reducing internal covariate shift. In: International conference on machine learning. PMLR, pp 448–456
18. He K, Zhang X, Ren S, Sun J (2016) Deep residual learning for image recognition. In: Proceedings of the IEEE conference on computer vision and pattern recognition, pp 770–778
19. Maldonado-Bascón S, Lafuente-Arroyo S, Gil-Jimenez P, Gómez-Moreno H, López-Ferreras F (2007) Road-sign detection and recognition based on support vector machines. IEEE Trans Intell Transp Syst 8(2):264–278
20. Girshick R, Donahue J, Darrell T, Malik J (2014) Rich feature hierarchies for accurate object detection and semantic segmentation. In: Proceedings of the IEEE conference on computer vision and pattern recognition, pp 580–587
21. He K, Zhang X, Ren S, Sun J (2015) Spatial pyramid pooling in deep convolutional networks for visual recognition. IEEE Trans Pattern Anal Mach Intell 37(9):1904–1916
22. Girshick R (2015) Fast R-CNN. In: Proceedings of the IEEE international conference on computer vision, pp 1440–1448
23. Abadi M, Barham P, Chen J, Chen Z, Davis A, Dean J et al (2016) Tensorflow: a system for large-scale machine learning. In: 12th (USENIX) symposium on operating systems design and implementation (OSDI'16), pp 265–283
24. Gulli A, Pal S (2017) Deep learning with Keras. Packt Publishing Ltd.
25. Oliphant TE (2006) A guide to NumPy, vol 1. Trelgol Publishing, USA, p 85

Densely-Populated Traffic Detection Using YOLOv5 and Non-maximum Suppression Ensembling

Raian Rahman, Zadid Bin Azad, and Md. Bakhtiar Hasan

Abstract Vehicular object detection is the heart of any intelligent traffic system. It is essential for urban traffic management. Recent state-of-the-art methods apply R-CNN, Fast R-CNN, Faster R-CNN, and YOLO for this task. However, region-based CNN methods have the problem of higher inference time which makes them unrealistic to use the model in real-time. YOLO on the other hand struggles to detect small objects that appear in groups. In this paper, we propose a method that can locate and classify vehicular objects from a given densely crowded image using YOLOv5. We apply non-maximum suppression ensembling of 4 different models of YOLOv5 trained on different setups. The performance of our proposed model was measured on the Dhaka AI dataset which contains densely crowded vehicular images taken from both top view and side view of the street in both day and night settings. Our experiments show that our model achieved mAP@0.5 of 0.458 with an inference time of 0.75s outperforming other state-of-the-art models on performance. Hence, the model can be implemented in the street for real-world traffic detection which can be used for traffic control and data collection.

Keywords Real-time object detection · Ensemble learning · YOLOv5 · Non-maximum suppression

R. Rahman (✉) · Z. Bin Azad · Md. Bakhtiar Hasan
Department of Computer Science and Engineering, Islamic University of Technology, Gazipur, Dhaka, Bangladesh
e-mail: raianrahman@iut-dhaka.edu

Z. Bin Azad
e-mail: zadidbinazad@iut-dhaka.edu

Md. Bakhtiar Hasan
e-mail: bakhtiarhasan@iut-dhaka.edu

M. S. Arefin et al. (eds.), *Proceedings of the International Conference on Big Data, IoT, and Machine Learning*, Lecture Notes on Data Engineering and Communications Technologies 95, https://doi.org/10.1007/978-981-16-6636-0_43

1 Introduction

An increasing number of vehicle types in urban areas pose many problems like traffic congestion, long queue in toll and parking sites. To solve traffic problems in megacities and to monetize traffics in areas like toll booths, parking lots, and analyzing types of vehicles in a city more efficiently and effectively, an intelligent system is required. As an indispensable part of the intelligent traffic monitoring system, accurate vehicle detection and real-time performance is the most challenging part which is gaining the attention of researchers all over the world. Efficient vehicle detection and classification in densely populated areas can facilitate automated toll collection, smart parking systems, and the identification of vehicles related to crimes.

The task of vehicle detection can be formulated as a multi-object detection problem. In simple terms, object detection is the task of locating the objects in an image with a bounding box and detecting the class of that object. For this, convolutional neural network (CNN) based methods have been widely used in the recent past. The prominent state-of-the-art methods utilize R-CNN [1], Fast R-CNN [2], Faster R-CNN [3] to achieve this task. But the problem with these two-stage-based models is that training happens in multiple phases and the network is too slow at inference time, which impedes real-time detection of vehicles. To solve this problem, recently You Only Look Once (YOLO) [4] introduced a faster way of real-time object detection making it usable in real-life applications. However, this architecture struggles to detect small objects that appear in groups [4].

To solve these issues, we trained 4 separate models that utilize the ensemble technique to aggregate the separate predictions using Non-Maximum Suppression. Our contributions are as follows:

- Trained a total of 4 YOLOv5 [5] models using different image sizes and hyper-parameters.
- Aggregated the prediction of 4 models using an ensemble model that facilitates faster detection of vehicles.
- Introduced additional difficulty by adding low-light nighttime images and top-view images with densely crowded vehicles to training samples to improve the accuracy and robustness of the model.

These steps resulted in a solution that can be used in real-time and low light situations even in densely populated streets. Besides it also ensured that our solution outputs a result with acceptable accuracy which makes our model usable in congested and complex scenes.

2 Related Works

The traditional approaches [6, 7] for vehicle detection apply common machine learning algorithms like the histogram of oriented gradient (HOG) to extract features from vehicle images. After extracting the features, the vehicles are then classified using Support Vector Machine (SVM). Other approaches use Deformable Part Model (DPM) [8] to detect vehicles. Even though these approaches provide comparable accuracy, they involve handcrafted feature designing that requires human intervention.

Recent advances in deep learning facilitated by the availability of large datasets and big compute have made them a viable option for vehicle detection. Earlier approaches [9–11] utilize Convolutional Neural Network (CNN) to perform feature extraction and softmax function for classification. Later, more efficient models like R-CNN [1] and fast R-CNN [2] and Faster R-CNN [3] models were proposed. All these models utilize a Region-based Convolutional Neural Network, which uses a technique called Selective Search [12] to select a small number of candidate regions among all possible regions. As a result, the model requires running an image classification algorithm for a smaller amount of region making the model run faster. R-CNN is comparatively slower among all three models as it generates lots of candidate regions. Fast R-CNN [2] addressed this issue by feeding the input image to a CNN to generate a convolutional feature map. Then, the candidate regions are proposed using an RoI pooling layer and feeding it into a fully connected network. The number of candidate regions proposed by Fast R-CNN is less than that of R-CNN, hence it requires less time for inference. But the Selective Search algorithm, used by Fast R-CNN, is not a machine learning algorithm, so it cannot learn from the context, and often proposes a bad candidate for the region. Later, Faster R-CNN [3] was proposed with the idea of replacing selective search as it is a time-consuming process. Faster R-CNN provides the fastest running time compared with R-CNN and Fast R-CNN. However, it is still not fast enough to detect objects in real-time. Additionally, all these three models require huge computation due to having a complex model containing a large number of parameters.

Recently, YOLO is being used for vehicle detection [13–15]. Instead of using the region selection method, YOLO uses Convolutional Neural Network that predicts the bounding boxes as well as the class for these boxes. It divides the image into an $S \times S$ grid where S is a constant value. For each grid, YOLO generates a constant number of bounding boxes. Then if a bounding box has confidence greater than a certain threshold, the bounding box is selected to locate the object within the image. YOLO is by far the fastest algorithm for vehicle detection and its speed is helpful to implement real-time vehicle detection systems.

3 Proposed Methodology

3.1 Overview

As shown in Fig. 1, our proposed method consists of 3 main modules. First, we acquired and preprocessed the dataset. During preprocessing, we applied augmentation, resized the images into uniform shapes, and created training and testing folds.

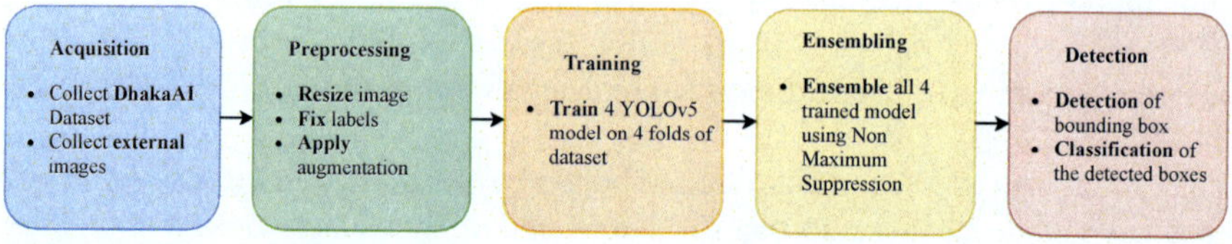

Fig. 1 The pipeline of our proposed solution: First, we acquired the training data. During preprocessing, images were resized, relabeled before creating four different folds of the dataset. Different augmentation techniques were applied to these folds. During training, these folds were trained independently with the YOLOv5 model with different setups. All four of our trained models were then ensembled using Non-Maximum Suppression. The last stage of our work is the images with bounding boxes surrounding the vehicle objects of an image with its class

Then, four different models were trained with these different training folds. After training, we ensembled the models using Non-Maximum Suppression [16] for final inference.

3.2 *Dataset Acquisition*

For this experiment, we used the "DhakaAI" [17] dataset developed under the "Dhaka AI 2020 challenge". The dataset consists of 3000 annotated images of traffic objects. The training dataset consisted of 21 classes. The most challenging part about the dataset is it contains images of vehicles from a different point of view. There were images from the front view, back view, side view, and top view of streets. We also added around 200 new images for training to increase the sample of rare class vehicles. These new images were hand-annotated using the LabelImg tool [18]. Most of the newly added images were top-view nighttime images.

3.3 *Preprocessing*

For generalizing a model for object detection using deep learning architecture, a prerequisite is to have enough training examples for each class so that the model can learn properly. But after exploring the DhakaAI dataset [17], we found that it has a huge class imbalance. The number of labels for each class is shown in Table 1. Here, some of the classes have less than 50 samples in the training dataset.

To resolve this issue, we used augmentation using tools from Roboflow[1] and Albumentations library [19] for image augmentation. Although augmentation did not provide decent results in the case of densely populated images, it improved the result in the case of night images.

[1] Available at:https://roboflow.com/.

Table 1 Sample distribution per class

Class name	Label count	Class name	Label count
Ambulance	76	Pickup	1178
Army vehicle	25	Police car	33
Auto rickshaw	465	Rickshaw	3495
Bicycle	465	Scooter	30
Bus	3340	SUV	667
Car	5574	Taxi	59
Garbage van	8	Three wheeler (CNG)	2982
Human Hauler	170	Truck	1475
Minibus	100	Van	682
Minivan	815	Wheelbarrow	251
Motorbike	2252		

(a) (b) (c)

Fig. 2 The figure shows an anomaly in the DhakaAI dataset. In **a**, we can see that a wheelbarrow is labeled as a rickshaw. In **b**, **c**, we can see that the same vehicle, which is a pickup is labeled as a truck and on the next image it is labeled as pickup

During the exploration of the dataset, we found lots of mislabeled images in the DhakaAI dataset training data. We also found that two images had different labeling of class for the same car in the same frame (illustrated in Fig. 2). So, we hand-annotated all 3000 images and labeled all the mislabeled objects as well as fixed labeling of wrongly labeled objects in the image.

Another challenge in the dataset is it does not have uniform image quality. Some of the images are in landscape mode while some of the images are in portrait mode. So, we resized the images to 1024×1024 pixels.

For the train and validation set split, we used the k-fold Cross-Validation technique so that our model could learn from the complete dataset. While creating the fold, we tried to make sure that images from the same frame in the train split do not occur in the validation split. Count of train-validation split for each fold is given in Table 2.

3.4 Model Selection

Although the key priority of our work was to localize and classify the vehicular objects on a street image, we also had to look into the inference speed so that it could be implemented in real-time. We had to discard R-CNN, Fast R-CNN, and

Table 2 Image resolution and applied augmentation for different folds of training dataset

Fold No.	Train set image count	Validation set image count	Augmentation
1	2506	600	Sharpened
2	2321	785	Sharpened
3	2400	706	Sharpened
4	1200	400	Darkened and Sharpened

All images had 1024 × 1024 resolution

Faster R-CNN as they could not compete with YOLO models in both performance and inference time. YOLO on the other hand, YOLO had a much less inference time with better accuracy. Among different versions of YOLO, we chose YOLOv5 [5] due to its simple architecture compared to R-CNN-based models. Even YOLOv5 is faster and more robust than other members of the YOLO family.

Even if the author of YOLOv4 [20] got official approval of YOLO, the version of YOLOv5 [5] developed by the Ultralytics LLC team did not get any acknowledgment from the original author of YOLO. Still, YOLOv5 provides much better performance compared to other models of the YOLO family [21]. YOLOv5 inherits the advantages of YOLOv4 [20] by adding SPP-NET [22] along with some enhancement techniques. YOLOv5 has become the new state of the art for object detection [23]. YOLOv5 was mainly developed to balance real-time performance and detection accuracy.

YOLOv5s, YOLOv5m, YOLOv5l, YOLOv5x [5] are the four versions of YOLOv5 where YOLOv5s being the lightest model and YOLOv5x being the heaviest model, respectively. Among all these four versions, there is a trade-off between the detection speed and real-time performance. The key differences among these versions are the number of feature extraction modules and convolution kernel in a specific location of the network.

The network consists of three networks. These are backbone network, neck network, and detect network. The backbone network is a convolutional neural network for aggregating fine-grained images and forming image features. The neck network is responsible for combining the image features collected by the backbone network and transmitting the feature map to the detect network. The detect network is responsible for the detection and classification part of the model. It applies anchor boxes on the feature map from the neck network. It also contains a softmax layer which predicts the probability of the class of the bounding box surrounding the object.

For image enhancement, YOLOv5 uses mosaic data augmentation to solve the small dataset problem. It applies operations like random inversion, zooming, cropping on four images and then combines them into a single image.

In traffic detection, the core priority was to improve the performance, so we chose YOLOv5x for our training model. It contains 607 layers with 88, 568, 234 trainable parameters. The model was pre-trained using Common Object in Context (COCO) dataset [24] to detect 80 classes. For our task, we changed the final layer to detect only 21 classes corresponding to the 21 vehicle classes available in the DhakaAI dataset.

Table 3 Training specification for each model

Model	Training data	Image size	Number of epochs	Batch size
1	Fold 1	1024 × 1024	80	4
2	Fold 2	1024 × 1024	80	4
3	Fold 3	1024 × 1024	80	4
4	Fold 4	640 × 640	120	16

All 4 models had Stochastic Gradient Descent optimizer with a learning rate 0.01 and momentum 0.937

3.5 Ensemble Learning

To ensure the robustness and accuracy of our model, we trained 4 separate models using different sets of images. Each of the models proposes multiple bounding boxes to specify candidate regions for vehicle detection. We used Non-Maximum Suppression [16] to aggregate these bounding boxes to select the ones having the most confidence. The way it works is the system takes all the bounding boxes proposed by all four models and puts them in a priority queue sorted based on the confidence of the models predicting them. It then selects the box with the highest confidence from the queue and calculates the Intersection over Union (IoU) with the rest of the boxes. If the IoU value exceeds a certain threshold for any of the remaining boxes, that box is discarded. It then removes the bounding box with the highest confidence from the queue and adds it to the selected box list. This process is repeated until there is no bounding box remaining in the priority queue. Finally, the boxes in the selected box list are returned.

4 Result and Analysis

4.1 Experimental Setup

As shown in Table 3, during training we had to train four different models and ensembled these four models for final output. All four of our model was trained on Google Colab[25]. Google Colab provides a cloud-based training utility with free GPU access for a limited amount of time. For each fold of our dataset, we trained a model. The first three models were trained with an image resolution of 1024 × 1024 pixels. But the fourth model was trained with an image resolution of 640 × 640 pixels. The first three folds of our dataset contained all the images, while the fourth fold contained only the night images. On the dataset, it was seen that the night images themselves were quite distorted noisy. So. we decided to train the night images on a lower resolution as it might then focus on larger objects of the night images. As a result, we could train our model for a longer time.

Table 4 Image augmentation parameters during training

Hyperparameter	Value
Image HSV—hue augmentation	0.015
Image HSV—saturation augmentation	0.7
Image HSV—value augmentation	0.4
Image rotation	5.0
Image translation	0.1
Image scale	0.5
Image flip left-right—probability	0.5
Image mosaic—probability	1.0
Image mixup—probability	0.0

All four of our models were trained with Tesla T4 GPU which comes with 16 GB of video memory. All four of our model was trained for around 12 hours each.

For training, we used a YOLOv5 implementation by the Ultralytics.[2] We used Stochastic Gradient Descent as our optimizer. Image augmentation parameters used for each of the models are given in Table 4.

For inference, we ensembled the weight of all four models we trained. We used Non-Maximum Suppression during the ensemble and the confidence threshold for each predicted bounding box was set to 0.3.

4.2 Evaluation Metrics

To evaluate our performance, we used mean Average Precision (mAP) over training epochs following. The formula for calculating mean average precision for object detection is

$$maP = \frac{1}{n} \sum_{k=1}^{k=n} AP_k \tag{1}$$

where n is the number of classes and AP_k is the average precision for class k. Average precision (AP) is a way of summarizing the precision-recall curve into a single value representing average of all precision. The formula for calculating AP is

$$AP@n = \sum_{k=0}^{k=n-1} [\text{Recall}(k) - \text{Recall}(k+1)] \times \text{Precision}(k) \tag{2}$$

where n is the number of thresholds and $\text{Recall}(n) = 0$ and $\text{Precision}(n) = 1$. We used the checkpoint where the model had most *mAP@0.5*.

[2] Available at https://github.com/ultralytics/yolov5.

Table 5 Comparison of performance and inference time with other models on the test set

Model name	mAP@0.5	Inference time (s)
Faster R-CNN	0.356	0.39
YOLOv3	0.266	0.18
YOLOv4	0.313	0.28
YOLOv5x	0.372	0.14
YOLOv5 with NMS ensembling (ours)	0.458	0.75

Here, Faster R-CNN. YOLOv3. YOLOv4 and YOLOv5x show performance trained on a single fold of train dataset, while YOLOv5x with NMS ensembling model shows the result of our four combined models

4.3 Result Discussion

We used all four training model's weights during the final inference. We ran an inference on test data—2 provided by DhakaAI. We hand-annotated 450 test images and executed inference. On that test, our model achieved $mAP@0.5$ value of 0.458. We also conducted inference on one of our validation sets. During validation set inference, our model achieved $mAP@0.5$ value of 0.883 and $mAP@0.95$ value of 0.677.

We compared the result of our model with other models of the YOLO family as well as the Faster R-CNN model. Table 5 shows the comparison between these models. For comparison, we compared our model's performance as well as the inference time on a single image with YOLOv3, YOLOv4, and Faster R-CNN. We trained each of these models for 12 hours on Google Colab in a similar environment. The table shows that our model has achieved the most $mAP@0.5$. As our proposed method ensembles 4 different models during inference, the inference time of our solution is a little bit higher compared to other models. Still, the precision performance of our model outperforms the other models. Our ensembling technique allowed us to emphasize on different aspects of the dataset equally, especially for the night time images, without sacrificing generality.

Output of our model for different scenario is illustrated in Figs. 3 and 4. Our model was able to localize and detect most of the vehicular objects for a given image taken from a different view of the street. It also performed well in the case of night images. Fig. 3 illustrates the performance of our model on night images. It could locate most of the vehicles as well as properly classify those vehicles in both densely populated images and less populated images. However, as seen in Fig. 3c, our model could not detect most of the vehicles in a very low-light noisy image.

In Fig. 4, we illustrated our model's performance on images taken from a different view of the streets which shows it can locate and detect the objects properly. As seen in the figure, the model also performs well in the case of occluded objects in the image.

Our model could run inference within 0.75 s per image. So, it could also be implemented in real-time vehicular traffic detection applications.

(a) (b) (c)

Fig. 3 Performance on night image in densely populated image Fig. 3a and less densely populated image Fig. 3b. In Fig. 3c, it can be seen that the model performs poorly in very low light sample

(a) (b) (c)

Fig. 4 Performance on images taken from different views of street

5 Conclusion

This paper proposed a new method of traffic object detection using YOLOv5. To improve the performance and robustness of our method, we ensembled 4 different models using Non-Maximum Suppression ensembling. We also tried to incorporate dataset modification by adding night images from different view-angles. Our experiments compared the performance of our model with other state-of-the-art models on the Dhaka AI dataset. Results show that our model had better mean average precision. Due to limited resources, we could not test our model's performance on other baseline datasets. For further experimentation, our work could be expanded on how we can use better ensembling methods like weighted ensembling or voting mechanisms for faster inference time.

Acknowledgements We would like to extend our gratitude to Mr. Redwan Karim Sony, Department of Computer Science and Engineering, Islamic University of Technology, and Mr. Mohammad Sabik Irbaz, Pioneer Alpha Limited for their continuous support and suggestions throughout the work. We would also like to thank the organizing committee of Dhaka AI 2020 for organizing the competition.

References

1. Girshick R, Donahue J, Darrell T, Malik J (2014) Rich feature hierarchies for accurate object detection and semantic segmentation. In: 2014 IEEE conference on computer vision and pattern recognition, pp 580–587. https://ieeexplore.ieee.org/document/6909475
2. Girshick R (2015) Fast r-cnn. In: 2015 IEEE international conference on computer vision (ICCV), pp 1440–1448. https://ieeexplore.ieee.org/document/7410526
3. Ren S, He K, Girshick R, Sun J (2017) Faster r-cnn: towards real-time object detection with region proposal networks. IEEE Trans Pattern Anal Mach Intell 39(6):1137–1149. https://ieeexplore.ieee.org/document/7485869
4. Redmon J, Divvala S, Girshick R, Farhadi A (2016) You only look once: Unified, real-time object detection. In: 2016 IEEE conference on computer vision and pattern recognition (CVPR), pp 779–788. https://ieeexplore.ieee.org/document/7780460
5. Jocher G, Stoken A, Borovec J, NanoCode012, Chaurasia A, TaoXie, Changyu L, V, A., Laughing, tkianai, yxNONG, Hogan A, lorenzomammana, AlexWang1900, Hajek J, Diaconu L, Marc, Kwon Y, oleg, wanghaoyang0106, Defretin Y, Lohia A, ml5ah, Milanko B, Fineran B, Khromov D, Yiwei D, Doug, Durgesh, Ingham F (2021) ultralytics/yolov5: v5.0 - YOLOv5-P6 1280 models, AWS, Supervise.ly and YouTube integrations (Apr 2021). https://github.com/ultralytics/yolov5
6. Laopracha N, Sunat K (2018) Comparative study of computational time that hog-based features used for vehicle detection. In: Meesad P, Sodsee S, Unger H (eds) Recent advances in information and communication technology 2017. Springer, Cham, pp 275–284
7. Cao X, Wu C, Yan P, Li X (2011) Linear svm classification using boosting hog features for vehicle detection in low-altitude airborne videos. In: 2011 18th IEEE international conference on image processing. IEEE, pp 2421–2424. https://ieeexplore.ieee.org/document/6116132
8. Pan C, Sun M, Yan Z (2018) The study on vehicle detection based on dpm in traffic scenes. In: Yen NY, Hung JC (eds) Frontier computing. Springer Singapore, Singapore, pp 19–27. https://link.springer.com/chapter/10.1007/978-981-10-3187-8_3
9. Tang Y, Zhang C, Gu R, Li P, Yang B (2017) Vehicle detection and recognition for intelligent traffic surveillance system. Multimedia Tools Appl 76(4):5817–5832. https://link.springer.com/article/10.1007/s11042-015-2520-x
10. Gao Y, Guo S, Huang K, Chen J, Gong Q, Zou Y, Bai T, Overett G (2017) Scale optimization for full-image-cnn vehicle detection. In: 2017 IEEE intelligent vehicles symposium (IV). IEEE, pp 785–791. https://ieeexplore.ieee.org/document/7995812
11. Huttunen H, Yancheshmeh FS, Chen K (2016) Car type recognition with deep neural networks. In: 2016 IEEE intelligent vehicles symposium (IV). IEEE, pp 1115–1120. https://ieeexplore.ieee.org/document/7535529
12. Uijlings JR, Van De Sande KE, Gevers T, Smeulders AW (2013) Selective search for object recognition. Int J Comput Vis 104(2):154–171. https://link.springer.com/article/10.1007/s11263-013-0620-5
13. Kasper-Eulaers M, Hahn N, Berger S, Sebulonsen T, Myrland Ø, Kummervold PE (2021) Short communication: detecting heavy goods vehicles in rest areas in winter conditions using yolov5. Algorithms 14(4). https://www.mdpi.com/1999-4893/14/4/114
14. Sang J, Wu Z, Guo P, Hu H, Xiang H, Zhang Q, Cai B (2018) An improved yolov2 for vehicle detection. Sensors 18(12). https://www.mdpi.com/1424-8220/18/12/4272
15. Asha C, Narasimhadhan A (2018) Vehicle counting for traffic management system using yolo and correlation filter. In: 2018 IEEE international conference on electronics, computing and communication technologies (CONECCT). IEEE, pp 1–6. https://ieeexplore.ieee.org/document/8482380
16. Neubeck A, Van Gool L (2006) Efficient non-maximum suppression. In: 18th international conference on pattern recognition (ICPR'06). vol 3, pp 850–855. https://ieeexplore.ieee.org/document/1699659
17. Shihavuddin A, Rashid MRA (2020) DhakaAI. https://doi.org/10.7910/DVN/POREXF

18. Tzutalin: Labelimg. Git Code (Dec 2018). https://github.com/tzutalin/labelImg
19. Buslaev A, Iglovikov VI, Khvedchenya E, Parinov A, Druzhinin M, Kalinin AA (2020) Albumentations: fast and flexible image augmentations. Information 11(2). https://www.mdpi.com/2078-2489/11/2/125
20. Bochkovskiy A, Wang C, Liao HM (2020) Yolov4: optimal speed and accuracy of object detection. Comput Res Repos (CoRR) abs/2004.10934. https://arxiv.org/abs/2004.10934
21. Liu Y, Lu B, Peng J, Zhang Z (2020) Research on the use of yolov5 object detection algorithm in mask wearing recognition. World Sci Res J 276–284
22. He K, Zhang X, Ren S, Sun J (2015) Spatial pyramid pooling in deep convolutional networks for visual recognition. IEEE Trans Pattern Anal Mach Intell 37(9):1904–1916. https://ieeexplore.ieee.org/document/7005506
23. Yan B, Fan P, Lei X, Liu Z, Yang F (2021) A real-time apple targets detection method for picking robot based on improved yolov5. Remote Sens 13(9). https://www.mdpi.com/2072-4292/13/9/1619
24. Lin TY, Maire M, Belongie, S, Hays J, Perona P, Ramanan D, Dollár P, Zitnick CL (2014) Microsoft coco: common objects in context. In: Fleet D, Pajdla T, Schiele B, Tuytelaars T (eds) Computer vision—ECCV 2014. Springer International Publishing, Cham (2014), pp 740–755. https://link.springer.com/chapter/10.1007/978-3-319-10602-1_48
25. Bisong E (2019) Google Colaboratory. Apress, Berkeley, CA, pp 59–64. https://link.springer.com/chapter/10.1007/978-1-4842-4470-8_7

Real-time Pothole Detection and Localization Using Convolutional Neural Network

Atikur Rahman, Rashed Mustafa, and Mohammad Shahadat Hossain

Abstract Pothole is a common problem in damaged roads and pavements. Vehicles get damaged, people get stumbled, drivers lose control over the car, and accidents take place. A system is required that can detect potholes as fast as possible and help to avoid them. In this paper, such a system is being described which is capable of detecting potholes in video frames as well as localizing and tracking them. The system can also make instantaneous signal and warn about the detected potholes. It works in real time even in mobile devices like Android smartphone. Convolutional neural network has been used as the basis of the model. A dataset was prepared where images having potholes were annotated by selecting the regions containing potholes. Following an approach of supervised transfer learning, a neural network model was developed. The model was then deployed as Android application for real-world testing purpose. Satisfactory result was found in both theoretical evaluation and practical real-world tests.

Keywords Pothole · Detection · Localization · Real time · ConvNet · CNN · Neural net · Fine-tuning · Transfer learning

1 Introduction

According to Cambridge Dictionary [32], a pothole is "a hole in a road surface that results from gradual damage caused by traffic and/or weather." Potholes cause huge trouble in regular transportation system. Vehicles may get damaged when they hit pothole. People who are visually impaired face a great problem due to potholes while navigating. Sometimes, drivers may lose their control over the car and result in accidents. Therefore, to get rid of this problem, either potholes must be repaired as soon as they appear or drivers and passers-by must be aware of them during navigation so that they can be avoided or passed safely.

A. Rahman · R. Mustafa (✉) · M. S. Hossain
University of Chittagong, Chattogram 4331, Bangladesh
e-mail: hossain_ms@cu.ac.bd

M. S. Arefin et al. (eds.), *Proceedings of the International Conference on Big Data, IoT, and Machine Learning*, Lecture Notes on Data Engineering and Communications Technologies 95, https://doi.org/10.1007/978-981-16-6636-0_44

Although the first one is very inefficient and troublesome, research has taken place to detect potholes and mark the location where they occurred [4, 7, 17, 23, 34]. The collected data then is sent to the authority so that the hole is repaired soon. Second one is also researched on [6, 30], to inform navigators about the potholes appearing on the road surface in front of them.

This research is also focused on the second one, i.e., detect potholes before you hit them and try to avoid if possible or hit safely. The research work targets to develop a system for automatic detection of potholes in real time and generate signal to warn about them. We have created a custom dataset, trained the model with it as well as built an Android application to deploy as a sample.

A supervised learning approach [18] was followed along with transfer learning [28] in this research to build the detector model. A dataset of annotated images was used to train and evaluate the model. Pre-trained MobileNet [11], a convolutional neural network model, was used as the feature extractor. Single-shot multibox detector (SSD) [22] was fine-tuned for detection and localization of potholes. Despite training on image dataset, the resulting model is capable of detecting, localizing, and tracking potholes in real time, analyzing the video frames.

An Android application has been developed using the model which can detect, localize, and track the potholes in the video coming from its camera. It also generates warning signal as long as pothole is detected in the video stream. Thus, it is very helpful for blind people to navigate. Localization capability of potholes in video stream can be very useful in case of automated vehicle driving.

In this paper, explanation and discussion of technical requirements, materials and methods, results and evaluation of our research are given following a review of existing and related works. After all, conclusion and future works are given followed by the bibliography.

2 Related Work

Mednis et al. proposed such a mobile sensing system for road [30] which was able to detect inconsistency with the help of a smartphone based on Android operating system [25, 30]. They took a 4.4 km long track for the testing purpose with ten consecutive laps. Using real-world data, their method presented around 90% true positive rate (TPR) [14].

This method could result in wrong information for the cases such as

- It detects hinges as well as joints of the road [14] as pothole event if it is not the case though.
- It fails to detect potholes which are in the middle of the lane.

Chang et al. showed that scanning as well as extracted focusing on some particular distress features was captured along with accurate 3D cloud points with their elevation by means of a grid-based approach [5]. Severity and coverage of the distress could be accurately and automatically calculated using this method [5, 14].

Li et al. presented an inspection system [23] which can be used to detect and identify distress features like potholes, shoving, and rutting with the help of a 3D transverse scanning technique [23], and it is a high-speed technology. The technique mentioned uses infrared waves-featured laser-line projector with a digital camera for detecting distress features as well as potholes [23].

Joubert et al. [4] proposed a low-cost sensor system using Kinect sensor and high-speed USB camera [4] to detect and analyze potholes. Some experiments have already taken place on using Kinect to examine potholes. This method is cost-effective which is a plus-point.

Buza et al. [4] proposed an unsupervised method using computer vision which does not require expensive equipment, filtering, or training phases [4, 31]. They used general image processing and clustering technologies for detecting and identifying potholes in the target image [14].

Their method consists of three steps as given below

1. Segmentation of target images
2. Shape and feature extraction
3. Detection and identification.

Using the method stated above, they reached an accuracy of 81% [14], and it could be used as a rough estimation for pothole repairs.

Lokeshwor et al. [12] proposed a method which could detect potholes, cracks, and patches of pavement by analyzing video frames. Using DFS algorithm [12, 14], they segmented video clips undoubtedly into two frames, namely stressed and distressed categories.

Jog et al. [13] showed a system for 2D recognition and 3D reconstruction [13] to detect and measure potholes along with their severity. They used video camera seated on the car to capture video of pavements.

They were able to find the depth, width, and number of potholes using this approach.

Koch et al. [16] proposed a method which was bound to single frame of the video coming form camera. It could not find the magnitude of potholes analyzing the frames of pavement video frames [14]. Koch et al. showed an updated composition signature for perfect pavement regions for pothole recognition. They also applied computer vision for tracking detected potholes in the all video frames[16].

3 Technical Requirements

The following technical requirements were identified for the research, preliminary system development, testing, and deployment

1. A dataset of images is required.
2. Dataset images must be annotated in such a way that the regions of interest, i.e., image areas having pothole must be selected, e.g., as bounding boxes [20].

3. For model train-up, a digital workstation [29] computer is recommended with "at least" following configurations

   - CPU: 2.2 GHz (6th Gen)
   - RAM: 16 GB (2400 MHz DDR4)
   - GPU:
     - VRAM: 12 GB
     - Computing [27] Capability: 5
     - CUDA [15] support.

   However, model developed in this research was trained on the Google Colaboratory [2].
4. Android [26] smartphone with speaker and camera support. Android OS version ≥ 6 (Marshmallow, API 22).

4 Materials and Methods

4.1 The Dataset

A portion (57.1%) of our dataset images was scrapped from Google image search. The rest (42.9%) was captured with an Android smartphone camera from some damaged roads.

The dataset contains total 665 images having a total of 1740 annotated potholes. Each image was annotated with bounding boxes around the regions having potholes. For this purpose, labelImg [19] was used as the annotation tool.

The training set contains 532 (80%) images, and the test set contains 133 (20%) images. The potholes were divided into three categories, namely small, medium, and large.

The sizes of the potholes were measured from the occupied pixels in the sample images. The number of pixels occupied was calculated after resizing the longer side of the images to 300 px keeping the aspect ratio of the shorter side.

The number of pixels occupied by the different categories of pothole is given in Table 1.

Distribution of different categories of potholes over the train and test data is shown in Fig. 1.

The dataset can be found in [1].

Table 1 Different categories versus number of pixels occupied

Pothole category	Occupied pixels
Small	area $\leq 32^2$
Medium	$32^2 <$ area $\leq 96^2$
Small	area $> 96^2$

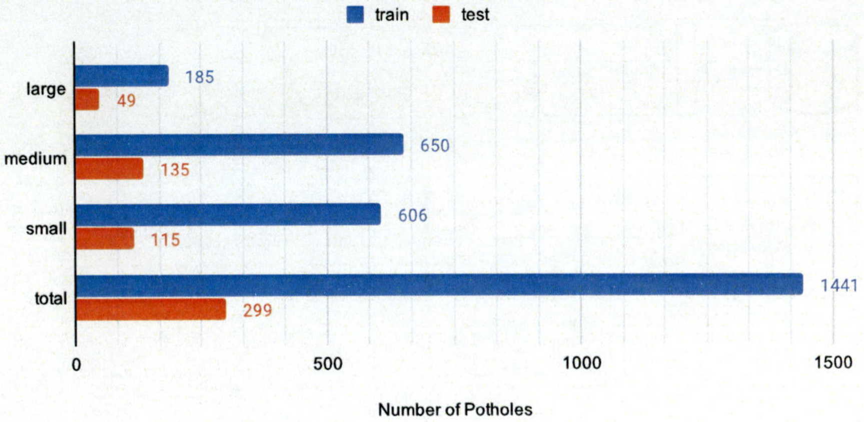

Fig. 1 Distribution of different categories of potholes

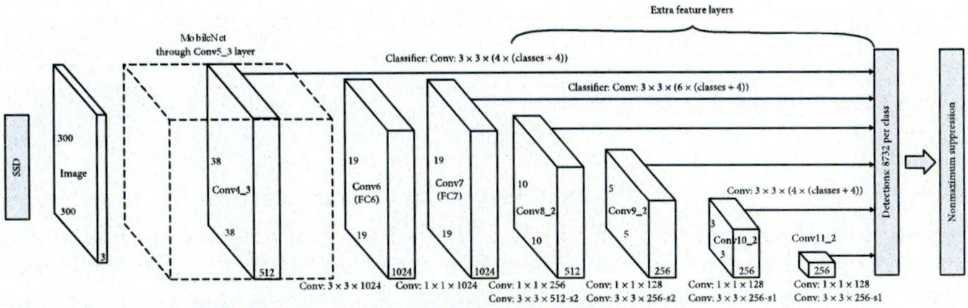

Fig. 2 Architecture of the SSD MobileNet model

4.2 Materials Used

Architecture of the SSD MobileNet [11, 22] is shown in Fig. 2.

With the help of TensorFlow [8] object detection API, the model was trained using the prepared dataset with some fine-tuned configurations. Figure 3 illustrates the interaction of different components during training. The whole model was trained and evaluated on the Google Colaboratory [2].

4.3 Training the Model

Data Augmentation The training images were augmented in different ways. The following augmentations were applied on the training images online, i.e., during training phase

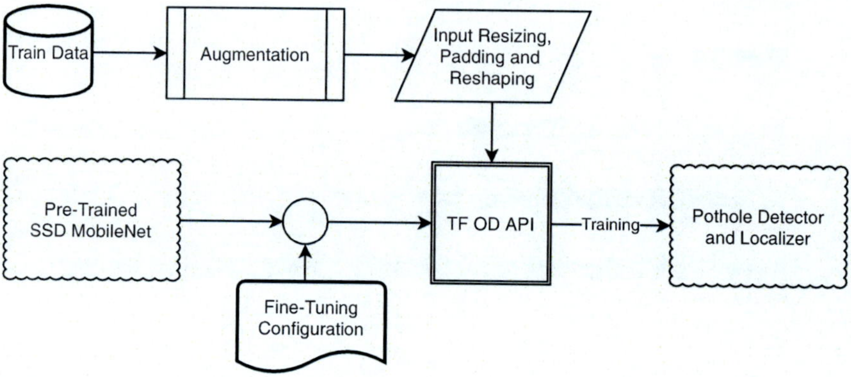

Fig. 3 Interaction of different training components

- Horizontal flip
- Random crop
- Resize keeping the aspect ratio
- Zero padding.

Input Resizing The shape of the input layer of our model is $300 \times 300 \times 3$; that is, it takes RGB images of 300 px width and 300 px height. We have resized the input such a way that the longer side was resized to 300 px, and the shorter size was resized keeping the aspect ratio intact. Remaining space of the shorter side was filled with zeros.

Preparing a Pre-trained Model We have followed transfer learning approach [28]. So, we required a pre-trained model ready to be used as a base model. Some of pre-trained models can be found in TensorFlow [8] Model Zoo [9].

We have used the SSD [22] MobileNet [11]V2 pre-trained model which was trained on Microsoft COCO dataset [21].

Our changes in different paramters are shown in Table 2.

Prepare TensorFlow Models Repository We reused the existing libraries, packages, and codes from TensorFlow "models" repository [10]. The repository contains almost everything we need for training and evaluation using TensorFlow API. The mentioned repository can be found in GitHub [10].

Run Training Process TensorFlow "models" repository provides necessary Python code for the whole training process. It saves the scalar and graphical values, i.e., the results of training and evaluation in "tfevent" files which can be monitored in TensorBoard [24] (Fig. 4).

Run Evaluation Process TensorFlow "models" repository provides the tools for the evaluation process too. The evaluation is triggered by the training tool automatically whenever it saves a checkpoint of the model [10].

Table 2 Fine-tuned configurations

Configuration option	Changed value
input_shape	(300, 300, 3)
num_classes	1
batch_size	24
image_resizer	keep_aspect_ratio_resizer
min_dimension	300
max_dimension	300
pad_to_max_dimension	True
initial_learning_rate	0.005
decay_steps	6000
decay_factor	0.85
quantization_delay	30000

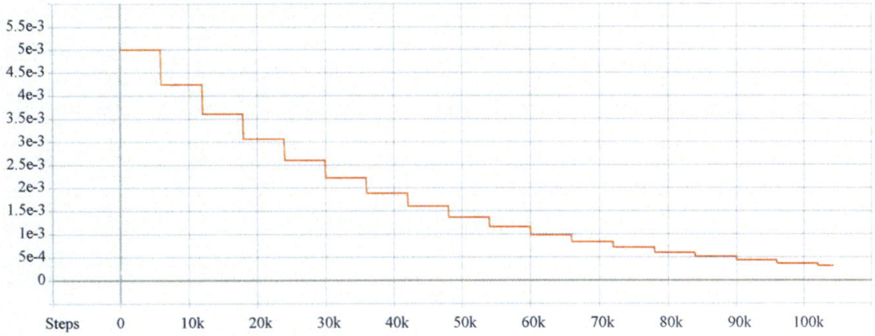

Fig. 4 Training steps versus decay of learning rate

4.4 Evaluation Methods

Intersection Over Union (IoU) To take a detected bounding box as true positive, different Intersection over Union (IoU) thresholds were considered (Fig. 5).

Following IoU thresholds were considered

- IoU@50%
- IoU@75%
- IoU@50%:5%:95%.

IoU@50%:5%:95% is a dynamic measurement where ten IoU thresholds are considered, and starting from 50% up to 95% with an interval of 5%, following thresholds are found—50%, 55%, 60%, 65%, 70%, 75%, 80%, 85%, 90%, 95%. Values at these ten thresholds were then averaged to get the value for IoU@50%:5%:95%.

Here, taking the above IoU thresholds into account,
TP = number of true positive predictions

Fig. 5 Measuring the Intersection over Union (IoU)

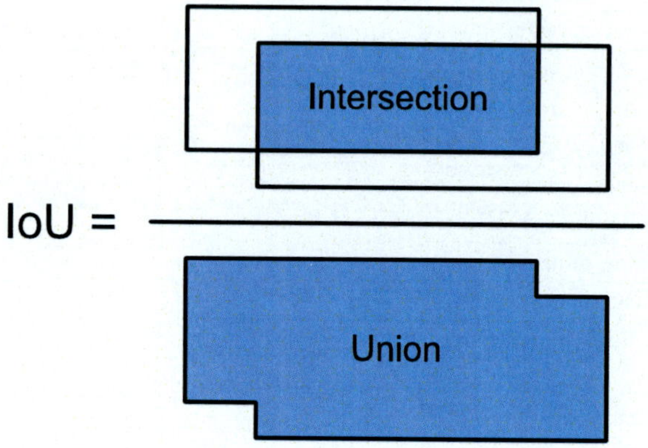

FP = number of false positive predictions
TN = number of true negative predictions
FN = number of false negative predictions.

Precision Calculation Precision shows how much the predictions are correct [3]. That is, percentage of correct predictions among all the positive-predicted values.

$$\text{Precision} = \frac{\text{TP}}{\text{TP} + \text{FP}} \tag{1}$$

Average precision is the area under precision–recall curve.

$$\text{AP} = \int_0^1 p(r)\, \text{d}r \tag{2}$$

Here, r represents recall, and p represents precision as a function of r. Therefore, $p(r)$ means "precision at recall r."

In this research, only a single class is available, and it is labeled "pothole"; therefore, mean average precision (mAP) is same as average precision (AP). For multi-class, mAP is the mean of average precisions of all individual classes.

$$\text{mAP} = \frac{1}{N}\sum_{i=1}^{N} \text{AP}_i \tag{3}$$

Here, mAP is mean average precision, N is the number of class labels, and AP_i is the average precision for ith class.

We considered and calculated mean average precision for different sizes of potholes as well as for different IoU thresholds.

Recall Calculation Recall means true positive rate and also known as sensitivity, a well-known parameter (measurement) for model evaluation in the context of classification [33].

We considered and calculated average recall for different sizes of potholes as well as for different maximum detection levels.

Average recall is same as recall in this research because there is a single class label in the dataset, namely "pothole." All of these recall values were calculated for 50%:5%:95% IoU threshold, i.e., using MS COCO metrics [21].

$$\text{Recall} = \frac{\text{TP}}{\text{TP} + \text{FN}} \tag{4}$$

Average recall is the mean of the recall values of all individual classes.

$$\text{AR} = \frac{1}{N}\sum_{i=1}^{N} r_i \tag{5}$$

Here, AR is average recall, N is the number of class labels, and r_i is the recall for ith class.

4.5 *Deployment to Android*

After training the model, it is exported as a **Protocol Buffer file** (has a *.pb* extension). Then, it is optimized for small memory devices and converted to *.tflite* file format. Tools for this conversion are provided by TensorFlow [8].

Finally, a simple Android application was built (just like a camcorder) where the exported *model.tflite* was included as an asset file. The app uses device-camera to gather video frames and feeds them to the model for detection of potholes. **TF-Lite** provides a Java API which is used to load the model into memory and running inference on video frames.

5 Results and Discussion

The training process was run for more than 100,000 steps with a batch size of 24. Then, evaluation process was run with the help of the TensorFlow "models" repository. The evaluation process used the performance metrics following the Microsoft Common Objects in Context (COCO) [21].

Precision values found on the test dataset after more than 100,000 steps of training are shown here in tabular form as well as using bar charts.

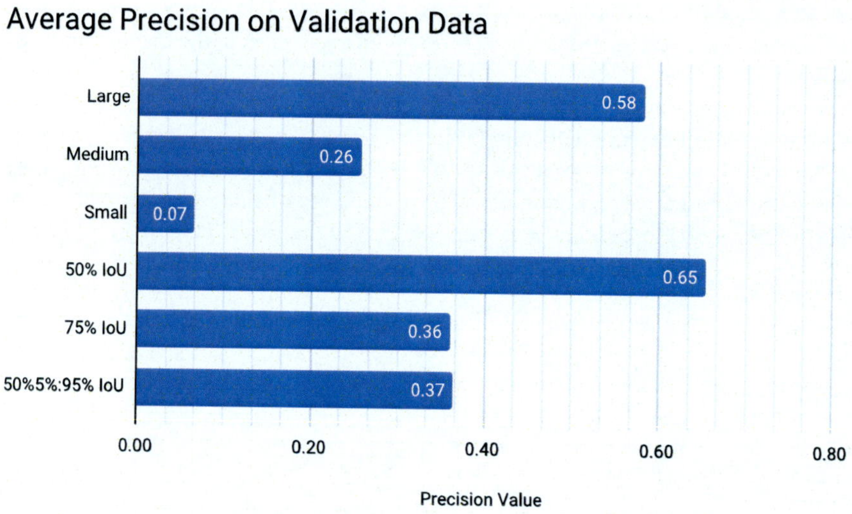

Fig. 6 Average precision on validation data

5.1 *Average Precision for Different IoU Thresholds*

In Table 3, the model's performance is shown as average precision considering different minimum IoU thresholds as the threshold for true positive detection.

5.2 *Average Precision for Different Pothole Sizes*

In Table 4, the model's performance is shown as average precision considering different categories of pothole sizes in the test dataset.

Figure 6 summarizes the mean average precision values considering different IoU thresholds and different pothole size categories.

Table 3 Precision at different IoU thresholds

IoU threshold	Precision on validation data
50%	0.65
75%	0.36
50%:5%:90%	0.37

Table 4 Precision for different pothole sizes

Area sizes	Precision on validation data
Small	0.07
Medium	0.26
Large	0.58

5.3 Average Recall at Different Detection Limits

We have taken the recall values at different limits of maximum detections. More specifically, we have taken the following limits for the calculation of recall values

- Average recall at maximum of 1 detection (AR@1)
- Average recall at maximum of 10 detections (AR@10)
- Average recall at maximum of 100 detection (AR@100).

In Table 5, average recall values are shown for different detection limits.

5.4 Average Recall for Different Pothole Sizes

In Table 6, average recall values are shown for different categories of pothole sizes. Here, all the values are calculated at the threshold of IoU@50%:5%:95%.

Figure 7, summarizes all the average recall values as measures of performance for the model. Here, all detections were calculated with IoU@50%:5%:95% threshold.

From Figs. 6 and 7, it is clear that the model performs better for larger potholes. Although the precision values are not very high, it is worth accepting because of such a lightweight model like MobileNet SSD which is built targeting the low-end mobile devices [11].

5.5 Shortcomings of Our Work

Main drawback of our proposed system is that it did not show a very high accuracy. This is a trade-off against speed. Because target was devices with small memory and processing resource, some accuracy was sacrificed. It is to make it working similar to real-time applications.

Table 5 Recall for different detection limits

Maximum detections	Recall on validation data
1	0.27
10	0.44
100	0.49

Table 6 Recalls for different pothole sizes

Area sizes	Recall on validation data
Small	0.18
Medium	0.44
Large	0.67

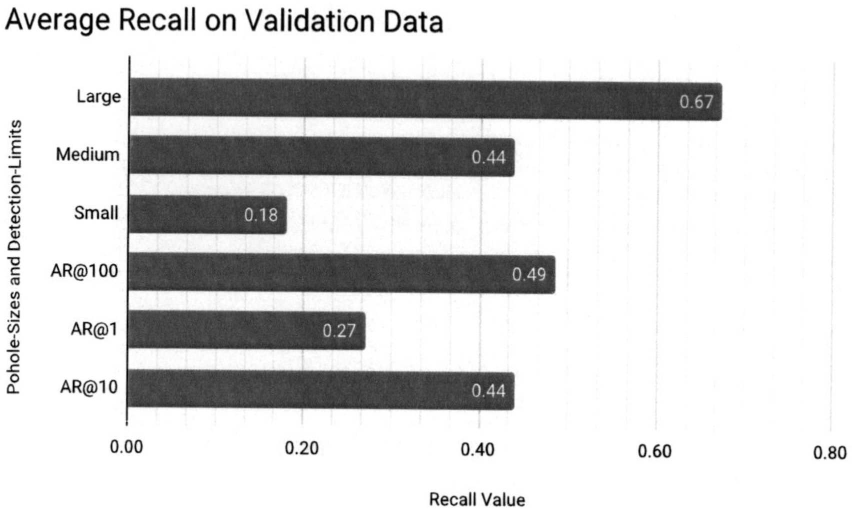

Fig. 7 Average recall on validation data

6 Conclusion

Considering the severity of potholes, a system was developed to automatically detect them in real time and generate warning signals to help for avoidance. An Android application was developed using the model which can detect, localize, and track potholes analyzing video frames. The app generates warning signal as long as it detects pothole in the video frames coming from its camera. The system can be used by visually impaired people to avoid potholes while navigating. It can also be used in automated vehicle driving.

7 Future Works

In the future, more research would be accomplished to make the system more accurate, more robust, more intelligent. Research for estimating the distance of the detected potholes would be done in the near future. Measuring the severity of the potholes would also be included in the future works. Making the model aware of different sizes of potholes may be included as future tasks. Also, measurement of dimension like area, depth of the detected pothole would be done in the future research.

References

1. Atikur R (2020) Annotated potholes image dataset. https://www.kaggle.com/chitholian/annotated-potholes-dataset. [Online]. Accessed 12 Mar 2020
2. Bisong E (2019) Google colaboratory. In: Building machine learning and deep learning models on Google cloud platform. Springer, pp 59–64
3. Buckland M, Gey F (1994) The relationship between recall and precision. J Am Soc Inf Sci 45(1):12–19
4. Buza E, Omanovic S, Huseinnovic A (2013) Stereo vision techniques in the road pavement evaluation. In: Proceedings of the 2nd international conference on information technology and computer networks, pp 48–53
5. Chang K-T, Chang J, Liu J-K (2005) Detection of pavement distresses using 3D laser scanning technology, pp 1–11
6. Danti A, Kulkarni J, Hiremath P (2012) An image processing approach to detect lanes, pot holes and recognize road signs in Indian roads. Int J Model Optim 2:658–662
7. De Zoysa K, Keppitiyagama C, Weerathunga S (2007) A public transport system based sensor network for road surface condition monitoring, p 9
8. Dillon JV, Langmore I, Tran D, Brevdo E, Vasudevan S, Moore D, Patton B, Alemi A, Hoffman M, Saurous RA (2017) Tensorflow distributions. arXiv preprint arXiv:1711.10604
9. Google (2020) Tensorflow model zoo. https://github.com/tensorflow/models/blob/master/research/object_detection/g3doc/detection_model_zoo.md. [Online]. Accessed 25 Feb 2020
10. Google (2020) Tensorflow models. https://github.com/tensorflow/models. [Online]. Accessed 25 Feb 2020
11. Howard AG, Zhu M, Chen B, Kalenichenko D, Wang W, Weyand T, Andreetto M, Adam H (2017) Mobilenets: efficient convolutional neural networks for mobile vision applications. arXiv preprint arXiv:1704.04861
12. Huidrom L, Das LK, Sud SK (2013) Method for automated assessment of potholes, cracks and patches from road surface video clips. Procedia-Soc Behav Sci 104:312–321
13. Jog GM, Koch C, Golparvar-Fard M, Brilakis I (2012) Pothole properties measurement through visual 2D recognition and 3D reconstruction. In: International conference on computing in civil engineering, pp 553–560
14. Kim T, Ryu S (2014) Review and analysis of pothole detection methods. J Emerg Trends Comput Inf Sci 5:603–608
15. Kirk D et al (2007) NVIDIA CUDA software and GPU parallel computing architecture. ISMM 7:103–104
16. Koch C, Jog G, Brilakis I (2013) Pothole detection with image processing and spectral clustering. J Comput Civ Eng 27:370–378
17. Koch C, Brilakis I (2011) Pothole detection in asphalt pavement images. Adv Eng Inform 25:507–515
18. Kotsiantis SB, Zaharakis I, Pintelas P (2007) Supervised machine learning: a review of classification techniques. Emerg Artif Intell Appl Comput Eng 160:3–24
19. LabelImg Tzutalin. Git code (2015)
20. Lempitsky V, Kohli P, Rother C, Sharp T (2009) Image segmentation with a bounding box prior. In: 2009 IEEE 12th international conference on computer vision. IEEE, pp 277–284
21. Lin T-Y, Maire M, Belongie S, Hays J, Perona P, Ramanan D, Dollár P, Zitnick CL (2014) Microsoft COCO: common objects in context. In: European conference on computer vision. Springer, pp 740–755
22. Liu W, Anguelov D, Erhan D, Szegedy C, Reed S, Fu C-Y, Berg AC (2016) SSD: single shot multibox detector. In: European conference on computer vision. Springer, pp 21–37
23. Li Q, Yao M, Yao X, Xu B (2009) A real-time 3D scanning system for pavement distortion inspection. Meas Sci Technol 21:015702
24. Mané D et al (2015) Tensorboard: tensorflow's visualization toolkit

25. Mednis A, Strazdins G, Zviedris R, Kanonirs G, Selavo L (2011) Real time pothole detection using android smartphones with accelerometers. In: 2011 international conference on distributed computing in sensor systems and workshops (DCOSS), June 2011, pp 1–6
26. OS Android. Android. Retrieved 24 Feb 2011
27. Owens JD, Houston M, Luebke D, Green S, Stone JE, Phillips JC (2008) GPU computing. Proc IEEE 96(5):879–899
28. Pan SJ, Yang Q (2009) A survey on transfer learning. IEEE Trans Knowl Data Eng 22(10):1345–1359
29. Pirkle FL (1988) Computer workstation. US patent 4,717,112, 5 Jan 1988
30. Rao A, Gubbi J, Palaniswami M, Wong E (2016) A vision-based system to detect potholes and uneven surfaces for assisting blind people, pp 1–6
31. Ryu S-K, Kim T, Kim Y-R (2015) Image-based pothole detection system for its service and road management system. Math Probl Eng 2015
32. Walter E (2008) Cambridge advanced learner's dictionary. Cambridge University Press
33. Wikipedia (2020) Precision and recall—Wikipedia, the free encyclopedia. https://en.wikipedia.org/w/index.php?title=Precision_and_recall&oldid=941734342. [Online]. Accessed 25 Feb 2020
34. Yu B, Yu X (2006) Vibration-based system for pavement condition evaluation, pp 183–189

Analysis of EEG Signal Classification for Application in SSVEP-Based BCI Using Convolutional Neural Network

Md. Saiful Islam Leon, Jarina Akter, Nazmus Sakib, and Md. Kafiul Islam

Abstract A Brain-Computer Interface (BCI) is a combination of hardware and software system that establishes a direct communication and real-time interaction between the human brain and external devices. For the non-invasive BCI applications, we have chosen Steady-State-Visual-Evoked Potential (SSVEP) and analyzed their behavior for multi-class classification as the signal provides high performance and reliable communication. Some of the applications of SSVEP-based BCI are neural rehabilitation, biometric authentication, word and letter recognition, digital gaming, and wheelchair control for Locked-in syndrome or ALS patients. The main aim of the project is to improve validation accuracy by tuning different hyper-parameters of the classifier. For classification, we have used a convolutional neural network, as there are scopes for improvement in classification accuracy. Several hyper-parameters of the classifier were tuned to achieve the highest validation accuracies through a large number of experimental trials. We have achieved 80.83% accuracy and 69.75% for LOSO methods through the stochastic gradient descent with momentum solver. The complexity of designing a general model for everyone is very high and our model is perfectly suitable if the user takes the initiative to pre-train the model.

Keywords EEG · CNN classifier · SSVEP signal · BCI applications

1 Introduction

The EEG of the human brain has been used primarily to determine neurological conditions in the clinical setting and to analyze brain functions in the laboratory after the first electroencephalography (EEG) studies on humans in 1929. An idea has increasingly emerged that brain activity could be used as a channel of communication. Considering the sophistication, distortion, and variability of brain signals, the likelihood of understanding a single message or instruction seemed to be highly remote.

Md. S. I. Leon · J. Akter · N. Sakib · Md. K. Islam (✉)
Department of Electrical and Electronic Engineering, Independent University, Bangladesh, Dhaka, Bangladesh
e-mail: kafiul_islam@iub.edu.bd

M. S. Arefin et al. (eds.), *Proceedings of the International Conference on Big Data, IoT, and Machine Learning*, Lecture Notes on Data Engineering and Communications Technologies 95, https://doi.org/10.1007/978-981-16-6636-0_45

However, EEG shows direct associations with user intentions, thus allowing a direct contact channel for the brain-computer interface (BCI). In conjunction with SSVEP, the BCI technology will potentially make the home environment more intelligent and assistive, offering additional means of communication to help the independent lives of elderly people affected by disabilities. The use of BCI and SSVEP-based assistive technology [1] will benefit the quality of life of people suffering from serious motor disabilities.

The authors of [2] and [3] have executed their experiments on the MAMEM SSVEP dataset experiment I same as ours. They have applied various classifiers such as SVM, CNN, K-NN, LDA to acquire highest validation accuracy. SVM and CNN classifier yields 79.47% [2] and 69.03% [3] accuracy, respectively, based on leaving one subject out (LOSO) experiment. Several stages are required such as pre-processing, feature extraction, feature selection in order to classify through SVM and choosing the best combination of the stages is a very time-consuming process. Accuracies vary with different feature extraction methods, whereas CNN does not require such complexity. CNN classifier achieved 69.03% accuracy on the MAMEM dataset [3] therefore we were motivated to bridge the research gap.

In our work, we have considered only the CNN classifier as it is one of the most advanced algorithms for classification problem. CNN does not require feature extraction layer as the classifier itself extracts features from the signal. Other than the classifier, we had to take into consideration few other stages such as data segmentation, signal pre-processing, 2-D image conversion from 1-D signal through spectrogram. Our unique contribution was to understand and tune the hyper-parameters of CNN classifier to achieve the highest validation accuracy. We have also filtered the raw signal with different frequency ranges to see the impact on the accuracy in the signal pre-processing. While converting the 1-D signal to 2-D image, we considered various configuration of spectrogram. After acquiring the best configuration in each stage, we combined it for further experiments.

Furthermore, we trained the classifier in two different methods. One of them was to segment the dataset class-wise after combining all the 11 subjects' data. The classifier was trained by 75% of the whole dataset and the rest was used as validation data. Another method was to segment the dataset subject-wise followed by the classes of signals. Here, the classifier was trained by the 10 subjects' data and one subject's data was used as validation data (LOSO—Leaving one subject out). We have proposed two SSVEP-based BCI model. Our best model performed at 80.83% accuracy while training the classifier by 75% of the whole dataset and in case of LOSO method, we have achieved 69.75% validation accuracy. To acquire such accuracies, we had to go through a large number of experimental trials.

2 Literature Review

The authors in [2] used MAMEM dataset experiment I which is available both in PhysioNet and MAMEM.eu (open sources). They classified the SSVEP signal using machine learning algorithms such as SVM, LDA, Naive Bayes. They have got the highest accuracy through SVM classifier which is 79.47%.

The authors in [3] also used the same dataset which is mentioned in [2]. They have used both machine learning and deep learning algorithms to get better results. They have got the highest accuracy of 69.03% by using CNN classifier.

The rest of the researchers in [4–10] used different datasets from [2] and [3] and achieved higher accuracies. By using CNN classifier validation accuracies were reached to 85.75%, 96%, 73.74%, and 92.33% in [4, 7, 9], and [8], respectively.

After surveyed literature, the following challenges are found. The classification accuracy requires further improvement and not much exploration on CNN-based classification was found for MAMEM dataset experiment I. The summary of the literature review in terms of classifier, dataset and respective classification accuracy is provided in Table 1.

Figure 1 shows the proposed SSVEP-based BCI model where several steps are taken into consideration to execute the experimental trials. The dataset has been extracted class-wise and segmented time-wise. Afterward, the signal has been filtered to remove the unnecessary noises. The input of CNN classifier has to be images thus the signal is converted to image through spectrogram. Our main contribution was on tuning the classifier's hyper-parameters to achieve the highest validation accuracy.

Table 1 The SSVEP-based BCI-related papers with the main characteristics referring to the classification method, source of the dataset, and its respective accuracy rate

Ref. no.	Classification algorithms	Source of dataset	Accuracy
[2]	SVM, LDA, K-NN, Naive Bayes	MAMEM SSVEP dataset experiment I	79.47%, 64.11%, 49.40%, 35.46%
[3]	K-NN, k-NN SFS, C4.5, C4.5 SFS, AdaBoost, LDA, SVMG, CNN, LSTM	MAMEM SSVEP dataset experiment I	46.17%, 51.03%, 49.41%, 52.36%, 66.67%, 58.59%, 65.13%, 69.03%, 66.89%
[4]	CNN	An offline SSVEP personal dataset	85.75%
[5]	SVM along with CCA	MAMEM SSVEP dataset experiment II	93.11%
[6]	K-NN	An offline SSVEP personal dataset	99.3%
[7]	CNN	An offline SSVEP personal dataset	96%
[8]	CNN along with CCA	An offline SSVEP personal dataset	92.33%
[9]	CNN	MAMEM SSVEP dataset experiment II	73.74%
[10]	SVM	MAMEM SSVEP dataset experiment III	88.3%

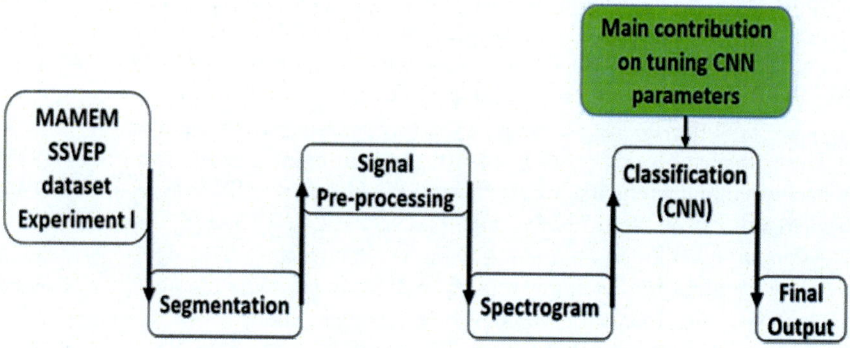

Fig. 1 Our proposed SSVEP-based BCI model

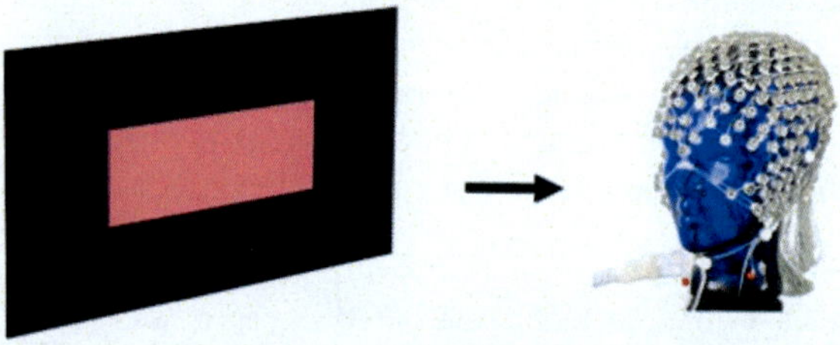

Fig. 2 A sample SSVEP-based BCI experiment setup where the subject is shown a flickering window in a screen while EEG recording is performed

3 Materials and Method

3.1 Dataset Description

We have downloaded the MAMEM SSVEP dataset of experiment 1 from MAMEM.eu (Open source). There were 11 subjects' SSVEP data where each subject has identical 5 sessions. And each session contains 23 trials of stimuli of different frequencies including adaption period. The sampling frequency of the recorded data is 250 Hz and the stimulating frequencies are 6.66, 7.50, 8.57, 10, and 12 Hz. The data was recorded through the device EGI GES 300 which has 256 channels [2]. The experiment setup is shown in Fig. 2.

The box on the black screen is flickering at different frequencies such as 6.66, 7.50, 8.57, 10, and 12 Hz during the visual simulation.

3.2 Software and Hardware

All the EEG data were processed offline for our experiment through MATLAB software 2020 version from MathWorks® Incorporation (US). A MATLAB toolbox titled EEG-processing-toolbox-master from Git Hub was used for extracting SSVEP

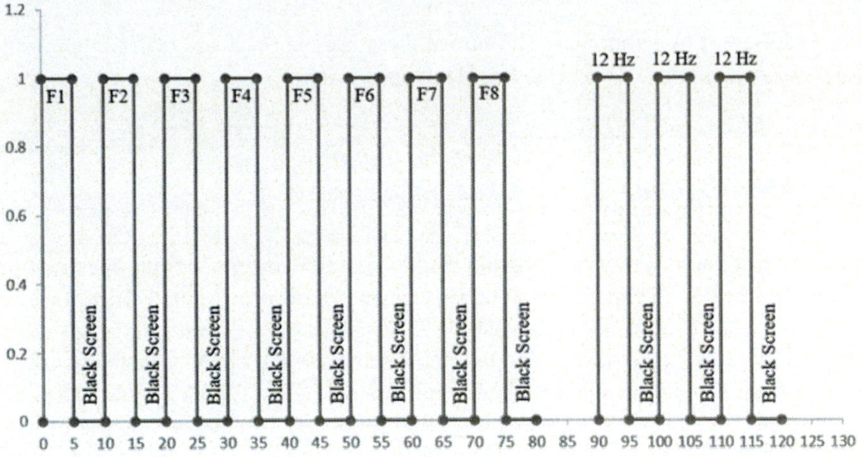

Fig. 3 Experimental setup for adaptation period and a main trail during visual stimulation

EEG data channel-wise. Besides, signal processing, filtering, spectrogram, classification, data analysis, visualization, etc., all the parts of programming are done using MATLAB software. The graphic card is used for our experiment is Ge Force® GTX 1050ti and the RAM size of the personal computer is 8 gigabytes.

3.3 Segmentation

Each stimulating frequency has been recorded for 5 s which is identified as one trial. In the same manner, there are 23 trials including the adaption period and all the trials are recorded at once one by one followed by 5 s resting period in each interval. For our experiments, we have separated each trial by half to increase the training and validation dataset for the classifier [2]. The experimental protocol is illustrated in Fig. 3.

3.4 Signal Pre-processing

The raw signal is corrupted by various kind of noises such as power line noise, muscle movement, eyeball movement and eye blinking noises. To remove such noises, we have implemented several filters on the raw signal and extracted the desired signal.

The Steady-State-Visual-Evoked Potentials (SSVEPs) are mainly active in the range of 5–50 Hz frequencies. We have applied several Butterworth filters to extract the required signals from the raw data. A high pass Butterworth filter was applied at 5 Hz followed by a low pass Butterworth filter at 50 Hz. There were power grid noises followed by harmonics. To remove these noises, two band stop Butterworth filters were applied at 50 Hz and 100 Hz, respectively.

3.5 Spectrogram

Most of the natural signals are aperiodic and the notion of time varying spectrum is proposed to study the aperiodic signals' behavior. Short time Fourier Transform is one of the many approaches to obtain time varying spectrum. Spectrogram uses short time Fourier transform to reveal the Fourier spectrum of the signal as it changes over time. Firstly, the signals are separated into equal segments (Chunks or frames) with overlapping the adjacent sides to minimize the artifacts at the boundary then applying Fourier transform on each segment and displaying the power spectral density over time by squaring the magnitude [11]. The conversion of a 1-D EEG epoch to a 2-D image through spectrogram is illustrated in Fig. 4.

$$\text{STFT}\{x(n)\}(m, w) \equiv x(m, w) = \sum_{-\infty}^{\infty} x[n]w[n-m]\text{e}^{-i\omega n} \tag{1}$$

where $m = \{1, 2, \ldots, N\}$ and $w[\cdot]$ are discrete variable and preselected window function, respectively.

$$\text{Spectrogram}\{x(n)\} = |x(m, w)|^2 \tag{2}$$

Here, in Fig. 4, the 1-D signal is representing the power spectral density where x-axis is frequency (Hz) and y-axis is power per frequency (dB/Hz). The 2-D image is showing time varying spectrum where x-axis is time (s) and y-axis is frequency (Hz). The color-bar is power per frequency (dB/Hz) where blue means lowest power and yellow means highest power.

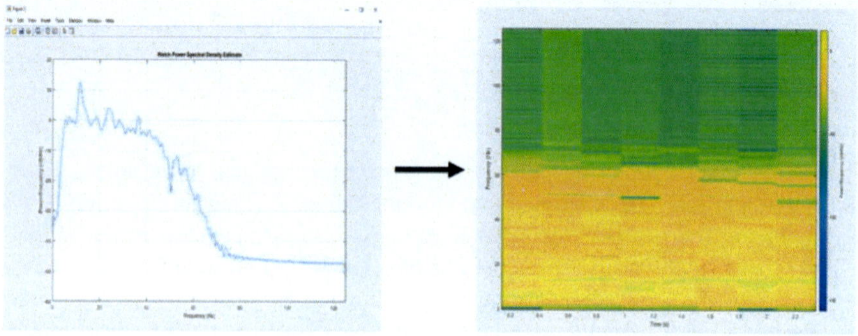

Fig. 4 Conversion of 1-D signal (PSD) to 2-D image through spectrogram

3.6 Convolutional Neural Network

Convolutional neural networks are the most used classifier in deep learning to study visual imagery. It is a powerful family of neural network that have been proposed to analyze the spatial structure of the pixels from image data. Mathematical operation convolution takes place between images and filters followed by down sampling before fetching them to neural networks. Some of the applications of CNN are image and video recognition, recommender systems, image classification, image segmentation, medical image analysis, natural language processing, brain-computer interfaces, and financial time series. The basic architecture of CNN consists of an input layer, hidden layers and an output layer. The input layer determines the format of the image and the pixels. In any feed-forward neural network, all the layers between the input and output layers are considered hidden layers. Layers such as convolutional layer, batch normalization layers, activation layers, pooling layers, fully connected layers and SoftMax layers are the hidden layers. These layers execute several mathematical operations as the image goes through. The output layer determines the final output that is associated with the image [12].

The convolutional neural network trains a system by several forward and backward propagation. There are various parameters, which are randomly initialized before training the classifier. The images are fed into the input layer in the form of numbers that denotes the intensity of pixels in the image. After executing the mathematical operations in the hidden layers, the system predicts an output and compares with the actual value. The difference between the actual and predicted value determines the error of the system and then update the parameters in the back propagation to minimize the error in each iteration to reach the optimization [12]. The finally selected CNN architecture used for this study is shown in Fig. 5.

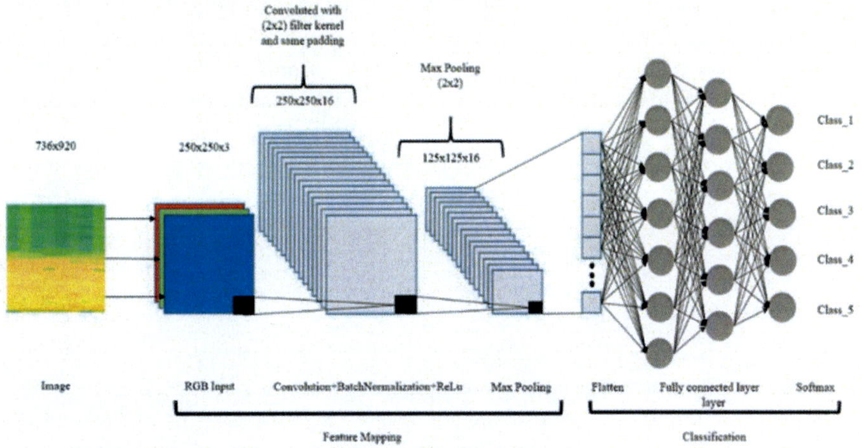

Fig. 5 Architecture for our best convolutional layer

3.7 CNN Hyper-Parameters

The hyper-parameters of the Convolution Neural Network used in this study are as follows:

The size and the number of filters of convolutional layer. It determines the height, the width and the depth of the filter (number of the filters). The values in the filters are generated randomly and during the training, these values are optimized by several iteration for lowest loss between the predicted value and the real value [12].

Padding and stride in convolutional layer. As the convolutional layer changes the output array's size, some of the pixel value is discarded from the input. To overcome this problem, zero padding is introduced. The input is attached with an extra border on every side with zero value so that the output array size remains the same as the input [12]. Stride is the factor that controls the sliding of the filter across the input vertically and horizontally [12].

Batch size, epochs iteration. Bigger data set needs more memory to compute the training. To overcome this problem, batch size (mini-batch size) is introduced. One forward propagation and one back propagation of the whole data in the training is known as one epoch. It is the number that is needed to complete the whole batch wise date per epoch. All these parameters are associated with neural network.

Gradient decent optimizer. Three optimization algorithms: Stochastic Gradient Decent with Momentum (SGDM), Root means square propagation (RMSprop), and Adaptive moment estimation (Adam).

Learning rate and dropout. The learning rate determines the arbitrary step sizes, which needs to be taken into consideration while training the network. This preselected value decides the amount that is used in the increasing and decreasing the parameters value. Different optimizer uses different method for selecting the learning rate. It can be used as a constant or as a variable. In case of variable learning rate, it will decrease over time and the decreasing factor will be determined through another option known as drop factor. Drop period is the value that decides after how many epochs the learning rate will drop [12]. Learning rate and optimizer is used in the neural network.

4 Results and Discussions

To achieve the best model, we have executed a large number of experimental trials. Firstly, we had to find out the best channel for the SSVEP signals from the headset. In the signal pre-processing, the frequency bands for SSVEP signals are determined by analyzing various bands. The conversion of the 1-D signal to 2-D images through spectrogram, uses various parameters and we have changed those parameters to see the impact on accuracy. There were exactly 1104 trails from all the volunteers. Each trail is of 5-s simulation for a specific frequency out of five frequencies, which was separated into 2.5 s windows before converting to images. To train convolutional neural networks, a huge amount of data is required. Separation of the trails was done solely to support the CNN training more accurately. Convolutional neural network has many tunable hyper-parameters such as learning rate, solver, drop factor, kernel size and no. of filters, etc. All these hyper-parameters are taken into consideration to increase the validation accuracy. Most of the researchers are working on leaving one subject out (LOSO) experiment to design a general model for everyone. We have also created a model for LOSO experiment. In the rest of the trials, we have considered 75% of all the subjects' data as training set and 25% as validation set. There are some challenges which we have to overcome in order to achieve the best model. Few of the challenges are stated below,

- To choose the kernel size and number of filters
- To achieve highest validation accuracy
- Data segmentation
- To avoid the overfitting of data while training
- To use different stimulus for a specific command.

4.1 Different Channels

The EGI 300 Geodesic EEG device has 257 electrodes including the reference one and each electrode is considered as one channel. Occipital channels are responsible for capturing Steady-State-Visual-Evoked Potentials (SSVEPs) and there are 40 channels lying around the occipital region. We have considered 15 channels to do our experiments in order to find the best one and all these channels are lying around the center of the occipital region. Channel 138 is yielding the highest validation accuracy of 77.50%.

4.2 Different Filtered Frequencies

Convolutional neural networks can train a system without any kind of data pre-processing or feature extractions and this fact is considered as the biggest advantage of CNN over other classifier. As to see how our CNN system works on raw and filtered data, we have executed several filtering processes before training the CNN classifier. We have compared the classifier's validation accuracy on both the raw and filtered data. The highest validation accuracy was 76.67% while we have filtered the raw data from 5 to 50 Hz.

4.3 Configuration of Spectrogram

We have applied different configurations of spectrogram for the conversion to see the impact on the validation accuracy and compared with the default settings. In the default settings, the whole data is separated into eight equal segments with 50% overlapping and 256 frequency points are taken to execute discrete Fourier transform. There are several windowing processes such as Hamming window, Blackman window and Kaiser Window. We have considered only Hamming and Blackman windows for our experiments. We have tried several combinations on our data but the highest validation accuracy was 77.50% for the default settings.

4.4 Kernel Size and No. of Filters in Convolutional Layer

The convolutional layer in CNN architecture extracts features from the input layer which is just an array of numbers known as tensor. The mathematical operation convolution is a special type of linear operation for feature extraction, which is done by a small array of numbers (known as Kernel/filter) across the input tensor. To see the impact of different filter size on performance, we have tried several filter sizes with changing number of filters. The best kernel size and no. of filters is [$2 \times 2 \times 16$] that yields 77.92% validation accuracy.

4.5 Multiple Convolutional Layer

Based on different applications, the image data can be passed through multiple convolutional layers for down sampling. Higher resolution images as an input, has a huge number of pixels. Multiple layers are used to minimize the points in the images. This totally depends on the application. For an image with 1000×1000, has 1,000,000 pixels value, to train such images computational time will be very high. So, multiple layer's concept is used to overcome such problems. Multiple layers are used to extract desired features from the images. We have tried several convolutional layers to see the impact on the validation accuracy for our image data. We have found the higher validation accuracy while using one convolutional layer.

4.6 Different Solver with Various Drop Factor

While executing our most of the experiments, we have faced the problem of overfitting. A model is supposed to learn the signal's properties rather than memorizing

Table 2 Solver experiments with drop factor

Filter size and no. of filters	Solver and learning rate	Drop factor and period	Validation accuracy (%)	Training accuracy (%)	Total accuracy (%)
[2 × 2]_16	Sgdm_0.01	0.1_5	77.92	79.86	79.38
[2 x 2]_16	**Sgdm_0.01**	**0.2_5**	**80.83**	**89.03**	**86.98**
[2 x 2]_16	**Adam_0.01**	**0.001_5**	**79.38**	**79.44**	**79.43**
[2 × 2]_16	Sgdm_0.01	0.5_5	81.25	80.90	80.99
[5 × 5]_7	Sgdm_0.01	0.1_5	81.25	72.92	75.00
[5 × 5]_7	Sgdm_0.01	0.2_5	78.96	80.56	80.16
[5 × 5]_7	Sgdm_0.01	0.5_5	79.58	81.18	80.78

the signal. Overfitting refers to a situation where the model learns statistical regularities such as irrelevant noises. This particular situation leads to poor performance on a subsequent new dataset. We have overcome such problem by using different drop factors in the training option. The highest validation accuracy is 80.83% while using stochastic gradient decent with momentum (SGDM) solver and drop factor of 0.2. The second-best accuracy is 79.38% with adaptive moment estimation (Adam) solver and drop factor of 0.001. The classification performance for different solvers and drop factors and periods is summarized in Table 2.

4.7 *Leaving One Subject Out (LOSO)*

In all of the above experiments, 75% of the whole dataset is used as training data while 25% as validation data. Now the following experiments are for the unseen validation data to evaluate the performance of the model. Here, one of the subjects is used as validation data whose data will not be used in training the model. Thus, the unseen testing experiments have been executed. We have trained the model by considering every subject's data as validation dataset. The highest mean accuracy is 69.75% in our experiments while using drop factor. The results are summarized in Tables 3 and 4.

4.8 *Comparison with Other Related Works*

The other related work was based on LOSO experiment and the highest validation accuracy is 69.03% in [3] for CNN, 79.47% for SVM. And in our project, we have achieved 69.75% validation accuracy through CNN classifier. While we trained 75% of the whole data, we acquired 80.83% accuracy. This indicates that our model will be working at 80.83% when the user pre-trains the model. The comparison is shown in Table 5.

Table 3 LOSO experiments with filter [2 × 2]_16

Filter size and no. of filters	Subject	Solver and learning rate	Validation accuracy (%)	Training accuracy (%)	Mean validation accuracy
[2 × 2]_16	1	Sgdm_0.01	92.75	100	**66.65%**
[2 × 2]_16	2	Sgdm_0.01	80.43	100	
[2 × 2]_16	3	Sgdm_0.01	44.93	100	
[2 × 2]_16	4	Sgdm_0.01	57.61	100	
[2 × 2]_16	5	Sgdm_0.01	27.39	100	
[2 × 2]_16	6	Sgdm_0.01	70.43	100	
[2 × 2]_16	7	Sgdm_0.01	61.30	100	
[2 × 2]_16	8	Sgdm_0.01	28.26	100	
[2 × 2]_16	9	Sgdm_0.01	96.09	100	
[2 × 2]_16	10	Sgdm_0.01	78.26	94.29	
[2 × 2]_16	11	Sgdm_0.01	95.65	100	

Table 4 LOSO experiments with drop factor

Filter size and no. of filters	Subject	Solver and learning rate	Drop factor and period	Validation accuracy (%)	Training accuracy (%)	Mean accuracy
[2 × 2]_16	1	Sgdm_0.01	0.2_5	96.38	78.36	**69.75%**
[2 × 2]_16	2	Sgdm_0.01	0.2_5	89.57%	74.97%	
[2 × 2]_16	3	Sgdm_0.01	0.2_5	42.75%	89.66%	
[2 × 2]_16	4	Sgdm_0.01	0.2_5	67.40	86.91	
[2 × 2]_16	5	Sgdm_0.01	0.2_5	27.83	86.50	
[2 × 2]_16	6	Sgdm_0.01	0.2_5	73.91	89.03	
[2 × 2]_16	7	Sgdm_0.01	0.2_5	66.09	82.91	
[2 × 2]_16	8	Sgdm_0.01	0.2_5	33.33	81.11	
[2 × 2]_16	9	Sgdm_0.01	0.2_5	97.83	79.12	
[2 × 2]_16	10	Sgdm_0.01	0.2_5	75.22	78.82	
[2 × 2]_16	11	Sgdm_0.01	0.2_5	96.96	89.28	

Table 5 Comparison with other related works

Ref.	Classification algorithms	Accuracy
[2]	**SVM**, LDA, K-NN, Naive Bayes	**79.47%**, 64.11%, 49.40%, 35.46%
[3]	K-NN, k-NN SFS, C4.5, C4.5 SFS, AdaBoost, LDA, SVMG, **CNN**, LSTM	46.17%, 51.03%, 49.41%, 52.36%, 66.67%, 58.59%, 65.13%, **69.03%**, 66.89%
Our work	**CNN (LOSO method)**	**69.75%**
Our work	**CNN**	**80.83%**

5 Conclusion and Future Work

Our study presented an investigation of an SSVEP-based BCI using the MAMEM database which can be applied in various application such as wheelchair control. From the literature review, it is noticeable that most of the researchers focus on improving the accuracy and trying to lessen the processing time of SSVEP-based BCI system for applying it in real life as well as trying to make it popular to the people. The complexity of designing a general model for everyone is very high. Our model will work perfectly at around 80% accuracy if the user takes the initiative to pre-train the model. While considering 75% of the whole data as training set, we have reached 80.83% accuracy with the solver stochastic gradient decent with momentum (SGDM), drop factor of 0.2, whereas with Adam solver and drop factor of 0.001, we have gained 79.38% accuracy. Subject 3, 5 and 8 yield the lowest accuracy while we have overcome the overfitting problem. All these subjects have thick hair and it can be a reason for such lowest accuracy. They might have been prone to more eye blinking. Subject 3 consistently gave the lowest accuracy in all of our experiments. Unlike other researchers, we have executed all of our LOSO experiments without any major feature extraction and gained 69.75% mean accuracy. So, our designed model offering a great result with higher accuracy for classification using CNN classifier. In our future work, we will be focusing on different filtering in pre-processing, applying different pooling layers in CNN and execute LOSO experiment with different solver. Our own generated data will also be taken into consideration for classification. After acquiring the best model, we will be implementing it in hardware for different applications such as wheelchair control.

References

1. Cincotti F, Mattia A et al (2008) Non-invasive brain-computer interface system: towards its application as assistive technology. Brain Res Bull **75**(6):796–803 (2008)
2. Oikonomou VP, Liaros G, Georgiadis K, Chatzilari E, Adam K, Nikolopoulos S, Kompatsiaris I (2016) Comparative evaluation of state-of-the-art algorithms for SSVEP-based BCIs. arXiv preprint arXiv:1602.00904
3. Thomas J, Maszczyk T, Sinha KT, Dauwels J (2017) Deep learning-based classification for brain-computer interfaces. In: 2017 IEEE International conference on systems, man, and cybernetics (SMC), Banff, AB, Canada, pp 234–239
4. Bevilacqua V et al (2014) A novel BCI-SSVEP based approach for control of walking in virtual environment using a convolutional neural network. In: 2014 International joint conference on neural networks (IJCNN), Beijing, China, pp 4121–4128
5. Chatzilari E, Liarios G, Georgiadis K, Nikolopoulos S, Kompatsiaris Y (2017) Combining the benefits of CCA and SVMs for SSVEP-based BCIs in real-world conditions. In: Proceedings of the 2nd international workshop on multimedia for personal health and health care, MMHealth'17, Mountain View, pp 3–10
6. Sendesi SFT (2018) Selecting and extracting effective features of SSVEP-based brain-computer interface. J Artif Intell Electr Eng 7(26):25–33
7. Nik Aznan NK, Bonner S, Connolly J, Al Moubayed N, Breckon T (2018) On the classification of SSVEP-based dry-EEG signals via convolutional neural networks. In: 2018

IEEE International conference on systems, man, and cybernetics (SMC), Miyazaki, Japan, pp 3726–3731
8. Ravi A, Beni NH, Manuel J, Jiang N (2020) Comparing user dependent and user-independent training of CNN for SSVEP BCI. J Neural Eng 17(2):026028
9. Nouri A, Azizi K (2020) Introducing a convolutional neural network and visualization of its filters for classification of EEG signal for SSVEP task. Front Biomed Technol 7(3):151–159
10. Rashid M, Sulaiman N, Mustafa M, Bari BS, Hasan MJ (2020) Five-class SSVEP response detection using common-spatial pattern (CSP)-SVM approach. Int J Integr Eng 12(6):165–173
11. Oppenheim AV, Schafer RW, Buck JR (1999) Discrete-time signal processing, 2nd edn. Prentice Hall, Upper Saddle River, NJ
12. Yamashita R, Nishio M, Do RKG et al (2018) Convolutional neural networks: an overview and application in radiology. Insights Imaging 9:611–629

Security Detection and Countermeasures

A Blockchain-Based Approach to Detect Counterfeit Drugs in Medical Supply Chain

Shabnam Sabah, A. S. M. Touhidul Hasan, and Apubra Daria

Abstract The production and circulation of counterfeit drugs in the supply chain are an earnest and progressively vital issue. However, the existing supply chain management system fails to guarantee genuine medicines to the patient. To ensure authentic medicine and to mitigate supply chain issues, we propose a blockchain-based approach to ensure authentic drugs for the patients. The blockchain-based distributed system will empower all the stakeholders, including patients, to know and trace the authenticity of the medicine. To make the process faster, we have adopted the Hyperledger Fabric platform to develop peer-to-peer distributed applications for drugs supply chain. Besides, smart contracts make the supply chain management system automated, more robust, and transparent to detect counterfeit drugs so that patients can get original products produced by a legitimate manufacturer. The experimental analysis demonstrates that the proposed system runs smoothly on a Hyperleadger Fabric platform, and each transaction can handle efficiently with the distributed smart contracts.

Keywords Blockchain · Supply chain · Hyperleadger fabric · Smart contracts

1 Introduction

Supply Chain Management (SCM) is the progression of goods and information through numerous providers like manufacturers, distributors, retailers, and clients. It helps to check the traversal of products and information without any difficulties.

S. Sabah · A. S. M. T. Hasan
Department of Computer Science and Engineering, University of Asia Pacific, Dhaka 1205, Bangladesh

S. Sabah · A. S. M. T. Hasan (✉) · A. Daria
Institute of Automation Research and Engineering, Dhaka 1205, Bangladesh
e-mail: touhid@uap-bd.edu

A. Daria
e-mail: apubra@iar-e.com

M. S. Arefin et al. (eds.), *Proceedings of the International Conference on Big Data, IoT, and Machine Learning*, Lecture Notes on Data Engineering and Communications Technologies 95, https://doi.org/10.1007/978-981-16-6636-0_46

However, the existing pharmaceutical supply chain cannot ensure authentic medicine to patients. Counterfeit drugs are sold across all the distribution channels from drug stores to top hospitals in Bangladesh. Nowadays, in developing countries, counterfeit drug production and circulation have become a significant concern and cause a severe threat to public health. In developing nations, 30% of medicines sold are found counterfeit [17]. In 2020, a total of 18 individuals were charged for manufacturing and selling counterfeit drugs. Fake drugs are manufactured in the capital's Uttara, Fakirapool, Chawkbazar, and Tongi of Gazipur and then are supplied to different known medicine shops [15]. World Health Organization (WHO) specified counterfeit drugs as falsely and intentionally mislabelled with identity [12].

Numerous strategies have already been used to detect counterfeit drugs in Bangladesh. A startup company called "Panacea" proposed an approach where a pharmaceutical company prints a unique code on each medicine strip to verify the originality of the product. Reneta, the fourth largest pharmaceutical company in Bangladesh applied Panacea's technology to Maxpro and Rolac to ensure the authenticity of the medicine [9]. In 2018, the Directorate General of Drug Administration (DGDA) of Bangladesh has launched six international standard mini-labs in different districts to enhance its capacity to check the sale of fake medicines [4].

However, from the existing method, a patient cannot know whether the drugs are produced precisely by following the actual drug manufacturing code or not from the existing pharmaceutical drug supply chain. At the production time, a manufacturer can use harmful/inactive ingredients or active ingredients with a small/large amount. Besides this, drugs can be produced with unsafe substances or mislabeled by an invalid manufacturer. Moreover, it fails to trace the drug as medicine ownership changes from time to time.

To mitigate the supply chain issue and ensure authentic and genuine medicine, we propose a blockchain-based approach to detect counterfeit drugs in the pharmaceutical supply chain. In the proposed system, we have integrated a hash function and digital signature for user validation and message authentication so that all authentic people can join in the transaction, and it will ensure the trustworthiness of the ledger. A pharmaceutical laboratory has been introduced to detect original or fake drugs. An observer (i.e., independent pharmaceutical laboratory, public, or organization) witnessed the supply chain to observe the medicine after its distribution into the market and challenge its authenticity. A tracking system has been proposed to detect real or fake drugs based on unique QR code verification. We have applied Hyperledger Fabric-based [3] smart contracts to assure an efficient, protected, and trusted environment for authentic and genuine medicine supply activities for the general people.

The paper is organized as follows: Sect. 2 provides the related work. Section 3 presents the proposed blockchain based drug supply chain model. Section 4 presents the tracking system. Section 5 discusses the experimental results. Section 6 presents the conclusions and future works.

2 Related Work

An automated supply chain management system is a process where a person can track and trace products' life cycles efficiently. Numerous approaches were proposed for detecting fake medicine in the medical supply chain with the integration of blockchain technology as a decentralized database [2, 5, 11, 13]. A blockchain and machine learning-based supply chain management and recommendation system (DSCMR) were proposed to monitor and track the drug delivery process [1].

Sylim et al. [16] proposed the pharmacosurveillance blockchain system, which can just distinguish drug movements that follow official circulation chains known to the regulatory agency but cannot track distorted medications that are distributed through routes outside of official conveyance chains. Numerous blockchain-based techniques are introduced to manage the records of the supply of drugs in a secure way [7, 18]. Huang et al. [6] proposed a practical blockchain system called "Drugledger" for drug traceability and regulation which ensures the both authenticity and privacy of traceability data and meanwhile achieves finally stable blockchain storage with time going by.

Jangir et al. [8] introduced a new structure by applying Etherium based distributed ledger technology and smart contract for the pharmaceutical supply chain management that help in achieving user privacy, data transparency, immutability, high availability, no single point of failure, non-repudiation, real-time tracking of the drug, and demand-supply management. Kumar et al. [10] resolved the issue of drug safety utilizing blockchain, which depends on PKI and digital signature and encrypted QR (quick response) code security.

However, the above mentioned approaches have remarkable drawbacks in terms of detecting fake drugs. Authors assumed that drugs are made properly by a legitimate manufacturer. Authors used only QR code verification, Near Field Communication (NFC) tags, and machine learning methods along with blockchain to detect the fake drug. However, from all of these an end-user cannot know whether the drugs are produced precisely by following the actual drug manufacturing code or not. Besides this, medicines can be produced with unsafe substances or mislabeled by an invalid manufacturer.

3 Methodology

This section presents the blockchain-based secure pharmaceutical drug supply chain model. Figure 1 shows the proposed secure pharmaceutical drug supply chain model based on blockchain. The proposed system is divided into four parts: Ingredient Verification, Drug Sample Verification, Drug Delivery and QR Code Verification, and Observation And Revoke.

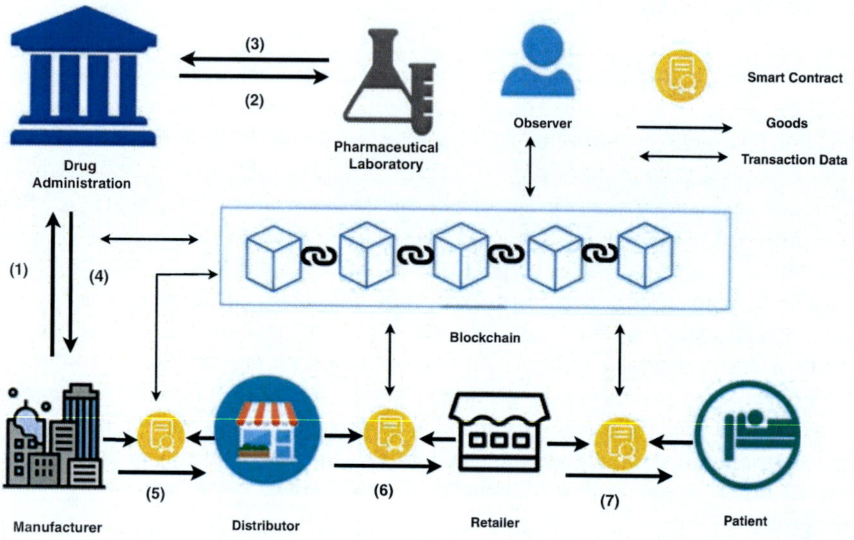

Fig. 1 Secure pharmaceutical drug supply chain based on blockchain

3.1 Ingredient Verification

The ingredient verification process ensures that the ingredients are authentic and original. Figure 2 shows the ingredient verification process. In this process, In this process, the manufacturer requests permission from the drug administration to produce a drug in a batch (e.g., a batch has *N* drugs.) and sends drug ingredients to the drug administration for verification. After getting those ingredients, drug administration adds the ingredients information into the blockchain. For ingredient verification, drug administration selects one pharmaceutical laboratory from several pharmaceutical laboratories that work under them and sends the ingredients to the selected pharmaceutical laboratory. After the ingredient checking, the laboratory gives the ingredients test result to drug administration. Afterward, drug administration adds the ingredient test result into the blockchain. Based on the ingredient test result, drug administration permits the manufacturer to produce the drug and generates a unique QR code for each drug of that specific batch and includes them into the blockchain. The manufacturer then produces *N* units of a drug in a batch after getting approval from the drug administration and appends the unique QR code to every drug packet of that batch at the time of packaging and labeling. In our proposed system, to add each transaction into blockchain, we have used the Practical Byzantine Fault Tolerance (PBFT) consensus algorithm which is permissioned voting based that allows our distributed system to reach a consensus even when a small number of nodes demonstrate malicious behavior (such as falsifying information).

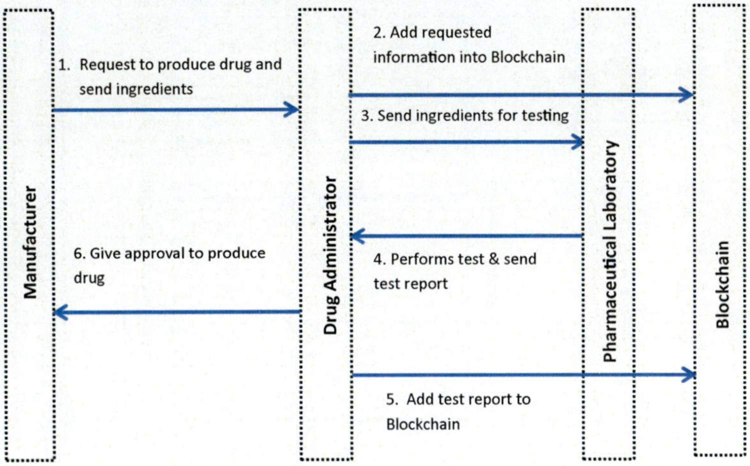

Fig. 2 Ingredient verification process

3.2 Drug Sample Verification

The drug sample verification process ensures that drugs are authentic. Figure 3 shows the drug sample verification process. In this process, the manufacturer sends the particular batch of a drug to the drug administrator (DA) for sample verification, and after the verification, DA includes the information into the blockchain. DA selects a pharmaceutical laboratory from a pool of laboratories for drug sample verification. The laboratory performs various types of tests on that particular batch of a drug to detect the counterfeit. There are three types of tests to detect counterfeit drugs which are given below.

- Thin Layer Chromatography (TLC) distinguish the counterfeit drug, and it is one of the efficient ways to determine the material in the drug, quantity of substances and adulterations [12].
- Analytical Techniques might be applied when counterfeit drugs are more sophisticated and require more sensitive tools to detect active ingredients in medicine. This technique includes near-infrared spectrophotometer, nuclear magnetic resonance, and mass spectrometry [12].
- Visual inspection is another rapid and straightforward strategy to recognize counterfeit drugs. It compares the original drug in terms of drug packaging and labeling. If there is no original drug to analyze, features such as altered/diverse packaging and non-uniform coloring of the drug can show that it may be counterfeit. In this way, authentic makers should give an exact depiction of the drug's physical qualities, and its materials to simplify the visual assessment [12].

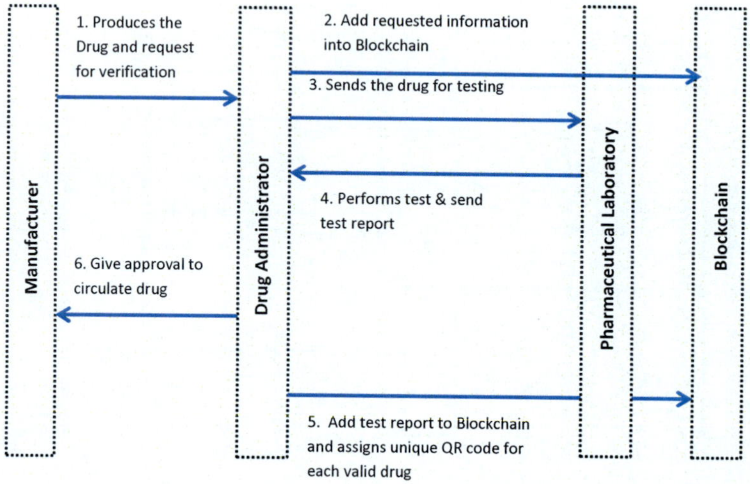

Fig. 3 Drug sample verification process

After the test, the laboratory gives the sample test report to the DA, and the DA appends the sample test report into the blockchain. If the sample test report returns that the specific batch of a drug is counterfeit, the DA does not allow the manufacturer to circulate the drug in the supply chain. Otherwise, the DA acknowledges the manufacturer to distribute the authentic drug's in the supply chain and stores the unique QR code of each drug into the blockchain as a valid QR code.

3.3 *Drug Delivery and QR Code Verification*

The manufacturer sells the drugs to distributors after getting approval from the drug administration. Afterward, distributors sell the drugs to retailers, and the retailer sells the drugs to patients. At the time of purchasing, the user scans the unique QR code by using drug verification application. If the drug verification application returns authentic, then it indicates that the drug is authentic. Thus, in this way, a user gets an authentic drug from the supply chain. Every entity links up the new transaction as a new block in the blockchain ledger, as shown in Fig. 4. The information of the delivered drug is stored in the blockchain by a smart contract. Every entity communicates with each other by sending digitally signed messages. The proposed structure utilizes blockchain capabilities and receives the original medicines and drugs' traceability from manufacturers to patients.

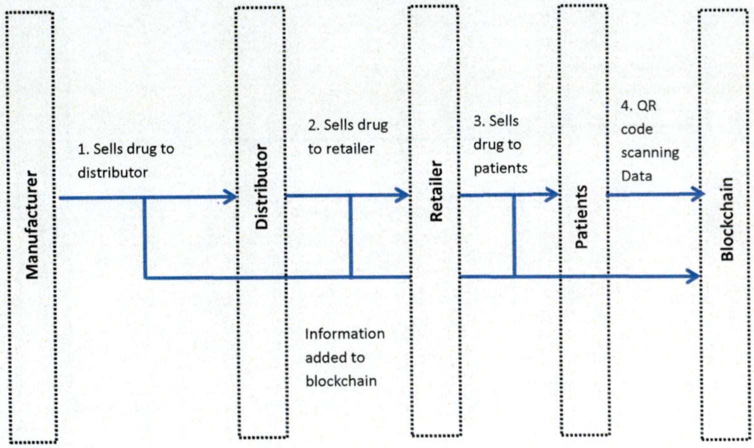

Fig. 4 Process of circulating the drug in the market

3.4 *Observation and Revoke*

Drugs must be observed after their distribution into the market. To observe a drug any time from the market, the observer plays an important role. Figure 5 shows the observation and revoking process where after the distribution of the valid drug, any observer takes the valid drug from the market and their information from blockchain and tests the drug again from their pharmaceutical laboratory. Afterward, the observer compares the information stored in the blockchain with their test result. If any problem is found in that after comparison, then the observer claims to drug administration that the drug is counterfeit. If the number of this claim is maximum, then the drug administration will be automatically notified about the claim by the proposed system. Thereafter, drug administration re-tests the drugs by any pharmaceutical laboratory, which is selected from several pharmaceutical laboratories. Following the re-testing of the drugs, drug administration compares that claim with their re-test result. If that claim is proved correct by re-testing the drugs, drug administration will revoke the drug from the pharmaceutical drug supply chain. Subsequently, after the revoking, a unique QR code of a previously valid drug will be destroyed from the smart contract of drug administration.

4 Tracking System

In this section, we describe the tracking system. Drugs are tracked as they move along the supply chain, first when produced, and afterward, each time they are provided to a buyer working at the next stage. The proposed mechanism can track whether a drug is valid or not based on unique QR code verification. The proposed tracking

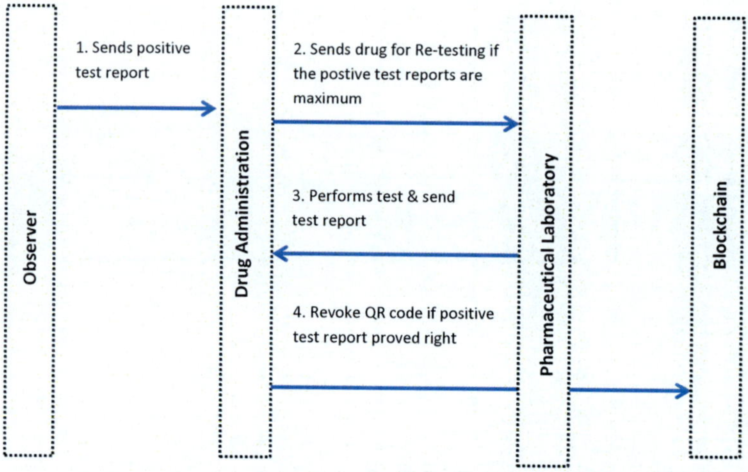

Fig. 5 Process of Observation and Revoke

system can track the information on valid drugs and their ingredients. It can track the locations of verified drugs and track the locations where people are getting the maximum number of counterfeit drugs. All the supply chain entities can keep track of the verified drug's information, who is the manufacturer of that drug, where the drug is sold, and to whom it is sold. All this tracking information is stored in the blockchain as transactions to ensure they are immutable and available to every entity in the supply chain.

5 Experimental Analysis

In this section, we discuss about the experimental tools that we have used to develop the proposed system and also evaluate the performance of the proposed system and compared the results with Etherium.

5.1 Experimental Setup

We have developed the proposed system on Hyperledger Fabric. The proposed system is conducted on a desktop computer with the following specifications.

- CPU: Intel Core i5-3517U 1.90 GHz
- Physical memory: 8 GB
- Operating System: Ubuntu 16.04.e

In order to assess the performance of our system, we have collected the data for each transactions as follows.

- Transaction deployment time (t_1): The time when transaction was deployed.
- Transaction end time (t_2): The time when transaction was confirmed by the blockchain.
- Transaction number.

We have evaluated the performance of the proposed system in terms of three metrics: execution time, average latency and average throughput according to Pongnumkul et al. [14] by varying number of transactions from 1 to 1000. We have also compared the results with Ethereum. A detailed description of these metrics is given below.

- *Execution Time*: It is the aggregate sum of time (number of seconds) that our system took to execute and confirm all the transaction in the dataset, for each set of transactions which is shown in Eq. 1 where n is the total number of transactions.

$$\text{Execution Time} = \sum_{i=1}^{n} (t_2 - t_1) \tag{1}$$

- *Average Latency*: The average latency can be specified as the average of latency of all transaction in a dataset, for a set of transactions, which is shown in Eq. 3. Latency can be defined as the difference between finishing time and deployment time for each transaction which is shown in Eq. 2.

$$\text{Latency} = t_2 - t_1, \text{ for each transaction} \tag{2}$$

$$\text{Average Latency} = \frac{\sum_{i=1}^{n} (t_2 - t_1)}{n}, \text{ for a set of transactions} \tag{3}$$

- *Average Throughput*: The average throughput can be determined as an average of throughput over the execution time, shown in Eq. 5. Throughput can be estimated as the number of successful transactions per second which is shown in in Eq. 4.

$$\text{Throughput} = \frac{n}{\sum_{i=1}^{n} (t_2 - t_1)} \tag{4}$$

$$\text{Average Throughput} = \frac{\text{Throughput}}{n} \tag{5}$$

5.2 *Experimental Results*

This section describes the result of evaluating the proposed system in three ways: evaluating execution time, evaluating average latency, and evaluating average throughput.

5.2.1 Evaluating Execution Time

We investigate the distinctions in execution time by varying the number of transactions in Fig. 6 with Hyperledger Fabric and Etherium. The x-axis shows the number of transactions, (ranging from 1 to 1000) and the y-axis shows the execution time (in seconds) for each set of transactions. The scale is linear. The execution time increments as the number of transactions in the data set grow. But Etherium fails to execute 1000 transactions, it only executes 980 transactions. The result shows that Hyperledger Fabric's execution time is consistently lower than Ethereum in all data sets. The gap between the execution time of Hyperledger Fabric and Ethereum also grows larger as the number of transactions increase.

5.2.2 Evaluating Average Latency

In Fig. 7, we evaluated the average latency by differing the number of transactions with Hyperledger Fabric and Etherium. The x-axis shows the number of transactions (varying from 1 to 1000), and the y-axis shows average latency (in seconds) for each set of transactions. We observe that Ethereum fails to execute 1000 transactions, it

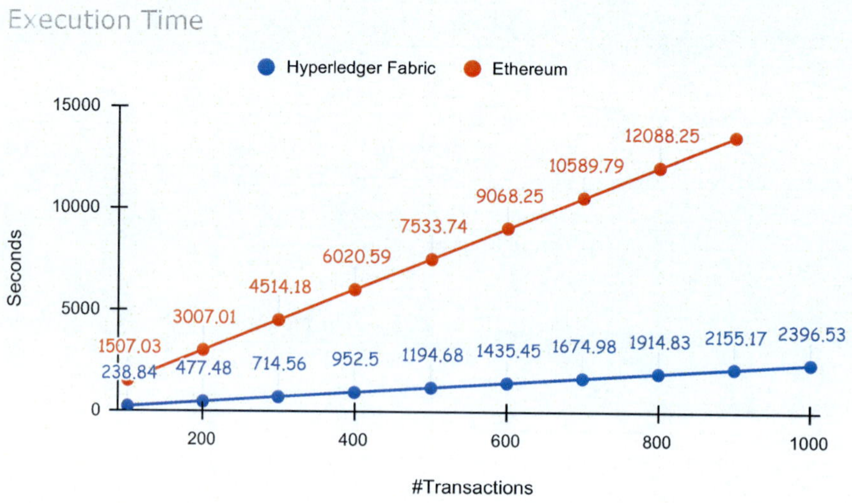

Fig. 6 Execution time

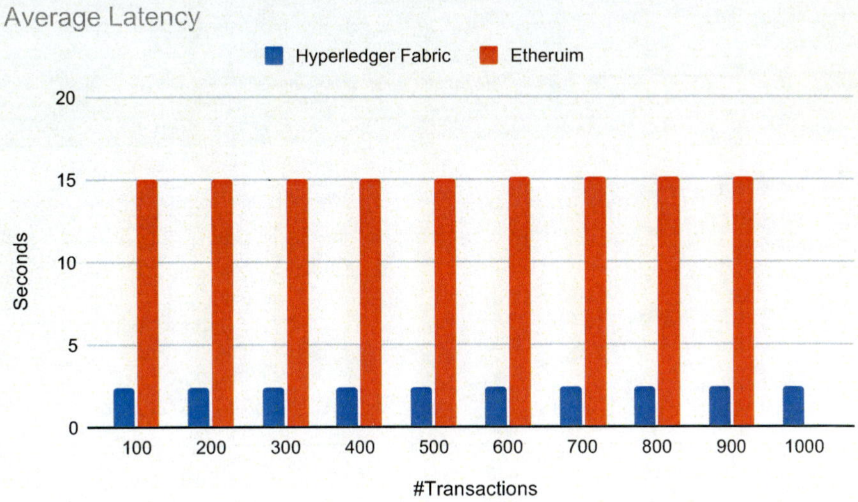

Fig. 7 Average latency

only executes 980 transactions. The result shows that Hyperledger Fabric's average latency is consistently lower than Ethereum in all data sets. As the average latency is lower, it proves that each transaction takes less time in Hyperledger Fabric. On the other-hand Ethereum takes more time for each transaction.

5.2.3 Evaluating Average Throughput

We evaluated the average throughput by changing the number of transactions in Fig. 8 with Hyperledger Fabric and Ethereum. The x-axis shows the number of transactions (ranging from 1 to 1000), and the y-axis shows average throughput (in transaction per second (tps)) for each set of transactions. We observe that to Ethereum fails to execute 1000 transactions, it only executes 980 transactions. The result shows that Hyperledger Fabric's average throughput is consistently higher than Ethereum in all data sets. In Hyperledger Fabric, the average throughput decreases as the number of transactions in the data set to grows. We observe that in Ethereum when the number of transactions varies from 500 to 900, the average throughput remains the same.

On a whole, we can say that our proposed system gives better performance in Hyperledger Fabric compared to Ethereum in terms of execution time, average latency and average throughput.

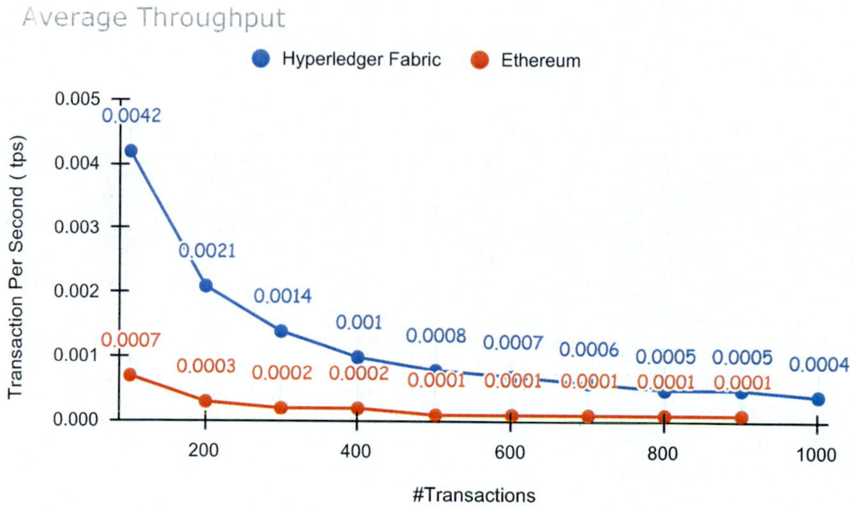

Fig. 8 Average throughput

6 Conclusion

This paper proposed a blockchain-based approach by integrating pharmaceutical laboratories to detect counterfeit drugs in the pharmaceutical supply chain. The proposed supply chain can verify the ingredients of the medicine before and after production. Moreover, an observer of the supply chain can re-check medicine and challenge its authenticity from the blockchain's data. The supply chain is deployed on the Hyperledger Fabric for its faster transaction process. We compared the performance of the proposed system with Ethereum, and it shows that Hyperledger Fabric makes the system more robust and faster. In future, we will build an integrated IoT device with a smart contract for the medical supply chain to verify and update each valid transaction.

References

1. Abbas K, Afaq M, Ahmed Khan T, Song WC (2020) A blockchain and machine learning-based drug supply chain management and recommendation system for smart pharmaceutical industry. Electronics 9(5):852
2. Alzahrani N, Bulusu N (2018) Block-supply chain: a new anti-counterfeiting supply chain using nfc and blockchain. In: Proceedings of the 1st workshop on cryptocurrencies and blockchains for distributed systems, pp 30–35
3. Androulaki E, Barger A, Bortnikov V, Cachin C, Christidis K, De Caro A, Enyeart D, Ferris C, Laventman G, Manevich Y et al (2018) Hyperledger fabric: a distributed operating system for permissioned blockchains. In: Proceedings of the thirteenth EuroSys conference, pp 1–15

4. Foisal Ahmed M (2019) Mini-labs in six districts. https://www.thedailystar.net/backpage/news/minilabs-six-districts-1683232
5. Haq I, Esuka OM (2018) Blockchain technology in pharmaceutical industry to prevent counterfeit drugs. Int J Comput Appl 180(25):8–12
6. Huang Y, Wu J, Long C (2018) Drugledger: a practical blockchain system for drug traceability and regulation. In: 2018 IEEE international conference on Internet of Things (iThings) and IEEE green computing and communications (GreenCom) and IEEE cyber, physical and social computing (CPSCom) and IEEE smart data (SmartData). IEEE, pp 1137–1144
7. Jamil F, Hang L, Kim K, Kim D (2019) A novel medical blockchain model for drug supply chain integrity management in a smart hospital. Electronics 8(5):505
8. Jangir S, Jaiswal A, Chandel S, Muzumdar A, Modi CN, Vyjayanthi C (2019) A novel framework for pharmaceutical supply chain management using distributed ledger and smart contracts. In: 2019 10th international conference on computing, communication and networking technologies (ICCCNT). IEEE, pp 1–7
9. Kabir K (2016) A panacea for counterfeit medicines. https://www.dhakatribune.com/feature/health-wellness/2016/06/22/panacea-counterfeit-medicines
10. Kumar R, Tripathi R (2019) Traceability of counterfeit medicine supply chain through blockchain. In: 2019 11th international conference on communication systems & networks (COMSNETS). IEEE, pp 568–570
11. Kumari K, Saini K (2019) Cfdd (counterfeit drug detection) using blockchain in the pharmaceutical industry. Int J Eng Res Technol (IJERT) 8:1–4
12. Organization WH (1999) Counterfeit drugs: guidelines for the development of measures to combat counterfeit drugs. World Health Organization, Tech rep
13. Pandey P, Litoriya R (2020) Securing e-health networks from counterfeit medicine penetration using blockchain. Wirel Personal Commun 1–19
14. Pongnumkul S, Siripanpornchana C, Thajchayapong S (2017) Performance analysis of private blockchain platforms in varying workloads. In: 2017 26th international conference on computer communication and networks (ICCCN). IEEE, pp 1–6
15. Rahman Rabbi A (2020) 18 held for selling counterfeit medicine in dhaka. https://www.dhakatribune.com/bangladesh/dhaka/2020/07/10/18-held-over-counterfeit-medicines-in-dhaka
16. Sylim P, Liu F, Marcelo A, Fontelo P (2018) Blockchain technology for detecting falsified and substandard drugs in distribution: pharmaceutical supply chain intervention. JMIR Res Protocols 7(9):e10163
17. (WHO) WHO (2010) Growing threat from counterfeit medicines. Bulletin of the World Health Organization 88(4)
18. Zhu P, Hu J, Zhang Y, Li X (2020) A blockchain based solution for medication anti-counterfeiting and traceability. IEEE Access 8:184256–184272

Enhanced Steganography Technique via Visual Cryptography and Deep Learning

Tasfia Seuti, Md. Al Mamun, and A. H. M. Sarowar Sattar

Abstract Steganography is one kind of information hiding technique where a file is hidden within a transferable medium, such as an image, video, or file. Many steganography methods have been proposed and implemented over the decades among which image-steganography is very popular. In image steganography, one of the most popular techniques is the Least Significant Bit (LSB) technique. However, there are certain security drawbacks to this method, such as the fact that anyone who knows where the information is concealed may simply recover it. In this paper, a new approach is proposed by integrating steganographic technique with deep learning and visual cryptography to solve the problem where a secret image can be hidden in a cover picture using steganography but the content of the secret image is embedded by both deep learning and visual cryptography first. The secret picture is initially sent into the autoencoder, which consists of an encoder and a decoder. It compresses the image and renders it unrecognizable. The picture is then subjected to visual cryptography by conducting an exclusive OR (XOR) operation on it with a randomly generated image named mask1. Using the LSB technique, the encrypted secret picture is then concealed within the carrier (cover) image. All of the encoding stages are reversed for the decryption of the secret picture. The consistency of the stego picture was assessed using image quality matrices, putting the experimental findings to test. The values of the image quality metrics indicate the enhancement of security. A comparison study was also conducted with various current tools, and our technique was shown to be superior to the majority of them.

Keywords Visual cryptography · LSB · Steganography · Autoencoder · Deep learning

T. Seuti (✉) · Md. Al Mamun · A. H. M. Sarowar Sattar
Department of Computer Science and Engineering, Rajshahi University of Engineering & Technology, Rajshahi, Bangladesh

M. S. Arefin et al. (eds.), *Proceedings of the International Conference on Big Data, IoT, and Machine Learning*, Lecture Notes on Data Engineering and Communications Technologies 95, https://doi.org/10.1007/978-981-16-6636-0_47

1 Introduction

In the modern communication system, people have become concerned about the safety of the data they post on the internet as the internet has grown in popularity and speed. They do not want their data to be exploited by third parties. This is where the concept of cryptography and steganography got introduced. Cryptography, also known as secret writing, is a method of converting a secret message into ciphertext and sending it to another person who decrypts it into plain text [1]. Steganography, on the other hand, is a method of concealing confidential information inside a file. It may be an image, audio, video, text, or HTML file. Image steganography enables different parties to secretly exchange image files [2, 3]. It is a method of concealing information from the adversary to create an unobservable communication channel [2, 4, 5].

Efficient cryptographic techniques are critical for protecting digital images from intruders. Since image data is in graphic form, visual cryptography is the most useful technique for image encryption [6], which was introduced by Naor and Shamir [7] in 1994. The aim was to preserve the privacy of people's hidden pictures. On the other hand, autoencoders are a form of artificial neural network which is trained with images to create the same image as output. Encoder, code, and decoder are the three parts of an autoencoder. The encoder compresses the input and generates the code, which the decoder then uses to recreate the input. Until executing both visual cryptography and steganography, an autoencoder is used to render the hidden picture unrecognizable.

Over the last few decades, several steganographic methods have been proposed [8–10]. The most popular method is to replace the hidden message by replacing the LSB of the pixels of the cover image. The key aim of image steganography is to keep the hidden details confidential by ensuring that the stego image is not manipulated by the cover image. While several steganographic approaches have been attempted to solve both problems, the methods have been discovered to be vulnerable. Some researchers have solved one case but are unable to continue on the other. In this paper, Image steganography, visual cryptography, and an autoencoder are all part of our core philosophy, where the key goals of this approach are to enhance the quality of stego images and the security of the secret images. After using an autoencoder to make the secret image unrecognizable, to make it much more difficult to discern, a randomly generated image called mask1 was introduced, and that image was XORed with the hidden image. The LSB system is used for steganography, where hidden information is typically stored in the particular location of LSB of a cover image [2, 8]. Therefore, this paper aims to develop a reliable tool that could solve both problems (ensuring that the secret information remains secure and that the stego picture is not affected by cover image). The main contributions of this paper are listed below.

- Combination of visual cryptography with image steganography
- Leveraging the compression technique of autoencoder to enhance image security and carrier capacity

- Employment of 1-LSB to preserve data (cover image) integrity

The rest of the paper is structured as follows. Following the introduction in Sect. 1, Sect. 2 examines the recent research works. Section 3 explains our proposed approach while the experiment results are discussed in Sect. 4. At last, our findings and observations are concluded in Sect. 6.

2 Literature Review

In the year 1994, Naor and Shamir were the first to present the notion of visual cryptography [7]. They also introduced $VCS(k, n)$ which was a new technique of using cover-based semi-group to enhance the contrast of an image [11]. An analysis of various recent research works in some of the digital watermarking approaches by combining visual cryptography mechanisms was included [12]. To ensure copyright security, Z. Tijedjadjine introduced a visual cryptography scheme in watermarking [13].

Among the recent studies, a paper published in 2021 proposed an adaptive fuzzy inference method for color image steganography that considers image complexity factors such as pixel similarity, pixel brightness, and color sensitivity [14]. Around the same time, a hybrid data transmission scheme incorporating Cryptography and Steganography was proposed and also an application incorporating Cryptography and Steganography was developed for hiding data by Gupta and Saxena [15, 16]. A little before that in 2020, a steganography algorithm for hiding data, as well as images in another images using the LSB technique was proposed by Shekhawat, Tiwari, and Patel [17].

One of the most frequent image steganography approaches is to encode the secret message in each pixel's LSB. This is because modifying the LSB has the least impact on the carrier picture, so human vision cannot detect the difference between the original cover image and the altered cover image. Chandramouli [18] investigated various LSB-based steganography methods wanting to see how a person could tell the difference between the stego image and the actual cover image. Paper [19] introduced a new LSB-based scheme in which they chose the cover image layer for hidden image concealment using a secret key. The secret picture was applied to the 1D bitstream after the stego key was converted into a 1D circular array bitstream. The system then executes the code on the red layer's first pixel LSB and the stego key's first bit. The system selects the green layer for the cover image when the resultant bit is 1, and the blue layer for the hidden image's 1-bit concealment if the resultant bit is 0. For the next bit of the hidden image, the method directs to the next red layer pixel and then the stego key's next bit. This procedure is repeated until the entire hidden image bitstream has been completed. An XOR operation was done between the red layer's LSB and the secret key pixel to decode the secret image. The secret information is in the green layer's LSB if the resultant bit is 1, and the secret information is in the blue layer's LSB if the resultant bit is 0. They then reshaped it into a 2D binary image matrix to retrieve the hidden image successfully.

Finally, M. Hossain proposed three steganography approaches in [8]. For estimating the smooth and edged areas, a bit's reliance on its neighborhood and psycho visual redundancy were used. In smooth regions, three bits are embedded, while in edged regions, variable-rate pixels are embedded. Even though their methods produce high-quality images, they did not have any security procedures for their work.

Also, there are some existing steganography tools. Hera Arif in [20] surveyed existing steganography tools. They surveyed S-Tools 4.00, VSL 1.1, Open Puff 4.00, CryptaPix 3.10, and Quick Crypto. But in this paper, the proposed system is compared with Online Image Steganography, East-Tec Invisible Secrets 4, and Openstego tools as these are the latest and three of the most useful steganography tools.

3 Proposed Methodology

Our proposed framework is built on a combination of visual cryptography, image steganography as well as a deep learning based autoencoder. The main goal of this approach is to improve the quality of the stego images and the protection of secret images.

3.1 Embedding and Encryption Technique

The image that is intended to be concealed is not explicitly hidden inside the chosen cover image for security concerns. To secure the hidden image, it is first passed through an autoencoder, which compressed and rendered it unrecognizable (i.e., secret image + autoencoder (encoding) = compressed secret image), as illustrated in Fig. 1. Afterward, a randomly created image called mask1 is introduced. The mask1 image was used to perform the Visual Cryptography part. And it is a must when it comes to extracting the secret image.

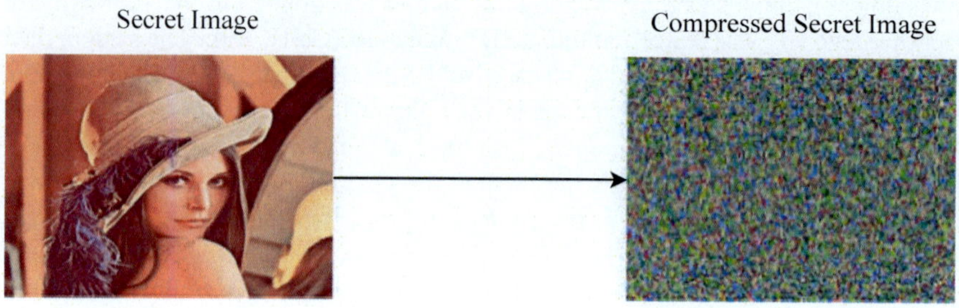

Fig. 1 Formation of the compressed secret image

Table 1 Red (R), Green (G), and Blue (B) channels production of mask2 image

Mask1 Image	Operation	Secret Image	Mask2 Image
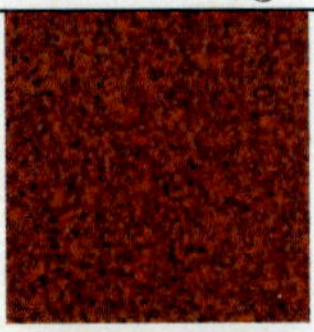	XOR	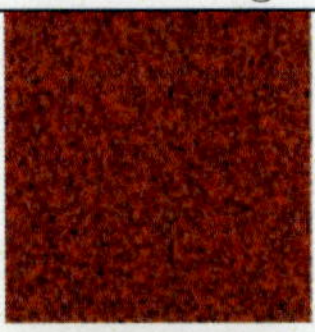	
	XOR		
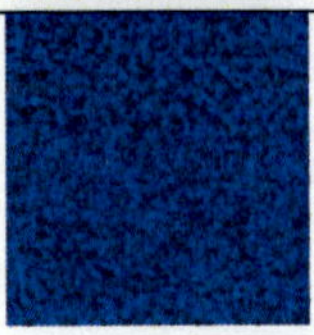	XOR		

Table 2 Secret to mask2 conversion output

Secret Image	Mask1 Image	Mask2 Image
	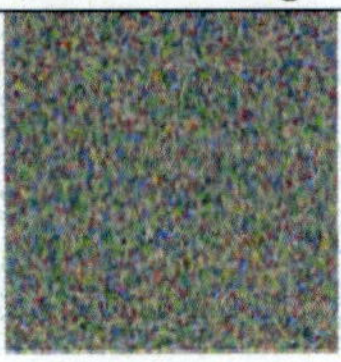	

The secret image and the mask1 image are both 24-bit color images of the same dimensions. Both the secret image, which is compressed and unrecognizable already, and the mask1 image are converted to binary matrices and divided into R, G, and B layers. After that pixel by pixel, an XOR operation was conducted on the matching layers between the compressed secret picture and mask1 image. The R, G, and B channels of the mask2 image were then established (Table 1).

When all three layers of the mask2 image matrix are stacked together to create a 24bit mask2 image, that looks nothing similar to the secret image. It appears to look similar to random noise. Table 2 shows the contrast between the hidden image and the mask2 image. It is reasonable to say that the secret picture and the mask2 image bear no resemblance, satisfying the fundamental premise of visual cryptography, which is to encrypt visual information in an unidentifiable manner. This mask2 picture is now concealed within a cover image, rather than the secret image.

Table 3 Creation of 3 channels of stego image

Cover Layer	Mask2 Layer	Stego Layer
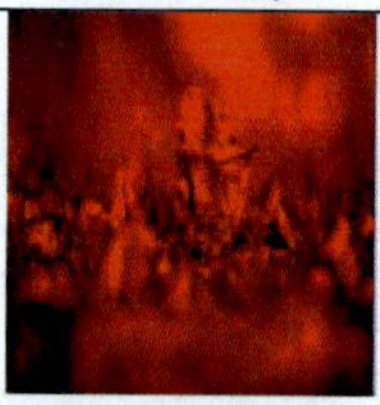	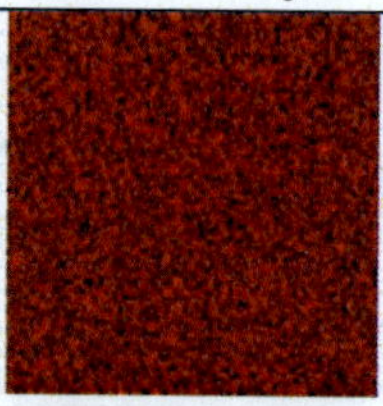	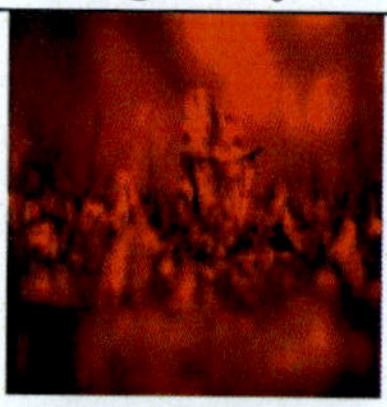
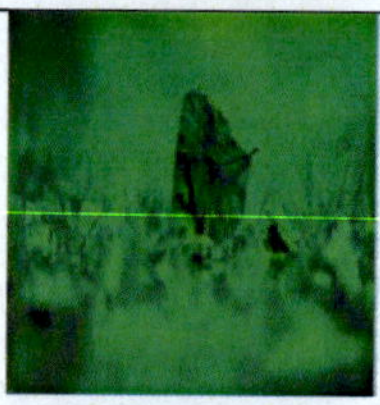	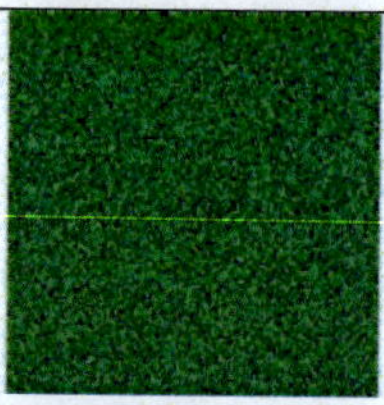	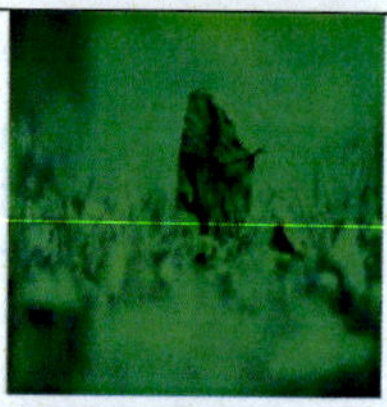
		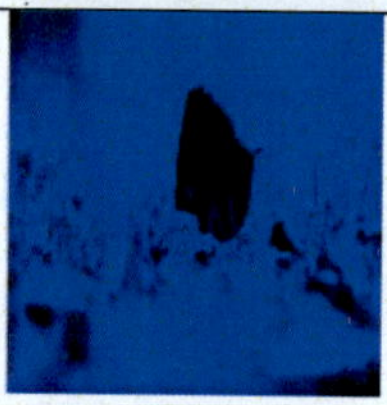

The cover image is also divided into R, G, and B layers after being converted to a binary image. Using the LSB approach, every layer of the mask2 image is then disguised into the corresponding layer of the cover image. In this step, each bit of the mask2 image matrix is serially changed by the LSB bit of a pixel of the cover image matrix. The transformation of the cover image's three layers into the stego image's three layers is seen in Table 3.

The same concealment technique is applied to hide the mask1 image matrix after removing all pixels in the mask2 image. Then, these three layers are combined to create the final stego image, which is identical to the cover image. The final stego picture is shown in Table 4. In this case, an XOR operation is performed between secret and mask1 pixel, yielding mask2. By performing this process, the R, G, and B channels of the mask2 images are created from the secret image.

The flow chart in Fig. 2 shows the entire encoding procedure for the secret image step by step. The initial step was to compress the secret picture using the autoencoder and obtain the image's binary matrix. The next step was to use a random picture named mask1 to conduct Visual Cryptography on the compressed image. To achieve this, both the compressed secret and mask1 pictures were first separated into layers of red (r), green (g), and blue (b). Then, the binary matrices of the compressed secret picture and the random image were then XORed. The next step was merging the three layers to get the color image mask2. This mask2 image was finally hidden inside the cover image using the popular steganography method LSB, and the stego image is formed.

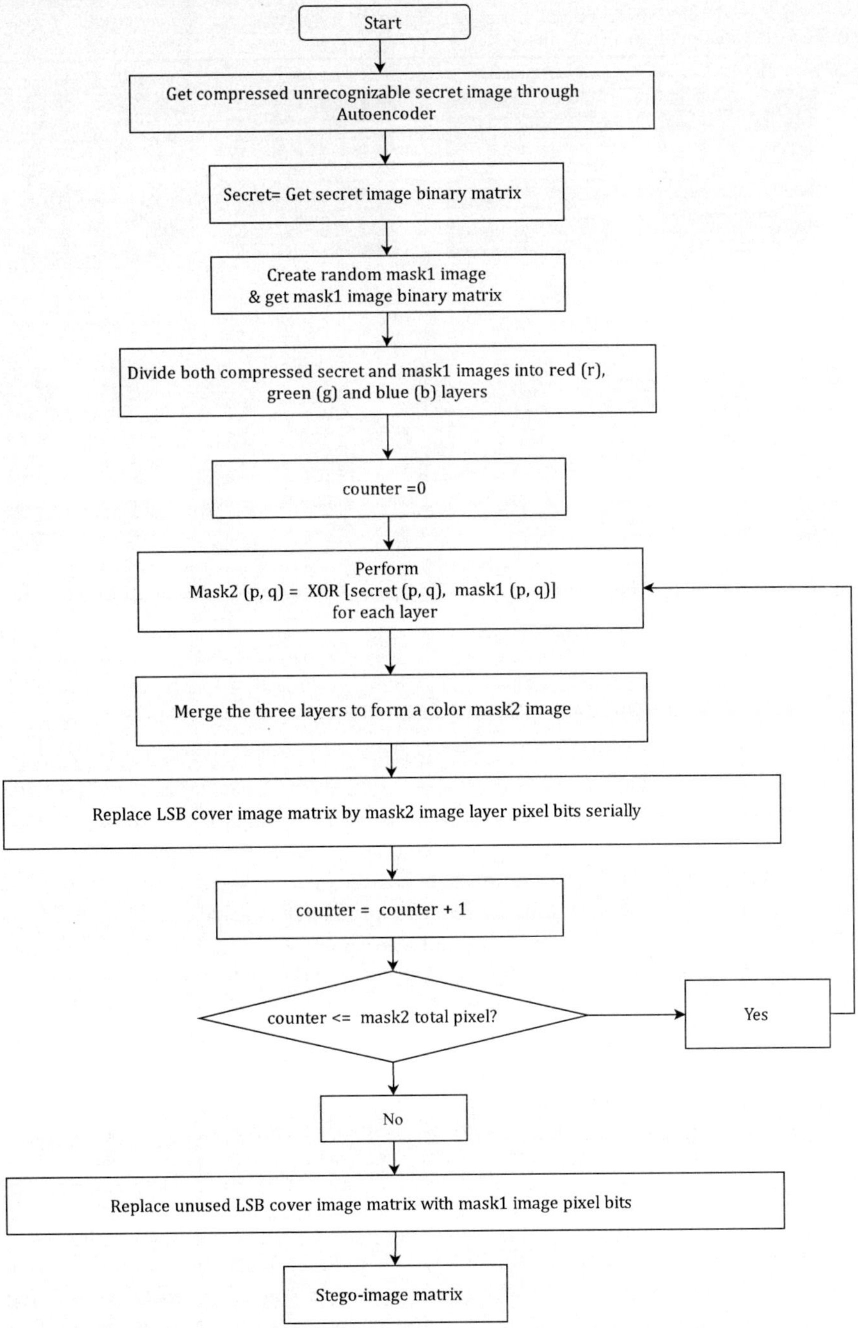

Fig. 2 Flowchart of Embedding and Encryption Approach of the Proposed Method

Table 4 Formation of final stego image

Input1	Input2	Input3	Output
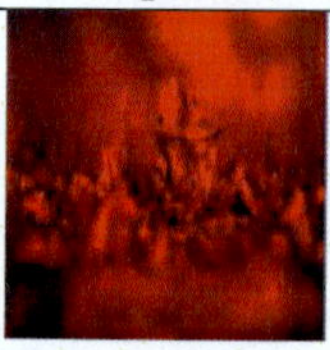		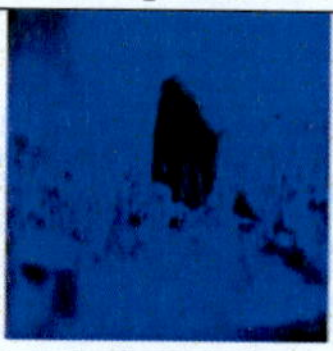	

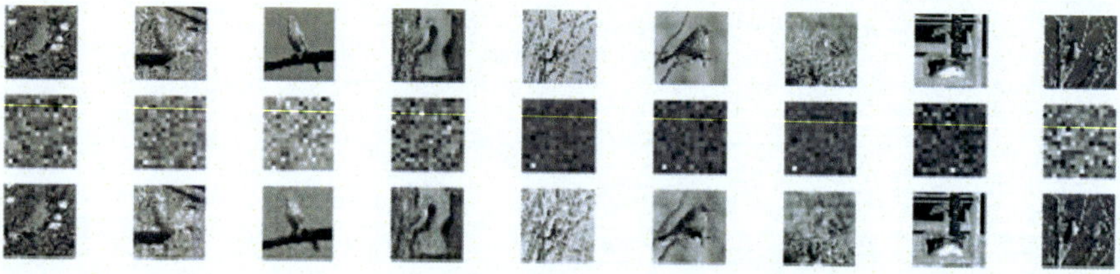

Fig. 3 Original images (first row) vs compressed images (second row) vs decoded Images (third row)

3.2 *Extraction and Decryption Technique*

To extract the concealed picture, the mask1 and mask2 images are decoded from the stego image. The stego image is first transformed into a binary image matrix. Then, it is divided into layers of red, green and blue image matrix layers. Since our secret information is hidden in three layers, the data must be decoded from each layer separately. For the mask2 image, three empty image matrices are formed. Every matrix corresponds to one of the 24-bit color image layers. The LSB bits from each pixel in each layer of the stego image are extracted and serially transferred to a new image matrix array. An empty image matrix was used for the stego image's red layer, and the stego image's red layer was used for the mask2 image's red layer. After that is done, the empty matrix's initial pixel was copied with the first eight LSB bits from the stego picture. This is repeated until the mask2 full red layer is obtained. In the same way, the green and blue layers are recovered. After obtaining three layers from the mask2 image, these three layers are merged to create the image. The mask1 image is decoded using the same tool.

Then, between each layer of the mask1 and mask2 images, an XOR operation is performed bit by bit. As a result, the red, green, and blue layers of the hidden image have been discovered. Our compressed secret image was achieved by combining them, which was achieved through the autoencoder's encoding process in the first place, This compressed noise-like image is then fed into the autoencoder's decoding portion yielding the final secret image. A comparison is provided of the original and decoded images from the dataset in Fig. 3.

In Fig. 3, the first row depicts some of the secret pictures that we wish to keep hidden. The compressed hidden images that were achieved after the autoencoder's encoding portion are shown in the second row. Finally, the third row displays the hidden photos that are recovered

4 Experiment Results

To assess the quality of stego pictures subject to cover images, five image quality metrics were used: SNR, PSNR, SSIM, MSE, and SC. As stated in Table 5, one cover image and two hidden images were used for the study. So far, two stego images have been discovered. The values for different performance measure metrics are provided in Table 6. The SNR rating ranges from 100 to 4.77 from greatest to worst. The proposed system's SNR values vary from 45 to 46, which indicates the appropriate result. The PSNR value for two identical images should be 100, and for completely dissimilar images, it should be 0. PSNR values range from 52 to 54 for various sizes of the hidden image. As a result, the PSNR values found here are also appropriate. The scale of best to worst values for SSIM is 1–0. The findings range from 0.983 to 0.996 in this case. As a result, this is also appropriate. The optimum to worst value range for MSE is somewhere between 0 and 60000+. The values of our suggested tool, on the other hand, vary from 0.340 to 0.500, showing that it performs well. Finally, the best value for SC is 1.00, while the worst value is infinity. 1.00 SC values have been identified for this proposed system, indicating that our proposed system achieves a satisfactory result. As a result, this approach produces stego images that are unnoticeable to human eyesight and the image quality measurements for the system are excellent.

Without the mask1 image, nobody can extract the hidden image from the stego image. People who know the extraction method for the mask2 image, which is very different from the actual hidden image, may extract the secret image. As a result, this proposed method can guarantee stego image quality as well as secret image protection.

5 Comparison with Other Tools

The same five image quality measures used to measure the quality of stego images in our proposed system were also used to measure the quality of stego images in existing tools: Online Image Steganography, East-Tec Invisible Secrets 4, and Openstego, and the results were compared to our proposed system. One cover image and three secret images were used for the research as shown in Table 7.

Tables 8, 9 and 10 show all the metric values of each of the steganography tools that were used in this experiment namely SNR, PSNR, SSIM, MSE, and SC. And then Table 11 shows the metric values of our proposed system for the data of Table 7.

Table 5 Image data set details for analysis

Cover Images	Secret Images	Stego Images

Cover Image (1) Size: 2.10 MB Dimensions: 720*720	secret image (1) Size: 65 KB Dimensions: 180*140	Stego image (1) Size: 2.02 MB Dimensions: 720*720
		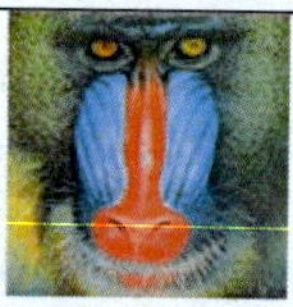
Cover Image (2) Size: 2.10 MB Dimensions: 720*720	secret image (2) Size: 135 KB Dimensions:310*140	Stego image (2) Size: 2.01 MB Dimensions: 720*720

Table 6 Our proposed approach's metric values

Cover Images	Stego Images	SNR	PSNR	SSIM	MSE	SC
Cover Image (1)	Stego image (1)	44.311	53.003	0.987	0.347	1
Cover Image (2)	Stego image (2)	45.551	53.704	0.992	0.480	1

And lastly, Table 12 shows the average metric values of three secret images for each tool, making the comparison analysis significantly easier.

The average SNR value of the proposed system here is better than two existing tools namely Online Image Steganography & East-Tec Invisiblesecret4. The same goes for PSNR, SSIM, MSE. And the SC value is the same for all the tools. After a comparison of five Image Quality Metrics, it can be concluded that the proposed tool produces an undoubtedly good result and performs better than two of the reputed existing steganography tools. This can be written as

OpenStego > Proposed System > Invisible Secret 4 > Online Image Steganography.

Table 7 Image data set details for analysis

Cover Images	Secret Images	Stego Images

Cover Image Size: 3400 KB Dimensions: 1512*850 Format: BMP	secret image (1) Size: 289 KB Dimensions: 310*324 Format: BMP	Stego image (1) Size: 3280 KB MB Dimensions: 1512*850 Format: BMP

Cover Image Size: 3400 KB Dimensions: 1512*850 Format: BMP	secret image (2) Size: 14.5 KB Dimensions: 70*70 Format: BMP	Stego image (2) Size: 3280 KB Dimensions: 1512*850 Format: BMP

Cover Image Size: 3400 KB Dimensions: 1512*850 Format: BMP	secret image (3) Size: 69 KB Dimensions: 180*128 Format: BMP	Stego image (3) Size: 3280 KB Dimensions: 1512*850 Format: BMP

Table 8 Metric values of online image steganography

Cover Images	Stego Images	SNR	PSNR	SSIM	MSE	SC
Cover Image	Stego image (1)	38.52	38.87	0.84	8.45	1
Cover Image	Stego image (2)	38.50	38.86	0.85	8.41	1
Cover Image	Stego image (3)	38.47	38.89	0.84	8.46	1

Table 9 Metric values of East-Tec Invisiblesecret4

Cover Images	Stego Images	SNR	PSNR	SSIM	MSE	SC
Cover Image	Stego image (1)	50.65	51.00	0.90	0.52	1
Cover Image	Stego image (2)	50.77	51.12	0.90	0.55	1
Cover Image	Stego image (3)	50.80	51.11	0.92	0.52	1

Table 10 Metric values of Openstego

Cover Images	Stego Images	SNR	PSNR	SSIM	MSE	SC
Cover Image	Stego image (1)	61.00	61.35	0.97	0.05	1
Cover Image	Stego image (2)	76.57	76.85	1.00	0.00	1
Cover Image	Stego image (3)	61.83	62.19	0.98	0.04	1

Table 11 Metric values of the proposed method

Cover Images	Stego Images	SNR	PSNR	SSIM	MSE	SC
Cover Image	Stego image (1)	50.33	52.35	0.92	0.32	1
Cover Image	Stego image (2)	67.91	66.85	0.99	0.01	1
Cover Image	Stego image (3)	58.83	59.35	0.98	0.10	1

Table 12 Average metric values of each system

Tools Name	SNR	PSNR	SSIM	MSE	SC
Online Image Steganography	38.497	38.874	0.843	8.44	1
East-Tec Invisiblesecret4	50.74	51.077	0.91	0.53	1
Openstego	66.46	66.79	0.98	0.03	1
Proposed System	59.03	59.52	0.97	0.14	1

6 Conclusion

A new steganography approach is introduced that combines a visual cryptographic scheme with a deep learning-based autoencoder technique. The cover image, like the stego image, exhibits very minimal distortion, according to the experimental results. As a result, normal human eyes are unable to detect the stego image as a carrier of concealed material. The LSB method was used for the steganography process. For security reasons, a method other than the widely used cryptographic methods has been used. The method involves using an autoencoder to convert the hidden image to a noise type image (mask2 image), XORing a randomly generated image called mask1, and then hiding it as the hidden data inside the cover image. This approach would provide good protection for the secret image and distinguish it from other image steganography methods since the secret image is rendered unrecognizable first by both autoencoder and the XOR operation between the compressed secret image and mask1 image. In future, this system's compatibility and performance can be tested on more recent datasets. Also, there are many more useful image quality metrics namely Average Difference (AD), Maximum Difference (MD), Mean Absolute Error (MAE), Signal Noise Ratio (SNR), Structural Dissimilarity (DSSIM), etc., which can also be used to measure the quality of the stego image subject to the cover image. Measuring the quality with all these metrics.

References

1. Uddin MP, Marjan MA, Sadia N, Islam MR (2014) Developing a cryptographic algorithm based on ASCII conversions and a cyclic mathematical function. In: 2014 international conference on informatics, electronics & vision (ICIEV). IEEE Press, pp 1–5
2. Kutte M, Hartung JF (1999) Information hiding-a survey. In: Proceedings of The IEEE: special issue on identification and protection of multimedia content, vol 87, no 7, pp 1062–1078
3. Uddin MP, Saha M, Ferdousi SJ, Afjal MI, Marjan MA (2014) Developing an efficient solution to information hiding through text steganography along with cryptography. In: 2014 9th international forum on strategic technology (IFOST), pp 14–17
4. Islam MR, Siddiqa A, Uddin MP, Mandal AK, Hossain MD (2014) An efficient filtering based approach improving LSB image steganography using status bit along with AES cryptography. In: 2014 international conference on informatics, electronics vision (ICIEV), pp 1–6
5. Sultana S, Khanam A, Islam MR, Nitu AM, Uddin MP, Afjal MI, Rabbi MF (2018) A modified filtering approach of LSB image steganography using stream builder along with AES encryption. HBRP recent trends in information technology and its applications 1(1):1–10
6. Bhatia S, Khatri SK, Singh AV (2018) Digital image security using hybrid visual cryptography. In: 2018 7th international conference on reliability, Infocom technologies and optimization (trends and future directions) (ICRITO). IEEE Press, pp 570–576
7. Naor M, Shamir A (1994) Visual cryptography, Eurocrypt'94. In: Lecture notes in computer science, vol 950
8. Hossain M, AL Haque S, Sharmin F (2009) Variable rate steganography in gray scale digital images using neighborhood pixel information. In: 2009 12th international conference on computers and information technology. IEEE Press, pp 267–272
9. Simmons GJ (1984) The prisoners' problem and the subliminal channel. In: Advances in cryptology. Springer, Berlin, pp 51–67

10. Nabavian N (2007) CPSC 350 data structures: image steganography. nabavl00@ chapman. edu
11. Naor M, Shamir A (1996) Visual cryptography II: improving the contrast via the cover base. In: International workshop on security protocols. Springer, Berlin, pp 197–202
12. Vyas C, Lunagaria M (2014) A review on methods for image authentication and visual cryptography in digital image watermarking. In: 2014 IEEE international conference on computational intelligence and computing research. IEEE Press, pp 1–6
13. Tifedjadjine Z, Atamna N, Dibi Z, Bouridane A (2005) Halftone image watermarking based on visual cryptography. In: 2005 12th IEEE international conference on electronics, circuits and systems. IEEE Press, pp 1–4
14. Tang L, Wu D, Wang H et al (2021) An adaptive fuzzy inference approach for color image steganography. Soft Comput 1–18
15. Shree R, Swami D (2021) Hybrid secure data transfer scheme using cryptography and steganography. In: Proceedings of the second international conference on information management and machine intelligence. Springer, Berlin, pp 577–583
16. Gupta Y, Saxena K (2021) Application developed on data hiding using cryptography and steganography. In: Innovative data communication technologies and application. Springer, Berlin, pp 107–119
17. Shekhawat VS, Tiwari M, Patel M (2021) A secured steganography algorithm for hiding an image and data in an image using LSB technique. In: Computational methods and data engineering. Springer, Berlin, pp 455–468
18. Chandramouli R, Memon N (2001) Analysis of LSB based image steganography techniques. In: Proceedings 2001 international conference on image processing (Cat. No. 01CH37205), vol 3. IEEE Press, pp 1019–1022
19. Karim SM, Rahman MS, Hossain MI (2011) A new approach forlsb based image steganography using secret key. In: 14th international conference on computer and information technology (ICCIT 2011). IEEE Press, pp 286–291
20. Arif H, Hajjdiab H (2017) A comparison between steganography software tools. In: 2017 IEEE/ACIS 16th international conference on computer and information science (ICIS). IEEE Press, pp 423–428

Developing a Framework for Credit Card Fraud Detection

Yeasin Arafath, Animesh Chandra Roy, M. Shamim Kaiser, and Mohammad Shamsul Arefin

Abstract Credit card is one of the most popular online or manual payment methods, and credit card fraud is increasing that can cause enormous financial damage. Protective action must therefore be taken to stop the credit card. Several new technologies can be utilized to detect frauduleous transactions based on artificial intelligence, data mining, and machine learning. Several new techniques may be used to identify artificial intelligence, data mining, machine learning, sequence alignment, genetic programming, etc. This article provides a new paradigm for credit card fraud detection based on the characteristics of previous user credit card transactions. Detecting fraudulent purchases from past credit card transactions is a difficult challenge, as it depends on a number of factors, like timing, amount, etc. The data on credit card transactions is rising at a huge rate every day. This constant influx of new data is also challenging to handle and to construct new models to determine if a transaction is fraudulent or not. To do this, we use the PCA data type of user transformation since user data supply sensitive information about users and user transactions. We utilize a tweaked fraud detection model utilizing a RandomizedSearchCV hyperparameter tweaking for detecting fraudulent transactions. Our mechanism provided can determine whether a transaction is fraudulent based on the data patterns of prior transactions of the user.

Keywords Logistic regression · K-neighbors classifier · Random forest classifier · LinearSVC · GaussianNB · Decision tree classifier

Y. Arafath · A. C. Roy · M. Shamim Kaiser · M. S. Arefin (✉)
Department of Computer Science Engineering, Chittagong University of Engineering and Technology, Chattogram 4349, Bangladesh
e-mail: sarefin@cuet.ac.bd

M. Shamim Kaiser
Institute of Information Technology, and Applied Intelligence and Informatics (AII), Wazed Miah Science Research Centre (WMSRC), Jahangirngar University, Savar, Dhaka 1342, Bangladesh

M. S. Arefin et al. (eds.), *Proceedings of the International Conference on Big Data, IoT, and Machine Learning*, Lecture Notes on Data Engineering and Communications Technologies 95, https://doi.org/10.1007/978-981-16-6636-0_48

1 Introduction

Credit card theft is the fraudulent use of credit card details without the consent of the cardholders. The credit card can be used in person or online. Cardholders use their keys at the end of the dealership in the event of physical use. The fraudster must procure the card in physical form by deception and then use it to commit fraud. To commit fraud, information such as card verification value (CVV code), expiry date, card number, and pin code are required for Internet card transactions. Fraudsters collect card data by intercepting e-mails, phishing, and skimming victims' online transactions.

Several modern approaches [1] focused on artificial intelligence, data processing, deep learning, sequence matching, genetic programming, and other technologies that can be used to detect fraudulent transactions. Many algorithms such as logistic regression, K-neighbors classifier, random forest classifier, linearSVC, GaussianNB, decision tree classifier, and others [2] can be used to detect defect fraud in machine learning.

Fraud detection algorithms that are based on machine learning are widely recognized as one of the most prominent and successful study areas in fraud detection [3]. Classification is the process of dividing a given set of data into groups based on their characteristics. It is possible to accomplish this with both structured and unstructured data.

Predicting the class of provided data points is the first step in the process. The groups are also known as the objectives, marks, or divisions. Classification algorithms include linear classifiers, support vector machines, decision trees, and others. One of the most studied fields of fraud detection is credit card fraud detection, which relies on automatic analysis of recorded transactions to determine fraudulent activity. Fraud detection systems are vulnerable to a range of issues and obstacles, which are described below. A successful fraud detection method should be capable of dealing with these issues to provide the best results. Credit card fraud detection can be used in a variety of situations. Fraud detection from this process may be used for a wide range of applications, including online banking systems, online payment systems using credit cards, and online transactions. The vast majority of transactions are now conducted electronically, necessitating the use of credit cards and other online payment services.

Both the business and the customer prosper from this approach. Consumers save time by not needing to go to the supermarket to place their orders, and businesses save money by not having to buy physical stores to save costly rent costs.

The new era seems to have incorporated some very useful features that have changed how companies and consumers interact with one another, but at a cost. Businesses must hire trained software engineers and penetration testers to ensure that all transactions are genuine and not fake. These individuals are building the company's servers in such a manner that the customer does not influence vital transaction components such as payment numbers. Most (if not all) of the issues can be avoided with proper design, but even the architecture used to construct the server is not flawless. In this application, a hybrid feature extraction approach that blends various features is used to detect credit card fraud. The key stages are to obtain the generalized format of the dataset, preprocessing steps are carried out, baseline models are trained and score using the training and testing dataset, among the baseline models, the best accurate models are set up for random hyperparameter tuning with RandomizedSearchCV, best params of the tuned model are evaluated. Imbalanced data, the evidence on credit card fraud is distorted, meaning that only a small percentage of all credit card transactions are fraudulent. Because of this, identifying illegal transactions is complex and imprecise. Overlapping data, many transactions may be deemed fraudulent while still appearing to be normal (fake positive), and a fraudulent transaction may appear to be real (fake negative). As a result, achieving a low false positive and false negative rate is a significant challenge for fraud detection systems. Lack of adaptability, the dilemma of adaptability often confronts classification algorithms. Both controlled and unsupervised fraud detection mechanisms are inefficient at identifying new patterns of ordinary and fraudulent conduct. Fraud detection cost, the method should consider both the importance of the detected fraudulent transaction and the expense of stopping it. Stopping a dishonest sale of a few dollars, for example, yields little money. Lack of standard metrics, there is no common assessment criteria for evaluating and comparing the outcomes of fraud detection systems to determine which is the most efficient. Research work is done to accomplish a particular set of targets, such as introducing a hybrid approach.

2 Related Work

Several approaches to bringing strategies to detect fraud have been proposed in previous research, ranging from regulated approaches to unsupervised approaches to hybrid approaches, making it possible to review the technologies associated with credit card fraud detection and to get a clearer understanding of the forms of credit card fraud. As time progressed, fraud patterns evolved, introducing new forms of fraud, making it a popular research subject. The rest of this section goes into individual machine learning algorithms, machine learning models, and fraud detection systems that are used in fraud detection. The issues found during the study have been researched in order to later implement an efficient machine learning model. Researchers suggested a vast range of credit card fraud prevention algorithms, the majority of which focused on neural network and data mining approaches. Bahnsen et al. [4] discussed a comparison study of credit card fraud detection: supervised versus unsupervised. Delamare et al. [5] suggested a paradigm that operates in two stages: preparation and detection. The credit card holder's shopping behavior is evaluated using the k-means clustering algorithm during the training process, and the sequence is assembled during the detection phase. If the present transaction fits the chain, it is considered legitimate; otherwise, it is considered fraudulent. Awoyemi et al. [6] suggested a two-stage method for detecting credit card fraud.

In the first step, using sequence alignment, a successful score is determined based on true cardholder transaction history and transaction behavioral changes. In the second point, the bad score is calculated by using the fraudulent transaction signature provided by the previous fraudulent transaction. If the difference between the good and poor scores exceeds a predetermined threshold, the transaction is illegal; otherwise, it is legal.

Zojaji et al. [7] suggested a BLAHFDS hybridization of BLASTA–SSAHA for detecting credit card fraud using a two-stage model. These are profile analyzer and deviation analyzer. Profile analyzer stage is used to compare the time and sequence of the current transaction to the transactional record, whereas the deviation analyzer performs the comparison of the deviated time-amount series to the fraud history index. It computes the cumulative variance between the profile score and the deviation score. The overall discrepancy is used to spot fraud. Xuan et al. [8] suggested GASS, a combination of two common algorithms, genetic algorithm (GA), and scatter search (SS). GASS incorporates certain SS components into the GA steps. The proposed method aims to reduce classification costs. Pozzolo et al. [9] discussed the idea for credit card fraud detection using the decision tree.

Patidar et al. [10] proposed a paradigm based on transaction aggregation. They aggregated the transaction and developed the customer's buying behavior in this model. These behaviors are used to detect fraudulent credit card transactions. Quah et al. [11] extended the purchase aggregation approach to observe customers' periodic purchasing actions. They improved fraud prevention by using feature processing and cost awareness. Kou et al. [12] used an interaction rule to create natural behavior patterns from a false transactional database. These trends are used to spot fraud. For fraud prevention, Carcillo et al. [13] used the self-organizing map (SOM). SOM is used for previous transactional data classification and clustering, deriving secret patterns from previous data, and as a filtering tool. Singh et al. [14] suggested an innovative solution that combines network-based and inherent functions. The intrinsic function determines how the latest requested transaction differs from previous transactions in terms of the card's regency–frequency–monetary parameters (RFM). The merchant-card relationship is a network-based mechanism that produces a time-dependent suspicious score for each merchant. D. Sá [15] proposed a cost-sensitive decision tree method for identifying fraudulent credit card transactions and reducing the cost of misclassification.

As several authors [16–19] have stated, one of the most significant problems associated with credit card fraud detection is the lack of datasets from which researchers may conduct a study. The explanation for the lack of real-world data is that banks and financial institutions are unable to disclose confidential consumer activity data for privacy purposes. Credit card fraud databases contain highly distorted results, with far more legitimate transactions than fraudulent transactions, and the lawful and fraudulent transactions differ by at least a hundred times. In practice, 98% of transactions are legitimate, while just 2% are fraudulent. According to Ogwueleka et al. [20], millions of credit card transactions are processed every day. Analyzing such large numbers of transactions necessitates highly qualified methods that scale well, as well as a considerable amount of computational power. It places some restrictions on the researchers. According to detailed research done by Clifton Phua and his colleagues, approaches used in this field include data mining applications, automated fraud detection, and adversarial detection. Suman, Research Scholar, GJUS&T at Hisar HCE, explored approaches such as supervised and unsupervised learning for credit card fraud detection in another article. Despite surprising success in some areas, these approaches and algorithms were unable to provide a long-term and consistent answer to fraud detection.

Unusual methods, such as hybrid data mining/complex network classification algorithms, can detect illegal occurrences in real-world card transaction data. Based on network reconstruction methods, these strategies enable the development of representations of a single instance's departure from a reference group. Multiple supervised and semi-supervised machine learning techniques [21, 22] are used for fraud detection, but our goal with card fraud datasets is to overcome three major challenges: strong class imbalance, the inclusion of labeled and unlabeled samples, and the ability to process a large number of transactions. To detect fraudulent transactions in real-time datasets, many supervised machine learning techniques, including decision trees, naive Bayes classification, least squares regression, logistic regression, and SVM, are utilized. To understand the behavioral characteristics of regular and anomalous transactions, two random forests methods are utilized. On severely skewed credit card fraud data, the performance of logistic regression, K-nearest neighbor, and naive Bayes is explored. It is also being studied how to use meta-classifiers and meta-learning methods to cope with severely unbalanced credit card fraud data. Fraudsters with complex behavior alter their behavior over time in order to avoid detection by new detection systems and adapt fraud types. As a result, fraud is becoming more difficult and advanced, to the point that human experts are unable to forecast it.

Detecting fraudulent transactions from credit card transactions, the major limitation is the lack of real-world data. As every credit card transaction contains user credentials like user account information, amount of many, etc., so that no organizations are willing to expose their user's data. So for developing new machine learning or deep learning model, it seems not to able to provide real-world transactions data. Besides this, in most cases, a fraud detection model is developed on the baseline model, baseline models are not tuned properly for the best suitable parameters. Some works proposed models on the synthesis dataset, which is highly mismatched with the real-world dataset.

The primary goal of this work is to increase the identification accuracy of credit card fraud detection. The primary contributions of this research are:

- Preprocessing the imbalanced dataset.
- Compare between different machine learning models and evaluate the best models based on accuracy.
- Set up random hyperparameter models with RandomizedSearchCV.
- Evaluate the best params of the tuned model and score the hyperparameter tuned model with the best params.

3 System Architecture and Design

Credit card misuse is the illegal use of a credit card number without the permission of the cardholder. Credit cards can be used both physically and online. Anomaly detection strategies for credit card fraud are categorized as regulated or unsupervised.

The use of supervised methods has several drawbacks. If a suspicious transaction occurs and is not conformed to the database, these transactions are considered normal, while anomaly events are found by new transactions and happened reports of unsupervised approaches. There are 492 illegal transactions in the sample of 284,807 transactions.

Since the amount of fraudulent transactions is very limited, once machine learning models are trained using data, models become overfit. In our research work, we have minimized overfitting and increased the model's accuracy based on hyperparameter tuning the best parameters.

The basic steps of the suggested protocol for the credit card fraud identification process are depicted in Figs. 1 and 2. Figure 1 shows data preprocessing such as evaluating the data and correcting missing data after loading the dataset from CSV format to data frame. Following data preprocessing, the entire dataset is divided into two

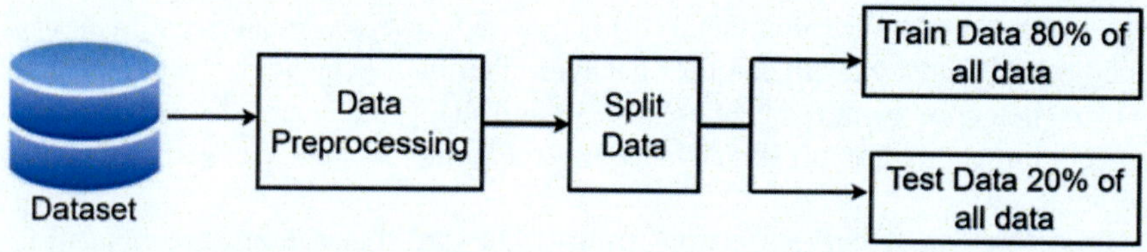

Fig. 1 Splitting of dataset

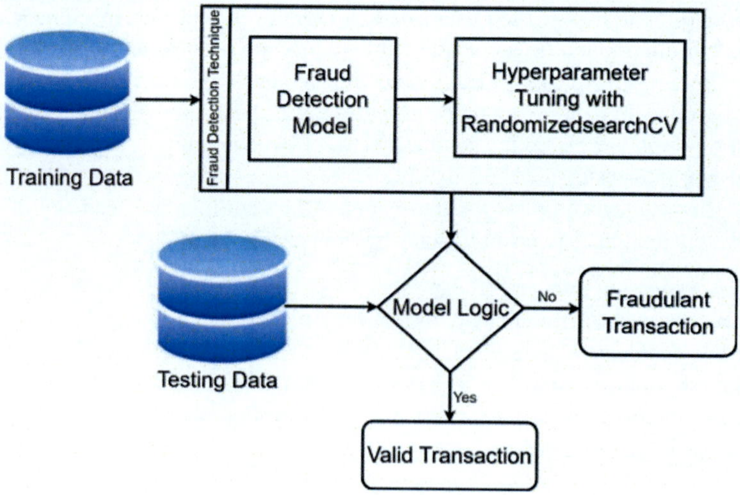

Fig. 2 Steps of the proposed framework

collections, with 80% of the data processed as training data and the other 20% stored as testing data. Figure 2 shows how, after training the machine learning algorithm with training data, the dependent model improves by hyperparameter tuning with RandomizedSearchCV, and model logic is built up. After refining the simple model with the right parameters, the model is checked by analyzing data and determining whether the transaction is true or fraudulent. If the model rationale is yes, it implies that the transaction is valid; otherwise, it indicates that the transaction is fraudulent.

3.1 Dataset Description

We collected the dataset used in this paper from Kaggle [23]. The dataset contains the transactions of credit card holders of Europe in September 2013. There are 492 fraud transactions out of 284,807 transactions under consideration.

From Table 1, we can find that the accepted values are all numerical values. Unfortunately, due to the data confidentiality, we cannot include any more background information. The characteristics are V1, V2, …, V28 which is the major component created using PCA; the only characteristics not transformed using PCA are "Time" and "mount." The "Time" feature is used to record elapsed time in second between each transaction and the first transaction in the dataset. The feature "Amount" represents the transaction amount. This information is helpful for example-dependent cost-sensitive learning.

A credit card transaction is any amount paid to a merchant by a consumer at a given time. As a consequence, the following are the six key features that summarize a transaction: the transaction ID, transaction date and time, the customer ID, the terminal ID, the transaction amount, and the fraud level. The "fraud level" is a binary variable having either of the two outcomes: 0 for a legitimate transaction and 1 for a fraudulent transaction.

Table 1 Dataset example

	Time	Amount	Class
Count	284,807.00	284,807	284,807
mean	94,813.86	88.35	0.0017
Std	47,488.15	250.12	0.0415
min	0.000000	0.0000	0.0000
25%	54,201.50	5.6000	0.0000
50%	84,692.00	22.000	0.0000
75%	139,320.0	77.165	0.0000
max	172,792.0	25,691.2	1.0000

3.2 Data Preprocessing

The credit card fraud identification dataset was collected from Kaggle in CSV format. The CSV dataset is first loaded into the module, which tests the entire dataset as a data frame. After reviewing the entire dataset, searching for missing data, and restoring any missing data, the data can be divided into training and analysis data. After the initial data preprocessing, split the whole data into input variables X and output variables y. And using the train test split method, data of input variable and output variable, X and y, is split into train and test data, where test size is assumed as 20%. The credit card fraud monitoring system considers 80% of all data to be training data, while the other 20% is believed to be study data.

3.3 Model Training and Testing

We considered six separate machine learning models in our research work: logistic regression, K-neighbors classifier, random forest classifier, linearSVC, GaussianNB, and decision tree classifier. The fit and score system is used to train the models. As model parameters, models, train data, and test data are passed. The NumPy random seed value of 42 is used to set the random shown in NumPy. Get the name and model of each model by looping through the model objects. The input training data and output training data are used to match models. Models are scored on the research results after they have been developed using the training data. Inside the suit and score process, testing data of input variables and output variables are used to score the model's results. The model scores for each model are saved in a list, and the model scores list is eventually returned by the suit and score process.

3.4 Model Hyperparameter Tuning with RandomizedSerachCV

Hyperparameter tuning with RansomizedSearchCV is used for tuning to boost the baseline models. Build a hyperparameter grid for each model for hyperparameter tuning with RandomizedSerachCV. A hyperparameter grid is the values of several estimators, such as max-width, min samples break, min sample leaf, and so on. After building the baseline model's hyperparameter grids, the models are ready for tuning with RandomizedSerachCV. NumPy random seed is initially set to 42. Randomized-SearchCV employs the RandomizedSearchCV approach from the sklearn model collection. The RandomizedSearchCV system parameters passed are models, parameter distributions as hyperparameter grid, cv, number of items, and verbose. This creates a setup for random hyperparameter quest for models. For models with input and output training results, fit the random hyperparameter search model. The best params are evaluated from the model's grid system using the best params attribute after fitting the random hyperparameter check for models. The best params attributes indicate the best-valued parameters for the random hyperparameter search model. Rate the hyperparameter tuning algorithm by evaluating the RandomsearchCV model score using input and output testing data and the RandomsearchCV grid process.

4 Implementation and Experimental Result

4.1 Experimental Setup

The proposed system has been implemented on a machine having Windows 10, Core i7 2.4 GHz with 8GB RAM. Python is used for developing it.

4.2 Results and Discussion

NumPy, Pandas, Matplotlib, and Seaborn are used for routine exploratory data processing and plotting in this research work. NumPy is the foundational Python package for scientific computation. Pandas is an open-source data analysis and manipulation framework that is quick, efficient, scalable, and simple to use. It is designed on top of the Python programming language. Matplotlib is a Python library that allows you to create static, animated, and immersive visualizations. Seaborn is a Matplotlib-based Python data visualization library. Scikit-learn models such as logistic regression, K-neighbors classifier, random forest classifier, linearSVC, GaussianNB, and decision tree classifier are used in this research work. Logistic regression is a data processing technique for describing and explaining the relationship between one dependent binary variable and one or more independent nominal, ordinal, interval, or ratio-level variables.

K-neighbors classifier implements classification by voting by the target point's closest k-neighbors, while radius neighbors classifier implements classification by voting by all neighborhood points within a set radius, r, of the target point. Random forest is a versatile, user-friendly machine learning algorithm that delivers excellent results much of the time even without hyperparameter tuning. Because of its simplicity and variety, it is perhaps one of the most widely used algorithms (it can be used for both classification and regression tasks). A linearSVC (support vector classifier) goal's is to match the data you have by returning a "best fit" hyperplane that separates or categorizes the data.

After obtaining the hyperplane, you can then feed some features to your classifier to determine the "predicted" class. Implementation of Gaussian–naive Bayes We built a GaussianNB classifier. To make it simpler to grasp, the decision tree acquires information in the form of a tree, which can also be rewritten as a series of distinct laws.

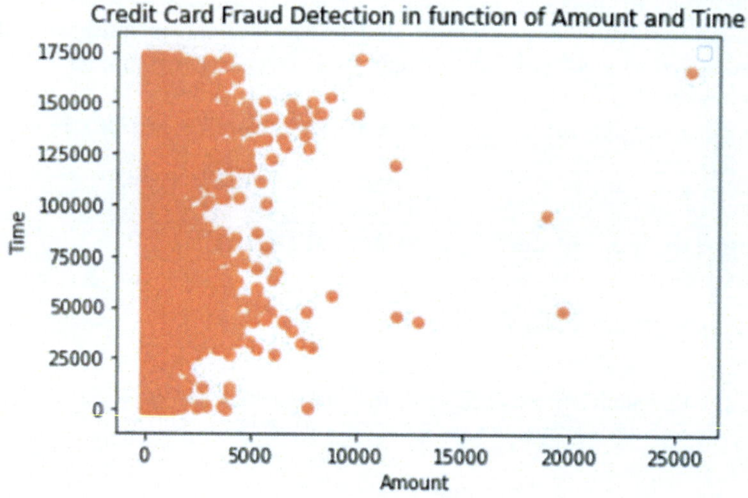

Fig. 3 Credit card fraud detection in function of amount and time

Model assessment is a critical phase in the model development process. For model evaluation in this research work, the train–test break process, cross-val score method, RandomizedSearchCV, confusion matrix, classification report, precision score, recall score, F1-score, and plot ROC curve are used. The very first step in this research work is to load the dataset into the data frame. Using the Pandas library, a dataset in CSV format is loaded as a data frame, and the dataset volume is calculated to be 284,807. And there are 31 attributes in the dataset. There are two kinds of data in the dataset. Class 1 denotes legitimate data, while Class 0 denotes fraudulent data. There are 284,315 fraud data and 492 valid data from the 284,807 data. Figure 3 describes the function between the amount and time feature of the dataset. In dataset, two important features of the user transactions are time and amount, and Fig. 3 shows that transactions between the 0 and 5000 amount happens most cases and the density of the data in between this range both for the amount and the time feature is very high.

Correlation matrix Fig. 4 is nothing more than a table that shows the correlation coefficients for various variables. The matrix illustrates the relationship between all potential pairs of values in a table. It is a valuable method for summarizing a broad dataset as well as identifying and visualizing trends in the data. Correlation matrix is made up of rows and columns that represent the variables. The correlation coefficient is contained in each cell of a table. Furthermore, the correlation matrix is used in conjunction with other mathematical research approaches.

After splitting data into input variables X and output variables y. Output variables y consist of class attributes, and other attributes are consist of input variables X. Set NumPy random seed as 42. Split the whole dataset of input variables X and output variables y into X train, X test, y train, and y test sets, where test size is assumed as 20% of all data.

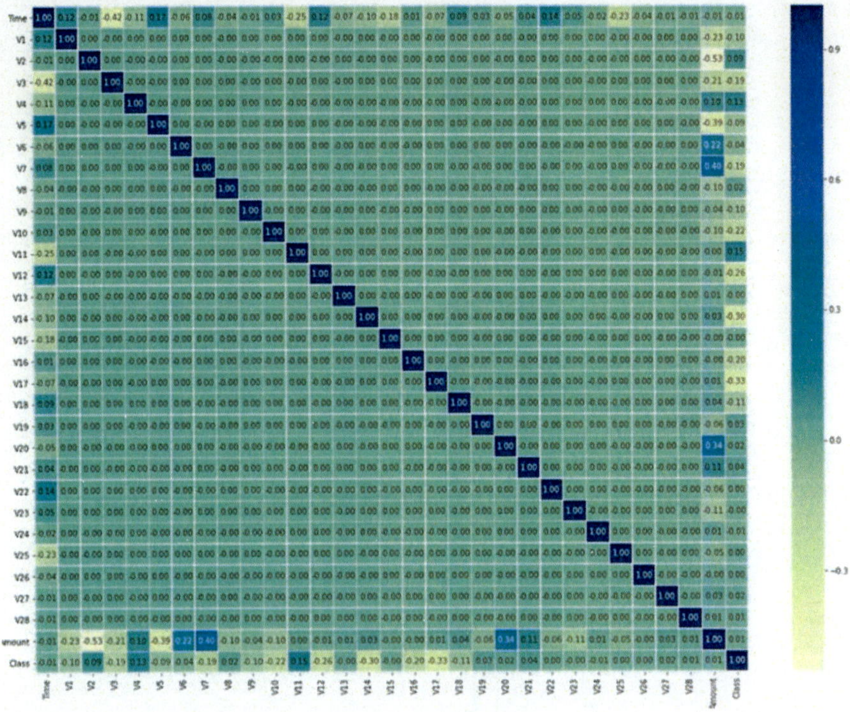

Fig. 4 Correlation matrix

To measure the performance of the models, a fit and score approach is built in this research work, with parameters including a list of models, X train data, X test data, y train data, and y test data. Set the NumPy random seed to 42 in the fit and score process, and model scores are saved in a list. Figure 5 shows the performance between the six baseline machine learning models, they are logistic regression, KNN, random forest, linearSVC, Gaussian–Naive Bayes, and decision tree classifier [24]. Performance between these models are evaluated on the accuracy based on test case dataset. Among the six baseline models, for our case on the dataset, we have considered, random forest classifier model perform the best accuracy, and we have selected this model for our hyperparameter RandomSearchCV model tuning. To address this problem, we look at hyperparameter tuning with RandomizedSearchCV. Random forest classifier outperforms all baseline models in terms of precision, and for this purpose, hyperparameter tuning with RandomizedSearchCV is considered for random forest classifier. For hyperparameter tuning with RandomizedSearchCV, create a hyperparameter grid for random forest classifier. For tuning the random forest classifier, set the NumPy random seed as 42. Setup hyperparameter search for random forest classifier using RandomizedSearchCV with the parameter of random forest classifier method, param distribution as random forest grid, cv as 5, number of items is 20, and verbose as True.

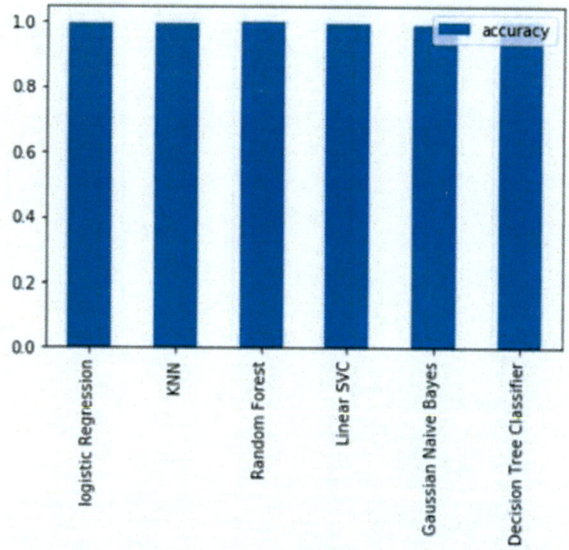

Fig. 5 Model comparisons

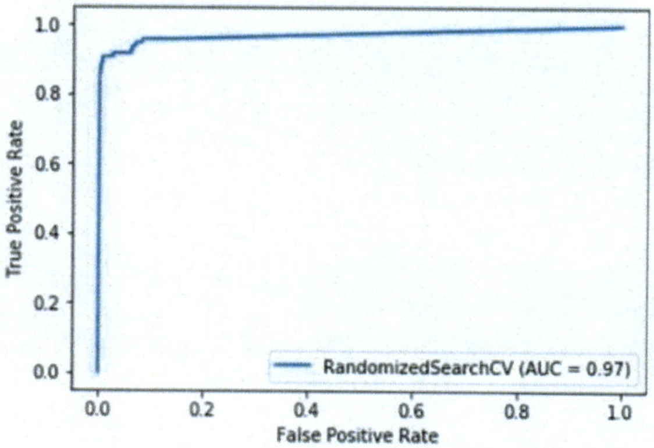

Fig. 6 ROC curve

After setting up the random hyperparameter search model for random forest classifier, fit the model with the training data. After fitting the model, the best params of the fit model of random hyperparameter search model for random forest classifier is evaluated and the evaluated value for fit model is, number of estimators is 510, min samples split is 14, min samples leaf is 1, and max depth is none. Random forest classifier hyperparameter tuned model is evaluated on ROC curve, recall score, and precision score. ROC curve Fig. 6 provides the relation between the true positive rate vs false positive rate of the hyperparameter tuned model, which seems that the AUC value of the RandomizedSearchCV is almost 0.97. Confusion matrix Table 2 provides the evaluation of RandomizedSearchCV hyperparameter tuned random forest classifier model, where the number of true negative is the lowest, also false positive and false negative values are very low, and most of the prediction cases are true

Table 2 Confusion matrix

0	5.7e+04	3
1	24	74

Table 3 Classification report

	0	1	Accuracy
Accuracy	1.00	0.96	–
Recall	1.00	0.76	–
F1-score	1.00	0.65	1.00
Support	56864	98	56962

positive. Classification report Table 3 of the RandomizedSearchCV hyperparameter random forest classifier model evaluates the tuned model, where precision value for the valid cases is 1.00 and the false cases are 0.96; recall value for the valid cases is 1.00 and for the false cases is 0.76; F1-score for the valid cases is 1.00 and the false cases are 0.85. The evaluation value for the RandomizedSearchCV hyperparameter tuned model performance is very high which indicates the performance on the highly imbalanced dataset which is better than existing works or baseline models.

5 Conclusion

Divide the total data into train data and test data, with a test scale of 20. Sckikit learn models are educated using 80% of all results. In our study, we looked at six different model training algorithms. Logistic regression, K-neighbors classifier, random forest classifier, linearSVC, GaussianNB, and decision tree classifier are among them. After training all models, rate them using the testing results, which represent the remaining 20% of all data. Based on the 284,807 transactions, baseline models estimate the score. Random forest classifier does comparatively well than the other baseline models. So, in our project work, we tuned our random forest classifier model, and for this, we used RandomizedSearchCV to set up a random hyperparameter tuning model. We trained our new hyperparameter tuning model for random forest classifier once more, and this time, we evaluated the model's score which was higher than before.

In this research work, a hybrid feature extraction approach that blends various features is used to detect credit card fraud. A dataset that is used is highly imbalanced, so it is a tough task to work this highly imbalanced dataset model building a highly imbalanced dataset is being processed and extract the features according to the model evaluation. Different machine learning baseline models are evaluated and compared in this work. The best proficient model is selected for hyperparameter RandomizedSearchCV tuning. The baseline model tuned and evaluated the best params for the high performance to detect credit card fraud detection. This research work does not make use of any of the user's material. Collecting more and more characteristics from consumer purchases would have greater precision in detecting fraud. Aside from that, the amount of fraud data in the Kaggle dataset is very limited, which is a major issue for model testing. Models are constructed using a vast volume of legitimate data, so models do well for this particular forecast, but increasing the number of fraud data would yield more reliable results. By resolving these constraints, a solid architecture for detecting credit card fraud can be developed.

References

1. Raj SBE, Portia AA (2011) Analysis on credit card fraud detection methods. In: 2011 international conference on computer, communication and electrical technology (ICCCET). IEEE, pp 152–156
2. Shen A, Tong R, Deng Y (2007) Application of classification models on credit card fraud detection. In: 2007 International conference on service systems and service management. IEEE, pp 1–4
3. Chaudhary K, Yadav J, Mallick B (2012) A review of fraud detection techniques: credit card. Int J Comput Appl 45(1):39–44
4. Bahnsen AC, Aouada D, Stojanovic A, Ottersten B (2016) Feature engineering strategies for credit card fraud detection. Expert Syst Appl 51:134–142
5. Delamaire L, Abdou H, Pointon J (2009) Credit card fraud and detection techniques: a review. Banks Bank Syst 4(2):57–68
6. Awoyemi JO, Adetunmbi AO, Oluwadare SA (2017) Credit card fraud detection using machine learning techniques: a comparative analysis. In: 2017 International conference on computing networking and informatics (ICCNI). IEEE, pp 1–9
7. Zojaji Z, Atani RE, Monadjemi AH et al (2016) A survey of credit card fraud detection techniques: data and technique oriented perspective. arXiv preprint arXiv:1611.06439
8. Xuan S, Liu G, Li Z, Zheng L, Wang S, Jiang C (2018) Random forest for credit card fraud detection. In: 2018 IEEE 15th international conference on networking, sensing and control (ICNSC). IEEE, pp 1–6
9. Dal Pozzolo A, Boracchi G, Caelen O, Alippi C, Bontempi G (2017) Credit card fraud detection: a realistic modeling and a novel learning strategy. IEEE Trans Neural Netwo Learn Syst 29(8):3784–3797
10. Patidar R, Sharma L et al (2011) Credit card fraud detection using neural network. Int J Soft Comput Eng (IJSCE) 1(32–38)
11. Quah JT, Sriganesh M (2008) Real-time credit card fraud detection using computational intelligence. Expert Syst Appl 35(4):1721–1732
12. Kou Y, Lu CT, Sirwongwattana S, Huang YP (2004) Survey of fraud detection techniques. In: IEEE international conference on networking, sensing and control, 2004, vol 2. IEEE, , pp 749–754
13. Carcillo F, Le Borgne YA, Caelen O, Kessaci Y, Oblé F, Bontempi G (2019) Combining unsupervised and supervised learning in credit card fraud detection. Inf Sci
14. Singh A, Jain A (2019) Adaptive credit card fraud detection techniques based on feature selection method. In: Advances in computer communication and computational sciences. Springer, Berlin, pp 167–178
15. de Sá AG, Pereira AC, Pappa GL (2018) A customized classification algorithm for credit card fraud detection. Eng Appl Artif Intell 72:21–29

16. Jain Y, NamrataTiwari S, Jain S (2019) A comparative analysis of various credit card fraud detection techniques. Int J Recent Technol Eng 7(5):402–407
17. Varmedja D, Karanovic M, Sladojevic S, Arsenovic M, Anderla A (2019) Credit card fraud detection-machine learning methods. In: 2019 18th international symposium INFOTEH-JAHORINA (INFOTEH). IEEE, pp 1–5
18. Save P, Tiwarekar P, Jain KN, Mahyavanshi N (2017) A novel idea for credit card fraud detection using decision tree. Int J Comput Appl 161(13)
19. Niu X, Wang L, Yang X (2019) A comparison study of credit card fraud detection: supervised versus unsupervised. arXiv preprint arXiv:1904.10604
20. Ogwueleka FN (2011) Data mining application in credit card fraud detection system. J Eng Sci Technol 6(3):311–322
21. Maniraj S, Saini A, Ahmed S, Sarkar S (2019) Credit card fraud detection using machine learning and data science. Int J Eng Res 8(09)
22. Mahmud M, Kaiser MS, McGinnity TM, Hussain A (2021) Deep learning in mining biological data. Cognit Comput 13(1):1–33 Jan
23. Credit card fraud detection | kaggle (June 2013), https://www.kaggle.com/mlg-ulb/creditcardfraud
24. Mahmud M, Kaiser MS, McGinnity TM, Hussain A (2021) Deep learning in mining biological data. Cognit Comput 13(1):1–33

Automatic Malware Categorization Based on K-Means Clustering Technique

Nazifa Mosharrat, Iqbal H. Sarker, Md Musfique Anwar, Muhammad Nazrul Islam, Paul Watters, and Mohammad Hammoudeh

Abstract The android operating system is a popular operating system for mobile phone applications. This is also known as an open-source operating system so that the developers can easily update and add new features to it. However, it poses significant challenges related to malicious attacks or cyberattacks because of its open system design philosophy. Nowadays, the number of malware applications is increasing rapidly and proportionally with safe android applications. As a result, it has become very challenging to identify their behaviors or signatures or categorizes them to implement protection in the android system. In this research work, we propose an automated system for malware categorization using the K-means clustering method that automatically chooses the cluster number. In our method, we have categorized malware into an optimum number of different cluster families by using a real-time malware dataset. We also compare our automated model with the traditional clus-

---

N. Mosharrat
Department of Computer Science and Engineering, East Delta University, Chittagong 4209, Bangladesh

I. H. Sarker (✉)
Department of Computer Science and Engineering, Chittagong University of Engineering & Technology, Chittagong 4349, Bangladesh
e-mail: iqbal@cuet.ac.bd

M. M. Anwar
Department of Computer Science and Engineering, Jahangirnagar University, Dhaka, Bangladesh

M. N. Islam
Department of Computer Science and Engineering, Military Institute of Science and Technology, Dhaka 1216, Bangladesh

P. Watters
Department of Security Studies and Criminology, Macquarie University, Sydney, Australia

M. Hammoudeh
Department of Computing & Mathematics, Manchester Metropolitan University, Manchester M1 5GD, UK

M. S. Arefin et al. (eds.), *Proceedings of the International Conference on Big Data, IoT, and Machine Learning*, Lecture Notes on Data Engineering and Communications Technologies 95, https://doi.org/10.1007/978-981-16-6636-0_49

ter selection technique with Elbow and Silhouette method. Experimental results demonstrate that our model determines the optimal cluster number with less user intervention for malware categorization.

Keywords Cybersecurity · Machine learning · Clustering · K-means · Malware detection · Malware categorization · Android applications

1 Introduction

Malware comes from malicious software which is a harmful program that hackers use to steal sensitive information or destroy it. Malware has many forms, such as Rootkits, Viruses, Trojans, Worms, Spyware, and so on [1, 2]. Android applications are rapidly growing three times greater than other mobile applications in the phone market [3]. According to a recent report in the US, the number of recent data breaches is 1, 001, and more than 155.8 million individuals were affected by data exposures in the year 2020 [4].

Sensitive information can be damaged or stolen or misused by a simple malware attack [5, 6]. In consequence, protection of the android operating system from being attacked by malicious software should be one of the preliminary tasks for the application developers and phone manufacturers. The malware detection technique is categorized in *static* and *dynamic* analysis that was introduced in the region of the mid-'90s [7, 8]. The *static* method works on the application's source code to categorize them without having the application being run. It can only identify the existing malware and fails against the unseen variants of malware [9]. This static analysis includes some approaches such as signature-based approach, permission-based analysis, etc. [2]. On the other hand, dynamic analysis works, while the program is running. But unfortunately, most of these techniques can not identify new types of malware [10]. Shijo et al. [11] proposed a method of an integrated approach that uses both static and dynamic methods which evaluates that the integrated method has better accuracy than the individual method. However, this also sometimes fails to detect new types of malware.

Malicious files have increased in recent times and the pattern of malware has been changing day by day. In 2020, 36,000 new malicious files were detected per day by Kaspersky lab [12]. Therefore, this new type of malware needs a reliable and new technique for detecting them and classifying them for protection. Instead of using traditional signatures-based malware detection, an alternative technique such as a characteristic-based method is needed to detect malware by observing the statistics and categorize them into different batches.

In recent years, data science and machine learning have exemplified outstanding capabilities in many real-world application areas including the cybersecurity field, summarized briefly in Sarker et al. [13, 14]. Machine learning-based models can be trained to find infected applications and also categorize them according to their behavior into the malware family. Several popular machine learning-based algorithms such as Naive Bayes (NB), Support Vector Machine (SVM), Logistic Regression (LR), K-means, Linear Discriminant Analysis, Random Forest, Stochastic Gradient Descent, Linear regression, Polynomial Regression, etc., are used in the area [9, 15, 16].

In this paper, we propose an automated system for malware categorization using machine learning methods. The main advantage of our model is that it does not need any static analysis or manual feature selection for categorizing. It will extract the most relevant and important features using Random Forest and Decision Tree [13] and eliminate the non-important features and finally, categorize the malware using the K-means algorithm.

The main contributions of our work are as follows:

- Our proposed method dynamically finds the optimum cluster number and feeds that cluster number into the K-means clustering model.
- We have categorized the malware family automatically into different batches that have almost similar characteristics based on their behaviors.
- We have compared our model with different existing systems using a real-time malware dataset.

The remaining sections are distributed as follows. In Sect. 2 we review the related work and associated methods. In Sect. 3, we explain the step-by-step procedure. Section 4 explicates the result and also compares them with the existing methods. Section 5 summarizes this paper and highlights the proposed work.

2 Related Work

The use of K-Means for malware detection or categorization is not new; however, the accuracy of the result and less user interaction during detection or categorization has been the goal for most researchers. The results pave the way for further explorations of novel variants of K-Means. Chumachenko et al. [17] have studied malware detection and classification using machine-learning-based methods where they work with 1156 malware files of 9 families of different types with additionally 984 benign files. The proposed model is evaluated with the different classifiers in which Random Forest achieved the highest accuracy of 95.69%.

Shhadat et al. [15] analyzed machine learning techniques for behavior-based malware detection. They also compared the performance of their proposed approach with K-Nearest Neighbor (KNN), Naïve Bayes, SVM, Decision Tree (DT), MLP, and they state that the best performance was achieved by using Decision Tree with a true positive rate and a false-positive rate of 95.9% and 2.4%, respectively, along with the positive predictive value and accuracy of 97.3% and 96.8%, respectively. The authors in [18] compared the performance of K-means and mini-batch K-means clustering where 800 samples are being used out of 1260 and then evaluated the accuracy. In [19], the authors defined the points of K for centroids that have been chosen randomly and takes each data point, and enumerate its distance from the centroid.

However, all the above technical need static analysis and are also time-consuming and it may fall in many aspects so this static system needs some modification to work with less delay and also less involvement. In this work, a method has been introduced to choose the cluster number which will find out the inertia for each feature, and then, it will compare them with the weighted mean of the total inertia. Finally, the model will choose the closest inertia from the average inertia to choose the optimum cluster. This system is a simple generalized solution that will work for any type of dataset. Additionally, an automatic feature extraction method has been formulated with RF (Random Forest) and DT (Decision Tree) algorithms for a better solution.

3 Methodology

In this work, we mainly aim to develop an automatic malware categorization model with less user involvement. It will extract important features from the given dataset and all irrelevant features will be removed in order to feed the dataset with the most relevant and important feature into the classifier model. In our approach, we use the K-means clustering technique to categorize malware. Additionally, it will choose the cluster number and also feed that cluster number into the K-means model without any external data given or user involvement [20]. All the steps of the process, followed by the model have been summarized in Algorithm 1.

3.1 Data Input

In this work, a malware dataset, collected by Canadian Institute for Cybersecurity [21], consists of 2883 values and 84 features have been used which come from 42 unique malware families. This dataset has been created from 10,854 samples from several resources. The categorical values contained in this dataset are shown in Figs. 1 and 2.

Algorithm 1: K-means Clustering based model for Malware Categorization

Input: DS, a dataset containing n malware instance
Output: Optimum number of clusters in different malware family's
Data: Initial dataset and fill all null field with 0

```
                                                // dtype =data types
1  if DS.dtype ≠ numeric then
2  |  Convert data into numeric
3  Feed DS into the RF model and check importance.     // RF = Random Forest
4  if Feature importance < minimum threshold then
5  |  eliminates features.
6  Feed the DS with important features into the DT model      // DT = Decision Tree
7  repeat
8  |  4
9  until 5
10 for each DS feature do
11 |  calculate the inertia.
12 calculate the weighted mean.
13 for for each inertia do
14 |  if the inertia ≈ weighted mean then
15 |  |  cluster ← cluster number for closest inertia
16 feed the updated cluster number into the k means model. Plot the k model with
   categorized malware
```

Figure 1 illustrates the outline of the whole process of our proposed method.

3.2 *Data Preprocessing*

The dataset that is used in this work has a combination of different types of columns including objects. Hence, some preprocessing is needed to ensure that all the value of the dataset is numerical for further manipulation. In this regard, we have created a structure that will find out the non-numerical columns and then, dynamically encode them via the hot encoding technique without the user's involvement.

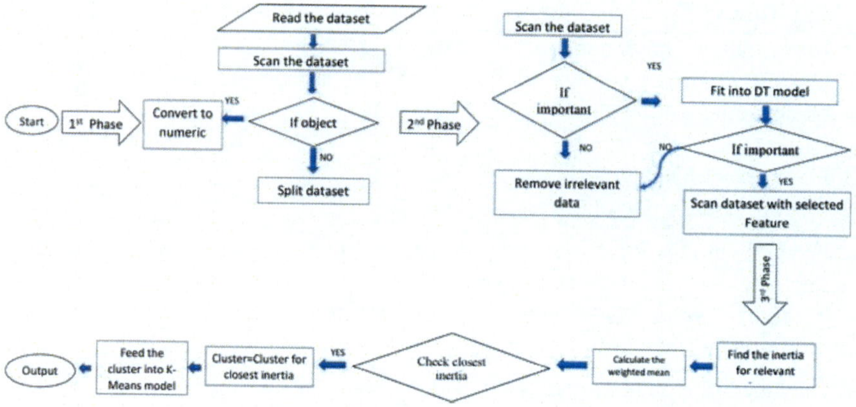

Fig. 1 Structural outline for the proposed method

	Flow ID	Source IP	Source Port	Destination IP	Destination Port	Protocol	Timestamp	Flow Duration	Total Fwd Packets	Total Backward Packets	Total Length of Fwd Packets	Total Length of Bwd Packets	Fwd Packet Length Max	Fwd Packet Length Min	Fwd Packet Length Mean
0	157.240.2.20-10.42.0.211-443-46559-6	157.240.2.20	443	10.42.0.211	46559	6	11/07/2017 10:47:35	57	2	0	42.0	0.0	42.0	0.0	21.0
1	157.240.2.20-10.42.0.211-443-46560-6	157.240.2.20	443	10.42.0.211	46560	6	11/07/2017 10:47:40	46	2	0	42.0	0.0	42.0	0.0	21.0
2	157.240.2.20-10.42.0.211-443-46560-6	157.240.2.20	443	10.42.0.211	46560	6	11/07/2017 10:47:40	108842	1	4	0.0	42.0	0.0	0.0	0.0

Fig. 2 Sample data from malware dataset

3.3 *Feature Selection*

Feature selection is an important step to eliminate the irrelevant data from a big dataset because a large dataset is sometimes difficult to handle which may result in poor efficiency. In this work, we propose a strategy that adapts the dynamic process which runs the malicious file by executing a loop through the dataset to gather the importance of its characteristics and filter the less important features automatically. As a result, the model is able to keep only the relevant and important features, and hence, the accuracy and efficiency of the model improve.

There are a lot of techniques for feature selection such as chi-square, Baruta, Decision Tree, Random Forest [13]. But the chi-squared method is not suitable for those datasets with negative values. Baruta is also an automatic feature elimination system which is also nothing but an iterative RF technique. This technique is iterating the dataset and applying RF multiple times. So this technique is costly and time-consuming which is not suitable for large datasets.

In this work, we first apply a Random Forest for selecting the first 80% of its important features and then apply the Decision tree to get more optimum results for feature selection. We have created a structure that will loop through the dataset and will calculate the importance and will reject the less important feature without any user involvement. This way this model selects 35 features from 84 features of the actual dataset.

3.4 Categorization Model Creation

We apply the K-means clustering method [13] for malware categorization. Clustering is the process of separating a given set of patterns into unique clusters, where patterns in the same cluster have much more similarity than the patterns belonging to two different clusters. The efficiency of the K-means technique mainly depends on its cluster number [22].

In the K-means algorithm, the minimum cluster number is one, and the maximum cluster number is the total features number. That means every feature belongs to a unique cluster. But it's not an optimum model. For this reason, there is an algorithm *elbow* method has been introduced. The elbow method is used for determining the number of clusters in a dataset by plotting the variation of the clusters on the graph. In the elbow method, the range of centroid needs to be fixed in order to perform static analysis on the elbow graph.

Our proposed model uses an iteration process as shown in Fig. 1 that will choose the optimum cluster number and categorize malware based on the cluster number. This iteration process will calculate the inertia for all the relevant features and store those values as a list to find the weighted mean of the inertia list. It also finds the closest inertia from the weighted mean as shown in Fig. 4. The cluster number for the closest inertia is the optimum inertia as the change of the value of inertia later is ignorable as shown in Fig. 3.

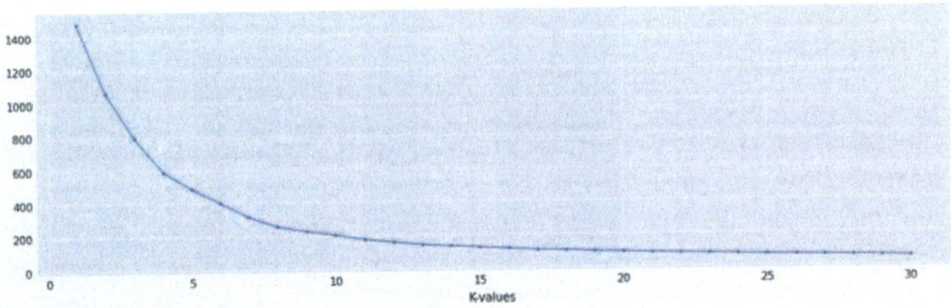

Fig. 3 Elbow plot for inertia

```
Average Inertia=  665.3506853013679
Closest Inertia =  590.4768527806027
Optimal  Cluster =  4
```

Fig. 4 Optimal cluster selection

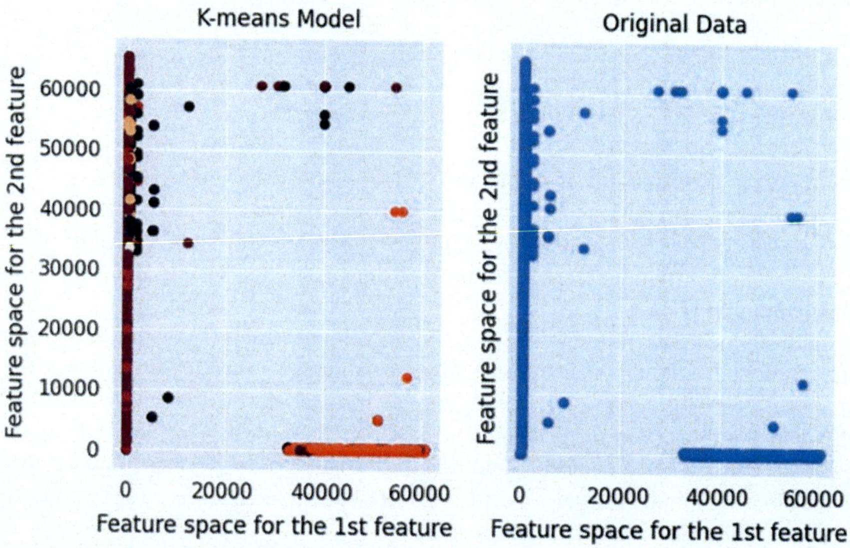

Fig. 5 Comparing cluster distribution

Next, the system will automatically feed the cluster number into the K-means model and plot the cluster distribution. The difference between original data distribution with our model is depicted in Fig. 5 in which our proposed model categorizes malware into different clusters automatically.

4 Result Analysis and Evaluation

In the above sections, we have discussed our methodology for finding the optimum cluster. We have experimented with our model with different numbers of clusters in the traditional method for better comparison. We use *Silhouette* analysis to study the distance between the resulting clusters. The silhouette plot displays a measure of distance between clusters which has a range of $[-1, 1]$. Silhouette coefficients near $+1$ indicate that the distance between the sample and the neighboring clusters is very big, while the value 0 (zero) indicates that the sample is near to the decision boundary [23].

We plot the dataset for different numbers (2, 3, 4, 5) of clusters shown in Figs. 6, 7, 8 and 9. We see that the silhouette model combines three subclusters into one large

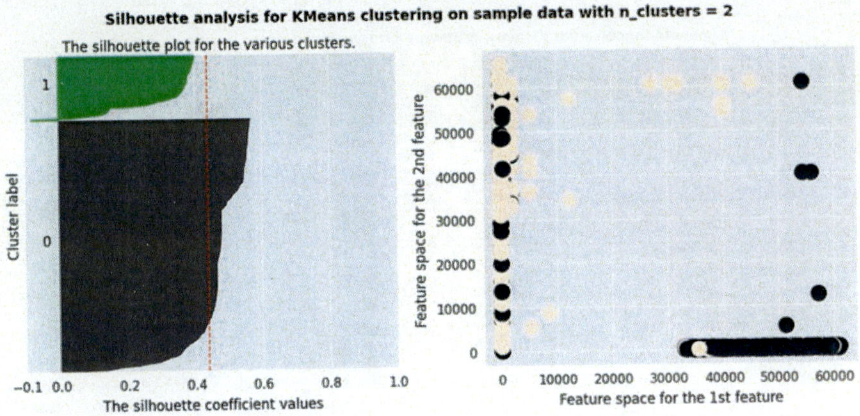

Fig. 6 Comparing cluster distribution for cluster number 2

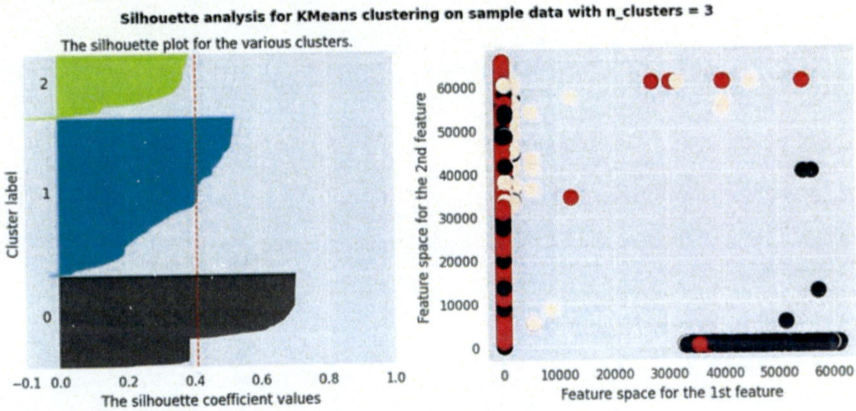

Fig. 7 Comparing cluster distribution for cluster number 3

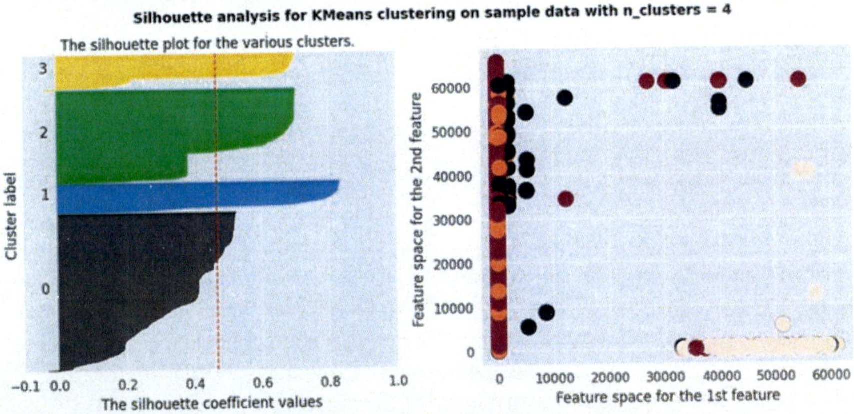

Fig. 8 Comparing cluster distribution for cluster number 4

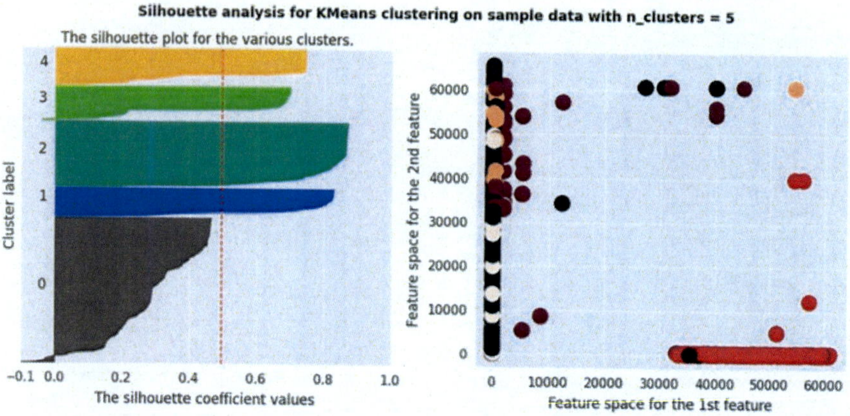

Fig. 9 Comparing cluster distribution for cluster number 5

Table 1 Cluster distribution and Silhouette scores for cluster number 2

Distribution of clusters		Silhouette scores
C-0	C-1	
2289	594	0.440

Table 2 Cluster distribution and Silhouette scores for cluster number 3

Distribution of clusters			Silhouette scores
C-0	C-1	C-2	
887	1415	581	0.411

Table 3 Cluster distribution and Silhouette scores for cluster number 4

Distribution of clusters				Silhouette scores
C-0	C-1	C-2	C-3	
1415	326	878	264	0.471

cluster when the cluster number is 2 as depicted in Fig. 6. Table 1 also proves that the $C - 0$ or the 1st cluster is bigger than the 2nd cluster.

From Figs. 7 and 9, it's quite apparent that the numbers 3 and 5 are wrong picks for cluster numbers because of the presence of clusters with below mean silhouette results and also, there is a large fluctuation in the size of the plots which are also reflected in Tables 2 and 4, respectively. On the other hand, when the cluster number is equal to 4, all the plots have an almost similar thickness as displayed in Fig. 8. Table 3 also shows the similar distribution of the clusters.

So from the above analysis, we can see that cluster number 4 should be chosen as the optimum cluster number for the given dataset. Our proposed model also chooses

Table 4 Cluster distribution and Silhouette scores for cluster number 5

Distribution of clusters					Silhouette scores
C-0	C-1	C-2	C-3	C-4	
798	1419	264	185	217	0.496

number 4 automatically as the optimum cluster number as shown in Figs. 4 and 5. It is easier to decide manually which cluster is optimum or better for 2, 3 or 10 variations. On the other hand, for the larger dataset with a large variation, it becomes difficult to decide the cluster number by just visualize the datasets and also the process is very time-consuming. Our proposed model is capable to automatically decide the optimum cluster number without any static analysis and can categorize the malware accordingly.

5 Conclusion

In this paper, we have proposed a model for malware categorization using the K-means clustering method that automatically chooses the cluster number. This model performs feature extraction using RF and DT and then categorizes malware using the clustering method. At first, the model pre-processes the dataset to extract the important features and eliminate the irrelevant features. It then chooses the optimum cluster for the given dataset and feeds that number without user interactions into the K-means model to categorize malware into different clusters. Experimental results show that our model determines the optimal cluster number with less user intervention for malware categorization. This model can also be generalized for other relevant datasets in the domain of cybersecurity.

References

1. Carlin A, Hammoudeh M, Aldabbas O (2015) Intrusion detection and countermeasure of virtual cloud systems-state of the art and current challenges. Int J Adv Comput Sci Appl 6(6)
2. Sarker IH, Hasan Furhad M, Nowrozy R (2021) AI-driven cybersecurity: an overview, security intelligence modeling and research directions. SN Comput Sci 2(3):1–18
3. Lookout app genome report. https://www.mylookout.com/appgenome, 2011. https://www.statista.com/statistics/273550/, Online accessed (14/07/2021)
4. Johnson J (2021) Annual number of data breaches and exposed records in the United States from 2005 to 2020, 3 Mar 2021. https://www.statista.com/statistics/273550/, Online accessed (14/07/2021)
5. Ghafir I, Prenosil V, Hammoudeh M, Baker T, Jabbar S, Khalid S, Jaf S (2018) BotDet: a system for real time botnet command and control traffic detection. IEEE Access 6:38947–38958

6. Belguith S, Kaaniche N, Hammoudeh M, Dargahi T (2020) Proud: Verifiable privacy-preserving outsourced attribute based signcryption supporting access policy update for cloud assisted iot applications. Future Gen Comput Syst 111:899–918
7. Shabtai A, Kanonov U, Elovici Y, Glezer C, Weiss Y (2012) Andromaly: a behavioral malware detection framework for android devices. J Intell Inf Syst 38(1):161190
8. Kephart J, Arnold W (1994) Automatic extraction of computer virus signatures. In: Proceedings of 4th virus bulletin international conference, pp 178–184
9. Amro B (2017) Malware detection techniques for mobile devices. Int J Mob Netw Commun Telemat 7. https://doi.org/10.5121/ijmnct.2017.7601
10. Ghafir I, Prenosil V, Hammoudeh M, Han L, Raza U (2017) Malicious ssl certificate detection: a step towards advanced persistent threat defence. In: Proceedings of the international conference on future networks and distributed systems
11. Shijo PV, Salim A (2015) Integrated static and dynamic analysis for malware detection, Procedia Comput Sci 46:804–811. ISSN 1877-0509. https://doi.org/10.1016/j.procs.2015.02.149
12. kaspersky (2021) The number of new malicious files detected every day increases by 5.2% to 360,000 in 2020, December 15, 2020. https://www.kaspersky.com/about/press-releases/2020_the-number-of-new-malicious-files-detected-every-day-increases-by-52-to-360000-in-2020, Online accessed (14/07/2021)
13. Sarker IH (2021) Machine learning: algorithms, real-world applications and research directions. SN Comput Sci 2(3):1–21
14. Sarker IH (2021) Data science and analytics: an overview from data-driven smart computing, decision-making and applications perspective. SN Comput Sci
15. Shhadat I, Bataineh B, Hayajneh A, Al-Sharif ZA (2020) The use of machine learning techniques to advance the detection and classification of unknown malware. Procedia Comput Sci 170:917–922. ISSN 1877-0509. https://doi.org/10.1016/j.procs.2020.03.110
16. Sarker IH (2021) CyberLearning: effectiveness analysis of machine learning security modeling to detect cyber-anomalies and multi-attacks. Internet of Things 14:100393
17. Chumachenko K et al (2017) Machine learning methods for malware detection and categorization
18. Feizollah A, Anuar NB, Salleh R, Amalina F (2014) Comparative study of K-means and mini batch K-means clustering algorithms in android malware detection using network traffic analysis. In: International symposium on biometrics and security technologies (ISBAST), pp 193–197. https://doi.org/10.1109/ISBAST.2014.7013120
19. Shameem MUS, Ferdous R (2009) An efficient K-means algorithm integrated with Jaccard distance measure for document clustering. In: Proceedings of the first Asian Himalayas international conference on internet, 2009, Kathmandu, Nepal, pp 1–6
20. Anwar MM, Liu C, Li J (2018) Discovering and tracking query oriented active online social groups in dynamic information network. In: WWWJ, pp 1–36
21. Lashkari AH, Kadir AFA, Taheri L, Ghorbani AA (2018) Toward developing a systematic approach to generate benchmark android malware datasets and classification. In: The proceedings of the 52nd IEEE international Carnahan conference on security technology (ICCST), Montreal, Quebec, Canada
22. Alsabti K, Ranka S, Singh V (1997) An efficient K-means clustering algorithm. Electrical Engineering and Computer Science. Paper 43. http://surface.syr.edu/eecs/43. Accessed 21 Jan 2016
23. Rousseeuw PJ (1987) Silhouettes: a graphical aid to the interpretation and validation of cluster analysis. J Comput Appl Math 20:53–65. ISSN 0377-0427. https://doi.org/10.1016/0377-0427(87)90125-7
24. Marques RS, Epiphaniou G, Al-Khateeb H, Maple C, Hammoudeh M, De Castro PAL, Dehghantanha A, Choo K-KR (2020) A flow-based multi-agent data exfiltration detection architecture for ultra-low latency networks

Improved Spam Email Filtering Architecture Using Several Feature Extraction Techniques

Priyo Ranjan Kundu Prosun, Kazi Saeed Alam, and Shovan Bhowmik

Abstract Research on spam email filtering is drawing experts from all over the world, as these junk email messages continue to affect people's daily lives, whether consciously or unconsciously. The overwhelming use of irritating, destructive, and misleading emails appears to have damaged the values of email which prompted us to perform this research to construct a model for spam filtering with faster training time and enhanced accuracy. We have proposed two voting architectures built upon machine learning models and ensemble classifiers, respectively. In our work, we have also analyzed the performance of several individually applied classifiers and ensemble techniques with various feature retrieval strategies. Additionally, we have compared the training time of the proposed models with the deep LSTM-CNN hybrid model. Both of our suggested models have performed adequately, while the ML-based voting model (Type 1) produces the most accurate filtering (98%) taking bag of words for feature extraction and can be trained above 200 times faster than the LSTM-CNN model.

Keywords Spam email · Junk messages · Voting model · Text classification · Ensemble techniques · Machine learning

1 Introduction

Email services are among the most widely used communication tools because of their speed and effectiveness. However, the virtues of email seem to be ruined by the widespread use of unpleasant, damaging, immoral, and deceptive email messages, which are frequently sent haphazardly by dishonest people who have no

P. R. K. Prosun · K. S. Alam
Khulna University of Engineering and Technology, Khulna, Bangladesh
e-mail: saeed.alam@cse.kuet.ac.bd

S. Bhowmik (✉)
Bangladesh Army International University of Science and Technology, Cumilla Cantonment, Cumilla, Bangladesh

M. S. Arefin et al. (eds.), *Proceedings of the International Conference on Big Data, IoT, and Machine Learning*, Lecture Notes on Data Engineering and Communications Technologies 95, https://doi.org/10.1007/978-981-16-6636-0_50

direct connection with the recipient. These types of email messages are commonly referred to as spam emails. The contents of spam emails can include commercials, online marketing, and scams including malware distribution, exposed data, and phishing [1].

For detecting spam in email messages, there are two main methodologies—knowledge engineering (KE) and machine learning (ML). The first method is to create a knowledge-based process [2] that uses predetermined laws to determine whether an inbound email message is genuine or not. The primary disadvantage of this strategy is that the list of rules must be maintained and updated on a regular basis by the client or another organization. The ML approach, on the other hand, does not demand predefined guidelines, but rather text messages that have been correctly pre-classified [3]. Sample messages can be used to build the training set that is utilized to fit the model's particular learning strategy. As a result, the computer algorithm learns from the data input and applies what it has learned to categorize fresh observations.

Due to the huge amount and variety of spam emails received on a daily basis, it is necessary to develop solutions that provide effective protection against it. In this article, we have designed two voting classifiers with ML and ensemble-based algorithms for specific features in order to increase accuracy. We have also performed the calculation of the training time for our proposed voting model and compared our result with a well-known word embedding-based LSTM-CNN hybrid model. Besides, a performance assessment has been demonstrated for various models along with voting classifiers.

2 Related Works

Spam email messages have been an issue for almost two decades, as spammers fight with filter developers by developing and perfecting novel hybrid and adaptable spam strategies.

Email spam and ham categorization was accomplished by either a machine learning or a non-machine learning strategy as shown in [4]. This paper provides a review of some prominent filtering techniques that use text categorization to determine whether or not an email is unsolicited. In [5], the authors introduced a comparative study where many algorithms such as Bayesian classification technique, k-NN, artificial neural network, SVM, and artificial immune system were implemented on the spam assassin dataset, and their capabilities were compared as well. But this review work mainly focused on the performance analysis of various classification techniques, but feature extraction techniques and model training time measurement were not examined. In [6], the authors constructed spam filtering technique by utilizing nonlinear SVM classifiers using specified kernel functions that were implemented on the Enron Dataset. Naïve Bayes spam detection approach was used on two distinct datasets that were studied for evaluation matrices like accuracy, precision, recall, and F1-score in [7]. Nevertheless, these works used only a few algorithms and did not showcase any comparative study based on the computational time of their proposed algorithm with other state-of-the-art methods.

An original spam text filter was proposed in [8], combining N-gram "TF-IDF" as text feature, improved balancing algorithm, and a regularized multilayer deep neural network model. Their model was checked out on four standard spam message datasets, and its performance was compared with other contemporary spam detection filters and various ML algorithms. But this article also does not show any usage of ensemble-based models. In [9], the authors assessed their framework that relies on a semantic feature selection along with a support vector machine classifier, achieving an accuracy of 94%, which is comparatively low. In [10], the authors proposed a unique spam message identification model in which a genetic algorithm is used as a feature extractor and random weight network (RWN) is employed as a classifier. This architecture recognizes the required features for spam email classification autonomously and obtained an accuracy of 96.70%.

However, these aforementioned models outperformed the most common spam email detection methods, date back to the early 2000s, and current spammer tactics are not taken into consideration. This evolution of non-spam and spam email messages across time is known as concept drift. This issue of concept drift was addressed in [11], where the researchers attempted to provide a multi-category summary as well. In brief, several recent studies have emphasized the growth of identifying mechanisms that can detect spam email messages. In this article, we attempted to model a voting classifier for the filtration of spam emails and also compared the obtained performance with other contemporary algorithms.

3 Proposed Spam Email Filtering Voting Model

For our suggested approach, we have worked with several well-known ML models and ensemble strategies to create the voting schemes. A popular spam filtering dataset is picked up for our task. An all-inclusive architecture of the whole spam filtering task based on our proposed context can be found in Fig. 1.

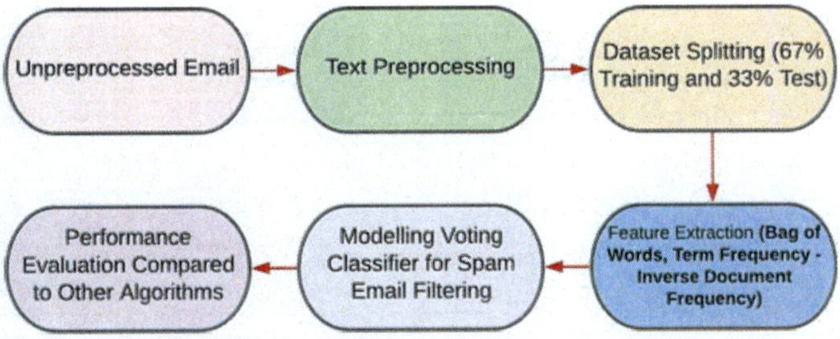

Fig. 1 Workflow of spam email filtering process

3.1 Dataset Description

As the digital environment has grown and technology has evolved, the spreading of spam email messages has become more prevalent, particularly among youth and mischievous people. There are several databases that incorporate various types of spam email messages. We have utilized a preexisting benchmark dataset namely "Spam Assassin dataset" [12] as we are focusing on recognizing spam messages from different emails. This dataset contains 15,736 emails with proper labeling. Labeling categorizes all the existing emails into two groups—"Spam" and "Ham." Among the 15,736 emails, 7850 (49.9%) are tagged as "Ham" (0) and the remaining 7886 (50.1%) emails are tagged as "Spam" (1).

3.2 Data Preprocessing

Since we have obtained unprocessed data from the dataset, we needed to clean it up before we can apply it for our task. As a result, we have refined our raw data using a variety of preprocessing techniques. If not deleted, URLs, contractions, digits, whitespaces, and punctuations produce noise while training the data. So, first of all, we have removed noise and punctuation from the data. After that, the entire document is transformed to a lower case in order to maintain consistency, before tokenization is performed. The next task is to eliminate all of the most commonly used words, sometimes known as "stop words." Stop words are mostly trivial terms that, if not discarded, will end up causing text categorization to become noisy. Following that, we have applied lemmatization using "WordNetLemmatizer" to change all of the words into their root form in order to reduce the variety of terms in our dataset. Detokenization is executed in the final step to obtain clean, polished, and acceptable data that can be used in the following stage of our work. In Fig. 2, the overall preprocessing procedures are depicted.

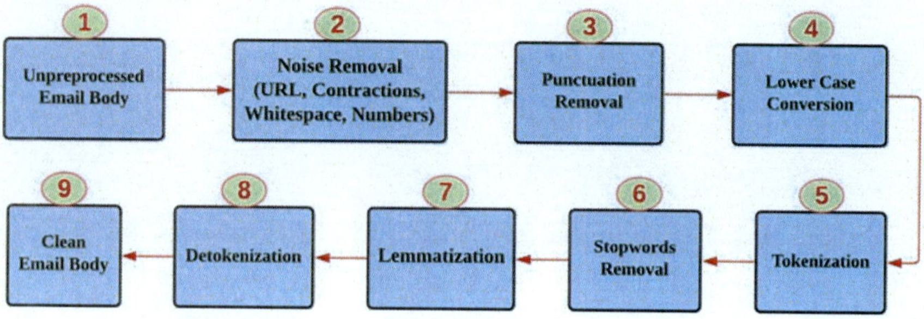

Fig. 2 Preprocessing steps

3.3 Feature Extraction

Due to the huge number of words, concepts, and phrases in text categorization, learning from a large amount of data is difficult. As a result, the entire operation becomes computationally expensive. Furthermore, any classification model's accuracy and performance are affected by irrelevant and repetitious features. Hence, to keep the classification process simple, precise, and less repetitive, it is better to work with only the significant features extracted from data. We have focused on two feature extraction methods of all the available techniques for our work: "bag of words" (BoW) and "term frequency-inverse document frequency" (TF-IDF). "Bag of words" is a simple and flexible way of text representation that describes the frequency with which words appear in a document. We only keep records of word occurrences and do not pay attention to grammatical subtleties or word arrangement. Since any information about the sequence and structure in the document is removed, it is referred to as a "bag" of words. Here, the number of appearances of each term in the document is represented as key-value pairs in a count vector format. However, in the TF-IDF method, term frequency is multiplied by the inverse document frequency. Counting the total occurrences of a word in a document can be used to determine term frequency, and the IDF is determined by dividing the total document number by the sum of documents in the corpus that contain the term. It is helpful for decreasing the weight of terms that appear most commonly in a group of documents. Finally, multiplying the log of this value with term frequency, we get the final product of TF-IDF. In the case of the LSTM-CNN hybrid model, we have customized an embedding layer where the related words are located near each other in a vector space measuring cosine similarity. In the embedding layer, 100 feature vectors have been taken measuring the highest length of the email.

3.4 Classification Process

Many traditional ML-based and ensemble-based strategies have been proposed for the specific task of spam mail detection [3, 13]. Also, several deep learning-based works have been done in this field. In addition to that, sublime performance and superiority over individual classifiers using voting classifiers for other text classification tasks have been shown by researchers which motivated us to carry out this research work [14, 15]. In our research work, we have analyzed the performance of voting-based models for spam filtering and compared the performance based on various evaluation metrics and the time required for training the model. We have also checked the performance of several individually applied ML models (MNB, LR, DT, LSVC) and few ensemble-based techniques (RF, Gboost, Adb, bagging) which are proven to be well performed for spam filtering in the previous works done [2].

For our work, we have focused on two types of voting models. The first type is comprised of only the ML algorithms, whereas the second type is built upon the

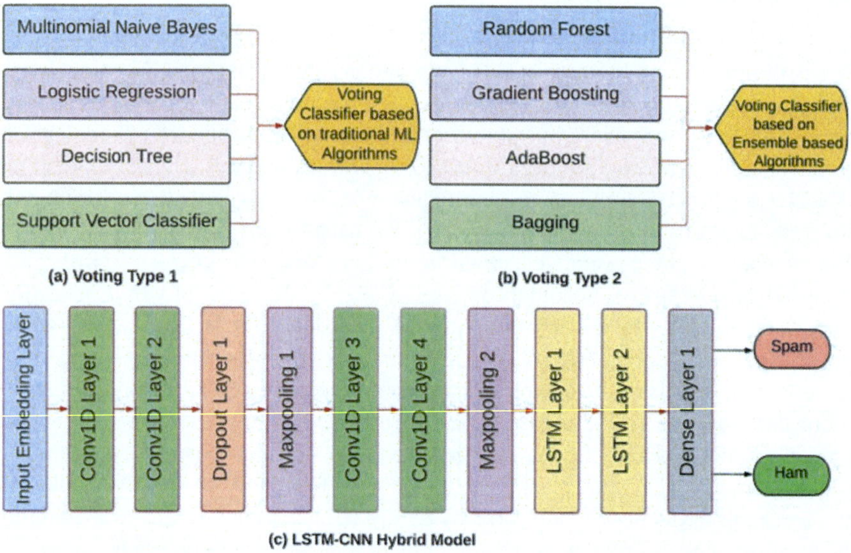

Fig. 3 Architecture of the proposed voting types

ensemble-based techniques. Figure 3 shows the detailed formation of two voting types and a deep LSTM-CNN hybrid framework.

For our suggested technique to be more confident to validate the result than individually used techniques, we have applied the hard voting technique. If most of the classifiers classify a mail to be "spam," the outcome of the voting model will be "spam." After cleaning the data as discussed in the previous sections, features are generated by BoW and TF-IDF. Extracted features are deployed to train the models illustrated above to finally predict any mail as "spam" or "ham."

As the LSTM-CNN hybrid model works remarkably for sequence data classification [16], we have created an LSTM-CNN combined architecture to benchmark our voting model along with ML and ensemble algorithms. One hot representation based embedding layer has been considered as the input layer which is followed by two convolutional layers of one dimension. Then, a dropout layer has been incorporated which is pooled by a max-pooling layer. After that, two convolutional layers have been added again with another pooling layer. Finally, two LSTM layers have been fed before adding the dense layer to classify the two email labels. This whole arrangement has been constructed to get the highest accuracy of the voting model. We have used taken 40, 20, 10, and 8 filters, respectively, for the Conv1D layers. LSTM layers have eight units each for training the model and storing the semantic information. "Relu" has been emphasized as the activation function for both CNN and LSTM layers, and "Sigmoid" has been employed in the final dense layer. To achieve the highest performance of the voting model, this LSTM-CNN hybrid model has taken 10 epochs with the "Adam" optimizer.

4 Experimental Results

In our research work, we have applied several individual ML models (MNB, LR, DT, LSVC) and ensemble models (RF, Gboost, Adb, bagging). After that, we have compared the performance with our proposed voting schemes. BoW and TF-IDF are used to retrieve features from cleaned data for the ease of training the models. We have compared and checked the performance of each individual and voting model based on three evaluation metrics: accuracy, AuC, and F1-score.

Overall, our voting type 1 (VT1) model has achieved the highest accuracy of 98% when BoW and TF-IDF ("word") are considered for feature extraction which surpassed all the individual models applied. VT1 works best among all models with all feature extraction techniques. On the other hand, the ensemble-based voting scheme (VT2) also works pretty well compared to singly applied models but slightly falls short of VT2. Among the ML models, LR performs the best, while the accuracy for DT is the worst in this case. Conversely, we have found "bagging" is the best ensemble technique to filter spam emails, whereas the accuracy obtained for AdB is not satisfactory. Both of our voting schemes have outperformed individual best models by a profound margin.

Turning to the feature extraction techniques, BoW, TF-IDF ("word") and TF-IDF ("character") perform extraordinary well for both of our voting schemes. After applying various n-gram techniques, we have found that the less the value of n in n-gram, the more the accuracy. The combination of TF-IDF ("unigram") with VT1 performs astonishingly well. For bigram, trigram, and N-gram (2:3), also a great performance by voting schemes with 93%, 85%, and 93% accuracy, respectively, is achieved which is adequate. Since these features hold more semantic information. Another thing is, the results are lesser than BoW, unigram, and character as spam emails do have uncommon words sometimes which have not enough significance. As we have worked with a balanced dataset, we need not perform any cross-validation analysis. The overall performance evaluation based on accuracy along with AuC and F1-score can be found in Table 1. Also, a comparison of training time in seconds for both the voting models is demonstrated in Fig. 4.

From Fig. 4, it can be visible that VT2 takes almost six times larger model training time than VT1. When trigram has been extracted, the models have faster output compared to character extraction since trigram has fewer amount features than character-based N-gram model because of its three-word similarity calculation.

To justify our voting model performance, we have compared model training time for VT1 and VT2 with respect to the popular LSTM-CNN model. Table 2 shows how much amount of time is needed for voting models compared to the LSTM-CNN Hybrid model measured in seconds.

The mean model training time for VT1 is around 8 s for all the features incorporated in the spam email classification task. This model execution time is more for VT2. Unfortunately, the LSTM-CNN takes quite an enormous amount of time for achieving the same level of accuracy. It indicates a good choice for using voting

Table 1 Performance comparison of various classifier models

Metric	Feature generation		ML algorithms				Ensemble techniques				Voting model	
			MNB	LR	DT	LSVC	RF	Gboost	Adb	Bagging	Type 1	Type 2
Accuracy	BOW		0.96	0.97	0.92	0.96	0.95	0.94	0.84	0.96	**0.98**	0.96
	TF-IDF	Word	0.97	**0.98**	0.92	0.97	0.95	0.93	0.83	**0.98**	**0.98**	0.96
		Character	0.94	0.96	0.87	**0.97**	0.91	0.94	0.85	**0.97**	**0.97**	0.96
		N-gram (2:3)	0.91	0.92	0.89	0.92	0.91	0.83	0.75	0.92	**0.93**	0.9
		Unigram	0.96	0.97	0.92	**0.98**	0.95	0.93	0.83	0.97	**0.98**	**0.98**
		Bigram	0.92	0.92	0.9	0.92	0.91	0.83	0.77	0.92	**0.93**	0.91
		Trigram	0.83	0.84	0.84	0.84	0.84	0.78	0.74	0.84	0.83	**0.85**
F1-score	BOW		0.95	0.97	0.92	0.96	0.95	0.94	0.84	0.96	0.98	0.96
	TF-IDF	Word	0.97	0.98	0.92	0.97	0.95	0.93	0.83	0.98	0.98	0.96
		Character	0.94	0.96	0.87	0.97	0.91	0.94	0.85	0.97	0.97	0.96
		N-gram (2:3)	0.91	0.92	0.89	0.92	0.91	0.83	0.77	0.92	0.92	0.9
		Unigram	0.96	0.97	0.92	0.98	0.95	0.93	0.83	0.97	0.98	0.97
		Bigram	0.92	0.93	0.9	0.92	0.91	0.84	0.78	0.92	0.93	0.91
		Trigram	0.83	0.84	0.84	0.84	0.85	0.79	0.76	0.84	0.84	0.85
AuC	BOW		0.96	0.97	0.92	0.96	0.95	0.94	0.86	0.97	0.98	0.96
	TF-IDF	Word	0.97	0.98	0.92	0.97	0.95	0.94	0.86	0.98	0.98	0.96
		Character	0.94	0.96	0.87	0.97	0.91	0.95	0.86	0.97	0.97	0.96
		N-gram (2:3)	0.91	0.92	0.89	0.92	0.91	0.85	0.82	0.93	0.93	0.91
		Unigram	0.96	0.97	0.92	0.98	0.95	0.93	0.86	0.97	0.98	0.97
		Bigram	0.92	0.92	0.9	0.93	0.91	0.86	0.81	0.93	0.93	0.92
		Trigram	0.86	0.86	0.87	0.87	0.87	0.84	0.82	0.87	0.86	0.87

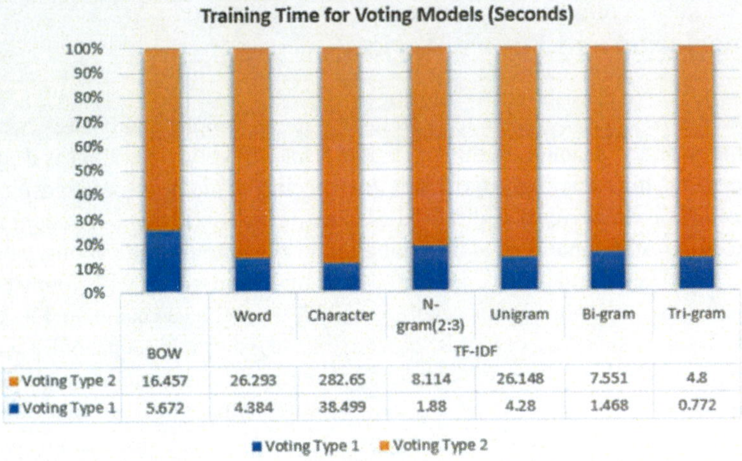

Fig. 4 Training time summary for voting models

Table 2 Average model training time (s)

Voting type 1	Voting type 2	LSTM-CNN hybrid
8.1364	53.1447	1920

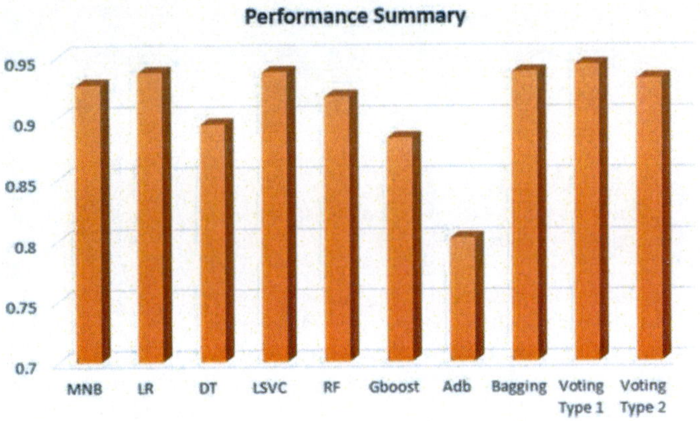

Fig. 5 Mean accuracy of all the models

models on a balanced dataset for increasing accuracy in sequence analysis works instead of using deep learning-based models.

A graph is constructed in Fig. 5 which depicts the average accuracy obtained for all feature extraction techniques. It gives a clear idea that both of our voting schemes are superior to all the individually applied ML and ensemble techniques.

5 Conclusion and Future Works

Spam email messages continue to have an impact on people's daily lives, whether deliberately or accidentally. In this work, our proposed voting classifier is designed with only machine learning algorithms for specific features in order to increase accuracy. We have performed different N-gram feature extraction for spam email classification. Additionally, model training time calculation for the voting model is evaluated, and a performance assessment of various models along with voting classifiers is demonstrated. The highest accuracy of 98% was achieved using the ML-based voting classifier using "bag of words" as an extracted feature. A couple of limitations regarding our work was—we implemented our voting classifier model solely on one dataset and although we designed a voting classifier using either ML algorithms or ensemble techniques, we did not employ a combination of those together. Moreover, as we took a balanced data, we did not apply cross-validation yet. In the future study, we wish to implement our voting classifiers on other benchmark datasets and showcase the experimental results. Also, we will try to employ "word2vec" as well as "seq2seq" techniques along with other feature extraction models. Furthermore, K-fold cross-validation will be applied in coming research regarding spam email classification. Eventually, in the current state of the art, our proposed technique can be simply implemented in software-based applications.

References

1. Colladon AF, Gloor PA (2019) Measuring the impact of spammers on e-mail and Twitter networks. Int J Inf Manage 48:254–262
2. Guzella TS, Caminhas WM (2009) A review of machine learning approaches to spam filtering. Expert Syst Appl 36(7):10206–10222
3. Awad WA, ELseuofi SM (2011) Machine learning methods for spam e-mail classification. Int J Comput Sci Inf Technol (IJCSIT) 3(1):173–184
4. Saab SA, Mitri N, Awad M (2014) Ham or spam? A comparative study for some content-based classification algorithms for email filtering. In: MELECON 2014–2014 17th IEEE Mediterranean electrotechnical conference, Apr 2014, pp 339–343. IEEE
5. Mujtaba G, Shuib L, Raj RG, Majeed N, Al-Garadi MA (2017) Email classification research trends: review and open issues. IEEE Access 5:9044–9064
6. Chhabra P, Wadhvani R, Shukla S (2010) Spam filtering using support vector machine. Int J Comput Commun Technol 1(2):322–341
7. Rusland NF, Wahid N, Kasim S, Hafit H (2017) Analysis of Naïve Bayes algorithm for email spam filtering across multiple datasets. IOP Conf Ser Mater Sci Eng 226(1):012091. IOP Publishing
8. Barushka A, Hajek P (2018) Spam filtering using integrated distribution-based balancing approach and regularized deep neural networks. Appl Intell 48(10):3538–3556
9. Bahgat EM, Rady S, Gad W, Moawad IF (2018) Efficient email classification approach based on semantic methods. Ain Shams Eng J 9(4):3259–3269
10. Faris H, Ala'M AZ, Heidari AA, Aljarah I, Mafarja M, Hassonah MA, Fujita H (2019) An intelligent system for spam detection and identification of the most relevant features based on evolutionary random weight networks. Inf Fusion 48:67–83

11. Ruano-Ordas D, Fdez-Riverola F, Méndez JR (2018) Concept drift in e-mail datasets: an empirical study with practical implications. Inf Sci 428:120–135
12. https://www.kaggle.com/nitishabharathi/email-spam-dataset
13. Saeedian MF, Beigy H (2012) Learning to filter spam emails: an ensemble learning approach. Int J Hybrid Intell Syst 9(1):27–43
14. Alam KS, Bhowmik S, Prosun PRK (2021) Cyberbullying detection: an ensemble based machine learning approach. In: 2021 third international conference on intelligent communication technologies and virtual mobile networks (ICICV), Feb 2021. IEEE, pp 710–715
15. Bhowmik S, Prosun PRK, Alam KS (2021) A novel three-level voting model for detecting misleading information on COVID-19. Paper presented at the 6th international conference on emerging applications of information technology (EAIT), Kolkata
16. Ajao O, Bhowmik D, Zargari S (2018) Fake news identification on twitter with hybrid CNN and RNN models. In: Proceedings of the 9th international conference on social media and society, July 2018, pp 226–230

Detecting Smishing Attacks Using Feature Extraction and Classification Techniques

Rubaiath E. Ulfath, Iqbal H. Sarker, Mohammad Jabed Morshed Chowdhury, and Mohammad Hammoudeh

Abstract Phishing scams via SMS have become a common phenomenon due to the widespread use of smartphones and the availability of mobile Internet technologies. Identifying a phishing SMS via analyzing unstructured short texts is a challenging issue in the domain of AI-driven cybersecurity. Machine learning-based techniques integrated with natural language processing have massive potentials to identify differentiating patterns between phishing and legitimate SMS. In this paper, we have experimented with several state-of-the-art machine learning algorithms on a benchmark dataset. Also, NLP-based feature extraction and feature selection steps are incorporated to build an automated phishing detection strategy. Support vector machine classifier when applied after feature extraction and feature selection has outperformed the tenfold cross-validation score of 98.27%, F1-score of 99.08% for legitimate SMS, and accuracy of 98.39%. The performance of the tested methods has been evaluated through popular evaluation metrics on a benchmark dataset.

Keywords Smishing · ANOVA test · TF-IDF · Natural language processing · Machine learning

---

R. E. Ulfath · I. H. Sarker (✉)
Department of Computer Science and Engineering, Chittagong University of Engineering and Technology, Chittagong 4349, Bangladesh
e-mail: iqbal@cuet.ac.bd

M. J. M. Chowdhury
La Trobe University, Melbourne, Australia

M. Hammoudeh
Department of Computing and Math, Manchester Metropolitan University, Manchester M1 5GD, UK

M. S. Arefin et al. (eds.), *Proceedings of the International Conference on Big Data, IoT, and Machine Learning*, Lecture Notes on Data Engineering and Communications Technologies 95, https://doi.org/10.1007/978-981-16-6636-0_51

1 Introduction

SMS phishing or smishing is a form of phishing attack which targets the users of mobile SMS messaging. Smishing attacks have become a common phenomenon nowadays [14, 15]. Due to the rapid advancement of smart and advanced technologies in the last decade, devices like smartphones have become available to people of all ages and classes.

Today, smartphones are one of the most essential parts of our daily life [24]. From the corporate world to home, almost everyone uses smartphone-like devices which provide the provision of Internet connectivity. This gives phishers a new room to start phishing through SMS. Fake text messages containing links that look real and legitimate but originally malicious are sent via SMS with the intention of stealing personal information, committing fraud, spreading malicious smartphone viruses.

SMS texts data are unstructured in the manner and difficult to process. Due to the nonlinearity involved in the processing and analyzing SMS text data, it is difficult to identify phishing and legitimate SMS efficiently. Extracting meaningful features from text data is also computationally expensive. This makes proper identification of phishing SMS a challenging problem in the domain of AI-driven security [20].

Proofpoint, a software security company, reported that SMS-based scams have risen 328% in the middle of 2020 alone [14]. It is because general smartphone users with less knowledge about the Internet and phishing scams easily get deceived by these phishing attacks via SMS. And, attackers are finding new strategies to deceive a user and get them into the trap of smishing attack to steal their sensitive personal information or spread malware in their mobile devices. The Bank of Ireland was forced to shell out €800,000 to over 300 bank customers in a single smishing scam in 2020 [5]. According to a 2018 survey by Wandera, a cloud-infrastructure company, 17% of its business customers were exposed to phishing connections on their mobile devices. In comparison, only 15% of users got a phishing note, and only 16% received phishing links via social media applications [10].

In this context, artificial intelligence researchers have devoted themselves to study the potential of machine learning and deep learning-based methods in combination with natural language processing to fight against smishing attacks [9]. Many state-of-the-art researches have been conducted with noteworthy outcomes which have contributed to the detection of phishing detection with AI methods [2, 4, 20]. Inspired by the previous research works and to overcome their limitations, we have conducted an extensive study to find a simple but efficient framework for smishing detection from raw SMS text data.

After empirical investigation, we have used machine learning algorithms [19] in association with natural language processing to detect smishing attacks in an efficient manner. The key contributions of our study are as follows,

- We have extracted features from raw text-based phishing data limiting the top 10,000 most significant vocabularies according to TF-IDF analysis with an N-gram range of 1–3.

- After feature extraction, feature selection is done using statistical significance analysis with ANOVA test.
- Finally, we have explored the potential of multiple machine learning algorithms for smishing detection and identified an efficient framework for automated detection of smishing attacks.

2 Literature Review

Boukari et al. [4] investigated the potential of a machine learning-based detection system for smishing attacks that provides users an early alarm. It can also be adapted for phishing and vishing attacks. Statistical significance-based feature selections with multiple correlation algorithm were explored in [17, 21].

In [21], the authors have reduced the feature dimensions with correlation algorithms which have increased the performance of both tree-based and linear classifiers. Among all these classifiers, AdaBoost with Kendal's correlation coefficient method exhibited the best accuracy. Mishra and Soni [13] developed an efficient model which reduces false positives in analyzing contents of SMS and behaviors of URLs along with the integration of the naive Bayes algorithm. Deep learning-based models, including convolutional neural network (CNN), have the ability to perform better in feature extraction for text analysis and text classification-based problems due to the complex and state-of-the-art dense architecture. Long short-term memory (LSTM)-based architectures have potential for text classification-based problems because of their efficient encoding and sequence learning. In this context, Ghourabi et al. [7] proposed a hybrid CNN-LSTM architecture for smishing detection in Arabic and English messages. The complex structure of deep learning-based models sometimes makes it difficult to interpret underlying discriminating factors which improve classification performance. As deep neural network-based architectures are sometimes complicated to implement, machine learning-based models with natural language processing have gained popularity and acceptance by researchers for easy implementation and good performance. Thus, several researchers have proposed a novel and state-of-the-art smishing classifiers using machine learning algorithms [8, 22].

The researchers of this study have found that an automated framework containing feature extraction followed by feature selection leading to classification is lacking in the prior state-of-the-art research. The main motive of this study is to propose a simple but fast and efficient automated framework utilizing the massive potential of machine learning methods. We have conducted a thorough investigation to find an efficient pipelining technique for an automated detection strategy to defend against SMS phishing.

3 Methodology: Automated Smishing Detection Framework

To detect phishing SMS, we have built an automated smishing detection model, followed by certain data prepossessing steps. We have done feature extraction with N-grams and TF-IDF and feature selection with ANOVA test, leading to classification by state-of-the-art machine learning algorithms. In this section, we have described every individual component of our method. Also, the overall training procedure of our smishing detection model has been described in Algorithm 1.

3.1 Data Preprocessing

In phishing SMS analysis, preprocessing of raw data is a very crucial step that can contribute to improving the performance of the machine learning classifiers [16]. Here, we have followed some widely used and efficient data cleaning steps that include the removal of extra whitespaces and stop words from the corpus of the SMS dataset. There are also two commonly used preprocessing steps named stemming and lemmatization. In our study, we have not to employ stemming and lemmatization during our experimentation to avoid loss of significant information from the words.

3.2 Feature Extraction and Feature Selection

3.2.1 TF-IDF and N-Grams

Term frequency (TF) is a metric of text analysis that is used to describe the frequency of occurrence of a certain term across the entire SMS. Since SMS texts vary in size and volume, the term frequency effects are normalized by dividing them by the total number of words in the SMS. To state mathematically,

$$\mathrm{TF}(x) = \frac{N(x)}{n} \tag{1}$$

where $\mathrm{TF}(x)$ denotes the term frequency of a particular term x, $N(x)$ denotes number of times term x appears in the SMS, and n denotes total number of terms in the SMS.

$$\mathrm{IDF}(x) = \log_{\mathrm{e}} \frac{n}{N(x)} \tag{2}$$

where $\mathrm{IDF}(x)$ denotes the inverse document frequency of a particular term x, n denotes total number of instances in the SMS, and $N(x)$ denotes number of instances with term x in it. TF-IDF has been proved an efficient feature extraction method for

Table 1 Example of N-grams

Unigram	Bi-gram	Tri-gram
Bonus	Second time	Guaranteed 1000 cash
Club	Claim reward	Second attempt contact
Credits	Club credits	Bonus caller prize
Age 16	Horny guys	Cash 2000 gift
Announcement	Credits pls	Chances win cash

Table 2 Examples of some TF-IDF scores

Feature	TF-IDF score
Islands holiday await	0.9996621492
Jackpot txt	0.9996755409
Valued mobile customer	0.9996755409
Videophones 09063458130 videochat	0.9999886424
Wc1n3xx	0.9999533068

phishing detection in recent studies [3]. Considering the immense potentials of hand-crafted features in the domain of text analysis, we have performed vectorized term frequency and inverse document frequency (TF-IDF) analysis to generate useful features from the SMS dataset. In this regard, we have used the N-gram analysis method, a popular feature generation method in natural language processing. An N-gram is a contiguous sequence of *N* entities from a given sample of SMS texts. When *N* equals one, the result is referred to as a unigram. Similarly, an *N* value of two is referred to as a bi-gram, an *N* value of three is referred to as a tri-gram, and so on. Many state-of-the-art studies have found N-gram analysis as a valuable feature generation method in the field of phishing detection. Table 1 shows some of the examples of N-grams, and Table 2 shows some TF-IDF scores that have been used in our study.

In our study, we set the value of *N* in a range of 1–3, which covers the use of Unigram, bi-gram, and tri-gram-based features from the SMS corpus. After that, we limit the vocabulary to consider the top 10,000 terms ordered by term frequency-inverse document frequency across the entire SMS dataset. By the end of this step, we obtain the top 10,000 significant features from the SMS corpus according to their term frequency and inverse document frequency scores [1, 11].

3.2.2 ANOVA Test

The analysis of variance (ANOVA) test is a statistical analysis technique or approach that looks for a significant difference in data distribution between two or more groups to see if they are substantially different. It is a non-parametric hypothesis test that

Table 3 Number of features selected with ANOVA test

P-value threshold	Number of features selected
0.05	4329
0.01	4184
0.001	4123

extends the *t*-test beyond two means. ANOVA test generates *p*-values to indicate how statistically significant a numerical variable is against the class variable. Several studies have got profound evidence on the high importance of feature selection using statistical correlation methods in text classification [12]. To state mathematically, the test statistic F is,

$$F = \frac{B_v}{G_v} \tag{3}$$

where B_v denotes between group variance and G_v within group variance. The *p*-values of each feature vector can be determined from the ANOVA test. A *p*-value less than or equal to a certain threshold indicates that the certain feature exhibits prominent statistical significance. We have analyzed the statistical significance of our extracted 10,000 features using the ANOVA test. After performing the ANOVA test, we have selected statistically significant features referring to the *p*-value thresholds of 0.05, 0.01, 0.001 to evaluate the effect of different thresholds. The effect of different *p*-value thresholds is reflected through Table 3.

The algorithm for training process of our model is given in Algorithm 1.

3.3 Machine Learning Classifiers: Output Generation

In this study, we have experimented with five state-of-the-art machine learning classifiers [19], and these are; XgBoost, random forest, classification and regression tree (CART), support vector machine, and AdaBoost, and all of these are tree-based classifiers. Classification and regression tree which is popularly named as decision tree is a rule-based classifier that builds a decision tree with a certain depth and splitting criteria on the nodes and leaves. Gini impurity is the splitting criterion, which determines where and how to split the tree. Random forest creates a number of randomized decision trees, where each decision tree is created by considering a random number of samples and features from the training set. Then, following a majority voting criteria, the final classification is justified by aggregating the decisions of each individual decision tree.

AdaBoost creates a number of weak decision trees with a depth of 1, which is called a stump. These stumps are weak classifiers which ultimately result in a strong classification model by accumulating the individual performance of each weak classifier. When building trees, XgBoost uses gradient descent techniques to achieve

Algorithm 1 Training process of the SMS phishing detector

Input: $|\mathbb{X}|$ represents number of training instances, with input vectors in $\{x_1, x_2, \ldots, x_{|\mathbb{X}|}\}$, where x_i represents vectors of individual SMS text with a label associated with it, indicating whether the SMS is phishing or legitimate. $\mathbb{F}$ represents the set of vocabularies going to be used as features for training.

for Each x_i in $\mathbb{X}$ **do**
  for $n \in \{1, 2, 3\}$ **do**
    Extract word-based features using n gram analysis and add it to the feature set $\mathbb{F}$.
  end for
end for
for Each f in $\mathbb{F}$ **do**
  Compute TF scores for extracted feature f according to Eq. 1.
  Compute IDF scores for extracted feature f according to Eq. 2.
  Compute TF-IDF scores for the feature f.
  Add corresponding TF-IDF score with feature f and update $\mathbb{F}$.
end for
Select top 10,000 features according to max TF-IDF score and create a list of vocabularies $\mathbb{T}$
Compute statistical significance score for $\mathbb{T}$ using ANOVA test by Eq. 3 and store the p-values in a list $\mathbb{A}$.
Create a list $\mathbb{S}$ of a vector $\{s_1, s_2, \ldots, s_{|\mathbb{S}|}\}$, which will contain the most significant features according to ANOVA test values.
for Each a in $\mathbb{A}$ **do**
  if p-value of $a \leq 0.05$ **then**
    Include a in the list of finally selected vocabularies, $\mathbb{S}$.
  end if
end for
for Each machine learning classifier **do**
  Train the model with input data $\mathbb{X}$ by selecting features based on $\mathbb{S}$.
end for

Output: Smishing Detection Model

distinct decision trees, starting with an initial baseline and updating it over iterations by optimizing residuals. The tree remains unique after each iteration since the previous tree's flaws or weaknesses are minimized or regularized in the next trees to be built [6].

Support vector machine (SVM) is a supervised machine learning classifier that delineates a decision boundary along the data point used for training based on differences in data distribution along with different classes. The decision boundary of SVM is a hyperplane in an N-dimensional space which evidently classifies the data points of classes like phishing SMS or legitimate SMS. The category of SVM is decided based on the kernel function it uses to build the hyperplane. In our study, we have employed the radial basis function as the kernel function. Some example of input text and class detected by machine learning classifier is shown in Table 4.

In this study of classifying phishing and legitimate SMS, we have employed the aforementioned machine learning classifiers individually in our experimentation stage followed by the feature extraction with TF with N-grams and a statistical feature selection with ANOVA test (p-value 0.05).

Table 4 Examples of SMS classification

Input text	Class detected by our smishing model
Text 1—"Free entry in 2 a wkly comp to win FA cup final tkts May 21st, 2005. Text FA to 87121 to receive entry question (std txt rate) T&C's apply 08452810075 over 18's"	Phishing SMS
Text 2—"As per your request 'Melle Melle (Oru Minnaminunginte Nurungu Vettam)' has been set as your callertune for all callers. Press *9 to copy your friends callertune"	Legitimate SMS
Text 3—"You have won ?1000 cash or a ?2000 prize! To claim, call 09050000327"	Phishing SMS
Text 4—"Our mobile number has won £5000, to claim calls us back or ring the claims hot line on 09050005321"	Phishing SMS
Text 5—"Even my brother is not like to speak with me. They treat me like aids patent"	Legitimate SMS
Text 6—"WINNER!! As a valued network customer you have been selected to receive a £900 prize reward! To claim call 09061701461. Claim code KL341. Valid 12 hours only"	Phishing SMS
Text 7—"I know! Grumpy old people. My mom was like you better not be lying. Then again I am always the one to play jokes…"	Legitimate SMS
Text 8—"I call you later, don't have network. If urgnt, sms me"	Legitimate SMS
Text 9—"For the most sparkling shopping breaks from 45 per person; call 0121 2025050 or visit www.shortbreaks.org.uk"	Phishing SMS
Text 10—"You will be in the place of that man"	Legitimate SMS

According to the automated framework proposed in our study, the text of an SMS would be given to our pipeline, and it will be classified into the category of phishing or legitimate SMS after going through feature extraction and feature selection steps followed by a machine learning classifier. The hyperparameters of each individual machine learning classifier in our study are stated in Table 5. The following hyperparameters were chosen with random searching on a trial and error basis.

4 Experimental Results

In this section, we evaluated the performance of five machine learning classifiers; extreme gradient boosting (XgBoost), random forest, classification and regression tree (decision tree), support vector machine (SVM), and AdaBoost, followed by the

Table 5 Hyperparameters of classifiers

Classifier	Hyperparameters	Values	Definition
XgBoost	booster	gbtree	Algorithm to use
	max_depth	5	Maximum depth of a tree
	n_estimators	5000	Number of trees
	learning_rate	0.01	Learning rate for weight update
	importance_type	gain	Splitting criterion
Random forest	n_estimators	1000	Number of trees
Decision tree	criterion	gini	Splitting criterion
SVM	C	1	Regularization parameter
	kernel	rbf	Kernel trick to be used
AdaBoost	n_estimators	50	Number of trees
	learning_rate	1	Learning rate for weight update

several experimental combination of mentioned preprocessing, feature extraction, and feature selection steps to identify the most suitable classifier for this problem domain.

4.1 Dataset Description

For this study, we collected our dataset [23] from the UCI machine learning repository which is a popular repository for datasets in machine learning researches. It contains 5572 data instances from which we labeled 4825 instances as "legitimate" (legitimate SMS) and 747 instances as "phishing" (fake or phishing SMS). For employing machine learning classifiers, we split our dataset into training and testing sets with a hold-out validation ratio of 80:20 using stratified sampling.

4.2 Evaluation Metrics

To justify the efficiency of a machine learning model, it is important to evaluate the performance with well-designated evaluation metrics. The performance of our model is evaluated over the below-mentioned measures:

$$\text{Accuracy} = \frac{\text{TP} + \text{TN}}{\text{TP} + \text{FP} + \text{TN} + \text{FN}} \tag{4}$$

Recall: It is defined as the ratio of the number of positive samples that have been correctly predicted as legitimate corresponding to all legitimate samples in the data. It can be defined as

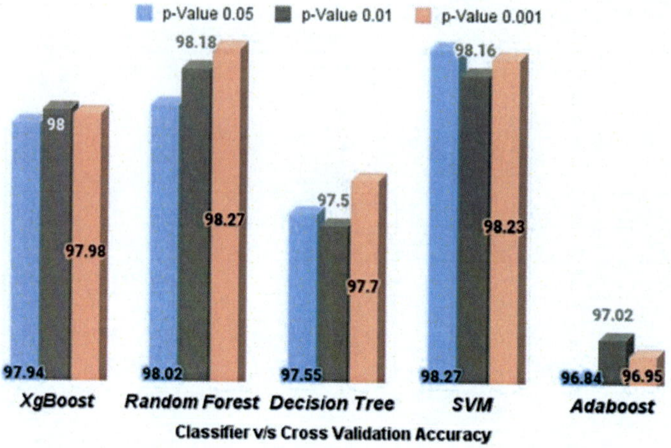

Fig. 1 Tenfold cross-validation accuracy scores

$$\text{Recall} = \frac{\text{TP}}{\text{TP} + \text{FN}} \tag{5}$$

Precision: It is defined as the ratio of the number of positive samples that have been correctly predicted as legitimate corresponding to all samples predicted as legitimate. It can be defined as

$$\text{Precision} = \frac{\text{TP}}{\text{TP} + \text{FP}} \tag{6}$$

Here, TP refers to true positive, TN refers to true negative, FP refers to false positive, and FN refers to false negative.

F1-score: It is defined as the term that balances between recall and precision. It can be defined as

$$\text{F1-score} = 2 \times \frac{\text{Recall} \times \text{Precision}}{\text{Recall} + \text{Precision}} \tag{7}$$

In this study, we have performed tenfold stratified cross-validation on training data considering accuracy scores. Also, we have done tuning with three p-value thresholds for ANOVA test in terms of feature selection to identify the threshold that the best complements the machine learning algorithms. Here, from Fig. 1 according to log-scale comparison, we can see that support vector machine outperforms all other classifiers with a p-value of 0.05 and 0.01, whereas with a p-value 0.001, the random forest performs the best.

In Table 6, the confusion matrix values of each classifier have been depicted from which the classification accuracy of each individual classifier on the test dataset can be justified.

Table 6 Confusion matrix

Classifier	True positive	False positive	True negative	False negative
XgBoost	964	23	126	2
Random forest	966	25	124	0
Decision tree	957	22	127	9
SVM	964	16	133	2
AdaBoost	961	34	115	5

Table 7 Evaluation metrics

Metrics	XgBoost	Decision tree	Random forest	SVM	AdaBoost
Precision (phishing)	98.44	93.38	98.99	98.52	95.83
Recall (phishing)	84.56	85.24	83.89	89.26	77.18
F1-score (phishing)	90.98	89.12	91.24	93.66	85.5
Precision (legitimate)	97.67	97.75	97.58	98.37	96.58
Recall (legitimate)	99.79	99.06	98.99	99.79	99.48
F1-score (legitimate)	98.72	98.41	98.77	99.08	98.01
Accuracy	97.76	97.22	97.85	98.39	96.5

In Table 7, we have done a thorough investigation of classifiers' performance by evaluating the precision, recall, and F1-scores of each class label, naming phishing and legitimate.

5 Discussion

The scores of evaluation metrics clarify that support vector machine (radial basis kernel) followed by the aforementioned feature extraction and feature selection steps outperforms all other classifiers with a cross-validation score of 98.27% and an accuracy of 98.39%. The better performance of SVM in this certain scenario is well-justified as naturally, and SVM is more effective in high-dimensional spaces where the number of features is relatively large and a clear margin can easily be drawn between classes. In contrast, the tree-based algorithms are sensitive to overfitting issues with high-dimensional feature space. Moreover, the best performing model of our study has outperformed the recent state-of-the-art studies of Sonowal and Kuppusamy [22] by 2.24% and Amir Sjarif et al. [3] by 1% in the domain of smishing detection with machine learning-based methods.

According to Table 4, phishing SMS typically contains numbers, tempting terms, for example, free entry, winner, reward, cash, prize, call back and also contains Web site links, contact number, claim codes, etc. Legitimate SMS, on the other hand, is subject-oriented, comparatively well-organized and human-understandable.

As a part of the future work, we would like to solve the class imbalance problem by collecting more data to support better classification, and also, we will do experiments on additional real-world datasets, by extracting robust features using deep learning techniques [18], e.g., CNN to identify smishing more efficiently. The goal of the future model is to discover the optimal feature set in the shortest amount of time. The primary difficulty in our study is imbalanced data, and no deep architecture was employed in the feature selection procedure. We would like to tackle the problem of class imbalance and employ deep feature selection architecture in the near future.

6 Conclusion

Smishing messages are rapidly growing, and they dominate cyber-attacks in cyberspace. Despite the fact that most researchers are introducing various advanced ways to slow down the pace of these attacks, they have yet to achieve more. In this study, we have found an automated strategy that efficiently differentiates between legitimate and phishing SMS. We experimented with several state-of-the-art machine learning algorithms. We have performed intensive feature extraction followed by a statistical feature selection process. After careful analysis of the performance of these machine learning classifiers, we have found that SVM (radial basis kernel) is outperforming all other classifiers. Moreover, feature selection with the ANOVA test evinced a good accuracy with reduced feature dimensions.

References

1. Aiyar S, Shetty NP (2018) N-gram assisted YouTube spam comment detection. Procedia Comput Sci 132:174–182. https://doi.org/10.1016/j.procs.2018.05.181, https://www.sciencedirect.com/science/article/pii/S1877050918309153. In: International conference on computational intelligence and data science
2. Alam MN, Sarma D, Lima FF, Saha I, Ulfath RE, Hossain S (2020) Phishing attacks detection using machine learning approach. In: 2020 third international conference on smart systems and inventive technology (ICSSIT), pp 1173–1179. https://doi.org/10.1109/ICSSIT48917.2020.9214225
3. Amir Sjarif NN, Mohd Azmi NF, Chuprat S, Sarkan HM, Yahya Y, Sam SM (2019) SMS spam message detection using term frequency-inverse document frequency and random forest algorithm. Procedia Comput Sci 161:509–515. https://doi.org/10.1016/j.procs.2019.11.150. https://www.sciencedirect.com/science/article/pii/S1877050919318617. In: The fifth information systems international conference, 23–24 July 2019, Surabaya
4. Boukari BE, Ravi A, Msahli M (2021) Machine learning detection for smishing frauds. In: 2021 IEEE 18th annual consumer communications networking conference (CCNC), pp 1–2. https://doi.org/10.1109/CCNC49032.2021.9369640

5. Burke-Kennedy E, Brennan J, Taylor C (2020) Bank of Ireland does U-turn after refusal to reimburse 'smishing' victims. https://www.irishtimes.com/business/financial-services/bank-of-ireland-does-u-turn-after-refusal-to-reimburse-smishing-victims-1.4326502
6. Chen T, Guestrin C (2016) Xgboost: a scalable tree boosting system. In: Proceedings of the 22nd ACM SIGKDD international conference on knowledge discovery and data mining, pp 785–794
7. Ghourabi A, Mahmood MA, Alzubi QM (2020) A hybrid CNN-LSTM model for SMS spam detection in Arabic and English messages. Future Internet 12(9). https://doi.org/10.3390/fi12090156. https://www.mdpi.com/1999-5903/12/9/156
8. Goel D, Jain AK (2017) Smishing-classifier: a novel framework for detection of smishing attack in mobile environment. In: International conference on next generation computing technologies. Springer, pp 502–512
9. Kumar S, Pal AK, Islam SH, Hammoudeh M (2021) Secure and efficient image retrieval through invariant features selection in insecure cloud environments. Neural Comput Appl 1–26
10. Martens B (2021) 11 facts + stats on smishing (SMS phishing) in 2021. https://www.safetydetectives.com/blog/what-is-smishing-sms-phishing-facts/
11. Mathew NV, Bai VR (2016) Analyzing the effectiveness of n-gram technique based feature set in a Naive Bayesian spam filter. In: 2016 international conference on emerging technological trends (ICETT), pp 1–5. https://doi.org/10.1109/ICETT.2016.7873648
12. Meesad P, Boonrawd P, Nuipian V. A chi-square-test for word importance differentiation in text classification
13. Mishra S, Soni D (2020) Smishing detector: a security model to detect smishing through SMS content analysis and URL behavior analysis. Future Gener Comput Syst 108:803–815. https://doi.org/10.1016/j.future.2020.03.021https://www.sciencedirect.com/science/article/pii/S0167739X19318758
14. Mobile phishing increases more than 300% as 2020 chaos continues | Proofpoint US (2021). https://www.proofpoint.com/us/blog/threat-protection/mobile-phishing-increases-more-300-2020-chaos-continues
15. Saleem J, Hammoudeh M (2018) Defense methods against social engineering attacks. In: Computer and network security essentials. Springer, pp 603–618
16. Sarker IH (2021) Data science and analytics: an overview from data-driven smart computing, decision-making and applications perspective. SN Comput Sci
17. Sarker IH (2021) Cyberlearning: effectiveness analysis of machine learning security modeling to detect cyber-anomalies and multi-attacks. Internet Things 14:100393
18. Sarker IH (2021) Deep cybersecurity: a comprehensive overview from neural network and deep learning perspective. SN Comput Sci 2(3):1–16
19. Sarker IH (2021) Machine learning: algorithms, real-world applications and research directions. SN Comput Sci 2(3):1–21
20. Sarker IH, Furhad MH, Nowrozy R (2021) AI-driven cybersecurity: an overview, security intelligence modeling and research directions. SN Comput Sci 2(3):1–18
21. Sonowal G (2020) Detecting phishing SMS based on multiple correlation algorithms. SN Comput Sci 1(6):1–9
22. Sonowal G, Kuppusamy KS (2018) SmiDCA: an anti-smishing model with machine learning approach. Comput J 61(8):1143–1157. https://doi.org/10.1093/comjnl/bxy039
23. UCI machine learning repository: SMS spam collection data set (2012). https://archive.ics.uci.edu/ml/datasets/sms+spam+collection
24. Walker-Roberts S, Hammoudeh M, Aldabbas O, Aydin M, Dehghantanha A (2020) Threats on the horizon: understanding security threats in the era of cyber-physical systems. J Supercomput 76(4):2643–2664

InterPlanetary File System-Based Decentralized and Secured Electronic Health Record System Using Lightweight Algorithm

Sanjida Sharmin, Iqbal H. Sarker, M. Shamim Kaiser, and Mohammad Shamsul Arefin

Abstract The electronic health record (EHR) system is a cloud-based patient health record in digital format that often includes contact information about the patient, test reports, medical history, and current and previous prescriptions. However, data breaches in cloud-based EHRs pose significant privacy and security concerns for a variety of health care organizations. Cryptographic techniques currently in use are inadequate to secure EHR data in the cloud from data breaches. Blockchain technology is a new technology that can be used to address security and privacy problems with EHR data on the blockchain in a decentralized manner. We have created a stable decentralized medical blockchain in this paper to address privacy and security concerns when sharing patient data on health care between medical organizations. The health care data is encrypted using Advanced Encryption Standard-based lightweight authenticated encryption algorithm before being uploaded to a cloud-based blockchain and Solidity smart code built on Ethereum to restrict access to EHR data in the cloud. We have used an InterPlanetary file system to store data because it is distributed and ensures record immutability. The medical blockchain also ensures that patient EHR data is interoperable, traceable, and anonymous across organizations. The stable cloud-based blockchain of medical records visualizes patient care data in a distributed and immutable environment with enhanced protection. AQ1

Keywords IPFS · Blockchain · Smart contract · Ethereum

S. Sharmin · I. H. Sarker · M. S. Arefin (✉)
Department of Computer Science Engineering, Chittagong University of Engineering and Technology, Chattogram 4349, Bangladesh
e-mail: sarefin@cuet.ac.bd

M. Shamim Kaiser
Institute of Information Technology, and Applied Intelligence and Informatics (AII), Wazed Miah Science Research Centre (WMSRC), Jahangirngar University, Savar, Dhaka 1342, Bangladesh

M. S. Arefin et al. (eds.), *Proceedings of the International Conference on Big Data, IoT, and Machine Learning*, Lecture Notes on Data Engineering and Communications Technologies 95, https://doi.org/10.1007/978-981-16-6636-0_52

1 Introduction

AQ2

In response to the widespread adoption of the Internet of Things (IoT) and the growing need for patient-centered health care, health care providers are increasingly implementing electronic health-related services (called e-health) [5, 10]. As a result of these IoT sensors, there is a large influx of health-related data. The electronic health record (EHR) database is where these records are kept [3, 11]. This data is converted into actionable knowledge through the use of cloud computing, big data, and machine learning algorithms [17]. This information is critical for health care professionals in the detection of anomalies as well as the construction of patient treatment plans [9, 15].

The advantages of EHR include the provision of exact, up-to-date, and vast data on patients, as well as faster access to patients' data from any location in the world, which aids in the diagnosis of problems and the updating of patient records. Distributed computing is another breakthrough that may be used to safely store and update EHR information, preventing access to potentially dangerous areas of the system without first obtaining authorization from the patient. Cloud-based specialist organizations are in charge of storing and exchanging adaptive clinical information with patients and clinics on a global scale [15].

With its decentralized and trustworthy existence, blockchain has shown enormous promise in a variety of e-health sectors, including safe exchange of EHRs and data access management through multiple medical agencies [1, 12]. To store EHR in the cloud, blockchain technology maintains a decentralized health care data management ledger. When joining a peer-to-peer network, the EHR information is linked to retaining the immutability character. The anonymity of the patient is also reserved for the privacy of the patient. The blockchain network ensures the integrity, secrecy, authenticity, interoperability, and accountability of EHRs between two groups. As a result, blockchain adoption has the potential to offer promising strategies for facilitating health care delivery and revolutionizing the health care industry. On a cloud platform, we proposed a new EHR sharing architecture that incorporates blockchain and the decentralized InterPlanetary system (IPFS).

We created a reliable access control system based on smart contracts to ensure safe EHR sharing between patients and medical service providers. The performance improvements in lightweight access control architecture and minimal network latency are also demonstrated in the system evaluation and security review.

The rest of the sections are organized as Sect. 2 includes literature review; the proposed method is discussed in Sect. 3. The implementation of the proposed security scheme is included in Sect. 4. The performance evaluation is presented in Sect. 5, and the research is concluded in Sect. 6.

2 Literature Survey

Many researchers are working for ensuring the security of EHR. Shi et al. [18] have presented a comprehensive assessment of blockchain-based approaches for EHR implementation, with an emphasis on security and privacy. They began by integrating prior information relevant to and associated with both health record systems and blockchain before delving into blockchain applications in EHRs.

Decentralization concepts were used by Arunkumar and Kousalya [2] to achieve data protection in peer-to-peer networks while maintaining the appearance of pseudo secrecy for patient data.

Bhavin et al. [4] suggested a way for securing a normal encryption scheme against quantum attacks by using quantum computing. During the hyperledger fabric blockchain block construction, the quantum blind signature is used.

Jiang and Guo [8] suggested a dynamic data sharing system with re-encryption and encryption. The cloud service re-encrypts the encrypted data before it is exchanged with the intended consumers, and this method manages users in a cloud system without modifying the keys when new users are added or removed.

Interworking of patient EHR data between hospitals and other private organizations is advocated by Gordon and Catalini [6]. Through exchanging EHR data across multiple health care networks using a blockchain network, patient-driven intractability addresses additional security and safety concerns.

According to Omar et al. [16], blockchain technologies may be used to protect health care data management. The decentralization of EHR data in the cloud, which decreases the risk of cyber-attacks, is discussed in this article.

Performing attribute-based authentication and ciphertext feature policy encryption, Huang et al. [7] achieve data confidentiality and data access security. Using fog nodes, this paper achieves faster encoding, decryption, and signature protocol execution and processing times undefinable.

Kaur et al. [13] recommend using blockchain technology to store heterogeneous medical data in the cloud. This paper covers a variety of EHR formats as well as problems of accessibility in cloud and blockchain environments.

Li et al. [14] proposed a blockchain-based accessible encryption technique for EHRs. The file for EHRs is constructed utilizing modern rationale articulations and put away on the blockchain, permitting an information client to look through the record utilizing the articulations.

Thus, decentralized and lightweight algorithms are required for ensuring the security of the EHR system.

3 Proposed System

The current system of cloud-based medical services information portrayed in (Fig. 1) stores different EHR from different information sources. Most hospital uses a centralized computerized repository where authorities handle all the data. While

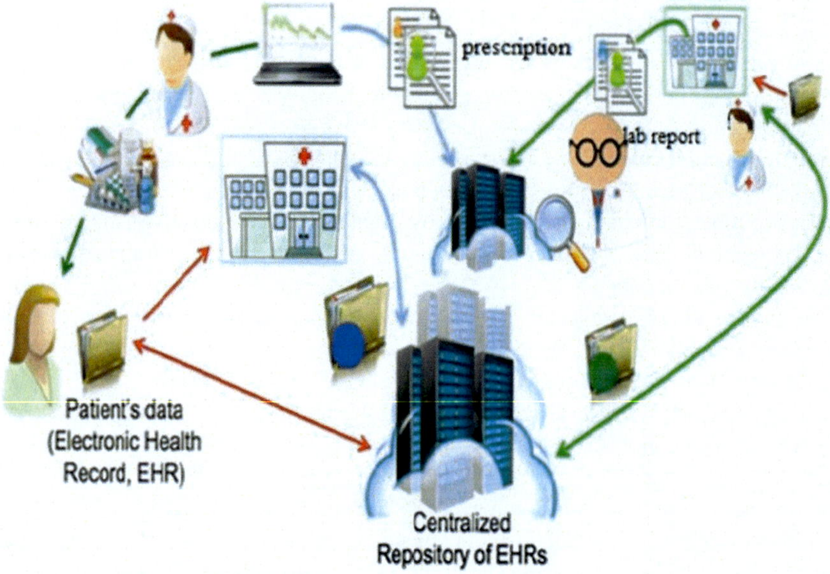

Fig. 1 EHR system using centralized repository

this solves the dilemma of paper-based records, patients do have concerns about medical record reliability, consumer control of data, data privacy, and other problems. Also, a central repository has to handle a large volume of data which increases the maintenance cost. It also had to deal with data replication and replication as a result of the fact that the patient's data was not authenticated.

Using IPFS and a lightweight authentication encryption technique (ALE), we proposed a blockchain-based secure decentralized health care system (see Fig. 2). We present the ALE algorithm, which stands for authenticated lightweight encryption. The AES round transformation and the AES-128 key scheduling are the foundations of ALE. The ALE is a single-pass online authenticated encryption technique with optional related data support. Its security is based on the use of nonces.

Here, we show a portion of ALE's benefits as far as execution:

1. AES equipment or programming executions can be reused with a couple of little changes, like the utilization of Intel AES guidelines.
2. Side-channel attack countermeasures, for example, edge executions in equipment to thwart first-request power have been made for the AES [15].
3. For long associations, ALE, as ASC1, requires just around four AES rounds to scramble and validate a square of a message. AES-256 GCM, then again, needs around ten AES adjusts.
4. ALE only requires the AES encryption engine for both encryption and authentication, as well as decryption and verification. AES-256 GCM requires both encryption and decryption engines to complete these duties.

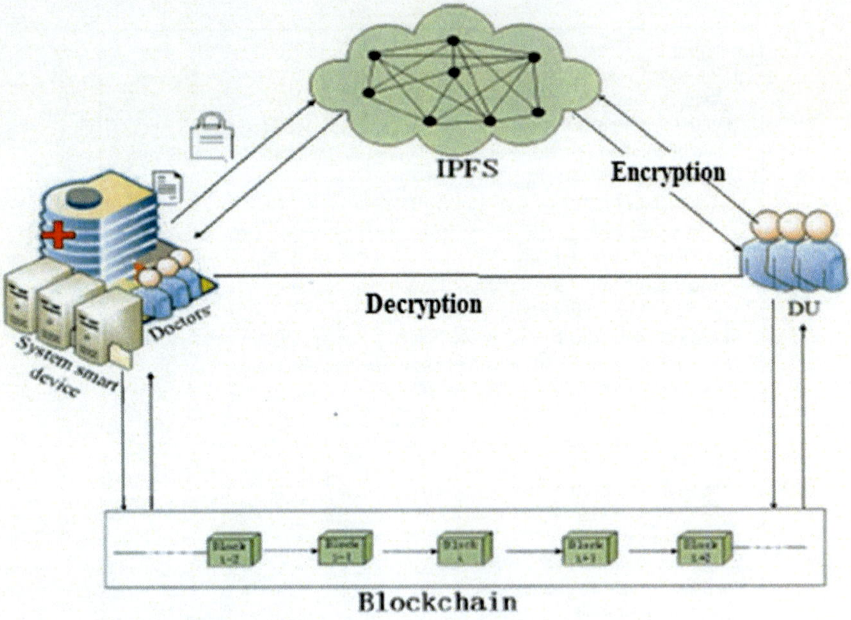

Fig. 2 Proposed IPFS-based blockchain

Medical data is encrypted using the patient's private key and processed on the blockchain in blockchain for health care. This information can be decrypted using the patient's public key, which is shared with users who have the patient's agreement, such as hospitals, diagnostic centers, and research organizations. This is diametrically opposed to our method, which provides patients with complete control over who can access their information. Additionally, we use IPFS to store data rather than the blockchain. This method is employed in the IPFS network by storing the patient's public key to hash. This map's hash is saved in a smart contract. When a doctor or a patient demands a document's hash, the smart contract must first collect the document's hash. They will then enter the record hash with the patient key. The primary benefit of this approach is that the hash of the map is the only thing the smart contract stores. The proposed architecture uses a decentralized and immutable blockchain database to address the privacy and security concerns associated with outsourcing EHR to the cloud. This design uploads the encrypted EHR to a decentralized cloud-based blockchain using a lightweight authentication encryption algorithm.

4 Implementation

Data encryption, data uploading to blockchain using IPFS, smart contract formation, data downloading, and proof of work validation are all part of the implementation process (Fig. 3).

Setup(n)
1:The nonce value is created by the data owner.
N={0,1}n and encryption secretkeys k->{0,1}n
2: With the support of public and private addresses, the secret keys nonce and encryption secret key aresafely shared with the data sender utilizing Elliptic wave cryptography.
KDC public key A (Nonce, Secretkey)→DataSender

EHR Encryption (ALE/RSA)
1:With the support of the ALE algorithmand Nonce parameters and secrete key value, the Data Sender encrypts the EHR file.
ALENonce, Secret(EHR)→S2: Using Elliptic wave cryptography and public and private addresses, the secret keys nonce and encryptionkey are safely exchanged with data senders.
E(n) , sk (EHR)→patient(ID)

EHR Uploading Phase
1:During the EHR process, each user uploads an encrypted file to a cloud Block Chain Network block by signing the file
signature(msg,privatekey):
private_key =web3.toBytes(hexstr=privatekey)
message = encode_defunct(text=msg)
signed_message = web3.eth.account.sign_message(message, private_key=private_key)
2: The Proof-of-Work or Proof-of-State consensusAlgorithm will verify the block and add to the Blockchain Network Using IPFSmessage = encode_defunct(text=msg)
return message,signed_message.signature
signature_verify(messagehash,signature)
a=web3.eth.account.recover_message(messagehash)
signature=signature

Storing the IPFS hash in the smart contractInput: IPFS Hash of the record
1: ihash → IPFS hash of the Data
2: p1 → ihash.stringh0; 24
3: p2 → ihash.stringh25; 48i
4: p1 r1→ p1+randCharact
5: p2 r2 → p2+randCharact
6: fpart1 → convtoBytes32(p1 r1)
7: fpart2 → convToBytes32(p2 r2)

(1) EHR Downloading Phase:
The smart contract is used to validate the transactionand pass it to the specific user-based rules written onthe smart contract during the downloading process. Step of EHR Downloading
(2) During the downloading process, the smart contractis utilized to validate the transaction andpass it on to the appropriate rules,which are user-based and written on the smart contract.

Fig. 3 Algorithm used for uploading and downloading data in EHR

4.1 Data Encryption Phase

During the data encryption method, to encrypt the patient's health record data, the ALE algorithm is used. The KDC produces two random numbers, N and K, which it sends over a protected communication channel. Per transaction in a block will generate a smart contract during the data uploading process in the patients' encrypted record. The patients' health record will produce a private key and a public key using ECDSA to validate the encrypted record transaction in a blockchain after adding a smart contract to each transaction in a block of health record info. The KDC generates a patient's public and private keys and sends them to them through secure channel communication.

$$\text{(Pulickey, Privatekey)} \rightarrow \text{ECDSA (Patient health record)} \quad (1)$$

4.2 Data Uploading Phase

During the data uploading step, a smart contract will be created of the patients' encrypted EHR for every transaction in a block. The patient then signs the encrypted EHR with his or her private key and uploads it to IPFS. After data being signed by the patient, data is uploaded to the IPFS. It will return a hash value and upload the data to the blockchain network. The blockchain network, where the miner verifies it using the patient's public key (Fig. 4).

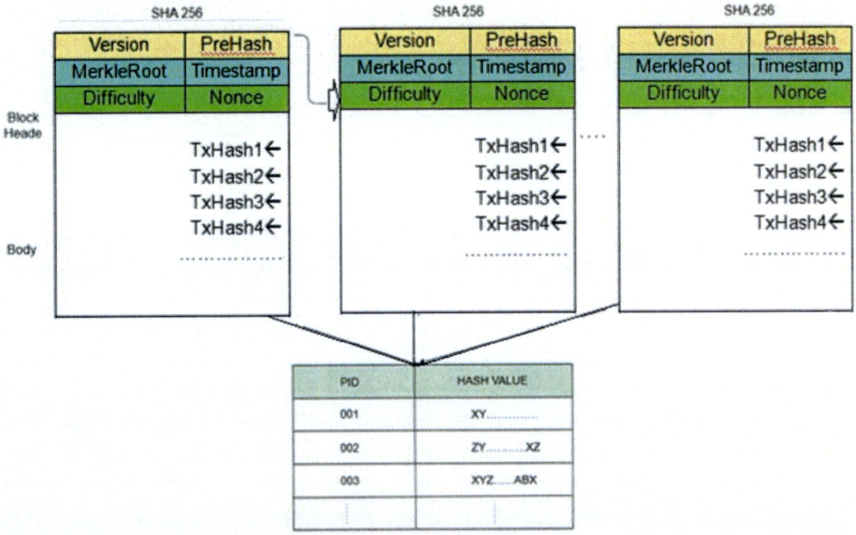

Fig. 4 Blockchian creation in EHR system

$$S = \text{SignPrivate key (EncryptedEHR)} \rightarrow \text{IPFS} \rightarrow \text{BlockChain} \tag{2}$$

4.3 Data Downloading Phase

During the data downloading process, the smart contract is applied to verify the transaction and transfer it to the relevant rules based on the user is written on the smart contract. A smart contract is used in a blockchain network to give permissions for each transaction in a block.

4.4 Proof of Work

The initial consensus algorithm in a blockchain network is proof of work (PoW). The algorithm confirms the transaction and adds another block to the blockchain. The cloud-based blockchain is built on the idea of a public blockchain network, with new blocks added using a consensus process called proof of work, which validates the block by solving a hard mathematical problem. After the block is solved, the new block is broadcasted to the blockchain network, where each transaction in the newly added block is validated using the SHA-256-bit and EDCSA (Fig. 4).

POW (proof of work) = double (SHA-256 hash) (previous block hash, nonce, Merkle root, EHR transaction)
Hash H = Proof of work $(N - 1)$.

5 Performance Analysis

The proposed method was tested on a PC with an Intel operating system and a 2.60 GHz i7-4510U processor with 6 GB RAM. We built a cloud-based public blockchain using Ethereum and IPFS and validated the blocks using two separate consensus algorithms: proof of work (PoW) and proof of stake (PoS). Using the MIRACL cryptography library, encrypt the patient's health record file and upload it to a cloud-based blockchain.

Table 1 lists the different operations performed on a cloud-based blockchain, as well as their cryptographic cost and execution time in milliseconds. According to the findings, the lightweight AES algorithm would result in the shortest processing and setup period for uploading health care data to the cloud-based blockchain Table 2. In addition, when compared to the previous study, our approach produces better results, as shown in Table 2.

Figure 5 illustrates the execution time before and following the use of IPFS in the blockchain. The time required to process the block progressively grows. Figures 6

Table 1 Cost for single-round cryptographic operation

Method	Setup	Keygeneration	Encryption	Sign/ver	Dec
ALE	0.004	0.0007210	0.0000638	0.0001336	0.00040
ECDSA	0.005	0.021399	0.0223669	0.0256238	0.00035
RSA	0006	0.1849945	0.0023114	0.0005830	0.00284

Table 2 Comparative analysis for single cryptographic operation of AES

Features	Arunkumar [10]	Proposed method
Setup	0.005	0.004
Keygen	0.0017	0.0007210
Enc	0.0015	0.0000638
Dec	0.0030	0.00040

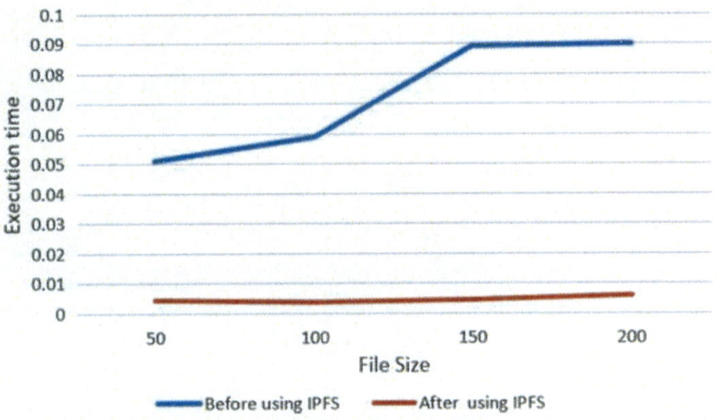

Fig. 5 Execution time before and after using IPFS

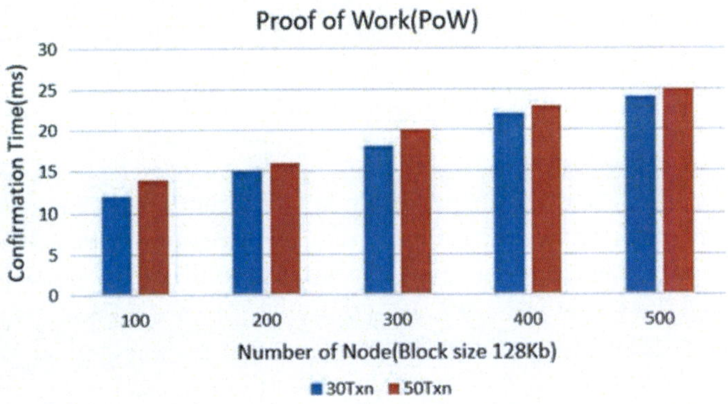

Fig. 6 Confirmation time of consensus algorithm (PoW) for different transaction

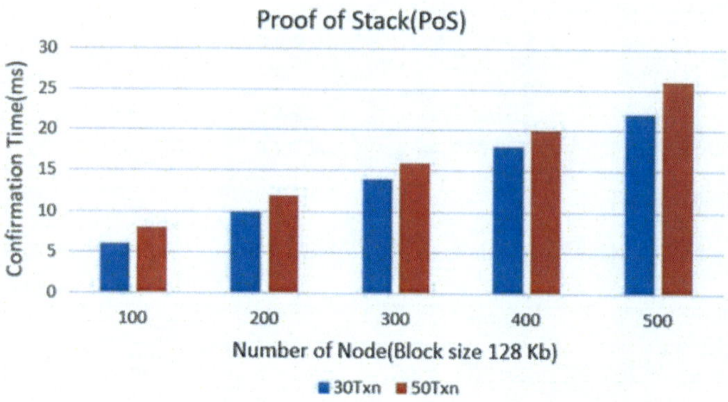

Fig. 7 Confirmation time of consensus algorithm (PoS) for different transaction

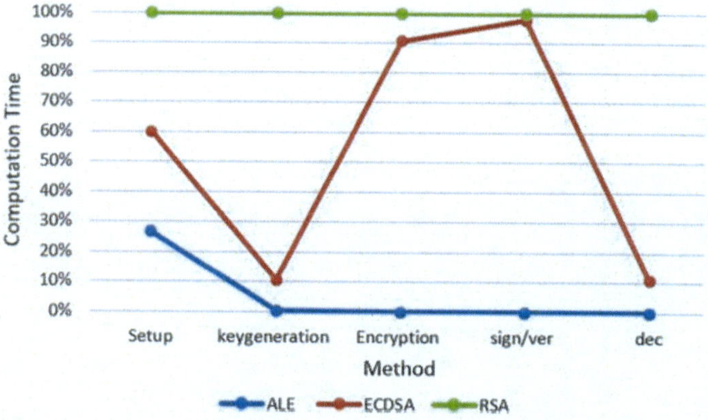

Fig. 8 Cryptographic cost for various operations in IPFS-based blockchain

and 7 illustrate the difference between proof of work (PoW) and proof of stack validation (PoS). We used a block size of 128 KB. The confirmation time against various nodes is between 100 and 500 ms, and the number of transactions is between 30 and 50 trillion. Confirmation time is longer when there are a large number of nodes involved in several transactions. Figure 8 illustrates the cryptographic costs associated with various operations in an IPFS-based blockchain using various methods. Cryptographic costs are determined based on the volume of transactions sent to a cloud-based clinical blockchain. Cryptographic costs are the highest when decoding EHR data and the lowest when key sharing is used. As can be seen, ALE has the lowest computing cost when compared to the other methods.

6 Conclusion

The cloud-based blockchain architecture utilizes a separate key processing center to establish and exchange public and secret keys over an insecure medium to encrypt and decrypt electronic health record information. Additionally, a cloud-based peer-to-peer storage architecture based on IPFS is being developed to enable decentralized data storage and data access management for EHR sharing. As a result, the suggested architecture makes use of cloud-based medical blockchain technology to ensure the longevity, decentralization, and traceability of EHR data, as well as its anonymity. The proposed methodology reduces security and processing complexity by utilizing blockchain-based cloud operations on electronic health record data. This is achievable due to the proposed algorithm's small weight.

References

1. Arifeen MM, Al Mamun A, Kaiser MS, Mahmud M (2020) Blockchain-enable contact tracing for preserving user privacy during COVID-19 outbreak
2. Arunkumar B, Kousalya G (2020) Blockchain-based decentralized secure lightweight e-health system for electronic health records. In: Intelligent systems, technologies and applications. Springer, New York, pp 273–289
3. Asif-Ur-Rahman M et al (2019) Toward a heterogeneous mist, fog, and cloud-based framework for the internet of healthcare things. IEEE Internet Things J 6(3):4049–4062
4. Bhavin M, Tanwar S, Sharma N, Tyagi S, Kumar N (2021) Blockchain and quantum blind signature-based hybrid scheme for healthcare 5.0 applications. J Inf Secur Appl 56:102673. https://doi.org/10.1016/j.jisa.2020.102673. https://www.sciencedirect.com/science/article/pii/S2214212620308255
5. Biswas S, Anisuzzaman, Akhter T, Kaiser MS, Mamun SA (2014) Cloud based healthcare application architecture and electronic medical record mining: an integrated approach to improve healthcare system. In: 2014 17th international conference on computer and information technology (ICCIT), pp 286–291. https://doi.org/10.1109/ICCITechn.2014.7073139
6. Gordon WJ, Catalini C (2018) Blockchain technology for healthcare: facilitating the transition to patient-driven interoperability. Comput Struct Biotechnol J 16:224–230. https://doi.org/10.1016/j.csbj.2018.06.003. https://www.sciencedirect.com/science/article/pii/S200103701830028X
7. Huang Q, Yang Y, Wang L (2017) Secure data access control with ciphertext update and computation outsourcing in fog computing for internet of things. IEEE Access 5:12941–12950. https://doi.org/10.1109/ACCESS.2017.2727054
8. Jiang L, Guo D (2017) Dynamic encrypted data sharing scheme based on conditional proxy broadcast re-encryption for cloud storage. IEEE Access 5:13336–13345. https://doi.org/10.1109/ACCESS.2017.2726584
9. Kaiser MS, Al Mamun S, Mahmud M, Tania MH (2021) Healthcare robots to combat COVID-19. Springer Singapore, Singapore, pp 83–97
10. Kaiser MS, Zenia N, Tabassum F, Al Mamun S, Rahman MA, Islam MS, Mahmud M (2021) 6G access network for intelligent internet of healthcare things: opportunity, challenges, and research directions. In: Proceedings of international conference on trends in computational and cognitive engineering. Springer, pp 317–328
11. Kaiser MS et al (2021) iWorksafe: towards healthy workplaces during COVID-19 with an intelligent pHealth app for industrial settings. IEEE Access 9:13814–13828. https://doi.org/10.1109/ACCESS.2021.3050193

12. Kaiser MS et al (2018) Advances in crowd analysis for urban applications through urban event detection. IEEE Trans Intell Transp Syst 19(10):3092–3112. https://doi.org/10.1109/TITS.2017.2771746
13. Kaur H, Alam MA, Jameel R, Mourya AK, Chang V (2018) A proposed solution and future direction for blockchain-based heterogeneous medicare data in cloud environment. J Med Syst 42(8):1–11
14. Li J, Liu Z, Chen L, Chen P, Wu J (2017) Blockchain-based security architecture for distributed cloud storage. In: 2017 IEEE international symposium on parallel and distributed processing with applications and 2017 IEEE international conference on ubiquitous computing and communications (ISPA/IUCC). IEEE, pp 408–411
15. Mamun AA, Hasan SR, Bhuiyan MS, Kaiser MS, Yousuf MA (2020) Secure and transparent KYC for banking system using IPFS and blockchain technology. In: 2020 IEEE region 10 symposium (TENSYMP), pp 348–351. https://doi.org/10.1109/TENSYMP50017.2020.9230987
16. Omar AA, Bhuiyan MZA, Basu A, Kiyomoto S, Rahman MS (2019) Privacy-friendly platform for healthcare data in cloud based on blockchain environment. Future Gener Comput Syst 95:511–521. https://doi.org/10.1016/j.future.2018.12.044. https://www.sciencedirect.com/science/article/pii/S0167739X18314201
17. Sarker IH (2021) Machine learning: algorithms, real-world applications and research directions. SN Comput Sci 2(3):1–21
18. Shi S, He D, Li L, Kumar N, Khan MK, Choo KKR (2020) Applications of blockchain in ensuring the security and privacy of electronic health record systems: a survey. Comput Secur 101966

Text Mining and Education 4.0

Multi-label Emotion Classification of Tweets Using Machine Learning

Simon Islam, Animesh Chandra Roy, Mohammad Shamsul Arefin, and Sonia Afroz

Abstract Twitter is one of the biggest social media network in the world. It has 330 million active monthly users. Users of Twitter use micro-blogs called tweets to express their opinions. This generates a huge amount of textual data every minute. Analyzing this data to search for emotions in it will lead us to understand the emotions currently presiding over the Internet. Emotion classification has a long history. Based on the approach, three ways of solving the emotion classification problem are devised. Binary classification detects whether an emotion is present or not. Multi-class classification classify tweets into one of the many available classes. In this paper, we proposed multi-label emotion classification of tweets. In multi-label classification, it is possible to label a tweet with more than one emotion. Multi-label classification methods follow two approaches to solve emotion classification problems. A total of 13 advanced multi-label classification methods was used to train and evaluate a tweet dataset containing 8501 tweets. 10 of them were problem transformation methods, and 3 were algorithm adaptation methods. We found that, although all the classifiers performance are pretty close, problem adaptation method like binary relevance and label powerset performs better that other multi-label classifiers. We have also found that random forest classifier works better than support vector machine as base classifier in problem transformation methods for multi-label classification. We achieved micro F-score up to 0.91 and subset 0/1 loss of 0.28. We also showed that, Senticnet5 can be used to improve the accuracy of the models. Considering our dataset contains tweets from various incidents, this represents a statistically significant improvement.

Keywords Multi-label emotion classification · Binary relevance · Classifier chains · MLkNN · Twitter

S. Islam (✉) · A. C. Roy · M. S. Arefin
Chittagong University of Engineering & Technology, Chattogram 4349, Bangladesh
e-mail: animesh_roy@cuet.ac.bd

M. S. Arefin
e-mail: sarefin@cuet.ac.bd

S. Afroz
University of Information Technology & Sciences, Dhaka, Bangladesh
e-mail: sonia.afroz@uits.edu.bd

M. S. Arefin et al. (eds.), *Proceedings of the International Conference on Big Data, IoT, and Machine Learning*, Lecture Notes on Data Engineering and Communications Technologies 95, https://doi.org/10.1007/978-981-16-6636-0_53

1 Introduction

In this current age of information technology, Twitter is one of the most prominent names. It is a micro-blogging Web site. Micro-blogs are usually small and compact. Twitter demonstrates this by capping the blog size to 140 characters. But in 2017, they increased the size to 280 characters. The micro-blogs on Twitter are generally known as 'tweets'. The topics of tweets can range from various topics, pictures, GIFs, or even sharing links to other Web sites. Twitter wants its reach to be global. For this reason, an unregistered user can see what a registered user has posted. And two users also need not be mutually connected to follow each other. Twitter has achieved this global reach by committing to this one-way follower relationship.

With the rapid advancement of technology, the world is advancing at a hectic pace. Current news and affairs are shifting constantly. Twitter keeps up with this rapid pace with its trending tab. It determines the trends by calculating the volume of a keyword over a period of time. Since it is a global platform, it is an excellent place for people to express their opinions and emotions. From average people to global leaders, many use Twitter to express their opinion on current topics. In return, the followers of the people also reply to the tweets with their own opinions. By extracting emotions from these active exchanges, it will be beneficial to understand the mindset of the people participating in the conversation. Multi-label emotion classification can be used to understand emotions in tweets.

Multi-label classification problem shows relation to multi-output classification problem where a problem can have multiple labels assigned to it. It is not similar to multi-class classification. In it, only one label among multiple labels can be assigned to it. Due to a tweet having multiple emotions embedded into it, multi-label classification is necessary over multi-class classification. The complex human nature can simply not be expressed by one label. In this work, we aim to perform multi-label emotion classification of tweets using machine learning. We hope to classify the tweets into eight classes. They are joy, sadness, anger, disgust, admiration, surprise, interest, and fear. We will collect tweets from Twitter using Twitter API. Then we will label the tweets with the appropriate emotions. Since our work is multi-label classification, we will assign more than one appropriate label if necessary to the tweets. Then the labeled tweets will be used to train our model. And finally, we wish to evaluate our model.

2 Motivation

We live in an age of social media. It has been ingrained into our daily lives. Peeking into social media is a very good way to stay up to date with current trends. For this reason, analysis of tweets can give us a good glimpse at current affairs and how people are being affected by them. Due to the variety of topics, it may not provide us with a totally accurate analysis but it can give us the gist of it.

Many previous works done in the past regarding the classification of tweets have been on sentiment analysis. Some of the works also looked for a multi-class classification solution. But as we are aware, human emotion is very complex. People do not think in one emotion at a time. Human thought can be directed in multiple direction. Thus, it consists of multiple emotions. For this reason, it is safe to assume that a tweet posted by a human may also contain multiple emotions. This is why sentiment analysis or multi-class classification of tweets may not be enough to represent the human psyche. We need multi-label emotion classification for this purpose.

3 Related Works

3.1 Sentiment Classification

Sentiment analysis is a natural language technique that determines if the given data asserts positive sentiment or negative sentiment. It is also referred to as opinion mining. It has been extensively used to study various product reviews and micro-blogs like Twitter [1]. Since it deals with classification from negative to positive, sometimes a neutral class is also introduced. The task of analyzing the sentiment polarity of work can be divided into three approaches. There are lexicon-based approaches [2]. They consider the sentiment polarity of words in a document to get the polarity of the text. Then there is the machine learning-based approaches [3], where one or multiple models are trained on given datasets to build classifiers to determine the polarity of text. And finally, there is a hybrid method that combines aspects of both of the previous ways together.

Since tweets used in this work can display many emotional states, a sentiment analysis approach will not be enough. For us, it is not enough to know whether the polarity of a tweet is negative or positive, we also have to know what emotions does it display.

3.2 Multi-class Classification

An improvement over sentiment analysis is multi-class classification. In this approach, the text can be classified into one of three or more classes, thus classifying it as part of one concrete emotional class. But in a dynamic world like ours, it is considered to be a drawback of this approach. Lin et al. [4] proposed an approach to classify based on readers emotions. Liang et al. [5] developed a system that will recommend emoticons to readers based on the content of what they are typing. And [6] proposed a scalable approach to quantify tweets into different sentiment classes. They used seven different sentiment classes which consist of three pairs of different sentiments and one neutral class. All of these works assign only one concrete sentiment to a text. Thus, they neglect other sentiments that might be present in it.

3.3 Multi-label Classification

The problem of one text having multiple emotions leads us to multi-label classification. In this process, one text can have more than one class assigned to it. Previous works of multi-label classification are divided into two groups based on the approach. One of them is problem transformation methods, and the other one is algorithm adaptation method.

In problem transformation methods, the given problem is transferred to a sentiment classification problem or multi-class classification problem. Methods like binary relevance (BR) [7] change the problem of detecting multiple labels into multiple binary classification problems. And at the end, the result of all the classifiers is combined again to produce multi-label output. On the other hand, label powerset (LP) [8] transforms the problem into multi-class classification. Each unique combination of labels is considered as a separate class. The models are then trained and tested on that assumption. Random *k*-Labelsets (RAkEL) [8] builds a collection of different LP classifiers. Here, different random subsets of labels are used to train each of the classifiers. It is then combined to generate the output of multi-label prediction.

The advantages of problem adaptation methods are that they are easy to implement and understand. But they fail to adjust and understand the interdependencies between models. Classifier chain (CC) [9] transforms the problem of classification into a group of binary classification problems that resembles a chain. Here, the length of this chain is determined by the amount of unique labels in the dataset. Each binary classification problem is solved by one classifier chain. But their knowledge is augmented by previous models which were trained on different emotions.

Algorithm adaptation methods convert prominent algorithms of machine learning to handle multi-label data. One of the main examples of this is the multi-label k-nearest neighbor algorithm (MLkNN) [10]. Here, *k*-nearest neighbor algorithm is modified to work with multi-label data. It uses the maximum a posteriori rule to make a multi-label prediction. Other examples of this include multi-label decision tree, Rank SVM, predictive clustering trees. Liu and Chen [11] used 11 multi-label classification-based approaches to train and predict 2 micro-blog datasets collected from different incidents. They also did a comparison of various multi-label classification approaches. Cabrera-Diego et al. [12] used issue trackers comments on Stack Overflow and JIRA to perform multi-label classification of the texts. They used two multi-label classifiers (RAkEL and HOMER) to train and test their models.

In this paper, we present a framework to perform multi-label emotion classification of tweets using machine learning. Our contribution in this paper can be summed up as follows:

- Development of a tweet dataset with properly labeled emotions.
- Training various multi-label classification models with the features extracted from the dataset.
- Using Senticnet5 to boost the accuracy of the models.

4 Methodology

4.1 Outline of the Methodology

To collect and label tweets in real time, we will use Twitter API. The collected tweets will be sent to our multi-label emotion classifier to classify it as various emotions. But before that, we need to train and develop our emotion classifier models. The first part is to collect and label tweets which will be used to train the models. After the collection, we will develop models that can perform multi-label emotion classification. The tweets that will be used to train classifiers were selected from *Sentiment140* [13] dataset. The tweets will then be labeled by hand and preprocessed before being used to train the classifiers. The labeling power of the classifiers will be augmented using *Senticnet5* [14] (Fig. 1).

4.2 Tweet Preprocessing

Before training our models, we needed to process our text data to be able to pass it through the training models. The purpose of this preprocessing is to keep features that are related to their labels. We also needed to discard or trim features that complicate the training phase. We started the preprocessing by removing links and pictures from the tweets. The sentence was then separated into multiple sentences. Each sentence was evaluated separately. The hashtags present in the tweet were processed by removing the hash symbol and reading the word to the sentence. If the tweet had any punctuation in it, the flags for question marks and exclamation points were turned on. We detected words that have negation (Ex. not, n't) before them. We removed stopwords from the sentence. We then tokenized the words and performed lemmatization. This will group the infected forms of words so that we can process them as a single item instead of differentiating between them. Finally, we implemented tf-idf over the analyzed tweet dataset. This transforms raw texts in the dataset to a matrix representation of tf-idf words. A table representing the process is shown in Table 1 (Fig. 2).

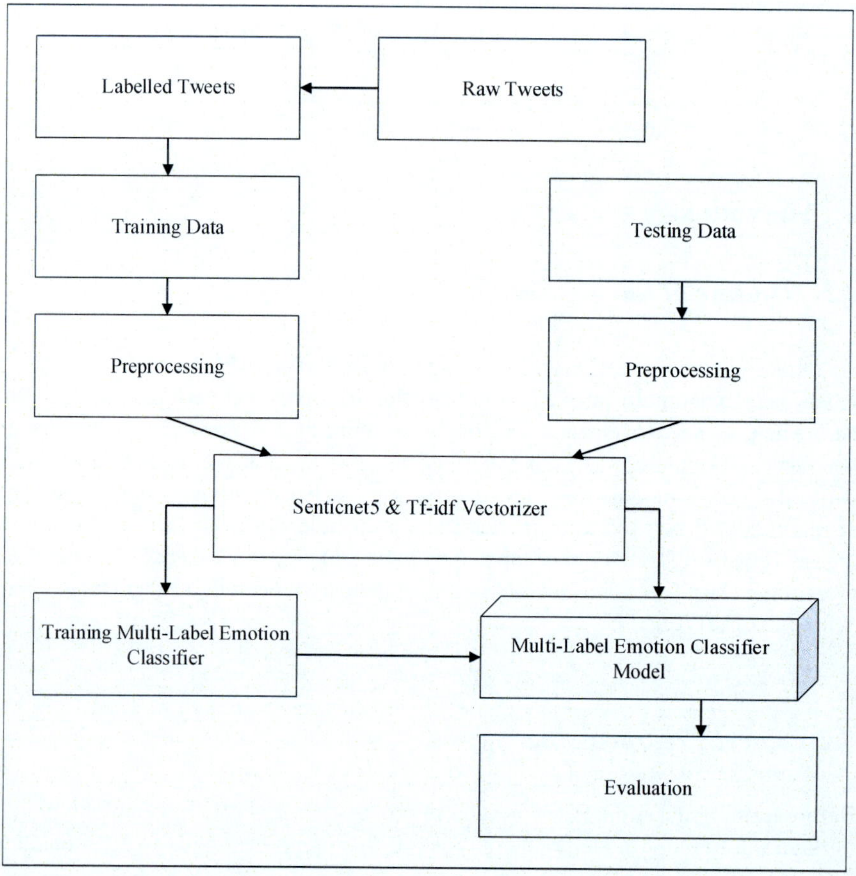

Fig. 1 Multi-label emotion classifier framework

4.3 Senticnet5

Senticnet is an initiative that began in the MIT media laboratory in 2009. It is used for emotion-aware intelligent applications in various fields spanning from natural language processing to human–computer interaction. It has many applications. In this work, we used it as a concept-level knowledge base to enhance the training phase of our model. In *Senticnet5* [14], various words are labeled according to their primary and secondary emotions. We transformed this information suitable to our dataset and added it during the training phase. Since compared to the vast amount of tweets out there, our dataset only covers a tiny fraction. This outside help is needed to understand various words it encounters which may not be present in our dataset.

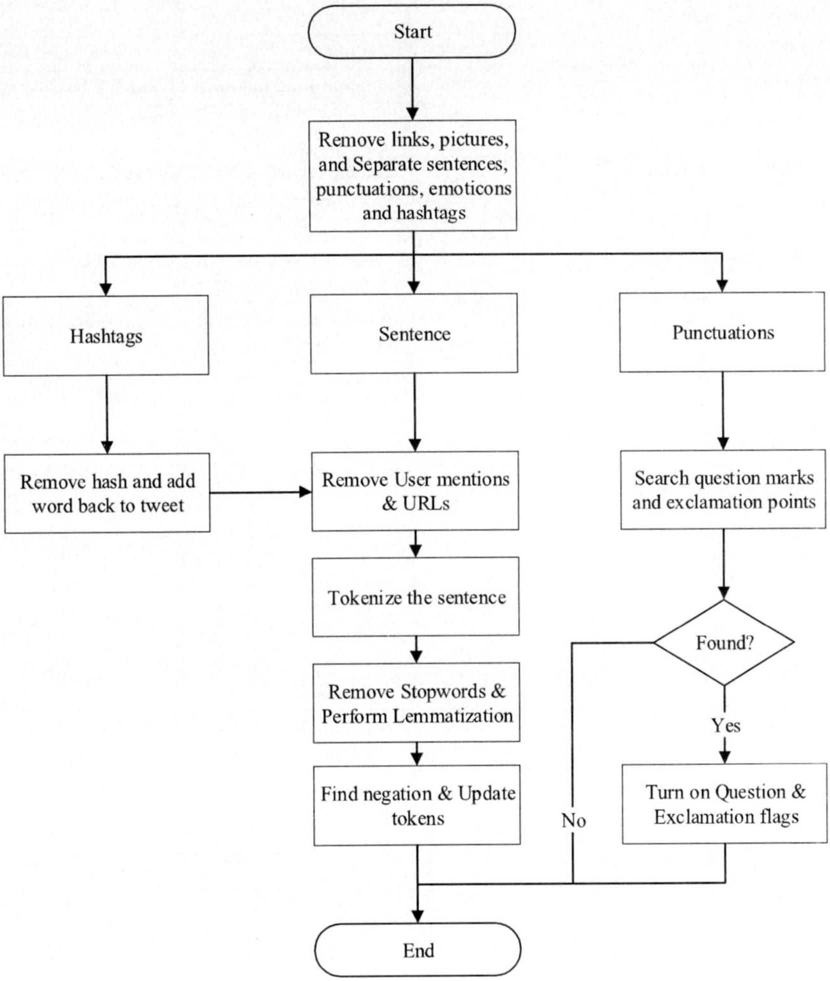

Fig. 2 Tweet preprocessing

4.4 *Tf-idf*

Tf-idf is based on numerical statistics. It determines the importance of a word in a document. In our case, it will indicate the significance of words to a tweet. The value of tf-idf is correlated to the occurrence of a word in the document. Tf-idf is a combination of two-part. And the other one is inverse document frequency. As understood from the name, term frequency is derived from the count of a term in the document. Inverse document frequency measures the amount of information provided by the word. It indicates the rarity of the word. It is logarithmically scaled.

Table 1 Tweet processing

Process	Example
Raw tweet	@alamin I did not have a good day but @refat had a blast. #Mixed #day Watch it here: https://youtu.be/dqw4w9wgxcq
Remove user mention	I did not have a good day but had a blast. #mixed #day watch it here: https://youtu.be/dqw4w9wgxcq
Remove hyperlinks	I did not have a good day but had a blast. #mixed #day watch it here:
Process hashtags	I did not have a good day but had a blast mixed day watch it here
Tokenization and remove stopwords	[‘i’, ‘did’, ‘not’, ‘have’, ‘good’, ‘day’, ‘had’, ‘blast’, ‘mixed’, ‘day’, ‘watch’, ‘here’]
Lemmatization	[‘i’, ‘do’, ‘not’, ‘have’, ‘a’, ‘good’, ‘day’, ‘but’, ‘have’, ‘a’, ‘blast’, ‘mixed’, ‘day’, ‘watch’, ‘it’, ‘here’]
Negation words	[‘have’, ‘a’, ‘good’, ‘day’]
POS tagging	[(‘i’, ‘NN’), (‘did’, ‘VBD’), (‘not’, ‘RB’), (‘have’, ‘VB’), (‘a’, ‘DT’), (‘good’, ‘JJ’), (‘day’, ‘NN’), (‘but’, ‘CC’), (‘had’, ‘VBD’), (‘a’, ‘DT’), (‘blast’, ‘NN’), (‘mixed’, ‘JJ’), (‘day’, ‘NN’), (‘watch’, ‘VB’), (‘it’, ‘PRP’), (‘here’, ‘RB’)]
Post POS tagging	[‘i’, ‘did’, ‘not’, ‘have’, ‘good’, ‘day’, ‘had’, ‘blast’, ‘mixed’, ‘day’, ‘watch’, ‘here’]
Final output	I do not have good day have blast mixed day watch here

We use tf-idf to perform the vectorization of our tweet dataset. It will convert the raw tweets to a matrix of tf-idf features. As our models cannot understand the meaning of written English words, performing tf-idf and converting the dataset to a matrix will help it to evaluate one term against other terms. This will help it to assign terms with emotion and improve its classification process.

4.5 *Training and Testing*

To train and test our multi-label models, various popular multi-label classification techniques were used. During the training phase, *Senticnet5* was used to enhance the training set.

Binary relevance (BR) [7] transforms the problem of multi-label classification into a group of single label binary classification problem. It is very easy to implement, and it is faster than many later methods. The number of models needed for the binary relevance solution is equal to the number of models. It divides the multi-label problem into numerous independent binary polarity detection tasks (one per label). All the models are trained parallelly, independent from each other. Since the tasks are independent of each other, it ignores the correlation that may be present between the labels.

Label powerset (LP) [8] transforms multi-label classification problem into multi-class classification. Each unique combination of labels is assigned a class. Thus, for a dataset with N labels, there are 2^N possible classes. Although it simplifies a multi-label problem into a multi-class problem, it also increases the runtime complexity of training the model. Also, there may not be enough data for some classes to train it properly. It will lead to training loss. And our model will only be capable of finding classes with high enough samples.

Classifier chains (CC) [9] is somewhat similar to BR. But instead of training the binary models parallelly, it trains them sequentially. And the result of previous training is used to augment the results of the next training session. It maintains the efficiency of binary relevance while still maintaining the correlation between different labels. The problem arises on deciding how the chain is formed. Since for a dataset with N labels, there are $N!$ possible unique combinations. Thus truly determining the best chain combination actually negates the advantages of using a classifier chain. An improvement of the classifier chain is **Ensemble of Classifier Chains (ECC)** [9], where several CC classifiers are trained with a random combination of chains. And after training all of the models, the labels are predicted based on a user-given threshold. We used majority voting for prediction. It means a tweet is labeled with emotion if at least half of the classifier chains voted for it.

Random k-Labelset (RAkEL) [8] is an improvement over label powerset. Instead of dividing N labels into 2^N classes, it breaks the starting set of labels into a small subset of labels. The subsets are created randomly. Then after the training phase, they are ensembled to produce the final prediction. In RAkEL, the size of the subset is specified by k. If N labels are divided into L subset of size k, the complexity will go down from 2^N to $L \times 2^k$ where $k < N$. In documents with large number of labels, it reduces the training time and complexity while maintaining the effectiveness of label powerset. It is implemented using the scikit multilearn package that uses the implementation of [15].

MLkNN [10] which stands for multi-label K-nearest neighbors is one of the algorithm adaptation methods to solve multi-label classification problems. Here, K-nearest neighbors algorithm is modified to work with a multi-label dataset. In MLkNN, for every iteration, its nearest k neighbors are calculated. After identifying, maximum a posteriori (MAP) principle is used to determine the labels for unidentified data. MAP principle calculates the probability of a label appearing in the unidentified tweet.

BRkNN [16] combines approaches from both binary relevance and K-nearest neighbor. It adopts a lazy learning approach. In this method, the k nearest neighbors are needed to be found only once. This is a big improvement from binary relevance with kNN as the base classifier. Instead of running kNN $|L|$ number of times where $|L|$ represents the number of labels, it is run only once. Thus, it is $|L|$ times faster than traditional BR methods. Finally, the resulting nearest neighbors are evaluated using BR to find multi-labels. There are two extensions to BRkNN. They are BRkNN-a and BRkNN-b.

Both BRkNN-a and BRkNN-b are based on calculating the confidence score for each of the labels. The confidence of a label is calculated by considering the k-nearest neighbors where the label is present. Formally, the confidence of label j can be written as:

$$cf_j = \frac{1}{k}\sum_{i=1}^{k} y_j^i \tag{1}$$

Now, the first extension BRkNN-a outputs an empty set if a label is not present in at least half of the k-nearest neighbors. If that happens, then BRkNN the labels with the highest confidence is selected. In BRkNN-b, the average size of the nearest k neighbors label set s is calculated by taking the average of positive labels present in k neighbors. Finally, the model outputs top $[s]$ labels with the highest confidence.

5 Experimental Results

5.1 Dataset Description

The tweets used to train multi-label classifiers was selected from *Sentiment140* [13] dataset. It contains 1.6 million tweets extracted using Twitter API. We randomly selected 8500 tweets from this. These tweets were then analyzed and properly labeled according to the emotions present in them. The emotions used are joy, sadness, anger, disgust, admiration, surprise, interest, and fear. According to [17], there are six basic emotions. They are anger, disgust, fear, joy, sadness, and surprise. We added the emotions interest and admiration to our dataset because while labeling we found some tweets hard to classify with the six basic emotions. A sample of the dataset is shown in Table 2.

Multi-label datasets are not like other datasets. Hence, we need new parameters to properly describe them. They are label cardinality (LC) and label density (LD). Let, Y_i to be the number of samples for ith tweet in the dataset, N be the number of tweets present in the dataset, and L be the number of emotions present in the dataset. We can describe LC and LD using,

$$\text{LC} = \frac{1}{N}\sum_{i=1} N|Y_i| \tag{2}$$

Table 2 Sample of the dataset

Text	Emotion
So much for sleeping in.	Fear
College days are loooong days.. 3 more hours #tired	Sadness, Interest
@daihard I'm headed to Kentucky this time. Never been so it should be fun!? http://blip.fm/xgqz1	Interest
hella tired.. where is gilbert for the usual basketball talk?!	Interest
Not as dry this morning as would have liked lot of moisture on the dune grass this am meant me and the dogs came home soaking wet!	Sadness, Disgust
@lil_laura_loo Really? I think we have some! I've taken Piriteeze but only works for a little while and can only take 1 a day! xo	Sadness, Fear

Table 3 Dataset properties

Property	Value
Total number of tweets	8500
Label cardinality	1.5
Label density	0.1875
No. of tweets with 'Joy' label	2406
No. of tweets with 'Sadness' label	4126
No. of tweets with 'Anger' label	989
No. of tweets with 'Disgust' label	1380
No. of tweets with 'Admiration' label	578
No. of tweets with 'Surprise' label	624
No. of tweets with 'Interest' label	2134
No. of tweets with 'Fear' label	674
No. of tweets with zero labels	323
No. of tweets with one label	4247
No. of tweets with two labels	3142
No. of tweets with three or more labels	789

$$\text{LD} = \frac{1}{N} \sum_{i=1} N \frac{|Y_i|}{|L|} \tag{3}$$

Various properties of the dataset are described in Table 3. We also generated association rules from the dataset. Interesting associations and relationships were discovered during this process. Although it is mostly used in market transactions, we used it here to show the co-relations between the emotion labels. Python 'Apyori' package which has a method for implementing apriori algorithm was used to generate the rules. Some of the generated rules are shown in Table 4.

Table 4 Association rules generated from the dataset

Association rules	Support	Confidence	Lift
Admiration → Joy	0.089	0.682	1.904
Disgust → Anger	0.087	0.394	1.252
Disgust → Sadness	0.231	0.732	1.277
Joy → Interest	0.161	0.464	1.296
Joy → Surprise	0.046	0.131	1.005
Admiration, Interest → Anger	0.0017	0.005	1.188
Admiration, Joy → Interest	0.011	0.029	1.028
Admiration, Sadness, Interest → Anger	0.0002	1	2.792

5.2 *Evaluation of Performance*

A total of 26 experiments was conducted on our dataset in which we used different multi-label classifiers. Thirteen of them were on raw tweets and thirteen of them were on the processed dataset with Senticnet5 added to improve the learning of the models for negation and missing words. Eight different multi-label classifiers were used. Five of them used problem transformation methods, and the other three used the algorithm adaption methods. Problem transformation methods used were binary relevance (BR), label powerset (LP), classifier chains (CC), random k-labelset (RAkEL), and ensemble of classifier chains (ECC). Algorithm adaption methods used were MLkNN, BRkNN-a, and BRkNN-b. For classifiers that used problem transformation methods, a base classifier was needed. For this purpose, both random forest classifier [18] and support vector machine [19] were used. All of the models were trained with K-fold cross-validation with $k = 10$. The results shown here are taken as the average of the results achieved from k-fold cross-validation.

We based our evaluation on five criteria. They are selected from the sixteen metrics used by Madjarov et al. [20] in their extensive experimental comparison of multi-label classification methods. Three of them are example-based metrics, and two of them are label-based metrics. Ranking-based metrics were not considered since the emotions are not ranked in contrast to each other. They are Hamming loss (HL), subset 0/1 loss, macro F_1, micro F_1, and average accuracy.

I. **Hamming loss** (HL) is used to calculate the dissimilarity between test cases and predictions. It is the fragment of labels that was falsely determined for a sample. It has a value between 0 and 1. Low value of hamming loss indicates high performance.

$$\mathrm{HL} = \frac{1}{|N||L|} \sum_{i=1}^{|N|} \sum_{j=1}^{|L|} T_{ij} \oplus P_{ij} \tag{4}$$

II. **Subset 0/1 loss** considers a label wrong if even one of the label is predicted wrong. Thus, it can be classified as absolute accuracy of the model. Bigger value of subset 0/1 loss means better prediction capability. It is also known as subset accuracy.

$$\text{Subset } 0/1 \text{ Loss} = \frac{1}{|N|}\sum_{i=1}^{|N|} T_i \oplus P_i \tag{5}$$

III. **Macro *F*-score** evaluates the prediction accuracy of labels. Here, the proportion of every label class is taken into account. It informs how well the classifier performs over the dataset.

$$\text{Macro } F\text{-Score} = \frac{2}{|L|}\sum\nolimits_{j=1}^{|L|} \frac{\sum_{i=1}^{|N|} T_{ij} \times P_{ij}}{\sum_{i=1}^{|N|} T_{ij} + P_{ij}} \tag{6}$$

IV. **Micro *F*-Score** calculates average label and instances prediction accuracy. If there is label imbalance, micro f-score is more preferable than macro f-score.

$$\text{Micro } F\text{-Score} = \frac{2}{|L|}\frac{\sum_{j=1}^{|L|}\sum_{i=1}^{|N|} T_{ij} \times P_{ij}}{\sum_{j=1}^{|L|}\sum_{i=1}^{|N|} T_{ij} + P_{ij}} \tag{7}$$

V. **Average Accuracy** evaluates average prediction score for each labels.

$$\text{Average Accuracy} = \frac{1}{|L|}\sum\nolimits_{j=1}^{|L|} \frac{|N| - \sum_{i=1}^{|N|} T_{ij} \oplus P_{ij}}{|N|} \tag{8}$$

The evaluation of the models on the dataset without any preprocessing is shown in Table 5 As we can see from Table 5, without any preprocessing binary relevance with SVM as base classifier has the lowest hamming loss at 0.162. It also has the best performance of macro F-score, micro F-score, and average accuracy. The result obtained in these three criteria is 0.893, 0.906, and 0.837, respectively. But label powerset with random forest classifier has shown the highest subset 0/1 loss with a value of 0.272. BRkNN-b has shown significantly worse performance in all the criteria. Evaluation of models trained with data preprocessing and Senticnet5 boosting is shown in Table 6.

In Table 6, binary relevance with random forest classifier shows the lowest hamming loss at 0.160. It also has the best performance of macro F-score, micro F-score, and average accuracy. The results obtained in these three criteria are 0.895, 0.907, and 0.839, respectively. Classifier chains with SVC as base classifier shows the highest 0/1 subset loss at 0.281. BRkNN-b has the worst performance in all the criteria. Since the output parameter of BRkNN-b is related to the average size of the k-nearest label sets, it needs a more balanced dataset. As our dataset is unbalanced, it can be considered as the reason for poor performance. The comparison between Tables 5 and 5 shows noticeable improvements over all of the evaluating criteria. A compar-

Table 5 Evaluation of the models using raw data

Evaluation metric	BR		LP		CC		RAkEL		ECC		MLkNN	BRkNN	
	RFC	SVM	RFC	SVM	RFC	SVM	RFC	SVM	RFC	SVM		*a*	*b*
Hamming loss	0.164	0.162	0.172	0.18	0.163	0.175	0.168	0.167	0.165	0.174	0.166	0.172	0.284
0/1 Subset loss	0.192	0.197	0.272	0.253	0.23	0.261	0.24	0.228	0.227	0.247	0.207	0.183	0.063
Macro F-score	0.892	0.893	0.865	0.84	0.888	0.866	0.878	0.875	0.885	0.86	0.888	0.883	0.790
Micro F-score	0.905	0.906	0.897	0.893	0.904	0.895	0.9	0.901	0.903	0.896	0.902	0.901	0.822
Average accuracy	0.835	0.837	0.827	0.819	0.836	0.824	0.831	0.832	0.834	0.825	0.833	0.827	0.715

Table 6 Evaluation of models trained with preprocessed data

Evaluation metric	BR		LP		CC		RAkEL		ECC		MLkNN	BRkNN	
	RFC	SVM	RFC	SVM	RFC	SVM	RFC	SVM	RFC	SVM		*a*	*b*
Hamming loss	0.16	0.161	0.169	0.172	0.162	0.17	0.165	0.167	0.162	0.171	0.169	0.17	0.255
0/1 Subset loss	0.223	0.213	0.277	0.278	0.263	0.281	0.249	0.262	0.26	0.265	0.202	0.169	0.094
Macro F-score	0.895	0.894	0.877	0.867	0.888	0.875	0.884	0.875	0.885	0.869	0.887	0.891	0.821
Micro F-score	0.907	0.906	0.899	0.897	0.904	0.898	0.903	0.901	0.904	0.897	0.901	0.902	0.843
Average accuracy	0.839	0.838	0.83	0.827	0.837	0.829	0.835	0.832	0.837	0.828	0.83	0.829	0.744

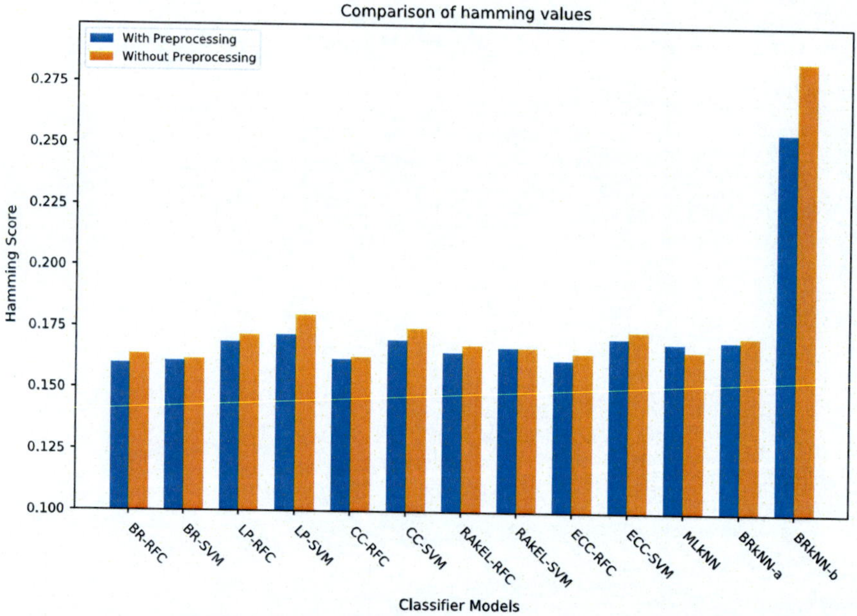

Fig. 3 Comparison of hamming loss over datasets

ison of hamming loss and subset 0/1 loss over preprocessed and raw data are shown in Figs. 3 and 4, respectively.

As shown from the figures, preprocessing the tweets and using Senticnet5 for handling unknown words and negation during the training phase leads to a better hamming loss in 12 out of 13 models. The models show an average improvement of 2.25%. Same observations can be made for subset 0/1 loss. Here, the results of 10 models out of 12 are improved by using preprocessing and senticnet5. The average improvement in subset 0/1 loss is 10.1%. But the unsatisfactory results in MLkNN and BRkNN show that our framework is finding it hard to adapt to algorithm adaption methods.

6 Conclusion

A multi-label emotion classification approach to classify tweets is proposed in this thesis. This prototype can be divided into two parts. The parts are processing of the tweet dataset and training and evaluation of multi-label classifiers. *Senticnet5* was used as a booster to improve the classifiers during the training phase. Various multi-label classifiers are applied. Among them, binary relevance (BR) with random forest classifier as base classifier has shown the best result. An average hamming loss of 0.16 is pretty impressive considering that the dataset was chosen at random. It means

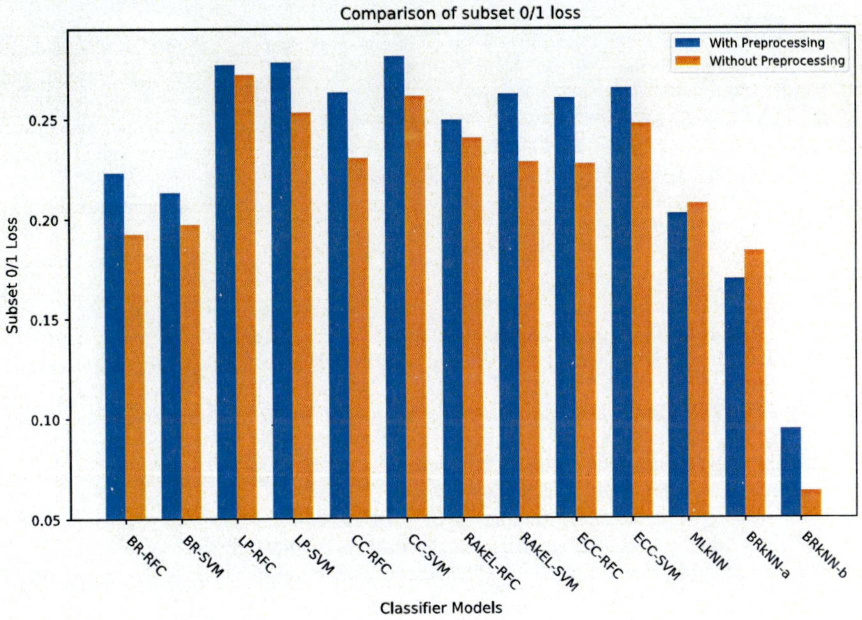

Fig. 4 Comparison of subset 0/1 loss over datasets

the correlation between tweets is pretty low. Thus, even with this disadvantage, our classifiers are able to find correct emotion labels 85% of the time. In classifiers that use the problem transformation method, we used both random forest classifier and support vector classifier as base classifiers. In all of the cases, random forest classifier outperformed support vector classifier.

In the future, this work can be extended in three directions. Enriching the dataset with more labeled tweets is the obvious way to go. We can also add image and emoji analysis along with text analysis to get a better understanding of the tweet's emotions. We believe that combining this research with other new technologies like neural networks will vastly improve its performance and will contribute to the development of various open-source emotion classifier software.

References

1. Tang H, Tan S, Cheng X (2009) A survey on sentiment detection of reviews. Expert Syst Appl 36(7):10760–10773
2. Taboada M, Brooke J, Tofiloski M, Voll K, Stede M (2011) Lexicon-based methods for sentiment analysis. Comput Linguist 37(2):267–307
3. Pang B, Lee L (2008) Opinion mining and sentiment analysis. Found Trends Inform Retrieval 2(1–2):1–135

4. Lin KHY, Yang C, Chen HH (2007) What emotions do news articles trigger in their readers? In: Proceedings of the 30th annual international ACM SIGIR conference on research and development in information retrieval, pp 733–734
5. Liang WB, Wang HC, Chu YA, Wu CH (2014) Emoticon recommendation in microblog using affective trajectory model. In: Signal and information processing association annual summit and conference (APSIPA). IEEE, Asia-Pacific, pp 1–5
6. Bouazizi M, Ohtsuki T (2017) A pattern-based approach for multi-class sentiment analysis in twitter. IEEE Access 5:20617–20639
7. Boutell MR, Luo J, Shen X, Brown CM (2004) Learning multi-label scene classification. Pattern Recogn 37(9):1757–1771
8. Tsoumakas G, Katakis I, Vlahavas I (2010) Random k-label sets for multilabel classification. IEEE Trans Knowl Data Eng 23(7):1079–1089
9. Read J, Pfahringer B, Holmes G, Frank E (2011) Classifier chains for multi-label classification. Mach Learn 85(3):333
10. Zhang ML, Zhou ZH (2007) ML-KNN: a lazy learning approach to multi-label learning. Pattern Recogn 40(7):2038–2048
11. Liu SM, Chen JH (2015) A multi-label classification based approach for sentiment classification. Expert Syst Appl 42(3):1083–1093
12. Cabrera-Diego LA, Bessis N, Korkontzelos I (2020) Classifying emotions in stack overflow and JIRA using a multi-label approach. Knowl-Based Syst 195(105):633
13. Go A, Bhayani R, Huang L (2009) Twitter sentiment classification using distant supervision. CS224N project report, Stanford 1(12)
14. Cambria E, Poria S, Hazarika D, Kwok K (2018) SenticNet 5: discovering conceptual primitives for sentiment analysis by means of context embeddings. In: Proceedings of the AAAI conference on artificial intelligence, vol 32
15. Tsoumakas G, Katakis I, Vlahavas I (2011) Random k-label sets for multilabel classification. IEEE Trans Knowl Data Eng 23(7):1079–1089. https://doi.org/10.1109/TKDE.2010.164
16. Spyromitros E, Tsoumakas G, Vlahavas I (2008) An empirical study of lazy multilabel classification algorithms. In: Proceedings of the 5th hellenic conference on artificial intelligence (SETN 2008)
17. Ekman P (1992) An argument for basic emotions. Cogn Emot 6(3–4):169–200
18. Ho TK (1995) Random decision forests. In: Proceedings of 3rd international conference on document analysis and recognition, vol 1. IEEE, pp 278–282
19. Cortes C, Vapnik V (1995) Support-vector networks. Mach Learn 20(3):273–297
20. Madjarov G, Kocev D, Gjorgjevikj D, Džeroski S (2012) An extensive experimental comparison of methods for multi-label learning. Pattern Recogn 45(9):3084–3104

Bangla News Classification Using GloVe Vectorization, LSTM, and CNN

Pallab Chowdhury, Ettilla Mohiuddin Eumi, Ovi Sarkar, and Md. Faysal Ahamed

Abstract Text mining has gained considerable popularity over the last few years. Since the awareness of mobile media, streaming media, print media, and many other outlets are now accessible to consumers. Due to the large availability of text in many respects, research experts registered many unstructured data and found various ways in the literature of converting this dispersed text into a given, organized volume. Compared to the short text, the emphasis on complete classification (complete news, big records, long text, etc.) is prevalent. We addressed in this paper the process of text classification, grading, and various methodologies for feature extraction in short texts, i.e., news classification based on their headlines. Existing classification is compared and their operating methodologies presented efficiently. This work serves the purpose of classifying different types of Bangla Newspaper articles into 10 specific categories. The classification task is performed on Bengali text from three renowned newspapers of Kolkata. We have used advanced data tokenization techniques and unsupervised 'GloVe' vectorization for better classification performance. We applied LSTM and CNN as our main feature extractors. Comparing with other models like binary SVM classifier, standard LSTM, BiLSTM, CNN, or ANN, this proposed work gives better accuracy of 87%.

Keywords GloVe · LSTM · CNN · Binary SVM classifier · BiLSTM · ANN

1 Introduction

Natural language processing (NLP) is a part of machine learning that empowers computers to comprehend characteristic discourse. NLP advances innovative basic reading comprehension tools, including text and speech. In simple words, NLP is talking about the structured handling of normal human pronunciations, along with speech or code. However, it is presented in such a way that perhaps the concept itself

P. Chowdhury (✉) · E. M. Eumi · O. Sarkar · Md. F. Ahamed
Department of Electrical and Computer Engineering, Rajshahi University of Engineering & Technology, Rajshahi, Bangladesh

M. S. Arefin et al. (eds.), *Proceedings of the International Conference on Big Data, IoT, and Machine Learning*, Lecture Notes on Data Engineering and Communications Technologies 95, https://doi.org/10.1007/978-981-16-6636-0_54

is capturing, the true motivation behind this invention comes through impact [1]. NLP will assist us with a lot of tasks, and the use of domains tends to be increasing steadily. NLP helps to analyze the impression of speech and text, enabling us to anticipate a certain point by evaluating details or by labeling messages.

In our research work, we use authoritative Bangla news texts from many newspapers and online journals to understand the emotions of these articles, where we will classify the texts into 10 groups. These 10 types of data will be recognized while our model can be successfully learned. We are accomplishing this work altogether on Bangla messages. We have developed a system that can differentiate easily between the data provided in Bangla. The ten categories are divided into 'Entertainment', 'National', 'Kolkata', 'State', 'International', 'Sport', 'Nation', 'World', 'Travel'. This work is implemented by using convolutional neural network [2] and long short-term memory [3]. As we use the hybrid model, our working accuracy is much better than another single model. For model training, we also preprocess our Bangla texts with some methods, like removing stop words, punctuation, word embedding.

There is a significant result, in actuality, utilizing NLP. Building this model, effectively usable in Web applications and it is benevolent for the Bangladesh working zone area. In general, people's behavior analysis is being easier for understanding [4]. Any individual comments throughout this respect, as it takes a bunch of costs including countless personnel to perform such assessments, may be completely unprecedented. This puts a cover on a gadget that can get a handle on the attitude of the commentators' editorials in numerous social stages or diaries, given the remarkable ascent in guests just as clients' details. The main aim in this situation is to decide what customers are thought with the word assessment of those goods.

The key goal of the research of emotion is to split the task into negative or positive amplitude so that it would distinguish parental attitudes or details. This research is used to increase consumer penetration and revenue, branding strategies, and several areas including certain spam detecting, banking, economy, stock exchange, selling and buying products, as well as many other companies. Effective intuition analyzes could have an immense effect on numerous fields such as policy, governance or organization, campaigns, and corporations, as they can respond efficiently and allow individuals to profit from the behavior or decision-making needed. Neural networks can easily acquire for a lower cost. There are thousands of evaluations, commentaries, e-mails, and many more. Text categorization approaches should be expanded to cover all major or small enterprises. There are many urgent circumstances in which businesses must recognize and take decisive steps when quickly and efficiently as possible. Computer information retrieval should often and in real time imitate the designer labels so that can recognize vital details and respond quickly. In the realm of natural language processing, text categorization is not a new concept. However, work on the Bangla text has begun in recent years. The categorization of online news covers a large range in this sector. People rely on this problem in the age of Internet news sources. This classification is the goal of the proposed study, which is based on the Bangla language. Some Bangla dataset is utilizing some examination work that is represented in our literature survey segment. Comparatively, our approached hybrid model is being more efficient than any machine learning approach.

2 Literature Review

Text classification for analyzing emotions is a type of information wrenching from the content of emerging research and commercial interest. Various researchers aimed to examine this field, and there are substantial quantities of research papers particularly on this subject. But using the Bengali language to determine the sentiments through text classification is not enriched in this field. Several researchers have tried many ways to achieve a good result with good dataset.

In restaurant reviews, Sharif et al. [5] introduced an automatic sentiment analysis technique for Bengali's text to classify a positive or negative impression. To categorize the sentiment from the Bengali review text, three algorithms are developed, such as the decision tree, random forest, and multinominal Naive Bayes classification.

In the evaluations or comments on this Web site, Adnan et al. [6] intended to seek good or negative judgments and assessed the results of the process, especially in 'Surabaya' restaurant. These observations or reviews are in the form of text or word information. The text data were then categorized by decision Tree-J48 into negative or positive judgments.

Tabassum and Khan [7] enhanced a sentimental experimental framework, creating a dataset of 1050 Bengali Facebook and Twitter comments. In order to offer a more accurate result of around 85%, the proposals include an unigram, a POS tag, denial handling, and the random forest classification.

Banik and Rahman [8] conducted a comparison of machine learning algorithms on Bengali emotional text interpretation. They made use of two distinct datasets. The first is the parts-of-speech (POS) tagset, which has 3000 Bengali sentences, 42,000 words, and 32 tagsets and is accessible for paid download on the Web. Another is their self-created dataset of 6314 Facebook comments. They trained 4700 data points, tested 940 data points, and obtained the maximum accuracy of 52.98%.

Pran et al. [9] attempted to use deep learning to evaluate the feelings or sentiments of Bangladeshi people under COVID-19 crisis scenarios. They evaluated 1120 data points divided into three categories. They used CNN and LSTM since they performed the best in terms of accuracy. This study sought to assist people in taking the essential actions to improve their position in the face of the epidemic.

3 Methodology

The proposed framework is consisting of three essential parts. They are data collection, data preprocessing, and applied model. In Fig. 1, the flowchart of the proposed framework is shown.

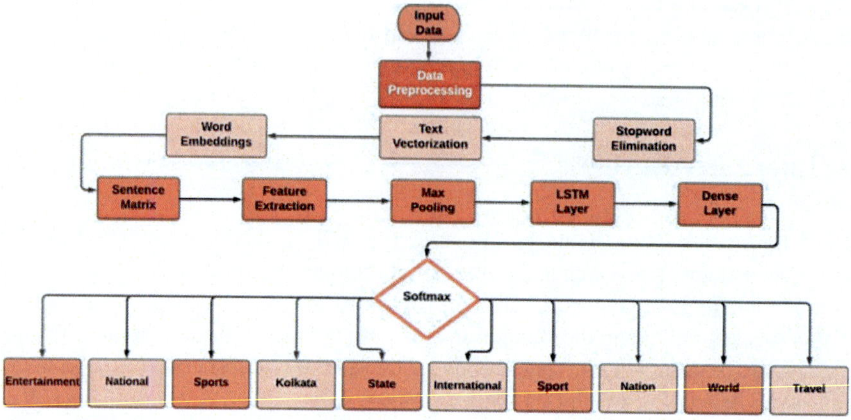

Fig. 1 Flowchart of the proposed framework

3.1 Data Collection

For classification purpose, the dataset is collected from 'Kaggle' [10], the world's largest data science community. The dataset contains 14k different news articles from three popular newspapers of Kolkata named 'Anandabazar', 'Ebela', and 'ZeeNews'. The entire dataset is categorized into two subsections, train and val containing 11k and 3k data, respectively. The dataset is divided into three major columns named 'title', 'article', 'label'. The label columns comprise of 10 different categories with respect to different news articles. 'Entertainment', 'National', 'Kolkata', 'State', 'International', 'Sport', 'Nation', 'World', and 'Travel'.

3.2 Data Preprocessing

The dataset needed to be cleaned for classification and summarizing purposes. For this purpose, the common expressions of Bengali text like , : , ' ', ?, etc. were removed. Mentioned ten categories were again labeled by some sequential numeric values. Keras tokenization is used for these purpose, and then tokens are converted into sequences with the help of Keras pad sequence. Tokenization is done on the basis of maximum feature value and pad sequence works on maximum sequence length. For vectorization purpose, an advanced approach named 'GloVe' [11] is used which increased the accuracy of the model. The maintenance of text sequence, a type of recurrent neural network, LSTM is used. For working on text data using neural networks, Keras embedding layer is implemented here. The input data was encoded such that it can be represented by unique integer values. SpartialDropout1D drops entire 1D feature maps. In Fig. 2, the changes have been made after preprocessing of the data has been presented.

Original	Cleaned	Category
এবার হুমকি অধ্যাপককে। বন্ধ পুলিশকর্তাও নামলেন আক্রমণে	এবার হুমকি অধ্যাপককে বন্ধ পুলিশকর্তাও নামলেন আক্রমণে	state
কালো পতাকার বিক্ষোভের 'জবাব', প্রেসির জন্য ১১৮ কোটি টাকা বরাদ্দ করলেন মুখ্যমন্ত্রী	কালো পতাকার বিক্ষোভের জবাব প্রেসির জন্য ১১৮ কোটি টাকা বরাদ্দ করলেন মুখ্যমন্ত্রী	kolkata
সোশ্যাল মিডিয়ায় আক্রান্ত জাহ্নবী, কেন ট্রোলড হতে হল	সোশ্যাল মিডিয়ায় আক্রান্ত জাহ্নবী কেন ট্রোলড হতে হল	entertainment
বেজিং-নীতিকে আক্রমণ কংগ্রেসের	বেজিং নীতিকে আক্রমণ কংগ্রেসের	national

Fig. 2 Original and processed data

3.3 *Applied Model*

Convolutional Layer In this framework, convolutional operations are performed on input matrixes to create feature maps. The filters slide across the matrix without padding the edges of the narrow convolution. The activation function determines whether or not the neuron will be triggered. Using vocabulary word indexes, the matrix generator creates a vector representation of each word in a sentence that is transformed into the convolution layer's input matrix. With 250 filters in the first convolution layer, Conv1D is generated. Three filters of size 4, 3, 2 were presented in our example, where a row defines a word in a sentence, and each column defines each letter in a word. We have modeled a column size equivalent to the maximum length of the word. Each convolution procedure of the words in a sentence for classification with a feature extraction architecture produces a feature map with different shapes. To create feature vectors, we have applied max pooling to the featured maps. We have concatenated the vectors thereafter to construct a large vector. In Fig. 3, the architectural design of the convolutional model is shown.

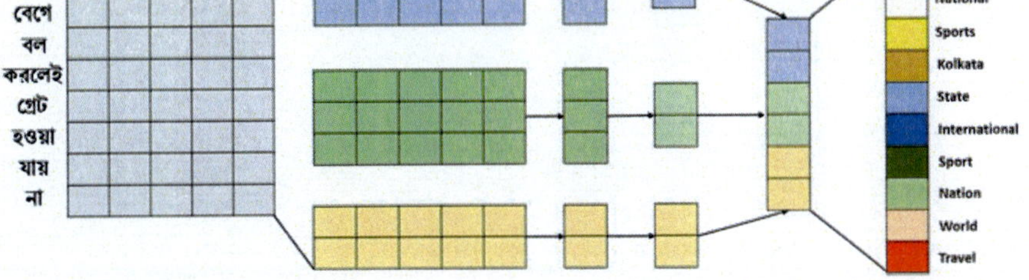

Fig. 3 Architectural design of the convolutional model

LSTM Layer The concatenated vector that results from the max pooling layer is the input of this layer. This layer uses three gates to filter the data, which eliminates the unwanted words or feature vectors by the delete gate and preserves the chronological order of the words in a sentence. This layer's output is fed to the fully connected layer.

Softmax After CNN, combined with LSTM, extracts high-level features from the text, the features are sent for classification to the softmax classifier in a fully linked way. Softmax is a special functional form. As a result of the prediction, it can map neuron output and pick the class with the largest probability value.

4 Statistical Analysis of Bangla Text

We have provided a statistical review of the Bangla text for classification in this section. The quantitative values are represented in Table 1.

5 Experimental Result and Discussion

The implementation of our proposed model was done on 14k dataset consisting of Bangla news articles on 10 different categories from 3 different Indian Newspaper. The existing paper in different research sectors basically works on dataset of Bengali comments, tweets, magazines or political, sports news. Bengali comments, text, etc., are quite unofficial data as far need much moderation [12]. Again, for classifying political news, problem will arise dealing with limited terms [13]. Comparing those,

Table 1 Quantitative data of Bangla text

Properties	Values
Total number of words	13978
Number of words in category-1 (entertainment)	1450
Number of words in category-2 (national)	1765
Number of words in category-3 (sports)	1589
Number of words in category-4 (kolkata)	5764
Number of words in category-5 (state)	2710
Number of words in category-6 (international)	650
Number of words in category-7 (sport)	20
Number of words in category-8 (nation)	15
Number of words in category-9 (world)	14
Number of words in category-10 (travel)	1

our proposed model works much efficiently as it deals with 10 categories, preprocessed data with 'GloVe', filter-based analogy using advanced features of LSTM.

5.1 Model Comparison

Working on Bangla news dataset using different classifiers, like binary SVM classifier, single LSTM, CNN, ANN, BiLSTM, etc., gave relatively poor results on the test dataset compared to our proposed model. The proposed model gives a higher training accuracy of 98.75% and test accuracy87% compared with other works on Bangla dataset. Table 2 shows this.

5.2 Model Accuracy and Loss

Our proposed model works with 'GloVe' in the word embedding section. This improves accuracy of the traditional CNN-LSTM model. Table 3 shows this.

In Fig. 4, the accuracy and the loss curves of our model have also been shown after 10 epochs, where Y-axis represents accuracy or loss with respect to epochs on X-axis.

Table 2 Accuracy comparison of different models on Bangla news dataset

Model name	Training accuracy (%)	Test accuracy (%)
Binary SVM classifier [14]	93.39	66.23
LSTM [14]	96.97	74.74
CNN [14]	89.03	60.49
ANN ADAM [15]	70.94	71.01
ANN RMS [15]	70.73	70.76
BiLSTM [15]	85.14	80.69
Proposed model	98.75	87.00

Table 3 Comparison of training accuracy and test accuracy of the proposed model with and without GloVe

Model name	Training accuracy (%)	Test accuracy (%)
CNN-LSTM without GloVe	96.00	79.00
CNN-LSTM with GloVe	98.75	87.00

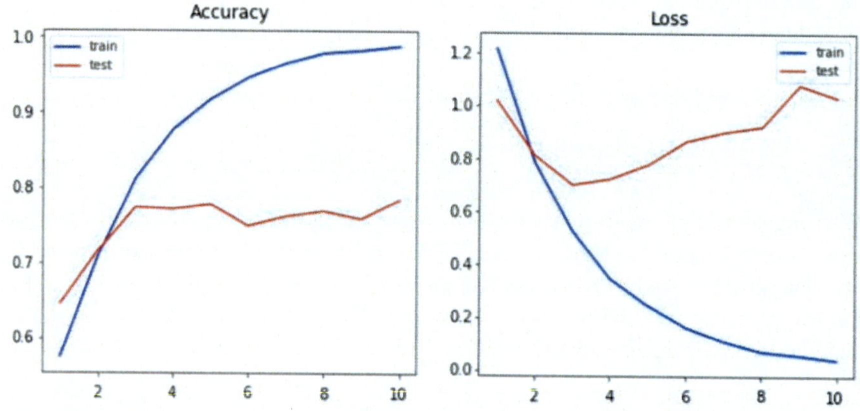

Fig. 4 Accuracy and loss curves of our proposed model

ইডেনে অনুশীলনে গরহাজির আফ্রিদি, কারণ শরীর নাকি মন খারাপ।

Fig. 5 Sentence as an example

5.3 *Error Analysis*

For the testing purpose, each article of our dataset is treated individually with respect to its sentences. Some errors are reported in classification, especially dealing with famous people name, popular places, and news related to Kolkata or state, like a sentence shown in Fig. 5 as an example. This sentence is classified as Kolkata news rather than Sports news. Because the model categorizes it based on the name of the place mentioned rather the name of the player. This problem basically arises due to the ambiguity of data as the mentioned first word in the sentence which is the name of a certain place in Kolkata is mentioned in several places in an article or different sentences of an article. Another major error is that, the model faces overfitting after 20 epochs as it closely fits to its limited set of text data.

6 Conclusion

In this research paper, Bangla news articles are classified into ten categories. The primary theme is to analyze efficiently each article's subjects individually. We also construct several other machine learning algorithms on the dataset we have used here for testing whether or not our proposed model performs correctly. We observed a better accuracy using GloVe compared with the same model without using GloVe. This comparison is shown in Table 3. For further accuracy and model modification, we may focus on increasing the size of the dataset. For this, our main focus will be developing our own balanced dataset and using other advanced NLP models like transformer in our future works.

References

1. Surabhi MC (2013) Natural language processing future. In: 2013 International conference on optical imaging sensor and security (ICOSS), pp 1–3
2. Johnson R, Zhang T (2014) Effective use of word order for text categorization with convolutional neural networks. arXiv preprint arXiv:1412.1058
3. Yao L, Guan Y (2018) An improved LSTM structure for natural language processing. In: 2018 IEEE International conference of safety produce informatization (IICSPI). IEEE, pp 565–569
4. Alam MH, Rahoman MM, Azad MAK (2017) Sentiment analysis for Bangla sentences using convolutional neural network. In: 2017 20th International conference of computer and information technology (ICCIT). IEEE, pp 1–6
5. Sharif O, Hoque MM, Hossain E (2019) Sentiment analysis of Bengali texts on online restaurant reviews using multinomial naïve bayes. In: 2019 1st International conference on advances in science, engineering and robotics technology (ICASERT). IEEE, pp 1–6
6. Adnan M, Sarno R, Sungkono KR (2019) Sentiment analysis of restaurant review with classification approach in the decision tree-j48 algorithm. In: 2019 International seminar on application for technology of information and communication (iSemantic). IEEE, pp 121–126
7. Tabassum N, Khan MI (2019) Design an empirical framework for sentiment analysis from Bangla text using machine learning. In: 2019 International conference on electrical, computer and communication engineering (ECCE). IEEE, pp 1–5
8. Banik N, Rahman MHH (2018) Evaluation of naïve bayes and support vector machines on Bangla textual movie reviews. In: 2018 International conference on Bangla speech and language processing (ICBSLP). IEEE, pp 1–6
9. Pran MSA, Bhuiyan MR, Hossain SA, Abujar S (2020) Analysis of Bangladeshi people's emotion during covid-19 in social media using deep learning. In: 2020 11th International conference on computing, communication and networking technologies (ICCCNT). IEEE, pp 1–6
10. Chatterjee S (2019) Classification: Bengali news articles (IndicNLP). https://www.kaggle.com/csoham/classification-bengali-news-articles-indicnlp
11. Pennington J, Socher R, Manning CD (2014) Glove: global vectors for word representation. In: Proceedings of the 2014 conference on empirical methods in natural language processing (EMNLP), pp 1532–1543
12. Amin R, Sworna NS, Hossain N (2020) Multiclass classification for Bangla news tags with parallel CNN using word level data augmentation. In: 2020 IEEE Region 10 symposium (TENSYMP). IEEE, pp 174–177
13. Saha BN, Senapati A, Mahajan A (2020) LSTM based deep RNN architecture for election sentiment analysis from Bengali newspaper. In: 2020 International conference on computational performance evaluation (ComPE). IEEE, pp 564–569
14. Ashik M, Shovon S, Haque S (2019) Data set for sentiment analysis on Bengali news comments and its baseline evaluation, pp 1–5
15. Shahin MMH, Ahmmed T, Piyal SH, Shopon M (2020) Classification of Bangla news articles using bidirectional long short term memory. In: 2020 IEEE Region 10 symposium (TENSYMP), pp 1547–1551

An Ensemble Method-Based Machine Learning Approach Using Text Mining to Identify Semantic Fake News

Fahima Hossain, Mohammed Nasir Uddin, and Rajib Kumar Halder

Abstract Fake news is a frequent problem that is having a massive influence on our social lives, especially in the political arena. Social media provides a forum for people to publicly share their thoughts and feelings, and it has made conversation easier. It also allows people to manipulate the power to spread misleading facts intentionally. Misleading or unreliable information is widely disseminated via prominent social media sites such as Facebook and Twitter in the form of videos, tweets, blogs, and URLs. Anyone can produce and spread fake news content for personal or professional benefit. In these circumstances, detecting and flagging certain material on social media is an emerging task in the recent era. We have proposed an ensemble method-based machine learning approach to identify semantic fake news directly from the text using text mining. Natural Language Processing technique is applied for data preprocessing on LIAR dataset collected from PolitiFact and convert it into a vector form. Univariate Selection, Select Percentile, Select from Model, Linear SVC, Extra Trees Classifier, and Chi-Square feature selection techniques are used to select the effective features directly from the text data. To combine outputs from multiple classifiers, namely Multinomial Naïve Bayes (MNB), Random Forest (RF), K-Nearest Neighbor (KNN), and Gradient Boosting (GB), an ensemble method is constructed. The ensemble method allows producing better prediction compared to a single classifier. Our proposed model obtained an accuracy of 45.48% for multi-class, and an accuracy of 71.8424, AUC-ROC score of 0.6351 better than the previous studies.

Keywords Fake news detection · NLTK · TF-IDF vectorizer · Text mining algorithm tuning · Ensemble machine learning

---

F. Hossain (✉) · M. N. Uddin · R. K. Halder
Jagannath University, Dhaka 1100, Bangladesh

M. S. Arefin et al. (eds.), *Proceedings of the International Conference on Big Data, IoT, and Machine Learning*, Lecture Notes on Data Engineering and Communications Technologies 95, https://doi.org/10.1007/978-981-16-6636-0_55

1 Introduction

The convenient use of internet or World Wide Web (WWW) has expanded the use of social media such as Facebook, Instagram, Telegram, Twitter, Google Plus, etc. This social media usage is growing expeditiously by people for online discussions, business promotions, online seminars, knowledge sharing, etc. As a result, social media has decreased the need for Television, Radio, Magazine or written newspapers in need of national and global news. These analog news media has been converted to digital media such as online news platforms, blogs, social media feeds, etc., [1]. According to the report mentioned in [2], 36% of people use Facebook, 21% use YouTube, 16% use WhatsApp, and 12% use Twitter for getting news. Though the news on traditional media is reviewed and controlled [3], the veracity of the news contents on digital media are hardly verified. Therefore, fake news is being diffused by abusers at a very fast rate [4]. These fake news are transmitted due to several reasons fake news is normally generated for defaming a public figure, political person and destroying a business. It is a very strenuous task to recognize the fake news as these news are generated intentionally to cover up the truth behind lies [5]. It is also impossible to manually separate the real news from fake news as there will be a need of proper and in-depth knowledge to identify the fake news [6]. So, we need to define which words are consolidated to create fake news [7]. Building an automatic fake news detection system requires the application of various Natural Language Processing techniques and Artificial intelligence (AI) techniques. Natural Language Processing techniques are used to summarize the news content into a vector of word count. Artificial Intelligence techniques are used to extract the relationship between two words to predict presence of fake news [8]. Various Natural Language Processing techniques are used such as CountVectorizer, Bag-of-Words (BoW), Word Embeddings, etc. The main objective of this research work is to design an efficient feature selection mechanism based on ranking of the features collected from six existing feature selection techniques.

The main contributions of this research work are:

- Proposed an efficient feature selection mechanism based on ranking of the features collected from six existing feature selection techniques to select the features needed to identify fake news more effectively.
- Performed a comparative analysis of the proposed architecture for both multi-class classification and binary class classification.

The rest of the paper is organized as follows: Sect. 2 gives details of the existing works; Sect. 3 illustrates details about data and methodology. In Sect. 4, we describe the evaluation, validation, experiments. And the last section is about conclusion and future work.

2 Literature Review

Brasoveanu and Andonie [9] proposed a fake news detection model using integrated machine learning approach. In this work, they experimented the proposed architecture for LIAR dataset. In data preprocessing, sentiment, named entities or facts were extracted from both structured (e.g., Knowledge Graphs) and unstructured data and then word embedding is performed using GLOVE. Several classic machine learning and deep learning techniques, namely Multinomial Naive Bayes, SGDClassifier, CapsNet, BasicLSTM, etc., are implemented to classify the fake news.

Hakak et al. [10] built an ensemble classification model for detection of the fake news. ISOT and LIAR fake news dataset was used in this research. Stopwords, punctuation marks, html tags, url, emojis, etc., were eliminated in the preprocessing section using NLTK toolkit. Stemming and tokenization are also performed using the NLTK toolkit. 26 features were used to build the predictive model to improve accuracy and reduce training time. Random Forest, Extra Tree Algorithm and Decision Tree are used for classification using ensemble techniques to build more accurate predictive model. Parameter tuning is applied to increase accuracy using Random search hyper-parameter tuning method to choose optimal hyper-parameters.

Liu et al. [11] proposed a two-stage model based on BERT for fine-grained fake news detection. In this work, BERT to extract the features from LIAR dataset. BERT is constructed using multi-layer bidirectional Transformer encoder. BERT is used for reduction of time and cost. BERT works in two steps: the text data is converted to a vector and the numerical data is then classified using BERT. BERT is implemented both as a feature extractor and classifier.

Balwant [12] proposed a hybrid model based on Bidirectional LSTM and POS tags. The NLTK POS tagger is applied to the LIAR dataset for word tagging and word embedding is used to transform it to a vector format. Then, CNN creates a feature vector. This feature vector is then sorted in alphabetical order and used to perform classification. Bidirectional LSTM is used to train and test model. In Bidirectional LSTM, softmax activation function is used.

Goldani et al. [13] built a fake news detection model based on Convolutional Neural Networks with loss. Both LIAR and ISOT dataset is used to conduct experiments. Glove.6B.300d is used to perform word embedding on a large volume of text to convert it to a vector. Then, CNN is used for classification task by utilizing softmax activation function and margin loss.

Most of the authors didn't reduce dimensionality of the feature vector which is obtained after extracting features. No optimized reduction technique is used for feature selection. Moreover, algorithm tuning has a major impact on machine learning algorithms which is not implemented in most of the existing works. In our research work, we have mitigated all of above limitations to propose a model that improves predictive performance for both multi-class and binary class classification.

3 Methodology

The proposed methodology completes the process in four steps: (1) Fake News Data Collection, (2) Data Preprocessing, (3) Feature Selection, and (4) Data Splitting and Ensemble Classification as displayed in Fig. 1.

3.1 Fake News Data Collection

In this work, the LIAR dataset was used which is collected from PolitiFact [11]. The dataset contains 10,269 number of instances as training set, 1283 as testing and 1284 as validation set [14]. There are six labels on the target column of this dataset such as: pants_fire, false, barely_true, half_true, mostly_true, and true. The columns of the dataset are json_id (the ID of the statement), statement (news content), subject_data (the subject), speaker, speaker_job_title (the speaker's job title), state_info (the

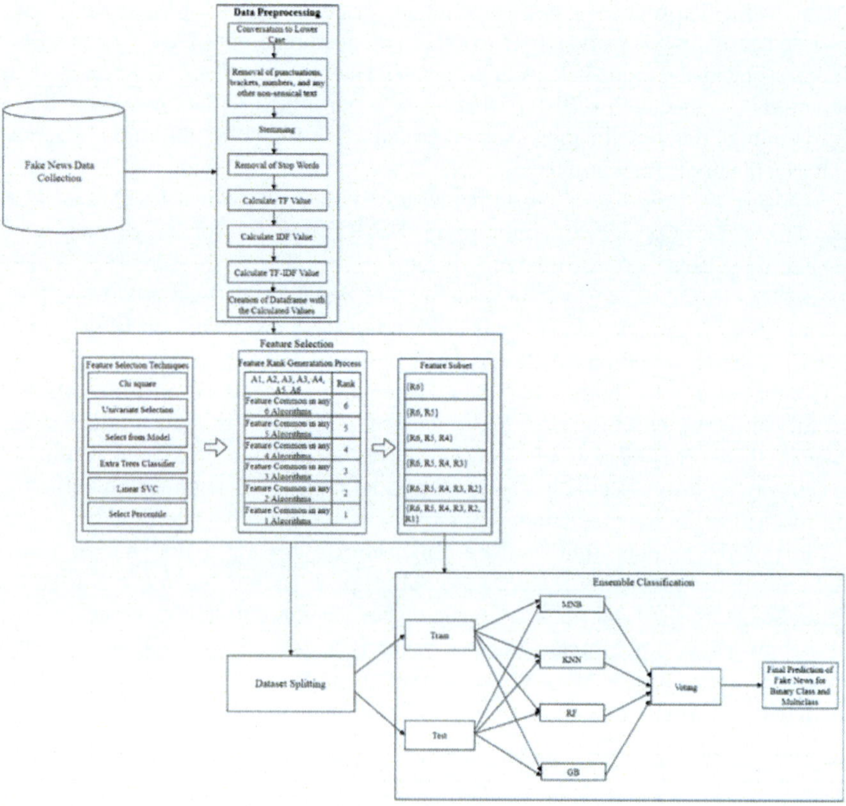

Fig. 1 Proposed architecture for fake news detection

state information), party_affliation (the party affiliation), context (value/location) (venue/location of the speech or statement), and the other columns are the total credit history count, including the current statement [15].

3.2 Data Preprocessing

In data preprocessing, json_id is removed from the collected dataset as this feature is an irrelevant attribute and this feature won't have any effect for prediction. Then, data cleaning and vectorization (text mining) is performed for the columns that are string type in the collected dataset such as statement, subject_data, and context (value/location). Regular Expression is used for data cleaning. Term Frequency-Inverse Document Frequency (TF-IDF) is used for vectorization in this work. In data cleaning, the text data is first converted from uppercase to lower case letters. Only the alphabetical words in the text data file are kept. Then, NLTK tool performs stemming of the words. Stemming removes the suffixes and prefixes from a word and returns only the root word. The stop words are also eliminated. TF and IDF value is calculated where TF is number of occurrences of a specific words and commonness of that word in different texts. A word with high TF value will get more relevance and a low TF value will get less relevance as this word is more common in the texts. TF value is calculated with IDF value and a final data frame is created with the calculated TF-IDF values. These data frames of each of the string type features are merged with other numeric features to make a preprocessed dataset. This process is performed to create a numerical dataset from the text dataset as no predictive models using Artificial Intelligence (AI) can be created with the text data.

3.3 Feature Selection

Six conventional feature selection techniques such as Univariate Selection, Select percentile, Select from Model, Linear SVC, Extra Trees Classifier, and Chi-Square are used to select the effective features in this proposed work. Chi-Square selects 12 features, Extra Tree Classifier selects 11, Linear SVC selects 11, Univariate Selection selects 19, Percentile selects 64, and Select From Model selects 63 features from the processed dataset. Then, rank generation process is introduced in this work. A feature selected by all the six algorithms will be having a rank of 6, a feature selected by five algorithms will be having a rank of 5 and so on till the generation of rank 1. A subset of these ranked features is created such as features of rank 6, features of both rank 6 and 5, features of rank 6, 5 and 4, and so on (Table 1).

Table 1 Rank wise features

Rank	Features name
6	_senator, others, false_count, half_true_count, mostly_true_count, fire_count, party, state.3, speaker_1
5	barely_true_count
4	health_care, president, social_media_posting
3	former_governor, congresswoman, radio_host, presidential_candidate, labor_state_budget, religion.1
2	candid_biographi_obama_birth_certif, elect.1.1, iraq.1, interview.1, speech.1, educ.1, economi.1, speaker_of_the_house_of_representatives, political_commentator, ad.1, health_care_medicar, foreign_polici_terror, feder_budget, state_budget_tax, candid_biographi
1	advocacy_group, blogger, judge, maryland_governor, ohio_supreme_court_justice, radio_talk_show_host, spokesman_for_the_60_plus_association, state_house_representative, bipartisanship_vote_record, candid_biographi_elect_messag_machin, censu.1, children.1, civil_right_crime_gun, corpor_tax, crime_crimin_justic, deficit_economi, drug.1, economi_immigr, economi_job_stimulu, fake_news, foreign_polici_histori_militari, health_care_medicar_messag_machin, health_care_messag_machin_tax, health_care_vote_record, immigr_terror, job_women_worker, messag_machin, patriot.1, ohio.1, onlin.1, post.1, speech.1, sunday.1

3.4 Data Splitting and Ensemble Classification

Data splitting techniques produce two subsets of the dataset, namely training and testing set. In this research, 85:15 splitting ratio is used. The datasets with the subset of features obtained from feature selection are used for classification. All the six subsets of features were used for classification to determine the best subset of features for detection of fake news. Ensemble classification algorithm is used to classify the fake news. It is a meta-algorithm instead of algorithm. This method joins the outcomes from single classifiers to improve the predictive performance of a model. It improves performance by making corrections of the mistakes done by the single classifiers. This method sometimes provide generalization to new and unseen data [16]. Stacking produces a meta-level (higher level) classifier by combining multiple single classifiers to get improved performance [17]. In majority voting, each single classifier makes its own prediction and final prediction is the one with the highest number of votes [18]. In this work, both the performance of stacking and voting ensemble method is analyzed and the voting method outperforms stacking method. The base classifiers used in the ensemble method are: Multinomial Naïve Bayes, K-Nearest Neighbor, Random Forest and Gradient Boosting. Algorithm tuning is also performed to find an optimal solution for identifying fake news.

4 Result Analysis

Various evaluation measures were performed to check the performance of the proposed model. There are six classes in multi-class. True and mostly true are considered true and the others are considered false for binary class classification problem. The model is experimented based on different rank of features for both multi-class and binary class classification as illustrated in Tables 2 and 3. The model's performance is then analyzed using different subset of features for multi-class and binary class classification explained in Tables 4 and 5. The best performance is obtained by the feature subset created by joining the features from rank 1, 2, 3, 4, 5 and 6 and subtracting two irrelevant features state.3 and speaker_1 as this feature is different for each of the news created. This best feature subset is then used for the rest of the experiments.

Tables 6 and 7 show the results for multi-class and binary class classification after applying voting ensemble integration method results, and Table 8 and 9 show the results for multi-class and binary class obtained after applying stacking ensemble integration method with the help of algorithm tuning to obtain satisfiable. It is easy to interpret from these results on Tables 6, 7, 8 and 9 that voting method performs clearly better for multi-class classification and stacking method performs well for binary class classification for the best feature subset.

The performance of the classification model is also tested by performing feature selection and without performing feature selection as shown in Table 10. The highest

Table 2 Performance analysis based on different rank of features for multi-class classification

Feature subset	Result		
	Accuracy	Training time	Testing time
Rank 6	27.6941	3.0572	0.0782
Rank 5	21.7844	2.6822	0.0974
Rank 4	19.3511	2.3336	0.0338
Rank 3	19.2932	2.5283	0.0319
Rank 2	22.3059	131.6442	2.2157
Rank 1	21.7845	4.0078	0.2936

Table 3 Performance analysis based on different rank features for binary class classification

Feature subset	Result		
	Accuracy	Training time	Testing time
Rank 6	63.6153	0.6156	0.0570
Rank 5	63.3256	0.3801	0.0408
Rank 4	62.5145	0.3990	0.1631
Rank 3	64.4264	0.4273	0.2129
Rank 2	64.7161	0.5303	0.3638
Rank 1	64.1947	0.6343	0.2335

Table 4 Different combination of features for multi-class classification

Feature subset	Voting ensemble method			Stacking ensemble method		
	Accuracy	Training time	Testing time	Accuracy	Training time	Testing time
Rank 6	23.3488	3.1888	0.3150	27.2885	24.2111	0.2609
Rank 6, 5	28.5052	3.8716	0.0553	23.5226	30.1836	0.3032
Rank 6, 5, 4	32.6767	3.2640	0.0501	26.6802	25.6466	0.2773
Rank 6, 5, 4, 3	42.0626	3.0108	0.4202	37.4275	24.0860	0.8454
Rank 6, 5, 4, 3, 2	24.5655	4.2347	0.4149	25.8691	32.5749	0.8128
Rank 6, 5, 4, 3, 2, 1	34.0093	5.7167	0.3948	24.5655	50.8789	0.8266

Table 5 Different combination of features for binary class classification

Feature subset	Voting ensemble method			Stacking ensemble method		
	Accuracy	Training time	Testing time	Accuracy	Training time	Testing time
Rank 6	63.0649	0.5570	0.0628	68.2793	4.4376	0.2424
Rank 6, 5	66.1935	0.6702	0.0634	65.0058	5.5465	0.2683
Rank 6, 5, 4	64.7161	0.5835	0.0708	63.8760	4.6622	0.2438
Rank 6, 5, 4, 3	69.2352	0.4909	0.6222	72.4508	5.0161	0.7614
Rank 6, 5, 4, 3, 2	62.4565	0.6288	0.6304	66.6859	6.2945	0.7482
Rank 6, 5, 4, 3, 2, 1	65.7879	0.8512	0.6434	70.1043	9.8392	0.8537

accuracy is achieved using feature selection techniques for both multi-class and binary class classification.

Confusion Matrix for binary class and multi-class classification for the best subset of features:

	0	1
0	1013	144
1	342	227

	0	1	2	3	4	5
0	174	78	46	3	14	13
1	56	221	60	7	17	2
2	32	81	180	6	15	8

(continued)

(continued)

	0	1	2	3	4	5
3	38	91	66	66	13	5
4	56	84	42	5	95	8
5	33	26	19	9	8	49

AUC-ROC Score for binary class = 0.6351. The model is lastly compared to the other existing systems that is illustrated in tabular format in Table 11.

5 Conclusion and Future Work

Fake news identification is gaining importance as sometimes innocent people becomes victim due to this fake news. The truth is covered by the lies. Some fraudulent people deceive innocent people by publishing misinformation about them or the other things related to them. In this research work, we have developed a model for fake news detection on linguistic features based on text mining and ensemble method in machine learning. Through this work, we have introduced a unique feature subset generation method for better classification of fake news. Algorithm tuning (parameter tuning) is also induced in this work to analyze the performance of this model. Our proposed model obtained accuracy of 45.4809 for multi-class classification, 71.84 for binary class classification which has outperformed the other existing systems. The AUC-ROC score for binary class classification is 0.6351. In the future direction, we will try to work with more benchmark fake news dataset. As the news content includes both linguistic and visual features, we will work with more linguistic features and visual features in the enhanced version of this research.

Table 6 Performance analysis of voting ensemble method for multi-class classification using algorithm tuning

Serial	Tuning parameters	Result		
		Accuracy	Training time	Testing time
1.	KNN = {n_neighbors = 23} GB = {n_estimators = 100, learning_rate = 1.0}	44.6118	13.5997	0.3854
2.	KNN = {n_neighbors = 110} GB = {n_estimators = 250, learning_rate = 0.7}	45.4809	18.5989	0.4647
3.	KNN = {n_neighbors = 210} GB = {n_estimators = 300, learning_rate = 0.7}	44.8436	114.8591	0.5306
4.	KNN = {n_neighbors = 240} GB = {n_estimators = 300, learning_rate = 0.6}	44.1113	22.0336	0.5594
5.	KNN = {n_neighbors = 90} GB = {n_estimators = 300, learning_rate = 0.6}	42.4598	51.3309	0.5246

Table 7 Performance analysis of voting ensemble method for binary class classification using algorithm tuning

Serial	Tuning parameters	Result		
		Accuracy	Training time	Testing time
1.	KNN = {n_neighbors = 125} GB = {n_estimators = 200, learning_rate = 0.6}	70.1622	3.6671	0.4039
2.	KNN = {n_neighbors = 95} GB = {n_estimators = 100, learning_rate = 0.6}	70.6257	899.7229	0.4175
3.	KNN = {n_neighbors = 126} GB = {n_estimators = 230, learning_rate = 0.6}	71.8424	3.9339	0.4278
4.	KNN = {n_neighbors = 120} GB = {n_estimators = 230, learning_rate = 0.7}	70.9154	693.4781	0.4161
5.	KNN = {n_neighbors = 110} GB = {n_estimators = 210, learning_rate = 0.7}	69.3511	28.6023	0.4357

Table 8 Stacking ensemble method for multi-class classification using algorithm tuning

Serial	Tuning parameters	Result		
		Accuracy	Training time	Testing time
1.	KNN = {n_neighbors = 126} GB = {n_estimators = 230, learning_rate = 0.8}	38.3256	41.9669	0.8112
2.	KNN = {n_neighbors = 110} GB = {n_estimators = 230, learning_rate = 0.8}	40.2955	41.9846	0.7866
3.	KNN = {n_neighbors = 110} GB = {n_estimators = 210, learning_rate = 0.7}	40.5272	38.0210	0.8073
4.	KNN = {n_neighbors = 90} GB = {n_estimators = 200, learning_rate = 0.8}	39.9189	37.7198	0.7999
5.	KNN = {n_neighbors = 160} GB = {n_estimators = 200, learning_rate = 0.6}	38.6153	38.1029	0.8597

Table 9 Stacking ensemble method for multi-class classification using algorithm tuning

Serial	Tuning parameters	Result		
		Accuracy	Training time	Testing time
1.	KNN = {n_neighbors = 126} GB = {n_estimators = 230, learning_rate = 0.8}	71.9003	8.1674	0.7858
2.	KNN = {n_neighbors = 110} GB = {n_estimators = 100, learning_rate = 0.6}	71.9873	8.3167	0.7841
3.	KNN = {n_neighbors = 150} GB = {n_estimators = 210, learning_rate = 0.6}	71.4948	7.8720	0.8443
4.	KNN = {n_neighbors = 170} GB = {n_estimators = 250, learning_rate = 0.6}	73.1460	8.6569	0.8074
5.	KNN = {n_neighbors = 180} GB = {n_estimators = 240, learning_rate = 0.5}	72.0451	9.0897	0.8629

Table 10 Performance analysis with and without performing feature selection

Parameter	Without feature selection method		With feature selection method	
	Multi-class classification	Binary class classification	Multi-class classification	Binary class classification
Accuracy	27.6362	64.2526	45.4809	71.8424
Training time	376.0949	33.2946	18.5989	3.9339
Testing time	2.3919	2.3406	0.4647	0.4278

Table 11 Proposed model's performance analysis with other author's model

Models	Accuracy
Adrian M. P. Brasoveanu [9]	32.60
Saqib Hakak [10]	44.15
Chao Liu [11]	40.58
Manoj Kumar Balwant [12]	41.50
Mohammad Hadi Goldani [13]	41.60
Proposed Model	45.48

References

1. Ahmad I, Yousaf M, Yousaf S, Ahmad M (2020) Fake news detection using machine learning ensemble methods. Complexity 2020:1–11
2. Vorhaus M, People increasingly turn to social media for news. https://www.forbes.com/sites/mikevorhaus/2020/06/24/people-increasingly-turn-to-social-media-for-news/?sh=5db8eabf3bcc
3. Agudelo G, Parra O, Velandia J (2018) Raising a model for fake news detection using machine learning in python. In: Lecture notes in computer science, pp 596–604
4. Ozbay F, Alatas B (2020) Fake news detection within online social media using supervised artificial intelligence algorithms. Phys A Stat Mech Appl 540:123174
5. Kesarwani A, Chauhan S, Nair A, Verma G (2020) Supervised machine learning algorithms for fake news detection. In: Lecture notes in electrical engineering, pp 767–778
6. Ahmed H, Traore I, Saad S (2017) Detection of online fake news using N-gram analysis and machine learning techniques. In: Lecture notes in computer science, pp 127–138
7. Ibrishimova M, Li K (2019) A machine learning approach to fake news detection using knowledge verification and natural language processing. In: Advances in intelligent networking and collaborative systems, pp 223–234
8. Thota A, Tilak P, Ahluwalia S, Lohia N (2018) Fake news detection: a deep learning approach. SMU Data Sci Rev 1:1–21
9. Braşoveanu A, Andonie R (2020) Integrating machine learning techniques in semantic fake news detection. Neural Process Lett
10. Hakak S, Alazab M, Khan S, Gadekallu T, Maddikunta P, Khan W (2021) An ensemble machine learning approach through effective feature extraction to classify fake news. Futur Gener Comput Syst 117:47–58
11. Liu C, Wu X, Yu M, Li G, Jiang J, Huang W, Lu X (2019) A two-stage model based on BERT for short fake news detection. In: Knowledge science, engineering and management, pp 172–183
12. Balwant M (2019) Bidirectional LSTM based on POS tags and CNN architecture for fake news detection. In: 2019 10th International conference on computing, communication and networking technologies (ICCCNT)
13. Goldani M, Safabakhsh R, Momtazi S (2021) Convolutional neural network with margin loss for fake news detection. Inf Process Manage 58:102418
14. Wang W (2017) "Liar, liar pants on fire": a new benchmark dataset for fake news detection. In: Proceedings of the 55th annual meeting of the association for computational linguistics. Volume 2: Short papers
15. thiagorainmaker77/liar_dataset. https://github.com/thiagorainmaker77/liar_dataset
16. Ensemble/voting classification in python with scikit-learn. https://www.stackabuse.com/ensemble-voting-classification-in-python-with-scikit-learn/
17. Su Y, Zhang Y, Ji D, Wang Y, Wu H (2013) Ensemble learning for sentiment classification. In: Lecture notes in computer science, pp 84–93
18. Mehanović D, Mašetić Z, Kečo D (2019) Prediction of heart diseases using majority voting ensemble method. In: IFMBE proceedings, pp 491–498

Fuzzy Logic-Based Assessment of Students Learning Outcome in Implementing Outcome-Based Education

Abdul Aziz and M. M. A. Hashem

Abstract Academic program accreditation is becoming more demandable all over the world among university students. But the degrees provided by most of the Bangladeshi universities are not accredited because of their deficiency of trained teachers, staff, etc. The Accredited Board of Engineering and Technology (ABET) works on accrediting academic programs and they have some criteria. On the other hand, the evaluation techniques practiced in most Bangladeshi universities are not up to the ABET standards but an absolute grading system. In this research paper, we have proposed a course learning outcome (CLO) and program learning outcome (PLO)-based student performance evaluation technique using the fuzzy logic system. We have considered semester final examinations (SFE) and continuous assessment (CA) consisting of class tests (CT), spot tests, home works, attendance, etc., as evaluation parameters. The course teachers and moderators set question papers assigning marks based on CLOs and the course teachers track the earned marks. Then, the ratios to the earned marks and the assigned marks considering by the CLOs in the SFE and CA are computed and fuzzified. The defuzzification stage returns the attainment of the CLOs of the courses. Finally, the PLOs are also demonstrated using the CLOs. This technique removes the biasedness, unfairness of absolute grading systems. We have case studied for five theoretical courses.

Keywords Attainment · Course learning outcome · Program learning outcome · Fuzzy logic · ABET · Accreditation

---

A. Aziz (✉) · M. M. A. Hashem
Khulna University of Engineering & Technology, Khulna 9203, Bangladesh
e-mail: abdulaziz@cse.kuet.ac.bd

M. M. A. Hashem
e-mail: hashem@cse.kuet.ac.bd

M. S. Arefin et al. (eds.), *Proceedings of the International Conference on Big Data, IoT, and Machine Learning*, Lecture Notes on Data Engineering and Communications Technologies 95, https://doi.org/10.1007/978-981-16-6636-0_56

1 Introduction

In recent times, institutional quality assurance has become one of the most vital issues for students all over the world universities. In most Bangladeshi universities, the evaluation methods that exist are not so much reliable and fair. So, the graduates from this unaccredited program face some problems especially in their higher education and career purposes all over the world. For those several problems, the departments of those universities are trying to accredit the program provided by them. So, we keep restricted our discussion for the department of Computer Science and Engineering (CSE) of Khulna University of Engineering & Technology (KUET). ABET works on four types of program accreditation [1] and sets some standards and criteria to follow. A fair student evaluation is one of the most important criteria for ensuring a program accreditation and implementing the outcome-based education (OBE). OBE is the education system where an academic program is opened knowing what is actually need for students and then designing the curriculum and an evaluation technique so that the outcome can be measured [2].

Normally in Bangladeshi universities a percentage-based grading is used which means if a student gets 80% or above marks then he/she will obtain the grade A^+, getting 75% to less than 80% marks will obtain the grade A, 70% to less than 75% marks return grade A^-, 65% to less than 70% marks return grade B^+, 60% to less than 65% marks return grade A^- and so on [3]. In this grading system, two students getting 80 and 99% marks are the same. On the other hand, getting 79 and 80% is not the same. This is totally unfair to the students. Ewing [4] discussed the relative grading where $x\%$ students are assigned to maximum grade, $y\%$ to second maximum grade and $z\%$ are assigned as failures which should not be happened. Solving these problems, we have presented in this paper a student's performance evaluation system based on CLO and PLO using the fuzzy inference system (FYS). For CLOs demonstration, we have distributed marks into two evaluation categories. One is SFE consisting of several questions, and the other is CA consisting of several items like class tests, attendance, homework, etc. The marks are assigned by the course teachers and moderators to the SFE questions and CA items considering the CLOs. Then, the ratio to the earned marks and assigned marks in the SFE and CA are fuzzified for the CLOs attainment of a course. Finally, using those CLOs, the PLOs of the program are also demonstrated.

This outcome-based evaluation method provides reliability and fairness for both the teachers and students. The following list has summarized the contributions of us.

- An OBE-based student evaluation model is designed theoretically and experimentally using fuzzy logic system.
- The marks are fuzzified for computing the students level of attainment and comparing with the existing method.

The presentation of this paper is done in several sections. Section 2 contains the literature review of different similar papers. The methodology with mathematical analyses, example, and case studies is shown in Sect. 3. Section 4 represents the result analyses and discussion of our system. The conclusion is discussed in Sect. 5.

2 Literature Review

In modern days, researchers all over the world are trying to develop more reliable, fair student performance evaluation systems for better educations. Aziz et al. [5] have proposed an evaluation system using the fuzzy logic system. They did not work for the accreditation process but a simple evaluation procedure. Shafi et al. [6] have introduced a hypothetical system for ABET accreditation for the computer science and computer information system departments. They provided generalized evaluation frameworks. Hussain et al. [7] presented the impact evaluation for engineering students based on ABET students outcome (SO) using regression analyses. They did not evaluate any course outcome (CO). Shanableh [8] evaluated for IT-facilitated students computing the course outcome and program outcome (PO) for the department of Civil Engineering. Hameed and Sorensen [9] demonstrated an evaluation technique using fuzzy logic. They have fuzzified the three parameters and generated the final grade to compare with traditional grading. Chandna [10] has developed evaluating method especially for weak students measuring the POs on three different levels such as high, medium, and low. They have also set a target of attainment so that the students can find their weakness for further developments. Varghese et al. [11] have presented an assessment technique based on the outcome using fuzzy logic. They computed Cos on five different levels and did not use all the fuzzy rules for computation. Abou-Zeid and Taha [12] describe the program accreditation process requirements and challenges toward the procedure such as inadequacy of faculty members, understaffing, data collection, preparation, and analyses. Lakshmi [13] demonstrated a CO- and PO-based education system for two different courses microwave and radar taking as examples, and they gave more precedence on COs evaluations. Akir et al. [14] have shown that the OBE is better for student's evaluation comparing the outcome-based evaluation and traditional grading. Rasha and Shatakumari [15] analyzed a trend and discussed the advantages and disadvantages, nature, origin, guidelines, etc., to implement the OBE. Ma and Zhou [16] proposed a fuzzy logic-based methods for student's performance evaluation turning the teachers-centered learning process to students centered which evaluate the performance from normal grades to fuzzy grades. Buragga et al. [17] made a rubric cube assessing the ABET standards SOs for computer science program at King Faisal University. Zaini et al. [18] designed a framework for online outcome-based education having a central main data storage system allowing concurrent transaction among multiple distributed users.

3 Proposed Evaluation Methodology

3.1 CLO- and PLO-Based Evaluations

The CLO- and PLO-based evaluations mean the determination of student's level of attainment measuring the CLOs of the courses and PLOs of the program for attaining the OBE. The statements which will be learned by the students at the end of the semester taught by the course teachers indicating the knowledge, skill, and attitude are called CLOs. And PLOs are the statements by which help to know what is learned by students during the time of their graduation and what is expected to know in terms of the cognitive, affective, and psychomotor domain [19]. The CLOs and PLOs should be predefined for both students and teachers by the department.

3.2 Challenges Have to Meet

Designing such an evaluation technique for accreditation, there have been several challenges to meet such as available, trained, and experienced teachers engaging with this process. Especially there need enough staffing for data collection, preparation, documentation, question papers setting, storing marks for each of the students, result preparation, etc. This is more challenging for a developing country like Bangladesh to provide enough employees.

3.3 Dataset Preparation

The dataset we have used is not a benchmark but randomly generated for our hypothetical evaluation technique. It has two parts of the same structure for each of the courses. Firstly, the SFE contains the q (no. of questions to be answered for evaluation) number of attributes and secondly the CA contains p (no. of items to be considered for evaluation) number of attributes. Both parts of the dataset contain the n (total no. of CLO of a course) number of rows.

3.4 Mathematical Background for CLO Evaluation with Example

The system is designed for Bangladeshi universities especially for the department of CSE of KUET. The evaluation system for this department has two parameters like SFE and CA. Assume for a theoretical course that, the SFE has q no. of questions of total k marks to be answered and the CA has p no. of items of total h marks to be considered and total n no. of CLOs for evaluation. Both in the question papers and items, marks are assigned according to the CLOs by the course teachers and moderators of the department and earned marks are stored and tracked by the course teachers.

Now, in the SFE, the assigned marks to each of the n CLOs and q questions are represented by the matrix S of dimension $n \times q$ shown in Eq. (1) where s_{ij} represents the assigned marks to CLO_i on question j in SFE.

$$S = \left[s_{ij}\right], n \times q. \tag{1}$$

Similarly, in the CA, the assigned marks to each of the n CLOs and p items are represented by the matrix C of dimension $n \times p$ shown in Eq. (2) where c_{ij} represents the assigned marks to CLO_i on item j in CA.

$$C = \left[c_{ij}\right], n \times p. \tag{2}$$

Then, the earned marks in SFE according to each of the CLOs are represented by the matrix A of dimension $n \times q$ shown in Eq. (3) where a_{ij} represents the earned marks on CLO_i of question j in SFE.

$$A = \left[a_{ij}\right], n \times q. \tag{3}$$

And the earned marks in CA according to each of the CLOs are represented by the matrix B of dimension $n \times p$ shown in Eq. (4) where b_{ij} represents the earned marks on CLO_i of item j in CA.

$$B = \left[b_{ij}\right], n \times p. \tag{4}$$

For evaluating the CLO attainment, the total earned marks and assigned marks in SFE based on CLOs and their ratios are calculated by the matrix X^T, W^T and R^T of dimension $1 \times n$ from Eqs. (5), (6) and (7), where x_i, w_i, and $r_i \in [0, 1]$ represent the total earned marks, assigned marks, and their ratios based on the CLO_i in the SFE. Here, X^T, W^T, and R^T are the transpose of matrix X, W, and R, respectively.

$$X = \left[x_i\right] = \left[\sum_{j=1}^{q} a_{ij}\right], n \times 1. \tag{5}$$

$$W = \left[w_i\right] = \left[\sum_{j=1}^{q} s_{ij}\right], n \times 1. \tag{6}$$

$$R = \left[r_i\right] = \left[\sum_{i=1}^{n} \frac{x_i}{w_i}\right], n \times 1. \tag{7}$$

Table 1 CLOs for theory of computation (CSE 2209) course for evaluation

CLO_i	Description
CLO_1	Identify the connection between language and computation using DFA and NFA
CLO_2	Prove the equivalence of languages described by finite state machines and regular expressions
CLO_3	Practice techniques of program design and development by using abstract machines
CLO_4	Design pushdown automata and the equivalent context-free grammars
CLO_5	Apply the equivalence of languages described by pushdown automata and context-free grammars
CLO_6	Explain the equivalence of languages described by turing machines and post machines

And the total earned marks and assigned marks in CA based on CLOs and their ratios are calculated by the matrix Z^{T}, Y^{T}, and T^{T} of dimension $1 \times n$ from Eqs. (8), (9) and (10), where z_i, y_i, and $t_i \in [0, 1]$ represent the total earned marks, assigned marks, and their ratios based on the CLO_i in CA. Here, Z^{T}, Y^{T}, and T^{T} are the transpose of matrix *Z*, *Y*, and *T*, respectively.

$$Z = [z_i] = \left[\sum_{j=1}^{p} b_{ij}\right], n \times 1. \tag{8}$$

$$Y = [y_i] = \left[\sum_{j=1}^{p} c_{ij}\right], n \times 1. \tag{9}$$

$$T = [t_i] = \left[\sum_{i=1}^{n} \frac{z_i}{y_i}\right], n \times 1. \tag{10}$$

Example: We have taken the "Theory of Computation" course code of CSE 2209 as an example for assessing the CLOs. Assume, CSE 2209 course has a total of 6 CLOs listed out in Table 1. The department of CSE of KUET considers 6 questions to be answered in SFE having 35 marks for each and a total of 210 marks. The CA considers 6 items such as CT_1, CT_2, CT_3, homework (HW), spot test (ST), and attendance (ATT) having 20, 20, 20, 10, 10, and 10 marks for each item, respectively, with total of 90 marks. So, number of CLOs, $n = 6$; questions in SFE, $q = 6$; items in CA, $p = 6$.

Now, the assigned marks matrix *S* and *C* from Eqs. (1) and (2) are generated by course teachers and moderators, and earned marks matrix *A* and *B* from Eqs. (3) and (4) are stored by course teachers.

$$S = [s_{ij}] = \begin{bmatrix} 6 & 12 & 4 & 13 & 8 & 0 \\ 5 & 0 & 9 & 0 & 4 & 13 \\ 7 & 6 & 10 & 6 & 0 & 4 \\ 5 & 7 & 5 & 0 & 11 & 9 \\ 4 & 5 & 7 & 10 & 5 & 3 \\ 8 & 5 & 0 & 6 & 7 & 6 \end{bmatrix}, \quad C = [c_{ij}] = \begin{bmatrix} 6 & 4 & 4 & 1 & 0 & 0 \\ 5 & 0 & 0 & 0 & 5 & 6 \\ 0 & 8 & 4 & 3 & 3 & 2 \\ 4 & 0 & 3 & 0 & 0 & 1 \\ 5 & 5 & 7 & 2 & 2 & 0 \\ 0 & 3 & 2 & 4 & 0 & 1 \end{bmatrix},$$

$$A = [a_{ij}] = \begin{bmatrix} 3 & 9 & 0 & 12 & 8 & 0 \\ 5 & 0 & 9 & 0 & 3 & 8 \\ 5 & 5 & 6 & 6 & 0 & 4 \\ 3 & 2 & 4 & 0 & 10 & 9 \\ 4 & 5 & 6 & 8 & 5 & 3 \\ 7 & 5 & 0 & 6 & 5 & 6 \end{bmatrix}, \quad B = [b_{ij}] = \begin{bmatrix} 4 & 2 & 3 & 1 & 0 & 0 \\ 5 & 0 & 0 & 0 & 3 & 5 \\ 0 & 7 & 3 & 2 & 3 & 2 \\ 3 & 0 & 3 & 0 & 0 & 1 \\ 5 & 2 & 5 & 2 & 2 & 0 \\ 0 & 3 & 2 & 2 & 0 & 1 \end{bmatrix}$$

Using the matrix S, C, A, and B, the matrices X^{T}, W^{T}, R^{T}, Z^{T}, Y^{T}, and T^{T} are demonstrated from Eqs. (5), (6), (7), (8), (9), and (10), respectively.

$$X^{\mathrm{T}} = [x_i] = [32\ 25\ 26\ 28\ 31\ 29], \quad W^{\mathrm{T}} = [w_i] = [43\ 31\ 33\ 37\ 34\ 32],$$

$$R^{\mathrm{T}} = [r_i] \quad = [0.744\ 0.806\ 0.788\ 0.757\ 0.912\ 0.906],$$

$$Z^{\mathrm{T}} = [z_i] = [10\ 13\ 17\ 7\ 16\ 8], \quad Y^{\mathrm{T}} = [y_i] = [15\ 16\ 20\ 8\ 21\ 10],$$

$$T^{\mathrm{T}} = [t_i] \quad = [0.667\ 0.812\ 0.850\ 0.875\ 0.762\ 0.800]$$

Here, $r_i, t_i \in [0, 1]$ represent the input values of CLO_i for fuzzification.

Fuzzification of Input Variables The evaluation system has two input variables r_i and t_i representing the marks ratio in SFE and CA, respectively, for fuzzification and one output variable is CLO_i. We have considered five levels ($v = 5$) of ratio or linguistic values for the fuzzy set. Using the fuzzy domain and marks ratios r_i and t_i, two matrices SF and CA of dimension $n \times v$ indicating the degree of membership for each of the CLOs in the fuzzy domain are demonstrated and shown in Eqs. (11) and (12) where sf_{ij} and ca_{ij} represent the degree of membership for CLO_i and ratio level j in SFE and CA.

$$\mathrm{SF} = [\mathrm{sf}_{ij}], n \times v. \tag{11}$$

$$\mathrm{CA} = [\mathrm{ca}_{ij}], n \times v. \tag{12}$$

The ratio levels $j = 1, 2, 3, 4$, and 5 represent the linguistic values "Very Low (vl)", "Low (low)", "Average (av)", "High (high)", and "Very High (vh)", respectively. The membership function we have used (combination of trapezoidal and triangular) is shown in Fig. 1. If the marks ratios in the matrix R^{T} and T^{T} are given as input to the function, we will get the degree of membership matrix from Eqs. (11) and (12).

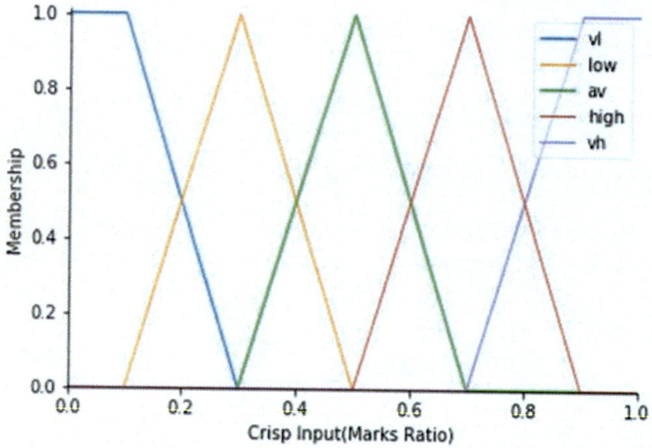

Fig. 1 Membership function used for fuzzification

$$\text{SF} = \left[\text{sf}_{ij}\right] = \begin{bmatrix} 0 & 0 & 0 & 0.780 & 0.220 \\ 0 & 0 & 0 & 0.470 & 0.530 \\ 0 & 0 & 0 & 0.560 & 0.440 \\ 0 & 0 & 0 & 0.715 & 0.285 \\ 0 & 0 & 0 & 0 & 1.000 \\ 0 & 0 & 0 & 0 & 1.000 \end{bmatrix}, \quad \text{CA} = \left[\text{ca}_{ij}\right] = \begin{bmatrix} 0 & 0 & 0.165 & 0.835 & 0 \\ 0 & 0 & 0 & 0.440 & 0.560 \\ 0 & 0 & 0 & 0.250 & 0.750 \\ 0 & 0 & 0 & 0.125 & 0.875 \\ 0 & 0 & 0 & 0.690 & 0.310 \\ 0 & 0 & 0 & 0.500 & 0.500 \end{bmatrix}$$

Here, $\text{sf}_{34} = 0.560$ means the degree of membership for CLO_3 to the ratio level high $(j = 4)$ in the SFE.

Fuzzy Rule Generation and Aggregation The Mamdani type min–max fuzzy inference system and IF–THEN rule bases are used in our system. As r_i and t_i are the input variables and CLO_i is the output variable, the rule bases are as "IF r_i is 'vl (1)' and t_i is 'vl (1)' THEN CLO_i is 'vl (1)"'. So, all the fuzzy rules for evaluating the CLO_i are generated in Table 2. For the aggregation of the rule bases, a matrix AG of dimension $n \times v$ is generated and shown in Eq. (13).

$$\text{AG} = \left[\text{ag}_{ij}\right], n \times v. \tag{13}$$

where ag_{ij} denotes the aggregated value for CLO_i of linguistic value j and ag_{ij} can be demonstrated using Eq. (14), where $(k, l)|\text{Rule}\ (k, l) = j$ represents the rules, for which the output CLO_i are j if the input r_i are k and t_i are l.

$$\text{ag}_{ij} = \max_{\{(k,l)|\text{Rule}(k,l)=j\}} \{\min(\text{sf}_{ik}, \text{ca}_{il})\} \tag{14}$$

For calculating ag_{14},

Table 2 Fuzzy rule base for CLO_i evaluation

$r_i \backslash t_i$	vl (1)	low (2)	av (3)	high (4)	vh (5)
vl (1)	vl (1)	vl (1)	vl (1)	low (2)	av (3)
low (2)	vl (1)	low (2)	low (2)	low (2)	av (3)
av (3)	low (2)	av (3)	av (3)	av (3)	high (4)
high (4)	av (3)	av (3)	high (4)	high (4)	vh (5)
vh (5)	av (3)	high (4)	vh (5)	vh (5)	vh (5)

$$\begin{aligned}\mathrm{ag}_{14} &= \max_{\{(k,l)\mid((3,5),(4,3),(4,4),(5,2))\}} \{\min(\mathrm{sf}_{1k}, \mathrm{ca}_{1l})\} \\ &= \max\{\min(0, 0), \min(0.780, 0.165), \min(0.780, 0.835), \min(0.220, 0)\} \\ &= \max\{0, 0.165, 0.780, 0\} = 0.780\end{aligned}$$

Using Eq. (14) and Table 2, the AG matrix is generated for all the CLOs.

$$\mathrm{AG} = \left[\mathrm{ag}_{ij}\right] = \begin{bmatrix} 0 & 0 & 0 & 0.780 & 0.220 \\ 0 & 0 & 0 & 0.440 & 0.530 \\ 0 & 0 & 0 & 0.250 & 0.560 \\ 0 & 0 & 0 & 0.125 & 0.715 \\ 0 & 0 & 0 & 0 & 0.690 \\ 0 & 0 & 0 & 0 & 0.500 \end{bmatrix},$$

Defuzzification The final crisp output values are produced by the method is called defuzzification. We have implemented the whole system through the Python scikit-fuzzy package. It uses center of gravitation (CoG) method for demonstrating the final crisp output or CLOs. The defuzzification uses the AG matrix and CoG method to compute the CLOs. In Fig. 2, CLO_1 for the CSE 2209 course is demonstrated. Similarly, the other CLOs of the CSE 2209 course are generated and represented by matrix CLO of dimension $1 \times n$ as follows.

$$\mathrm{CLO} = \begin{bmatrix} 0.733 & 0.782 & 0.807 & 0.844 & 0.881 & 0.873 \end{bmatrix}$$

3.5 *PLO Evaluation from the CLOs*

In the previous section, let n no. of CLOs evaluation are done. Now, assume that there is m no. of PLOs of a program to evaluate. The mapping between the CLOs and PLOs is represented by the matrix G of dimension $n \times m$ shown in Eq. (15). We have considered five levels of mapping such as very highly (5), highly (4), averagely (3), lowly (2), and very lowly (1). The zero (0) value indicates the no relation between

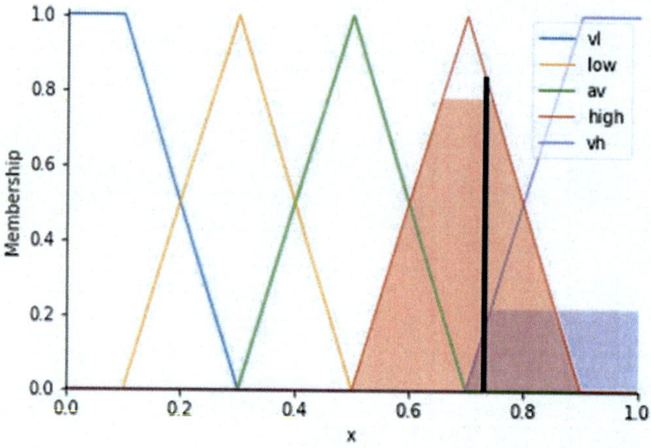

Fig. 2 Defuzzification to evaluate CLO_1 for CSE 2209 course

the CLOs and PLOs.

$$G = \left[g_{ij}\right], n \times m. \tag{15}$$

where g_{ij} represents the mapping level between CLO_i and PLO_j.

Our system is designed for an Engineering program and the Board of Accreditation for Engineering and Technical Education (BAETE), Bangladesh has listed 12 PLOs from the Washington accord for this type of program partially shown in Table 3 [20]. So, the degree provided by the department of CSE, KUET has 12 PLOs. As the course CSE 2209 has 6 CLOs, the mapping of the CLOs and the PLOs is set by the course teachers and accrediting committee. Then, we get the G from Eq. (15).

$$G = \begin{bmatrix} 0 & 0 & 1 & 5 & 0 & 3 & 0 & 0 & 0 & 0 & 2 & 0 \\ 2 & 0 & 3 & 0 & 4 & 0 & 0 & 0 & 0 & 0 & 3 & 0 \\ 4 & 5 & 0 & 1 & 0 & 0 & 0 & 0 & 0 & 0 & 0 & 0 \\ 0 & 3 & 5 & 0 & 5 & 0 & 0 & 0 & 0 & 0 & 0 & 0 \\ 0 & 2 & 3 & 0 & 0 & 0 & 0 & 0 & 4 & 0 & 0 & 0 \\ 0 & 0 & 5 & 0 & 2 & 0 & 0 & 0 & 5 & 0 & 0 & 0 \end{bmatrix}$$

Now, the PLO matrix of dimension $1 \times m$ is demonstrated using Eq. (16).

$$\text{plo}_i = \begin{cases} \frac{\sum_{j=1}^{n} g_{ji} \times \text{CLO}_j}{\sum_{j=1}^{n} g_{ji}}, & \text{if } \sum_{j=1}^{n} g_{ji} \neq 0. \\ 0, & \text{if } \sum_{j=1}^{n} g_{ji} = 0. \end{cases} \tag{16}$$

For demonstrating PLO_1,

Table 3 PLOs for engineering program by Washington accord

PLO_i	Description
PLO_1	Apply knowledge of mathematics, natural science, engineering fundamentals, and an engineering specialization, respectively, to the solution of complex engineering problems
PLO_2	Identify, formulate, research literature, and analyze complex engineering problems reaching substantiated conclusions using first principles of mathematics, natural sciences, and engineering sciences
…	…
PLO_{12}	Recognize the need for, and have the preparation and ability to engage in, independent and life-long learning in the broadest context of technological change

$$\begin{aligned} \text{plo}_1 &= \frac{0*0.733+2*0.782+4*0.807+0*0.844+0*0.881+0*0.873}{0+2+4+0+0+0} \\ &= 0.799 \end{aligned}$$

Similarly, from Eq. (16) and CLO matrix, we get other PLOs and the matrix of PLO as follows.

$$\text{PLO} = \begin{bmatrix} 0.799 & 0.824 & 0.821 & 0.807 & 0.807 & 0 & 0 & 0 & 0.877 & 0 & 0.762 & 0 \end{bmatrix}$$

3.6 *Case Studies*

The degree provided by the department of CSE comprises 161 credit hours, 8 semesters of 4 academic years, total of 40 theory courses having 5 in each semester, and few more practical courses. We have implemented our system for 5 theoretical courses from 2nd year 2nd semester from where CSE 2209 is already discussed as an example. Let the remaining 4 courses are CSE 2201 (Algorithm Analysis and Design), EEE 2213 (Digital Electronics), HUM 2207 (Economics and Accounting), and MATH 2207 (Complex Variable, Vector Analysis and Statistics) having 7, 6, 8, and 9 CLOs, respectively. The CLO and PLO attainment of these courses is demonstrated using the same procedure and is shown in Tables 4 and 5, respectively.

4 Results and Discussions

Our outcome-based model through evaluating the CLOs and PLOs helps the engineering students to improve the opportunities for attaining the learning outcomes. OBE is helpful for both students and teachers where students are acknowledged

Table 4 CLO demonstration of five courses from the 2nd year 2nd semester

Course \CLO_i	CLO_1	CLO_2	CLO_3	CLO_4	CLO_5	CLO_6	CLO_7	CLO_8	CLO_9	Average
CSE 2209	0.733	0.782	0.807	0.844	0.881	0.873	–	–	–	0.820
CSE 2207	0.583	0.724	0.646	0.706	0.605	0.637	0.667	–	–	0.653
EEE 2213	0.728	0.816	0.728	0.668	0.763	0.575	–	–	–	0.713
HUM 2207	0.779	0.802	0.882	0.865	0.834	0.878	0.700	0.888	–	0.828
MATH 2207	0.805	0.881	0.892	0.794	0.774	0.862	0.681	0.778	0.745	0.801

Table 5 PLO demonstrations of five courses from the 2nd year 2nd semester

Course\PLO_i	PLO_1	PLO_2	PLO_3	PLO_4	PLO_5	...	PLO_9	PLO_{10}	PLO_{11}	PLO_{12}
CSE 2209	0.799	0.824	0.821	0.807	0.807	...	0.877	0	0.762	0
CSE 2207	0	0.622	0.657	0.667	0.667	...	0	0	0	0
EEE 2213	0.728	0.727	0.816	0.671	0	...	0	0	0	0
HUM 2207	0.788	0.877	0.808	0	0.858	...	0	0	0	0
MATH 2207	0.779	0.836	0	0	0	...	0	0	0	0
Average	0.774	0.777	0.776	0.715	0.777	...	0.877	0	0.762	0

**The values of PLO_6, PLO_7 and PLO_8 are zero (0) as PLO_{10}

whether they have achieved the target level of attainment or not and teachers can find their lacking during their teaching periods. We have initially case studied for 5 courses. Let us discuss the percentages-based absolute grading system where the teachers are simply collect the total earned marks and the assigned marks without considering CLOs. So, converting the 5 courses data into absolute grading and proposed fuzzy grading are shown in Table 6. We can see that the students in the CSE 2207 and EEE 2213 courses getting 64.67 and 69.67% marks which are very close to 65 and 70%. In the hard boundary of absolute grading system, if they got one more mark their grade would have changed to one upper grade from B and B^+ to B^+ and A^-, respectively, which is so unsatisfactory and not very much fair to the students. In these cases, the teachers may get biased or show kindness to the students. This problem can be easily solved by the proposed outcome-based evaluation using the fuzzy logic system. Even, the course teachers do not know what will be attained by the students in the soft boundary of the fuzzy system. If we convert the average CLO attainment of the courses to percentages of attainment (marks), we can see that the grade is upgraded to the next grade which is shown in Table 6. The comparison between this two evaluation is also shown in Fig. 3. Considering this issue and solving the problem discussed above, we have chosen fuzzy logic system to evaluate so that this provides no biasedness but reliability to both students and teachers. In OBE, we have also measured the attainment of PLOs using the CLOs

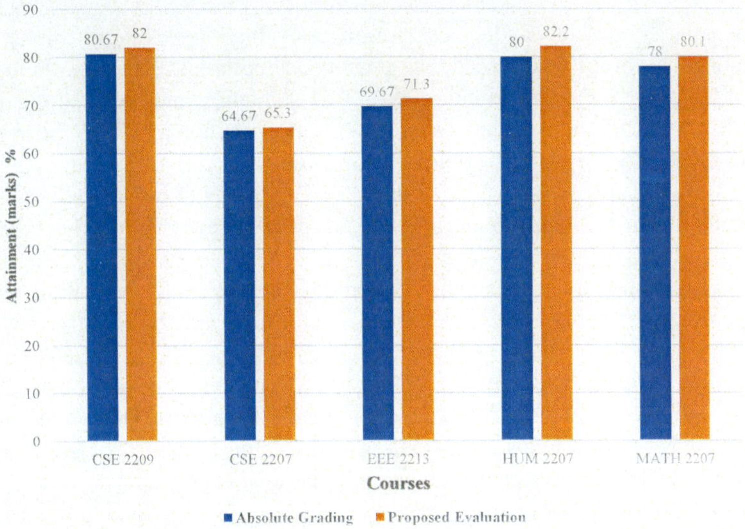

Fig. 3 Comparison between existing grading and proposed evaluation

Table 6 Absolute grading and proposed fuzzy grading of five courses

Courses	Earned marks	Assigned marks	Absolute marks %	Absolute grade	Proposed marks %	Proposed grade
CSE 2209	242	300	80.67	A^+	82.00	A^+
CSE 2207	194	300	64.67	B	65.30	B^+
EEE 2213	209	300	69.67	B^+	71.30	A^-
HUM 2207	240	300	80.00	A^+	82.20	A^+
MATH 2207	234	300	78.00	A	80.10	A^+

by five-level of mapping. The accreditation committee of the department of CSE of KUET provided the binary mapping (0 and 1) which means PLO attainment from CLO will be the same. So, we have done five-level of mapping for better evaluation. The average PLO attainment for each of the five courses is also measured in Table 5.

5 Conclusion

For improving the quality of students and academic programs, the institutions are applying for ABET accreditation. An outcome-based evaluation helps the students especially the engineering students to get acceptance in international standards. The main challenge here is to provide a sufficient no. of employees to prepare a well documentation, a reliable student's performance evaluation technique, implement a computer program for whole academic process accreditation. Recently, the department of CSE of KUET is working on ABET accreditation and prepared documentation. But they did not provide a reliable evaluation method and keep the absolute grading system as they are practicing this method. Regarding a proper evaluation method, in this paper, we have described a CLO- and PLO-based student performance attainment technique using fuzzy logic for adding to the documentation for ABET accreditation. We have considered the class tests, homework, spot tests, and attendance for continuous assessment and semester final exam for evaluation because the department of CSE of KUET has considered these parameters. Firstly, the CLOs are evaluated from the earned marks and assigned marks then PLOs are demonstrated from those CLOs. In this system, there are need some extra efforts of teachers for question setting and tracking the marks for each of the students. Though the system has some drawbacks it helps to generate a meaningful, reliable, fair grade attainment for students. We have cases studied only theoretical courses but it can also be implemented for practical courses.

References

1. Accreditation: setting the standard worldwide. Available: https://www.abet.org/accreditation/. Last Accessed: 22 Mar 2021
2. University Grants Commission of Bangladesh. Available: http://ugc.portal.gov.bd/sites/default/files/files/ugc.portal.gov.bd/notices/e4c1bdfd_8db9_4af8_a538_34dbc84ed2b0/OBE131019.pdf. Last accessed: 22 Mar 2021
3. Integrated University System, Khulna University of Engineering & Technology. Available: https://academic.kuet.ac.bd/public.php?page=grading. Last accessed: 22 Mar 2021
4. Ewing AM (2012) Estimating the impact of relative expected grade on student evaluations of teachers. Econ Educ Rev 31(1):141–154
5. Aziz A, Golap MAU, Hashem MMA (2019) Student's academic performance evaluation method using Fuzzy Logic system. In: 2019 1st International conference on advances in science, engineering and robotics technology (ICASERT). IEEE, pp 1–6
6. Shafi A, Saeed S, Bamarouf YA, Iqbal SZ, Min-Allah N, Alqahtani MA (2019) Student outcomes assessment methodology for ABET accreditation: a case study of computer science and computer information systems programs. IEEE Access 7:13653–13667
7. Hussain W, Spady WG, Khan SZ, Khawaja BA, Naqash T, Conner L (2021) Impact evaluations of engineering programs using Abet student outcomes. IEEE Access 9:46166–46190
8. Shanableh A (2011) IT-facilitated student assessment: outcome-based student grades. In: 2011 International conference on information technology based higher education and training. IEEE, pp 1–6
9. Hameed IA, Sorensen CG (2010) Fuzzy systems in education: a more reliable system for student evaluation, pp 978–953. ISBN
10. Chandna VK (2015) Course outcome assessment and improvement on weak student. In: 2015 IEEE 3rd International conference on MOOCs, innovation and technology in education (MITE). IEEE, pp 38–40
11. Varghese A, Sreedhar JP, Kolamban S, Nayaki S (2017) Outcome based assessment using fuzzy logic. Int J Adv Comput Sci Appl (IJACSA) 8(1):103–106

12. Abou-Zeid A, Taha MA (2014) Accreditation process for engineering programs in Saudi Arabia: challenges and lessons learned. In: 2014 IEEE Global engineering education conference (EDUCON). IEEE, pp 1118–1125
13. Lakshmi MV (2014) Outcome-based teaching: microwave and radar. In: 2014 IEEE International conference on MOOC, innovation and technology in education (MITE). IEEE, pp 227–231
14. Akir O, Eng TH, Malie S (2012) Teaching and learning enhancement through outcome-based education structure and technology e-learning support. Procedia-Soc Behav Sci 62:87–92
15. Eldeeb R, Shatakumari N (2013) Outcome based education (OBE)-trend review. IOSR J Res Method Educ 1:9
16. Ma J, Zhou D (2000) Fuzzy set approach to the assessment of student-centered learning. IEEE Trans Educ 43(2):237–241
17. Buragga KA, Khan AR, Zaman N (2013) Rubric based assessment plan implementation for computer Science program: a practical approach. In: Proceedings of 2013 IEEE international conference on teaching, assessment and learning for engineering (TALE). IEEE, pp 551–555
18. Zaini N, Latip MFA, Omar H (2011) Semantic-based online Outcome-based education measurement system. In: 2011 3rd International congress on engineering education (ICEED). IEEE, pp 218–222
19. Mapping of OUM Outcome Based Education (OBE) System. Available: http://library.oum.edu.my/repository/1149/1/library-document-1149.pdf. Last access: 09 Apr 2021
20. Board of Accreditation for Engineering and Technical Education. Available: https://www.baetebangladesh.org/poa.php. Last access: 04 May 2021

Performance Comparisons in Association Rule Mining Over Public Datasets

Jaher Hassan Chowdhury, Md. Billal Hossain, M. Shamim Kaiser, and Mohammad Shamsul Arefin

Abstract Association rule mining techniques are widely employed in a variety of applications, including stock analysis, log mining, medical diagnostics, consumer market analysis, bioinformatics, and many more. When it comes to uncovering interesting relationships between variables in huge databases, association rule learning is a rule-based machine learning method that can be used. With the help of some metrics of interestingness, it is intended to find powerful rules that have been identified in databases. Discovery and validation are two subprocesses of association rule mining, which is a data processing technique that is divided into two parts. The first method is known as finding frequent itemsets, and the second method is finding association rules from these frequent itemsets. The ideas associated with the use of frequently used itemsets are retrieved from the data during this subprocess. Numerous techniques and procedures for locating frequently occurring itemsets and association rules have been devised by academics throughout the years. This study presents the results of a comprehensive performance analysis of ten distinct association rule mining methods, which was carried out on five different datasets.

Keywords Data mining · Association rule mining · Support counts · FP-Growth · MNR algorithm

1 Introduction

The process of analyzing data and summarizing data into usable information is the main objective of data mining techniques [6, 7]. It is a method for analysis that allows a user to examine the data and to sum up the relationship between these variables. The

J. H. Chowdhury · Md. B. Hossain · M. S. Arefin (✉)
Department of Computer Science Engineering, Chittagong University of Engineering and Technology, Chattogram 4349, Bangladesh
e-mail: sarefin@cuet.ac.bd

M. S. Kaiser
Wazed Miah Science Research Centre (WMSRC), Institute of Information Technology, and Applied Intelligence and Informatics (AII), Jahangirnagar University, Savar, Dhaka 1342, Bangladesh

M. S. Arefin et al. (eds.), *Proceedings of the International Conference on Big Data, IoT, and Machine Learning*, Lecture Notes on Data Engineering and Communications Technologies 95, https://doi.org/10.1007/978-981-16-6636-0_57

work in [1] addresses the problem of first establishing the relationship between data which is considered as part of research connected to association mining. Association rules are used to find the links between frequently used objects together [2, 13]. The applications to association rules are basket analysis, classification, cross-marketing, classification, catalog design, and loss-leader analyzes [8]. The association rules apply two key elements, support and trust. Association rules are typically required to comply with the user's minimum support and the user's minimum trust.

In this study, we compared the performance of 10 distinct algorithms (Apriori, Fp-Growth; FP-Growth with Lift; RP-Growth; FP-Close; Indirect; MNR; Sporadic; Top K; IGB) used by five different datasets (Breast Cancer, Facebook Comment Prediction Dataset, High School Dataset, Sales Transaction Dataset, Zoo Dataset) in association rules mining. The detail of these algorithms is explained in Sect. 3. To the best of our knowledge, none of the work have considered mentioned diverse datasets for association rule mining employing 10 distinct algorithms.

The rest of the paper is organized as Sect. 2 provides a detailed review of data mining and association rules mining researches. The details of association rules mining methods employed in this paper are explained in Sect. 3. In Sect. 4, we discuss experimental data and analysis. Finally, in Sect. 5, we present the conclusion of the paper.

2 Literature Review

This section contains a survey of the literature on association rule mining algorithms. Apriori is the first algorithm for mining association rules introduced by Agarwal et al. [1]. The association rule fosters support and confidence which has two steps—the first step is to develop candidate itemsets, and the second step is to generate a large itemset based on the minimal support threshold values. Normally, if any k itemset is uncommon, the (k+1) super-itemsets are equally uncommon. Here, the algorithm mines the frequent itemsets using the candidate generation procedure. Two points are critical here. These two parameters are utilized to trim and determine the most precise association rules. However, the primary disadvantage of the Apriori association rule is that it is extremely time demanding and consumes a great deal of space. Additionally, it does multiple searches of the database to generate potential itemsets.

To address the Apriori's inadequacies, data scientists developed a new algorithm. That is the frequency pattern (FP)-Growth that Han et al. suggest. [5]. It is often tree-based and performs a single scan of the entire database twice. It extracts from the database a frequent pattern tree and a conditional pattern base that satisfy the minimum support. Due to its small construction, it requires less time and space for execution. Additionally, it is ideal for huge itemsets. Following that, the FP-Growth with lift was implemented. This is a variant of the algorithm used to extract all association rules from a transaction database. Traditionally, association rule mining has been conducted using two critical metrics: rule support and confidence in rule evaluation. Han et al. demonstrated how to employ another widely used metric known as the lift or interest. The researchers then sought to ascertain the various sorts of association rules.

Szathmary developed a new technique for discovering novel association rules in datasets. It is a closed association rule mining algorithm. A condensed representation of all association rules. Szathmary [12] was concerned in identifying rules of sporadic association in datasets. They developed a method for mining absolutely sporadic association rules. The technique begins by utilizing Apriori-Inverse to generate perfectly rare itemsets. Then, using these itemsets, it constructs the association rules.

Koh and Rountree [9] developed the zart method for determining Informative Generic Based (IGB) association rules. This algorithm extracts a subset of all association rules known as IGB association rules (Informative and Generic Basis of Association Rules) from a transaction database. This subset was originally proposed by Gasmi et al. [4]. This algorithm discovers the IGB association rules in two steps: (1) The approach begins by discovering closed itemsets and their related generators using the Zart technique. (2) Following that, association rules are built using closed itemsets and generators.

Kryszkiewicz [10] discovered a set of minimal, non-redundant rules for associations that is both underperforming and compact. In this implementation, he uses the Zart technique to discover closed itemsets and their associated generators. Following that, the minimal non-redundant association rules were built utilizing this information. The author then established an indirect association between the data by looking forward in time. As a result, they developed an approach called indirect association rule mining for detecting indirect links between items in transaction databases. This approach is significant since traditional association rule mining algorithms focus on direct relationships between itemsets. This method is capable of detecting indirect relationships, which is beneficial in domains such as biology. The analysis of indirect association rules is beneficial for a variety of applications, including stock market analysis and competitor product analysis.

Tan et al. [14] discovered the above-mentioned relationship. After working with these algorithms, the scientists coined a phrase to substitute support in order to boost the algorithms' performance. Top-K rules is a search algorithm for identifying a transaction database's top-k association rules. Other algorithmic approaches to association rule mining require the definition of a difficult-to-set minimum support (minsup) parameter (usually users set it by trial and error, which is time consuming). Top-K rules avoids this issue by allowing users to directly define k, the number of rules to discover, rather than using minsup.

Fournier-Viger et al. [3] offer RP-Growth, a fast technique based on FP-Growth for top-k mining of discriminative patterns that are highly relevant to the class of interest. In branch and bound search with antimonotonic upper limit values such as f-score 2, the RP-Growth algorithm produces a minimum support increase, a well-known and straightforward pruning approach for top-k mining. The branch pruning and bound search results in minimum aid increasing. Additionally, by introducing

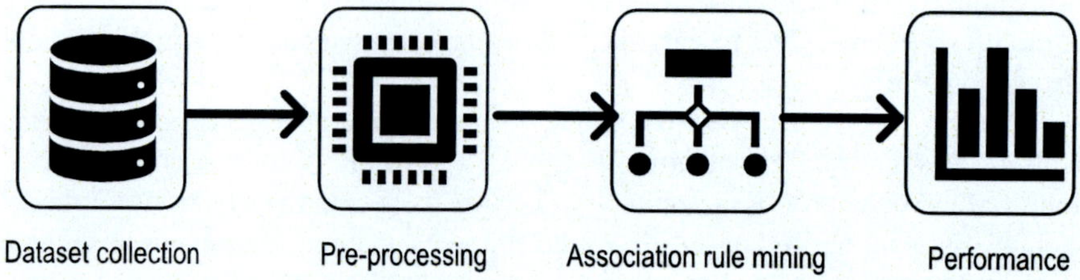

Fig. 1 Steps involved in the performance evaluation of association rule mining

Table 1 Description of dataset

Dataset name	Source	Data description
Breast cancer dataset	UCI	Instances: 286 Attributes: 09
Facebook comment prediction dataset	Kaggle	Instances: 40949 Attributes: 28
High school dataset	UCI	Instances: 3095 Attributes: 33
Sales transaction dataset	Kaggle	Instances: 811 Attributes: 107
Zoo dataset	UCI	Instances: 101 Attributes: 17

the concept of weakness and an aggressive pruning technique based on it, RP-Growth effectively discovers *k* patterns with a high degree of diversity and relevance to the class of interest.

3 System Architecture and Design

The paper is comprised of the following major steps: (1) dataset collection, (2) preprocessing of datasets, (3) applying association rule mining algorithms, (4) evaluating performance, and (5) the decision (Fig. 1).

3.1 Dataset Collection

The most critical aspect of this research is selecting the appropriate datasets. Association rule mining has a variety of real-world applications. Thus, to ascertain the relationship between datasets from various sectors, we selected five datasets. Table 1 has a complete description of the datasets. One dataset is organized by medical diagnosis. Thus, we can see how the association rule mining method might be used to diagnose a patient using this dataset. Additionally, we are interested in social datasets to ascertain how people communicate. As a result, we used the Kaggle dataset for Facebook comment prediction. On the other side, we used the high school dataset to assess the students. We used the sales transaction dataset for market basket analysis and the "zoo" dataset from the UCI machine learning repository for categorizing the animals in the zoo.

3.2 Data Preprocessing

Preprocessing is the stage at which data is converted or encoded in such a way that it can be easily parsed by a machine in any machine learning process. In other words, the algorithm can now read the data's features effortlessly. In this step, we will prepare our datasets for the algorithms' implementation. We were missing data in the Facebook dataset, which required us to instantly correct using the median method. Continuous numeric values in dataset columns may be supplanted in mode by the middle, the middle, or the remainder of the column values. This technique is capable of avoiding data loss. The two approximations above (medium, medium) are a statistical technique for dealing with missing values. The remaining four datasets are preprocessed, which saved us time. Additionally, the primary difficulty involved converting the dataset's values to numerical values. We did this to ensure that our algorithm runs more quickly. Additionally, with a minimal level of support and trust, we can have a huge number of association rules. Prior to data mining, data preprocessing is always critical.

3.3 Applying Algorithm

For our investigation, we used 10 different association rule mining algorithms: Apriori, FP-Growth, FP-Growth with lift, RP-growth, Minimum Non-redundant Association Rules (MNR), Indirect, Sporadic, IGB, FP-Close, and TOP-K. The association rule mining methods were built in the Java programming language. Because Java enables data scientists and programmers to grow their applications more easily. Due to the comparative nature of our work, we concentrated on the association rules generated and the algorithm's temporal complexity during implementation.

The Apriori technique is used to look for frequent itemsets inside a data collection that follows the Boolean association rule. We employ an iterative or level technique to locate $k + 1$ articles that include k-frequent itemsets. The Apriori characteristic is critical since it aids in reducing the search area and so increases the efficiency of frequent item production. All non-empty subsets that are commonly utilized must also be frequent (Apriori property). The Apriori algorithm's antimonotonic support measures are a fundamental notion. Apriori holds the belief that—Each subset of a common item must be frequent (Apriori property). If an object is uncommon, all of its supersets are uncommon. The Apriori technique has two major drawbacks: candidate sets must be built at each stage and the program must repeatedly scan the database for candidate sets.

Both of these traits unavoidably cause the algorithm to slow down. To circumvent these unnecessary stages, a novel association mining technique dubbed frequent pattern growth algorithm has been devised. It addresses the drawbacks of the Apriori method by storing all transactions in a trie data structure. This algorithm has improved the Apriori approach to a new level. Candidate creation is not required to generate a frequent pattern. For instance, the frequent pattern tree (FP tree) growth algorithm represents the database using an FP tree. Using this tree structure, itemsets will remain connected. Due to the presence of a single regularly occurring item, the database has become fragmented. That is the name of the pattern fragment. The fractured patterns' itemsets are analyzed. This strategy considerably reduces the time required to look for frequently occurring itemsets. To determine the most frequent pattern in a tree-like structure constructed from the database's initial itemsets, the FP tree must be employed. Each item in the itemset is represented as a node on the FP tree by this tree.

The nodes beneath the root node, which symbolizes null, represent itemsets. While constructing the tree, the nodes' relationships with the lower nodes are preserved, i.e., the itemsets' associations with one another.

A preliminary database scan is run to determine the presence of each itemset in the database. Apriori's second phase is identical to its first. The number of 1-itemsets in a database is referred to as the support count or frequency of 1-itemsets. In the second stage, the FP tree is created. To accomplish this, create the tree's root. Null is the root. Following that, the database will be scanned again and the transactions examined. The initial transaction should be reviewed to ascertain the itemset it contains. The itemset with the most items is displayed first, followed by those with less items, and so on. On the branch of the tree, transaction itemsets are organized in ascending order of count. As a result, the transaction's common itemset is linked to the newly created node of another itemset. Additionally, once transactions are completed, the itemset's count is incremented. As new nodes are established and associated with transactions, both the common node and new node counts grow by one. After that, it is time to begin mining your newly built FP tree! This is accomplished by first examining the lowest node and its relationships. The lowest node in the graph represents the frequency pattern with length 1. From this point on, follow the FP tree path. These are referred to together as a conditional pattern basis (or pattern base). The pathways to the prefixes in the pattern tree are maintained in a separate database called conditional pattern base (suffix). By counting the amount of itemsets along the route, a conditional FP tree is generated. The conditional FP tree evaluates just those itemsets that support the threshold. Frequent patterns are generated by the conditional FP tree.

In the FP-Growth with lift method, we included a new parameter called lift. If the lift equals one, the X and Y are independent under the association rule $X \rightarrow Y$. If the lift is more than one, the correlation between X and Y is positive. If the lift is less than one, the relationship between X and Y is negative. When calculating the conditional probability of $\{Y\}$ $\{X\}$ occurrence, subsequent lift controls (frequency).

FP-Close is a variant of the FP-Growth algorithm that is optimized for mining frequently closed objects. FP-Close is reputed to be one of the quickest closed mining algorithms. When a transaction database is extremely thick and the minimal support need is low, i.e., when the database contains a high number of large frequent itemsets. For a frequent itemset of size l, for example, all $2 * (l * l)$ non-empty subsets must be generated. Because all subsets of a frequent itemset are frequent, it suffices to discover only the most often occurring itemsets (MFIs). Because there is no frequent itemset Y that is identical to X, X can be considered to be maximum. One need just dig along the lattice's edge to mine commonly occurring itemsets. Those on either side of the boundary are uncommon, whereas those on the other side are all frequent. As a result, some existing algorithms can only mine the most frequently occurring itemsets. As a result, mining solely MFIs has the following disadvantages. Our knowledge of an MFI's support and subsets is limited. All frequently occurring itemsets must be considered while generating association rules. This difficulty is solved by a sort of frequent itemset called a closed frequent itemset. An algorithm for mining closed association rules returns a list of closed association rules. A closed association rule is an $X \rightarrow Y$ association rule that represents the closed itemset of the union of X and Y. The algorithm returns all closed association rules that satisfy the user's minimum support and confidence standards. Thus, FP-Close frequently returns closed itemsets. A frequent itemset is one that appears in at least minsup transactions in the transaction databases. A often closed item is a frequently occurring itemset that lacks the same support in a proper superset. As a result, the frequently closed array is a subset of the common array. The collection of frequently closed things is typically significantly smaller than the collection of common items, and no information is lost.

Closures and Galois connections are employed in more sophisticated procedures that generate only a subset of the total set of rules. FCA-based techniques provide a better trade-off between the size of the mining result and the amount of information communicated than frequent pattern algorithms. In this setting, generic thinking received less attention in comparison to the number of articles produced to describe it. Their primary concern was with syntactic techniques for deriving rules from generic bases. IGB set of association rules is a new, sound, and instructive general framework for association rules. The soundness characteristic is utilized to evaluate the "syntactic" derivation since it ensures that all association rules can be deduced from the generic basis. The informativeness of a derivable rule enables precise and accurate determination of its support and confidence. Once these generic bases are established, it is trivial to extract the remaining (duplicate) rules. This approach extracts a subset of association rules called IGB set. It is a generic and informative collection of association rules. This algorithm discovers the IGB association rules in two steps: (1) The closed items and their associated generators are located initially; (2) the association rules are produced using the closed objects and generators. We describe sporadic rules as those that have a low degree of support but a high degree of confidence, such as a rare connection of two symptoms indicating the presence of a rare disease. To discover such rules, Apriori's well-known minimal support setting is required, resulting in a huge number of trivial frequent itemsets. Our method, Apriori-Inverse, disregards all candidate itemsets with a support greater than a predefined threshold to uncover rules that occur at random. It is conceivable to have rules that are sporadic (items that fall below the maximum support) and rules that are imperfectly sporadic (items that do not fall below maximum support). Apriori-Inverse is significantly faster than Apriori at discovering all completely sporadic rules. Additionally, we propose that Apriori-Inverse be expanded to discover some imperfectly sporadic laws (but not necessarily all). An algorithm for mining precisely sporadic association rules is referred to as a sporadic association rule mining algorithm. To begin, the algorithm generates extremely rare goods. A rare itemset (alternatively referred to as a sporadic itemset) is an itemset that is not frequently utilized, and all its subgroups contain uncommon things. Additionally, assistance must be greater than or equal to the minimum criterion. Then, using these itemsets, association rules are formed.

Taxonomy information can be utilized to prune out insignificant rules, resulting in a 60% reduction in the number of rules. It is customary to quantify the level of interest in a rule. In [10], authors discussed two other ways for extracting useful rules from databases. If rules provide maximum prediction with the least amount of knowledge feasible, they are deemed intriguing in the first approach to rule-making. These forms of association rules will be referred to as minimum condition maximum consequence rules (MMR). As stated in [10], a second strategy comprises searching for the smallest set of association rules from which all other association rules may be derived without the usage of a database. The term "set of representative association rules" refers to this collection of rules (RR). They demonstrate the existence of a subset of RR called MMR. This implementation has closed items and their associated generators. This information is then utilized to generate the association's minimum non-redundant rules. The minimum set of non-redundant rules shall be specified as $P_1 \rightarrow P_2/P_1$, where P_1 is a generator of P_2, P_2 is a closed element, and the rule shall have at least as much support and confidence as the minimum support and confidence. The MNR algorithms generate a sequence of non-redundant association rules.

In [11], a novel pattern termed indirect association is introduced and its applicability in a variety of application fields is investigated. Apriori and other associative mining algorithms will only discover itemsets that have support above a user-defined threshold of support. Any itemsets with support levels less than the minimum support requirement are discarded. In our perspective, an uncommon pair of objects can be valuable if they are associated indirectly via another set of items. As a result of an algorithm, these patterns can be applied in the retail, literary, and stock market realms. Indirect is a database search algorithm that identifies indirect relationships between objects in transaction databases. An indirect association takes the form $(x, y) \rightarrow M$, where x and y are discrete elements and M is a set dubbed "mediator". The following conditions must be completed to establish an indirect association: The total number of transactions divided by the total number of transactions containing all $\bigcup_x M$ items must equal or exceed minsup. It must be bigger than or equal to the difference between the number of transactions containing all items from $\bigcup_y M$ and the total number of transactions. Subtract the total number of transactions (ts) from the total number of transactions containing (x, y). If x is more confident than y about M, then y must be more confident than minconf about M. Top-K rules is a technique for discovering associations that meet the highest standards in a transaction database. Other algorithms connected with mining rules require the setting of a minimum support (minsup) parameter (usually users set it by trial and error, which is time consuming). Top-K rules addresses this issue by allowing users to specify the number of rules to be discovered directly, rather than utilizing minsup.

Top-K rules is an algorithm for the discovery in a transaction database of the highest standards for associations. Other algorithms associated with mining rules require a minimum (minsup) support parameter to be set (usually users set it by trial and error, which is time consuming). Top-K rules solves this problem by letting users directly indicate k, instead of using minsup, the number of rules to be discovered.

In a transaction database, RP-Growth is an algorithm for the finding of itemsets (group of items) that occur seldom (rare itemsets). Apriori levelwise algorithms are used by all current algorithms for rare association rule mining. In RP tree, Tsang et al. [15] propose a method for mining a subset of rare association rules using a tree structure, as well as an information gain component that aids in identifying the more interesting association rules. Researchers have found that RP tree itemset and rule generation are faster than modified versions of FP-Growth and ARIMA, and that it uncovers 92–100% of all interesting, rare association rules in real-world datasets. A rare itemset is an itemset appearing within the minrare support and minimum support range. Then, each itemset is annotated with its support value. The number of times an itemset appears in the transaction database is the number of support value.

3.4 Evaluating Performance

Finally, we evaluate the performance of the algorithms mentioned above. As a result, after applying the algorithm to five datasets, we must determine which technique performs the best. Thus, we must examine the generated association rules, the time required by the algorithms to generate them, and the data support for each association rule mining algorithm. Additionally, we will analyze how these algorithms generate association rules in terms of their fixed confidence value.

4 Results and Discussion

In this section, we present the experimental setup, and performance evaluation in association rule mining over 10 public datasets.

4.1 Experimental Setup

This section presents the experimental setup and performance evaluation in association rule mining over ten public datasets.

The proposed system has been implemented on a machine having Windows 10, Core i5 2.4 GHz with 8 GB RAM. JAVA is used for developing it.

4.2 Results and Discussion

We have to compare the results of the algorithms one by one. We have taken two different criteria for comparing.

- Association generated by the algorithms in different datasets, and
- Time needed to generate this association rule

Association Rules Generation In this comparison, we are going to keep the confidence value fixed. We took that as 0.8. Moreover, as the support we took 0.7 and for K value we took 3. After that, we will apply the algorithm in different datasets.

In Tables 2, 3, and 4, we have observed that the RP-Growth algorithm is not performing well in this breast cancer dataset. Moreover, Apriori, FP-Growth, FP-Growth with lift are generating the same amount of association rules. However, Top-K is showing superior performance improvement in compared to the other algorithms for this breast cancer dataset.

In the Facebook dataset, we can see Apriori, FP-Growth, and FP-Growth with lift are generating the same number of association rules. However, RP-Growth and FP-Close algorithms also performed well with this dataset.

In the School dataset, we can found that the Sporadic and Top-K association rule mining algorithms are not performing well. Moreover, Apriori, FP-Growth, and FP-Growth with lift are generating the same number of association rules again. But, with these three algorithms, we can see that RP-Growth, MNR, and FP-Close algorithms also perform well in this dataset. So, the amount of non-redundant and closed association rules is a large amount in this dataset.

In the Sales Transaction dataset, we can observe that the RP-Growth is outperforming the other algorithms and generating a huge number of association rules. Other algorithms in this dataset with a high fixed confidence value do not perform at all. They failed to generate a high number of association rules in terms of high support and confidence.

Using the Zoo dataset, we can see that the algorithm exhibits the same performance that we have seen in the datasets of sales transaction data, Facebook data, and the breast cancer data. Apriori, FP-Growth, and Fp-Growth with lift are all executing the same role in order to construct association rules for the same dataset. RP-Growth, on the other hand, is a solid performer in this regard. Furthermore, Top-K performs admirably in this dataset as well.

While summarize the data from Tables 2, 3, and 4, we can see that Apriori, FP-Growth, and FP-Growth with lift all produce the same number of association rules. However, the performance of other algorithms varies depending on the dataset. A dataset is analyzed using the RP-Growth algorithm, which looks for patterns that are rare in the data. Thousands of uncommon patterns of human genes, as well as the proteins and amino acids from which the genes are derived, will be revealed in biology. In addition, there are millions of species, each with a unique genome and set of features, which makes classification difficult. As a result, when we apply RP-Growth to the Zoo dataset, we will find a large number of association rules.

Table 2 Association rules generated by different algorithms in different datasets for support value 0.5

Algorithms	Breast cancer	Facebook	Sales transaction	School	Zoo
FP-Growth	871	291103	331	60324	835011
FP-Growth with lift	871	291103	331	60324	835011
RP-Growth	2898	117249	139901	7043610	10980582
FP-Close	385	312715	318	62223	43803
Indirect	401	401	1008	21425	27913
MNR	338	31234	101	68435	13131
Sporadic	122	8	402	6	840
TOP-*K*	45	15	9	17	13
IGB	99	7548	0	542	1210

Table 3 Association rules generated by different algorithms in different datasets for support value 0.6

Algorithms	Breast cancer	Facebook	Sales transaction	School	Zoo
Apriori	692	28003	245	57581	778938
Fp-Growth	692	240032	245	57581	778938
FP-Growth with Lift	692	240032	245	57581	778938
RP-Growth	2333	105981	115082	670482	1024124
FP-Close	263	2713220	213	59416	41412
Indirect	288	288	874	20813	24581
MNR	258	274198	87	59260	1008)
Sporadic	89	3	265	1	558
TOP-*K*	28	9	2	11	8
IGB	57	4715	0	305	845

Furthermore, these datasets contain evidence of indirect relationships. FP-Close, on the other hand, is always on the lookout for data that has a close link. The number of these types of contacts is larger on social media platforms. It is performing well in this regard on Facebook or in the social connection dataset, among other places, as well. As opposed to other sources of information, school datasets contain accurate data. Direct relationships between students are at a minimum. It is common for MNR algorithms to discover non-redundant rules in a dataset; therefore, it is likely that the school dataset will have a large number of association rules. Sporadic refers to something that occurs just once or twice a year at random. In order to find the unusual and the isolated, irregular searches are conducted. It therefore works flawlessly with the Zoo dataset as well.

Table 4 Association rules generated by different algorithms in different datasets for support value 0.7

Algorithms	Breast cancer	Facebook	Sales transaction	School	Zoo
Apriori	401	271706	180	55942	714386
Fp-Growth	401	271706	180	55942	714386
FP-Growth with lift	401	271706	180	55942	714386
RP-Growth	1573	101259	109306	5986305	9930582
FP-Close	209	256801	100	55418	35150
Indirect	244	244	780	18989	21459
MNR	169	249985	60	55289	9975
Sporadic	51	1	180	0	447
TOP-*K*	19	4	0	3	3
IGB	33	3019	0	182	672

Time Taken to Generate the Association Rules It has been demonstrated that three algorithms perform nearly identically in the case of association rules generated by these algorithms, as shown in Table 4. As a result, we require another method of distinguishing between these algorithms. We applied ten algorithms to the breast cancer and Facebook datasets in order to determine how long it takes to construct the association rules. There are two of them, one of which is small and the other of which is very enormous. Following the application, we plotted the value in the table to show how it changed. Lift algorithms enable us to distinguish between the Apriori, FP-Growth, and FP-Growth via lift algorithms. The values are expressed in milliseconds, and we can see that the FP-Growth algorithm performs admirably across all of the datasets tested.

In a nutshell, Apriori, FP-Growth, and FP-Growth with lift are generating the same amount of rules in different datasets. But the time taken to generate the association rule is different. Because FP-Growth is using an FP tree to generate these association rules. Moreover, the pruning techniques are better than the Apriori algorithm. But, the lift value is an important measure of a rule. One can specify the desired lift field in the settings. The ratio of confidence of the rule and the expected confidence of the rule is a lift value of the association rule. We are using it to have some precise and desired association rules. This third value (lift) is used as a pruning technique and for a certain value of lift like 0.8, and the FP-Growth with lift is generating the same amount of rules as Apriori and FP-Growth. But, it takes an extra amount of time to generate an association rule rather than FP-Growth. That is why the FP-Growth wins.

On the other hand, association rules like IGB, sporadic, and TOP-*K* association rules take less time on some datasets than the above-mentioned three algorithms. But you have to keep in mind that they are also generating a lower number of association rules. That is why FP-Growth can be used in different kinds of datasets to get desired association rules in a shorter time.

Figure 2 shows the time taken to generate the association rules by different algorithms using different datasets (Table 5).

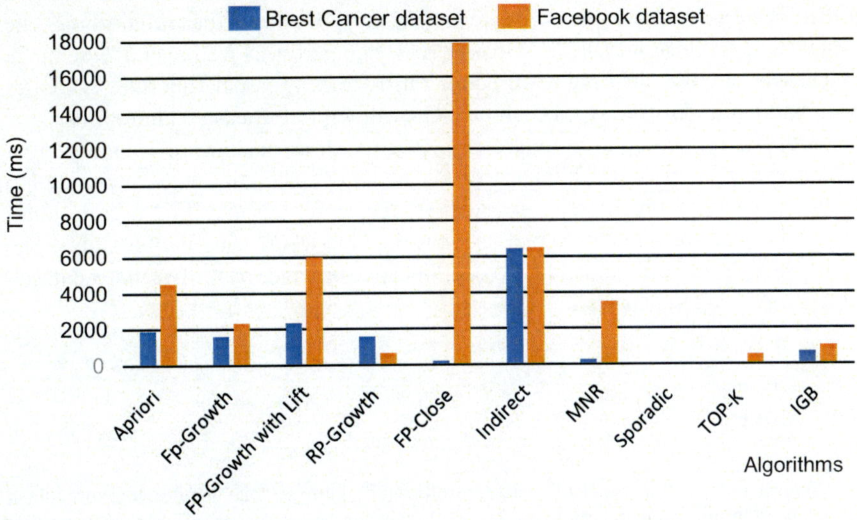

Fig. 2 Time taken to generate the association rules by different algorithms using different datasets

Table 5 Comparative association rules generation times by different algorithms for support value 0.7

Algorithms	Breast cancer dataset	Facebook dataset
Apriori	1898 ms	4517 ms
Fp-Growth	1609 ms	2310 ms
FP-Growth with Lift	2329 ms	6017 ms
RP-Growth	1573 ms	648 ms
FP-Close	198 ms	17846 ms
Indirect	6424 ms	6451 ms
MNR	255 ms	3459 ms
Sporadic	3 ms	1 ms
TOP-*K*	4 ms	514 ms
IGB	672 ms	1013 ms

5 Conclusion

In this study, we have used ten techniques and five datasets to accomplish our goals. Furthermore, we provide variety in the method so that our work may be distinguished from others. On the basis of the association rules generated and the time it takes to construct the association rules, we evaluate the performance of these algorithms. The comparison revealed that the FP-Growth approach performs well over a wide range of datasets, and that the time required to construct these association rules is shorter than that required by the Apriori and FP-Growth with lift measure algorithms.

Furthermore, using the Sporadic and RP-Growth association rule mining algorithms, we can discover the hidden sporadic and rare association rules in transaction datasets that were previously unknown. Furthermore, we may find a large number of closed associations in social network datasets when using the FP-Close. The MNR algorithm also demonstrated exceptional performance on the Facebook dataset, which we discovered as well.

References

1. Agrawal R, Srikant R et al (1994) Fast algorithms for mining association rules. In: Proceedings of the 20th international conference on very large data bases, VLDB, vol 1215. Citeseer, pp 487–499
2. Farhin F, Sultana I, Islam N, Kaiser MS, Rahman MS, Mahmud M (2020) Attack detection in internet of things using software defined network and fuzzy neural network. In: 2020 icIVPR, pp 1–6
3. Fournier-Viger P, Wu CW, Tseng VS (2012) Mining top-k association rules. In: Canadian conference on artificial intelligence. Springer, Berlin, pp 61–73
4. Gasmi G, Yahia SB, Nguifo EM, Slimani Y (2005) IGB: a new informative generic base of association rules. In: Pacific-Asia conference on knowledge discovery and data mining. Springer, Berlin, pp 81–90
5. Han J, Pei J, Yin Y (2000) Mining frequent patterns without candidate generation. ACM Sigmod Rec 29(2):1–12
6. Kaiser MS, Al Mamun S, Mahmud M, Tania MH (2021) Healthcare robots to combat covid-19. In: COVID-19: Prediction, decision-making, and its impacts. Springer, Berlin, pp 83–97
7. Kaiser MS et al (2021) iWorksafe: towards healthy workplaces during covid-19 with an intelligent Phealth app for industrial settings. IEEE Access 9:13814–13828. https://doi.org/10.1109/ACCESS.2021.3050193
8. Kaiser MS et al (2018) Advances in crowd analysis for urban applications through urban event detection. IEEE Trans Intell Transp Syst 19(10):3092–3112. https://doi.org/10.1109/TITS.2017.2771746
9. Koh YS, Rountree N (2005) Finding sporadic rules using apriori-inverse. In: Pacific-Asia conference on knowledge discovery and data mining. Springer, Berlin, pp 97–106
10. Kryszkiewicz M (1998) Representative association rules and minimum condition maximum consequence association rules. In: European symposium on principles of data mining and knowledge discovery. Springer, Berlin, pp 361–369
11. Prithiviraj P, Porkodi R (2015) A comparative analysis of association rule mining algorithms in data mining: a study. Open J Comput Sci Eng Surv 3(1):98–119
12. Szathmary L (2006) Symbolic data mining methods with the Coron platform. Ph.D. thesis, Université Henri Poincaré-Nancy 1

13. Tabassum F, Nazrul Islam AKM, Kaiser MS (2021) Performance evaluation of fuzzy-based hybrid MIMO architecture for 5G-IoT communications. In: Ray K, Roy KC, Toshniwal SK, Sharma H, Bandyopadhyay A (eds) Proceedings. ICDSA. LNNS, Springer, Singapore, pp 289–297
14. Tan PN, Kumar V, Srivastava J (2000) Indirect association: mining higher order dependencies in data. In: European conference on principles of data mining and knowledge discovery. Springer, Berlin, pp 632–637
15. Tsang S, Koh YS, Dobbie G (2011) RP-tree: rare pattern tree mining. In: International conference on data warehousing and knowledge discovery. Springer, Berlin, pp 277–288

Students' Satisfaction with Virtual Interaction Mediated Online Learning: An Empirical Investigation

Md. Hafiz Iqbal, Md. Masumur Rahaman, Md. Shakil Mahamud, Serajum Munira, Md. Armanul Haque, Md. Amirul Islam, Md. Abdul Mazid, and Md. Elias Hossain

Abstract Virtual interaction offers a diversified and smart learning environment. It works as an effective knowledge hub for the students to increase their satisfaction. This study assesses students' satisfaction with virtual interaction-mediated learning and explores relevant contributors to the level of students' satisfaction. Mixed methods were used for proper empirical investigation. Survey and semi-structured and open-ended interview questions ($n = 385$) through random sampling technique were used to capture cross-sectional data for measuring the students' satisfaction with virtual interaction-mediated online learning by a simple linear regression model. Most of the students at the undergraduate and graduate levels in different educational institutions of Pabna district have ambivalent feelings about online learning. The flipped classroom, asynchronous discussion, interaction of teachers and students, timely feedback, personal needs, infrastructural support, uninterrupted internet facility, and online technology-mediated communication are significant contributors to virtual mediated online learning and these attributes are responsible for increasing students' satisfaction. This study highlights the necessary contribution to the prevailing studies on students' satisfaction with online learning in Bangladesh and other South Asian Countries.

Md. H. Iqbal (✉) · Md. A. Mazid
Government Edward College, Pabna, Bangladesh

Md. M. Rahaman
Bangladesh Embassy, Bangkok, Thailand

Md. S. Mahamud
Bangladesh High Commission, Ottawa, Canada

S. Munira
Green University of Bangladesh, Dhaka, Bangladesh

Md. A. Haque
Wuhan University, Wuhan, China

Md. A. Islam
Pabna University of Science and Technology, Pabna, Bangladesh

Md. E. Hossain
University of Rajshahi, Rajshahi, Bangladesh

M. S. Arefin et al. (eds.), *Proceedings of the International Conference on Big Data, IoT, and Machine Learning*, Lecture Notes on Data Engineering and Communications Technologies 95, https://doi.org/10.1007/978-981-16-6636-0_58

Keywords Online learning · Knowledge management · Sustainable learning · Massive open online course

1 Introduction

Electronic gadgets and access to repositories and bibliographic databases make students' learning attractive and interesting. Students get more learning benefits from the virtual interaction-mediated online learning because it promotes students' cognitive level, expertise, imagination power, concept development, and self-confidence by exchanging essential learning tips and class notes or assignment-making strategies across students and teachers [1, 2]. Technology-mediated learning is the principal catalyst to manage diversified learning. Educational institutions in advanced countries give more importance to such technology-mediated learning for online learning and knowledge management [3]. A sustainable learning practice requires a perfect match between students' satisfaction and online technology. Empirical studies suggest that Zoom, Webex, Meet, Kortext, Teams, Google Classroom, Dropbox, Google Drive, YouTube, Chat Room, Skype, and Wikis are essential and significant components of online technology that promotes online learning and students' satisfaction [4]. These online technologies have greater scope to make online learning popular because of its interaction capacity among teachers and students [5]. The rising demand for online learning in any situation, time, and geographical location can change traditional teaching–learning practice and bring greater satisfaction to students [6, 7]. A greater portion of students at the undergraduate and graduate levels in the USA are motivated by the greater benefits of online learning and prefer online-mediated courses [8]. This learning environment allows students to interact with teachers, mentors, and coaches and upload class tasks, and assignments. They can also download teachers' feedback and other reading materials within a very short time.

Teachers' role is active in the virtual interaction-mediated online learning [9]. A teacher can provide feedback on an assignment, the direction of making an assignment, asynchronous online discussion and academic session, timely assessment, and meet other academic supports such as referencing and citation technique, development of the conceptual framework, and note-taking strategies [10]. The popular form of virtual interaction-mediated online learning is synchronous or asynchronous online discussions [11]. Asynchronous online academic discussions are suitable for the students who have engaged themselves with part-time work or household work because of their greater flexibility in terms of place and time [12]. Timely assessment and academic support are another part of virtual interaction-mediated learning. For instance, timely assessment brings more attractiveness to online learning [13]. Similarly, academic support through online technology can bring a diversified learning strategy for the learners [14].

Recently, within the literature, there is widespread consensus regarding a range of impacts of digital interaction-mediated online learning on students' academic grades.

In addition, there has been a growing acknowledgement regarding the effectiveness of online learning. While concepts related to students' satisfaction with online learning have been unexplored. Valuation of the effectiveness of online learning and satisfaction of students are highly required for teachers, students, administrators, policymakers, development partners and the government to implement, plan, and modify digital platform-mediated teaching and learning for a better academic atmosphere. Therefore, the general objective of our study is to assess students' satisfaction with virtual interaction-mediated online learning. The specific objectives of our study are to explore students' attitudes toward virtual interaction-mediated online learning and identify the attributes contributing to the level of students' satisfaction with online learning.

The remainder of the paper is organized as follows. Section 2 covers the literature review. Section 3 of the paper outlines the theoretical framework and methodology. The results, findings, and discussion are presented in Sect. 4. Section 5 concludes.

2 Literature Review

Students' satisfaction is a principal catalyst of online learning [15–17]. A study by Alqurashi [18] presents that online learning has greater potentiality in better learning and satisfaction. Students can enjoy and gain knowledge from such a learning atmosphere at any time, geographical location, hazardous condition, and age [19, 20]. For instance, a large number of students worldwide are habituated with online learning in the present COVID-19 pandemic situation [21]. Due to the prevalence of COVID-19, high-ranking universities launched technology-based online learning for their international students in different countries [22].

Online learning is essential for distance mode learning [23]. Learners get more satisfaction from this learning practice because of its easy access to learning [24]. Under this learning practice, students can share their experience, view, and knowledge with each others and teachers [25]. Accessibility and availability of student-friendly online technology can increase capacity building on online learning with students' satisfaction at all locations, like remote areas, isolated islands, rural, and hilly regions [26]. The common forms of online learning are flipped classrooms, asynchronous discussion, uninterrupted internet facility, online technology-mediated communication, and seating and classroom management [27].

Teachers are the main actors of online learning. Without their active participation, it is difficult to make online learning popular [28]. Having no timely feedback and more interaction opportunities from a course teacher or instructor make online learning insignificant [29]. With the assistance of teachers, technology-mediated online learning is essential for fulfilling students' personal needs like information about the job market, internship opportunities, a summer course in other universities, and guidelines for making resumes or curriculum vitae (CV) [30].

Many universities in Bangladesh have recently introduced online learning during the COVID-19 pandemic situation for continuing academic session. For instance, the National University of Bangladesh has conducted academic sessions through the Zoom platform since last year. Every student of this university gets access to online learning through YouTube. Bangladesh Open University (BOU) depends on distance mode online learning for its students aiming to cover inclusiveness and equitability to all levels and areas. Online learning can promote a paperless campus and ensures a low-carbon society by reducing paper use [31]. Strong coordination among government and authorities of colleges and universities are essential to implement technology-mediated online learning. Special packages for the internet and gadgets and online learning supported curriculum and assessment are the preconditions of implementing online learning for all.

Assessment of the students' satisfaction with virtual communication and interacted learning is profoundly needed for instructors, school executives, the authority of the National University, Bangladesh, and the public authority to carry out the plan, and other online innovation learning for better education and learning environment. Existing literature perfectly highlights the effectiveness of online technology in learning as far as various countries' points of view. A very few studies focused on the assessment of students' satisfaction with virtual interaction-mediated learning. The extent to which current prosthetic assessments of students' satisfaction with virtual interaction-mediated learning is unclear. To gain further knowledge in this area, existing evidence gaps and method design issues must be identified, therefore, informing the design of future research. Our study tries to cover this issue significantly and attempts to reduce such a gap by assessing students' satisfaction with online technology-aided learning in Bangladesh and other South Asian countries.

3 Theoretical Framework and Methodology

Every society can follow an alternative option when it fails to reach optimum condition with its traditional or existing option. In this viewpoint, [32] first conceptualized the theory of second-best in their article entitled 'The General Theory of Second Best' followed by an earlier work by James E. Meade. The main focus of their theory is what happens when the optimum condition is not satisfied in a society.

Suppose there be some function $F(x_1 \ldots x_n)$ of the n variables$x_1 \ldots x_n$, which is to be maximized or minimized subject to a constraint on the variables $\varnothing(x_1 \ldots x_n) = 0$. This is a formalization of the typical choice situation in the analysis. Suppose the solution to this problem be the $n - 1$ condition $\varphi'(x_1 \ldots x_n) = 0, i = 1 \ldots n - 1$, then the following condition will be focused under the theory of second-best:

If there are additional constraints imposed of the type $\varphi' \neq 0$ for $i = j$, then the minimum or maximum of F subject to both φ' and the constraint $\varphi' \neq 0$ will be such that none of the still attainable by the Paretian optimality conditions $\varphi' \neq 0, i \neq j$, will be satisfied (Fig. 1).

Ox and Oy show the quantities of two goods x and y. The linear line AB highlights a transformation function (to be considered as a boundary condition) and CD presents a constraint condition. In the absence of the CD, the optimum position will be some

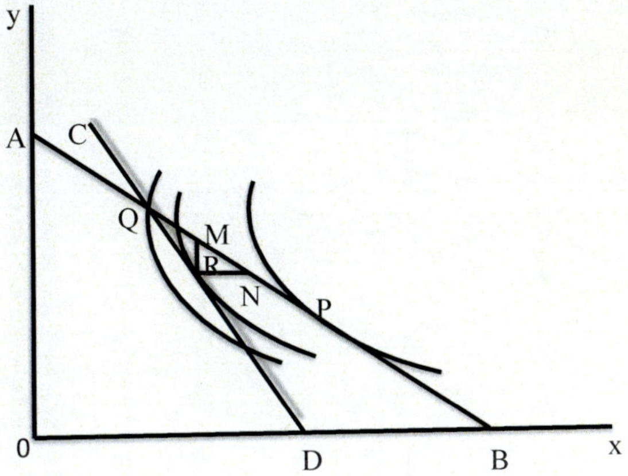

Fig. 1 Social welfare curve [32]

point, e.g., *P* takes a position on the transformation line at the point of its tangency with one of the contours of the welfare function. The constraint condition must be satisfied only points along *CD* can be preferred, and the optimum point *P* is no longer attainable. A point on the transformation line (*Q*) is still attainable. If the welfare contours and the constraint line are located in the diagram, then the second-best point will be at the point *R*, inside the transformation line.

Findings of [32] have important implications not only to trade policy and public finance but also to virtual interaction-mediated online learning- online technology-aided feedback on an assignment, asynchronous online academic discussion, and timely assessment through online technology because classroom-based teaching and learning practices fail to establish a desirable academic atmosphere in all situations or hazardous conditions. This theory helps us to design our methodology (Fig. 2).

3.1 Selection of Attributes, Data Collection Procedure, and Sampling Technique

Proposed attributes such as flipped classroom, asynchronous discussion, the interaction of teachers and students, timely feedback on students' assignments, infrastructural support, uninterrupted internet facility, and online technology-mediated communication were selected from [29, 33–38]. The surveys and face-to-face interviews were conducted to collect data in different educational institutions of Pabna, Bangladesh. These two instruments were used to assess our proposed attributes or explanatory variables concerning virtual interaction-mediated online learning and the level of students' satisfaction. For proper understanding, the questionnaire was constructed by the local language, Bangla.

A four-point Likert scale was applied to assess students' perception and satisfaction with online learning, where 1 indicates students were highly dissatisfied with

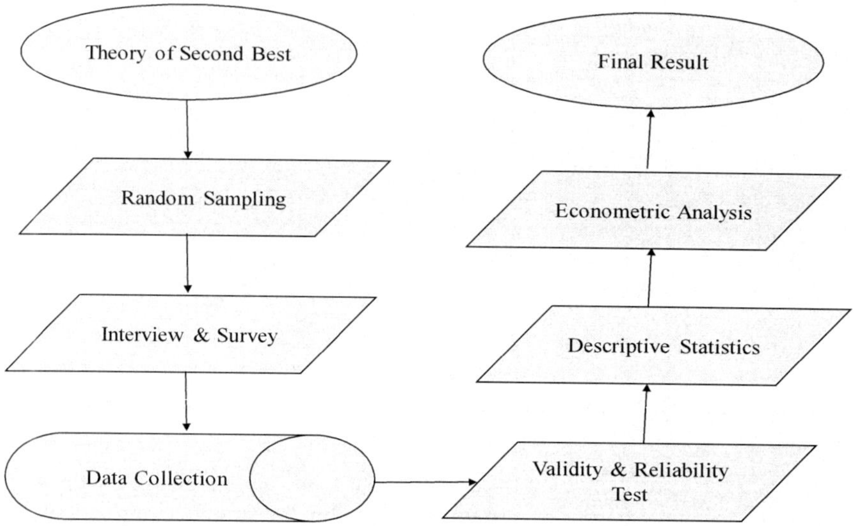

Fig. 2 Flow graph for data collection to the final result

online learning, 2 reveals students were dissatisfied with online learning, 3 expresses students were satisfied with online learning and 4 indicates students were highly satisfied with online learning. Furthermore, the dependent variable (students' satisfaction) was figured out by any value ranging from 8 to 32, where 8 to 16 indicates dissatisfaction with online learning and 24 to 32 indicates satisfaction with online learning. However, scores ranging from 17 to 23 are treated as the ambivalent zone. A simple linear regression model was used for proper empirical assessment.

Obtaining in-depth insight into *students' satisfaction* with virtual interaction-mediated *online learning* motivated us to carry out students' interviews ($n = 385$) through face-to-face and semi-structured interviews from 27 April to 15 June, 2019. The first category was expressed as "*Strongly disagree*" and figured out by 1, the second category was expressed as "Disagree" and figured out by 2, the third code was expressed as "*Agree*" and figured out by 3, and the fourth code was expressed as "*Strongly agree*" and figured out by 4. The random sampling technique was applied to the students having experience in online learning at Pabna University of Science and Technology and Pabna Government Edward College. 385 students gave consent to take part in the interview and survey process. Most of the respondents completed the survey form and interview within 15 min.

3.2 Ethical Consideration

The study was approved in line with guidelines from the ethical committee of the Department of Economics, Government Edward College, Pabna. The respondents (students) were assured that their responses would be handled confidentially, and they could withdraw themselves from the survey at any time. This form has also presented the details about the collection of data and offered a range of concepts and ideas in which consent was sought, and every respondent was requested to provide support, information, and data. Oral informed consent was obtained from all the respondents after the survey objectives and procedures were explained. They were also assured that they were not identifiable in any resulting presentations or publications that arose from the study. This study assigned a unique identification code to each respondent's data and separated personal identification information from the response data to maintain the data's high confidentiality and protect the respondents' anonymity. All sets of data were password protected and saved in different places.

4 Results, Findings, and Discussion

The average value of students' satisfaction associated with *online learning* is recorded at 21.01. It implies that students are indifferent between online learning and learning from the classroom because the recorded value lies in the ambivalent zone 17–23 (Table 1).

Students want more support from the authority of their university or college. During the survey and interview time, they argued that they will get online learning is more reliable and they get more benefit from the online learning like teachers if the university or college authority provides uninterrupted Wi-Fi facility in the campus and government and internet providers provide special internet package designing for the students. They also argued that timely feedback from the course teacher on their uploaded assignments may popularize online learning and enhance their satisfaction with the course teacher. The flipped classroom, asynchronous discussion, interaction of teachers and students, personal needs, Infrastructural support, and online technology-mediated communication may also contribute to online learning. Values of average and standard deviation for our proposed attributes for the surveys range from 3.07 to 2.09 and from 1.109 to 0.701. In contrast, these values range from 2.31 to 1.01 and 0.849 to 0.514 for the interview questions.

Identifying the significant contributors to students' satisfaction with online technology-mediated learning, the ordinary least squares (OLS) method supported simple linear regression model was applied using Statistical Package for the Social Sciences (SPSS) econometric software. Our econometric model was satisfied to pass a few diagnostic tests. For instance, our regression model is not affected by multicollinearity because all of our estimated predictors in the correlation matrix are less than 0.80. We also measured the Variance Inflation Factor (VIF) and tolerance statistics for further investigation of multicollinearity detection in our models. All estimated values of VIF lie below the cutoff of 10 as recommended by field [39] and tolerance statistics take the position in the value of 0.30–0.90 larger than the 0.20 recommended cutoff [40]. Most of our proposed attributes are qualified to predict significantly in the regression model (Table 2).

Table 1 Descriptive statistics for the attributes of the survey and interview questions

Attributes	Average	SD
Students' satisfaction	21.01	2.731
Flipped classroom	2.62	1.008
Asynchronous discussion	2.09	1.062
Teachers-students' interaction	2.92	0.922
Feedback in time	2.94	1.096
Personal needs (e.g. cover letter writing, CV writing, internship, etc.)	2.92	0.701
Infrastructural support (e.g. classroom management, seating arrangement, etc.)	2.50	1.109
Uninterrupted internet facility	3.07	0.778
Online technology-mediated communication	2.50	0.953
Online learning related interview questions		
Q1: Comfortability with online learning	2.14	0.639
Q2: Reliability with this learning	2.31	0.700
Q3: Higher workload with online learning	1.82	0.825
Q4: No opportunity of physically contact with the teacher through this learning	2.14	0.808
Q5: No opportunity to learn physically with this learning	1.82	0.514
Q6: Teachers are more active in communication with this learning	2.24	0.716
Q7: Proper knowledge management aided by online learning	1.98	0.769
Q8: Teachers are more enthusiastic under this learning atmosphere	2.29	0.768
Q9: Creativity develops in terms of the use of resources from online technology	2.16	0.842
Q10: Learning disturbances of online learning due to a technical problem	1.01	0.764
Q11: Capacity building and more opportunity gain under this learning	1.98	0.769
Q12: Longer time is required to prepare assignments under this learning	1.80	0.782
Q13: Get prompt feedback from a teacher in online learning	2.06	0.740
Q14: Easy access to online learning	2.02	0.742
Q15: Reduction of students' interest due to technical problem	1.94	0.849

Positive signs of all coefficients of our proposed attributes imply that these can increase students' satisfaction when we improve further increment of these attributes. More specifically, a one percent increase in any of our proposed attributes will increase students' satisfaction with online learning and vice-versa. The highest estimated value of asynchronous discussion in the standardized coefficients suggests that asynchronous discussion is a more effective and significant predictor of online learning because of its higher beta value. In contrast, the lowest value of personal needs is the weakest predictor of online learning. The value R^2 ensures that 37% of

Table 2 Regression model

Model	Unstandardized co-efficient		Standardized co-efficient	Z	Sig.
	β	Std. error	β		
Constant	0.751	1.070		0.708	0.487
Flipped classroom	0.900*	0.127	0.331	7.031	0.001
Asynchronous discussion	0.287*	0.118	0.497	11.127	0.002
Teachers-students' interaction	0.082*	0.136	0.360	8.407	0.001
Feedback in time	0.215*	0.112	0.482	10.730	0.001
Personal needs	0.423**	0.193	0.125	2.303	0.021
Infrastructural support	0.725*	0.105	0.300	6.939	0.000
Uninterrupted internet facility	0.861*	0.176	0.245	4.882	0.000
Online technology-mediated comm.	0.075*	0.127	0.377	8.530	0.001
Goodness of fit (R^2)	0.373				
Log-likelihood	−356.298				
Adjusted R^2	0.721				
Number of respondents (n)	385				

Outcome variable: students' satisfaction
*$P < 0.01$ indicates significant at the 1% level, **$P < 0.05$ indicates significant at the 5% level

the total variation of outcome variable can be interpreted by the variation of explanatory variables of the linear regression model. The estimated value of Cronbach's alpha ensures that our proposed attributes are reliable and valid.

Estimated results of descriptive statistics show that flipped classrooms, asynchronous discussion, the interaction of teachers and students, timely feedback, personal needs, infrastructural support, uninterrupted internet facility, and online technology-mediated communication are the influential attributes impinging on students' satisfaction. The mean and standard deviation of students' satisfaction with online learning are measured at 21.01 and 2.731, respectively and these values indicate a sense of indifference or ambivalence between traditional learning practice and online learning practice. Significant and reliable online learning always required more asynchronous discussion (Mean = 2.09 and Standard Deviation = 1.062), interaction of teacher and students (Mean = 2.92 and Standard Deviation = 0.922), timely feedback (Mean = 2.94 and Standard Deviation = 1.096), personal needs (Mean = 2.92 and Standard Deviation = 0.701), Infrastructural support (Mean = 2.50 and Standard Deviation = 1.109), Uninterrupted internet facility (Mean = 3.07 and Standard Deviation = 0.778), and Online technology-mediated communication (Mean = 2.50 and Standard Deviation = 0.953). Estimated results from the regression model suggests that all of our proposed attributes have positive relationship with virtual interaction-mediated online learning. For instance, a 1% rise in flipped classroom-based teaching will rise in students' satisfaction at 90%. Likewise, a 1% increase in asynchronous discussion will rise in students' satisfaction at 28%. Similar results were also obtained from the existing empirical studies. For instance, students' e-learning readiness is positively correlated with flipped classroom-based teaching and this teaching practice is a significant predictor of students' satisfaction [41–43]. Clarity of design of asynchronous-based courses, interaction with instructors, timely feedback, and active discussion significantly influenced students' satisfaction and improved learning quality [44].

5 Summary and Conclusions

Online learning can develop students' satisfaction. Without satisfaction in learning may hamper students' grades in the examination. Proper designing of online based learning curriculum can ensure sustainable, equitable, and inclusive learning for all. Perfect utilization of online learning may also reduce the dropout rate [45]. The flipped classroom, asynchronous discussion, interaction of teachers and students, timely feedback, fulfillment of personal needs, infrastructural support, uninterrupted internet facility, and online technology-mediated communication can enhance students' satisfaction with online learning. Students' ambivalent feelings on online learning imply that the current practice of online learning does not maintain an online-based better learning atmosphere. It fails to bring a new learning culture at the college and university level in Bangladesh [46]. Policymakers in the education sector of Bangladesh should give more emphasis on students' perceptions based on online learning. They should incorporate all of our proposed attributes in the curriculum of online learning. The government, the authority of National University, Bangladesh, and college authorities should give more support, motivation, and training to the teachers for implementing online learning. Assessment of students' satisfaction with online learning is rather rare in empirical studies. Our study tries to cover this issue. In this viewpoint, our study has contributed unique insights into the existing literature.

Despite the proper and careful design, our study is not free from certain limitations because of not included socio-economic-demographic (SED) characteristics and fewer educational institutions in a smaller study area. Thus, this study recommends further study to avoid such weakness and get a significant and desirable assessment of students' satisfaction with online learning.

Acknowledgements We are grateful to the respondents that participated in the study and formed our sample for collecting and analyzing the data.

References

1. Aydin S (2014) Wikis as a tool for collaborative language learning: implications for literacy, language education and multilingualism. Sustain Multimedia 5(1):207–236
2. Caruso SJ (2017) A foundation for understanding knowledge sharing: organizational culture, informal workplace learning, performance support, and knowledge management. Contemp Issues Educ Res 10(1):45–52
3. Greenhow C, Askari E (2017) Learning and teaching with social network sites: a decade of research in K-12 related education. Educ Inf Technol 22(2):623–645
4. Trelease RB (2016) From chalkboard, slides, and paper to e-learning: how computing technologies have transformed anatomical sciences education. Anat Sci Educ 9(6):583–602
5. Bere A, Rambe P (2016) An empirical analysis of the determinants of mobile instant messaging appropriation in university learning. J Comput High Educ 28(2):172–198
6. Elliott KM, Shin D (2002) Student satisfaction: an alternative approach to assessing this important concept. J High Educ Policy Manag 24(2):197–209
7. Kentnor HE (2015) Distance education and the evolution of online learning in the United States. Curriculum Teach Dialogue 17(1):21–34
8. Kaymak ZD, Horzum MB (2013) Relationship between online learning readiness and structure and interaction of online learning students. Educ Sci Theor Pract 13(3):1792–1797
9. Perrotta C, Gulson KN, Williamson B, Witzenberger K (2021) Automation, APIs and the distributed labour of platform pedagogies in Google classroom. Crit Stud Educ 62(1):97–113
10. Gaytan J, McEwen BC (2007) Effective online instructional and assessment strategies. Am J Distance Educ 1(3):117–132
11. Wang Q, Woo HL (2007) Comparing asynchronous online discussions and face-to-face discussions in a classroom setting. Br J Edu Technol 38(2):272–286
12. Sheail P (2018) Temporal flexibility in the digital university: full-time, part-time, flexi-time. Distance Educ 39(4):462–479
13. Lin HF (2010) An application of fuzzy AHP for evaluating course website quality. Comput Educ 54(4):877–888
14. Peters M, Romero M (2019) Lifelong learning ecologies in online higher education: students' engagement in the continuum between formal and informal learning. Br J Educ Technol 50(4):1729–1743
15. Semente E (2017) Student satisfaction and technology integration in teaching and learning: the case of University Education in Namibia. J Educ Pract 1(2):1–10
16. Arbaugh JB, Cleveland-Innes M, Diaz SR, Garrison DR, Ice P, Richardson JC, Swan KP (2008) Developing a community of inquiry instrument: testing a measure of the community of inquiry framework using a multi-institutional sample. Internet High Educ 11(3–4):133–136
17. Almaiah MA, Alismaiel OA (2019) Examination of factors influencing the use of mobile learning system: an empirical study. Educ Inf Technol 24(1):885–909
18. Alqurashi E (2019) Predicting student satisfaction and perceived learning within online learning environments. Distance Educ 40(1):133–148
19. Tran T, Ho MT, Pham TH, Nguyen MH, Nguyen KLP, Vuong TT, Nguyen THT, Nguyen TD, Nguyen TL, Khuc Q, La VP (2020) How digital natives learn and thrive in the digital age: evidence from an emerging economy. Sustainability 12(9):3819
20. Berg J, Ihlström J (2019) The importance of public transport for mobility and everyday activities among rural residents. Soc Sci 8(2):58
21. Lassoued Z, Alhendawi M, Bashitialshaaer R (2020) An exploratory study of the obstacles for achieving quality in distance learning during the COVID-19 pandemic. Educ Sci 10(9):232
22. Alsmadi MK, Al-Marashdeh I, Alzaqebah M, Jaradat G, Alghamdi FA, Mohammad RMA, Alshabanah M, Alrajhi D, Alkhaldi H, Aldhafferi N, Alqahtani A (2021) Digitalization of learning in Saudi Arabia during the COVID-19 outbreak: a survey. Inf Med Unlocked 100632
23. Ragusa AT, Crampton A (2018) Sense of connection, identity and academic success in distance education: sociologically exploring online learning environments. Rural Soc 27(2):125–142

24. Khalil R, Mansour AE, Fadda WA, Almisnid K, Aldamegh M, Al-Nafeesah A, Alkhalifah A, Al-Wutayd O (2020) The sudden transition to synchronized online learning during the COVID-19 pandemic in Saudi Arabia: a qualitative study exploring medical students' perspectives. BMC Med Educ 20(1):1–10
25. Rapanta C, Botturi L, Goodyear P, Guàrdia L, Koole M (2020) Online university teaching during and after the Covid-19 crisis: refocusing teacher presence and learning activity. Post Digital Sci Educ 2(3):923–945
26. Sugino C (2021) Student perceptions of a synchronous online cooperative learning course in a Japanese women's university during the COVID-19 pandemic. Educ Sci 11(5):231
27. Daugvilaite D (2021) Exploring perceptions and experiences of students, parents and teachers on their online instrumental lessons. Music Educ Res 23(2):179–193
28. Plump CM, LaRosa J (2017) Using Kahoot! in the classroom to create engagement and active learning: a game-based technology solution for eLearning novices. Manage Teach Rev 2(2):151–158
29. Martin F, Bolliger DU (2018) Engagement matters: student perceptions on the importance of engagement strategies in the online learning environment. Online Learn 22(1):205–222
30. Iqbal H (2020) E-mentoring: an effective platform for distance learning. e-mentor 84(2):54–61
31. Iqbal H, Ahmed F (2015) Paperless campus: the real contribution towards a sustainable low carbon society. IOSR J Environ Sci Toxicol Food Technol 9(8):10–17
32. Lipsey RG, Lancaster K (1956) The general theory of second best. Rev Econ Stud 24(1):11–32
33. Tseng MF, Lin CH, Chen H (2018) An immersive flipped classroom for learning Mandarin Chinese: design, implementation, and outcomes. Comput Assist Lang Learn 31(7):714–733
34. Croxton RA (2014) The role of interactivity in student satisfaction and persistence in online learning. J Online Learn Teach 10(2):314–325
35. Puška A, Puška E, Dragić L, Maksimović A, Osmanović N (2021) Students' satisfaction with E-learning platforms in Bosnia and Herzegovina. Technol Knowl Learn 26(1):173–191
36. Hamid R, Sentryo I, Hasan S (2021) Online learning and its problems in the Covid-19 emergency period. J Primary Educ 8(1):86–95
37. Faize FA, Nawaz M (2020) Evaluation and improvement of students' satisfaction in online learning during COVID-19. Open Praxis 12(4):495–507
38. Larbi-Siaw O, Owusu-Agyeman Y (2017) Miscellany of students' satisfaction in an asynchronous learning environment. J Educ Technol Syst 45(4):456–475
39. Field A (2013) Discovering statistics using SPSS. SAGE Publication, London
40. Suar D, Khuntia R (2010) Influence of personal values and value congruence on unethical practices and work behavior. J Bus Ethics 97(3):443–460
41. Yilmaz R (2017) Exploring the role of e-learning readiness on student satisfaction and motivation in flipped classroom. Comput Hum Behav 70:251–260
42. Nortvig AM, Petersen AK, Balle SH (2018) A literature review of the factors influencing e-learning and blended learning in relation to learning outcome, student satisfaction and engagement. Electron J E-Learn 16(1):46–55
43. Iqbal MH, Siddiqie SA, Mazid MA (2021) Rethinking theories of lesson plan for effective teaching and learning. Soc Sci Humanit Open 4(1):100172
44. Swan K (2001) Virtual interaction: design factors affecting student satisfaction and perceived learning in asynchronous online courses. Distance Educ 22(2):306–331
45. Bolliger DU, Wasilik O (2009) Factors influencing faculty satisfaction with online teaching and learning in higher education. Distance Educ 30(1):103–116
46. Häkkinen P, Järvelä S, Mäkitalo-Siegl K, Ahonen A, Näykki P, Valtonen T (2017) Preparing teacher-students for twenty-first-century learning practices (PREP 21): a framework for enhancing collaborative problem-solving and strategic learning skills. Teach Teach 23(1):25–41

Identification of the Resting Position Based on EGG, ECG, Respiration Rate and SpO_2 Using Stacked Ensemble Learning

Md. Mohsin Sarker Raihan, Muhammad Muinul Islam, Fariha Fairoz, and Abdullah Bin Shams

Abstract Rest is essential for a high-level physiological and psychological performance. It is also necessary for the muscles to repair, rebuild, and strengthen. There is a significant correlation between the quality of rest and the resting posture. Therefore, identification of the resting position is of paramount importance to maintain a healthy life. Resting postures can be classified into four basic categories: Lying on the back (supine), facing of the left/right sides and free-fall position. The later position is already considered to be an unhealthy posture by researchers equivocally and hence can be eliminated. In this paper, we analyzed the other three states of resting position based on the data collected from the physiological parameters: Electrogastrogram (EGG), Electrocardiogram (ECG), Respiration Rate, Heart Rate, and Oxygen Saturation (SpO_2). Based on these parameters, the resting position is classified using a hybrid stacked ensemble machine learning model designed using the Decision tree, Random Forest, and Xgboost algorithms. Our study demonstrates a 100% accurate prediction of the resting position using the hybrid model. The proposed method of identifying the resting position based on physiological parameters has the potential to be integrated into wearable devices. This is a low cost, highly accurate and autonomous technique to monitor the body posture, while maintaining the user's privacy by eliminating the use of RGB camera conventionally used to conduct the polysomnography (sleep Monitoring) or resting position studies.

Keywords Polysomnography · Resting position · ECG · EGG · Stacked ensemble learning · Decision tree · Random forest · XGBoost · Machine learning

---

Md. M. S. Raihan (✉) · M. M. Islam
Department of Biomedical Engineering, Khulna University of Engineering & Technology, Khulna 9203, Bangladesh
e-mail: mmi@bme.kuet.ac.bd

F. Fairoz
Department of Computer Engineering, Islamic University of Technology, Gazipur, Bangladesh
e-mail: farihafairoz@iut-dhaka.edu

A. B. Shams
Department of Electrical and Computer Engineering, University of Toronto, Toronto, ON M5S 3G4, Canada

M. S. Arefin et al. (eds.), *Proceedings of the International Conference on Big Data, IoT, and Machine Learning*, Lecture Notes on Data Engineering and Communications Technologies 95, https://doi.org/10.1007/978-981-16-6636-0_59

1 Introduction

For physical and mental fatigue removal, human beings take rest. Despite having some similarities with sleep, rest usually indicates a shorter period and the person remains awake during rest. Lying down is one of the prominent ways of taking a rest. There are 4 basic types of resting positions. They are: Lying on the back (supine), facing either of the left or right side and free-fall position [1]. There are many insights regarding healthy sleeping or resting postures. The absence of proper rest may eventually lead to a physical and mental breakdown. Also, incorrect body posture may potentially cause muscular strain. The interrelation between the quality of rest and resting posture is quite unavoidable. Researchers are trying to determine the best-suited position by observing the volunteers using a variety of processes. Resting posture observation is considered one of the key steps in determining the cause of diverse diseases. While observing and monitoring positions, there is always a trade-off between some constraints such as privacy, use of light, cost, and the accuracy of results.

In this paper, we considered the 3 states of resting position utilizing data collected through physiological examinations. The test parameters are Electrogastrogram (EGG), Electrocardiogram (ECG), Respiration Rate, and oxygen saturation (SpO_2). This study used easy and simple non-invasive processes to determine resting posture. All the data are classified by using a hybrid 2-layer stacked ensemble machine learning model. Each of the layers comprises three machine learning algorithms namely Decision tree, Random Forest, and XGboost. Our findings show a significant improvement in the prediction of posture after the use of the hybrid 2-layer stacked ensemble model which is quite promising. The highest possible accuracy gain implies the credibility of the analysis and this can be used in Polysomnography afterward. Easy implementation of the insight found from this study in wearable devices is possible and reliable as the study gained high accuracy. The cost-effectiveness is also an upvote for this study.

2 Related Works

There is a strong relationship between sleeping or resting position with the quality of rest. Uncomfortable posture significantly diminishes the sole purpose of taking rest. There are several studies on the determination of resting and sleeping position. A large spectrum of techniques is used in this sector to get optimal performance.

Many studies have used RGB cameras which have certain limitations like the invasion of privacy, environmental noise, etc. The use of 3d model [2], Kinect [3], and other sensors are also prominent in this field. The use of a single IR camera and image classification using CNN is also found [4] to analyze posture. Some studies show the use of pressure-sensitive mats [5].

In the determination of sleeping posture, the use of pressure sensors is costly. Result calculation becomes difficult if the patient is away from the central axis. The images developed by using such sensors may lead to complex ambiguity as the number of pixels is higher. Tang et al. used sensors in a different arrangement [5]. A special mat made of force sensing registers arranging in a 2D array is used to gather statistical information about posture. Later, the correct posture is determined using an artificial intelligence library "TensorFlow." The generated heat map from the sensors is used as the input and 200 images were used for each of the six positions. The accuracy of posture recognition was highest 100% in the case of an empty bed and lowest 80% for the right lateral.

Rasouli D. et al. employed only one depth sensor on a tripod for the determination of posture. The study also focused on the hand and leg positions in detail. The depth signals were processed using fast Fourier Transformation on scan planes [3] 14 volunteers were involved in the experiment. The extracted features are ranked by the T-test method. The machine learning algorithm that is used is the Support Vector Machine. Postures were divided into two main groups: side and supine. This experiment also considered the scenario with and without a blanket. The average accuracy is 97.96%, and the number of data to train the model was 1171.

Mohammadi et al. [4] studied 12 positions using IR images and neural networks. InfraRed images were captured using Microsoft Kinect depth camera but for simplicity depth data was removed. Later the 2D images were used for further analysis. CNN is used to classify the positions from 2D images. 103,68 frames were used to train the model. The average success rate of the method was 76% with a blanket and 91% without a blanket. This study aimed at recognizing 12 positions.

3 Methodology

The determination of the resting position from different physiological signals instead of direct visualization or video monitoring approach may be separated into different segments as shown in Fig. 1. The details of the methodology are described as follows:

3.1 Physiological Data Collection

This study used specific physiological data and each of them was collected using reliable and sophisticated technologies. All data were recorded at once in the morning hours after a light refreshment. Allocated time for every one of the positions was 15 min with 2 min interval in between for recheck the electrode placement and subject stabilization. The data collection procedure is thoroughly explained in the study Raihan and Islam with details [1]. Data was taken in the three position and they are right side, left side and supine side. Figure 2 shows the 3 resting positions we are considering for our study. Facing right, the supine position and facing left. The

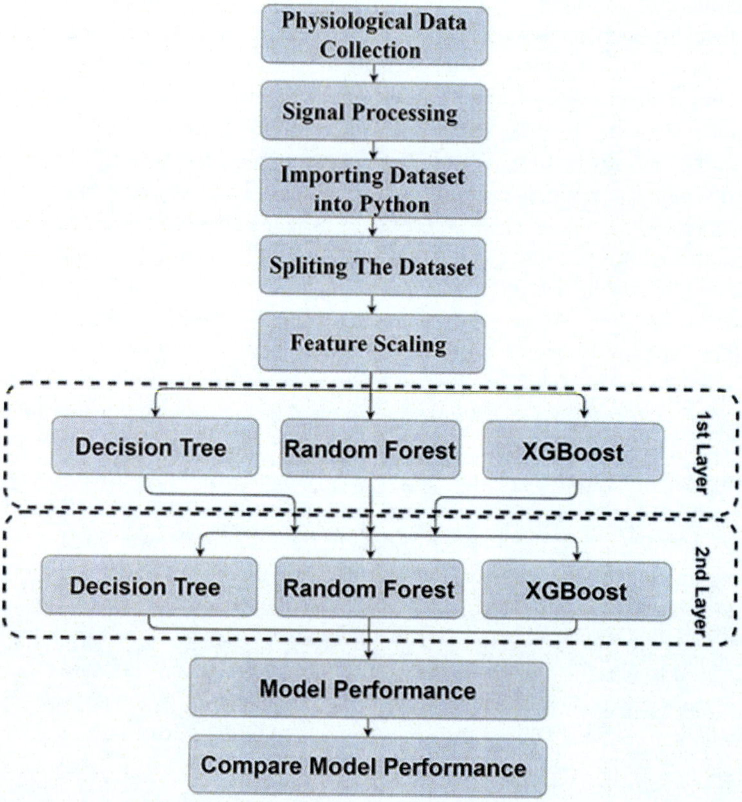

Fig. 1 Work-flow for the determination of the resting position from different physiological signals

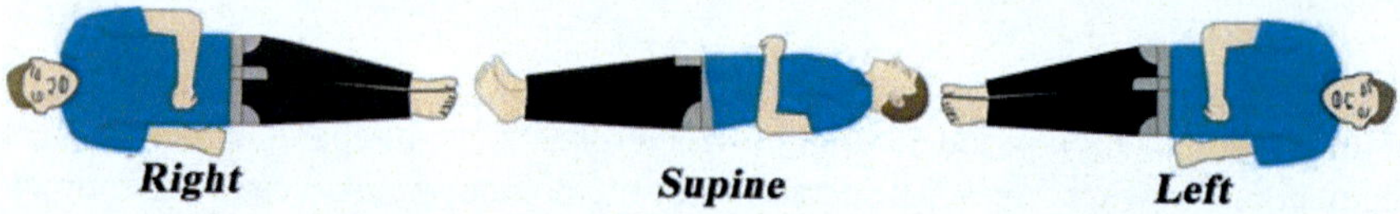

Fig. 2 Three types of resting positions

physiological signals collected were: Electrocardiogram (ECG), Electrogastrogram (EGG), Oxygen saturation (SpO_2), and Respiratory Rate (RR) at different resting positions. A signal acquisition system, MP150, Biopac Inc., USA, and AcqKnowledge applications mounted on a computer were used to capture all of the physiological signals. Figure 3 schematic view of different physiological signals collection with different sensors and electrodes placement.

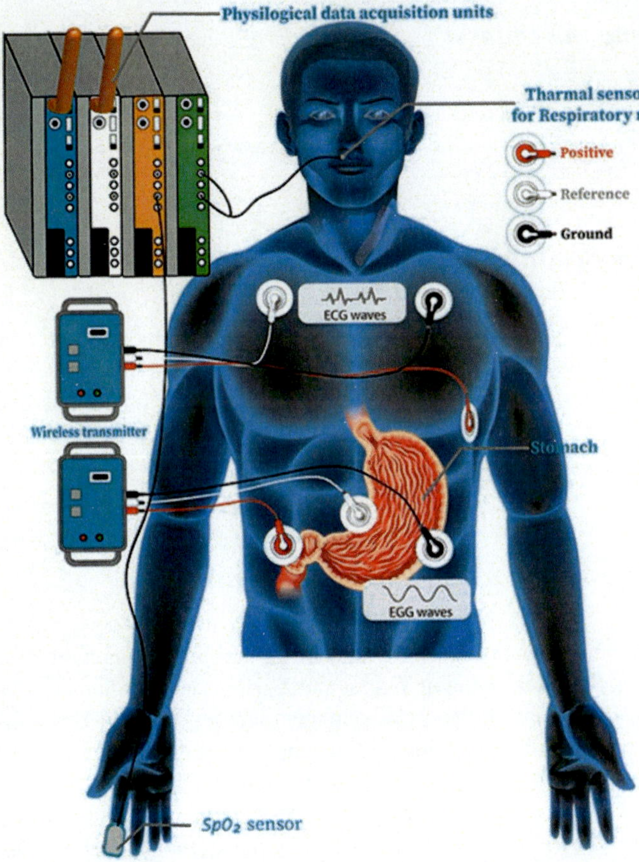

Fig. 3 Schematic view of different physiological signals collection with different sensors and electrode placement

3.2 *Signal Processing*

EGG signal processing, Heart rate calculation procedure from ECG signal, Respiration rate and SpO_2 calculation methodology are followed which mention in Raihan and Islam [1].

3.3 *Importing Dataset into Python*

After preparation of the dataset, we imported it which consists of EGG signal, Heart rate, Respiration rate, SpO_2, and resting positions.

3.4 *Splitting the Dataset*

The dataset had to be divided into two separate parts for training and testing purposes. 80% of data was used to train the model and the rest were used for testing [6].

3.5 *Feature Scaling*

The raw data we've gathered so far has a wide range of values, which could cause problems with calculations if they aren't normalized. If one of the features has a wide range of values, while the others are measured in a smaller range, this feature will manipulate the calculation. As a result, the range of all features should be normalized so that each contributes roughly the same proportion [7].

3.6 *Implemented Machine Learning Algorithms*

We have implemented a heterogeneous stacked ensemble learning method with base estimators (Decision Tree, Random Forest and XgBoost as a 1st layer) and Decision Tree, Random Forest and XgBoost as our meta learner. Stacked Ensemble works by aggregating the results of multiple algorithms by using another algorithms [8]. Initially, data is passed to multiple algorithms of our choosing, these are called base estimators. They classify directly using the dataset. The results of these base estimators are then passed to and subsequently aggregated by another algorithms which we call the meta leaner. We have designed a hybrid stacked ensemble learning algorithms with 2-layer. Three algorithms are used in both layers and the algorithms are described in short below.

Decision tree: A decision tree is a powerful yet simple machine learning algorithm. The basic induction method is recursively used to reach the prediction. As the name suggests, it creates a tree-like structure where branches reflect the outcome of the test and nodes denote any decision to be taken. The initial training set is tested and subdivided into small sets based on definite parameters. This process is done repeatedly and called "recursive partitioning" [9].

Random forest: This algorithm stands out because of its less training time, higher accuracy for large datasets, and estimation of missing data. Random Forest uses a set of decision trees for the training process. The prediction is considered to be right based on the majority of the trees. Random Forest is included in the division of supervised learning [10].

XGBoost: XGBoost stands for eXtreme Gradient Boosting. Through several improvements on Gradient Boosting Machines, the XGboost has evolved. Such optimization in both algorithmic and system utilization made this perfect combination of higher accuracy in a shorter time and lesser computation. XGBoost approaches the system of sequential tree building using parallelized implementation. XGBoost uses the "max-depth" parameter as specified rather than criterion first, and starts pruning trees backward. This algorithm ensures the efficient use of hardware resources by allocating internal buffers in each thread to store gradient statistics [10].

3.7 *Analysis of Model Performance*

The output gained from any model can easily be divided into four categories from the confusion matrix, and they are True positive, False-positive, True negative, False-negative.

Usually, this total scenario is captured by a confusion matrix. Multiple parameters can be determined from further processing from the matrix, and those are used to derive insights about the strength and effectiveness of a certain model.

The analysis of the machine learning algorithms is conducted here utilizing four derived standards from the confusion matrix. These are Accuracy, Recall, Precision, and F_1 score [6]. The performance of the study is monitored using some additional parameters along with overall accuracies, such as F_1 score, precision, and recall.

4 Results and Discussion

The correlation matrix helps to visualize the interrelationship between parameters of the dataset. It expresses the strength of variables within the range -1 to 1. Where the value close to $+1$ denotes a positive correlation. The closer it is to -1 it indicates the negative correlation. Neutral values or values closer to zero indicates an insignificant correlation between variables [6].

According to the matrix in Fig. 4, the correlation between position and EGG is 0.28, which indicates that they have a significant positive correlation. The inverse relationship is seen for SpO_2 as the value is -0.25, indicating a negative correlation. Respiration rate is also inversely related to the position but not as strong as SpO_2. The correlation value is -0.14. Heart rate has a weak correlation with position with the value 0.087. As the value is too close to the neutral position, it implies an insignificant correlation between heart rate and position.

Figure 5 shows the results and the performance of this study. In the first layer, the accuracy of the Decision tree is 80.56% with a precision of 84.44%. The recall and F_1 score were, respectively, 84.1% and 80.19%. While using the Random Forest, the output contained 83.33% of accuracy, precision was 81.54%, Recall 82.44% and F_1 score 81.56%. For XGBoost, the accuracy was 88.89%, with a precision of 88.99%. The value of Recall and F_1 score was 87.22% and 87.5%, respectively.

By the end of the first layer of analysis, all the algorithms predicted the output of the training set with a percentage of accuracy as written above. Stacked ensemble learning led to higher accuracy gain. The results showed a drastic change in the

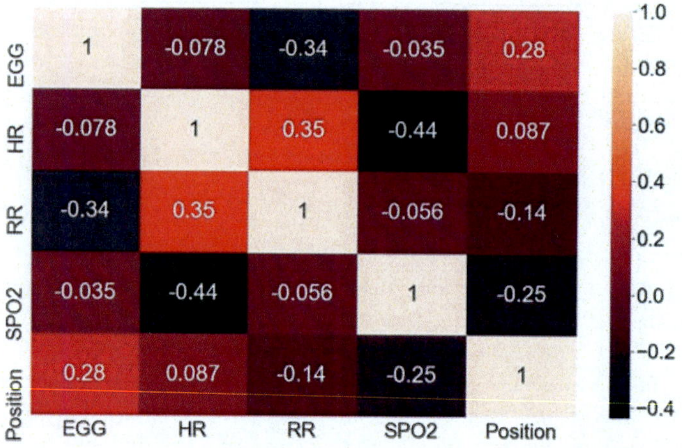

Fig. 4 Correlation matrix

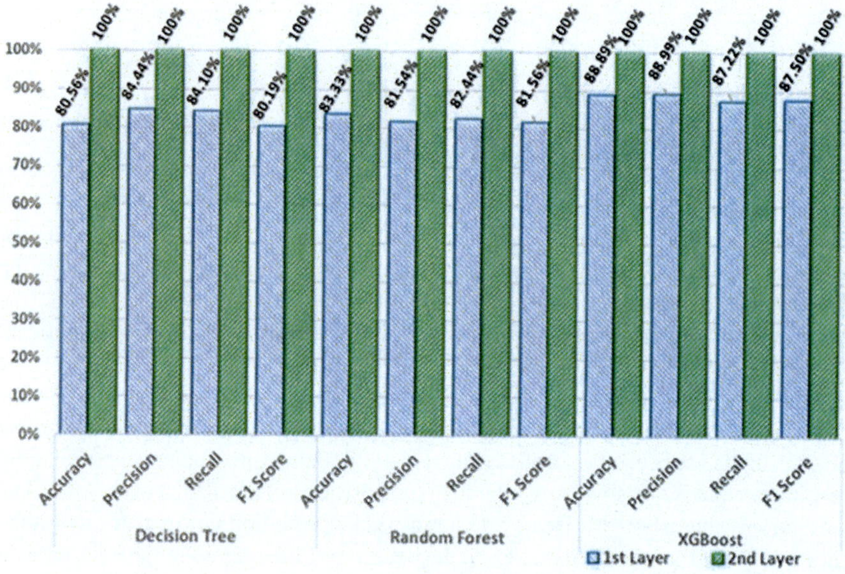

Fig. 5 Model performance result

Table 1 Comparisons with other existing systems

Ref.	Algorithm	Accuracy	Precision	F_1 score	Recall
[2]	CNN with transfer learning	90%	N/A	N/A	N/A
[5]	SVM	97.96%	N/A	N/A	N/A
[11]	CNN	With blanket: 76% Without blanket: 91%	N/A	N/A	N/A
[12]	fuzzy *c*-means clustering algorithm	88.05%	N/A	N/A	N/A
[13]	VGG 19 Tansor factorization	86%	N/A	N/A	N/A
This study	Decision tree Random forest XGBoost	100%	100%	100%	100%

second layer. In the case of stacked ensemble learning, the combination of multiple algorithms eliminates the drawbacks of each other resulting in a better gain.

As the primary predictions were collected from algorithms in the first layer, the training set along with the predictions of the previous layer was used to train the second layer. The accuracy, precision, recall, F_1 score for each of the three algorithms reached 100% after the second layer of training.

Comparison with the previous study: If we inspect the comparison table, we will find that none of the methods mentioned in the other studies performed as well as the method mentioned in this study. In our findings, we can classify with an accuracy of 100% with perfect scores in Precision, F_1 and Recall. Tang et al. [5] using SVM got 97.96% accuracy which is the nearest. Convolutional neural network with Transfer Learning got an accuracy of 90%. Other methods mentioned in the table got accuracies around 90%. It's quite evident from the table that our method shows a superior accuracy in comparison with the other methods (Table 1).

5 Conclusion

We looked at 3 resting positions using data from Electro-gastrogram (EGG), Electro-cardiogram (ECG), Respiration Rate, and oxygen saturation (SpO_2). In our study, we proposed a hybrid stacked ensemble model using Decision tree, Random Forest and Xgboost to identify the resting position. Our method achieved an accuracy of 100%. Unlike other studies this method doesn't invade people's privacy. Also because of the high accuracy, our method shows promise in being used in a cost-effective wearable device which would identify resting positions based on physiological parameters. As data was not collected via direct manual supervision, image, or video, the process was completely free from privacy violation. Such features make this study a dependable and sustainable source for future studies in a similar field of interest.

Acknowledgements The authors wish to thank all participants during this study and cordially grateful to the Department of Biomedical Engineering, Khulna University of Engineering & Technology for proving all facilities for this study.

References

1. Raihan M, Islam M (2020) Determination of the best resting position uing electrogastrography after having a light meal. In: 2020 IEEE Region 10 symposium (TENSYMP), pp 1684–1687
2. Boulay B, Brémond F, Thonnat M (2006) Applying 3d human model in a posture recognition system. Pattern Recogn Lett 27:1788–1796
3. Payandeh S et al (2019) A novel depth image analysis for sleep posture estimation. J Ambient Intell Humanized Comput 10:1999-2014
4. Mohammadi S, Alnowami M, Khan S, Dijk D, Hilton A, Wells K (2018) Sleep posture classification using a convolutional neural network. In: 2018 40th Annual international conference of the IEEE engineering in medicine and biology society (EMBC), pp 1–4
5. Tang K, Kumar A, Nadeem M, Maaz I (2021) CNN-based smart sleep posture recognition system. IoT 2:119–139
6. Raihan M, Shams A, Preo R (2020) Multi-class electrogastrogram (EGG) signal classification using machine learning algorithms. In: 2020 23rd International conference on computer and information technology (ICCIT), pp 1–6
7. Sklearn.preprocessing.standardscaler scikit-learn 0.23.1 documentation. Scikit Learn (2020). [Online] https://scikitlearn.org/stable/modules/generated/sklearn.preprocessing.StandardScaler.html
8. Wolpert D (1992) Neural Networks 5:241–259
9. Song Y, Ying L (2015) Decision tree methods: applications for classification and prediction. Shanghai Arch Psychiatry 27:130
10. Kabiraj S, Raihan M, Alvi N, Afrin M, Akter L, Sohagi S, Podder E (2020) Breast cancer risk prediction using XGBoost and random forest algorithm. In: 2020 11th International conference on computing, communication and networking technologies (ICCCNT), pp 1–4
11. Uçar M, Bozkurt M, Bilgin C, Polat K (2018) Automatic sleep staging in obstructive sleep apnea patients using photoplethysmography, heart rate variability signal and machine learning techniques. Neural Comput Appl 29:1–16
12. Hsiao R, Chen T, Bitew M, Kao C, Li T (2018) Sleeping posture recognition using fuzzy c-means algorithm. Biomed Eng Online 17:1–19
13. Mohammadi S, Kouchaki S, Sanei S, Dijk D, Hilton A, Wells K (2019) Tensor factorisation and transfer learning for sleep pose detection. In: 2019 27th European signal processing conference (EUSIPCO), pp 1–5

Author Index

M. S. Arefin et al. (eds.), *Proceedings of the International Conference on Big Data, IoT, and Machine Learning*, Lecture Notes on Data Engineering and Communications Technologies 95, https://doi.org/10.1007/978-981-16-6636-0